W9-DHO-166

PRINCIPLES OF

HUMAN GENETICS

A SERIES OF BOOKS IN BIOLOGY

Editors: Donald Kennedy
Roderic B. Park

THIRD EDITION

PRINCIPLES OF HUMAN GENETICS

Curt Stern
UNIVERSITY OF CALIFORNIA, BERKELEY

W. H. FREEMAN AND COMPANY
San Francisco

Library of Congress Cataloging in Publication Data

Stern, Curt, 1902–
Principles of human genetics.

1. Human genetics. I. Title.
QH431.S714 1973 573.2'1 72-4357
ISBN 0-7167-0597-4

Copyright © 1949, 1960, 1973 by Curt Stern

No part of this book may be reproduced
by any mechanical, photographic, or electronic process,
or in the form of a phonographic recording,
nor may it be stored in a retrieval system, transmitted,
or otherwise copied for public or private use
without written permission of the publisher.

Printed in the United States of America

International Standard Book Number: 0-7167-0597-4

2 3 4 5 6 7 8 9

To
Evelyn
and
Hilde,
Holly, and
Barbara

CONTENTS

PREFACE

The first edition of this book was published more than twenty years ago. When a second edition appeared in 1960, eleven years later, much had been added to our knowledge, but the principles established by the classical general geneticists remained intact. Now, after another twelve years, here is a new edition.

How did the second edition stand the test of time? Molecular genetics led to many refinements and now permits a satisfying approach to the structure of genes and their alleles and their actions in development and maturity. These achievements resulted in greatly deepened insights, but did not change the principles. Additional generalizations have been discovered since 1960, and innumerable individual facts have emerged and their interpretation become secured. The task of the present edition has been to expand the framework of the book to include general topics that had not yet been known or only very incompletely known, and to add to the treatment of the principles new examples that illustrate their generality and are noteworthy in their own right.

Throughout this third edition there have been numerous additions and deletions, making it considerably different from its predecessors. A new chapter whose subject is cytogenetics is included, and new sections are given to the single-active-X theory and to threshold theories of polygenic inheritance. The analysis of genetic mosaicism in humans furnished striking examples of participation of more than one egg nucleus and one sperm in the origin of an individual. Other additions were possible as medical genetics had become enriched by the invention of preventive treatment of Rh disease, by amniocentesis, and by leucocyte genetics as applied to transplantation compatibility. And to mention a few others of the topics introduced in the new edition: tests of Mendelian segregation by the method of "discarding the singles;" the enzymatic basis of testicular feminization; gene localization by means of man-mouse cell hybrids; final proof that there are genetic components in schizophrenia; and new aspects of selection in civilization.

Science is conducted by scientists. It is an understandable fact that many scientists not only want their discoveries cited but also their own names. Moreover the reader will gain some perspective when he becomes aware of the contributions of specific persons. I have named a good many investigators in the text, the figure and table legends, and the chapter references. No full justice has been done to the great number of human geneticists. This book is based on the findings of so many people that only a small number of investigators could be cited without overloading the text, and the choice of authors who are quoted and those not quoted cannot help having a considerable degree of arbitrariness. This I regret.

Several fellow geneticists were particularly helpful in the preparation of this edition. James Crow and Carter Denniston together commented on most of the content of the book, and Edward Novitski and Everett Lovrien together did likewise, as did Frederick Dill. Help on specific items came from Klaus Patau, Michael Lerner, Donald Shreffler, and Jack King.

Many excellent drawings which Dr. Aloha Hannah-Alava and Mrs. Emily Reid prepared for the first and the second edition have been used again and new drawings by Mrs. Reid have been added. The investigators who generously made new, original photographs available and the publishers and authors who kindly granted permission for the use of copyright material are acknowledged in the figure and table legends. The system of acknowledging borrowed material is explained in the introduction to the references at the end of Chapter 1.[1]

CURT STERN
August 1972

[1]The following acknowledgment is required by the publisher to be printed in the Preface: I am indebted to the Literary Executor of the late Sir Ronald A. Fisher, F.R.S., and to Oliver and Boyd, Edinburgh for their permission to reprint an abridged version of Table III from their book *Statistical Methods for Research Workers.*

PRINCIPLES OF HUMAN GENETICS

1

INTRODUCTION

The study of human inheritance is concerned with the existence of "inborn" characteristics of human beings: physical and mental, normal and abnormal. In its broadest sense, it deals with those qualities present in all human beings that distinguish them from nonhuman beings, as well as with those qualities that characterize only certain groups of man, certain families, or certain individuals. Thus, it is largely a study of hereditary similarities and differences among human beings. It is concerned with the causes of these similarities and differences and the way in which they are transmitted from generation to generation.

The science of inheritance is called *genetics,* a term derived from the Greek root *gen,* which means to become or to grow into something. It signifies that genetics deals not only with the transmission of hereditary factors but also with the ways in which they express themselves during the development and life of the individual.

The student of human genetics must draw on many sources of information. He obtains his material from studies of families or larger groups—from anthropological, psychological, medical, and sociological investigations. To evaluate the data, he needs the tools of the statistician. Human genetics is based on general principles derived originally from the study of plants and animals—principles that are as valid for man as they are for unicellular and other multicellular organisms. Since these general principles can often be demonstrated more clearly in organisms better adapted to experimental work than man, reference to such studies is helpful in many analyses of human inheritance.

An understanding of simple inheritance in man was not possible before Mendel's work on peas. Elucidation of the genetics of body size, intelligence scores, and other "quantitative" characters in man was not even begun until Nilsson-Ehle's and East's studies on shades of color in wheat and size of Indian corn paved the way. The peculiar mode of transmission of color blindness in man became clear only after similarly transmitted "sex-linked" traits were found in birds, moths, and flies. The basis of sex determination was not recognized until the cells of grasshoppers and plant bugs had been scrutinized. And no study contributed more toward the clarification of the heredity-environment problem than an investigation of the variability of size in beans. Work on nonhuman organisms will be cited, however, not only for historical reasons. Sometimes specific data on man are lacking and it is necessary to cite experiments with other organisms in order to estimate the situation in human beings.

Man as an Object of Genetic Study. At first sight, man appears to be an unfavorable object for genetic study. Ideally, the student of genetics works with groups of standardized organisms that are genetically identical or at least approach identity. He tries to breed and raise successive generations under similar environmental conditions. In man, however, the genetic diversity of individuals is great and uncontrolled, and biological and sociological environments vary greatly. The principal tool of the general geneticist is the experimental crossing of different genetic types. In man, on the other hand, parental unions are entered into with no intent to serve an experimental plan. Studies of inheritance are generally based on knowledge of a series of generations. Consequently, the preferred subjects for genetic study are organisms with rapid succession of parents and offspring. Annual plants and small mammals like mice and rats first served this purpose, with the still shorter-lived fruit flies, molds, and microorganisms gaining special favor later. More patience was required for work on fruit or forest trees or on horses and cattle—and relatively less scientific yield was expected. With man, the duration of a generation is alike in the observer and in the object of observation, and personal knowledge of successive generations is therefore restricted. Finally, many factors affecting transmission of hereditary traits obey statistical laws and are best studied when large numbers of offspring are available. In man, these numbers are always small; even large human families fall far short of the size desirable for statistical deductions.

Obstacles to the understanding of nature have always been challenging to the human mind. Even though genetic and environmental diversity in man is largely uncontrollable, the human geneticist can find groups that have similar heredity or similar environments. Even though he cannot arrange matings in accordance with a research plan, he can collect data from those marriages that happen to fit his schemes. Even though man's life cycle is longer than that of laboratory organisms, the scientist can devise special methods that enable him to get information from one or two generations, which normally only a longer series would provide. Even though the human family is small, pooling of data from many families may provide enough material for statistical analysis. The human population is large, and its millions of unions give the geneticist an immense amount of material from which to select what he needs.

The Principles of Human Genetics. This book will present the principles of human genetics, i.e., the general regularities of heredity in man that have been derived from the study of families, pedigrees, and large interrelated groups of individuals called populations. The various rules of transmission of hereditary traits and some of the methods that enable us to find out what kind of inheritance is operating in specific cases will be described. The results of human cytogenetics with its demonstration of the roles of normal and abnormal chromosome behavior in health and disease will be discussed. The way human genes act in the network of developmental and metabolic processes will be surveyed. Detailed discussions will be presented on the effects of environment on the expression of genes, the genetic aspects of sex, and the origin of new hereditary traits (mutation). Applications of these principles will be considered under such headings as Genetic Prognosis (Chap. 9), Medicolegal Applications (Chap. 13), The Genetic Hazards of Radiation (Chap. 24), Selection in Civilization (Chap. 29), Aspects of Medical Genetics (Chap. 30), and Genetic Aspects of Race and Race Mixture (Chap. 31). No attempt will be made to treat systematically the genetics of the seemingly infinite range of normal and abnormal human traits, but many of these will be used as examples to illustrate principles.

The Scope of Human Genetics. When, in the early twentieth century, the modern study of heredity in plants and lower animals resulted in the discovery of the laws of biological inheritance, enthusiastic men drew far-reaching conclusions about the hereditary nature of differences among human individuals and of the consequences for mankind of the transmission of these differences to future generations. These conclusions were used to project plans for manipulating the human gene pool through the legal prohibition of reproduction for certain large groups of humans and social incentives for the reproduction of other groups. It became evident later that the factual knowledge of man's inheritance was too narrow to justify such actions. The "eugenic movement" became discredited in the eyes of many, and interest shifted from hypothetical possibilities for changing the human gene pool to the accumulation of specific information and development of specific methods applicable to human genetics. The work of the past decades has greatly increased our knowledge, but the preliminary nature of much of it must still be emphasized. Human genetics can claim significant achievements, but, as in any growing science, future discoveries will not only add new facts, but may also invalidate apparently established views.

Knowledge of human genetics does not only satisfy our desire to know about ourselves; it must also form the basis of practical decisions. The physician and public health official need to understand the inheritance of diseases and abnormal characteristics. Every individual should know something about the genetic aspects of selecting a partner for marriage and about the kinds of children he may expect; he should also know that his prospects for physical and mental well-being and for a long or short life are influenced genetically. Unavoidably, though often obscurely, social measures affect the type of men and women who will people a country and the earth. In order to gauge such influences, an understanding is needed of the causes of human differences, individual and racial; of the part heredity and environment play in the determination of such differences; and of the role of social organization and education in molding men. If it is proposed to change a population hereditarily, the prospects of successful selection of favored types must first be understood in order to reach rational conclusions; if such a change is feared, the same knowledge is required to evaluate the future.

As will be shown, human genetics has already made important contributions to these practical problems, and there is every reason to believe that its usefulness will increase greatly in years to come.

REFERENCES

The references listed at the ends of chapters include important sources of the discussion as well as suggestions for further reading. Besides these references, other sources are also cited in the figure legends and table notes. If only the author's name is cited in a figure legend or table note, the full bibliographical information will be found in the references at the end of the chapter. References in the figure legends or table headings which cite the author's name and, in abbreviated form, the time and place of publication are not given again at the ends of chapters; these are references which should be of further help in tracing the origins and ramifications of the topics treated.

In the figure legends and table notes, an attempt has also been made to indicate whether the illustrative and tabular material borrowed from other authors is unchanged or is in modified form. For both modified and unchanged material, credit is acknowledged in the figure legends and table notes; but for modified material, the author's name is preceded by the word "after."

BOOKS ON GENERAL GENETICS

King, R. C., 1965. *Genetics.* 2nd. ed. 450 pp. Oxford University Press, New York

Srb, A. M., Owen, R. D., and Edgar, R. S., 1965. *General Genetics,* 2nd ed. 557 pp. W. H. Freeman and Company, San Francisco

Stent, G. S., 1971. *Molecular Genetics.* 650 pp. W. H. Freeman and Company, San Francisco

Strickberger, M. W., 1968. *Genetics.* 868 pp. Macmillan, New York

Watson, J. D., 1970. *Molecular Biology of the Gene,* 2nd ed. 662 pp. Benjamin, New York

BOOKS ON GENERAL HUMAN GENETICS

Boyer, S. H. (Ed.), 1963. *Papers on Human Genetics.* 305 pp. Prentice-Hall, Englewood Cliffs, N.J.

Burdette, W. J. (Ed.), 1962. *Methodology in Human Genetics.* 436 pp. Holden-Day, San Francisco.

Cavalli-Sforza, L. L., and Bodmer, W. F., 1971. *The Genetics of Human Populations.* 965 pp. W. H. Freeman and Company, San Francisco.

Cold Spring Harbor Laboratory of Quantitative Biology, 1964. *Human Genetics.* 492 pp. Cold Spring Harbor Laboratory, Cold Spring Harbor, New York. (Cold Spring Harbor Symposium on Quantitative Biology, **29.**)

Crow, J. F., and Neel, J. V. (Eds.), 1967. *Proceedings of the Third International Congress of Human Genetics.* 578 pp. Johns Hopkins University Press, Baltimore.

Emery, A. E. H. (Ed.), 1970. *Modern Trends in Human Genetics.* 379 pp. Butterworth, London

Ford, C. E., and Harris, H. (Eds.), 1969. *New Aspects of Human Genetics.* British Council, Medical Department, London. (British Medical Bulletin, **25.**)

Gates, R., 1946. *Human Genetics.* 2 vols., 1518 pp. Macmillan, New York.

Giblett, Eloise R., 1969. *Genetic Markers in Human Blood.* 629 pp. Blackwell, Oxford.

Haldane, J. B. S., 1938. *Heredity and Politics.* 202 pp. Norton, New York.

Hamerton, J. L., 1971. *Human Cytogenetics, Vol. I, General Cytogenetics* 412 pp. Academic, New York.

Hamerton, J. L., 1971. *Human Cytogenetics, Vol. II, Clinical Cytogenetics,* 545 pp. Academic, New York.

Harris, H., 1970. *The Principles of Human Biochemical Genetics.* 328 pp. North Holland Publ. Co., Amsterdam.

Hogben, L. T., 1933. *Nature and Nurture.* 143 pp. Norton, New York.
Huron, R., and Ruffié, J., 1959. *Les Méthodes en Génétique Générale et en Génétique Humaine.* 556 pp. Masson, Paris.
Lerner, I. M., 1968. *Heredity, Evolution, and Society.* 307 pp. W. H. Freeman and Company, San Francisco.
Levitan, M., and Montagu, M. F. A., 1971. *Textbook of Human Genetics.* 922 pp. Oxford University Press, New York.
Li, C. C., 1961. *Human Genetics.* 218 pp. McGraw-Hill, New York.
Maynard-Smith, Sheila; Penrose, L. S., and Smith, C. A. B., 1962. *Mathematical Tables for Research Workers in Human Genetics.* 74 pp. Little, Brown, Boston.
McKusick, V. 1969. *Human Genetics,* 2nd ed. 211 pp. Prentice-Hall, Englewood Cliffs, N. J.
Montagu, M. F. A., 1963. *Human Heredity,* 2nd ed. 432 pp. World, Cleveland.
Moody, P. A., 1967. *Genetics of Man.* 444 pp. Norton, New York.
Neel, J. V., and Schull, W. J., 1954. *Human Heredity.* 361 pp. University of Chicago Press, Chicago.
Penrose, L. S., 1963. *Outline of Human Genetics,* 2nd ed. 166 pp. Wiley, New York.
Race, R. R., and Sanger, Ruth, 1968. *Blood Groups in Man.* 5th ed. 599 pp. Blackwell Oxford.
Scheinfeld, A. 1965. *Your Heredity and Environment.* 830 pp. Lippincott, New York
Sutton, H. E., 1965. *An Introduction to Human Genetics.* 262 pp. Holt, Rinehart, and Winston, New York.
Vogel, F., 1961. *Lehrbuch der Allgemeinen Humangenetik.* 753 pp. Springer, Berlin.
Whittinghill, M., 1965. *Human Genetics and Its Foundations.* 431 pp. Reinhold, New York.
Wiener, A. S., 1961. *Advances in Blood Grouping.* 549 pp. Grune and Stratton, New York.
Wiener, A. S., with a section by M. Shapiro, 1965. *Advances in Blood Grouping II.* 452 pp. Grune and Stratton, New York.

BOOKS ON MEDICAL GENETICS

Becker, P. E. (Ed.), 1964– . *Humangenetik, Ein kurzes Handbuch in 5 Bänden.* Thieme, Stuttgart.
Carter, C. O., 1969. *An ABC of Medical Genetics.* 94 pp. Little, Brown, Boston.
Clarke, C. A. (Ed.), 1969. *Selected Topics in Medical Genetics.* 282 pp. Oxford University Press, Oxford.
Cockayne, E. A., 1933. *Inherited Abnormalities of the Skin and Its Appendages.* 394 pp. Oxford University Press, London.
Franceschetti, A., Francois, J. and Babel, J., 1963. *Les Hérédo-dégénérescences Chorio-Rétiniennes.* 1709 pp. Masson, Paris.
Francois, J., 1958. *L'Hérédité en Ophthalmologie.* 876 pp. Masson, Paris
Goodman, R. M. (Ed.), 1970. *Genetic Disorders of Man.* 1009 pp. Little, Brown, Boston.

Gottron, H. A., and Schnyder, U. W. (Eds.), 1966. *Vererbung von Hautkrankheiten. Handbuch der Haut- und Geschlechtskrankheiten,* VII. 1211 pp. Springer, Berlin

Kallmann, F. J., 1953. *Heredity in Health and Mental Disorder.* 315 pp. Norton, New York.

Knudson, A. G., Jr., 1965. *Genetics and Disease.* 294 pp. McGraw-Hill, New York.

Lenz, W., 1963. *Medical Genetics.* 218 pp. University of Chicago Press, Chicago. (English translation of *Medizinische Genetik,* 1961.)

McKusick, V., 1960. *Heritable Disorders of Connective Tissue,* 2nd ed. 333 pp. Mosby, St. Louis.

McKusick, V., 1971. *Mendelian Inheritance in Man: catalogs of autosomal dominant, autosomal recessive, and X-linked phenotypes,* 3rd ed., 738 pp. Johns Hopkins University Press, Baltimore.

Montagu, M. F. A., (Ed.), 1961. *Genetic Mechanisms in Human Disease.* 592 pp. Thomas, Springfield, Ill.

Penrose, L. S., 1963. *The Biology of Mental Defect,* rev. ed. 374 pp. Grune and Stratton, New York.

Robėrts, J. A. F., 1970. *An Introduction to Medical Genetics,* 5th ed. 296 pp. Oxford University Press, London.

Sorsby, A., 1951. *Genetics in Opthalmology.* 251 pp. Butterworth, London.

Sorsby, A. (Ed.), 1953. *Clinical Genetics.* 580 pp. Butterworth, London.

Stanbury, J. B., Wyngaarden, J. B., and Frederickson, D. S. (Eds.), 1972. *The Metabolic Basis of Inherited Disease,* 3rd ed. 1778 pp. McGraw-Hill, New York.

Thompson, J. S., and Thompson, Margaret W., 1966. *Genetics in Medicine.* 300 pp. Saunders, Philadelphia

Touraine, A., 1955. *L'Hérédité en Médecine.* 875 pp. Masson, Paris.

Waardenburg, P. J., Franceschetti, A., and Klein, D., 1961–1963. *Genetics and Opthalmology,* 2 vols., 1914 pp. Thomas, Springfield, Ill.

Witkop, C. J., Jr. (Ed.), 1962. *Genetics and Dental Health.* 300 pp. McGraw-Hill, New York.

SERIAL PUBLICATIONS DEVOTED WHOLLY OR IN PART TO HUMAN GENETICS

Acta Geneticae Medicae et Gemellologiae
Advances in Human Genetics
American Journal of Human Genetics
Annals of Human Genetics (formerly *Annals of Eugenics*)
Archiv der Julius Claus Stiftung für Vererbungsforschung
Behavior Genetics
Clinical Genetics
Copenhagen Universitet: Opera ex Domo Biologiae Hereditariae Humanae Universitatis Hafniensis.
Cytogenetics
De Genetica Medica

Eugenics Review (superseded by *Journal of Biosocial Science*)
Excerpta Medica, Section 22: Human Genetics
Human Biology
Human Heredity (formerly *Acta Genetica et Statistica Medica*)
Japanese Journal of Human Genetics
Journal de Génétique Humaine
Journal of Heredity
Monographs in Human Genetics
Oxford Monographs on Medical Genetics
Progress in Medical Genetics
Social Biology (formerly *Eugenics Quarterly*)
The Treasury of Human Inheritance

2

THE BIOLOGICAL BASIS OF MAN'S INHERITANCE

All the material a human being inherits from his two parents is contained in two cells, the egg and the sperm.

EGG AND SPERM

The Human Egg. The human egg (Fig. 1) is a spherical body about 1/7 of a millimeter, or about 1/175 of an inch, in diameter. Such small measurements are usually given in microns—a micron (μ) being one-thousandth of a millimeter; thus the diameter of the human egg is about 140μ. In spite of its relatively minute size, the egg is one of the largest cells of the human body. Its cellular character is easily recognizable, since it possesses a typical nucleus enclosed in a mass of cytoplasm.

The weight of the human egg has been estimated to be 0.0015 milligram, or approximately one twenty-millionth of an ounce. In this tiny bit of matter is contained the genetic contribution of the mother to her child.

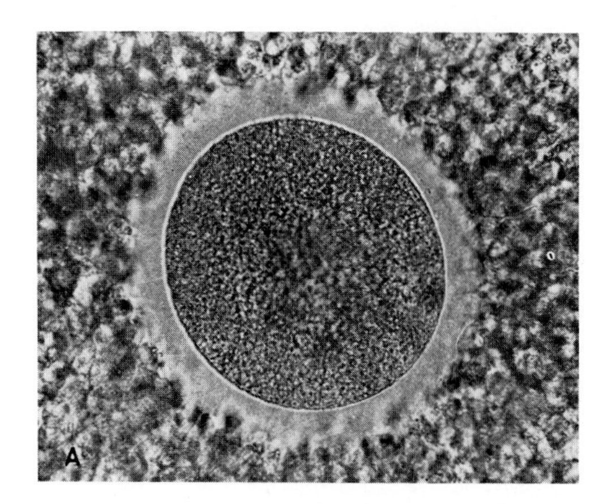

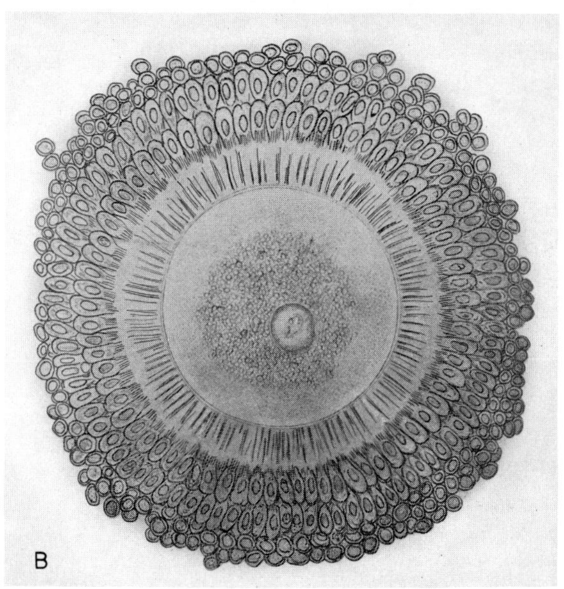

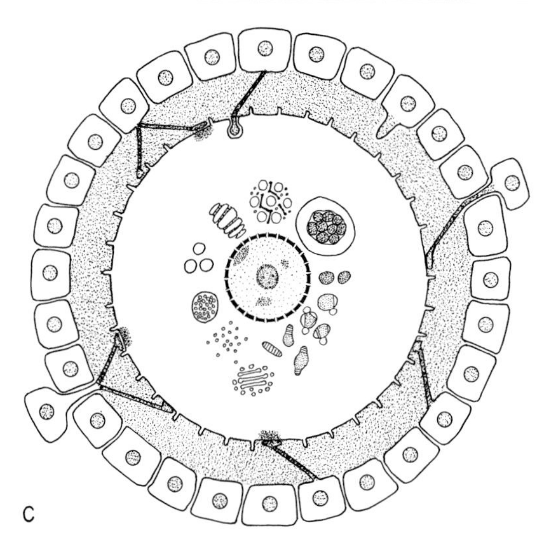

FIGURE 1
The human egg. **A.** An egg, surrounde by a noncellular layer, the zona pellucida, and corona cells, recovered from a large ovarian follicle. **B.** Drawin of an egg from a large ovarian follicle. The egg nucleus is visible. **C.** Simplifie diagram of an egg, based on light- and electron-microscopic studies. Note the "pores" of the nuclear membrane, the various types of cytoplasmic inclusion bodies, the microvilli of the cell membrane, and the processes connecting the follicle cells to an egg cell. (A, original photomicrograph by Dr. Warren H. Lewis; B, Nagel, *Arch. Mikroskop. Anat.*, **31,** 1888; C, Wartenberg and Stegner, *Zeitschr. Zellforschg.*, **52,** 470, Berlin, Heidelberg, New York: Springer, 1960

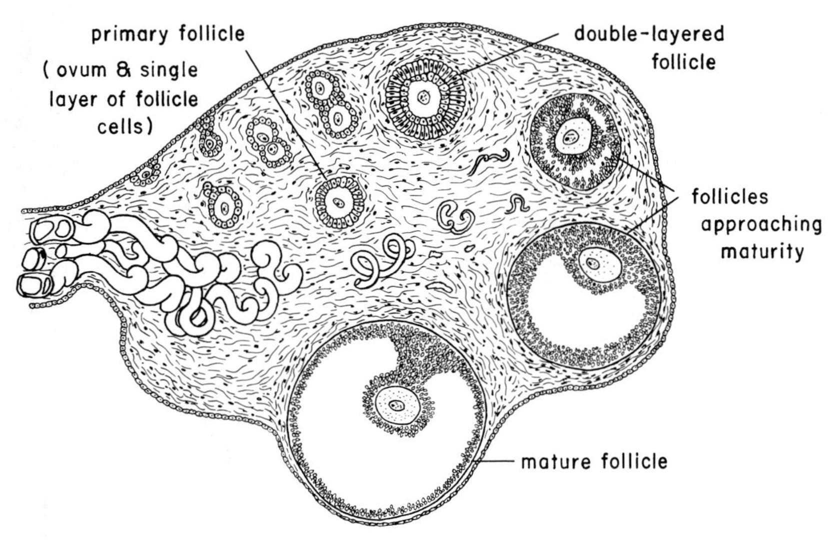

FIGURE 2
Diagram of a human ovary with eggs and follicles at various stages of development. (Adapted from Patten, *Embryology of the Pig*, Blakiston, 1931.)

The eggs are produced in two ovaries, organs about the size of walnuts, which are attached to the dorsolateral wall of the female abdominal cavity. In the ovaries, immature germ cells are found in various stages of growth (Fig. 2). Each egg cell is surrounded by a wall of *follicle cells*. Concurrent with the growth of an egg cell, the follicle cells multiply. Fluid-filled gaps appear between the cells of the follicular wall and so increase as to become a single fluid-filled space, splitting the wall into an external and an internal layer. Thus, a mature ovarian follicle originates. It consists of an outer layer of follicle cells, a large, fluid-filled center, and an egg cell located within an inner layer of follicle cells still connected on one side with the outer layer. As growth proceeds, the follicle bulges out on the surface of the ovary. The internal pressure of the follicular fluid stretches the thin sheets of tissue that separate the content of the follicle from the abdominal cavity until both the follicle and the wall of the ovary burst (Fig. 3). The egg, surrounded by the *corona*, or *crown*, the cover of follicle cells that had formed the inner layer, is released into the abdominal cavity, where it enters the funnel-like opening of the oviduct and starts its journey down the oviduct into the uterus.

If the oviduct is free of sperm, the egg cell disintegrates inside the uterus. If, however, as a result of a recent mating, live sperm are present, the egg may be fertilized in the oviduct and there begin its development. While undergoing the first steps of this process, the fertilized egg moves to the uterus, where it becomes embedded in the uterine wall and remains for the nine months of prenatal development.

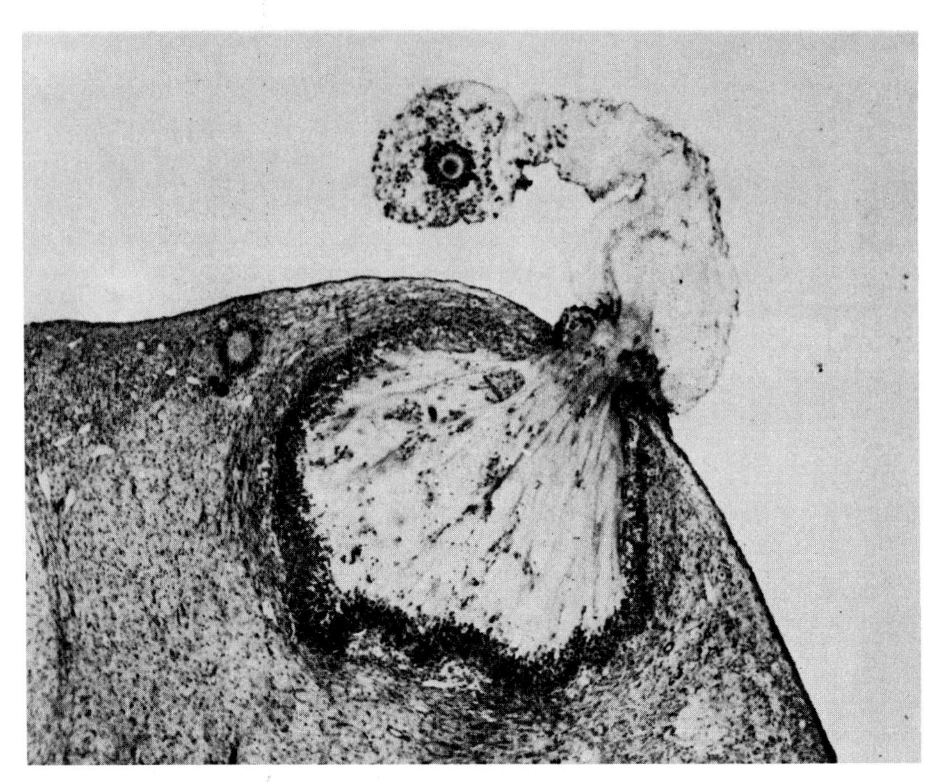

FIGURE 3
Ovulation of a rabbit follicle observed in the living. After ovulation the ovary was fixed and sectioned. (Original photomicrograph by Dr. R. J. Blandau.)

Normally, only one mature follicle develops during each monthly cycle. It may originate in either the right or left ovary: by and large, each has an equal share, but one ovary often has several successive ovulations. Occasionally, more than one follicle matures at the same time, either in the same ovary or in both. Consequently, more than one ripe egg may be ready for fertilization at the same time, and a multiple birth may result. Occasionally, too, one follicle may contain more than one egg cell, again setting the stage for a possible multiple birth.

The human egg was discovered in 1827 by Karl Ernst von Bär (1792–1876), the founder of modern embryology. The basic research that led to this discovery was done with dogs. Von Bär obtained a series of female dogs in various stages of pregnancy and succeeded in tracing the embryos back to very small specks of matter—much smaller than the large ovarian follicles that had been thought to be the eggs. Finally, he found that the unbroken follicles contained minute bodies that were identical with the egg and its corona, which at a later stage were present in the oviduct. To confirm that what was true of dogs was also true of man was only a short step.

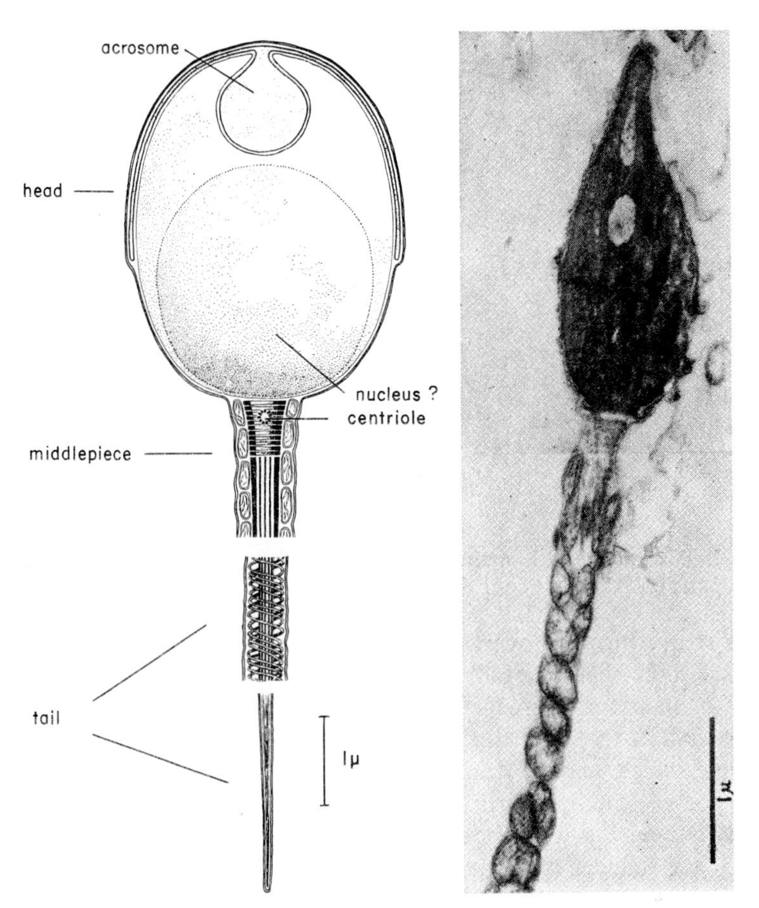

FIGURE 4
Human sperm. *Left:* Diagram of a "frontal" section (based on electron-microscopic studies). *Right:* Electronmicrograph of a somewhat tangential "longitudinal" section through head and middlepiece. (Left, Schultz-Larsen; right, Lord Rothschild, *Brit. Med. J.*, **1**, 1958.)

The Human Sperm. The father's genetic contribution to a child is contained in the *spermatozoon*, or *sperm* (Fig. 4), which is produced in the testes. The cellular structure of the sperm is less obvious than that of the egg. It consists of several parts, called *head, middlepiece*, and *tail*. Their dimensions are:

Head			*Middlepiece*		*Tail*	
	Width					
Length	(elliptical face view)	(side view anterior end)	Length	Width	Length	Width
3–5µ	1.6–4µ	1.8µ	2–4µ	0.4–0.8µ	30–50µ	0.08–0.5µ

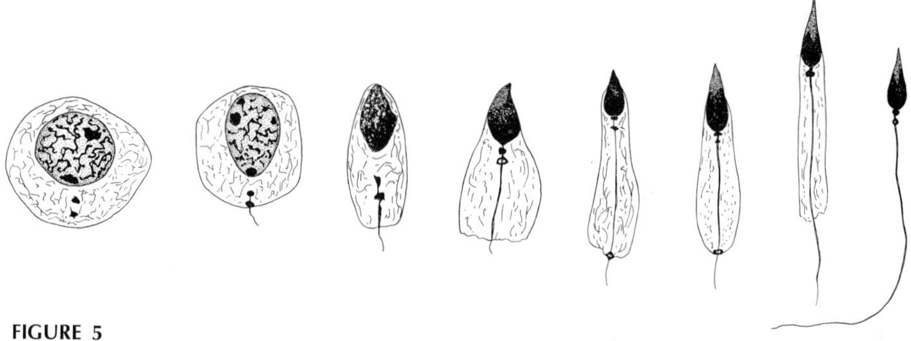

FIGURE 5
The transformation of a spermatid cell into a mature sperm. (After Stieve, *Hdbch. d. Mikroskop. Anat. d. Menschen,* **7,** 1930.)

The different types of immature germ cells, representing successive stages in the development of spermatozoa, are called *spermatogonia, spermatocytes,* and *spermatids.* Each consists of a mass of cytoplasm and a nucleus, and is thus a typical cell. In the course of the transformation of a spermatid into a spermatozoon, striking changes take place (Fig. 5). The nucleus becomes smaller and more compact. It forms part of the head of the mature sperm, which is somewhat pear-shaped in side view and oval in frontal view. A minute body in the cytoplasm, the *centriole,* sends out a bundle of fibers, which, embedded in a thin cylinder of cytoplasm, become the tail. Most of the original cytoplasm of the immature cell is cast off and disintegrates inside the testes: a part remains in the head, another develops into the conical middlepiece, and a third forms the outer covering of the tail. The head of the mature sperm consists largely of the nucleus and the acrosome, an organelle that contains several enzymes characteristic of the lysosomes of other cells. As will be described in the next section, these enzymes play a role in fertilization.

Inside the numerous fine tubes (Fig. 6) that constitute the greater part of the testes, many millions of cells are constantly being transformed into mature spermatozoa. They are stored in the ducts that lead from the testes to the outside. A discharge of human semen consists on the average of more than two hundred million spermatozoa suspended in the fluid secretions of glands of the male genital system.

Human spermatozoa were first seen in 1677 by a student named Ham. He reported his observation to the pioneer of microscopy, van Leeuwenhoek (1632–1723), who described the sperm in detail in some of his famous letters to the Royal Society of London. The sperm cells, swimming around in the seminal fluid with their motile tails, appeared not unlike the small animals that Leeuwenhoek had discovered in drops of pond water; he therefore called

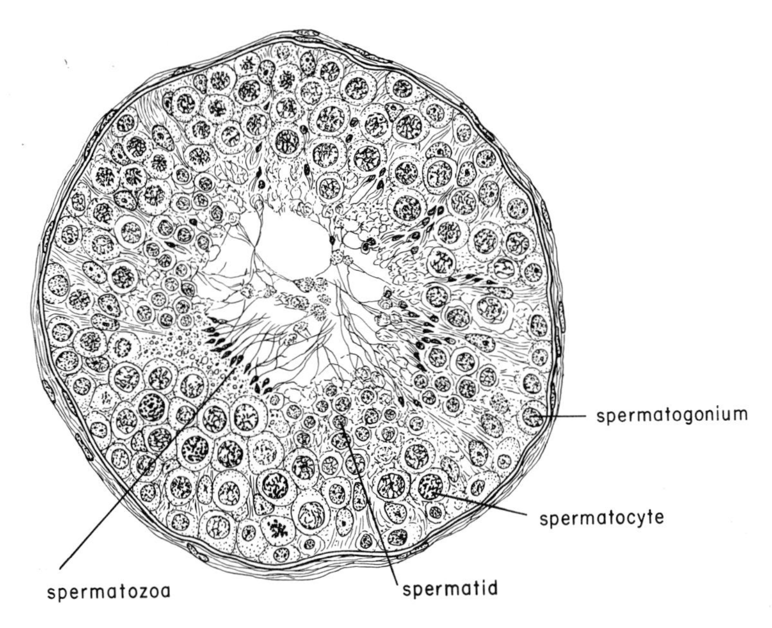

FIGURE 6
Cross section through a tubule of a human testis, showing cells in various stages of spermatogenesis: spermatogonia → spermatocytes → spermatids → spermatozoa. (After Stieve, *Hdbch. d. Mikroskop. Anat. d. Menschen,* **7,** 1930.)

them spermatozoa, or "seed animals." Although Leeuwenhoek himself believed that the spermatozoa were the essential elements of the male reproductive material, others doubted this and regarded them as independent organisms with no known function. A hundred years later, Spallanzani (1729–1799) tested the two conflicting theories by experiment. He filtered the seminal fluid of dogs and injected either unfiltered semen or a sperm-free filtrate into dogs' vaginas, thus inventing a method of artificial insemination. (A similar procedure had been practiced much earlier by Arabs in the breeding of horses.) Offspring resulted only from insemination with fluid containing sperm. Spallanzani did not draw the conclusion that the sperm takes part in the development of the egg, but believed that it in some way enabled the seminal fluid to stimulate development. Final proof for the theory that the sperm cell is the essential contribution of the male to his offspring was not obtained until 200 years after Ham's and Leeuwenhoek's discovery, when studies on fertilization of sea-urchin eggs showed the fusion of a single egg and a single spermatozoon at the beginning of each new life.

Fertilization. To fertilize an egg, a sperm must penetrate the corona and *zona pellucida* surrounding it. This process is facilitated by the action of the acrosomal enzymes. They include proteases and hyaluronidase, which, respectively, break down the proteins surrounding the egg and the hyaluronic acid that cements the cells of the corona together. As soon as a sperm fuses with an egg, the zona pellucida and the surface of the egg undergo changes that prevent the entry of other sperm.

After the egg and sperm have fused, the head of the sperm cell absorbs fluid from the cytoplasm of the egg, increases in size, and becomes spherical, resuming the appearance of a normal nucleus and resembling the nucleus of the egg, both in volume and morphological detail (Fig. 7). The two nuclei approach each other until they are in contact, and finally fuse. Thus nuclear fusion may be regarded as the ultimate process in fertilization. The fertilized egg cell is called the *zygote,* a term also frequently used to designate the individual who develops from the fertilized egg cell.

The female and male *gametes,* egg and sperm, contain all of the hereditary potentialities of a future child. The small space in which these potentialities lie can perhaps best be comprehended if one considers how large a volume would be required to contain the hereditary potentialities of all persons of the next generation. Mankind, at present, comprises approximately 3,500,000,000 individuals, and it is probable that at least 6,000,000,000 persons will be alive when all those now living have completed their reproductive period. These 6,000,000,000 people will have originated from 6,000,000,000 egg cells and an equal number of sperm cells. Calculating the volume of a sphere the size of a human egg and multiplying it by 6,000,000,000 yields a total volume of about two gallons (= about eight liters) in which the eggs could be compressed without space between them, or two and one-half gallons if the eggs were assembled touching one another but with space between them: in these small volumes could be contained the material contribution of all mothers of this generation to all beings of the next. By making a similar calculation for the volume of sperm, it is found that the total hereditary contribution of all fathers of this generation will be contained in a mass smaller than an aspirin tablet! Actually, these considerations can be refined by calculating the volumes not of the gametes but of the molecular hereditary material that they contain and that forms only a fraction of their mass. The amount of this material, deoxyribonucleic acid, or DNA, in a single human sperm weighs approximately 3/1000 billionth ($= 3 \times 10^{-12}$) of a gram and presumably there is the same amount in a human egg. If one estimates the volume occupied by the amount of DNA in one gamete and multiplies it by the number of gametes required to produce the next generation, all mothers and fathers of the present generation would jointly contribute one-fiftieth of a cubic centimeter (cm^3) or less than one-tenth of the volume of an aspirin tablet.

The Genetic Significance of the Nucleus. Egg and sperm are greatly different in size, but the hereditary influences of the mother and father are about equal. This is not only known from the general observation that children, on the whole, do not resemble the female more than the male parent, but it can be

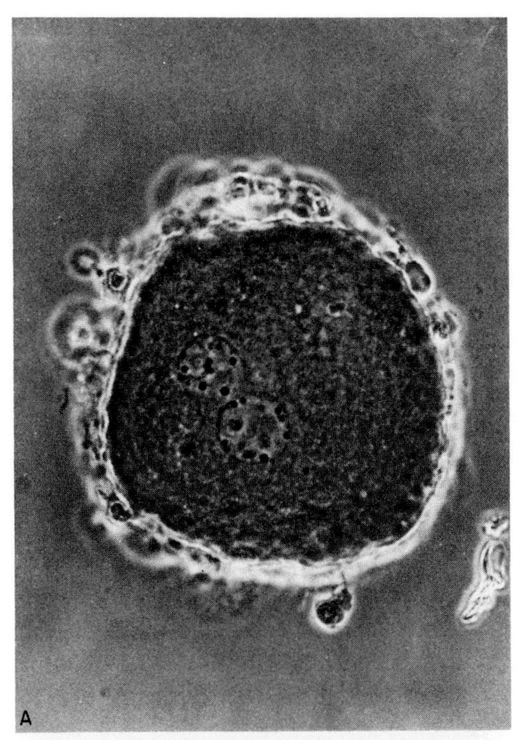

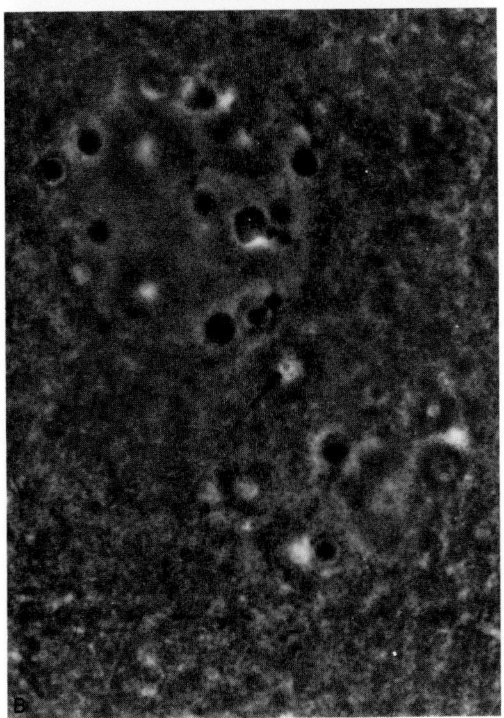

FIGURE 7
A human egg shortly after fertilization, fixed and stained. The zona pellucida was dissolved by the processing. **A.** Egg and sperm nuclei before fusion. **B.** The two nuclei photographed at a focal plane different from that in A. The tail of the fertilizing spermatozoon lies over one of the nuclei, which is almost out of focus. (Dickmann, Clewe, Bonney, and Noyes, *Anat. Rec.*, **152,** 1965.)

especially well demonstrated by the offspring of interracial crosses. The children from unions of whites and blacks are mulattoes, regardless of whether the mother is black and the father white or whether they have a white mother and a black father. If one considers that the sperm consists almost entirely of a nucleus, while the egg is made up both of a nucleus and a considerable amount of cytoplasm, it would perhaps have been expected that the children of black mothers and white fathers would be darker and, in general, more black-like than those resulting from "reciprocal matings." The observation that the children are *not* different from each other suggests that the cytoplasm of the egg does not transmit the specific properties that distinguish the two parents, but that the two nuclei, which are strikingly similar, are the material carriers of the hereditary contributions of the parents. It was this reasoning, first suggested by O. Hertwig (1849–1922) in the second half of the nineteenth century, that focused the attention of biologists on the nucleus as the bearer of hereditary material.

CHROMOSOMES

Mitosis. When it was realized that the nuclei are probably the essential agents in the transmission of hereditary differences, it was natural to seek more knowledge about them. Such knowledge was soon found to be obtainable from a study of cell division (Fig. 8).

When a cell divides, it forms two daughter cells, each of which derives its nucleus and cytoplasm from the nucleus and cytoplasm of the mother cell. Division of the cytoplasm is usually a relatively simple process: a furrow appears on the surface of the cell around its whole circumference and gradually grows deeper and deeper until it separates the cytoplasm into two halves.

While the division of the cytoplasm is under way, the nucleus is also dividing in two—a division that involves a much more elaborate series of events. Inside the nucleus, which in the undivided cell appears as a rather undifferentiated vesicle (Fig. 8, A), well-defined structures become visible (Fig. 8, B–D). They have various shapes: some are short rods and others long rods; some are V-shaped and others J-shaped. Although these bodies can be observed in living cells, they are more frequently studied in cells which have been fixed—that is, killed in such a manner as to cause very little structural change—and then treated with special stains. These stains are taken up more intensely by such nuclear structures than by the rest of the cell; some stains are exclusively absorbed by them. They have, therefore, been called *chromosomes* (from the Greek words *chromos* = color, *soma* = body).

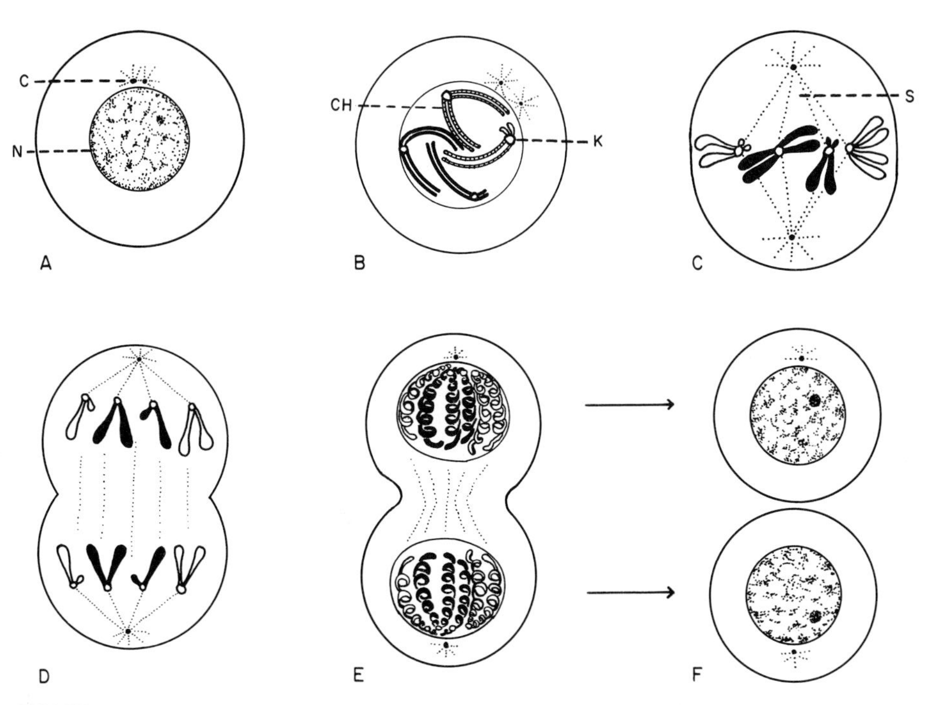

FIGURE 8
Cell division and mitosis. C = centriole; N = nucleus; CH = chromosome (showing two chromatids); K = kinetochore; S = spindle.

At about the time the chromosomes become visible in the dividing cell, the nuclear membrane and the nucleus disappear, and the chromosomes become associated with a new structure, the *spindle,* which now develops inside the cell (Fig. 8, C, D). (It derived its name from the spindle used in weaving, a familiar household object in earlier times.) In most animal cells, including those of man, the spindle is an elongated body that resembles two cones joined at their bases. It is formed as the centriole of the cell is duplicated and the two centrioles move apart (Fig. 8, B): these bodies assume positions on opposite sides of the cell, and the spindle forms between them. The centrioles are thus the poles of the spindle.

By the time the spindle is fully formed, the chromosomes have assembled in its equatorial plane (Fig. 8, C). At this stage, they show clearly the most important feature of nuclear division, which could be seen only indistinctly earlier. Each chromosome appears to be double (Fig. 8, B, C), with two identical strands (*chromatids*) lying nearly parallel to each other and joined at only one short region: close to the ends of the rod-shaped chromosomes and at the bend of the V- or J-shaped ones. This region has staining properties different from the rest of the chromosomes and is known as the *kinetochore,* or *centromere*. The kinetochores are anchored on the spindle; the rest of the chromosome may extend into the cytoplasm of the cell.

The final separation of the sister strands is initiated by a division of the kinetochores. They divide in such a way that one of the two sister kinetochores becomes connected with one of the sister chromosome strands, and the other kinetochore with the other chromosome strand. The sister kinetochores then move along the spindle to opposite poles, dragging behind them the sister strands. Thus, two groups of chromosomes move along the spindle, one group toward each pole. The two groups are identical in make-up, since each chromosome of one group has a sister chromosome in the other. As they near the poles, the chromosomes become less distinct, a nuclear membrane forms around each group, and, gradually, each daughter nucleus takes on the diffuse appearance that the mother nucleus exhibited before its division (Fig. 8, E, F).

The somewhat thread-like appearance of the chromosomes during the early stages has provided the name *mitosis* (from the Greek *mitos* = thread) for the whole process of nuclear division. Cell division and mitosis are so synchronized that the circular furrow that cleaves the cytoplasm into two halves, and that lies in the same plane as the equator of the spindle, separates the two new cells at the same time that each chromosomal group becomes transformed into a daughter nucleus. The whole process of mitosis and cell division ordinarily takes less than an hour, but wide variations in its duration are found.

It is known that the chromosomes never lose their identity in the nucleus. That they are indistinct during the period when the cell is not dividing is due to the fact that they are then long and extended threads which are so thin that they are almost invisible. The relatively thick chromosomes of the mitotic periods are these same long threads concentrated into tightly wound coils. Usually the coiled configuration of the condensed chromosomes is not apparent, but with special techniques it can be brought out clearly (Fig. 9).

Chromosome Reproduction. The continuous existence of the chromosomes in the nuclei accounts for the fact that in general each cell within the same organism has the same number of chromosomes. A human body is made up of thousands of billions of cells that are all derived, by division, from one fertilized egg. Even in the adult organism, division and mitosis continue to form new cells in order to replace old or lost ones. Typically, at a stage that precedes mitosis, each chromosome of a nucleus duplicates itself, and each daughter nucleus receives one of each pair of duplicates. Thus, mitosis supplies the nuclei of all cells with identical copies of the chromosomes of the fertilized egg.

The Significance of Mitosis. As soon as mitosis had been discovered, biologists began to speculate about its meaning. Why should the nuclear material

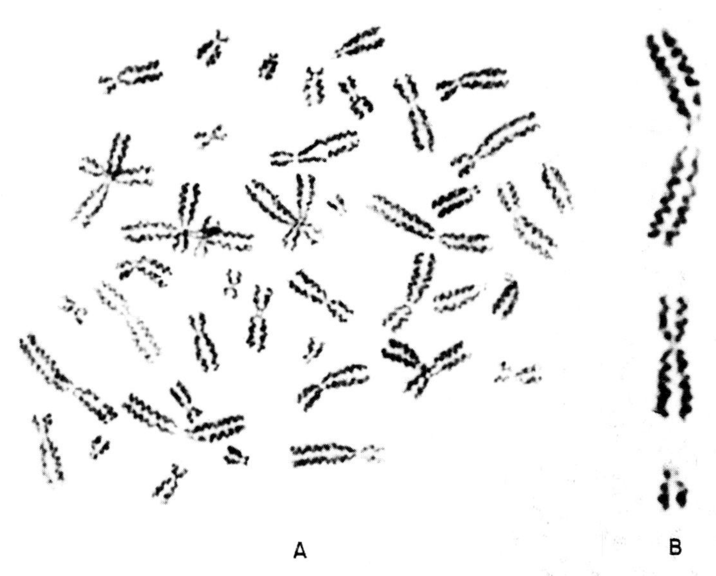

FIGURE 9
A. The chromosomes of a dividing human cell, showing coiling. **B.** Details of coiling of three chromosomes. (After Ohnuki, *Chromosoma*, **25**, 1968.)

be arranged in thread-like structures, and why should identical sister threads be transmitted to both daughter nuclei? The contrast between the extreme precision of assuring equality of nuclear matter to the daughter nuclei and the less accurate method of dividing the cytoplasm of the original cell between the daughter cells called for an explanation. It was suspected that the nucleus contained many different kinds of essential, self-reproducing substances and that, if these substances were present in a haphazard arrangement, simple constriction of the nucleus might not provide each daughter nucleus with part of every one of them. Would not the safest method for a qualitatively equal distribution of the nuclear content during division be the following? Arrange all these different substances in one or several linear structures, let these form identical linear sister structures, and have each double thread send its two representatives into opposite daughter nuclei.

Such reasoning treats cells as if they were constructed in order to achieve a specific purpose—in this case, to insure that every kind of hypothetical self-reproducing nuclear element became equally represented in the daughter cells. To inquire what purpose a biological phenomenon has is not always meaningful, for there are processes and structures that have no significance for the survival of the organisms of which they are a part and may even have slightly harmful effects. In mitosis, however, this inquiry was justified, since it was most unlikely that a process of such universal occurrence could have been retained during the evolution of nearly all species if it were not of great importance to them. The assumption of a linear arrangement of many qualitatively different essential substances was first proposed about ninety years ago, by Wilhelm Roux (1850–1924). He added the hypothesis that the essential substances were basic units representing hereditary properties. After many years of microscopic studies of chromosomes and after the twentieth-century science of genetics had greatly increased our knowledge, Roux's ideas were proven to be correct by Thomas Hunt Morgan (1866–1945) and Alfred H. Sturtevant (1891–1970).

The Chromosome Number of Man. As a rule the nuclei of the mature egg and the sperm each contribute the same number of chromosomes to the zygote. The number characteristic of the gametes of a particular species is called *haploid,* that characteristic of the zygote *diploid* (from the Greek *haploos* = single, *diploos* = double, *-id* from *eidos* = form; i.e., haploid, diploid = single or double number of formative genetic elements). It was easy to establish the haploid and diploid chromosome numbers in many species of animals and plants in which the chromosomes are relatively large and not numerous. In many other species the apparently simple task of counting chromosomes led different investigators to different results. In particular, the chromosomes of mammals proved to be not only rather small and numerous but also difficult to fix. Incomplete penetration of the fixing fluid into the tissues is common, resulting in conditions that lead chromosomes to clump together, making it difficult to decide whether an observed "chromosome" is really a single element or a conglomerate of two or even more actually separate chromosomes. These difficulties were at least in part responsible for many different chromosome numbers for man being reported by various early observers. Gradually, however, agreement seemed to have been reached. For about thirty years, most students of the subject were satisfied that in man the diploid number was 48 and the haploid number 24.

Unexpectedly, in 1956, the question became open again. Tjio and Levan, in Sweden, had prepared cultures of lung tissue from a number of aborted human embryos. Making use of a special method of handling the cells that leads to contraction of the individual chromosomes but spaces them widely apart, these cytologists obtained unusually clear preparations of fixed cells that, beyond any reasonable doubt, showed 46 chromosomes (Fig. 10). Some months later, Ford and Hamerton, in England, produced evidence that the germ cells in the testes of several men likewise possessed 46 chromosomes, present separately in the spermatogonia and as 23 pairs in the first of the two divisions of the spermatocytes (Fig. 10; see also Chap. 4). Since these discoveries, many cells from various organs or tissues have been studied. In nearly all, 46 chromosomes have been counted, most deviations from this number probably being caused by abnormal processes of chromosome distribution, which will be described below.

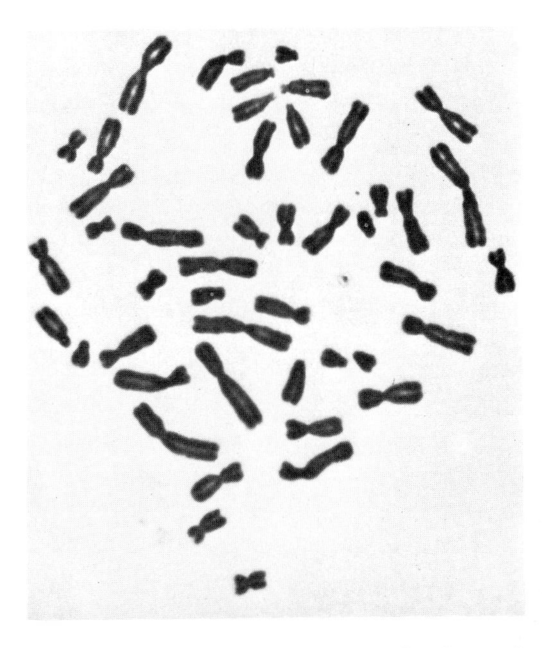

FIGURE 10
The chromosomes of a human male in a cultured white blood cell that is dividing. Each of the 46 chromosomes consists of two chromatids, held together by a kinetochore that has not yet divided. (Original photomicrograph by Dr. Margery Shaw.)

The chromosome numbers of other primates have also recently been restudied with modern methods. The numbers are determined in dividing cells from various tissues—particularly white blood cells, which can be obtained easily from blood samples of living individuals. The Rhesus monkey, whose diploid chromosome number had long been given as 48, is now known to possess only 42 chromosomes, but the original finding of 48 chromosomes in the chimpanzee has been confirmed and 48 is also the number in gorillas and orang-utangs. Other primate species have chromosome numbers as widely different as 42, 50, 60, 66, and 72 (Old World monkeys) and 34, 44, 46, 48, and 54 (New World monkeys).

Variations in Chromosome Numbers in the Individual. Normal mitosis during the development of a human being from the fertilized egg cell is to be expected to produce only cells with the diploid chromosome number characteristic of man, and all cells of his body should contain the same number. Apart from reservations concerning the accuracy of the earlier counts, it can be said that the diploid number of chromosomes has, indeed, been observed in many cells—for instance, in cells of the ovaries, testes, spleen, liver, lungs, and embryonic membranes; in white blood cells, bone-marrow cells, and cells of connective tissues.

There are, however, counts on record giving lower or higher numbers than the diploid for some cells. Some of these aberrant numbers are clearly the results of faulty technique or misinterpretation. Many, though, are correct: they are a result of the fact that mitosis, in spite of its marvelous accuracy, is a mechanism subject to mishap. It is known that occasionally two sister

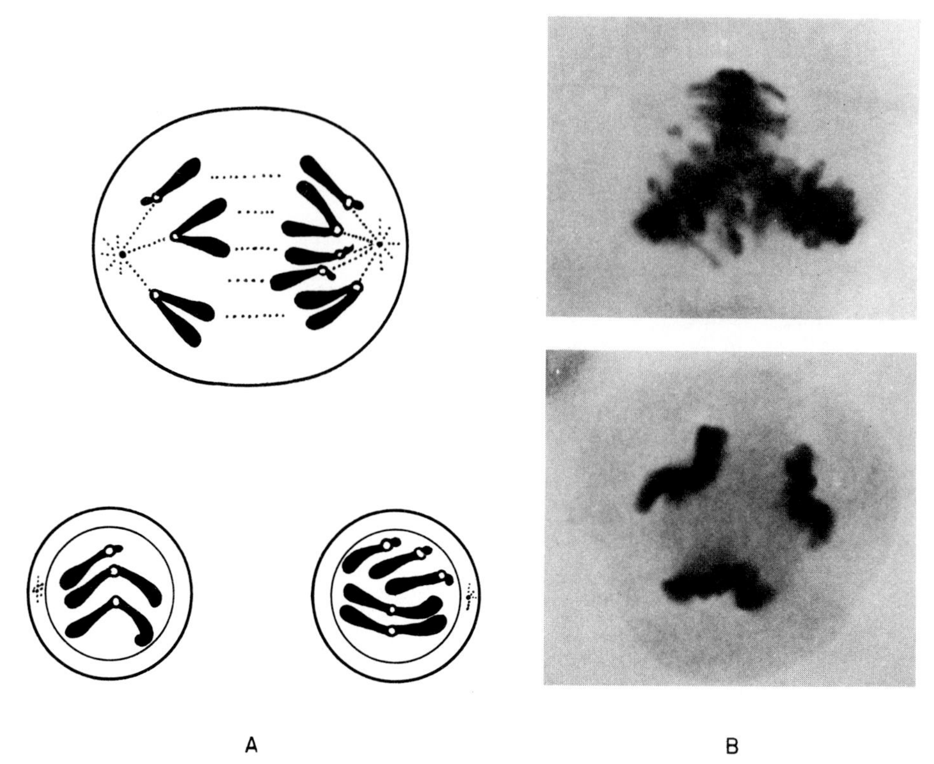

FIGURE 11
Abnormal distribution of chromosomes. **A.** Nondisjunction. *Above:* Two daughter chromosome strands go jointly toward the right spindle pole. *Below:* The resulting two daughter cells, one deficient for a chromosome, the other with an extra chromosome. **B.** Two cells with tripolar spindles. The chromosomes divide typically but are distributed irregularly to three instead of two daughter nuclei. (Original photomicrograph by Dr. R. Alava.)

chromosomes may fail to move to opposite poles (*nondisjunction*; Fig. 11,A). The result is a pair of unlike daughter nuclei, one with one chromosome more than the diploid number, and the other with one chromosome less. In a person who began life with 46 chromosomes, a cell with 47 and another with 45 would thus be formed. Once this has happened, the atypical chromosome numbers will be transmitted to each cell's subsequent generations; later mitosis in the cell with 47 chromosomes will result in two cells with 47 each, and mitosis in the cell with 45 chromosomes will result in two cells with this number. Further mitoses will keep constant the atypical chromosome numbers. It may also happen that one of the sister chromosomes is not included in either nucleus, but rather is left in the cytoplasm where it degenerates. The result is that one of the two nuclei will have 46 and the other only 45 chromosomes.

Another type of "accident" results when a nucleus that contains 46 chromosomes goes through mitosis only up to the stage in which the chromosomes

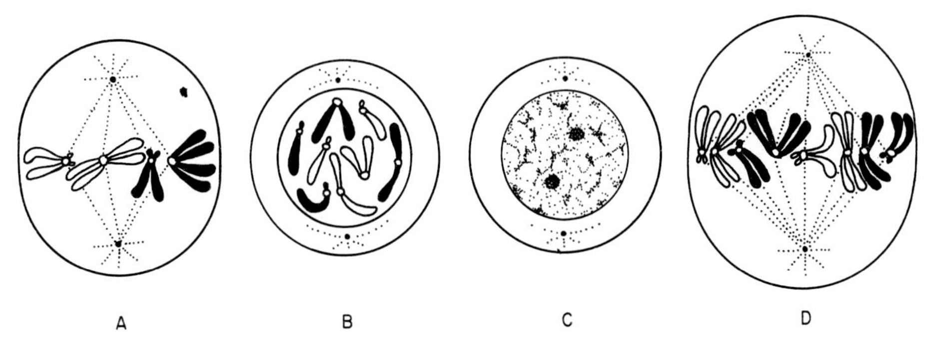

FIGURE 12
Formation of a cell with doubled chromosome number. **A.** 4 chromosomes on spindle. **B.** Breakdown of spindle; chromosomes have divided. **C.** Formation of resting nucleus. **D.** Subsequent mitosis; 8 chromosomes, ready to divide.

and their kinetochores become duplicated (see Fig. 12). Owing to breakdown of the spindle mechanism, separation of sister chromosomes does not take place; and when the new nuclear membrane forms, it encloses the 92 chromosomes. When mitosis occurs again, each of the 92 chromosomes behaves typically and duplicates itself (Fig. 13) so that the two daughter nuclei receive a "tetraploid" number of chromosomes. Tetraploidy is not always due to unusual accidental circumstances. For example, many cells of the liver regularly have 92 chromosomes but not as the result of breakdown of the spindle mechanism. Rather, in diploid liver cells mitosis may proceed normally, but be unaccompanied by cell division. The two sister nuclei, which have been left in the single cell, later fuse, producing a tetraploid nucleus. In turn, in such a tetraploid cell, omission of cell division at the time of mitosis results in a cell with 184 chromosomes. Such octoploid cells are also regularly found in normal livers.

In various tissues cells are occasionally seen that have considerably fewer than 46 chromosomes. One of the mechanisms that account for their origin depends on the occurrence of spindles having more than two poles (Fig. 11, B). If, for instance, a tripolar spindle is formed, the chromosomes in mitosis are distributed irregularly to three instead of two daughter nuclei; if the cell then divides into three instead of two daughter cells, each of these will have a hypodiploid chromosome number. Hyperdiploid numbers may result from multipolar divisions in tetraploid cells or other cells whose chromosome numbers are greater than the diploid.

Although reproduction of cells having aberrant chromosome numbers does occur, as has been mentioned in this section, it is likely that, except for tumor cells, most cells with aberrant chromosome numbers have low viability—that is, ability to survive—and either disintegrate without further division or, at most, grow and reproduce more slowly than cells with a normal chromosome

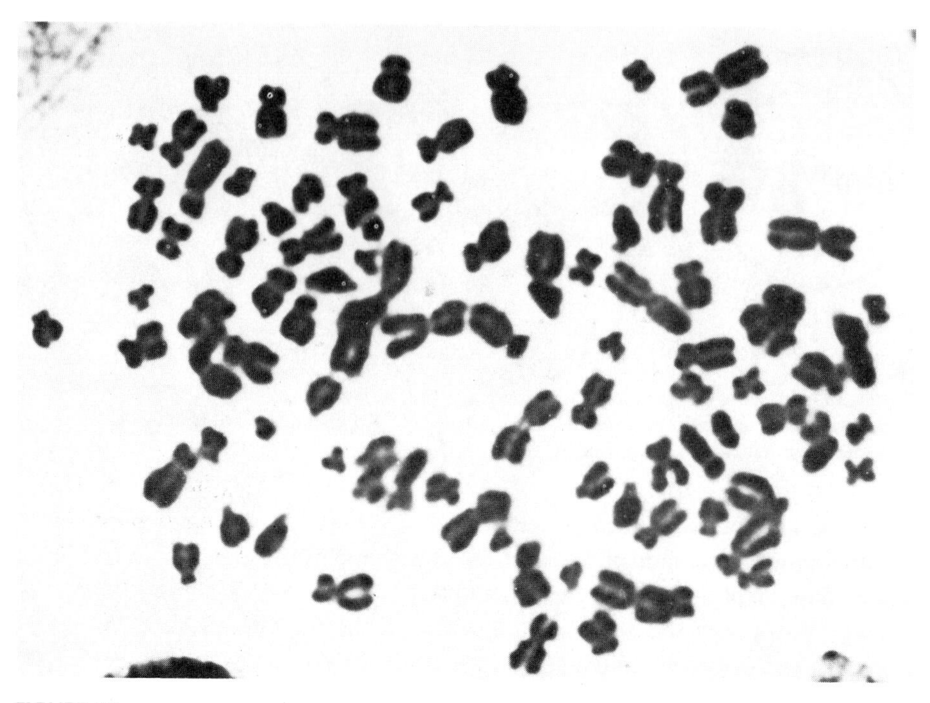

FIGURE 13
A tetraploid cell with its 92 chromosomes from culture of lung tissue of a male fetus. (Original photomicrograph by Dr. J. H. Tjio.)

complement. This view is supported by the results of an experiment in which human cells without nuclei, and hence without chromosomes, were produced in tissue cultures: they retained many normal features of cellular behavior for nearly two days, but then degenerated abruptly. Although this is admittedly an extreme example, the permanence of the body seems to depend on the basic constancy of the diploid chromosome constitution.

The Chromosome Set. A careful study of the diploid chromosomes in man shows (1) that they are of different sizes and shapes; and (2) that, with one exception there are two of each type of chromosome in each nucleus. Thus, in the cell whose chromosomes are arranged in pairs in Figure 14, each chromosome of the first pair is very large, and the arms of each are almost the same length. The chromosomes of the next pair are about the same length as the first, but with unequal arms; the third are the same shape as the first, but smaller; and so on. In good preparations it is not difficult to assign each chromosome to one of the seven groups, named A through G, which contain from 2 to 8 pairs. As summarized in Table 1, these groups are distinguishable by the length of their chromosomes and the position of the kinetochores, which are either more-or-less in the middle of the chromosome (median),

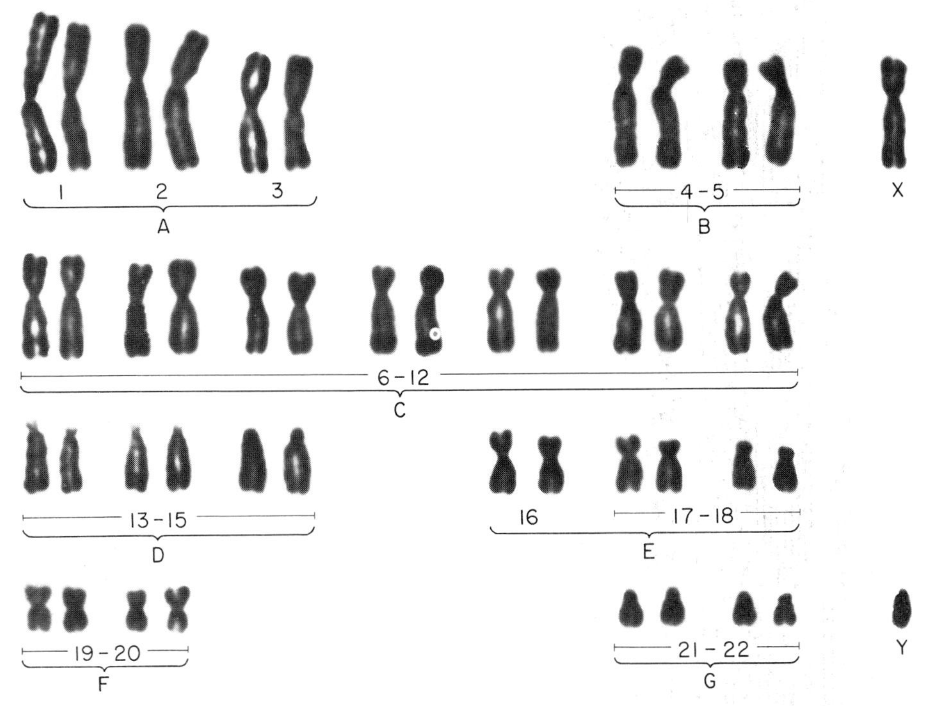

FIGURE 14
The chromosomes of the cell shown in Fig. 10, cut from the photomicrographic print, arranged in pairs, and grouped according to sizes and relative lengths of arms. Such an arrangement is called a *karyotype*. Designation of the seven groups by letter is conventional; within groups the numbering system is arbitrary except where there are clear differences in lengths of chromosome arms. (Original photomicrograph by Dr. Margery Shaw.)

somewhat off the middle (submedian), far off the middle (distal), or nearly terminal (acrocentric; the term acrocentric whose etymology would suggest that the kinetochore is *at* the extremity, does not do justice to the fact that a very small chromosomal section lies beyond the kinetochore.) Within some of the groups several of the chromosomes can be clearly shown to belong to pairs of "homologous" chromosomes. There is, however, a certain degree of variability from cell to cell in the lengths of the chromosomes and the relative lengths of their arms. Given the usual methods of preparing cells for observation, this variability makes the assignment of pairs difficult or impossible in those groups whose chromosomes are of very similar appearance. Thus, the four chromosomes in group B are so similar that the assignment to pairs is arbitrary. The same holds true for the large group C, consisting of chromosome pairs 6 to 12 plus the X-chromosome (see below); for group D, consisting of pairs 13 to 15; for group F, consisting of pairs 19 and 20; and for group G, consisting of pairs 21 and 22. It may be added that some chromosomes frequently exhibit minute terminal structures called satellites, or show specific

TABLE 1
Assignment of Human Chromosomes to Groups A through G.

Group	*Chromosome Numbers*	*Length*	*Approximate Position of Kinetochore*
A	1–3	long	median
B	4–5	long	distal
C	6–12, X	medium	submedian
D	13–15	medium	acrocentric
E	16–18	short	16, median; 17, 18, submedian
F	19–20	shorter	median
G	21, 22, Y	very short	acrocentric

Source: After Turpin and Lejeune.

constrictions apart from those caused by the presence of the kinetochore, which may help in their identification. Other means of identification are provided by experiments in which radioactive building blocks of DNA are incorporated earlier in the replication period by some chromosomes than by others.

Recently, Caspersson and his associates discovered that staining of the DNA of chromosomes with certain fluorescent compounds such as quinacrine mustard and quinacrine hydrochloride does not occur uniformly over a whole chromosome but affects preferentially specific regions of the different chromosomes. It results in patterns of cross-banding of the chromosomes that are different for each kind. This makes it possible to characterize individual chromosomes, not only by the conventional methods of staining, but also—more discriminatingly—by the patterns of fluorescent staining. Similar, but not identical, specific patterns of stained chromosomes have also been observed after a special pretreatment followed by Giemsa staining, one of the classical staining methods used in cytology (Fig. 15). These findings remove the uncertainties of identifying specific chromosomes within groups whose chromosomes are morphologically very similar. Each human chromosome and each of its arms can now be recognized individually and their involvement in chromosome abnormalities (which will be discussed later in this book) can be defined in unequivocal terms.

The exception—also encountered in many animals—to the rule that human chromosomes occur in pairs is regularly found in the male. In the cells of females, each chromosome can be matched with another, but in the cells of males, there are two that are unlike any others. The smaller of these, the *Y-chromosome,* is present exclusively in males. The other, the *X-chromosome,* occurs as a single chromosome in males, but twice—as a pair—in females. These two types of chromosomes are called *sex chromosomes,* and their

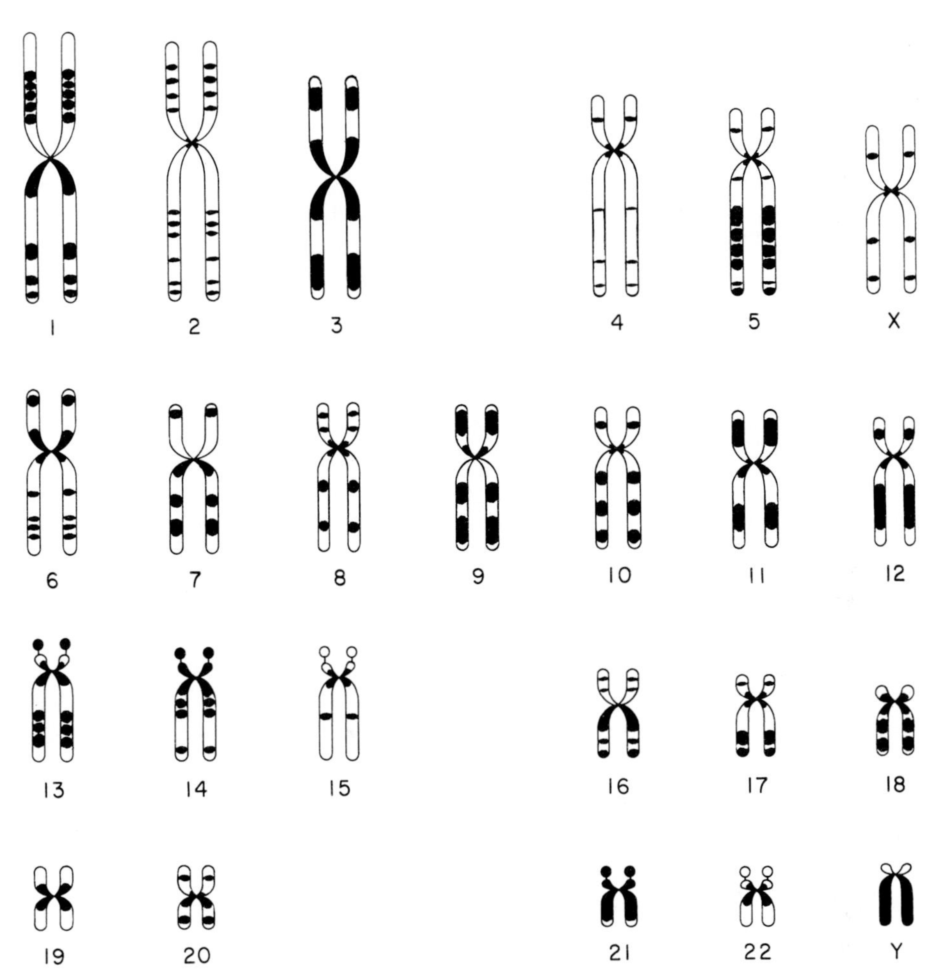

FIGURE 15
Diagram of most reliable banding patterns for each chromosome, seen after use of a special staining method. (Modified from Fig. 3, Drets and Shaw, *Proc. Nat. Acad. Sci.*, **68**, 1971.)

significance will be discussed in the chapters on sex linkage and sex determination: the "regular" chromosomes are called *autosomes.* In summary, there are 22 pairs of autosomes in human beings, and a twenty-third pair of sex chromosomes, which are alike in the female but unlike in the male.

We have seen that 23 of the 46 chromosomes in the cells of a human body originally come from the egg formed within the mother, and the other 23 from the sperm produced by the father. It should now be made clear that the 23 chromosomes which an egg or a sperm contains are not an arbitrary assembly of 23 chromosomes, but consist of one of each of 23 different kinds. Such an assortment is called a *chromosome set.* Every human being contains within his body (or somatic) cells two sets of 23 different chromosomes—one set derived from his mother, the other from his father.

Variations in Size and Structure of Chromosomes. The observed total length of chromosomes of the same kind and the ratio of their arm lengths vary considerably from cell to cell. This is usually the result of variable contraction of the chromosomes, which depends on incidental circumstances prevailing during cytological preparation. In addition to these random variations there is evidence for stable differences in the same kind of chromosome from person to person. This is most clearly established for the size of the Y-chromosome. Within an individual its length varies little from cell to cell and the same holds for the Y-chromosomes of various males of common male descent such as father and son or brothers. Different, unrelated males may, however, have Y-chromosomes of significantly different lengths (Fig. 16). Thus, there are persons whose Y-chromosome is only two-thirds as long as the mean length of the four group F chromosomes in the same cells and others fifty per cent longer than this mean. These size differences sometimes may reflect an intrinsic differential contraction of the Y-chromosomes, but the quinacrine-fluorescent-staining technique has shown in some cases that the size differences may actually result from there being different amounts of chromosomal substance.

The Y-chromosome does not contain genetic material necessary for the normal development or the well-being of human beings: it is absent in females. Whatever the biological function of much of its content, variations in its amount seem to be without obvious consequences. It is noteworthy that some racial groups differ from others in the average length of the Y-chromosome. In a study of five populations—Japanese, Jews, American blacks, Asian Indians, and Caucasian non-Jews—the mean length of the Y-chromosome in Japanese was significantly greater than that of the other groups and that of the Caucasian non-Jews significantly less than that of the others.

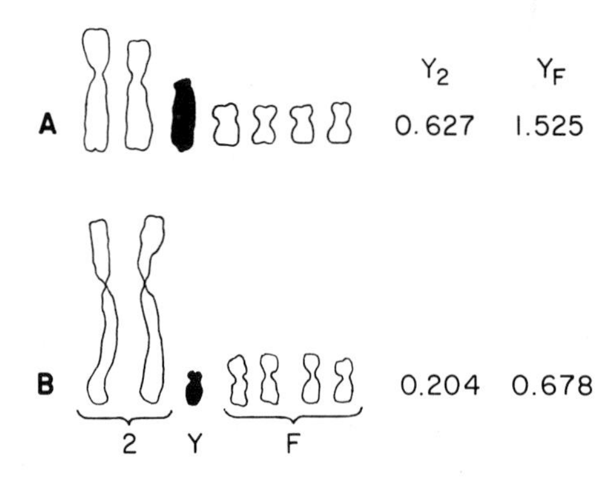

FIGURE 16
Variations in length of Y-chromosomes (black in the sketches) of two unrelated men. Each series shows the relative sizes of (left to right) chromosomes 2, chromosome Y, and those of group F (chromosomes 19 and chromosomes 20). Y_2 and Y_F are indices giving the lengths of the Y-chromosomes relative to the lengths of chromosomes 2 or to the mean lengths of chromosomes F, respectively. **A.** The longest Y-chromosome in the study population. **B.** The shortest Y-chromosome in the study population. (After Cohen, Shaw, and MacCluer, *Cytogenetics*, **5**, 1966.)

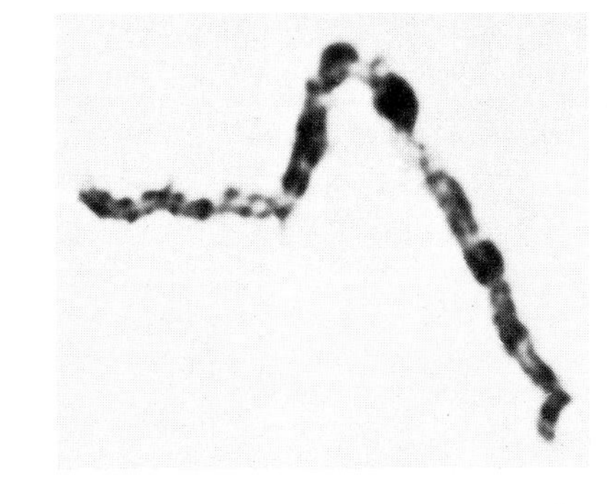

FIGURE 17
Human chromomeres. Two closely paired homologous chromosomes of an immature male germ cell. *Above:* Photomicrograph. *Below:* Diagram of the chromomere sequence. (Yerganian, *Amer. J. Hum. Genet.*, **9**, 1957.)

The Linear Order of Chromosomal Structures. A careful study of chromosomes in certain stages of the immature germ cells, in which they are relatively uncoiled long threads, has shown that these threads appear to be covered by a succession of fine beads, the *chromomeres* (Fig. 17). These chromomeres were found to be constant features of each chromosome. They are of different sizes and are arranged differently on each of the 23 pairs of chromosomes. Thus, a chromosome may have a small chromomere at one end, then a series of four slightly larger ones, then two smaller ones, and so on along the length of the chromosome. This is shown in the figure of a specific human chromosome in which it is also seen that the distances between successive chromomeres vary, as do the chromomeres' staining capacities (Fig. 17).

The patterns of chromomeres observed in fixed preparations may not represent well the actual structure of living chromosomes. But their constancy from cell to cell shows at least that constant and linearly arranged structural peculiarities exist in each chromosome and that these determine where, after fixation, a chromomere should appear and what its size and staining properties will be.

GENES

The most detailed insight into the linear arrangement of specific structures along the length of the chromosomes has been gained from study of the so-called giant chromosomes of the salivary glands of fly larvae of various species. These chromosomes, instead of resembling threads, look like rather wide

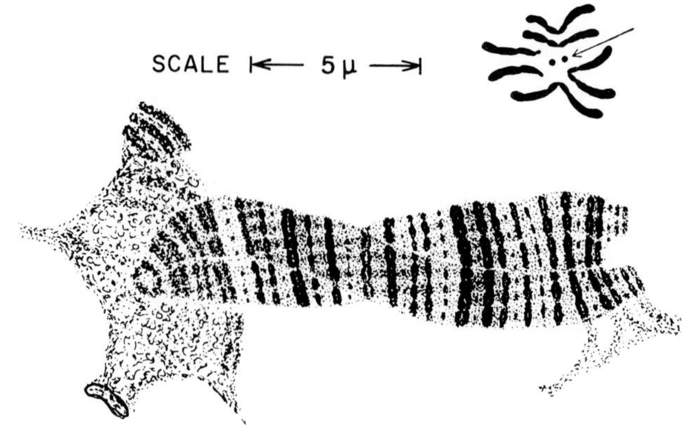

FIGURE 18
Chromosomes of the fly *Drosophila melanogaster. Above:* The 8 chromosomes of an ovarian cell. *Below:* The two chromosomes that are dotlike in the ovarian cell as they appear, at the same magnification, in the cell of a larval salivary gland. Note the identical type of cross banding of the two paired chromosomes. (Bridges, *J. Hered.*, **26,** 1935.)

cylinders marked by many crossbands or discs (Fig. 18). Just as the size of chromomeres and distances between them vary, so the bands of the salivary gland chromosomes may be thick or thin, close together or far apart; and all this in a manner constant and characteristic of each chromosome of a set, so that it is possible to number every band. Such microscopically visible linear differentiation of the chromosomes is highly suggestive of the linear arrangement of entities representing hereditary properties as postulated by Roux and demonstrated by Morgan. These entities are now called *genes.* It is natural to ask: Are the bands or the interband regions the genes themselves or are they, at least, associated with them? And the answer is: There is some evidence that the interband regions are associated with the genes; less evidence that they are identical.

The Localization of Genes. An introduction to human genetics is not the place to give a detailed account of the evidence for the existence of genes and the localization of each gene at a specific point on a specific chromosome. Such a task belongs to a general treatise on genetics. Since, however, the concept of genes underlies all discussions of human genetics, an example of the experimental, factual basis on which the concept is based will be given here. This example is only one of many investigations of plants and animals that, either by similar or by fundamentally different methods, have led to the recognition that the genes are localized in the chromosomes, and in an orderly manner.

Localization by Means of a Deficient Chromosome. H. J. Muller (1890–1967) discovered that X-rays may, besides producing other effects, permanently remove sections of chromosomes. In the fruit fly, *Drosophila melanogaster,* such sections seldom include the ends of a chromosome but are taken out somewhere between them. The remaining end pieces may fuse together at their breakage points, resulting in a chromosome that has a deletion. It is not possible to direct X-rays in such a fashion as to excise any desired section of a chromosome. Rather, the experimenter irradiates the whole fly, after which some of its cells show one or more deletions in different chromosomes or in different regions of the same chromosome. The deletions are of various lengths, some so short as to be hardly discernible under the highest magnification and some so large as to leave only the extreme end sections of the original chromosome.

Most offspring of an X-rayed fly are normal, but a few are not: some of the latter have an abnormality called Notch wing. Fruit flies normally have wings whose outline is smooth and continuous (Fig. 19, A); in Notch flies the wings look as if notches had been cut out of them (Fig. 19, B). This Notch character is transmitted to later generations. Apparently, the irradiation of the parent flies produces a change in some of the gametes, which causes loss of ability to produce normal-winged individuals. It is possible to study the chromosomes of the Notch flies under the microscope. There is no visible change in the chromosomes of some; the X-rays presumably caused alterations of the hereditary entities located in the chromosomes at a submicroscopic level. In many,

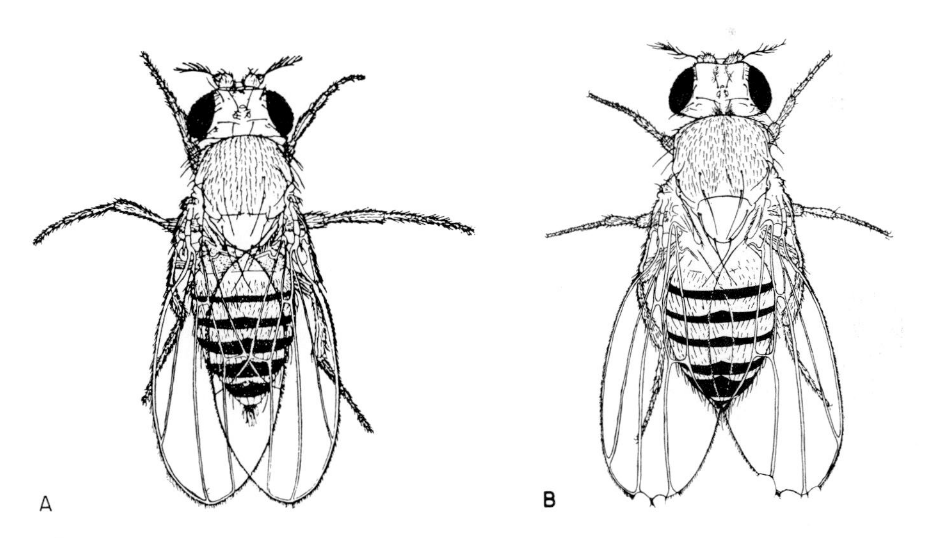

FIGURE 19
Drosophila melanogaster. **A.** Normal female. **B.** Notch female. (A, Morgan, Bridges, and Sturtevant, *Bibliograph. Genet.*, **2,** 1925; B, Mohr, *Zeitschr. Abst. Vererb.*, **32,** 1923.)

however, a specific rod-shaped chromosome shows absence of a specific region (Fig. 20, A). From this, it can be concluded that the ability to cause the development of a normal wing depends on the presence of an entity in the region that is missing from the deficient chromosome. This entity is called a "gene for normal wing outline," or a "gene for not-Notch wings." It may be represented by the symbol N^+ (+ = normal).

Genes and Chromosome Bands. In the giant chromosomes of the salivary gland cells of *Drosophila,* the location of the gene N^+ may be narrowed down to a very short section. Whenever a visible loss of chromosomal material accompanies the loss of the gene for not-Notch, at least one band in the chromosome, referred to as the 3C7 band, is absent. Some of the deficient chromosomes show loss of this band only (Fig. 20, B), others the loss of a larger number of bands; but all have lost the 3C7 band (Fig. 21). It follows that either this band or its immediate interband neighborhood is the seat of the gene for not-Notch. In similar ways, other genes have been localized, as indicated in Figure 21. Thus, the gene w^+ ("not-white"), whose loss causes the flies to have white, instead of their normal red, eyes lies to the left of that for not-Notch, within the region indicated by the bands 3C1 to 3C3. Between the genes for red eye and normal wing outline, at the 3C4 band, lies the gene rst^+ ("not-roughest"), whose absence leads to an irregular rough appearance of the eye surface. To the right of N^+ is the gene dm^+ ("not-diminutive"), which must be present if bristles are to develop to normal size. Its location in the chromosome is somewhere between 3C7 and 3D3.

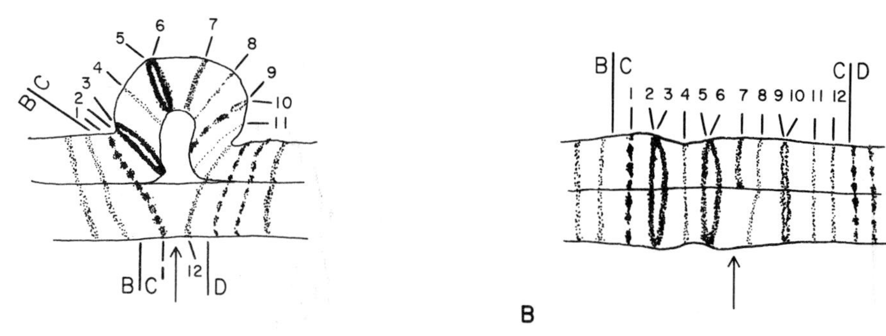

FIGURE 20
Notch deficiencies in the chromosomes of two different strains of *Drosophila melanogaster*. A short section of a chromosome pair in a salivary gland nucleus is shown. (The chromosomes of *Drosophila* have been divided into 102 main sections [1 to 102], each of which is divided into six subsections [A to F]. Within each subsection, a variable number of individual bands are labeled 1, 2, etc. In these figures, subsection C of section 3 is presented ["3C"]). **A.** The upper chromosome is normal, containing all bands, 3C1 through 3C12. In the lower chromosome the bands in sections 3B and 3D and bands 3C1 and 3C12 are present, so that pairing has taken place. However, bands 3C2 through 3C11 are absent. **B.** Only band 3C7 is missing from the deficient chromosome.

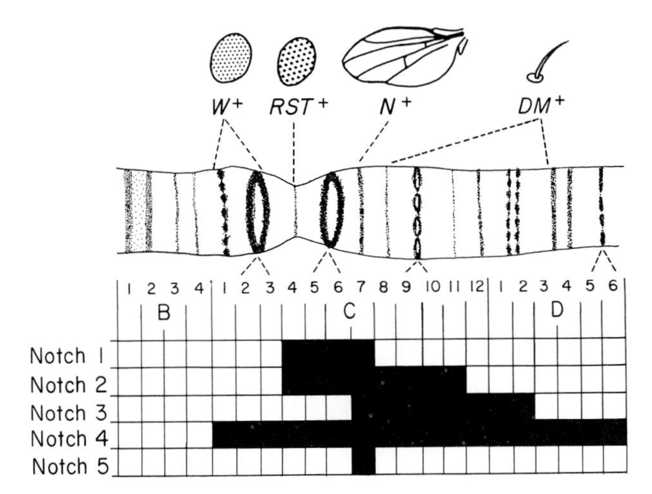

FIGURE 21
Above: Linear arrangement of four genes of *Drosophila* in relation to bands in chromosome sections 3B–D. *Below:* The extent of five different Notch deficiencies is indicated by the length of the five black bars. The bands located in these sections are absent in the Notch strains; band 3C7 is absent in the chromosomes of all five strains. See text also. (After Slizynska, *Genetics,* **23,** 1938.)

In *Drosophila,* hereditary differences, such as Notch and normal wing, or rough and smooth eye, can by no means always be traced back to presence or absence of a region or band in a chromosome. Cases of visible loss have been selected for the foregoing analysis, but most inherited abnormalities are found in flies whose chromosomes have all bands present. Furthermore, each different type of abnormality is not distinguished by the presence or absence of some genic content of the chromosomes; rather, all genes are usually present, though in modified form. Notch wings can be caused by the absence of the 3C7 band, but an invisible modification of the gene associated with 3C7 can also cause them. Successful localization of genes has not been restricted to *Drosophila* and other experimental organisms. In man, too, a beginning has been made in the determination of the linear arrangement and approximate location of genes (see Chap. 15).

We have seen that an individual possesses a pair of each kind of chromosome. Since each chromosome contains a series of *loci* (from the Latin *locus* = place), occupied by different genes, and since the two chromosomes of a pair have the same genes, it follows that each individual has two of each kind of gene. Just as each of the 23 kinds of chromosomes has a partner, so has each gene, the two forming a pair of genes occupying homologous loci on a pair of homologous chromosomes. The two partner genes are called *alleles,* a term that (in the singular) means "the other one." They may be compared to twins, and the genes making up the hereditary endowment of an individual may be compared to a population in which every individual has a twin.

At every genic locus in a pair of autosomes, an individual may have either two identical alleles or two different ones. Given two different alleles A^1 and A^2 for the locus, there are three possibilities for genetic makeup: an individual may carry the allele A^1 in both his chromosomes; he may carry A^1 in one chromosome and A^2 in the other; or he may carry A^2 in both. We speak of the first and third types of person, symbolized as A^1A^1 and A^2A^2, as *homozygotes,* since the alleles are alike; of the second type of person, A^1A^2, as a *heterozygote,* since the alleles are different. An individual may be homozygous at some loci and heterozygous at others.

Genes and Alleles. The different genes that occupy different loci either in the same chromosome or in different chromosomes produce different effects and so do the different alleles of a given gene. This may be illustrated by the effects of genes that direct the synthesis of chains of amino acids—polypeptide chains—that are the constituents of enzymes and other proteins. Different genes produce widely different polypeptides. Thus, one gene leads to the synthesis of the human growth hormone, which consists of a chain of about 188 amino acids, while another gene leads to the synthesis of the beta chain of hemoglobin, which consists of 146 amino acids. Moreover, the detailed compositions of the amino acid chains organized by different genes differ greatly. If, for example, we list the first ten amino acids of the growth hormone and of the normal beta-hemoglobin chains, we find no similarity between the two sequences (Fig. 22, Ib, IIb_1). In contrast, the amino acid chains synthesized by different alleles of the same gene are often very similar. Many alleles of the beta-hemoglobin gene, for instance, lead to polypeptide chains that are identical in type and sequence for 145 out of the 146 amino acids. The differences in the chains produced by different alleles usually consist in the substitution of one single amino acid somewhere along the sequence by another (Fig. 22, IIb_1, b_2, b_3, b_4). For example, the normal beta-hemoglobin possesses the amino acid histidine in position 2 of its sequence and glutamic acid in position 6. (Fig. 22, IIb_1). The rare allele $Hb^{Tokuchi}$ causes the substitution of histidine in position 2 by tyrosine (b_2). Two other alleles affect the amino acid in position 6. Instead of glutamic acid Hb^S places valine (b_3) and Hb^C lysine (b_4) in this position.

In the foregoing discussion, genes and alleles have been characterized by the kinds of polypeptide chains that are the products of their activities. It would be desirable to define genes and alleles in terms of their actual molecular structure. It is known, of course, that genes are macromolecules of DNA

I. Human Growth Hormone

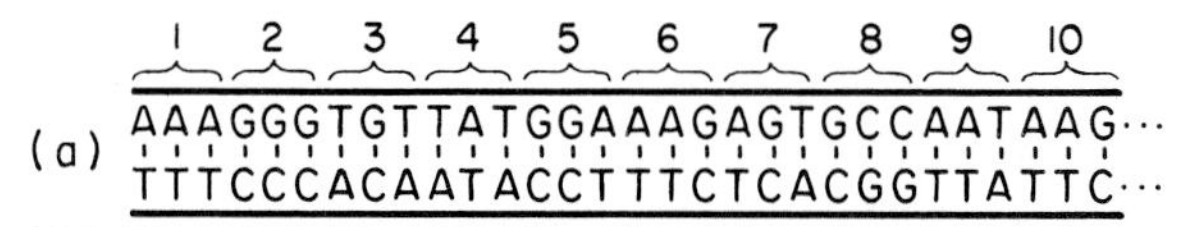

II. Beta Hemoglobin

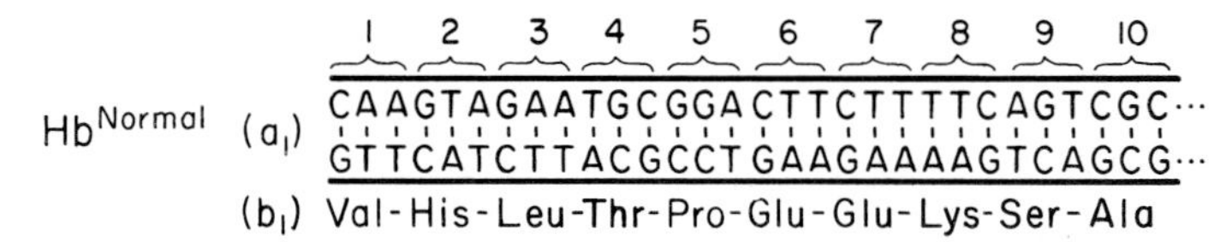

2

Hb^{Tokuchi} (a_2) ···ATA···
···TAT···

(b_2) -Tyr-

6

Hb^{S} (a_3) ···CAT···
···GTA···

(b_3) -Val-

6

Hb^{C} (a_4) ···TTT···
···AAA···

(b_4) -Lys-

FIGURE 22
A short terminal section of each of two different genes and of four alleles of one of the genes, and the corresponding terminal sections of the respective polypeptide chains. See text for discussion of some arbitrariness in base assignments. In the diagrams of base-pair sequences the upper strand of bases is coded for amino acids. I. Growth hormone. (a) DNA sequence of 30 base pairs (= 10 triplets); (b) polypeptide sequence of 10 amino acids. II. Beta hemoglobin. (a_1, a_2, a_3, a_4) DNA sequences of 30 base pairs in four alleles; (b_1, b_2, b_3, b_4) polypeptide sequences of 10 amino acids determined by the four alleles. a_1, a_2, a_3, and a_4 are identical except for base-pair substitutions in triples 2 and 6, respectively. Correspondingly, b_1, b_2, b_3, b_4 are identical except for amino acid substitutions in positions 2 and 6, respectively. The four beta polypeptides result in hemoglobins A_1, Tokuchi, S, and C, respectively.

and that their specificity is encoded in the sequence of their building blocks, the base pairs A-T, T-A, C-G and G-C, where A and G stand for the purine bases adenine and guanine, and T and C for the pyrimidine bases thymine and cytosine. A gene consists of two, helically twisted, phosphate-sugar backbone chains between which the base pairs are arranged like rungs on a ladder. (Since each base is paired with one specific other base – A with T and C with G – it is sufficient to designate a sequence of base pairs by a sequence of single bases. Moreover, in the assembly of polypeptide chains the sequence of bases along only one of the phosphate-sugar chains is utilized, without reference to the complementary sequence along the other chain. For present purposes, however, we will usually refer to base pairs).

The number of base pairs varies from gene to gene from less than one hundred to more than one thousand with perhaps an average of several hundred. Genes are distinguished from one another not only by the numbers of base pairs but even more significantly by their sequences.

There is a class of genes for which, in a slightly indirect way, these sequences have been derived by analytical chemical procedures. These are the genes concerned with the transfer of specific amino acids to their places within a polypeptide chain. The transfer DNA genes in the chromosomes are "transcribed" into transfer RNA molecules, whose single-stranded sequences of bases are complementary to the sequences of the transcribed strand of the DNA double helix. Therefore, knowledge of the sequence of bases in a transfer RNA molecule indirectly provides knowledge of the sequence of base pairs in its transfer gene.

For most other genes the sequence of base pairs is not known unequivocally, but some possibilities of genic structure can be deduced from the amino acid sequences of the polypeptides produced by a gene. These deductions do not lead to unique assignments of genic structure but leave open a variety of alternatives. The reason for this lies in the properties of the genetic code. It is known that each amino acid is represented in the DNA molecule by a sequence of three base pairs and that a sequence of amino acids is "colinear" with a sequence of base pair triplets, i.e., that the sequence of amino acids is a reflection of a corresponding sequence of groups of three consecutive base pairs. Since within a triplet each base pair may be represented by any one of the four kinds of base pairs, there are a total of $4 \times 4 \times 4 = 64$ different triplets. This is more than the number of different amino acids present in polypeptides, which amount to only about twenty. The result of this discrepancy between the number of possible base-pair triplets and the number of amino acids is that most amino acids are specified not by one specific triplet sequence but by two, four, or even six different sequences. Thus, the amino acid tryptophane is represented uniquely by the triplet bases ACC; but phenylalanine is represented by either AAA or AAG, and leucine by any one of the six triplets

AAT, AAC, GAA, GAG, GAT, and GAC. It is the existence of more than one triplet coding for the same amino acid—the "degeneracy" of the triplet code—that makes it impossible at present to derive a uniquely specific sequence of base pairs for given genes from knowledge of their polypeptide products. Nevertheless, for purposes of illustration of the differences between different genes and differences between different alleles of a gene, specific base-pair sequences have been assigned in Figure 22 (Ia, IIa_1–a_4) to the first ten triplets of the genes for growth hormone and beta hemoglobin, and for several alleles of the latter gene. These sequences are compatible with our knowledge of the relation between triplet types and amino acids but other sequences are likewise compatible. It is seen that the genes for growth hormone and hemoglobin are quite different but that the alleles of the hemoglobin gene differ from one another in one or another single triplet only. This sharp picture of the molecular constitutions of genes and their alleles becomes somewhat blurred when many genes are studied. Some genes, although at different loci, show certain similarities in base sequences, and some alleles of genes differ from one another by more than a single triplet. These exceptions find their explanations in terms of evolution, indicating common ancestry of now widely divergent genes or indicating divergence of alleles.

The Number of Genes in Man. More than 1800 different human genes are listed in McKusick's catalogs of genes. New ones, responsible for inherited normal and abnormal traits, are discovered each year. Such data can place a lower limit for the total number of genes but not an upper limit. In other organisms attempts have been made to obtain more specific information regarding number of genes. In *Drosophila,* the bands or interband regions in the salivary gland chromosomes have been counted, under the assumption that these numbers are close to those of the chromosomal genes. This assumption is based on genetic evidence concerning the numbers of genes in small sections of chromosomes and in whole chromosomes. Or, by a less direct approach, investigators have determined (a) the average frequency of genic changes (mutations) of a single gene and (b) that of the sum of all genes. The ratio b : a then gives an estimate of the total number of genes. These methods led to estimates of from 5,000 to 15,000 for the haploid set of genes in *Drosophila.* A different approach has been used in the bacterium *Escherichia coli.* Here, several thousand different enzymes and proteins are known, which, according to the rule "one gene–one polypeptide," suggests the existence of several thousand genes. Still another approach that at first sight seems the most direct one, rests on the determination of the amount of DNA in a cell. A simple calculation can convert this amount into the number of base pairs that are represented by it. Dividing the number by three yields the number of base pair triplets and dividing this number by that of the triplets required to code for an average length of polypeptide chains yields an estimate of the number of genes. When the amounts of DNA per cell were determined in different organisms wide variation from species to species was observed. Assuming an average polypeptide chain to consist of 200 amino acids, the DNA in a haploid nucleus of *E. coli* would suffice to code for 8,000 polypeptides, that of a haploid nucleus of *Drosophila melanogaster* for 65,000, of a toad for 7,000,000 of a frog for 14,000,000, of a lungfish for 95 million, of a mouse for 4,000,000, and of man for 5,000,000.

These calculated numbers cannot be indicative of the actual numbers of different genes in each species. Why should a frog nucleus have twice as many genes as that of a toad and why should both kinds of amphibians have more genes than mouse and man? Why should the amount of the lungfish's DNA surpass that of the toad by more than 13-fold and that of man by even more? And is it likely that the number of polypeptide chains that make up the proteins and enzymes of the multicellular animals listed is from 8 times to 12 thousand times greater than in the bacterium? On the contrary, it has been argued that the number of genes in the chromosome set of man and the number in other organisms are of the same order of magnitude. It may be objected that the human body has a higher organization than that of a fly or a bacterium and that man therefore must have a greater number of genes. But it is difficult to rank the complexity of two different organisms and, considering the great similarity in structure and function of cells from the most diverse groups of animals and plants, it may very well be that there are only minor differences between them in numbers and kinds of genes. The striking external differences between diverse species may largely be the developmental outcome of different interactions of similar genic reagents. To use an analogy: the same 26 letters of the alphabet may serve to compose either a nursery rhyme or a philosophical treatise.

What then is the meaning of the striking differences in amount of DNA in the nuclei of diverse species? Britten and Kohne have provided some data on this problem. They found by means of physical-chemical studies that the DNA of many, and perhaps all, species can be separated into two fractions: one fraction consisting of "families" of more-or-less like base-pair sequences that are repeated many times, and another fraction consisting of unique sequences. The latter may be made up of genes that are represented once or at most a few times only. The former would be genes that are present in many repetitions, some repeated perhaps a few thousand times, others more than a hundred thousand times. Within each family of repeated genes the degree of similarity of the individual genes may vary widely. Some of the repetitive families of genes may consist of rather closely similar genes, others of less similar ones. Although these findings account for some of the striking differences in DNA content of the nucleus in different species, it seems that the amount of redundant DNA is not enough to account for all differences. Moreover, there also seems to be within a given species an excess of unique-sequence DNA over that estimated from other considerations regarding gene numbers.

Another way of accounting for differences in total DNA of different species rests on the discovery, in microorganisms, of genes that do not code for the production of polypeptide chains but whose function is to produce molecules that are effective in regulating the activity of other genes. We may distinguish the *structural* genes, those that code for the proteins and enzymes of the structural organization of the cell, from the *regulator* and *operator* genes, which "turn the structural genes on and off." Little is known yet about nonstructural genes in man, but it has been suggested that such genes are numerous in multicellular organisms, where—in contrast to microorganisms—regulation of gene activity presumably is of great importance in differentiation and development in general. Nevertheless, it is difficult to envisage the presence of a hundred times as many regulatory genes in vertebrates as in *Drosophila* or a many-fold difference in the number of regulatory genes from species to species within the vertebrates.

In summary, we have no reliable information enabling us to estimate the number of genes in man. We may, rather arbitrarily, assume that the number of structural genes lies somewhere in the range from 10,000 to 100,000 (and refrain from any attempt to divine the number of nonstructural genes). Of course, this is a wide range, but it should be realized that these numbers represent, after all, a *relatively* limited variation. Without knowledge of various facts such as those found by Mendel and, later, in combined studies of inheritance and chromosomes, the question regarding a subdivision of the hereditary material might have been answered by as low an estimate as one—that is, no subdivisibility—or as high an estimate as hundreds of millions of structural genes.

A human individual receives, from his parents, a complete assortment of all genic loci in two sets of chromosomes, those of the egg and the sperm. Thus, the cells of his body harbor two assortments of genes or, assuming the correctness of the foregoing estimate, between 10,000 and 100,000 pairs.

PROBLEMS

1. The nucleus of an unfertilized human egg is approximately 25 microns in diameter. Calculate the total volume of the nuclei of all eggs from which the present generation of mankind originated.
2. If a newborn baby weighs three kilograms, how many times heavier is the child than the egg from which it originated?

3. Calculate the total volume occupied by the eggs that produced all the children born in a given year in the states of
   (a) California,
   (b) New York,
   (c) Nevada.
   (For number of births per year, consult *Vital Statistics of the United States, 1, Natality.* U.S. Dept. of Health, Educ., and Welfare, Public Health Service, Washington.)
4. During an abnormal mitosis of a cell with 46 chromosomes, the daughter chromosomes of one of the long chromosomes are included together in the same daughter nucleus. If nondisjunction also affects one of the short chromosomes, how many chromosomes may there be in the two daughter nuclei?
5. In a cell with 46 chromosomes an abnormal tripolar spindle is formed with three poles: I, II, and III. Eighteen chromosomes send daughter chromosomes to poles I and II, fifteen to poles I and III, and thirteen to poles II and III. What are the chromosome numbers in the resulting three daughter nuclei?

REFERENCES

Austin, C. R., 1961. *The Mammalian Egg.* 183 pp. Thomas, Springfield.

Austin, C. R., 1965. *Fertilization.* 145 pp. Prentice-Hall, Englewood Cliffs, N.J.

Austin, C. R., 1968. *Ultrastructure of Fertilization.* 196 pp. Holt, Rinehart, and Winston, New York.

Hartman, C. G., 1939. Ovulation, fertilization and viability of spermatozoa. Pp. 630–719 *in* Allen, E. (Ed.), *Sex and Internal Secretions,* 2nd ed. Williams and Wilkins, Baltimore.

Meyer, A. W., 1939. *The Rise of Embryology.* 367 pp. Stanford University Press and Oxford University Press, Stanford, California, and London.

Schultz-Larsen, J., 1958. *The Morphology of the Human Sperm.* 121 pp. Munksgaard, Copenhagen.

Shettles, L. B., 1960. *Ovum Humanum.* 79 pp. Urban and Schwarzenberg, Munich.

Spuhler, J. N., 1948. On the number of genes in man. *Science,* **108:**279–280.

Stern, C., 1959. The chromosomes of man. *J. Med. Educ.,* **34:**301–314 (also *Amer. J. Hum. Genet.,* **11,** 2 Part 2: 301–314).

Swanson, C. P., 1969. *The Cell,* 3rd ed. 150 pp. Prentice-Hall, Englewood Cliffs, N.J.

Turpin, R., and Lejeune, J., 1969. *Human Afflictions and Chromosomal Aberrations.* 392 pp. Pergamon, Oxford. (English translation of *Les Chromosomes Humaines,* 1965.)

Unnérus, Viveca, Fellman, J., and de la Chapelle, A., 1967. The length of the human Y chromosome. *Cytogenetics,* **6:**213–227

Wilson, E. B., 1925. *The Cell in Development and Heredity.* 1232 pp. Macmillan, New York.

3

GENIC ACTION

In the final analysis, all gene-dependent differences among human beings must be based on differences in the physiological processes of their cells. Since genes produce their effects by chemical interaction with other cellular constituents, much of the study of genic action belongs to the area of biochemical genetics.

Many genetic differences among human beings are described by reference to traits that can be distinguished by simple observation: tallness versus shortness, six-fingeredness versus five-fingeredness, straight hair versus curly hair, normal blood clotting versus bleeder's disease, color vision versus color blindness, and idiocy versus normal mentality. Sometimes simple observation constitutes an acknowledgement of obvious chemical differences (e.g., pigmentation versus albinism, the chemical distinction being the presence or absence of the pigment melanin). Often, however, the chemical basis of differences in appearance is far removed from the observable trait and requires special studies for its determination.

Genetic Control at the Molecular Level. Genes may produce their effects at different levels: within individual cells, in the specific morphology and physiology of organs, and in mental attributes that depend on the structure and the functioning of the central nervous system. The primary action of the structural genes, with which we will be mostly concerned, is to initiate a biochemical sequence that has been called the central dogma of molecular genetics, DNA→ m-RNA→ polypeptide chain. This scheme states that the DNA genes, which are chromosomal polynucleotides, serve as templates for the synthesis of molecules of messenger ribonucleic acid (m-RNA), which contain in somewhat differently coded form, or *transcribed,* the same information in their sequence of nucleotide bases that the DNA genes possess. Unlike the DNA genes, which remain as parts of the chromosomes, the RNA molecules separate from the chromosomes and pass from the nucleus into the cytoplasm. There, the message carried by them is "translated" into a sequence of amino acids that form chains of specific polypeptides, so called after the *peptide bonds* that join amino acids together.

A given kind of polypeptide chain may pair or join in higher multiples with polypeptides of the same kind or of other kinds, or may become part of other complex molecules. Thus, the proteins of a cell are derived expressions of the genes that determine the constitution of their component amino acid chains. An example is furnished by the normal human hemoglobins. Leaving aside types of hemoglobins produced only before birth, four different genes code for four different hemoglobin chains, α, β, γ, and δ. Each of these forms pairs (dimers) with its own kind, producing α_2, β_2, γ_2, and δ_2. Furthermore, the α_2 pairs may join with any one of the other pairs so that three types of hemoglobin are formed, $\alpha_2\beta_2$, $\alpha_2\gamma_2$, and $\alpha_2\delta_2$.

The assembly of hemoglobin molecules requires not only the various polypeptide chains that constitute the globin constituents but also an iron complex of porphyrin called heme. The cellular synthesis of heme consists of a series of steps under the control of genes other than those specifying the synthesis of the α, β, γ, and δ chains. The existence of genes for heme synthesis has partly been proven by genetic analysis and in addition is deduced from the discoveries of specific enzymes that catalyze different steps in the metabolic pathways which lead to heme.

The hemoglobins have been of fundamental importance in clarifying the relation between genes and their effects at the molecular level. At the same time, they have led to an understanding of the molecular basis of certain blood diseases. In many populations, particularly those of African stock, there are numerous individuals whose red blood cells take on a sickle-like shape when they are exposed to low oxygen tension outside the body (Fig. 23, A, B). Such persons fall into two classes. The majority are healthy and inside their bodies less than one per cent of the blood cells are abnormally shaped: they are said to possess the "sickle-cell trait." A small minority, however, has a severe

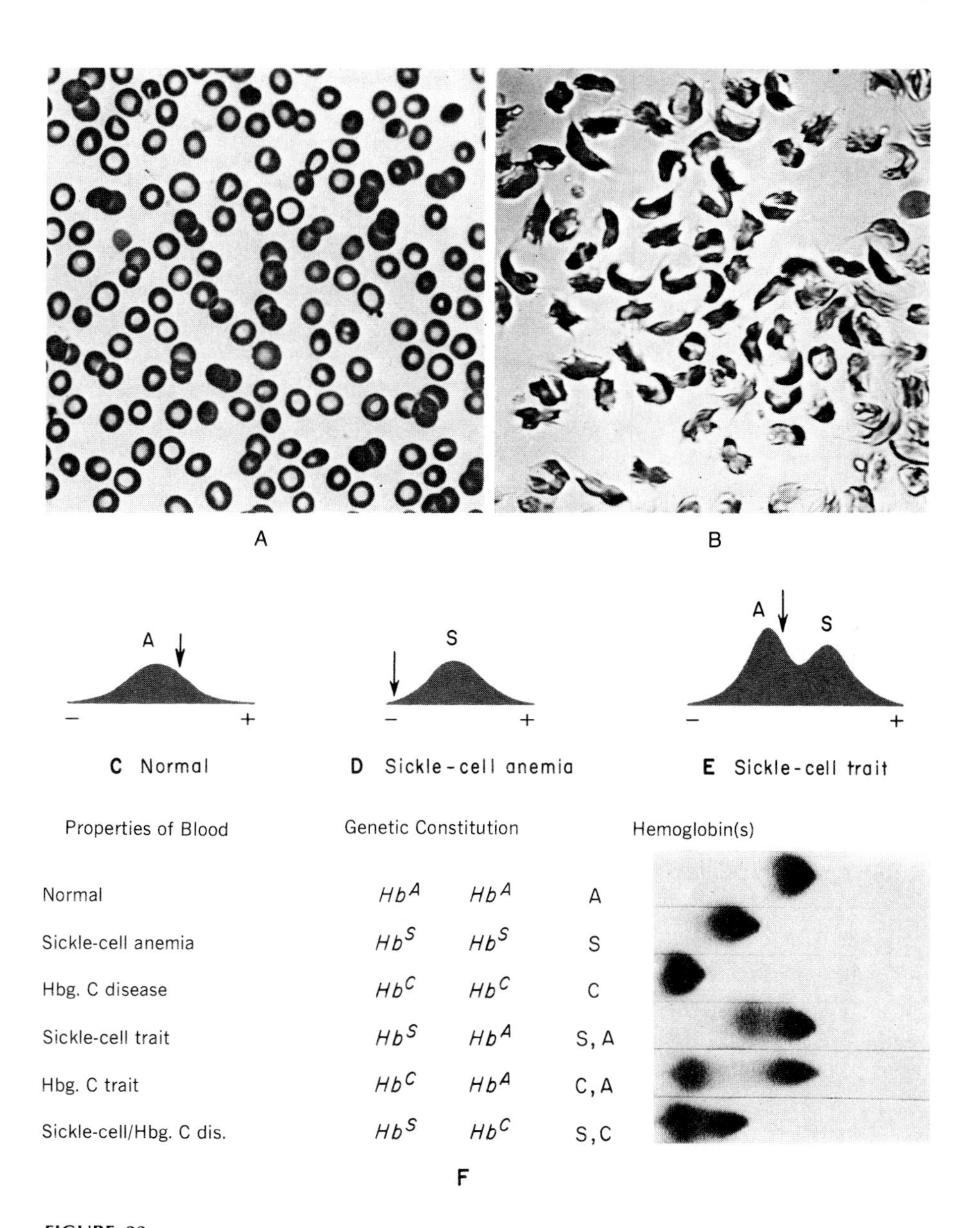

FIGURE 23
Red blood cells from **(A)** normal and **(B)** sickle-cell homozygotes. Movements of hemoglobins in an electric field **(C, D, E)**. The hemoglobins were dissolved and an electric current passed through the fluid. The black areas correspond to the position and amount of hemoglobin after its migration in the electric field. By using the original vertical axes (arrows in figure) as points of reference, it is seen that normal hemoglobin, A, has moved toward the negative pole, sickle-cell hemoglobin, S, toward the positive pole, and that the hemoglobin of the heterozygous individual became separated into two parts: about 60 per cent A and 40 per cent S hemoglobin. **F.** Movement of hemoglobins A, S, and C in an electric field on filter paper. The hemoglobins were initially placed on the edge at left. Different hemoglobins move with different speed. (A, B, originals from Dr. C. L. Conley; C, D, E, after Pauling, Itano, Singer, and Wells, *Science,* **110,** 1949; F. Conley and Smith, *Trans. Ass. Amer. Phys.,* **57,** 1954.)

anemia which often becomes fatal before those afflicted reach the age of reproduction, and more than one-third of their blood cells have abnormal shapes: they are called "sickle-cell anemics." The sickle-cell character is controlled by a pair of alleles. Individuals whose cells are not sickle-shaped are homozygous for an allele Hb^A ($Hb^A Hb^A$); possessors of the trait are heterozygous for the allele Hb^S ($Hb^A Hb^S$); and sickle-cell anemics are homozygous $Hb^S Hb^S$.

The molecular basis of the sickle-cell character began to be understood when it was discovered that a special kind of hemoglobin (S) different from that of normal individuals (A) is present in the blood cells of persons with the sickle-cell trait or sickle-cell anemia. The two kinds of hemoglobin may under certain conditions possess opposite electrical charges, so that if an electric current is passed through a solution that contains both types one type moves toward the negative and the other toward the positive pole ("electrophoresis"—Fig. 23, C-E). An $\alpha_2\beta_2$ hemoglobin molecule consists of two like halves, $\alpha\beta$, made up of about 4,000 atoms arranged to form 141 amino acids in the α chain and 146 in the β chain. A difference in a single one of the amino acids in the β chain constitutes the chemical difference between sickle-cell and normal hemoglobin: the latter contains glutamic acid; the former, valine (Fig. 24). The presence of a third allele at the *Hb* locus, Hb^C, causes another amino acid, lysine, to replace either glutamic acid or valine in the hemoglobin molecule. Ingram, the discoverer of the molecular difference between hemoglobins A and S, has remarked that "a change of one amino acid in nearly 300 is certainly a very small change indeed and yet this slight alteration can be fatal to the unfortunate possessor of the errant hemoglobin." There are, of course, many compounds synthesized in nature or in laboratories that have ill effects on man, and their chemical structure is often only slightly different from that of harmless or beneficial substances. Hence it is not surprising that the molecular entities called genes are at times responsible for alterations of molecular cellular material with far-reaching effects in development or functioning.

The difference between normal and sickle-cell hemoglobin resides in the β chains. More than 60 other abnormal β chains have been discovered as well as many abnormal α chains. They are all dependent on particular genes, and are, in homozygotes or in heterozygotes for two abnormal genes, responsible for anemias of varying degrees of severity.

Inborn Errors of Metabolism. Enzymes are proteins and as such are coded by genes. This has been shown in many cases in which different alleles of a given gene determine the presence or absence—or amount—of activity of specific enzymes. The most extensive evidence for the relations between genes and enzymes was obtained beginning in the 1940's from work with microorganisms such as the bacterium *E. coli* and, particularly, the bread-mold fungus *Neurospora crassa*. Essentially similar, fundamental discoveries had been made

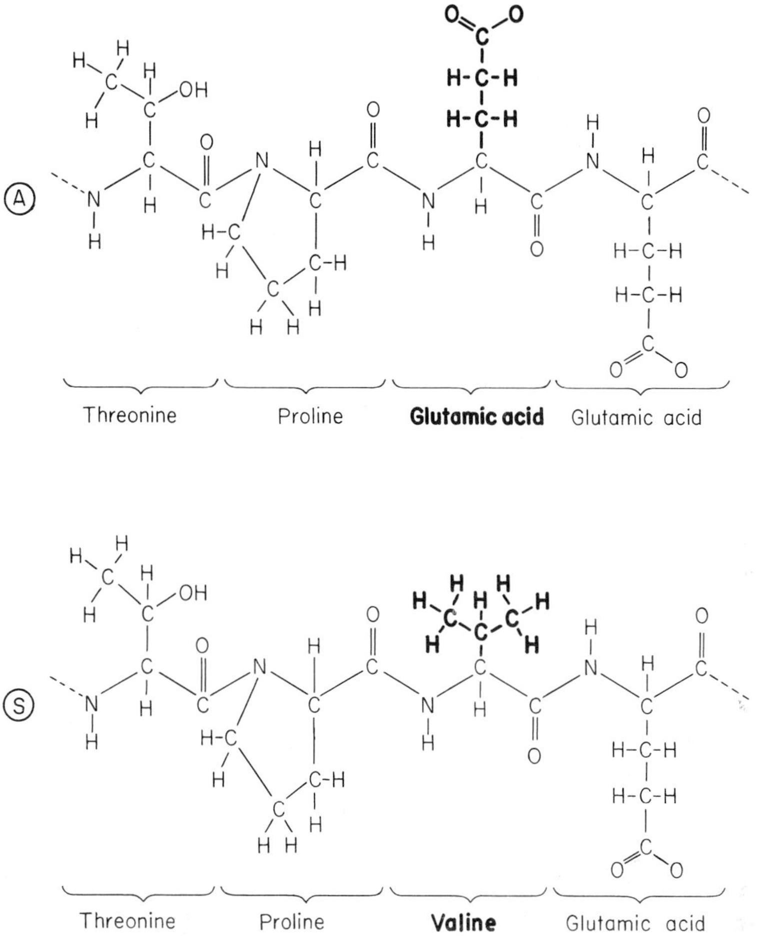

A: Threonine · Proline · **Glutamic acid** · Glutamic acid

S: Threonine · Proline · **Valine** · Glutamic acid

FIGURE 24
Short corresponding sections of the molecules of hemoglobin A and S. The amino acid units are the same in both hemoglobins except that glutamic acid in A is replaced by valine in S. (After Ingram, "How Do Genes Act?" Copyright © 1958 by Scientific American, Inc. All rights reserved.)

much earlier in human material by the English physician Garrod (1858–1936), whose work in biochemical genetics dates back to the earliest years of the twentieth century, immediately after the establishment of Mendelian genetics in 1900. Garrod's method consisted in the comparison of the biochemistry of normal and hereditarily abnormal individuals. He reached the conclusion that certain abnormalities were the consequences of "inborn errors of metabolism."

One of these errors leads to albinism, the absence of the melanin pigments that are present in the hair, skin, and the iris of nonalbinotic persons (see p. 162). Garrod showed that albinism is caused by homozygosity of a gene, *a*, responsible for the absence or inactivity of an enzyme that participates in

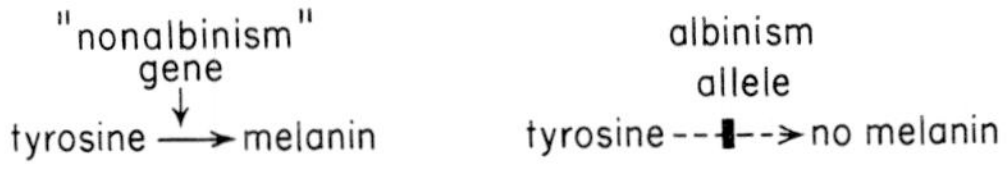

"nonalbinism"
gene
tyrosine → melanin

albinism
allele
tyrosine --▮--> no melanin

FIGURE 25
Simplified scheme of metabolic effects of alleles at the albino locus.

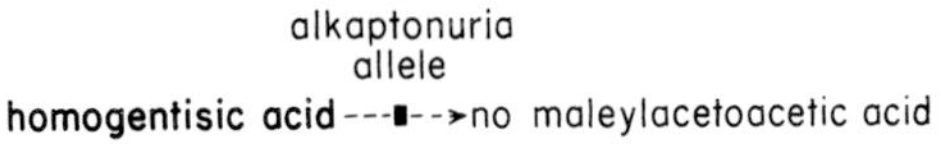

alkaptonuria
allele
homogentisic acid ---▮--> no maleylacetoacetic acid

FIGURE 26
Simplified scheme of metabolic effect of the alkaptonuria gene. The heavy print indicates accumulation of homogentisic acid. *Note:* The entry "no maleylacetoacetic acid" applies only to that fraction of acetoacetic acid whose source is homogentisic acid.

transforming the amino acid tyrosine into a precursor of melanin. In non-albinos the enzyme is present and catalyzes the reaction; in albinos the absence of enzymatic activity blocks the reaction (Fig. 25).

A similar situation was found by Garrod in regard to the rare inherited abnormality alkaptonuria. In normal individuals, a gene, al^+, is responsible for an enzyme in the blood that accelerates the breakdown of a normal product of metabolism, homogentisic acid (also called alkapton). In alkaptonurics homozygosity for the allele *al* is responsible for absence of this enzyme, and the homogentisic acid, instead of being degraded to maleylacetic acid and ultimately being metabolized into carbon dioxide and water, is excreted in the urine (Fig. 26). Since alkapton is a substance that darkens when exposed to air, the diapers of alkaptonuric infants and the urine of affected persons turn black after prolonged standing (Fig. 27).

The example of alkaptonuria illustrates two different though related phenomena. The metabolic block results (1) in the absence of the degradation product of homogentisic acid, maleylacetic acid, as well as (2) in the accumulation of the homogentisic acid itself. The gene thus has two-fold secondary effects. In alkaptonuria the presence of the accumulated precursor of the blocked reaction product is compatible with a long period of health. Nevertheless, the oxidation of alkapton gradually leads to dark pigmentation of cartilage and other body parts, externally noticeable in the ears, the sclera of the eyes and some other regions. Disease of the joints (arthritis), sometimes incapacitating, is the ultimate consequence of the presence of the abnormal gene.

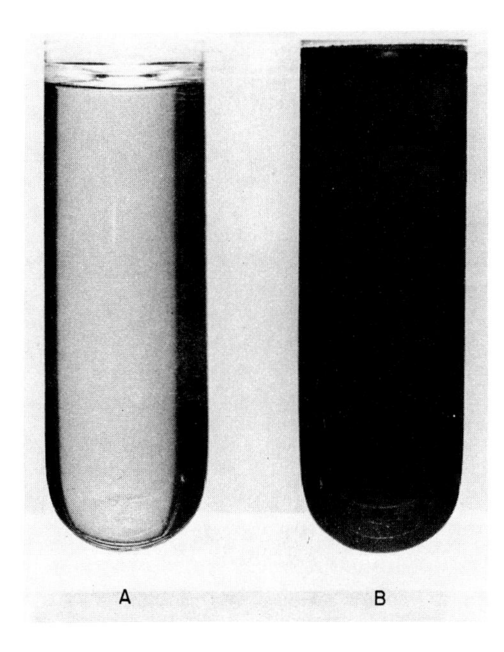

FIGURE 27
Two test tubes of urine of a person affected with alkaptonuria. **A.** Freshly voided, of same appearance as normal urine. **B.** After prolonged standing, blackened by oxidation. (Original from Dr. R. A. Milch.)

Absence of other enzymes resulting in blocking of reactions and accumulation of precursor substances may have still more serious effects. In normal individuals the enzyme phenylalanine hydroxylase catalyzes the conversion of phenylalanine into the very similar amino acid tyrosine. There are individuals who, for genetic reasons, lack the enzyme. Since they cannot convert phenylalanine in the regular way, large concentrations of it are accumulated in the blood, the cerebro-spinal fluid, and the urine. In normal persons some phenylalanine is converted into phenylpyruvic acid, but in persons lacking phenylalanine hydroxylase excessive amounts of this substance are produced and excreted in the urine (Fig. 28). Because the condition was first discovered by the presence of the ketone phenylpyruvic acid in the urine, it is known as phenylketonuria or PKU. Important biochemical insights have been gained in the study of this condition which has been given impetus by the fact that many phenylketonurics are afflicted with severe feeblemindedness. In some still unknown way the high level of phenylalanine in the body fluids causes damage to the developing brain of affected infants. It is likely that it is not the phenylalanine itself that is harmful but some derived metabolic substance.

Gene-controlled enzymes are not limited to those which participate in the metabolism of amino acids and their derivatives. On the contrary, all enzymes are regarded as ultimately gene determined. Three examples of genes concerned with carbohydrate metabolism will be given, as well as examples of their participation in other types of metabolism. One of the inborn errors of carbohydrate metabolism is caused by absence of an enzyme that acts on the

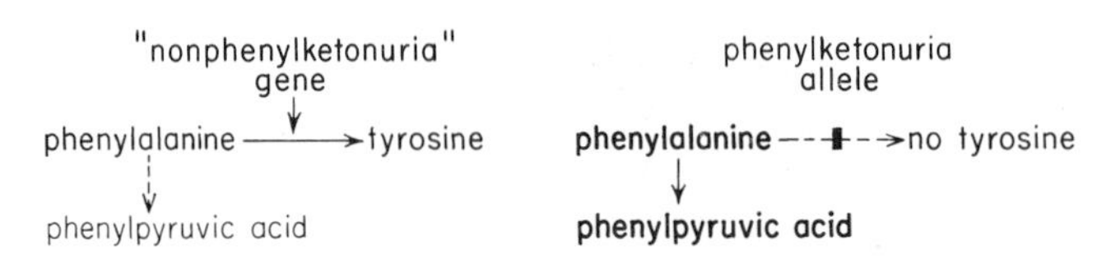

FIGURE 28
Simplified scheme of metabolic effects of alleles at the phenylketonuria locus. *Left:* In the presence of the normal allele only small amounts of phenylalanine are converted into phenylpyruvic acid, much of it being converted into tyrosine. *Right:* In the presence of an abnormal allele, considerable amounts of phenylalanine are converted into phenylpyruvic acid. *Note:* The entry "no tyrosine" applies only to that fraction of tyrosine whose source is phenylalanine. Bold-face type indicates accumulation of precursor or its derivative.

milk sugar galactose. Absence of this enzyme is responsible for a rare disease of young children, galactosemia. Individuals who possess the gene for presence of the enzyme galactose-l-phosphate-uridyl transferase convert derivatives of galactose into derivatives of the similar sugar glucose. Children who are homozygous for an abnormal allele of the gene lack the enzyme and accumulate galactose compounds, which lead to liver damage and mental defect soon after birth.

The second example of an abnormal gene in carbohydrate metabolism is that leading to a glycogen storage disease (Type I, von Gierke's disease). Normally, glycogen is stored in moderate amounts in the liver and kidneys, where it forms the body's main reserve of material that can easily be converted to glucose sugar. This conversion is one link in a chain, accomplished under the influence of the enzyme glucose-6-phosphatase. Some individuals lack the enzyme. They suffer from low blood sugar concentrations and accumulate glycogen in enormous amounts in the liver and, to a somewhat lesser extent, in the kidneys. Usually the condition leads to death during childhood.

A third example refers to a different glycogen storage disease (Type II, Pompe's disease) in which the enzyme α-1,4-glucosidase is deficient. Ultrastructurally, cells from various organs of affected fetuses and infants are characterized by abnormal lysosomes (Fig. 29).

Another genetically controlled biochemical condition is one that affects the utilization of copper in the body. Wilson's disease, characterized by degeneration of certain parts of the nervous system, of the liver, and various other symptoms, has long been recognized as a peculiar entity, and its dependence on an abnormal gene has been established. Bearn has shown that the diverse symptoms of the condition are the results of an inability to synthesize the normal amount of a copper-containing blood protein called ceruloplasmin. As a result of this inability, the copper atoms taken up in food are deposited in the brain, the liver, and other tissues, and there produce their multiple clinical effects.

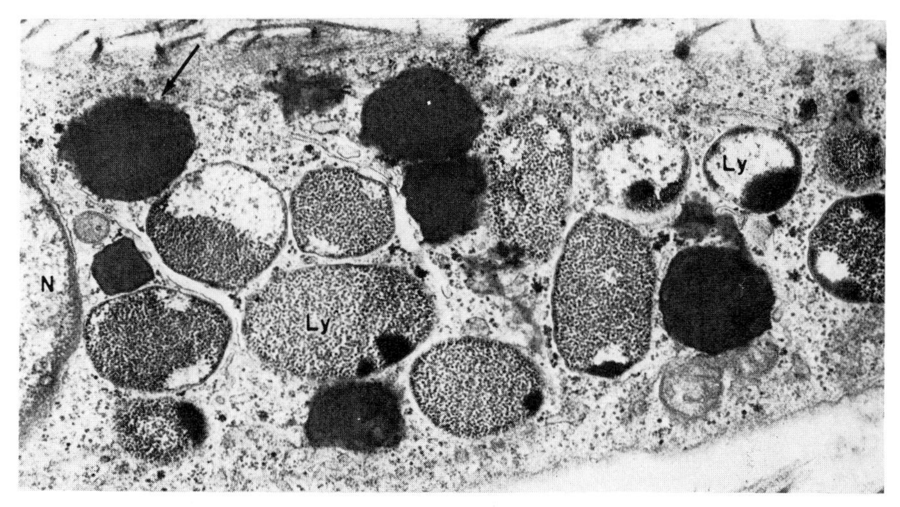

FIGURE 29
Part of a fibroblast from a culture of cells collected in a skin biopsy on an infant with Pompe's disease. There are numerous vacuoles filled to varying degrees with glycogen particles. The abnormal lysosomes (Ly) are the ultrastructural hallmark of the disease; they do not occur in normal fibroblast cultures. The arrow points to one of several lipid droplets that are not unusual. N = nucleus. (Original from Dr. G. Hug.)

The recognition of differences between normal, sickle-cell, and C hemoglobin molecules was made possible by their different electrical charges, which result in different degrees of mobility under the influence of an electric current. Other hemoglobins cannot be distinguished by electrophoresis but differ in their diffusive properties as revealed by chromatography. The molecular properties of the different chemicals determine the distance a solvent will carry each of them from the place on a paper or a starch gel where they had first been deposited. In this way they will be separated from one another, and they can then be made visible by certain other reagents. This and similar methods have been applied to urine of different individuals, and a variety of amino acids and other substances have been identified. There is evidence of genetic control of the excreted amounts of these metabolic products within groups of normal individuals. In most cases it still remains to be determined where the primary genic effect lies—whether in some general metabolic processes or in the degree to which the kidney controls excretion.

Most normal individuals excrete only six amino acids in easily discernible amounts (Fig. 30, A). Cystinuric persons, characterized by homozygosity for an abnormal gene, α, excrete unusually large amounts not only of cystine, as has been known for a long time, but also of arginine, ornithine, and lysine (Fig. 30, B). Some of these individuals do not show any symptoms of disease, but others form a special type of kidney stone composed of cystine and suffer damage to the kidney. An even more generalized "aminoaciduria" occurs in

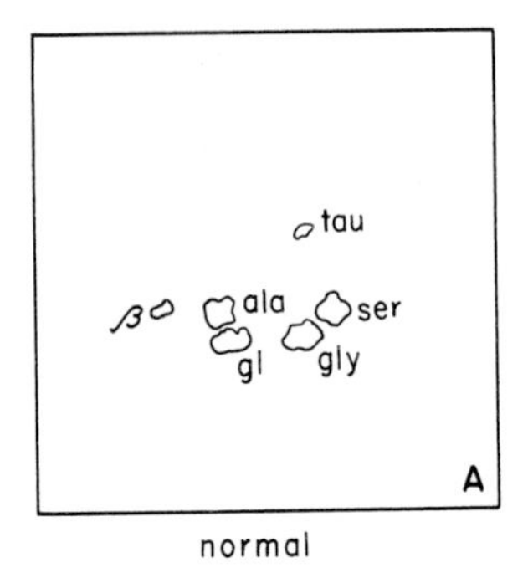

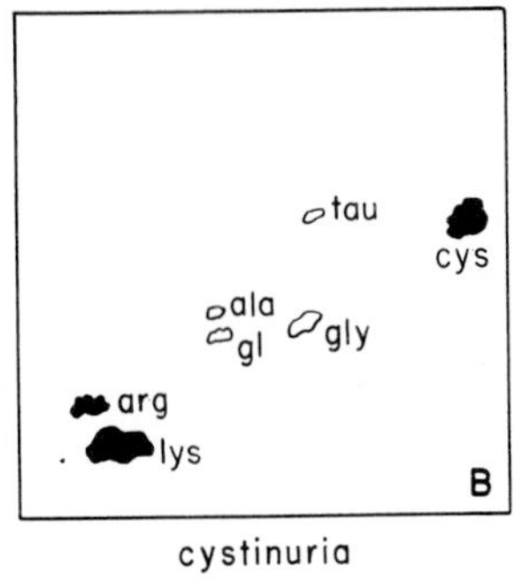

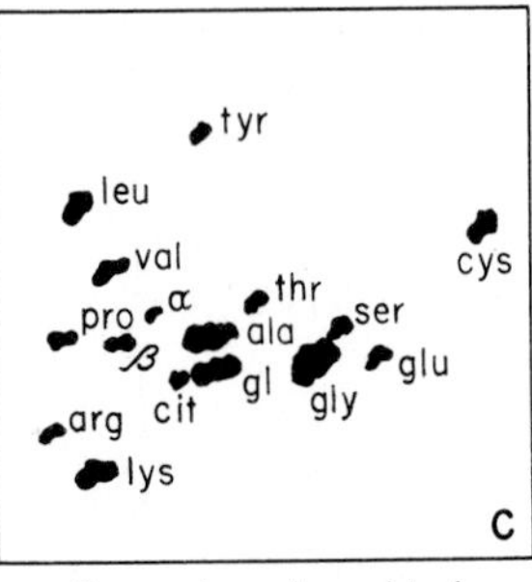

FIGURE 30
Paper chromatograms of urine from (**A**) a normal individual, (**B**) a cystinuric individual, and (C) an individual with Fanconi disease. The square represents the piece of filter paper to whose lower-right corner the urine sample had been applied. Solvents were then run from right to left and from bottom to top, thus separating the various amino acids. (tau = taurine; ser = serine; gly = glycine; ala = alanine; gl = glutamine; β = β-amino-iso-butyric acid; cys = cystine; arg = arginine; lys = lysine; glu = glutamic acid; thr = threonine; leu = leucine, iso-1; val = valine; α = α-amino-iso-butyric acid; cit = citrulline; pro = proline.) The normal individual excretes significant though small amounts of only six amino acids. A cystinuric individual is primarily characterized by excessive excretion of cystine, arginine, and lysine. (In this figure no separation of ornithine, which is also characteristic, is indicated.) A patient with Fanconi disease excretes large quantities of many amino acids. (Dent and Harris, *Ann. Eugen.*, **16**, 1951.)

infants with the rare and eventually fatal Fanconi disease, which affects bone formation and causes various physiological disturbances; it, too, is caused by an abnormal gene. Affected infants excrete large quantities of the majority of amino acids (Fig. 30, C). Initially it was assumed that cystinuria was caused by a metabolic block in amino acid metabolism. Later work, however, showed that the defect is one of renal transport. In normal individuals the tubular cells of the kidney possess an enzyme-like transport system that reabsorbs cystine, arginine, ornithine, and lysine out of the glomerular filtrate. Individuals homozygous for a specific abnormal allele do not reabsorb these substances but rather excrete them in the urine. The same sort of system seems to be defective in the intestine of cystinurics, as judged by their slower-than-normal absorption of amino acids from the diet. Garrod's original concept of inborn errors of metabolism can easily be broadened to include errors affecting functional substances other than enzymes.

Most biochemical analyses of gene effects have been concerned with rare defects. The biochemistry and the genetics of many rather common defects is less well understood, largely because most rare defects are caused by practically complete absence of a normal metabolic process, while the more common defects may be the consequences not of the absence of a biochemical reaction, but only of a quantitative abnormality. Obviously, it is easier to discover which genes result in a reaction's absence than to find which genes lead to quantitative variations in reactions. Yet there is every reason to assume that genes, by regulating biochemical processes, determine not only striking abnormalities but also less striking ones and, above all, much of the variability among normal human beings.

Insight into the biochemistry of genetic variation has provided tools for overcoming genic deficiencies. Infants with galactosemia can be helped by removing galactose from their diet; the mental development of phenylketonuric infants can be improved greatly by giving them food containing only the minimal necessary amount of phenylalanine; and the gene-controlled sugar diabetes, which is often due to a deficiency of the pancreatic hormone, insulin, can be controlled by regular injection of this hormone. We may even, with Pauling, "foresee the day when many of the diseases that are caused by abnormal enzyme molecules will be treated by the use of artificial enzymes. . . . It may be possible, for example, to synthesize a catalyst for the oxidation of phenylalanine to tyrosine. A small amount of this catalyst could be attached to a framework inside a small open-ended tube, which could be permanently placed within the artery of a new-born child who had been shown by chemical tests to have inherited the mental disease phenylketonuria. Through the action of the artificial enzyme, the child could then develop in a normal way." Other methods of overcoming genic deficiencies, often referred to as genetic engineering, will be cited in the last pages of this book.

The metabolism of an organism does not consist of biochemical processes that proceed side-by-side but independent of one another. On the contrary, many of these processes form an interconnected network of reactions. In a small, partial way this is indicated by Figure 31, which presents several of the interrelated metabolic events of amino acid metabolism. It shows five genetic blocks known at present, three of which—those leading to albinism, phenylketonuria, and alkaptonuria—were artificially isolated in earlier discussion in this section.

Other examples of interrelated, genetically controlled, biochemical events are provided by chemical properties of human blood apart from hemoglobins. Blood was the basis of some of the earliest studies of human biochemical genetics. The discovery by Landsteiner (1868–1943), in 1900, that individuals can be assigned different groups according to whether their red blood cells clump or remain separate when mixed with the blood of different persons was the beginning of a long series of studies. The goal of these studies, still incompletely attained, is the chemical characterization of the substances participating in the reactions and an analysis of the role that genes play in them. There are three main alleles of the gene I which control the presence or absence of (1) specific substances, called *antigens*, in the red blood cells and (2) very similar antigens in body fluids, including the fluid part of the blood, the serum. I^{A} leads to antigen A, I^{B} to B, and I^{O} to absence of A and B.

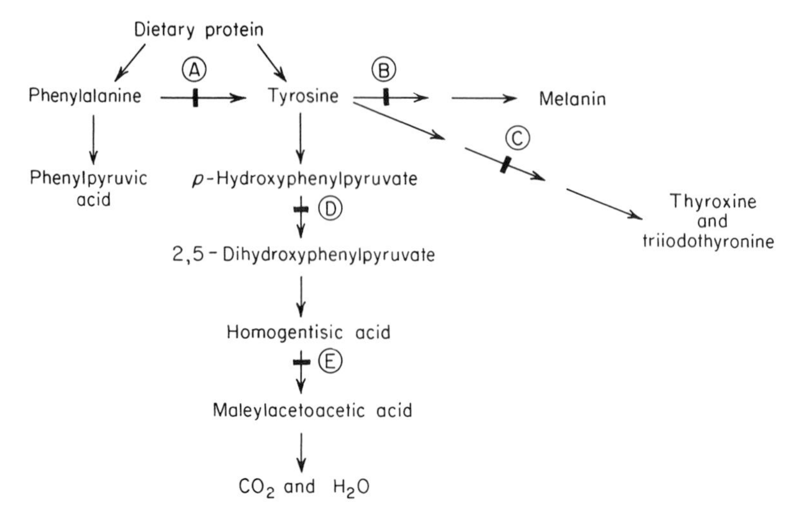

FIGURE 31
Interrelations in the metabolism of phenylalanine and tyrosine in man. Five different genes are known, each of which controls a different reaction. In the presence of the normal alleles of these genes, the reactions proceed as indicated by the arrows. If certain abnormal alleles are present, the reactions do not proceed. Such abnormal alleles of gene *A* lead to phenylketonuria, of *B* to albinism, of *C* to goiterous cretinism, of *D* to tyrosinosis (abnormal metabolism of tyrosine) and of *E* to alkaptonuria. (After Harris, *Human Biochemical Genetics,* Cambridge University Press, 1959.)

The antigens are complex macromolecules and the relation between the *I* alleles and their antigens is not a primary one; that is, genes other than the *I* alleles are responsible for the formation of large precursor molecules that in the body fluids are glycoproteins (consisting of carbohydrate and amino acid chains) and in the red blood cells glycolipids (consisting of carbohydrates linked through sphingosine to fatty acids). In presence of the Lewis (*Le*) gene a precursor substance is transformed into a Le^a substance, which in turn is transformed by the *H* gene into an H substance. It seems that the *H* gene controls an enzyme—a transferase—that adds a sugar at a specific position to the carbohydrate chains of the Le^a precursor molecule and thus transforms it into an H substance (Fig. 32). The alleles I^A and I^B control other transferases that add specific sugars to chains of the H substance. In contrast, the allele I^O does not lead to changes in the H substance. The proposed structures of the terminal sugars in the antigenic H, A, and B macromolecules are pictured in Figure 33. Watkins, who together with Morgan and Kabat, is mainly responsible for the elucidation of the structure of the A and B antigens summarizes her work as follows: "The picture that emerges from the chemical, serological and genetical analysis of the AB . . . substances is one in which patterns of synthesis resulting from the sequential action of the products of

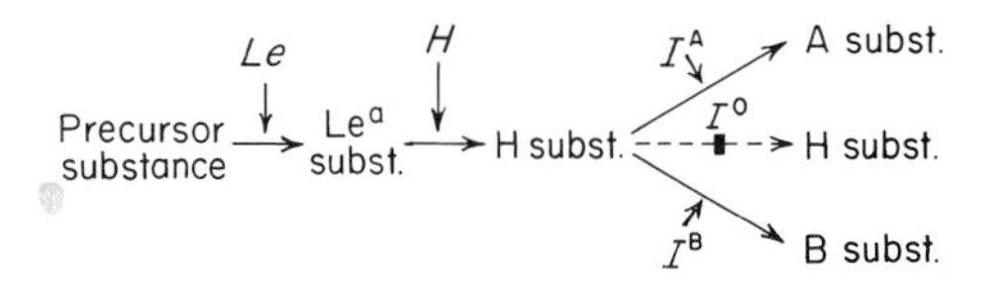

FIGURE 32
Simplified scheme of sequences of reactions controlled by the *Le*, *H*, and I^A, I^B, I^O genes, resulting in the appearance of A, B, and H blood-group substances.

different blood group genes on a common precursor . . . lead to the three-dimensional patterns of sugar residues responsible for the serological specificity."

The clumping of red blood cells in mixtures of the blood of appropriate individuals is caused by interactions of the AB antigens, which we have just discussed, with other substances, called *antibodies*, in the fluid part of the blood. Individuals with different kinds of antigens controlled by the *I* locus are distinguished by specifically different kinds of antibodies (see Chap. 11). In addition to their role in antigen formation, the blood-group alleles of the *I* locus determine, in an indirect way that is not yet fully understood, the type of antibodies present.

The group of blood antigens defined by the *I* alleles is only one of at least 15 different groups, each of which is controlled by a different genetic locus. The genetics of these other blood groups are on the whole well known and will be dealt with later. The biochemical facts are much less clarified at present. It is impressive to consider that all these different antigenic groups of the blood develop side-by-side in each single person.

A variety of substances constitute normal blood serum. It may be assumed that all of them are under genic control. Definite proof for this is available in a number of cases. Here, we shall single out two examples. One of these refers to a type of serum protein, the α_2-globulins or haptoglobins, which are concerned with the binding of hemoglobin from aged and broken-down red blood cells. The haptoglobin molecules are composed of two kinds of polypeptide chains, α and β (not related to the α and β chains of hemoglobin). A number of different though similar haptoglobins have been found in different individuals and it has been shown that their differences are based on variations in the α chains, with the β chains being everywhere the same. Three alleles at the *Hp* locus are responsible for the three main kinds of α chains found in human populations. Two of the alleles, Hp^{1F} and Hp^{1S}, control α chains that differ from each other by a single specific replacement of one amino acid. Both α chains move relatively fast in electrophoresis tests, Hp^{1F} faster than Hp^{1S}. The third allele, Hp^2, differs from both Hp^1 alleles in a much more striking way. The Hp^2 α chain moves considerably slower than either Hp^{1F} or Hp^{1S}. It is nearly twice as long as the Hp^1 α chains and its amino acid sequence shows it to be a combination of two nearly complete Hp^1 chains arranged in tandem sequence. We shall return to this remarkable phenomenon of gene duplication.

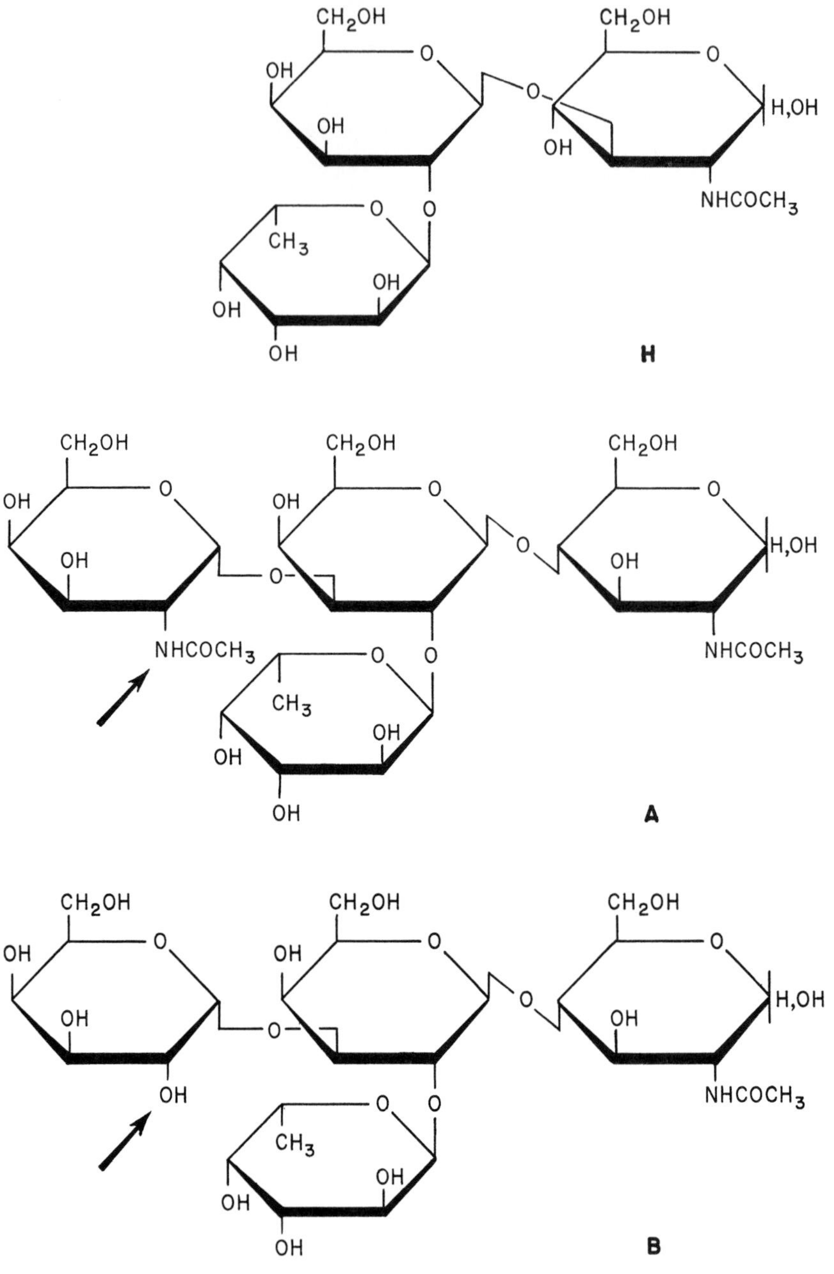

FIGURE 33
Proposed structures. *Top:* The three terminal sugars of the H substance, which give it its antigenic specificity. *Middle and bottom:* The four terminal sugars of the A and B substances, which give them their antigenic specificities. Note that these substances differ from the H substance by the possession of an additional terminal sugar. The arrows point to the only groupings that distinguish the A and B structures. The structures shown in these diagrams represent only one of two similar chains, both of which are present in the macromolecules. The similar chains differ in the linkages between the sugar on the right and its neighbor on the left. Shown is the 1-3 linkage; not shown is the 1-4 linkage. (After Watkins, *Science*, **152,** 172, 1966. Copyright 1966 by the American Association for the Advancement of Science.)

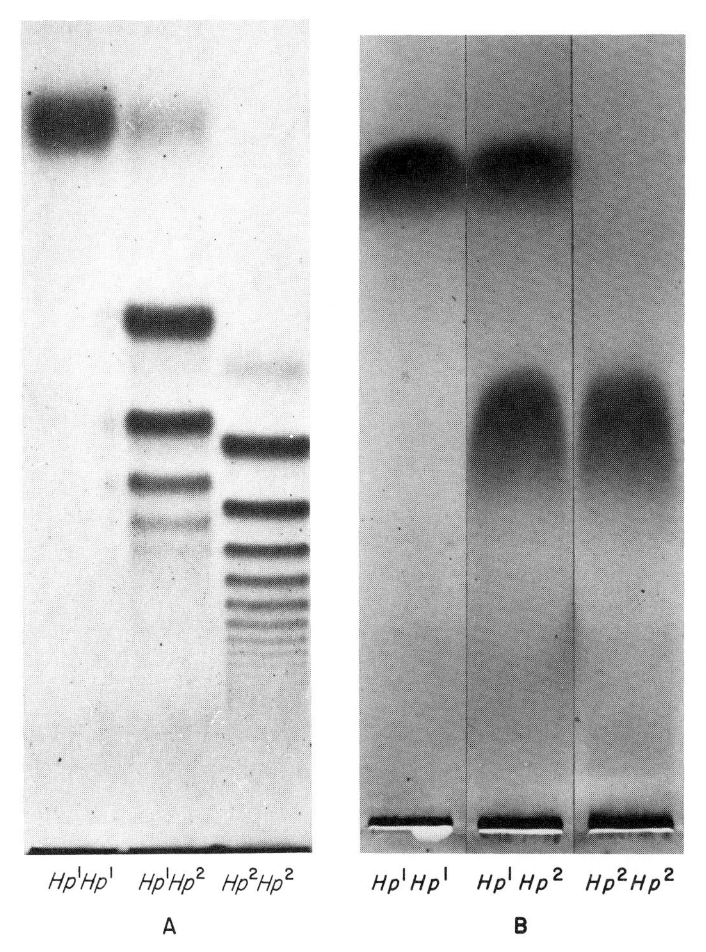

FIGURE 34
Electrophoresis of the haptoglobins Hp^1 and Hp^2 of the three genotypes Hp^1 Hp^1, Hp^1 Hp^2, and Hp^2 Hp^2. (No distinction has been made between Hp^{1F} and Hp^{1S}.) **A.** Purified haptoglobins. **B.** Alpha chains. Electrophoresis was carried out in starch gels in which the speed of migration depends both on charge and molecular size. (A, Connell and Smithies, *Biochem. J.*, **72,** 1959; B, original from Drs. O. Smithies and G. E. Connell.)

The haptoglobins furnish an interesting example of polymerization. Electrophoresis of the α chains of Hp^1 and Hp^2 homozygotes yields single bands only and electrophoresis of Hp^1Hp^2 heterozygotes yields two bands corresponding to those of the two homozygotes (Fig. 34, B). In contrast electrophoresis of the purified haptoglobins, while still resulting in a single band for Hp^1 homozygotes, yields a whole series of bands for Hp^2 homozygotes and Hp^1Hp^2 heterozygotes (Fig. 34, A). These facts are compatible with the "one gene–one polypeptide" concept if one assumes that the polypeptide chains have a tendency to unite in pairs and in higher polymers. The single Hp^1 band then represents a polypeptide that forms only one type of α haptoglobin; the multiple Hp^2 bands are indicators of different degrees of polymerization; and the multiple bands of the Hp^1Hp^2 heterozygotes are the result of different Hp^2 polymers plus other polymers of hybrid Hp^1/Hp^2 molecules.

Other gene-controlled chemical properties of the blood include those of components of the fluid that are concerned with the formation of antibodies in general. Thus there exists a rare gene that makes its bearers extremely susceptible to bacterial infections by causing the near absence of that blood fraction called gamma globulin, a protein material present in normal individuals which provides the antibodies that attack the foreign material produced by bacterial invaders. This hypogammaglobulinemia has been traced back to the almost total lack of a special type of cells in the lymph nodes, the spleen, and some other organs. These "plasma cells" are apparently essential in the production of antibodies.

A Diagram of Genic Interaction within a Single Cell. As we have seen, the primary products of structural genic activity are polypeptide chains. These enter into the formation of various molecules that interact in manifold ways within individual cells and between cells. A very simplified diagram of some of these interactions within a cell is given in Figure 35. There we have listed

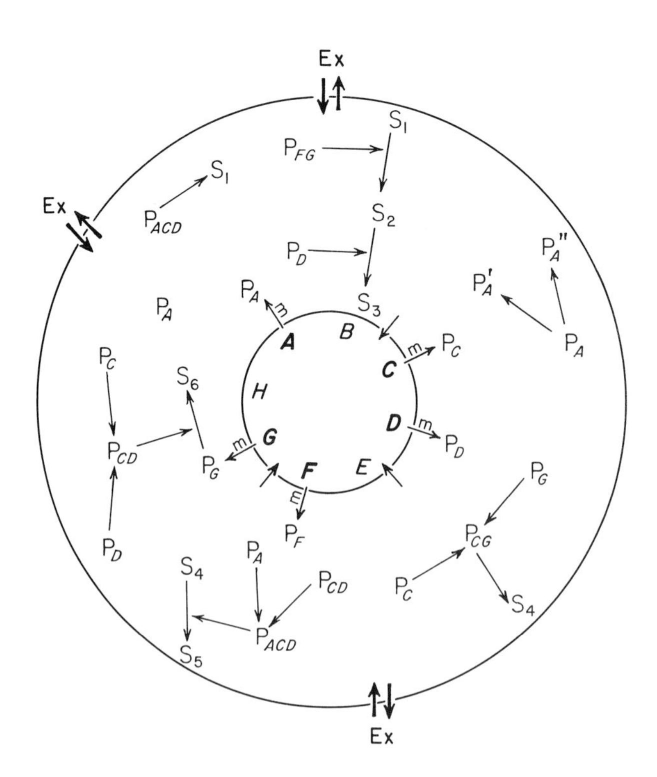

FIGURE 35
Genic action within a cell. See text.

eight genes (A, B, C, D, E, F, G and H) inside the nucleus of a cell. Three of these, B, E, and H, it is assumed, are inactive in the cell shown, but the other five send out messenger RNA molecules, which in the cytoplasm organize the formation of the primary gene products P_A, P_C, P_D, P_F, and P_G. The formation of these products depends on biochemical interactions between the messenger RNA's and other types of RNA—some types that are bound to protein and others that carry the amino acids that are to be assembled into polypeptide chains. The interactions are controlled by enzymes and the enzymes themselves are synthesized by the cell under the influence of genes, as are certain amino acids. (Other amino acids in the human body are obtained from the enzymatic degradation of proteins taken in as food.)

The diagram indicates further how various products P are changed into new products as, for example, $P_A \rightarrow P_A^1$ or in the interaction $P_C + P_D \rightarrow P_{CD}$ or in the action of gene products on substances S_1, S_2, etc., which are part of the cytoplasm. In addition to these intracellular events there are exchanges between the cell constituents and the environment, Ex, as indicated by the arrows on the cell membrane. While these reactions take place many more go on, controlled by still other genes, to form an interweaving network of sequences that all depend on one another and that, together, make up the life processes of the cell.

In a multicellular organism such as man, individual cells are not isolated from one another. They interact in both general and specific ways: in competing for oxygen and food carried by the circulatory system, and by discharging carbon dioxide and waste products into it; by release and uptake of hormones; and by being subjected to stimuli which originate in the nervous sytems and responding to them. These intercellular processes too are dependent on genic control.

Genic Action and Differentiation. A special cellular interaction is that which leads to the development of a complex, multicellular individual from the single-celled egg. After fertilization the egg cleaves by repeated divisions into more and more cells. Diagrams of 2-, 4-, and 8-cell stages, as found in the development of many animals, are shown in Figure 36. Externally, the cells of these early embryos are all very similar to one another, but some differentiation is apparent in the 8-cell stage. Here, as a result of slightly unequal divisions, the 4 upper cells are smaller than the 4 lower ones.

There is also an internal difference between the upper and lower cells. The cytoplasm of the original fertilized egg is not uniform throughout. A stratification can sometimes be observed, with more yolk granules in the lower half than in the upper half, and it has been determined experimentally that other differences in the cytoplasmic composition exist along an axis which extends from the upper to the lower pole of the egg. The cleavage of the egg, which separates its different regions from one another by assigning them to different

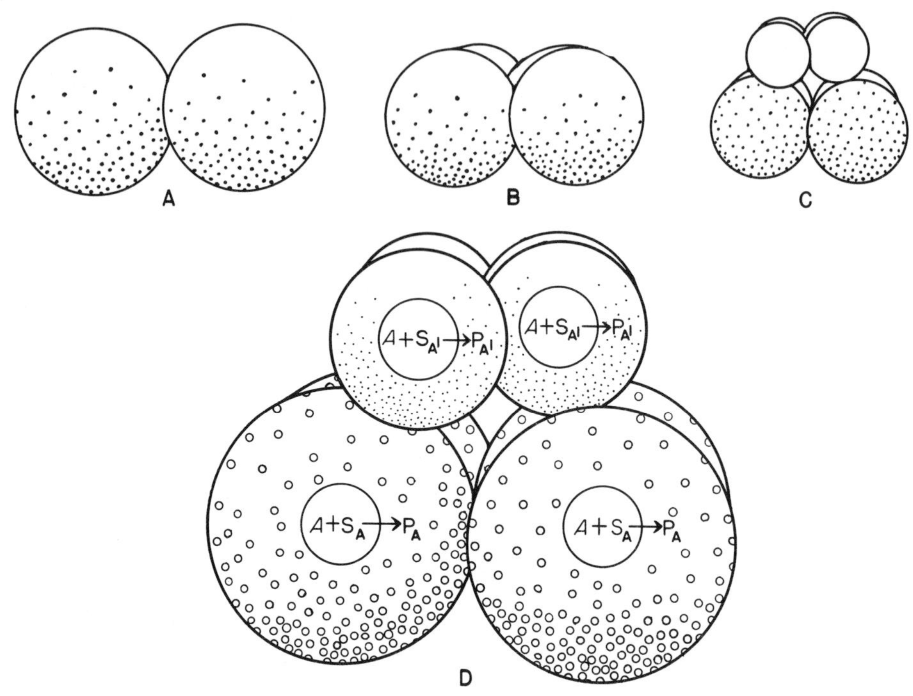

FIGURE 36

Cleavage and differentiation. **A.** Two-cell stage. A difference in the cytoplasm of the upper and lower halves of the cells is indicated. **B.** Four-cell stage. **C.** Eight-cell stage. The upper four cells are differentiated from the lower four cells. **D.** Eight-cell stage, enlarged. Interaction between the gene *A* in the nuclei with substrate S_A in the lower cells and substrate $S_{A'}$ in the upper cells leads to formation of different products P_A and $P_{A'}$. Contact between upper and lower cells results in further cytoplasmic differentiation, as indicated by fine stippling in upper cells.

cells, thus leads to a *differentiation* of the cytoplasmic contents of the cells. Once such a differentiation has begun, it will lead to others. In the 8-cell stage, for instance, the upper group of 4 cells is in contact with the lower group. This contact, however, is restricted to the basal parts of the upper cells, and it seems possible that these basal parts become differentiated by interactions with the underlying cells, as indicated by the stippling in Figure 36, D.

How do the identical genes present in the nucleus of each cell react to these different cytoplasmic "environments"? It seems, in analogy with insights gained from microbial molecular genetics, that different genes respond differently to different environments. In environment X genes *A, C, D, F, G, . . .* may be activated and *B, E, H, . . .* may be repressed. In environment Y genes *A, C, F, H, . . .* may be activated and *B, D, E, G, . . .* may be repressed; and in environment Z still other assortments of active and inactive genes may emerge. As development proceeds, further differentiation of the embryo occurs as various parts of the total genic sets are selectively called into action. The different fates of different cells of the embryo depend on the following sequence:

1. Initial regional differences in the cytoplasm of the egg.
2. Subdivision of the egg into differentiated cells.
3. Production of further differentiation by specific interactions between the identical gene sets and the differentiated cell contents.

The differentiation and organization of the embryo depend on the action of genes in all cells and interaction among the cells, among whole embryonic regions, and among the parts of the embryo and its surroundings.

The details of activation and inactivation of genes—or to use terms derived from microbial studies, of derepression and repression—are not well understood. It is believed that the different components of the cytoplasmic "environments" of different cells produce their effects in part by interacting with regulator genes which control the state of function of the structural genes.

It has been suggested that the production of hemoglobin chains during the prenatal and early postnatal development of a normal child may be an example of regulation of the activity of structural genes. No detectable amount of hemoglobin is formed during the earliest stages of embryonic life. During the second month, however, at least four different chains, α, β, γ, and ϵ are synthesized (Fig. 37). By the end of 3 months of prenatal life the α chains constitute the same proportion of the total amount of hemoglobin that they do in the adult. The β chains are produced more slowly, not reaching their ultimate propor-

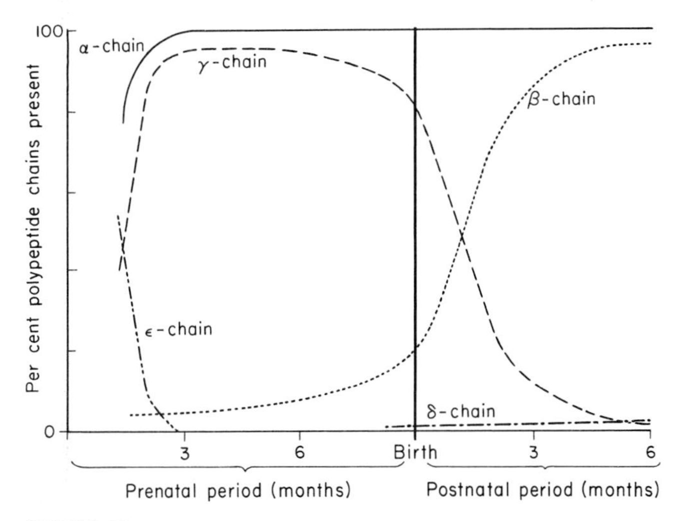

FIGURE 37
Percentages of hemoglobin polypeptide chains present during prenatal and early postnatal life. (After Huehns, et al., Human Embryonic Hemoglobins, *Cold Spring Harbor Symp. Quant. Biol.*, **29**, 1964.)

tion until several months after birth. The relative quantity of γ chains is inversely related to that of the β chains: they form almost as rapidly as the α chains until the end of the sixth month of prenatal life, but then begin to disappear, and by the third month of postnatal life have declined to about one per cent of the hemoglobin. In contrast to the α, β, and γ chains, all of which are produced throughout prenatal life and beyond, the ϵ chains are produced for only a short period, and disappear from the embryo before the end of the first three months of development. The δ chains, finally, do not appear until shortly before birth and never constitute more than about one per cent of the total amount of hemoglobin. It has been proved that some hemoglobin chains are determined by different structural genes, and it is probably true that all are. The striking changes in relative amounts produced during prenatal and postnatal development may well depend on the great changes that take place in the "internal environment" that surrounds the hemoglobin genes during the embryonic, fetal, and infantile stages: components of the internal environment of the cells interact with the regulating genes that control the activity of structural genes. This hypothesis finds support in the fact that a gene is known that causes persistent production throughout adult life of 10–20 times the number of γ chains ordinarily formed: since these chains differ in no way from those produced by a normal person, the gene that leads to their excessive production may be regarded as a regulating rather than a structural gene.

Regulation of the activity of hemoglobin genes is correlated with stage of development. Another type of regulation is that in which specific gene-dependent products are restricted to specific types of cells. Thus, the enzyme glucose-6-phosphatase involved in glycogen metabolism is found in the liver and kidneys but not elsewhere, and the same is true of the homogentisic oxidase whose absence results in alkaptonuria; the hormone insulin is produced exclusively in the Langerhans cells of the pancreas; and hemoglobin is synthesized by blood-forming cells only. Other gene-dependent substances, however, seem to be produced in all cells; examples are galactose-1-phosphate-uridyl transferase, which catalyzes conversion of galactose to glucose, and the catalase that breaks down hydrogen peroxide.

Regulation of enzyme activity during development of young rats and at different sites in adult rats may serve as another possible example of the regulation of genic activity. Three liver enzymes were studied by Kretchmer during development (Fig. 38). Each has its specific pattern of changing activity: the activity of phenyalanine transaminase rises from one-half that characteristic of the adult to nearly twice that of the adult before slowly falling again, while the tyrosine transaminase rises to a steep peak of high activity within two days and then falls steeply to the level characteristic for adult rats. Four enzymes produced by cells of the intestinal wall were then studied in adult rats (Fig. 39). Each shows a specific pattern of activity for particular levels

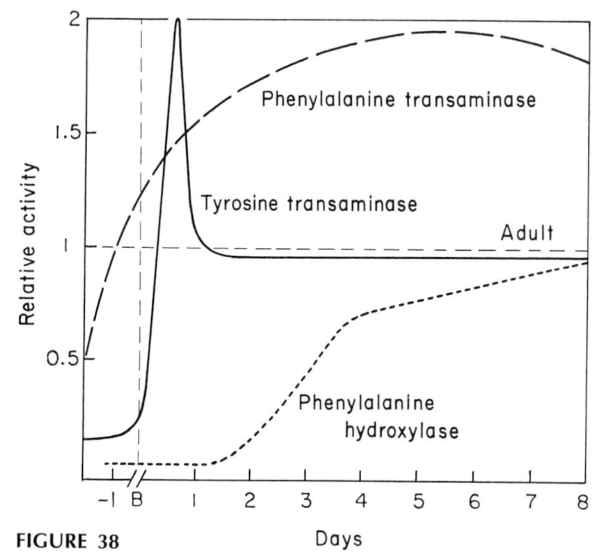

FIGURE 38
Patterns of enzymatic activity in rat liver during development from one day before birth to eight days after birth. The activities of the three enzymes are plotted relative to their adult activities. (Kretchmer, *Pediatrics,* **46,** 1970.)

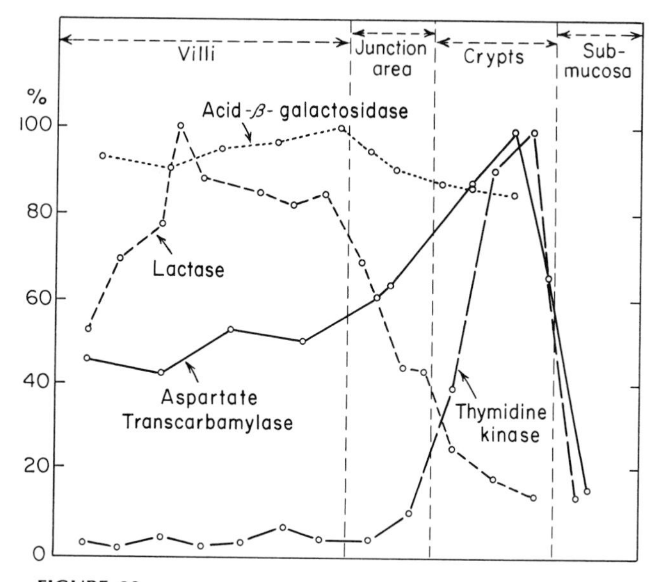

FIGURE 39
Patterns of enzymatic activities in rat intestinal wall at different histological levels: villi, junction area, cripts, and submucosa. The activities of four enzymes are plotted relative to 100, the maximum activity for each enzyme. (Fortin-Magana, Hurwitz, Herbst, and Kretchmer, *Science*, **167,** 1627, 1970. Copyright 1970 by the American Association for the Advancement of Science.)

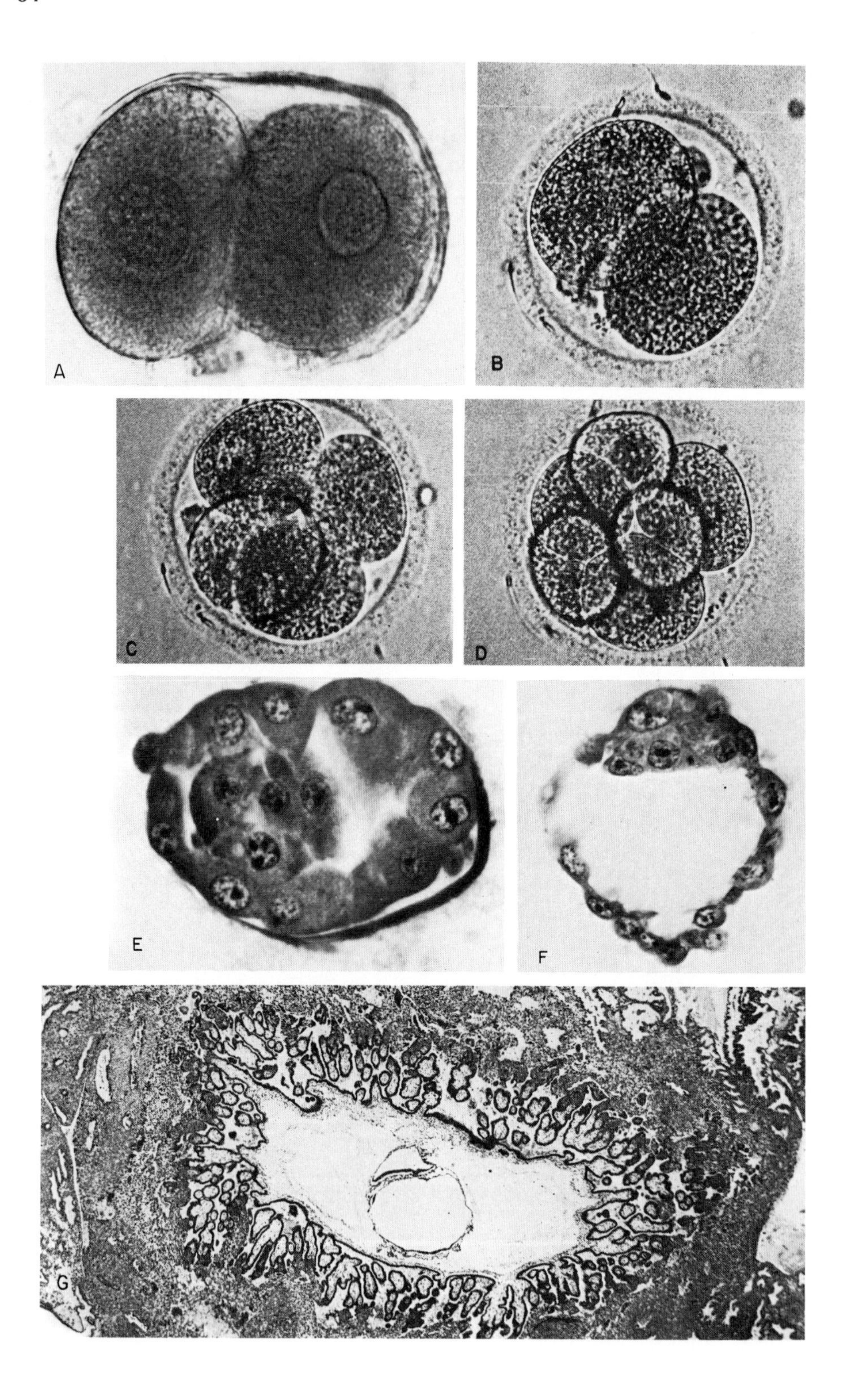
A
B
C
D
E
F
G

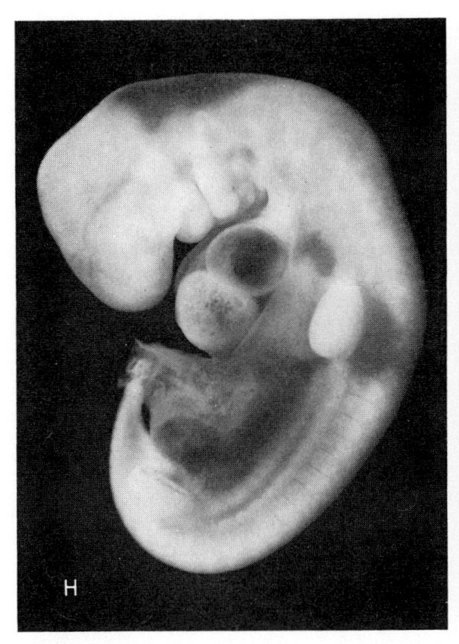

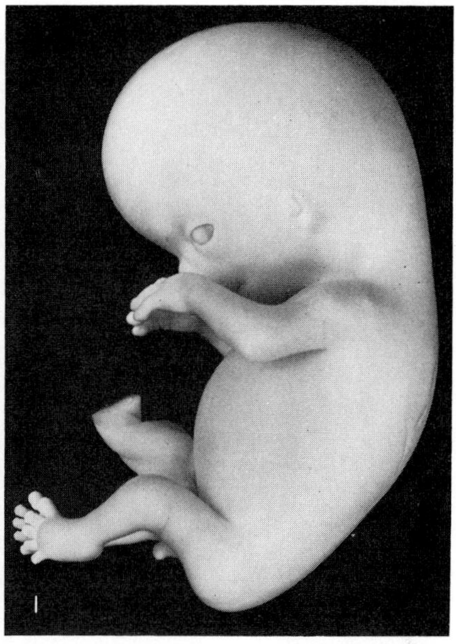

FIGURE 40
States of human development; supplemented by showing stages from the Rhesus monkey. **A.** A human egg at the two-cell stage (× 512). This egg, which was recovered from the oviduct of a woman during an operation, shows the earliest known human embryonic stage after fertilization. Fixed and stained. **B–D.** Early cleavage in the Rhesus monkey (× 288): Fertilized eggs after or during their first, second, and third divisions. Note some of the spermatozoa outside the developing egg cells. **E–F.** Blastocyst sections, fixed and stained (E × 560, F × 450). The blastocyst shown in E consisted of 58 cells, that in F of 117 cells. This latter blastocyst was at a stage shortly before implantation in the uterine wall. **G.** Section through an $18\frac{1}{2}$-day-old embryo implanted in the uterine wall (× 22). Fixed and stained. Compare with Figs. 41 and 42, A. Note the sectioned outgrowths of the chorion; the extra-embryonic coelom which corresponds to the large hollow space in the blastocyst shown in F; the cross section of the embryonic shield, above which is the small amniotic cavity and below which is the yolk sac. **H.** Embryo 28–30 days old, shown without extraembryonic parts (× 9). Note limb buds and tail. **I.** Embryo ± 47 days old, shown without extraembryonic parts (× 2.5). (A, E, F, Hertig, Rock and Adams; B, C, D, Lewis and Hartman, *Carnegie Inst. Publ.*, **525**, 1941; G, Jones and Brewer, *Carnegie Inst. Contrib. Embryol.*, **29**, 1941; H, I, Carnegie Numbers 8141, 4570.)

in the intestinal wall: Lactase activity is between 50 and 100 per cent in the villi but drops to less than 20 per cent in the crypts; thymidine kinase activity is nearly absent in the villi but rises sharply to nearly 100 per cent in the crypts. If enzyme activity, at least in parts, is an expression of genic activity these studies indicate its regulation.

Human Embryology. Several of the preceding pages have dealt rather schematically with problems of genic action. It may be appropriate, therefore, to diverge briefly and present an outline of observable, morphological details of human embryology. The development of the fertilized human egg follows in general the same course as that of other mammalian eggs. In particular, the embryology of man is similar to that of other primates, among which that of the Rhesus monkey is best known (see Fig. 40, B–D).

The cleavage of the egg begins while it is still in the oviduct. At first, repeated divisions of the egg result in the formation of a solid cell ball (Fig. 40, A–E). Within this structure, a space filled with fluid soon develops, so that a stage is reached in which the egg resembles a hollow ball (Fig. 40, F): this is equivalent to the blastula of lower vertebrates and is called the *blastocyst*. Four days after fertilization the blastocyst still consists of less than 100 cells, but it has moved down the oviduct and entered the uterine cavity. At this time an important differentiation becomes visible: the wall of the hollow sphere is of uneven thickness (Fig. 40, F). Most of it consists of a single layer of rather small cells but, at the upper pole, a mass of larger cells has accumulated on the inside. This is the first indication of the very divergent fate of different parts of the blastocyst. Most of its cells do not participate in the formation of the embryo proper, but produce the external embryonic membrane, the *chorion*. The embryo itself develops from the inner cell mass—but again, only from part of it.

During the sixth day the blastocyst attaches itself to the inner surface of the uterus and gradually becomes embedded in its wall. Within its inner cell mass, two new hollow spaces filled with fluid originate: the *amniotic cavity* near the periphery of the embryonic sphere, and the *yolk sac* toward the center (Figs. 40, G; 41; 42, A).

Between the amniotic cavity and the yolk sac, forming the floor for the former and the ceiling of the latter, lies a plate-like layer of tissue, the *embryonic shield* or *disc* (Figs. 40, G; 41). About the twelfth day, an embryonic streak becomes visible along part of one of its diameters. During the next three weeks the cells of the disc that lie adjacent to the embryonic streak differentiate into the true embryo with skin, brain, neural tube, intestine, heart, body segments, and many other organs and tissues (Fig. 40, H).

Early in this period the amniotic cavity, which is surrounded by a cellular membrane, the *amnion*, enlarges in such a way that all of it but a narrow stalk gradually becomes separated from the external embryonic layer of tissue (Fig. 42, B–D). The amnion, the inner embryonic membrane, thus comes to lie within the chorion.

Still another three weeks are required to transform the embryo, which in many ways resembles not only that of other mammals but also that of lower vertebrates, into a truly human fetus, with a face that foreshadows that of a child, with fingers and toes, and with a progressively smaller tail (Fig. 40, I). Even then, eight weeks after the beginning, the crown-rump length of the

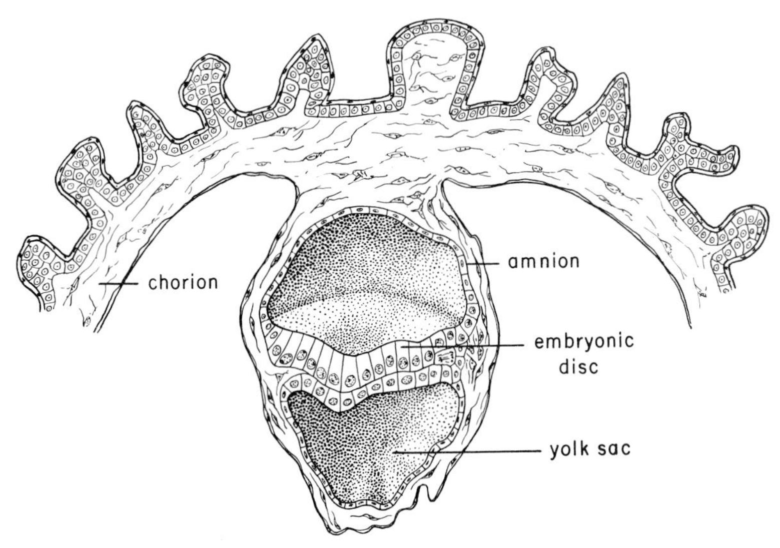

FIGURE 41
Three-dimensional diagram of the embryonic disc, amnion, yolk sac, and part of the chorion of an embryo slightly younger than the one in Figure 40, G. (After Arey, *Developmental Anatomy*, Saunders, 1965.)

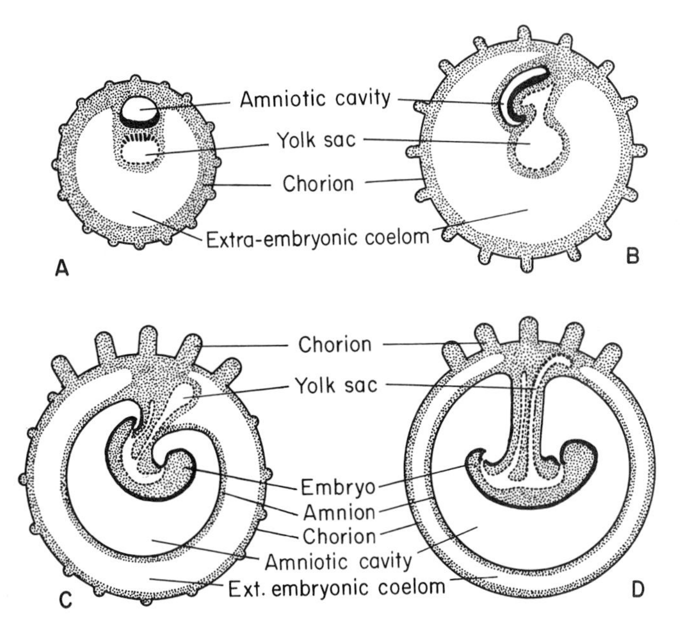

FIGURE 42
Diagrams of four stages in the early development of a mammalian embryo. (After Dodds, *The Essentials of Human Embryology*, Wiley, 1946.)

fetus is hardly more than 25 mm (one inch), and its weight is less than 2 gm. During the approximately thirty remaining weeks of pregnancy many details of organ formation are elaborated, and general growth increases the length and weight of the fetus to about 340 mm and 3,400 gm.

The first two months, counted from the time of fertilization, represent undoubtedly the most important period in the development of the new human being. As much as later events may still be important in his being born well and progressing after birth, the fundamental organization is laid down very early. Any significant deviation at this stage from the normal course of the differentiation of cells and organs is likely to have far-reaching consequences. Such deviations may be the result of actions of abnormal genes, of abnormal conditions in the surroundings of the embryo, or of combinations of the effects of genes and environment.

Genes and Characters. The general picture of the role of genes in development outlined earlier in this chapter helps us gain an insight into the interrelation between genes and "characters," which is an important aspect of human genetics. A *character,* or *trait,* may be defined as any observable feature of the developing or the fully developed individual: a biochemical property, a cellular form or process, an anatomical structure, an organ function, or a mental characteristic. The genic content of the nuclei of a given individual and his appearance are obviously different things: characters are derived from genic action. They are removed from the genes by at least one and often many steps. For the genetic constitution, the term *genotype* was coined; for the external appearance, the term *phenotype*. The genotype of a given individual is constant, fixed at the time of his origin as a fertilized egg; the phenotype is potentially variable, the result of interaction between the genotype and its nongenetic environment.

The general concepts of genic action lead us to expect (1) that no simple connection exists between most observable characters of the developed human being and a single gene; and (2) that a single gene, by being part of the network of developmental reactions, often influences more than a single character. We shall take up these two generalizations separately.

One character from many genes. Most characters are produced by the complex interaction of numerous genes; Figure 35 gave a glimpse of this interaction. There is no gene "for" eyes, or even for a part of an eye—no "one-to-one" relation between the genic units in the chromosomes and the different tissues and organs of the individual.

The recognition of this situation provides a guide for thinking about the effect of external or internal agents on the organs of the developing or fully developed individual and the possible effects of these agents on the genes themselves. If, for instance, exposure to the sun leads to darkening of the skin, should we assume that the many *genes* in the cells of the skin that enter into the interweaving processes of pigment formation have been changed? Or is it not more likely that some of the *reactions* have been altered under the influence of the new environment, "exposure to irradiation"? Or can we reasonably expect that the genes in the cells of the ovaries or testes of a suntanned individual were changed when his skin became browned? There are experimental ways of attacking these problems, and we shall return to them later, in Chapter 23, which deals with the causes of genetic changes.

Since most characters depend on many genes, it is obvious that changes in any one of many genes may result in a change in a character. Thus, hereditary blindness is known to be due to any one of very many different genes at different chromosomal loci. Some of these genetically different kinds of blindness are caused by morphologically different effects of different genes, as, for instance, an effect on the retina, on the lens, or on the general growth of the eyeball. Within each kind, it is possible to find effects specific to different types of genes, for example, different processes leading to opacity of the lens (cataract).

Frequently, what appears to be the same abnormality can be shown to be due to different genic loci. Evidence for this is obtainable, for instance, in the disease retinitis pigmentosa, a progressive degeneration of the retina which is accompanied by deposition of pigment. In some pedigrees, the "sex-linked" inheritance (see Chap. 14) of the disease proves that the gene responsible is located in the X-chromosome; in other pedigrees, the type of inheritance assigns the disease to a gene in an autosome. At present, medical examination of individuals with retinitis pigmentosa does not reveal specific differences by means of which the sex-linked and autosomal cases can be separated from each other. New methods of diagnosis, however, may someday show that the different genes control development of the disease by means of different mechanisms.

This is known to be true for the various "bleeder" diseases. Most of them have a property in common: impairment of the formation of fibrin in blood clotting. Fibrin is produced from fibrinogen by the action of an enzyme, thrombin, which itself is the end product of interactions of a whole series of substances, insufficiency of any one of which will produce a kind of bleeder phenotype. There is evidence of genetic control of at least eight of these substances. Among these the so-called antihemophilic factor is deficient in persons with "classical" sex-linked hemophilia (hemophilia A). Another type of sex-linked hemophilia (hemophilia B) is characterized by a deficiency in the plasma thromboplastin component, also called the "Christmas factor." Three different autosomal bleeder diseases are each characterized by insufficiency of a particular substance: in one, Ac globulin; in another, fibrinogen; in the third, proconvertin. In still another autosomal hemorrhagic condition, the blood seems to be normal, and the bleeding may be due to some defect in the capillaries.

The fact that different individuals who exhibit what appears to be the same disease entity or other phenotype actually often fall into genetically distinct groups is the natural result of the participation of many genes in the formation of a given trait. It is important to be aware of the possibility, perhaps the likelihood, that what appears to be a single phenotypic trait may be genetically heterogeneous.

It may appear to be a contradiction to speak first of the many genes on which a character depends, and then of "the" gene responsible for it. But this contradiction is only apparent. The starting of an automobile depends on the collaboration of many parts: the battery, the electrical wiring, the pistons, the transmission, and others. The character "not-starting" may be due to a disturbance in any one of the parts required for normal function. Therefore, the normal character, "starting," is controlled by many entities; but the difference between starting and not-starting, in any one instance, is usually controlled by a single entity. (Occasionally, inability of a car to start may be due to more than one cause, each of which alone would suffice to cause the effect. Similarly, a person may, in rare cases, be blind for more than one reason—for example, if he happens to possess both the genetic constitution for cataract and that for retinitis pigmentosa.)

There are further consequences of the interrelation of gene-initiated reactions. One of these is that individuals carrying identical genic constitutions may sometimes look quite different. The reason is that the gene-initiated reactions are subject to environmental influences, as are all other chemical or physical processes. Genes cannot be expected, under all environmental circumstances, to produce the same observable character. Just as a mixture of hydrogen and oxygen in a container will be stable if kept undisturbed but will explode if an electric spark is introduced, or as the speed of many other chemical reactions is low at one temperature and high at another, so a gene-controlled character may appear in one form under some circumstances and in another form under others. The problems of the relationship of heredity and environment in man form the subject of several chapters in this book (Chaps. 16, 17, and 25–27).

Another consequence of the interrelations of genic action is that the same gene at a given locus may, if other genes at other loci are not alike, lead to different effects in different individuals, even under the same external circumstances. If, for instance, a certain gene controls the presence of an enzyme which enters into some biochemical process, the activity of the enzyme may be influenced by the acidity of the cytoplasm, which may be under the control of another gene. Thus "genetic background" provided by other genes is often important in the study of any one gene. Examples in man will be presented in Chapter 16.

Many characters from one gene. We turn now to the second generalization derived from our concept of genic action: A single gene may often influence more than one character. Such multiple effects of a gene are called pleiotropic.

The primary products of genic action are polypeptide chains and it is assumed that primarily each structural gene has only a single effect, that of specifying the production of a certain polypeptide. The polypeptides themselves may enter more than one pathway of further reactions. Thus, since pairs of the α hemoglobin chains may join with pairs of β or γ or δ chains, participation in the production of three different hemoglobin types may be seen as a secondary effect of the Hb_α gene.

The operation of the genes in hemoglobin pleiotropy is still close to the primary gene effects. Some more distant relations have been described earlier in this chapter. The gene for alkaptonuria, for example, is responsible for lack of a functional enzyme that leads to the breakdown of homogentisic acid. Consequently, the degradation product maleylacetic acid is absent in homozygotes and homogentisic acid is present in increased amounts. Or, the gene for phenylketonuria is responsible for lack of a functional enzyme that converts phenylalanine to tyrosine. This results in excessive amounts of phenylalanine, a deficiency of tyrosine, and an excess of the alternate degradation product of phenylalanine, phenylpyruvic acid, in the tissue fluids and in the urine. Further consequences of these multiple biochemical abnormalities may be damage to the central nervous system expressed in impaired mental function and decreased production of melanin pigment, due to the tyrosine deficiency, which results in light hair color. In Caucasian patients having an excess of phenylpyruvic acid, the melanin-deficiency effect is obscured by the variety of pigmentation types encountered among normal persons. In populations with uniformly blackish hair color, such as Mongoloids, however, it is easy to separate on the basis of lighter pigmentation phenylketonuric defectives from those mental defectives whose deficiencies have other causes.

An instructive example of biochemical pleiotropism has been studied in *Drosophila*. Presence of a certain gene, called lethal-translucida, results in the accumulation of excessive amounts of blood fluid in the larvae, so that they become bloated and transparent. The chemical composition of their blood fluid differs greatly from that of normal larvae (Fig. 43): the amount of some substances in the abnormal larvae is much greater than in normal ones, and other substances they contain may be absent from normal larvae; still other substances are present in smaller amounts or are lacking completely in abnormal larvae. Undoubtedly, many other biochemical differences will be

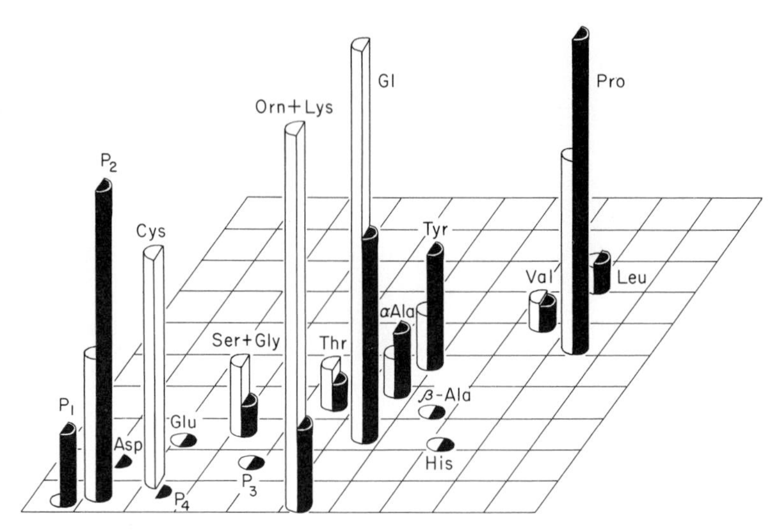

FIGURE 43
Differences in the concentrations of free amino acids, amides, and peptides in the blood fluid of normal (light columns) and lethal-translucida (black columns) larvae of *Drosophila melanogaster.* (P1 – P4 = peptides; Asp = aspartic acid; Cys = cystine; Glu = glutamic acid; Ser + Gly = serine and glycine; Orn + Lys = ornithine and lysine; Thr = threonine; Gl = glutamine; α-Ala = α-alanine; Tyr = tyrosine; β-Ala = β-alanine; His = histidine; Val = valine; Pro = proline; Leu = leucine, isoleucine.) The positions of the columns correspond to the positions of the substances on paper chromatograms (see p. 51). (Hadorn.)

discovered if tests for further substances in the blood fluid are made, and still more if other parts of the larvae are analyzed.

Inherited Syndromes. If several specific abnormal traits present in the same individual are transmitted to his offspring as a unit, as they often are, it can usually be assumed that they depend jointly on a single gene. In medicine, such a group of characters is called a *syndrome.* A well-known example is Marfan's syndrome, or arachnodactyly (spider-fingeredness), so-called because of the excessive length of the bones of fingers and toes (Fig. 44). This and other skeletal abnormalities are often accompanied by an abnormal position of the eye lens and by heart defects. Abraham Lincoln may have been afflicted with this syndrome. Another example is the Laurence-Moon-Biedl syndrome, in which mental deficiency, obesity, possession of extra fingers or toes, and subnormal development of the genital organs go together. It is not clear how one gene produces such diverse characters, but it is quite possible that they are all expressions of some single primary genic activity whose consequences are manifold. For example, the different symptoms of Marfan's syndrome, according to McKusick, may all be consequences of a defect in connective

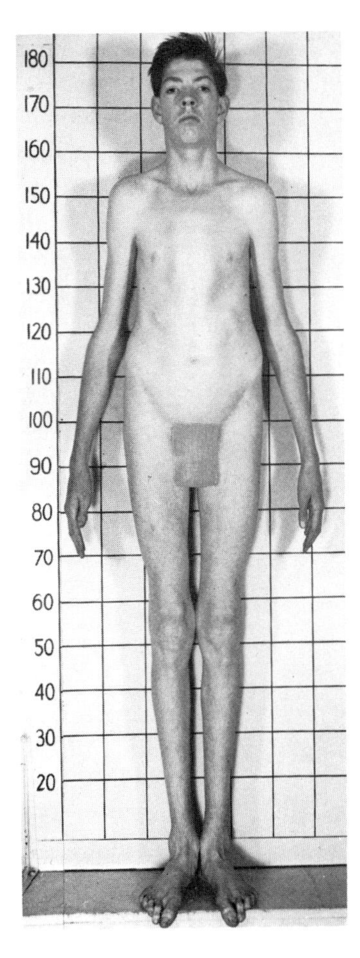

FIGURE 44
Arachnodactyly. *Above:* An affected individual; note length of legs and feet, and "pigeon chest." *Below:* Affected and normal foot for comparison. (Above, original from Dr. V. McKusick; below, Rados, *Arch. Ophth.*, **27**, 1942.)

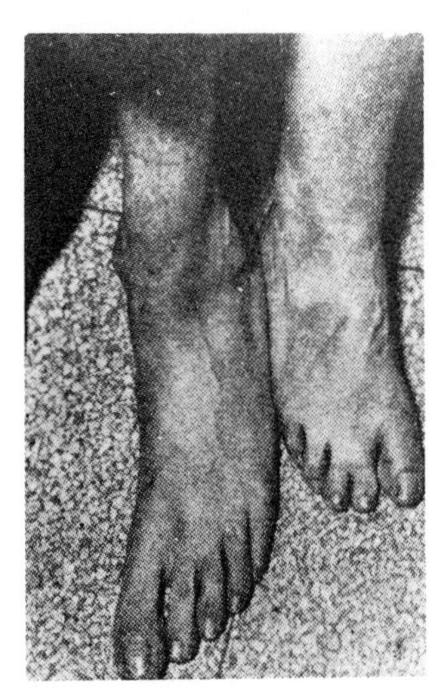

tissue. Grüneberg has been able to provide a far-reaching developmental clarification of an inherited syndrome in rats. In a special genetically atypical strain, individuals are produced who appear normal at birth, but within a few days their growth is retarded, breathing becomes abnormal, and death soon occurs. Anatomical studies show a variety of abnormalities, such as a narrowed windpipe, thickened ribs, bleeding into the lungs, abnormal position of the thoracic viscera, and high hemoglobin content of the blood—all characters that depend in some way on abnormal cartilage formation. Wherever cartilage is formed in the developing animal, it differentiates abnormally and thus leads to various defects, which, in turn, cause further abnormalities in other organs or functions. These abnormalities, singly or jointly, lead to the death of the individual.

In man, a particularly well understood syndrome is that of sickle-cell disease. As described earlier, the abnormal hemoglobin gene Hb^S has as its primary effect the production of an abnormal beta polypeptide chain, which differs from a normal chain in a single amino acid. This difference can lead "to a new isoelectric point, new ionic properties, new solubility for the haemoglobin, new red cell shape, new blood stream properties, haemolysis, jaundice, decreased partial pressure of O_2 in the blood, and so forth" (Fleischman, 1970, *Nature,* **225**:32). If not given special medical attention, many sickle-cell anemic patients die before reaching reproductive age. Numerous abnormalities of many parts and organs are observed, and death may be due to any of a variety of specific failures of the body. All of the abnormalities and failures can be traced back in a "pedigree of causes" to the primary effect of the sickle-cell hemoglobin allele (Fig. 45). The sickle-cell syndrome is a derived multiple consequence of the "molecular disease" that is characterized by hemoglobin S.

The Phenotypes of Heterozygotes. Persons having the genotype $I^{O}I^{O}$ belong to blood group O, and $I^{A}I^{A}$ persons belong to group A. What is the phenotype of the heterozygous $I^{O}I^{A}$ individuals?

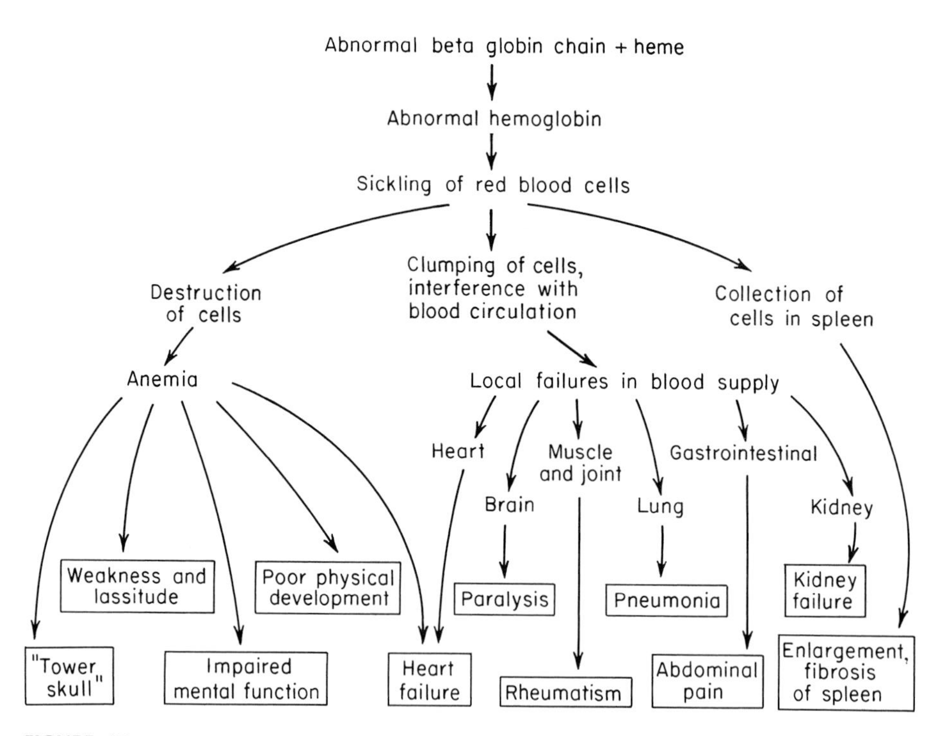

FIGURE 45
A "pedigree of causes" of the multiple effects of the abnormal beta chain of hemoglobin S. (After Neel and Schull, *Human Heredity*, University of Chicago Press, 1954.)

In heterozygosity, three main types of phenotypic expression have been encountered:

1. The heterozygote may exhibit only the properties typical of persons homozygous for one of the two alleles.
2. The heterozygote may exhibit properties intermediate between those of the two homozygotes.
3. The heterozygote may exhibit the properties of both of the homozygotes.

Dominance and recessiveness. The first type is represented by our example of the blood-group alleles I^O and I^A. The heterozygous individuals $I^O I^A$ belong to blood group A, just as the homozygotes $I^A I^A$ do. The allele I^A is said to be *dominant* over the allele I^O, the latter being *recessive* to the former (Table 2, A).

Dominance and recessiveness are obviously developmental phenomena resulting from genic action, since they refer to the effect of a combination of differing alleles as compared to the effects of a homozygous combination. Dominance of an allele in a heterozygote does not necessarily signify a competitive struggle between the dominant and the recessive allele in which the

TABLE 2
Traits of Homozygous and Heterozygous Individuals.

Trait	*Homozygous for One Allele*	*Heterozygous*	*Homozygous for the Other Allele*	*Allelic Relation*
A. Blood groups O, A (dominance involving 2 frequent alleles)	$I^A I^A$ a normal trait	$I^O I^A$ trait as in $I^A I^A$	$I^O I^O$ another normal trait	I^A dominant I^O recessive
B. Polydactyly (dominance of a rare allele)	A^1A^1 normal	A^1A^2 abnormal	A^2A^2 unknown, possibly more abnormal than A^1A^2	A^1 recessive A^2 dominant
C. Albinism (recessivity of a rare allele)	AA normal	Aa normal	aa abnormal	A dominant a recessive
D. Catalase in blood (intermediateness)	A^1A^1 normal	A^1A^2 intermediate between two homozygotes	A^2A^2 abnormal	A^1 } A^2 } intermediate
E. Blood groups A, B (codominance)	$I^A I^A$ a normal trait	$I^A I^B$ a normal trait showing properties of both homozygotes	$I^B I^B$ another normal trait	I^A } I^B } codominant

former is successful. In some instances, there may, indeed, be competition between two different reactions controlled by the two alleles:

$$\text{(1) Competition} \quad \left.\begin{array}{r} A \longrightarrow \\ \leftarrow a \end{array}\right\} \text{effect similar to } AA.$$

But in others it may be true that one allele is inactive while the other is active enough to produce an effect which seems as great as that produced in the homozygote by two alleles:

$$\text{(2) Inactivity of one allele} \quad \left.\begin{array}{l} A \longrightarrow \\ a \end{array}\right\} \text{effect similar to } AA.$$

In $I^{O}I^{A}$ individuals, this seems to be nearly true, the one I^{A} allele causing an effect similar to that of two in $I^{A}I^{A}$ homozygotes, and the I^{O} allele seeming to be inactive.

A dominant effect may even result from the joint action of the dominant and recessive allele since there are known to be recessive alleles whose effect is qualitatively identical or very similar to that of their dominant alleles, although quantitatively weaker than the latter:

$$\text{(3) Cumulative action of both alleles} \quad \left.\begin{array}{l} A \longrightarrow \\ a \rightarrow \end{array}\right\} \text{effect similar to } AA.$$

In such cases, the dominant effect of the heterozygote may be the outcome of the addition of a quantitatively strong effect of the dominant allele to the weaker, but similar, effect of the recessive allele.

Biochemically, dominant alleles are often characterized by their production of polypeptide chains that become parts of active enzymes. In recessive homozygotes it may be that either no active enzyme is formed or an enzyme with reduced activity is formed. Complete absence of an enzyme may be due to actual loss of the gene from the chromosome. More often it may be due to alteration of the base-pair structure of the gene, which then results in an abnormal polypeptide chain that in turn leads to an enzymatically incompetent protein. In experimental organisms it is sometimes possible to decide between the two alternatives since in some cases a protein can be shown to exist that is in some way similar to the active enzyme without being active itself. Alleles that do not lead to the production of active enzymes have been called *silent genes*. Alleles with partial activity are sometimes referred to as *leaky genes* since they do not completely block a given reaction. Other terms characterizing the activity states of alleles were proposed in 1932 by Muller on the basis of experiments with *Drosophila: amorph* for inactive alleles, *hypomorph* for alleles with qualitatively normal but quantitatively reduced activity, and *antimorph* for alleles whose activity in heterozygotes tends to decrease the effect of the allele with normal activity.

Detailed studies by means of enzyme assay, electrophoresis, and other physicochemical methods have frequently demonstrated the existence in a population of a variety of normal alleles. Instead of speaking of "the" normal allele one refers to one or the other of normal *isoalleles*. These then are alleles that each control phenotypes within the range of normality. Much normal variation has its basis in the presence of many normal isoalleles at many loci. Similarily, variations in the expression of abnormal alleles may be due to the existence of a variety of abnormal isoalleles. One of many examples is provided by hemophilia A, in which afflicted persons from different families may show differences in clotting time but afflicted persons within a family have more similar clotting times. Different families seem to carry abnormal isoalleles that differ in the severity of their abnormal effect.

From molecular considerations the existence of alleles with different degrees of activity including its absence is fully expected. If a certain sequence of base pairs in the gene leads to the production of a polypeptide chain with "normal" activity, a sequence that differs from it in a single base pair out of the hundreds or thousands present may lead to the production of a polypeptide having hardly any difference in activity from the "normal," or only a minor difference, or a profound difference depending on the type of base-pair change and, even more, on its position in the sequence. If, instead of the substitution of one base pair for another, the loss of a pair occurs, the transcription and translation of the base-pair sequence into an amino acid chain is likely to result in severely defective protein. Since "reading" of the base-pair code begins at one end of the gene, the modified polypeptide chain would be normal up to the point at which the base pair was lost. It would be abnormal for the rest of its length since the amino acids coded by the normal triplets

$$\underbrace{1\text{-}2\text{-}3}\text{-}\underbrace{1'\text{-}2'\text{-}3'}\text{-}\underbrace{1''\text{-}2''\text{-}3''}\text{-}$$

would not be formed. If, for instance, the base pair 1 were lost, the new triplets would be

$$\underbrace{2\text{-}3\text{-}1'}\text{-}\underbrace{2'\text{-}3'\text{-}1''}\text{-}\underbrace{2''\text{-}3''}\text{-},$$

resulting in a completely different sequence of amino acids in the polypeptide chain. Moreover, sooner or later, a triplet would be read that does not code for any amino acid, and the polypeptide chain would thus be terminated prematurely. Similar "shifts in the reading frame" of the gene would result from insertion of a base pair, and, in many cases, from losses or insertions of more than one base pair.

In human genetics, the term dominance is often used in situations to which it may not fully apply. In order to decide whether two alleles, A^1 and A^2, are related to each other as dominant and recessive or whether they produce an intermediate effect in the heterozygote, it is obviously necessary to compare the effects of all three combinations: A^1A^1, A^1A^2, and A^2A^2. In general, this comparison is not possible for relatively rare inherited traits. For example, we may take the trait polydactyly, the possession of extra fingers and toes (Table 2, B). Assuming that A^1A^1 is responsible for normal appearance—i.e., five-fingeredness—and A^1A^2 causes the development of extra fingers, it is important to know of what type A^2A^2 individuals are. Such individuals, however, can originate only if they receive the "abnormal" allele A^2 from both parents, and this will usually happen only if two abnormal individuals marry. The union of two individuals with such a rare abnormality as this—and striking abnormalities, by definition, are always rare—is very infrequent. Consequently, in most cases, the abnormal homozygote has not yet been found. There is reason to believe—by analogy with animals—that the homozygote would often be still more abnormal than the heterozygote, thus making the latter really intermediate. In practice, however, abnormal traits caused by the presence of a single allele are called dominants in man even if the homozygous expression of the abnormal gene is unknown, or, if known, is identical with, or more extreme than, the heterozygous expression. We may summarize the usage by the statement that rare alleles are called dominant if they cause an appreciable observable effect in the heterozygote.

Rare recessive alleles do not present a similar problem. Recessive homozygotes must receive an "abnormal" allele from each of their parents, most of whom, as will be shown later, are heterozygotes. Therefore, all three genotypes—homozygotes for the recessive, for the dominant, and heterozygotes—are available for comparison. An example of a relatively rare recessive allele is provided in albinism, which consists of the absence of pigment in skin, hair, and the iris of the eye. Individuals who are *aa* (homozygous for the allele *a*) are albinos; homozygotes who are *AA* and heterozygotes, who are *Aa*, are normally pigmented (Table 2, C).

An often quoted example of dominance in two alleles which both act on a normal trait is that of brown and blue eye color. It is usually stated that there is a pair of alleles, *B* and *b*, that control eye color, and that *BB* and *Bb* produce brown eyes, and *bb* blue. This description is only an approximation of the truth. The irises of some blue-eyed persons contain small spots of brown pigment which may not be noticeable without careful inspection. Two such seemingly blue-eyed individuals can have brown-eyed children. The genetics of eye color is complex and not yet fully understood.

Intermediateness. In heterozygotes properties intermediate between those of the two homozygotes are not rare if our consideration extends to biochemical features that presumably are not too distantly removed from the primary gene effects. If, for instance, an allele A^1 in the homozygous state leads to an active enzyme and an allele A^2 in the homozygous state to one of very low or no activity, then the combined activity of A^1 and A^2 in the heterozygote may lie between the activities found in the homozygotes. There is, of course, usually some variability in the measured effects of each of the three genotypes. Sometimes, in spite of this variability, there is no overlap between the data for each of them. Thus, a very rare gene in homozygous individuals leads to the absence in the blood of catalase, an enzyme found in considerable amounts in most individuals. The catalase content of heterozygotes is intermediate, without overlap, between those of the two homozygotes (Table 2, D; Fig. 301). In many other diseases or abnormalities there is some overlap between the heterozygote and either one or both homozygotes, which blurs the distinction between intermediateness and dominance or recessiveness. The rare disease galactosemia may serve as an example. Affected individuals seem to lack completely the enzyme galactose-1-phosphate-uridyl transferase, which is necessary for the normal metabolism of the sugar galactose. Heterozygotes generally have between 60 and 70 per cent of the enzymatic activity of normal homozygotes. As seen in Figure 299 (p. 804), the abnormal homozygotes are fully separable from the heterozygotes, but there is some overlap in activity between the heterozygotes and normal homozygotes.

The enzymatic expression of the genes for acatalasemia and galactosemia in heterozygotes lies between those of the two types of homozygotes. In a more general sense, that of the distinction between normal health and illness, the abnormal alleles for the two diseases are strictly recessive since the enzyme activity in the heterozygote is sufficient to assure normal health. This is true of many traits that are developmentally far removed from primary genic action. An important task, particularly for medical work in human genetics, is the recognition of the heterozygous "carrier state" for genes that at first may seem to fit the scheme of the simple alternatives, normal-dominant and abnormal-recessive. It should be added that reference to intermediateness in a heterozygote does not imply that an expressed character lies exactly in the middle between those of the homozygotes, as was just seen for galactosemia, in which the activity level of the heterozygote is significantly more than 50 per cent that of the normal homozygote. Obviously, regulatory processes influence the effect of the single normal allele in the heterozygote at some stage of the cellular activity initiated by it.

Clear cases of intermediate expression of heterozygotes are relatively rare in man as compared to experimental organisms. This is so in part because, as we have seen, genes that are responsible for abnormal traits in heterozygotes are liable to be classified as dominants, although their heterozygous effects may actually be intermediate between those of the normal and abnormal homozygotes. In part, the difficulty of establishing intermediateness concerns common traits. Such traits often depend on genetic differences at more than one locus—for reasons to be discussed in Chapter 7—so that it is difficult to make clear decisions concerning dominance or intermediateness of alleles at a single locus.

Codominance. For a well defined example of a heterozygote's exhibiting the properties of both homozygotes, we may refer again to sickle-cell hemoglobin. A normal homozygous Hb^{A} Hb^{A} individual forms only normal hemoglobin, and an anemic homozygous Hb^{S} Hb^{S} individual has only sickle-cell hemoglobin. The heterozygote, with the usually harmless sickle-cell trait, produces both types of hemoglobin, the abnormal one making up from one-quarter to nearly half of the mixture (Fig. 23, E, F).

Another example of codominance shows a different relation of the derived products of two codominant alleles. The blood-group gene I is one for which more than the two different alleles I^{A} and I^{O} are known. I^{A} leads to the presence of an "A" antigen in the blood. Another allele, I^{B} causes the presence of a "B" antigen. In combination with each other, I^{A} and I^{B} present a new relation (Table 2, E). The heterozygote $I^{A}I^{B}$ does not show dominance of either allele, nor does it show an intermediate expression. Instead, the specificities A and B are found together in the blood of the $I^{A}I^{B}$ individuals. It is noteworthy that a test-tube mixture of the A and B substances from A and B individuals can be separated experimentally into two kinds of macromolecules that carry one or the other of the A and B specificities, but that an AB individual does not carry separate A and B macromolecules. Instead, a third type of macromolecule is formed that carries the two specificities together.

The difference in the codominance effect in Hb^{A} Hb^{S} heterozygotes and in $I^{A}I^{B}$ heterozygotes reflects the difference in the level at which the gene effects are observed. The hemoglobin molecules are polymers of polypeptide chains whose synthesis is close to primary gene action. Thus the separate effects of the two Hb alleles express themselves in separate types of hemoglobin. The A and B antigens are the results of action of enzymes that are themselves controlled by the two I alleles or a macromolecular precursor. Thus the separate effects of the I alleles express themselves as minor differences at different sites of single molecules.

We have seen that a gene participates in the developmental network in such a way that it may be involved in the expression of more than one character. Though our procedure of naming a gene for its most conspicuous effect often causes us to forget its varied expressions, most genes seem to have multiple effects. Remembering this, it is not surprising that the same gene may be dominant in regard to one expression and recessive or intermediate in regard to another. For example, the skin disease xeroderma pigmentosum (Fig. 245, p. 599), characterized by a peculiar type of freckling, degenerating areas, and cancerous growth invariably leads to premature death, is inherited as a recessive abnormality as far as the serious pathological effects are concerned—i.e., the heterozygotes are normal people—but the freckling shows in many of them, and is thus a dominant expression of the gene.

Gene Symbols. Repeated use has already been made of gene symbols, and since their use will increase in later chapters, it may be helpful if they are explained here. It is customary to denote the genes of specific loci by one or a few letters of the alphabet printed in italics. Such a designation may be completely arbitrary—as is the use of the letters *A, B, M,* or *X*—or the letters chosen may be abbreviations of the names of specific traits or characters: for example, *I* for blood substances, called isoagglutinogens, or *Xe* for the locus concerned with xeroderma pigmentosum. Different alleles of a gene are all given the same letter with an additional distinguishing mark: for instance A and A', A^1 and A^2, or I^O and I^A. This makes it easy for the reader to recognize the allelic groupings in a genotype. In the formula $AA'I^OI^A$, it is clear that A and A' are a pair of alleles, and that I^O and I^A are a different pair. When one allele of a pair is recessive to the other, its symbol is not capitalized by many authors. *A* is used for the dominant and *a* for the recessive, or *D* for the dominant and *d* for the recessive (*not D* versus *r*!).

Sometimes, alleles with particularly similar action are distinguished by such symbols as I^{A_1} and I^{A_2}. Sometimes, also, the same base letters may be used for genes at different loci with distinguishing subscripts: Hb_α and Hb_β designate genes at two loci both of which are concerned with hemoglobin properties.

Genotype and Phenotype. The terms genotype and phenotype can be applied either to the totality of the constitution and its expression, or to part of it. They are often used in relation to a single pair of alleles and its expression. Thus, one may refer to the heterozygous genotype *Aa* and its dominant phenotype.

The simple statement "The phenotype is not always an indication of the genotype" summarizes important discoveries. We have seen that a genotypically phenylketonuric child can be made phenotypically normal by treatment with a special diet and that the phenotype blood group A may be the result of either of the two genotypes I^AI^A and I^AI^O. The two concepts, genotype and phenotype, form a frame of reference in many discussions of human genetics.

Autonomous and Dependent Characters. The existence of syndromes shows that a specific gene may produce effects in different parts of the body. The question arises whether (1) the gene is necessarily present in all tissues or organs that are affected, or whether (2) some basic effect at one stage of development in one particular primary organ may be the decisive factor that controls, secondarily, the production of the specific traits in other parts of the body. If the latter alternative were true, the appearance of the traits would be unrelated to the fact that the organs concerned also happened to contain the gene in their cell nuclei. Answers to this question are provided by work with animals, and two experiments conducted with mice will be described briefly.

In the first of these, color differences of the hair coat were studied. It is known that such differences in color are due to the presence of different kinds of genes. In this case, our question may be formulated in the following terms: Is the fur of a black mouse black because the pigment cells of the animal are genetically black, or because genic action in some other part of the animal sets up a physiological condition that affects the pigment cells and causes them to make the hair black? A decision is made possible by transplanting a piece of skin from a newborn or late embryonic genetically black mouse to a genetically nonblack one. At the time of transplantation, no pigment has been formed. Will the transplant obey its own genic commands or will it respond to the genic influences of the host? The answer is that the transplant forms black pigment, independently of the host, by *autonomous determination* (Fig. 46, A).

It may be wondered whether this autonomy of the transplant is really the result of genic action in its own cells, or whether the formation of black pigment is due to some processes that had been initiated in remote parts of the genetically black embryo before transplantation took place. This possibility has been tested in experiments in which two fertilized eggs at early cleavage stages are fused and then develop into single mouse individuals. When the two eggs were genetically different in their pigmentation potential, the resulting mice were phenotypic mosaics of the two pigmentation types showing that the genetic control of pigmentation is in the pigment cells themselves. Essen-

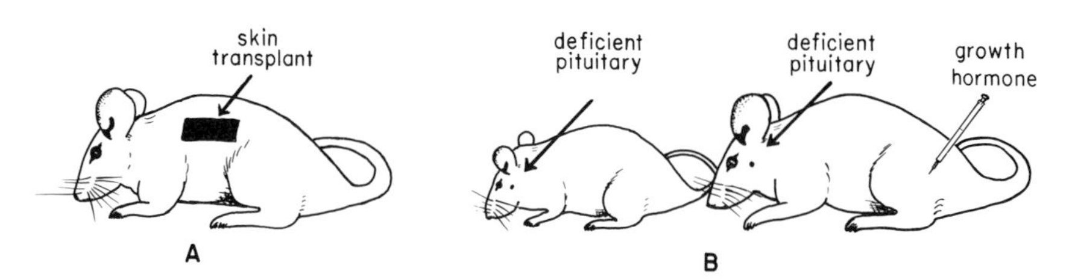

FIGURE 46
Autonomous and dependent development. **A.** Genetically pigmented piece of skin transplanted to an unpigmented host develops its intrinsic coloration. **B.** *Left:* A genetically dwarf mouse with deficient pituitary. *Right:* Growth to normal size of a genetically dwarf mouse dependent on injection of growth hormone.

tially similar observations in human mosaics, which have occurred "spontaneously"—i.e., without experimental manipulation—have confirmed that pigmentation is an example of *autonomous determination.*

The second experiment shows that genetic effects are not always brought about by autonomous determination. There is a class of mice in which the normal allele *D* is replaced by *d,* with the result that growth is retarded and the animals remain dwarfs. The pituitary gland of such a dwarf mouse is abnormal. The anterior lobe of this gland normally secretes into the blood stream hormones which are necessary for normal body growth, but none, or only insignificant amounts, of the hormones are produced by the pituitary of the dwarfs. This defect of the gland would be a sufficient explanation for the growth retardation of the animals. The question remains, however, whether the *d* allele, which, in the cells of the pituitary, is responsible for the deficient hormone production, may, in addition, by its presence in the cells of the growing parts themselves, inhibit their growth. A decision is reached by transplanting pituitary glands from normal mice (or rats) into the appropriate region of the brain, the hypothalamus, of dwarf littermates whose abnormal pituitaries have been removed, or by injecting mammalian growth hormone into them (Fig. 46, B). Such manipulations produce normal growth of the genetically dwarf mice. In contrast, when pituitary glands from dwarf animals are transplanted into normal littermates whose pituitaries have been removed, the animals are dwarfed. Here, then, is an example of *dependent determination.* The dwarf allele causes pituitary deficiency, and the easily observable character, dwarfness, is secondarily dependent on the insufficient production of the hormones.

As in mice, autonomous and dependent determination can be distinguished in inherited human traits, although clear-cut experimental proof is usually not obtainable. It seems likely, however, that autonomous genetic determination in man accounts for hair and eye colors, for the antigens that characterize various blood types, for certain diseases in which specific cells of the nervous system degenerate, for red-green color blindness, and for many other characters. Dependent determination in man probably accounts for the development of certain midgets with adult body proportions who are presumably pituitary-deficient dwarfs; for at least those types of sugar diabetes (diabetes mellitus) in which the abnormal carbohydrate metabolism is a consequence of defective pancreatic glands; and for genetic subnormal functioning of the thyroid in children, which may lead to stunted physical and mental development (cretinism). Dependent determination also brings about the development of those traits which are not mediated by hormones but by interrelated physiological processes comparable to some of those active in the development of the sickle-cell anemia syndrome.

Phenocopies. The skin of many people is heavily pigmented even in those regions of the body that are not exposed to sunlight. This is due to the action of specific alleles. Other people are only lightly pigmented in unexposed regions but become darkened where exposed. The latter carry alleles for light skin, but the processes leading typically to limited pigment formation can be reinforced by the external agent "sunlight." Sun-tanned, genetically light individuals thus are copies of genetically dark individuals. The term *phenocopy* has been coined to designate individuals whose phenotype, under the influence of nongenetic agents, has become like the one normally caused by a specific genotype in the absence of the nongenetic agents (Goldschmidt, 1878–1958).

Numerous examples can be cited to show that a given phenotype is either the result of a specific genotype or a phenocopy produced by the interaction of a nongenetic agent with a different genotype. Thus, cretinism can be due to the presence of certain alleles or to the lack of iodine in the diet of a child of any genotype, and cataract can be the result of specific genes or of damage to the lens by ionizing radiation. The existence of phenocopies is an aspect of genic and gene-dependent action which conforms to expectations based on the general concepts of the relation between gene and character.

In this chapter, we have tried to picture genic action in terms of intracellular reactions and have outlined the role of genic action in the development and the biochemistry of the individual. A few summarized experiments and some deductions from general principles have indicated the relationship between genes and specific traits. Important phenomena and concepts have been discussed and defined: dominance, recessiveness, intermediateness, and codominance; genotype and phenotype. The ground has been prepared for a more detailed treatment of many aspects of human genetics.

REFERENCES

Baumeister, A. A., 1967. The effects of dietary control on intelligence in phenylketonuria. *Amer. J. Ment. Defic.,* **71**:840–847.

Caspari, E., 1952. Pleiotropic gene action. *Evolution,* **6**:1–18.

Harris, H., 1963. *Garrod's Inborn Errors of Metabolism.* 207 pp. Oxford University Press, London. (A reprint of the first edition, 1909, with a supplement by H. Harris.)

Harris, H., 1970. *The Principles of Human Biochemical Genetics.* 328 pp. North Holland Publ. Co., Amsterdam.

Hertig, A. T., Rock, J., and Adams, E. C., 1956. A description of 34 human ova within the first 17 days of development. *Amer. J. Anat.,* **98**:435–493.

Hsia, D. (Yi-Yung), 1966. *Inborn Errors of Metabolism,* Part 1, Clinical Aspects. 2nd ed. 396 pp. Year Book Medical Publishers, Chicago.

Hsia, D. (Yi-Yung), 1968. *Human Developmental Genetics.* 400 pp. Year Book Medical Publishers, Chicago.

Huehns, E. R., and Shooter, E. M., 1965. Human haemoglobins. *J. Med. Genet.,* **2**:1–92.

Hunt, J. A., and Ingram, V. M., 1958. Allelomorphism and the chemical differences of the human hemoglobins A, S, and C. *Nature,* **181**:1062–1063.

Jervis, G. A., 1954. Phenylpyruvic oligophrenia (phenylketonuria). *Res. Publ. Ass. Nerv. Ment. Dis.,* **33**:259–282.

Kerr, C. B., 1965. Genetics of blood coagulation. *J. Med. Genet.,* **2**:254–303.

Kirk, R. L., 1968. *The Haptoglobin Groups in Man.* 77 pp. Karger, Basel. (Monographs in Human Genetics, **4**.)

Knox, W. E., 1958. Sir Archibald Garrod's "inborn errors of metabolism": I. Cystinuria, II. Alkaptonuria, III. Albinism, IV. Pentosuria. *Amer. J. Hum. Genet.,* **10**:3–32, 95–124, 249–267, 385–397.

Langman, J., 1969. *Medical Embryology,* 2nd ed. 386 pp. Williams and Wilkins, Baltimore.

Morgan, W. T. J., and Watkins, Winifred M., 1969. Genetic and biochemical aspects of human blood-group A-, B-, H-, Le^{a}- and Le^{b}- specificity. *Brit. Med. Bull.,* **25**:30–34.

Patten, B. M., 1968. *Human Embryology.* 3rd ed. 651 pp. McGraw-Hill, New York.

Stanbury, J. B., Wyngaarden, J. B., and Fredrickson, D. S. (Eds.), 1972. *The Metabolic Basis of Inherited Disease,* 3rd ed. 1778 pp. McGraw-Hill, New York.

Watkins, Winifred, 1967. The possible enzymatic basis of the biosynthesis of blood-group substances. Pp. 171–187 *in* Crow, J. F., and Neel, J. V., (Eds.), *Proceedings of the Third International Congress of Human Genetics.* Johns Hopkins University Press, Baltimore.

Witschi, E., 1956. *Development of Vertebrates.* 588 pp. Saunders, Philadelphia.

4

MEIOSIS

The cells that make up the body of a man or a woman contain two sets of 23 chromosomes, or 46 chromosomes in all. When, however, a mature egg is formed by a woman, or a mature sperm by a man, each of these germ cells contains only one set of 23 chromosomes. How is the reduction in number of sets from two to one—in number of chromosomes from 46 to 23—accomplished? This is clearly not just a problem of cellular detail but a question of great significance for human inheritance, since changes in the number of chromosomes entail, by necessity, changes in the number of genes.

It has taken nearly a half-century of the most painstaking study of many animals and plants to determine how chromosome reduction takes place. In the growing ovary or testis of a human, each cell, like the other cells of the individual, contains 46 chromosomes. After many cell divisions, with normal mitotic behavior of the chromosomes, numerous immature germ cells have been produced. These cells undergo a unique process called *meiosis* (from the Greek *meiosis* = diminution), which results in production of the mature germ cells.

A Preliminary Outline of Meiosis. The essential aspects of meiosis are not difficult to understand, but since several different processes go on simultaneously, it will be helpful to give a simplified description before a more complete account is rendered.

Figure 47, A and B, shows an immature germ cell in which the chromosomes become differentiated in a way characteristic of the early stages of nuclear division. As in the diagram of mitosis (Fig. 8), only two pairs of chromosomes are shown: a pair of long chromosomes with a kinetochore in the middle, and a pair of short chromosomes with a kinetochore near one end.

Figure 47, C, shows the *first* of the unique features of meiosis: homologous chromosomes attract each other and lie side by side. As the diagram indicates, this pairing brings together the homologous chromosomes in such a way that each chromomere pairs with its homologous chromomere, and kinetochore with kinetochore.

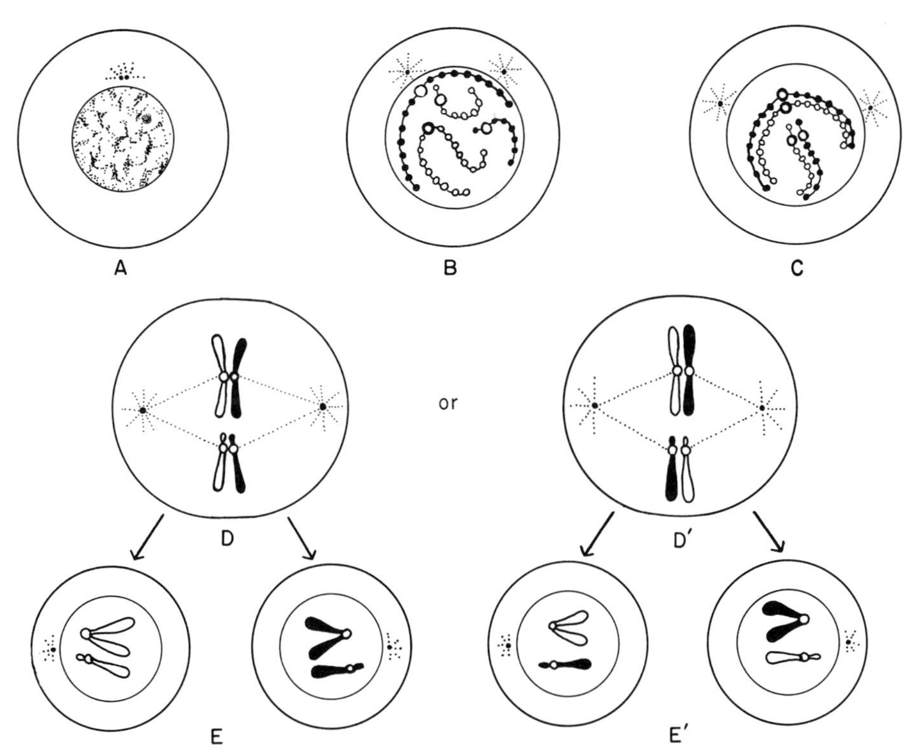

FIGURE 47
Meiosis, simplified: two pairs of chromosomes. *Dark:* Paternal chromosomes. *Light:* Maternal chromosomes. **A.** Nucleus in a premeiotic germ cell. **B–C.** Pairing of homologous chromosomes. **D–D′.** The two alternative arrangements of the chromosome pairs on a meiotic spindle. **E–E′.** The four different types of reduced chromosome constitutions of the gametes.

While the relatively uncoiled, paired chromosomes condense into shorter, heavier bodies, a spindle is formed on which they become arranged as in mitosis (Fig. 47, D and D′). At this stage, a *second* important feature of meiosis can be observed. In mitosis, each kinetochore lies apart from its homologue; in meiosis, it lies beside it. In mitosis, the kinetochores divide and lead their associated sister chromosomes to opposite poles. In the meiotic division pictured in Figure 47, D and D′, the paired kinetochores do not divide, but move apart, leading one of each pair of chromosomes to opposite poles. Thus, when cell division takes place, each daughter cell contains only one chromosome of each pair, or, stated differently, only one instead of two sets of chromosomes (Fig. 47, E and E′).

A *third* characteristic of meiosis is shown in the alternative diagrams of Figure 47, D and D′: there is no fixed pattern of arrangement of paternal and maternal chromosomes on the meiotic spindle. On the spindle of Figure 47, D, both maternal chromosomes, the long and the short, lie on one side, and the two paternal ones on the other. On the spindle of Figure 47, D′, the two maternal chromosomes lie on opposite sides, as do the paternal ones.

In microscopic preparations of cells, the paternal and maternal partners of a pair of homologous chromosomes are almost always indistinguishable from each other. Occasionally, however, slight permanent differences can be seen, and it has been observed that each pair of homologues acts as if it had been ordered to arrange itself on the spindle without paying attention to any other pair. The result is an independent arrangement of the different chromosome pairs. When division occurs and "reduced" daughter cells with only one set of chromosomes are formed, the meiotic process shown in Figure 47, D, produces one cell containing the long and short maternal chromosomes and another containing the long and short paternal ones (Fig. 47, E). The meiosis in Figure 47, D′, leads to one cell with the long paternal and the short maternal chromosomes and another cell with the long maternal and short paternal ones. Since the two arrangements, D and D′, are equally common, all four resulting gametes are also equally common. Meiosis thus entails the independent or free assortment of the maternal and paternal chromosomes during the reduction from two to one chromosome set.

The Two Meiotic Divisions. The preceding simplified description referred to a single division by means of which chromosome reduction is accomplished. In reality, meiosis always includes two successive divisions (Fig. 48). During the *first* meiotic division, the paired homologous chromosomes are present in duplicate, the duplicates being joined by a single kinetochore just as in the early stages of mitosis. Each pair, therefore, consists of four chromosome strands (*chromatids*), two of which are held together by a maternal and two by a paternal kinetochore. When the kinetochores move toward opposite

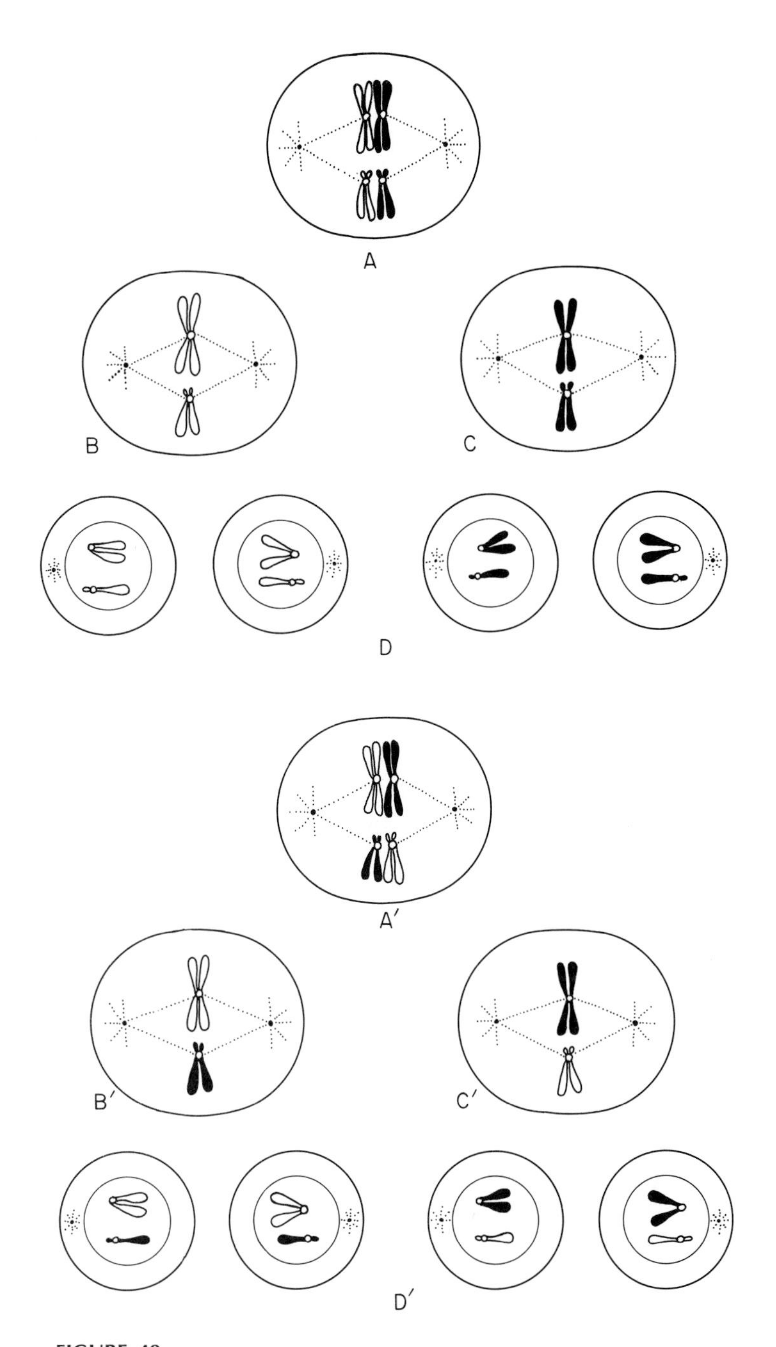

FIGURE 48
Meiosis, including the two meiotic divisions. **A – A′**. The two alternative arrangements of the chromosome pairs on the first meiotic spindle. **B–D**, and **B′–D′**. The second meiotic divisions and the different types of reduced chromosome constitutions of the gametes.

poles, each carries with it two chromosome strands. The two daughter cells which are formed as a result of the first meiotic division immediately divide again. During this *second* meiotic division, the kinetochores with their double strands assemble on the spindles of the daughter cells. As in mitosis, the kinetochores divide, and the daughter kinetochores, each joined to one chromosome strand, move to opposite poles. Thus, a premeiotic germ cell with two sets of chromosomes (Fig. 48, A, A′) forms four germ cells, each with a single set of chromosomes (Fig. 48, D, D′).

The two meiotic divisions are often called the reduction divisions. The first division reduces to one-half the number of kinetochores in a cell; the second reduces to one-half the number of chromosome strands. An answer to the question of why meiosis takes two instead of only one division will have to await a better knowledge of the molecular architecture of chromosomes and of the forces involved in duplication, pairing, and separation of kinetochores and chromosome strands.

The cells in which meiosis takes place, the oocytes in the female and the spermatocytes in the male, originate, by division involving ordinary mitosis, from oogonia and spermatogonia. After meiosis has been completed, the female germ cell is a mature egg, but the male germ cell, the spermatid, has still to transform itself into a mature sperm.

Crossing Over. Meiosis includes an important additional feature: *crossing over*. This is a process by which homologous segments of paired maternal and paternal chromosomes are exchanged. Its effect is that even genes that are "linked" in a single chromosome do not necessarily remain together from one generation to another, but are interchanged with their alleles in the homologous partner chromosome. If, for instance, the maternal chromosome contained, among other genes, the genes A and B, and the paternal chromosome the alleles A' and B', then, after meiosis, some gametes might contain the original chromosomes with AB or $A'B'$ (no crossing over) and others exchange-chromosomes with AB' or $A'B$.

Examples of crossing over are illustrated in Figure 49. Three pairs of loci have been indicated in a pair of chromosomes, the maternal alleles by M, N, and O, and the paternal alleles by M', N', and O' (Fig. 49, A). In the duplicated chromosomes, each locus is present twice: $MMM'M'$, $NNN'N'$, and $OOO'O'$ (Fig. 49, B). A crossover between the M, M' and N, N' loci is shown in Figure 49, C. Only two of the four chromosome strands are involved in the exchange. Two strands are noncrossovers and have retained the original maternal or paternal sequence of alleles, MNO or $M'N'O'$, respectively. The two other strands are crossovers. One of them begins with the maternal allele M, but continues with a section containing the paternal alleles N' and O'. The other begins with the paternal allele M', but continues with

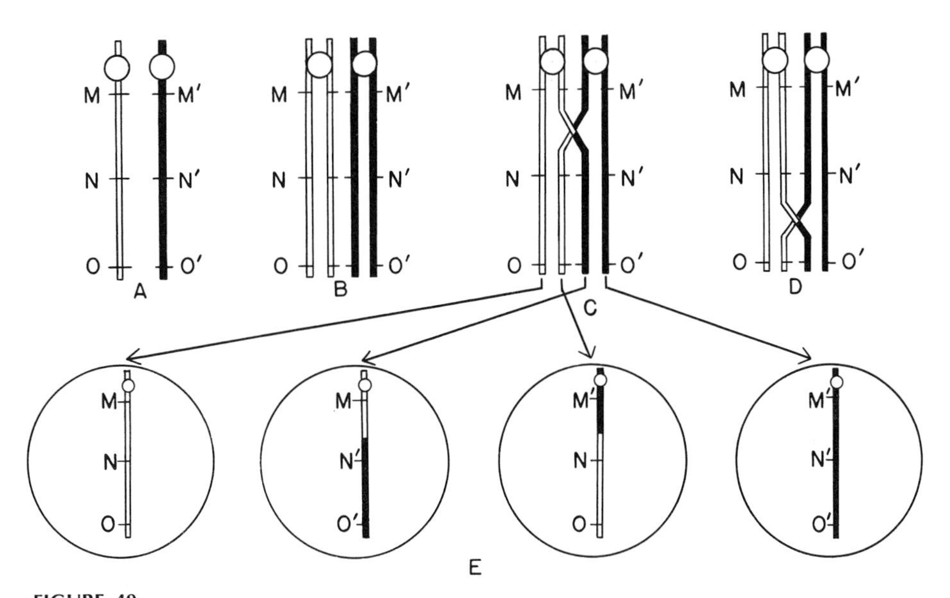

FIGURE 49
Crossing over. **A.** A pair of homologous chromosomes heterozygous for three pairs of loci *M, M′*; *N, N′*; and *O, O′*. **B.** Four-strand stage. **C.** Crossing over between two of the four strands in the region between *M, M′* and *N, N′*. **D.** Same in the region between *N, N′* and *O, O′*. **E.** The four types of reduced chromosome constitutions of the gametes resulting from the crossing over shown in C.

a section containing the maternal alleles *N* and *O*. It is as if the crossover strands originated by breakage of two original strands at exactly homologous locations and healed together after they had crossed over.

It has been found that the place of a crossover in a given pair of homologous chromosomes varies from one germ cell to another. Considering only the three pairs of loci—*M, M′*; *N, N′*; and *O, O′*—crossovers can occur between *M, M′* and *N, N′* in some cells, and between *N, N′* and *O, O′* in others. These crossovers are seen in Figure 49, C and D. Had more than three pairs of loci been shown in the diagrams, then more than two possible crossover sites could have been indicated. For instance, a pair of chromosomes with six pairs of loci, occupied by the alleles *MNOPQR* in the maternal chromosome and the alleles *M′N′O′P′Q′R′* in the paternal chromosome, could, in different cells, undergo crossing over between any two successive loci, that is, in five different regions.

Any single crossover between two homologous chromosomes affects only two of the four chromosome strands. Since these four strands are distributed into four separate cells by the two meiotic divisions, each of the four cells receives a different type of chromosome. If we follow, as an example, the crossover in Figure 49, C, we see (Fig. 49, E) that one cell receives the maternal noncrossover strand *MNO*, another the crossover strand *MN′O′*, a third the complementary crossover strand $M'NO$, and a fourth the paternal noncrossover strand $M'N'O'$.

Within a given cell a pair of homologous chromosomes may cross over at two or more places. This is shown in Figure 50 for those double crossovers that involve only two strands. Figure 51 shows that double crossovers may involve not only two strands (A) but also three (B) and all four strands (C) of the chromosome pair. In different oocytes of the same woman each one of the 23 chromosome pairs may undergo single crossing over, or some pairs may undergo double crossing over, or some may cross over three or even more

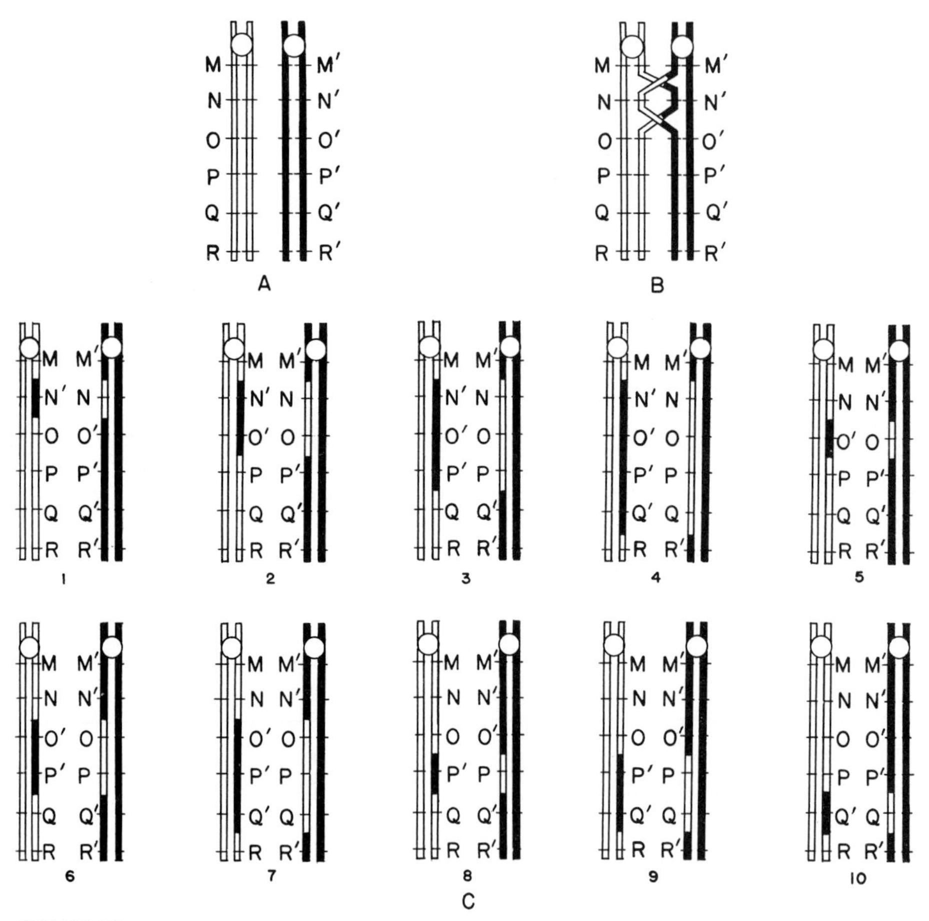

FIGURE 50
Double crossing over. **A.** A pair of homologous chromosomes heterozygous for three pairs of loci. **B.** A double crossover involving two of the four strands. **C, 1.** The four chromosome strands resulting from the double crossover shown in B. **C, 2–10.** Chromosome strands resulting from the outer 9 possible types of double crossovers involving two of the four strands in A.

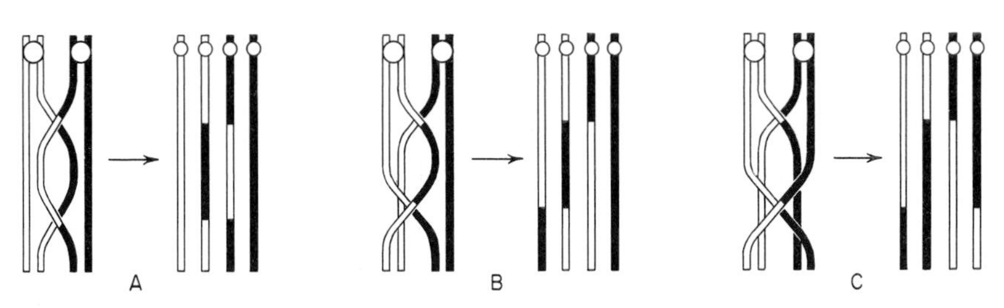

FIGURE 51
Double crossing over. **A.** Two strands. **B.** Three strands. **C.** Four strands. Note that the two-strand double crossing over results in two chromosomes that have undergone two exchanges and two chromosomes that have undergone no exchanges; that the three-strand double crossing over results in four chromosomes that had undergone 1, 2, 1, and no exchanges, respectively; and that the four-strand double crossing over results in four chromosomes, each of which has undergone one exchange only.

times; and the locations of the crossovers along the lengths of the chromosomes may vary from cell to cell. The same is true for crossing over in spermatocytes except for the pair of sex chromosomes (the X-chromosome and the Y-chromosome), which do not seem to cross over regularly, if at all.

Meiosis and the Formation of Egg and Sperm. Many of the important phenomena of meiosis are summarized in Figure 52, and a careful study of the figure is advisable. The formation of both mature eggs and sperm is illustrated. Essentially, meiosis is alike in both, but one difference is significant. In the male, the first meiotic division results in two cells of equal size; the second meiotic division, which occurs simultaneously in these two cells, leads to the formation of four equal-sized cells, each of which becomes a mature, functional sperm (Fig. 52, D-H, right). In the female, the two divisions do not yield four cells of equal size. The cytoplasm of an immature egg cell has been organized in a special way in preparation for development and has been loaded with reserve food material, so that its growth has made it one of the largest cells in the human body. The meiotic divisions occur in such a fashion that the structure and size of one of the four cells that are being formed are almost the same as the original cell. This is accomplished by the first meiotic spindle being near the cell's periphery, with its long axis lying along a radius of the egg (Fig. 52, D-H, left). Only a small bud of cytoplasm, including the peripheral daughter nucleus, is pinched off the main body of the cell. The second meiotic division occurs simultaneously in the large cell and in the pinched-off cell, which is called the first polar body. In the latter, the second division results in two equal-sized polar bodies, provided the division is completed, which is not always the case. In the large cell, the second meiotic spindle occupies the same asymmetrical position as the first meiotic spindle and, again, this results in an unequal cell division: a second polar body is separated from the now-mature

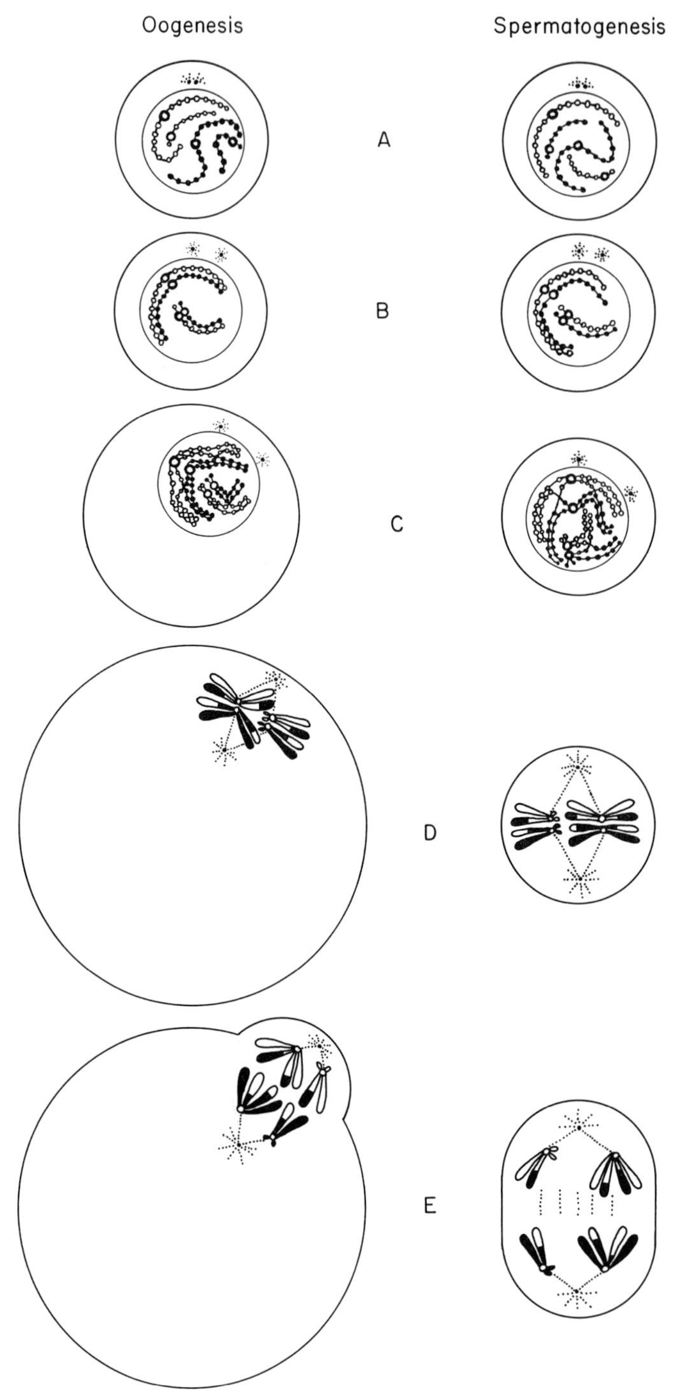

FIGURE 52
Meiosis and formation of gametes. **A–C.** Chromosome pairing and crossing over. **D–E.** First meiotic division.

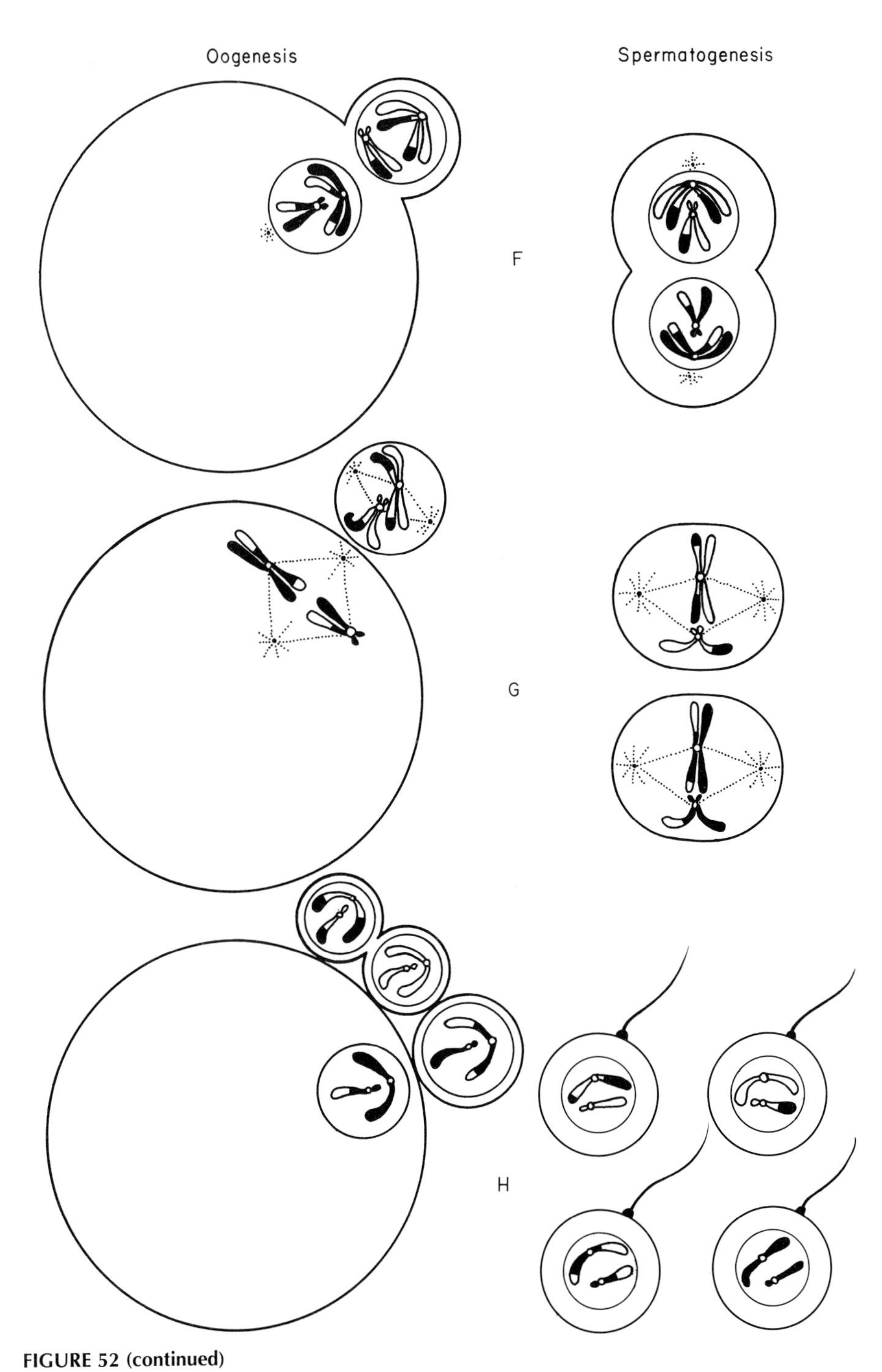

FIGURE 52 (continued)
F. Products of first division. **G.** Second meiotic division. **H.** Egg with polar bodies (*left*), sperm cells (*right*).

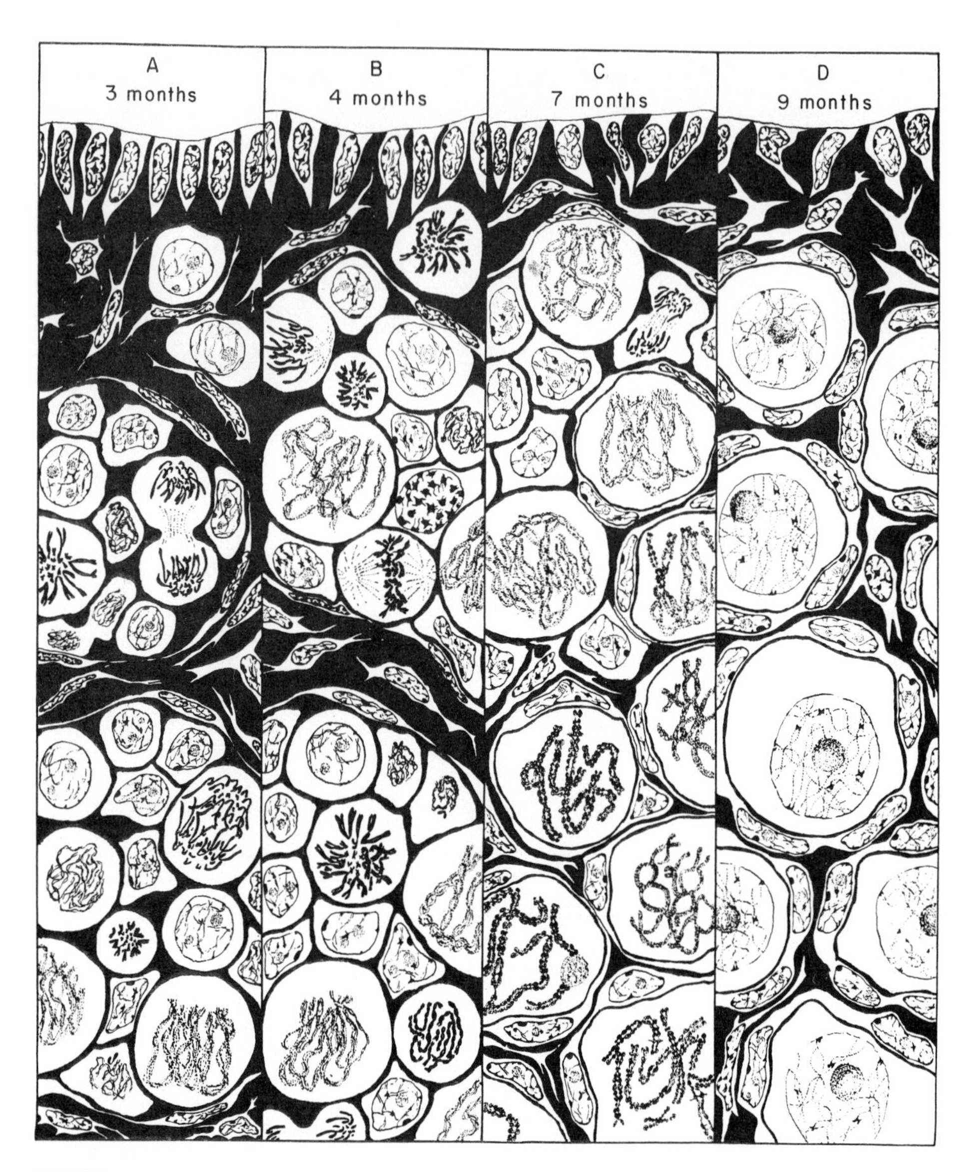

FIGURE 53
Sectors of the cortical areas of fetal human ovaries. **A.** Oogonia (large cells) and follicular cells (small) showing mitotic activity. In the lowest region, adjacent to the medulla, a few oocytes whose chromosomes are in early pairing stages. **B.** Greater numbers of oocytes in early pairing stages. **C.** Apparently, oogonia no longer present. Mitosis is restricted to follicular cells. At the superficial layer oocytes in early stages of pairing; in the deeper layers, at advanced stages. **D.** All oocytes have proceeded to the dictyotene stage, a stage in which the paired chromosomes are so extended as to give a diffuse appearance to the nucleus. (Woodcut by S. Ohno from Ohno, Klinger, and Atkin, *Cytogenetics,* **1,** 1962.)

egg, which has maintained a volume almost equal to that of the original egg cell. Four viable sperm cells but only one viable egg cell plus three degenerating polar bodies are thus formed by meiosis of a single cell.

Sections of parts of four fetal human ovaries with cells at various early stages of meiosis are shown in Figure 53. The paired chromosomes of a cell from a male at a stage shortly before the first meiotic division are seen in Figure 54.

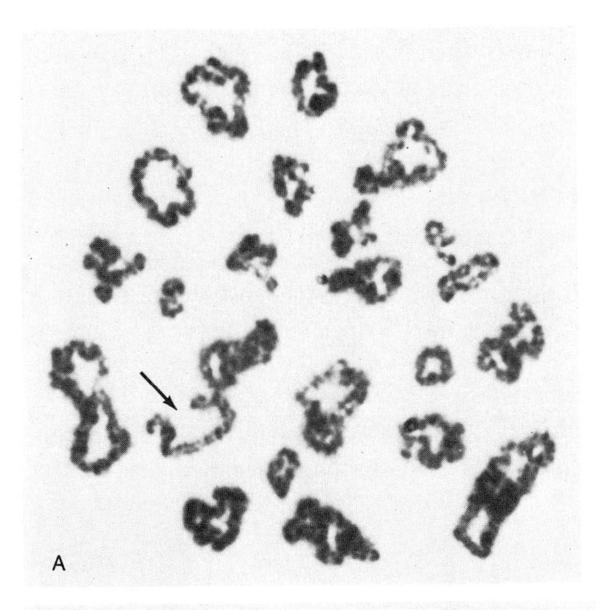

FIGURE 54
A. An immature male germ cell, shortly before the first meiotic division, showing pairing of the 23 homologous chromosomes.
B. The karyotype of the cell. The X- and Y-chromosomes, marked by an arrow, are in characteristic end-to-end association. (Court Brown, *Human Population Cytogenetics*, North Holland Publ. Co., 1967.)

GENETIC CONSEQUENCES OF MEIOSIS

The cellular details of meiosis will now be translated into the language of genetics. We shall first follow the fate of a single pair of alleles, then that of two or more pairs located in different pairs of homologous chromosomes, and, last, that of pairs of loci which are linked by being carried in the same pair of chromosomes.

Segregation of a Single Pair of Alleles. In the early stages of meiosis, every pair of alleles—A and A', for example—that is present in a pair of homologous chromosomes of the immature germ cells is duplicated when the chromosomes double. Therefore, the genotype is temporarily $AAA'A'$ (Fig. 55). The separation, during the two meiotic divisions of the four strands and their final enclosure within four cells, assigns one, and only one, of these alleles to each cell. This process, called *segregation,* results in mature germ cells that are *pure* for either the maternal or the paternal allele; or, as we might say, the alleles segregate themselves from each other, uninfluenced by the fact that they have resided together in the same nuclei since the individual began his existence as a fertilized egg.

Segregation of Two Pairs of Alleles. In order to understand the distribution of two pairs of alleles located in two pairs of chromosomes, it will be best to

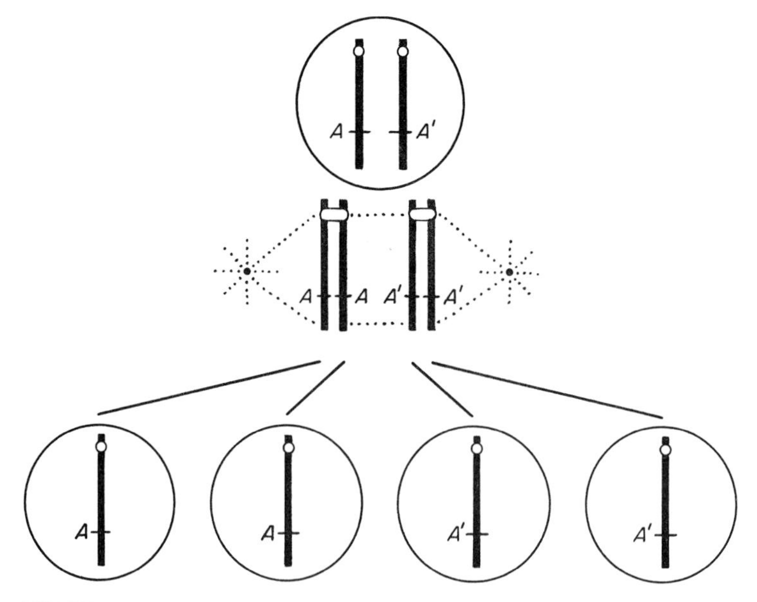

FIGURE 55
Meiotic segregation of a pair of alleles A, A'. Compare with Figure 48.

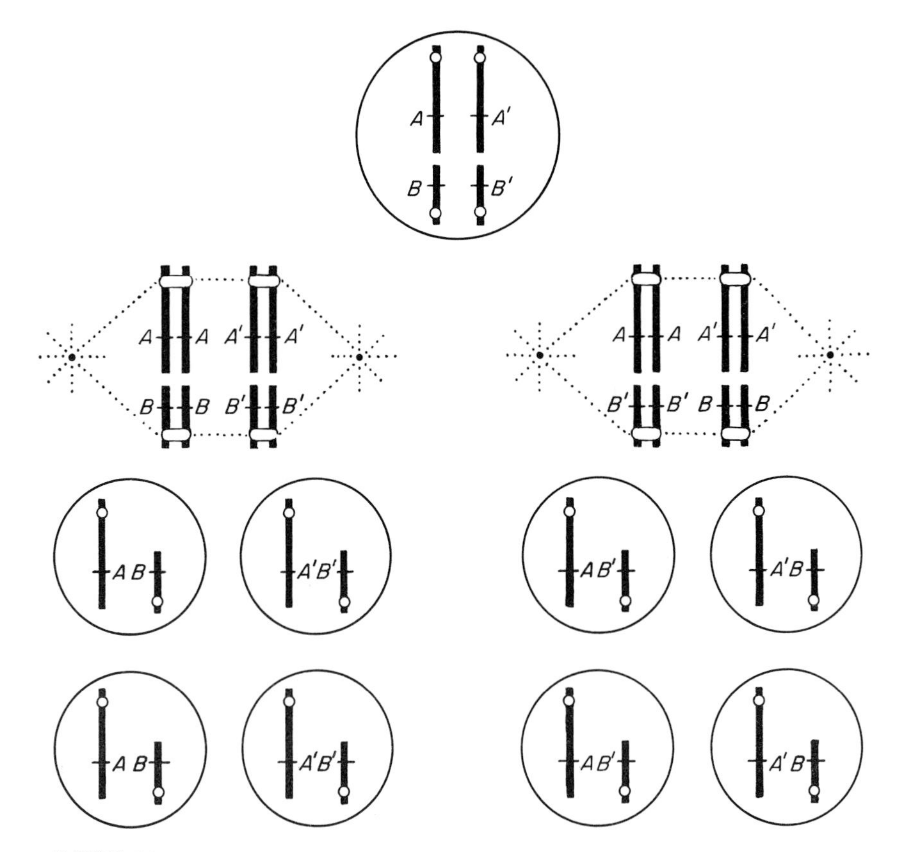

FIGURE 56
Meiotic segregation and the two alternative types of recombination of two pairs of alleles A, A' and B, B'. Compare with Figure 48.

assume at first that no crossing over takes place. We have seen (Fig. 48) that the pairs of homologous chromosomes take their positions on the spindle independently of each other, so that different, but equally frequent, arrangements of maternal and paternal partner chromosomes are found in different meiotic cells. In Figure 56, these arrangements are recognizable by the position of the alleles. A and B represent the maternal alleles in the two nonhomologous chromosomes, and A' and B' the paternal ones. The division at the left segregates the two original parental combinations, AB and $A'B'$, from each other; and the division at the right, the two new combination, AB' and $A'B$.

Four types of gametes are thus formed, as is illustrated by the following example. Assume that an individual received from his mother the allele P^2 for polydactlyly in one chromosome and the allele I^A for blood-group property A in another, and that the father had contributed the allele P^1 for five-fingeredness and the allele I^O for blood-group property O. At maturity, the individual will form gametes with one, and only one, of either P^2 or P^1 in equal proportion, and in addition, with one, and only one, of either I^A or I^O in equal proportion. Therefore, one-quarter of the gametes will be like that of the mother, carrying both P^2 and I^A; one-quarter, like that of the father, with both P^1 and I^O; one-quarter will contain the mother's P^2 and the father's I^O; one-quarter, the fathers P^1 and the mother's I^A.

Segregation of Many Pairs of Alleles. Independent assortment of maternal and paternal alleles belonging to genes located in different chromosome pairs occurs not only for two pairs of alleles but for as many pairs of alleles as there are pairs of chromosomes. It is easy to calculate how many different kinds of gametes may be formed by an individual whose 23 chromosome pairs are marked by a pair of alleles. If only one pair, A and A', is considered, two kinds of gametes can be produced, A and A'. If two heterozygous pairs of loci, A, A' and B, B', are assumed to exist, a gamete with A may have either B or B', a gamete with A' may have either B or B', and four kinds of gametes, AB, AB', $A'B$, and $A'B'$, would thus be formed. When a third pair of loci, C, C', is taken into account, each of the four kinds of gametes just mentioned may contain either the maternal (C) or the paternal (C') allele of the third pair of loci, and this leads to eight kinds of gametes: ABC, ABC', $AB'C$, $A'BC$, $AB'C'$, $A'BC'$, $A'B'C$, and $A'B'C'$. With each further pair of loci added, the number of possible kinds of gametes is doubled. This is shown graphically in Figure 57. The regularity can also be expressed in numerical form, as in Table 3, where it is seen that the number of kinds of gametes (which is equal to the number of combinations of maternal and paternal alleles) is 2^n, where n signifies the number of allelic pairs.

TABLE 3
Number of Potentially Different Gametes in Relation to Number of Pairs of Alleles.

No. of Heterozygous Pairs of Alleles	*No. of Kinds of Gametes*
1	$2 = 2^1$
2	$4 = 2^2$
3	$8 = 2^3$
4	$16 = 2^4$
⋮	⋮
23	$8{,}388{,}608 = 2^{23}$
⋮	⋮
n	$= 2^n$

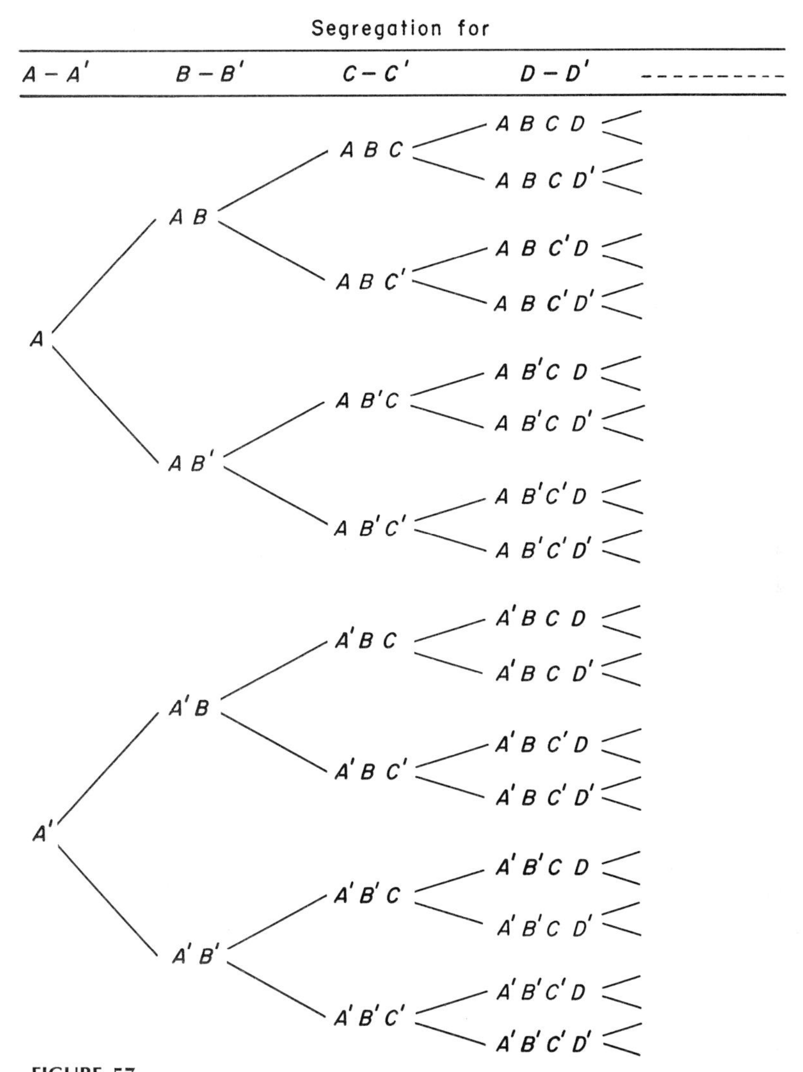

FIGURE 57
Meiotic segregation and random assortment of four heterozygous pairs of alleles, and the resulting different kinds of gametic constitutions.

When 23 allelic pairs, on the 23 pairs of chromosomes, are involved, 2^{23} or 8,388,608 kinds of gametes could be produced. This number may seem astoundingly large, particularly when we remember that it represents the combinations of only 23 elements, each of which occurs in only two forms.

In order to help the reader realize the significance of 8,388,608 possible combinations, we may cite a few of the individual combinations that are included in the total. First, there are two combinations with either all 23 maternal or all 23 paternal alleles. Second, there are combinations of 22 maternal and

1 paternal and of 22 paternal and 1 maternal alleles, each of which occurs in 23 different ways, since the 1 paternal or the 1 maternal allele of these combinations may be either the allele in the first pair of chromosomes, or that in the second pair; and so on, up to the twenty-third pair. Third, there are no less than 253 different ways in which 2 maternal and 21 paternal alleles, and 253 different ways in which 2 paternal and 21 maternal alleles, may be segregated and assorted as a result of meiosis. The two maternal alleles may belong to pairs 1 and 2, 1 and 3, 1 and 4, and so on to 1 and 23, adding up to 22 combinations. Or they may belong to pairs 2 and 3, 2 and 4, and so on to 2 and 23, adding up to 21 combinations. Similarly they may belong to pairs 3 and 4, etc., to 3 and 23, adding up to 20 combinations. The sum of all these combinations of 2 maternal and 21 paternal chromosomes cited up to this point is $22 + 21 + 20$.

Many more combinations are furnished by the maternal alleles of pair 4 with the maternal alleles of pairs 5, 6, . . . 23; of the maternal allele of pair 5 with the maternal alleles of pairs 6, 7, . . . , 23; and so on until, finally, the combination of the 2 maternal alleles of pairs 22 and 23 is reached. It should be apparent, then, that the sum total of all combinations that contain 2 maternal alleles (plus, necessarily, 21 paternal alleles) is $22 + 21 + 20 + 19 + 18 \ldots + 1$, which is 253. As the number of maternal alleles approaches that of paternal alleles, the number of combinations increases, until there are no less than 1,352,078 ways in which 12 maternal alleles of different loci may be combined with 11 paternal alleles of other loci, or 11 maternal with 12 paternal ones. (This number of combinations is obtained from the expression $23!/(12!\ 11!)$, which is explained on pages 198–199.)

The Number of Combinations. As large as the figure 8,388,608 may seem, it is by no means large enough to give a true description of the number of possible combinations of maternal and paternal genetic elements in the gametes of a single human being. A chromosome pair contains not only one pair of loci, but a great many. If crossing over did not occur, then the existence of many pairs of loci in a pair of homologous chromosomes would not add to the number of allelic combinations in the gametes, since the alleles in a maternal chromosome would always segregate as a unit from the paternal alleles. But crossing over, as we have seen, breaks up the original combinations and recombines maternal and paternal alleles in new chromosome strands. It thus increases greatly the number of hereditarily different kinds of gametes that one individual is potentially able to form.

This number depends, of course, on the degree of heterozygosity of the individual. All the gametes of a completely homozygous person are alike genetically, and only a heterozygote will produce different kinds of gametes. It is not known with any certainty how many genes are present in the heterozygous state in the average person. Estimates based on small samples of loci in English populations suggest that on the average at least 10 per cent of all loci are heterozygous. If we use the low estimate of a total of 10,000 loci in a chromosome set, we come to the conclusion that 9,000 of the genes supplied by the mother are exactly the same as those supplied by the father, but that the two parents contributed different alleles at 1000 loci. On the average, each of the 23 chromosome pairs would be heterozygous for nearly 40 different loci.

In different cells of the same individual, crossing over could occur in any one of the 39 consecutive regions delimited in each chromosome by 40 heterozygous loci. If only single crossovers occurred in each chromosome pair, the gametes could contain 80 different combinations of maternally and paternally derived alleles in one pair of chromosomes: namely, 2 noncrossover combinations and 78 different crossover combinations.

Since crossing over in one pair of chromosomes takes place largely irrespective of its occurrence in other chromosome pairs, and since segregation of the four strands of each original pair of chromosomes is independent of segregation in other pairs, the total number of possible combinations is the product of the combinations in each of the 23 pairs. This product, under the assumptions made—which do not include double or multiple crossovers—is $80 \times 80 \times 80 \times \ldots = 80^{23}$. The magnitude of this figure is beyond comprehension. It is, of course, equivalent to $8^{23} \times 10^{23} = (2^{23})^3 \times 10^{23}$, or our former figure of 8,388,608 cubed and followed by 23 zeros. A woman, during her reproductive years, produces only about 400 mature egg cells. Clearly, these 400 eggs are an infinitesimally small sample of the overwhelming variety of germ cells that she has the potentiality of forming. A man's total production of germ cells is much larger than of a woman: it has been estimated to be in the neighborhood of 1,000,000,000,000 (1,000 billion). Even this immense number is negligible if compared to 80^{23}, the total number of possible combinations of maternal and paternal alleles in his sperm; it is only about one sixty million trillion-trillionth of 80^{23}! It is hardly necessary to add that this number would be greatly larger if there were 100,000 instead of 10,000 loci on the human chromosome set.

The Shuffling of Genes. Because meiosis assorts and recombines the genetic contributions of an individual's parents, it has been compared to the shuffling of a deck of cards. The recombining of genes in meiosis explains why each child of a couple, excepting identical twins, has many genetic characters that the other brothers and sisters do not have. Even if individuals lived for geologic periods and had a litter of children each year, the chance of forming two identical gametes would be so small as to make it practically certain that no two children would have the same genotype.

We are prone to emphasize those genes that a child inherits from his parents. It is important not to forget the genes that he does *not* inherit. Of every pair of alleles a parent possesses, a child gets only one. As far as that child is concerned, the other allele is lost to the future, since segregation in meiosis excluded it from the germ cells that led to the child's being. Segregation explains some of his similarities to his parents, those due to his having half of the genes of each; and it also explains some of his dissimilarities, those due to his not having obtained the other half of their genes. Pride of ancestry is, at best, a questionable attitude, since an individual's value depends on himself rather than on the properties of others. If the pride is based on the assumption that one has the same genes as a distinguished ancestor, it is well to remember that half of a person's genes are not transmitted to his child, and that this process of halving takes places in each generation.

The argument may also be used in a different situation. Undesirable traits among ancestors are not uncommon, even though family tradition may seldom preserve their memory. Should these traits be based on specific genotypes, there may be consolation in the knowledge that only half of any ancestor's genetic material reaches the next generation.

PROBLEMS

6. A woman received from her father the five genes $ABCDE$ and from her mother the alleles $A'B'C'D'E'$. Which of the following combinations of alleles may be present in the eggs from which her children originate: $ABCDE$, $A'B'C'D'E'$, $ABCC'DE'$, $A'BCDD'$, $AB'CD'E$, $AB'DE$?
7. A man is heterozygous for ten pairs of genes, each pair in different chromosomes.
   (a) How many different types of gametes may be formed by him?
   (b) Had the ten pairs of genes been located in five pairs of chromosomes, with each of the chromosome pairs carrying two pairs of genes, how would you have answered Part (a)?
8. What kind of crossover (single, double, two-strand, etc.) has occurred in the rod-shaped chromosome pair drawn in Figure 52? In the V-shaped chromosome pair?
9. (a) Redraw Figure 52, D (left and right) with a different arrangement of the chromosome pairs. How many different arrangements are possible in each case?
   (b) On the basis of your redrawn stages D, make corresponding drawings of stages E–H. (Note that additional alternative arrangements are possible at the second meiotic division, giving a total of 16 possible kinds of gametes.)
10. Redraw Figure 52, C, assuming a three-strand double crossover in the V-shaped chromosome pair. Redraw D–H on the basis of your new Figure 52, C.

11. Assume that a man is heterozygous for five loci arranged in the order $ABCDE$ on one chromosome and $A'B'C'D'E'$ on the other.
    (a) What are all possible genotypes of the spermatozoa?
    (b) State for each genotype whether it was derived from a noncrossover or a crossover, and, if the latter, from what kind? Among multiple crossovers consider only two-strand multiples.

REFERENCES

Rhoades, M. M., 1961. Meiosis. Pp. 1–75 *in* Brachet, J., and Mirsky, A. E. (Eds.), *The Cell,* Vol. 3. Academic, New York.

Sturtevant, A. H., and Beadle, G. W., 1939. *An Introduction to Genetics.* Saunders, Philadelphia. (Chapters 5–7, pp. 63–126.)

Swanson, C. P., Merz, T., and Young, W. J., 1967. *Cytogenetics.* 194 pp. Prentice-Hall, Englewood Cliffs, N.J..

Wilson, E. B., 1925. *The Cell in Development and Heredity,* 3rd ed. 1232 pp. Macmillan, New York.

5

CYTOGENETICS

Homologous chromosomes are usually alike in form and size. Occasionally, variant chromosomes are found that can be traced from one generation to another. Such chromosomes provide visible evidence for segregation and hereditary transmission. Some of the variant chromosomes do not recognizably affect the phenotype of the individuals carrying them. Others may produce severely abnormal phenotypes. Even normal chromosomes may lead to the appearance of abnormal phenotypes if they are present in an individual in abnormal number. Cytogenetics deals with the interrelations of microscopically observable chromosomal phenomena and genetic transmission and function.

Pedigrees. Data of human genetics are frequently presented in the form of diagrammatic pedigrees (Fig. 58). In some countries, including the United States, females in these pedigrees are symbolized by circles and males by squares; in other countries, females by the sign ♀ and males by ♂. Symbols of parents are joined by a horizontal *marriage line,* and their offspring's symbols are placed in a horizontal row below a line to which they are connected by verticals. The horizontal line above the symbols for the children is itself

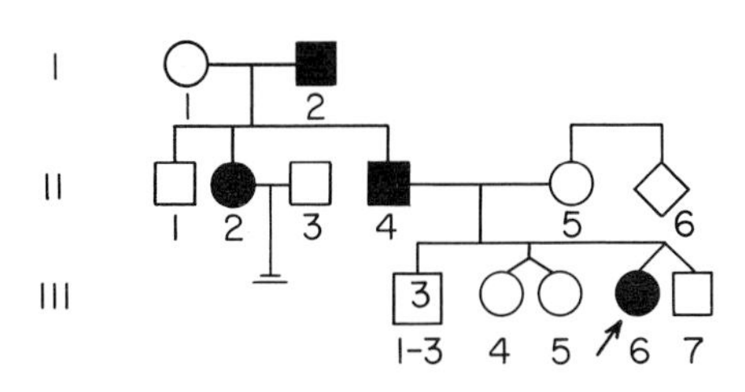

FIGURE 58
A sample pedigree. For explanation, see text. The marriage of II-2 and II-3 was childless. The twins, III-4 and III-5, are identical, as indicated by the short vertical line descending from the sibship line. The twins III-6 and III-7 are nonidentical, as indicated by their separate connection with the sibship line. II-5 and II-6 are sibs whose parents are not included in the pedigree.

connected to the parents' marriage line by a vertical line. The symbol for a single child is directly attached to this vertical line. The children of a parental pair form a *sibship,* and the individual children, regardless of sex, are called *sibs,* or *siblings.* Thus, sibs can be brothers, sisters, or brothers and sisters. In a pedigree the sibs are listed from left to right in order of birth. The term *family* is usually applied to a pair of parents and their children but may also refer to a larger circle of relatives, although this larger circle of persons who are interconnected by descent and marriage is generally called a *kindred.*

Often, each individual in a pedigree is designated by a number in order to facilitate reference to him. The numbering system may be consecutive from the earliest generation to the most recent, or each generation may be denoted by a Roman numeral and the individuals within a generation by Arabic numerals. Thus, II-4 identifies the fourth individual listed in the second generation of a pedigree.

Many additional symbols are used for special purposes. They are usually explained in the legend of the pedigree in which they occur. Some of these are used in Figure 58: a diamond denotes that the sex of the person was unknown to the recorder of the pedigree; a number enclosed in a large symbol indicates the number of sibs who are not listed separately; twins are represented by two symbols connected by lines to the same point on the sibship line. If an individual possesses the trait whose inheritance forms the subject of the pedigree, he is said to be "affected" and is designated by a black symbol. A hollow circle or square signifies absence of the trait ("not-affected"). Sometimes, to save space in extensive pedigrees, different generations are listed in consecutive concentric curves (Fig. 111, p. 188) rather than in horizontal rows.

Many pedigrees become known when a specific affected individual comes to the attention of a physician or some other investigator. The ascertainment of the family or kindred depends, therefore, on the discovery of the affected individual. He or she is called the *propositus* or *proposita* (or *proband* or *index case*) and is often marked in pedigrees by an arrow or a pointer. Many kindreds are ascertained more than once by independent discoveries of different propositi.

Transmission of a Variant Chromosome. The "affected" individuals in the pedigree Figure 59 carry a typical chromosome of a pair in the D group (chromosomes 13 to 15) and a homologous chromosome whose short arm is unusually long. Both the carriers and the noncarriers of the chromosome are normal individuals. The brother, I-3, of the woman I-2 was shown to carry the variant chromosome and, to judge from her offspring, she was also a carrier. The segregation of homologous chromosomes in meiosis should have resulted in one-half of her eggs receiving the typical chromosome and the other half receiving the variant. Actually, the three eggs that gave rise to her son, II-1, and her daughters, II-3 and II-5, all happened to carry the variant chromosome. The offspring of the affected persons represented both types resulting from segregation in a 1:1 ratio, III-2, III-4, and III-5, receiving the variant and III-1, III-3, and III-6 the typical chromosome. The pedigree demonstrates that the two homologous but morphologically different chromosomes, which were present together in the cells of heterozygotes, did not undergo a hypothetical "blending" to form two equal chromosomes with short arms of intermediate length but retained their original sizes and segregated into pure gametes.

Nondisjunction. It was pointed out earlier (p. 24) that an occasional error in mitosis may result in two sister chromosomes, instead of moving to opposite

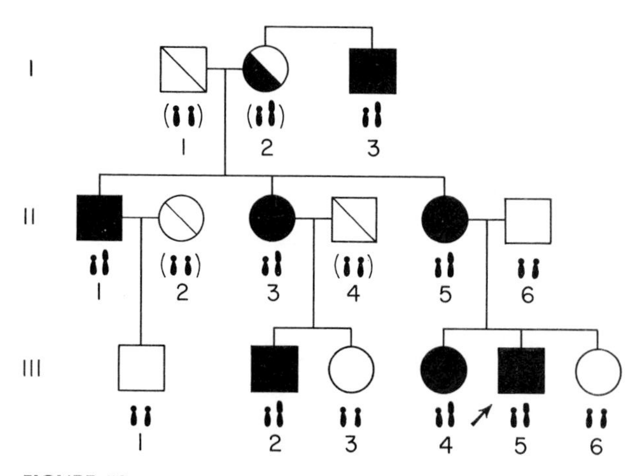

FIGURE 59
The transmission of a chromosome of the D group whose short arm was unusually long. Solid symbols = presence of the variant chromosome; empty symbols = absence of the chromosome. Half-filled female symbol = no direct cytological information, but the indication from offspring that the woman carried the variant chromosome. Diagonal line crossing an empty symbol = no cytological information but presumably normal chromosomes. (After Court Brown.)

spindle poles, being included in the same daughter nucleus. This nucleus thus receives one chromosome too many and the other daughter nucleus one too few. Another error in distribution occurs when one of two sister chromosomes so lags in its movement that it is not included in either daughter nucleus: one nucleus receives the normal number of chromosomes and the other one too few. It is often not possible to distinguish by genetic analysis which of the two errors in chromosome distribution is responsible for a given abnormal chromosome constitution. Both types of failure in distribution will therefore be referred to as nondisjunction.

When nondisjunction occurs during a meiotic division it results in gametes with abnormal chromosome content. It was first observed in the testes of plant bugs. In the males of these animals, as in man, a pair of visibly different sex chromosomes, X and Y, normally segregate to opposite poles, thus leading to the formation of X- and Y-sperm (Fig. 60, A, B, C.). Occasionally, however, the two sex chromosomes move together to the same pole (Fig. 60, D) so that the sperm cells originating from such nondisjunctional spermatocytes are of the abnormal types XY and "0" (zero). In man, direct cytological observations of nondisjoining chromosomes are lacking, but genetic evidence shows their occurrence to be a rare event for a variety of chromosome pairs both in spermatogenesis and oogenesis. In the following pages human autosomal nondisjunction will be discussed in some detail. Sex chromosomal nondisjunction will be treated in the chapters on sex linkage and on sex determination.

Autosomal Meiotic Nondisjunction: Down's Syndrome and Other Anomalies. Meiotic nondisjunction of an autosomal pair during oogenesis leads either to

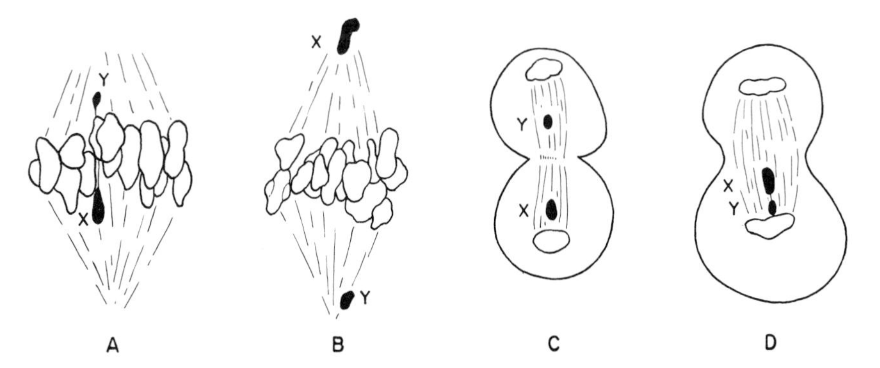

FIGURE 60
Normal disjunction of the X- and Y-chromosomes in (**A, B**) man, (**C**) the plant bug *Metapodius granulosus*. (**D**) Nondisjunction in *Metapodius femoratus*. (In man segregation of the X- and Y-chromosomes occurs at the first meiotic division, in *Metapodius* at the second. Note that the segregation of the sex chromosomes is out of step with that of the autosomes). (A, B, after Painter, *J. Exp. Zool.*, **37**, 1927; *C, D,* Wilson, *J. Exp. Zool.*, **6**, 1909).

a mature egg without a representative of the maldistributed chromosome or to an egg with two homologues of the pair (Fig. 61). When such eggs are fertilized by normal sperm, each of which has a single chromosome of each kind, two types of zygotes result, the first one having only one of the specified autosomes and the second having three. If meiotic nondisjunction of an autosome occurs during spermatogenesis, some mature exceptional sperm will be formed that correspond to the exceptional egg without a representative of the maldistributed autosome and others that correspond to the egg with two homologues (Fig. 62). Fertilization of normal eggs with nondisjunctional abnormal sperm will, as in the reciprocal case, result in *monosomic* and *trisomic* zygotes. Such *monosomic* and *trisomic* individuals have been known for more than half a century in *Drosophila* and in a whole series of plant species. They were found to differ from normal disomic individuals in a variety of morphological and physiological ways. Clearly, these abnormalities were not the result of abnormal genes but of the abnormal dosage of normal genes. Development and function of a diploid organism is based on a balance of gene action such that the activities of two alleles at many loci result in normality. If, however, most genes are present twice but some only once or three times then overall genic action is unbalanced and abnormal phenotypes are formed.

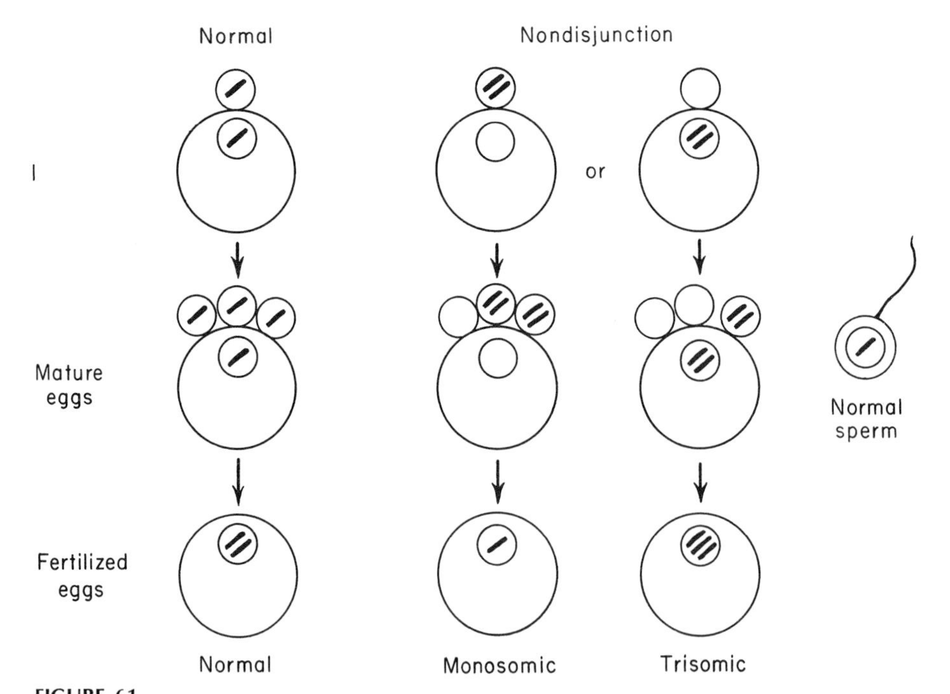

FIGURE 61
Meiotic nondisjunction of an autosomal pair in the female. The diagram represents nondisjunction at the first meiotic division.

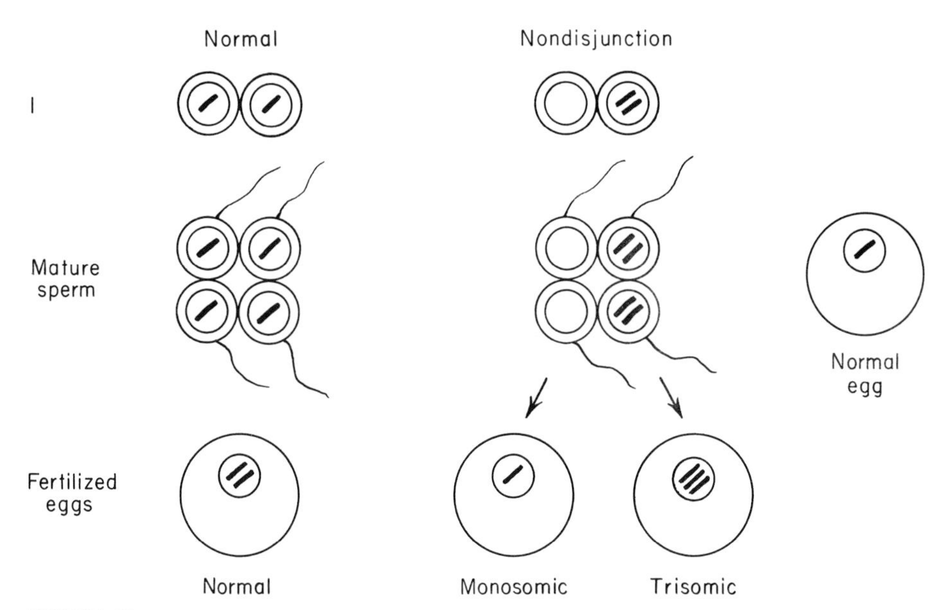

FIGURE 62
Meiotic nondisjunction of an autosomal pair in the male. The diagram represents nondisjunction at the first meiotic division.

Like the musical effect of an orchestra, which depends both on the presence of different instruments and on the number of each, genic effects are not the same if different quantities of alleles participate. As might be expected, some of the abnormal traits of monosomics deviate from the normal traits in the opposite direction of those of trisomics. Often the genic imbalance of monosomics and of trisomics leads to death at an early stage of development. In *Drosophila,* for instance, monosomy and trisomy of either one of the two large autosomal pairs of chromosomes is lethal. Monosomics for the small fourth, chromosome ("haplo-4") do survive until adulthood although they do poorly, whereas trisomics ("triplo-4") are nearly normal.

The occurrence of nondisjunction in experimental organisms and the resulting abnormal phenotypes should have suggested long ago that a search for similar situations be made in man. A few proposals of this kind were indeed made but no action was taken. That knowledge of the intricacies of chromosome behavior in a fly or a weed could be transferred to man seemed fantastic and, in any case, human cytological techniques were too primitive to serve as tools. Yet the time did come when one of the most remarkable and distressingly abnormal syndromes in man, Down's syndrome, or Down's anomaly, was shown to be the result of trisomy. Formerly the term mongolism was applied to this condition because many of the affected individuals show a fold of the eyelid that was erroneously taken to resemble the eyelid of members of the Mongoloid races.

Unfortunately, Down's syndrome is a relatively common anomaly, having an over-all incidence of about 0.15 per cent in all births in Caucasoid populations. (Early death of many affected individuals results in a much lower frequency of living Down's individuals.) The condition has also been observed in American Indians, Eskimos, Negroes, Orientals, and other Asians. It is probably present in all races.

The occurrence of the defect in the newborn is closely related to the age of the mother. More than two per cent of the children of women who became pregnant late in their reproductive years are affected, but only a very small fraction of one per cent of the children of young mothers. Among mothers eighteen years old, the frequency of affected babies is about 1 in 2500, but among mothers older than forty-five, 1 out of 50 babies are affected. In relative terms, the incidence of the trait rises more than fifty-fold with increasing age of the mother (Fig. 63). The age effect is also apparent from statistics on more than 1,700 affected children, nearly 40 per cent of whom were born to mothers forty years old or older, in contrast to normal children in control populations, of whom only between 3.5 and 5 per cent were born to mothers over forty.

It is generally true that older women have older husbands, and it might therefore be thought that affected births were caused by a direct or indirect

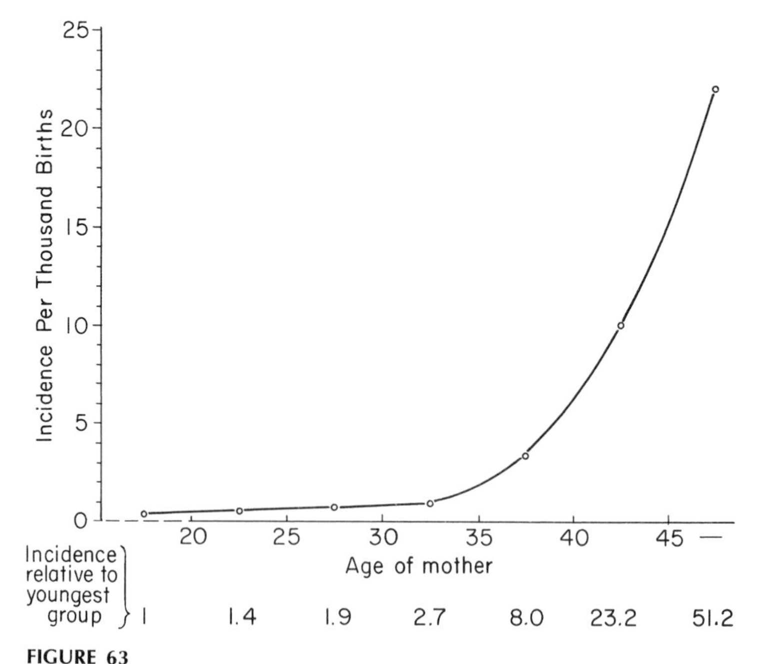

FIGURE 63

Frequencies, in relation to age of mothers, of 1,119 children with Down's syndrome born in Victoria, Australia, between 1942 and 1957 (After Collmann and Stoller, *Amer. J. Public Health*, **52**, 1962.)

influence of either the mother or the father. Penrose has shown, however, that the frequency of such births rises with the age of the mothers even if the age of the fathers is constant, and does not rise with the age of the fathers if the age of the mothers does not vary. In addition, it has been found that the age of the mother, not the number of preceding pregnancies, determines the frequency of affected births.

For many years it was usually assumed that the syndrome was produced by an unfavorable interaction between mother and fetus. It seemed that some kind of unknown physiological changes in the mother, correlated with her age, led to retarded development of affected embryos, perhaps as early as the second or third month of pregnancy. It was clear, however, that the cause of the damage was not inherent in the mother alone but must also reside in the egg. This was deduced from studies of Down's syndrome twins. In all cases in which it is certain that the twins are identical, both are affected, whereas with the great majority of nonidentical twins only one of them is affected and the other is normal. This important finding seemed to show that it is not the physiological condition of the mother alone, but its interaction with a specific condition of the embryo, that causes the anomaly. Since identical twins have identical genotypes, the most obvious explanation for both being affected is that a specific genotype of the embryo is involved in the defect—an explanation for which there seemed to be some independent evidence. Thus the frequency of affected sibs following the birth of an affected child is somewhat greater than in the general population, and, more significantly, there is a slightly increased frequency of affected among the offspring of relatives of both mothers and fathers of propositi. Further hints come from the few women with Down's syndrome who have given birth to a child: some of these children were normal, but others were affected with the syndrome. All of this made it seem that one or more specific genotypes might make an embryo susceptible to a damaging influence of the intra-uterine environment of an older mother, and that most other genotypes would be immune to such damage.

This point of view had to be abandoned. In 1959, Lejeune, Turpin, and Gautier in France, and soon after, investigators in various other countries, studied the chromosomes from tissue cultures of fibroblasts and bone marrow cells of a number of Down's individuals and found that one of the small chromosomes in the G group is present in triplicate instead of in duplicate. The affected thus have 47 chromosomes (Fig. 64; but see below). It must therefore be assumed that it is the imbalance of their genic content—the presence of three sets of genes in the triplicated chromosome—that is the cause of the specific maldevelopment of the embryos. Presumably the unusual trisomic chromosome constitution in Down's syndrome is the result of the fusion of a normal haploid gamete with an abnormal nondisjunctional one that carries two homologues of one small chromosome, although it is haploid for all others.

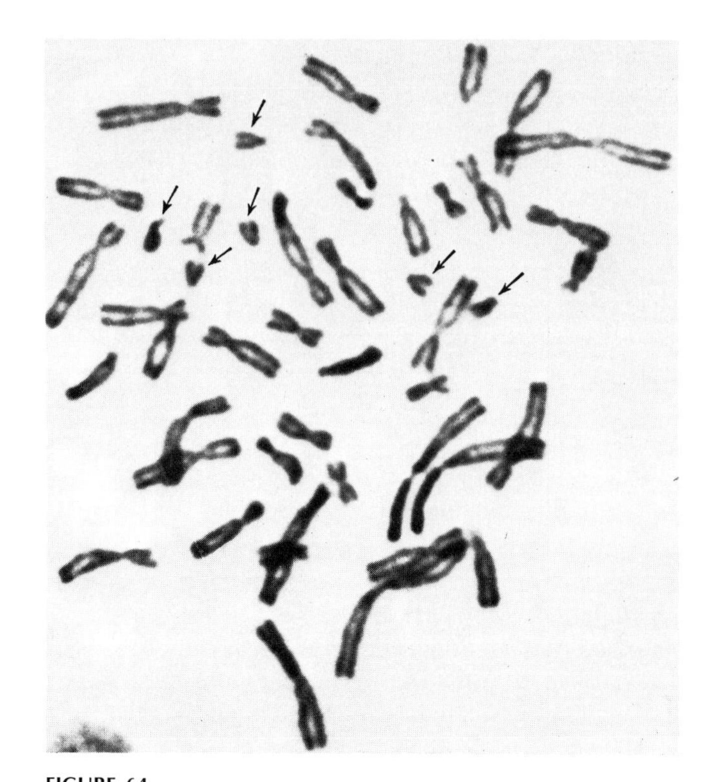

FIGURE 64
The 47 chromosomes of a boy with Down's syndrome. Among the normal diploid chromosome content of a male cell, there are *five* small chromosomes with approximately terminal kinetochores – the Y-chromosome and two pairs of autosomes numbered 21 and 22. The cells of Down's males contain *six* small chromosomes – the Y-chromosome, one pair of autosomes, and one trisomic autosomal assembly. The six small chromosomes are marked by arrows. (Lejeune, Turpin, and Gautier, *Ann. de Génét.*, **1,** 1959.)

It is possible that either parent may occasionally form such nondisjunctional gametes, but the relation between Down's syndrome and age of the mother indicates that the majority of affected individuals are derived from nondisjunctional eggs. Contrary to the earlier opinion, the syndrome is not caused by an unfavorable mother-fetus interaction, but by abnormal chromosome behavior of inactive eggs in the ovaries of the mothers or, after ovulation in the eggs produced by them. It will be shown later in this chapter that the nondisjunctional event may even occur as late as after fertilization of an egg and that nondisjunction is not the only source of Down's anomaly.

The chromosomal uniqueness of the eggs that produce Down's individuals explains the facts concerning affected identical and nonidentical twins. If a zygote which, as the result of nondisjunction, possesses 47 chromosomes gives rise to identical twins, both will have the abnormal chromosome number and both will be affected. If two separate zygotes lead to nonidentical twins, usually only one will be affected, since only very rarely will each of two eggs experience meiotic nondisjunction (or both be chromosomally abnormal descendants of a mitotic nondisjunction event having occurred in an early oogonial stage). The chromosomal uniqueness of affected individuals explains also the fact that normal and affected children appear in approximately equal proportion among the rare offspring of women with this affliction. These women form two kinds of eggs since during meiosis the chromosomes of a trisomic group are distributed in such manner that two chromosomes move to one pole and the third to the other, resulting in eggs with one and with two chromosomes of this trisomic group. This type of partial segregation has been called "inevitable" or "secondary" nondisjunction since it follows the primary nondisjunctional event that led to the disomic gamete responsible for the presence of the syndrome in the mother.

The two chromosome pairs numbered 21 and 22 in a normal chromosome group G are very similar in size and are not easily distinguishable using ordinary staining techniques. There has been uncertainty, therefore, which one is involved in the trisomy responsible for Down's syndrome. This uncertainty has finally been removed. The pattern of fluorescent and of Giemsa staining of cells of a Down's syndrome individual shows that it is the very slightly shorter of the two G chromosomes that is trisomic in affected individuals. This is in agreement with the results of study by Hungerford and associates of the clearly analyzable, elongated, paired meiotic chromosomes of another Down's person which shows that it is the shorter of the G chromosomes that is present three times. As far as size is concerned, it is thus established that chromosome number 22 is involved in Down's syndrome, and we should thus speak of 22-trisomy. Since, however, the term 21-trisomy has been extensively used for more than ten years, it appears advisable to retain it and define number 21 notwithstanding its size characteristic as that chromosome which if present thrice leads to Down's syndrome. The term 21-trisomy is sometimes preferred to the term Down's syndrome.

Most cases of Down's syndrome occur as isolated instances. This is as expected if we regard nondisjunction as an accident that hits meiosis in a random fashion. There is, however, evidence in a fraction of cases that there is a hereditary element that causes an increased frequency of additional affected individuals among the relatives of an affected propositus.

Some of the evidence is clear-cut, deriving from the existence of visibly abnormal chromosome constitutions superimposed on 21-trisomy. Other evidence is less direct. It consists of a variety of pedigrees in which more than one individual with typical, uncomplicated cytological 21-trisomy is observed. Such pedigrees may be no more than examples of the statistically unlikely coincidence of more than one random event of meiotic nondisjunction or of oogonial nondisjunction occurring in a single family, but they may also be the result of genetic predisposition to nondisjunction. In *Drosophila,* corn plants, and other organisms genes are known that interfere with the pairing and subsequent segregation of either specific chromosomes or all chromosomes, and it is probable that similar genotypes occur in man. It is suggestive in this respect that 21-trisomy is sometimes combined in the same individual with abnormal chromosome numbers for the sex chromosomes and that in some pedigrees relatives of a 21-trisomic person are monosomic or trisomic for the sex chromosomes or trisomic for a different autosome than number 21. Since the frequency of thyroid disease has been found to be greater among mothers and other close female relatives of Down's syndrome patients than in the female population at large, it seems that thyroid disease may be a predisposing factor to having children with Down's syndrome.

Striking proof for hereditary predisposition for Down's syndrome comes from the discovery of translocations between a chromosome 21 and some other chromosome or between two chromosomes 21. An example of the first type of translocation is shown in Figure 65. The pedigree includes three generations. The first two generations have normal phenotypes, but in the last generation two children were born with Down's syndrome. A study of the chromosomes showed that the woman in generation I had an unusual chromosome, which could be interpreted as being the fusion product of the long arm of a chromosome 21 and the long arm of a chromosome 14 since, instead of six long acrocentric chromosomes, representing the three pairs

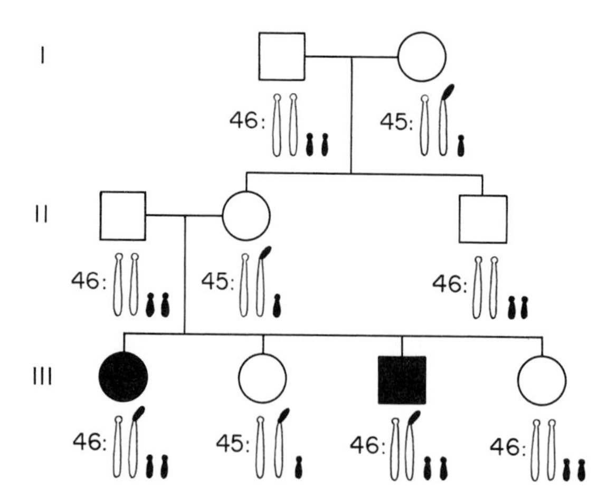

FIGURE 65
Two sibs having Down's syndrome due to translocation 21-trisomy. See text. (After Penrose and Delhanty, *Ann. Hum. Genet.,* **25,** 1961.)

13, 14, and 15, only five such chromosomes were present. Similarly, instead of the four short chromosomes 21 and 22 only three were present, the fourth constituting the short arm of the fusion product. Given these facts, the woman of generation I had only 45 chromosomes but she did possess an essentially normal genotype since she had two chromosomes of each kind even though there had been attachment of one chromosome to another. The woman's husband had the usual chromosome number of 46 with all chromosomes being normal. Their two children were normal in phenotype. The son was chromosomally typical, having obtained a normal chromosome 14 and a normal 21 from each parent and thus possessing 46 chromosomes. The daughter had the same genetically balanced chromosomal constitution as the mother, having 45 chromosomes of which the father had contributed one each of normal chromosomes 14 and 21 and the mother the fusion chromosome. The husband of the daughter had the typical 46-chromosome endowment. Their children were of three different kinds. The first and third were affected with Down's syndrome. They had two chromosomes 14, one free and one as part of the fusion product, and three chromosomes 21, two free and one as part of the fusion product. Notwithstanding their having the normal total number of separate chromosomes, 46, they actually were trisomic for most of chromosome 21 and therefore affected by the imbalance of their genes. The second child in the sibship had the same balanced, 45-chromosome constitution as her mother and grandmother and the last, fourth child was typical in having 46 chromosomes consisting of 23 normal pairs. Using the symbol $\widehat{14\text{-}21}$ to designate the fusion chromosome, it may be inferred that the mother of the four children had produced three kinds of mature eggs: those with chromosomes $\widehat{14\text{-}21}$, 21; with $\widehat{14\text{-}21}$; and with 14, 21. Obviously, meiotic pairing and segregation in a chromosomally abnormal 14, $\widehat{14\text{-}21}$, 21 constitution such as the mother's is disturbed and may lead to balanced and unbalanced gametes. The occurrence of the $\widehat{14\text{-}21}$, 21 mature eggs suggests that eggs of the complementary constitution, 14 alone, would also be produced. They would give rise, after fertilization with normal 14, 21 sperm, to 14, 14, 21 zygotes. Their monosomy for chromosome 21 would, presumably, lead to early embryonic death. When data concerning the offspring of many mothers of the 14, $\widehat{14\text{-}21}$, 21 constitution are pooled, it becomes apparent that the three types of viable offspring are produced with different frequencies. Down's syndrome affects about 10 per cent; the other 90 per cent of the children are all phenotypically normal, with about half being of normal chromosomal constitution and half chromosomally balanced carriers. Unexpectedly, pooled data on the offspring of fathers having the 14, $\widehat{14\text{-}21}$, 21 chromosomal constitution show that most offspring are normal in phenotype, somewhat less than one-half of them also being normal chromosomally and the others balanced carriers. Only very few, perhaps about 2 per cent, of Down's type children are produced by carrier fathers. It is not understood why the results of meiotic segregation in carriers are so different from one sex to the other.

It should be clarified that the term "fusion chromosome" does not fully describe the events that may lead to its origin. Cytogenetic knowledge in general provides the following interpretation (Fig. 66). Within the same nucleus a break occurs in each of one chromosome 14 and 21, one break near the kinetochore in the long arm of chromosome 14 and another in the short arm of chromosome 21. (The complementary situation of a break in the short arm of 14 and another in the long arm of 21 will not be discussed.) This results in the presence of four chromosome fragments, which have a tendency to fuse in any one of four possible ways. A first possibility would restore the original two chromosomes. A second possibility would lead to chromosomes without any or with two kinetochores, both types of which would probably be lost in later divisions. A third possibility would fuse the long arm of 14 without its kinetochore with the long arm of 21 with its kinetochore, and would entail the complementary fusion of the two short arms with their one kinetochore: This is the possibility that accounts for the origin of the $\widehat{14\text{-}21}$ chromosome with its long arms from two originally acrocentric chromosomes. The process of chromosome breakage and subsequent rearrangement is called reciprocal translocation and the $\widehat{14\text{-}21}$ chromosome is best called a translocation chromosome. Neither the term fusion chromosome nor the sometimes used term centric fusion (because fusion occurs near the centromere, or kinetochore) will be employed in our future discussions. (The classical theory of the origin of translocation chromosomes as due to reciprocal exchanges, described in the preceding paragraph, has recently been challenged. It has been suggested rather that translocations of chromosome sections may often depend on a single break in a cell, followed by nonreciprocal attachment of a fragment to an unbroken chromosome.)

A rare type of chromosome that results in Down's syndrome consists of a unit of two arms each of the length of the long arm of chromosome 21. Apparently this chromosome results either from translocation of the long arm of one chromosome 21 to the short arm of another, or from a misdivision of a single chromosome 21 such that its two long-arm chromatids remain permanently joined to a single kinetochore while the two short arms are lost. A person whose gonads carry this $\widehat{21\text{-}21}$ chromosome can only form two kinds of gametes, those with and those without the chromosome. After fusion with a normal gamete two kinds of zygotes can originate $\widehat{21\text{-}21}$, 21 and 0, 21. Presumably the latter do not develop beyond early stages, whereas the former result in Down's syndrome. Among studies of families with a $\widehat{21\text{-}21}$ translocation, one family has been described in which four children were born, and

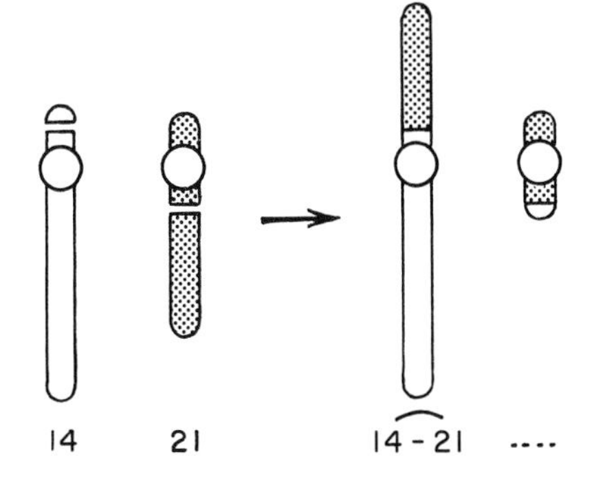

FIGURE 66
Origin of a 14-21 translocation chromosome after breakage near the kinetochore in the short arm of 14 and the long arm of 21.

a fifth medically aborted, all of whom in consequence of presence of a 21-21 chromosome had been afflicted with Down's syndrome!

The Age Effect in Down's Syndrome. The English geneticist Penrose (1898–1972), who contributed greatly to the understanding of Down's syndrome, tried to partition its causes into (A) those independent of the mother's age and (B) those dependent upon it. In class A are placed "built-in" mechanisms such as inherited translocations and other chromosome abnormalities that may result in abnormal segregation as well as—possibly—environmental disturbances of cell division. Class B causes are based on deterioration of the egg before fertilization due to aging in the mother. A very uncertain guess as to the proportions of the two classes in an unselected sample of affected children is 40 per cent A and 60 per cent B. Perhaps as many as 5 per cent of the whole might be of the inherited translocation types. If the cases of affected offspring are subdivided into maternal age classes, we expect to find a high proportion of the inherited types among the offspring of younger mothers and a low proportion among those of older ones. This would be so since segregation in the built-in cases presumably is not influenced by mother's age and would account for a constant low incidence of the syndrome over the whole age range; this low incidence would be close to the total incidence among the youngest mothers, but a relatively small part of that among the oldest. A young couple who have had an affected child are well advised to have their chromosomes studied for the possible presence of a heritable chromosome aberration. For an older couple to whom an affected child is born, the likelihood of complete chromosomal normality of the parents and "simple" 21-trisomy in the child, due to nondisjunction, is so great that in the majority of cases the offspring will indeed show the presence of three free chromosomes 21, i.e., the cause of their child's having Down's syndrome will be of the usually nonrecurrent type.

The nature of the age effect has been a matter of speculation. It has generally been assumed that the aging affects the ovarian oocytes, which from before birth until ovulation do not undergo cell division but remain for decades at an arrested stage of the prophase, the dictyotene. The age effect, it was thought, consists in some kind of metabolic damage to the "waiting" oocyte that reveals itself in a breakdown of chromosome disjunction during the meiotic divisions. This hypothesis may well be valid but a supplementary hypothesis has been proposed by German. It is based on the fact that an ovulated egg decays within the oviduct in a day or less unless it is fertilized. Fertilization toward the end of this brief period in other species is known often to lead to abnormal development. Might it then be, that Down's syndrome is the consequence of overripeness of an ovulated egg rather than of damage to an ovarian egg? Why, however, would the frequency of overripe ovulated eggs be correlated with age of mother? German's hypothesis makes use of the fact that frequency of sexual intercourse decreases with the age of a couple. Such decrease may well result in a higher frequency than among younger couples of the fertilization of eggs that are at a stage beyond their prime condition. It has been pointed out by several investigators that German's hypothesis can at best account for only part of the aging effect in Down's syndrome. Nevertheless, whatever the ultimate fate of the hypothesis, it may serve as an ever-needed reminder that association between two sets of data, such as those on age of mother and frequency of 21-trisomy is no proof of a direct causal relation.

Postzygotic Nondisjunction. Down's syndrome of the 21-trisomy type is not necessarily due to oogonial, or meiotic, nondisjunction. If a fertilized egg is of the normal 21-disomy type, subsequent nondisjunction at the first, mitotic, cleavage division could result in two cells, one with three chromosomes 21 and the other with only one. The latter cell, or its descendants, presumably would be nonviable while the trisomic cell could form an embryo that would develop into a child with Down's syndrome.

Should the mitotic nondisjunction event occur in a blastomere of an egg that consists of more than two cells, then three cell types would result: one with the normal disomic constitution, derived from cells with normal chromosome distribution in mitosis; one with a monosomic and another with a trisomic constitution, both derived from the same event of mitotic nondisjunction. Such an egg would comprise a mixture of two viable cell types, the normal disomic and the Down's type trisomic. Mosaics of this nature have indeed been discovered. The proportion of the two cell types varies greatly from one mosaic individual to another, as do their phenotypes in terms of morphology and mental capacity, which may approach either normality or Down's syndrome or be intermediate in a series of different ways. Mosaics for 21-disomy and 21-trisomy may also be the result of loss of a chromosome 21 in some cell of an initially trisomic egg. Whenever the trisomic condition is present in cells of the adult gonads—whether in all of the cells or as part of a mosaicism—disomic gametes may be produced, thus leading to the recurrence of 21-trisomy among the offspring.

A special instance of mosaicism for chromosome 21 is the unique case of "identical" one-egg twin boys, one of whom is normal and disomic while the other is 21-trisomic and afflicted with Down's syndrome: As far as is known, neither twin is a mosaic. The single developing egg, however, that gave rise to them must have been a mosaic, some of whose cells were disomic and others trisomics. Separation at an early cleavage stage led, as is usual in one-egg twins, into two independently developing embryos, each of which happened to have been derived from one or the other of the two different cell lines.

It is not known what proportion of 21-trisomy children from nonmosaic normal parents arises in consequence of oogonial, or meiotic, nondisjunction in the mother and what fraction in consequence of post-fertilization mitotic nondisjunction. The age effect is compatible with a disturbance of either type of chromosome distribution. If cytological or genetic markers were available for distinguishing the maternal and paternal chromosomes 21 of a trisomic child, an estimate might be made of the relative proportions of oogonial, or meiotic, nondisjunction and post-fertilization nondisjunction. The former would show two of the three chromosomes 21 to have been derived from the mother, the latter would probably involve in some cases nondisjunction of the maternal and in others of the paternal chromosomes.

Mosaicism for 21-trisomy may be one reason why certain physical traits that are usually part of Down's syndrome have an increased frequency among the parents of affected individuals. One of these is the so-called four-finger or simian crease in one or both hands (Fig. 67). It is present in approximately 50 per cent of the affected but only in considerably less than 10 per cent of control individuals. It seems to be more frequent in close relatives of affected persons than in controls. Another relevant trait is defined by the positions of several points characterized by specific ridges on the palm, as measured by the size of a certain angle between them (Fig. 68). An angle greater than 57° is found in more than 80 per cent of all affected, but in only 7–9 per cent of the general population. Among the parents and sibs of affected persons, the frequency of large angles is somewhat increased.

Infants trisomic for an autosome other than number 21 have been found, one such autosome being in the 17-18 group and another one in the 13-15 group. The anomalies are often referred to as 18-trisomy and 13-trisomy, implying that chromosomes 18 and 13 are those that give rise to the two syndromes, which are named after Edwards and Patau, respectively. The two syndromes are different, but trisomy for either chromosome 13 or for 18 causes

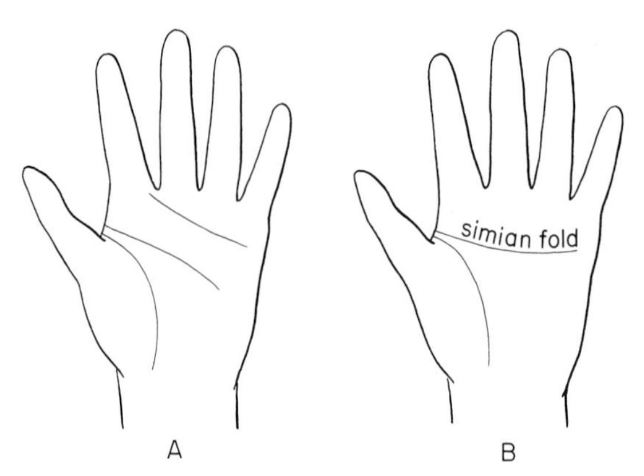

FIGURE 67.
A. The hand of a normal individual. **B.** The hand of an individual with Down's syndrome (s. f. = simian fold). (After Schiller, *Ztschr. menschl. Vererb.*, **25,** 1941.)

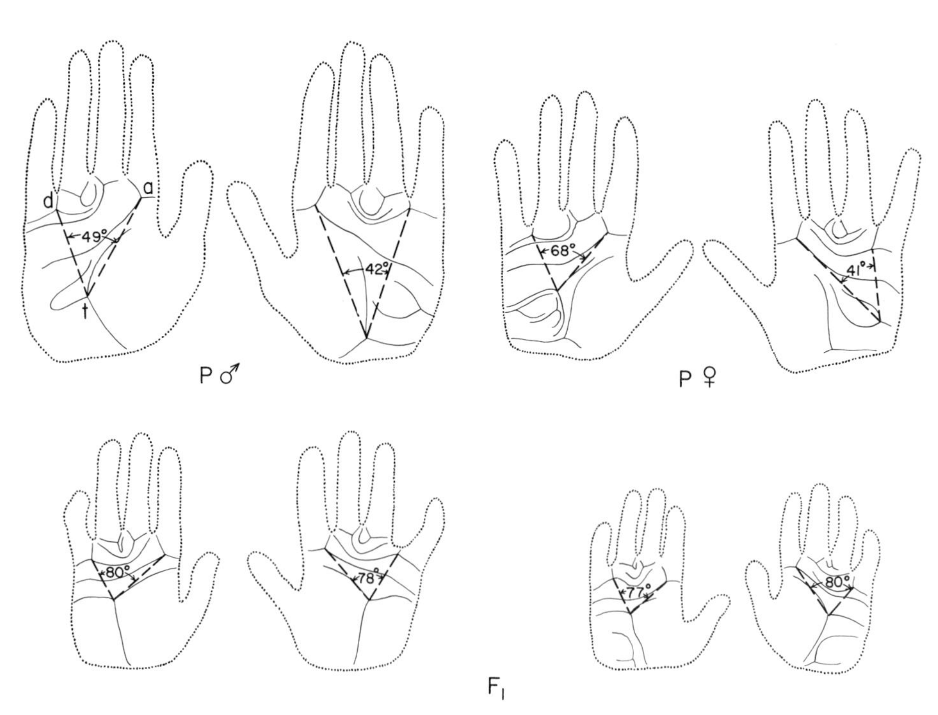

FIGURE 68
The main lines on hands of two normal parents and their two children with Down's syndrome. The numbers signify the degree of the maximal atd angle: a, t, and d being three points defined by the meeting of three lines each on the bases of the first and fourth finger and on the palm. (After Penrose, *Ann. Eugen.*, **19,** 1954.)

affected infants to show much more severely abnormal development than 21-trisomy; either leads to early death. In a general way this is probably the consequence of the larger size of chromosomes 13 and 18 than of 21 and the concomitantly larger number of genes brought into imbalance by the triple dose. No other autosomes have been found in a trisomic condition in any live-born children, except in mosaics, but trisomy for one or the other of many different autosomes has been observed in abortuses at early embryonic stages that were spontaneously expelled. The absence of trisomy in living persons for all but chromosomes 13, 18, and 21 is not due to nonoccurrence of nondisjunction among the other chromosomes but to inviability of the resulting zygotes.

Chromosome Deficiency: the Cri du Chat Syndrome. Among a variety of defective human types in which there is a chromosomal abnormality, a peculiar syndrome has been discovered by Lejeune and associates and named *cri du chat* (cat's cry). Affected individuals are severely abnormal physically and mentally but the condition is not lethal. The property of the syndrome that has given it its name is a plaintive continuous crying, particularly by the younger children, which resembles that of a cat. This syndrome is the result of a deletion of about one-half of the length of the short arm of a chromosome 5, as determined by autoradiography as well as by fluorescent or Giemsa staining. In most pedigrees the parents of the cri du chat propositus have normal chromosomes, meaning that one of them apparently formed a gamete newly deficient for part of chromosome 5, which gave rise to the abnormal child. In a few pedigrees a translocation of part of chromosome 5 to some other chromosome has been observed in one of the parents. A parent who possesses the two products of the reciprocal translocation is normal but may produce unbalanced gametes, which, after fertilization, cause abnormal development. One family having such a translocation is represented in Figure 69. A normal couple had seven children, of whom five were subject to study. One of the children is normal, the first and the fourth child are affected with the cat's cry syndrome, and the second and fifth child exhibit a different group of severe developmental abnormalities. Cytological studies show the chromosomes of the father to be normal but those of the mother to have one normal chromosome 5, one abnormally short-armed chromosome 5, one normal chromosome 13 and one abnormally long chromosome 13. Apparently part of the short arm of chromosome 5 had been translocated to chromosome 13. Eggs produced by the mother as judged from her offspring are of three chromosomal types (1) normal 5, normal 13 (child II-3) (2) abnormal 5, normal 13 (II-1, II-4) and (3) normal 5, abnormal 13 (II-2, II-5). The two defective phenotypes were caused by "partial monosomy" for chromosome 5 (cat's cry syndrome) and by partial trisomy for the translocated part of chromosome 5. The latter

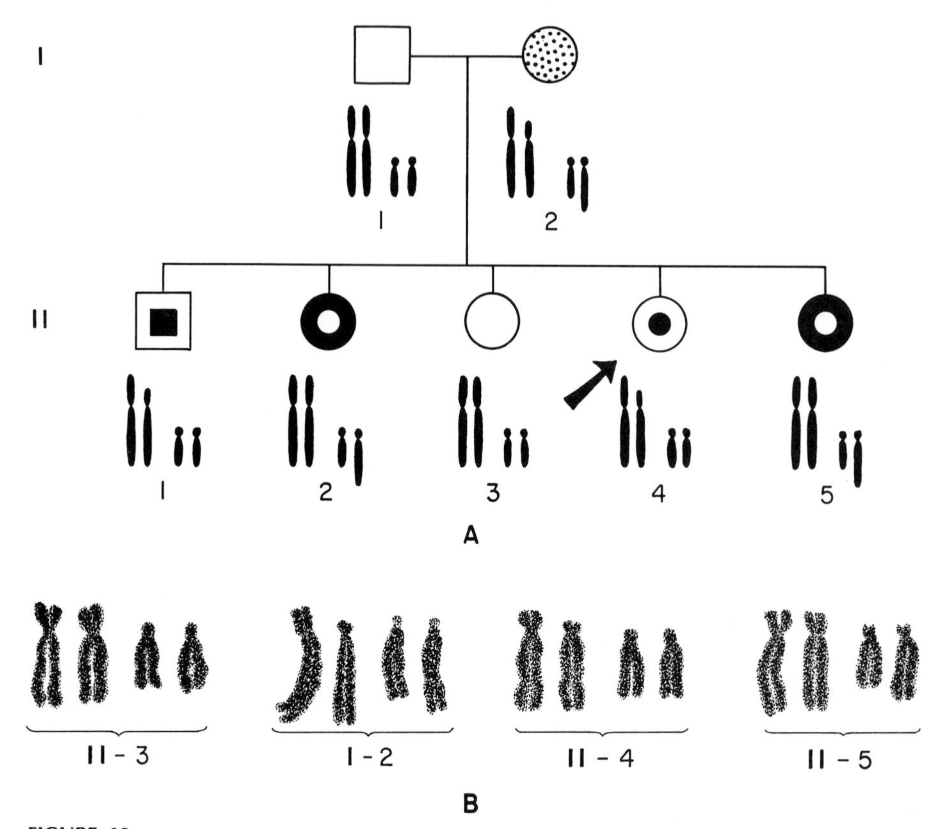

FIGURE 69
Cri du chat syndrome deriving from a reciprocal translocation. **A.** A pedigree showing hereditary effects of a reciprocal translocation between part of the short arm of chromosome 5 and part of the long arm of chromosome 13. The woman I-2 is normal phenotypically but carries the translocation. The children II-1 and II-4 are affected with the cat's cry syndrome, and II-2 and II-5 with its counter type. II-3 is normal. **B.** The chromosomes 5 and 13 of four persons shown in the pedigree. (After Lejeune, Lafourcade, Berger, and Turpin, *Comptes Rendus Acad. Sci. Paris,* **258,** 1964.)

chromosomal aberration has been called by Lejeune "le contre-type" of the partial monosomy. Phenotypically, the abnormal traits of its bearers deviate from normal traits in the opposite direction from those carrying the partial monosomy. A similar situation has been encountered in presumed deficiencies of chromosome 21 that result in partial monosomy in the heterozygous bearers of this chromosome. The abnormal features resulting from partial 21-monosomy result in an "anti-Down" phenotype.

The Origin of Abnormal Chromosomes. As described earlier, the 14-21 translocations in heritable Down's syndrome are the result of two breaks, one in chromosome 14 and another in chromosome 21, and the subsequent union such that the two chromosomes exchange terminal fragments. We shall now give a more general account of the origin of abnormal chromosomes.

When a chromosome breaks, the two fragments may either reunite later or remain permanently separate. If they reunite, there is no lasting effect on the genetic property of the chromosome, but permanent separation leads to important genetic consequences. The chromosomes of an overwhelming majority of species, including man, possess only one specific region—the kinetochore—that controls chromosome movement on the mitotic or meiotic spindle. And since only one of the two fragments will possess the kinetochore, while the other will be without it, this latter fragment or its daughter products will not be distributed by the spindle to the daughter cells in the normal way, but will be lost to the nuclei and degenerate in the cytoplasm (Fig. 70). The loss

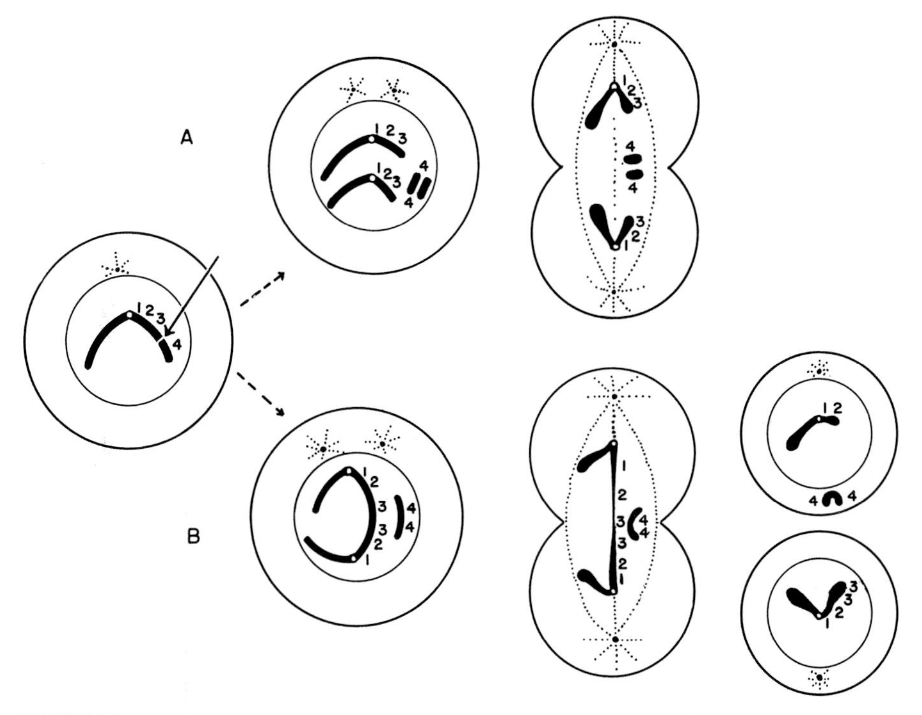

FIGURE 70
Chromosome breakage. A single break has occurred in a chromosome between loci 3 and 4. **A** and **B:** Alternative fates of the broken chromosome. **A.** The two parts of the broken chromosome reduplicate. The fragments with the kinetochore are distributed to opposite poles of the mitotic spindle. The fragments without the kinetochore remain in the equatorial region and will be eliminated in the cytoplasm. **B.** The two parts of the broken chromosome reduplicate and the broken ends of sister parts unite, thus forming a chromosome with two kinetochores and another chromosome without a kinetochore. During mitosis, the two kinetochores move to opposite poles and the chromosome section between them breaks. The two daughter cells receive different, unbalanced chromosome constitutions. The fragment without a kinetochore is eliminated in the cytoplasm.

of the genes located in the fragment will upset normal development and a new phenotype may appear, as it may when a whole chromosome is lost.

The fate of the chromosome fragment that retains the kinetochore varies. After some types of breakages, or in some tissues, this fragment behaves like a typical whole chromosome, and its daughter products are distributed normally in later divisions (Fig. 70, A). Under other circumstances, the fragment reduplicates to form two fragments with broken ends, and these ends unite, forming a long compound chromosome with two kinetochores (Fig. 70, B). Such chromosomes are not stable, since the two kinetochores may go to opposite spindle poles, with the result that the chromosome stretches and finally breaks at some point. Each newly created fragment is incorporated in a different daughter nucleus. As may be seen from Figure 70, B, one of the fragments may contain a certain doubled section that the other lacks completely. At the next division of the nuclei, the same cycle is repeated, namely, fusion of the broken ends of the sister chromosomes, formation of a "bridge" between the poles of each spindle, and a new break at some point on the stretched chromosome bridge. Repetition of the "breakage-fusion-bridge-breakage" cycle leads to more new genotypes for the cells involved, genotypes that are new not because they carry new alleles, but because of the quantity of alleles present—they may lack some alleles and possess others in multiple quantity.

The breakage-fusion-bridge cycle was discovered in 1938 by Barbara McClintock in highly technical studies of the genetics and cytology of maize. It is likely that the same processes also occur in man. As we shall see in a later chapter, knowledge of this phenomenon is probably of importance in the understanding of the effect of irradiation on man—a typical example of the unforeseeable results of scientific endeavor.

Chromosomes may also break under the effects of a variety of external agents. They may break at more than one place, or more than one chromosome may break in the same nucleus. If two breaks occur in the same chromosome, three fragments will be produced—two end pieces and a middle piece (Fig. 71). These may reunite in the original way without further consequences, or they may reunite with the middle piece inverted. This new *inversion* chromosome has lost no genetic material and behaves normally in later mitotic divisions: its meiotic fate will be discussed below. As a third alternative, the broken ends of the end pieces may unite, leaving the middle segment by itself. This segment, lacking a kinetochore, is lost (sometimes after having formed a ring chromosome, by fusion of its two broken ends); but the other reunited chromosome, in spite of lacking a middle section, behaves normally in future divisions. A new transmissible genotype, lacking the alleles located in the middle section, is thus created.

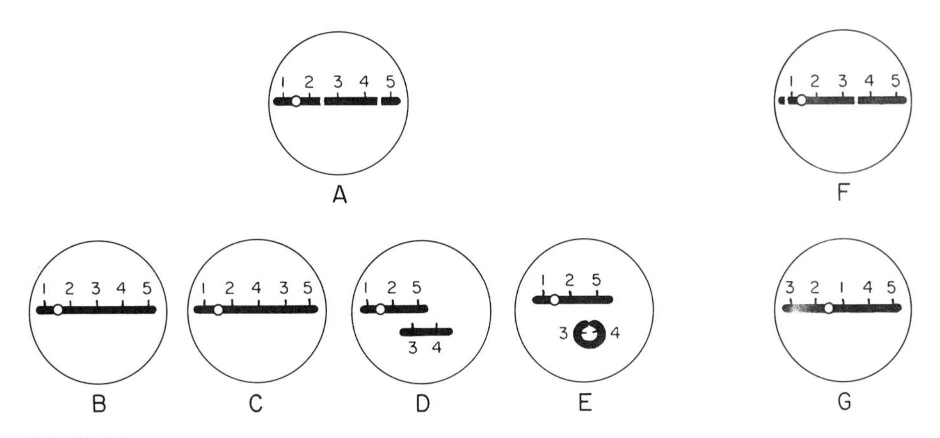

FIGURE 71
Chromosome breakage. **A.** Two breaks have occurred in the long arm of a chromosome, between loci 2 and 3 and between 4 and 5. **B–E.** Different consequences of the double break. **B.** Reunion of the fragments in the original order: 1–2–3–4–5. **C.** Reunion of the fragments with the middle fragment inverted (paracentric inversion): 1–2–4–3–5. **D.** Union of the two outside fragments: 1–2–5; the inner fragment, 3–4, which has no kinetochore, will be lost in a subsequent division. **E.** Same behavior as in D except the fragment 3–4 has formed a ring chromosome. **F.** The breaks in the chromosome are in different arms. **G.** The reunion of the fragments resulted in a pericentric inversion.

There are two different classes of inversions, distinguished by the positions of the original breakage points in relation to the kinetochore. If both breakage points lie in the same chromosome arm, leading to a *paracentric inversion,* no obvious rearrangement is observed under the microscope (compare Fig. 71, B and C). If the two breakage points lie on different sides of the kinetochore, the resulting *pericentric inversion* usually changes the relative lengths of the two chromosome arms and thus is apparent (compare Fig. 71, C and G).

Mitosis in cells heterozygous for an inversion proceeds normally since each chromosome replicates its specific sequence. In meiosis, however, inversion heterozygotes may initiate abnormal processes. As shown in Figure 72, A, two homologous chromosomes, one of which contains a paracentric inversion, segregate normally in meiosis, as long as no crossing over occurs within the inverted section. Thus, each gamete resulting from this type of meiosis has a whole chromosome of the pair considered. On the other hand, as shown in Figure 72, B, a crossover within the inverted section results in one chromosome with two kinetochores and a fragment without a kinetochore. The dicentric chromosome forms a bridge in the first meiotic division and may initiate a breakage-fusion-bridge cycle. The acentric fragment is lost. The resulting gametes are unbalanced and thus, after fertilization, cause incomplete development of the embryo. In heterozygotes for a pericentric inversion, other types of abnormal chromosomes are formed in consequence of crossing over within

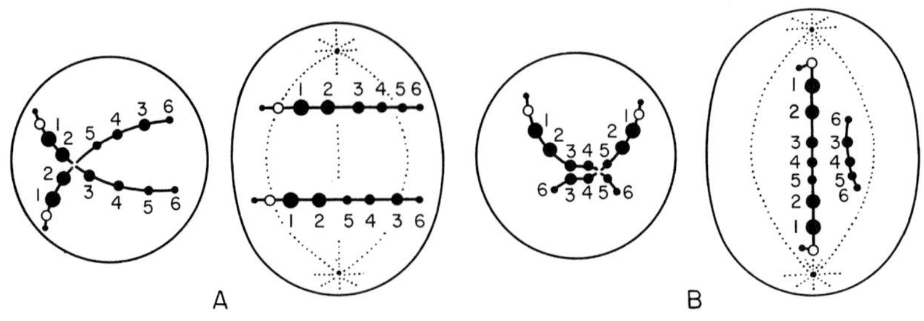

FIGURE 72
Production of balanced and unbalanced gametes during meiosis in paracentric inversion heterozygotes. The sequence of loci in one chromosome is 1–2–3–4–5–6, in the homologous chromosome 1–2–5–4–3–6. **A.** Crossing over outside the inverted section results in balanced chromosomes. **B.** Following pairing of the two chromosomes in the regions 3–4–5, crossing over within the inverted section results in a chromosome bridge and a fragment. For the sake of simplicity, only two of the four strands of the chromosome pair are represented in this figure.

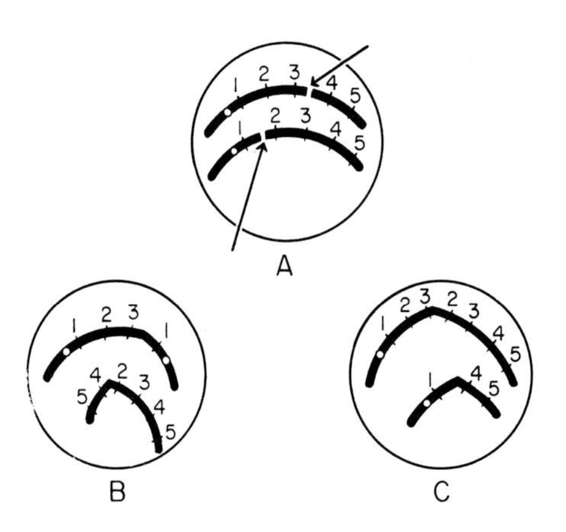

FIGURE 73
Chromosome breakage. **A.** Two breaks have occurred, at different places in two homologous chromosomes. **B, C.** Different consequences of the double break: **B.** Union of the two fragments with kinetochores, and of the two fragments without kinetochores. **C.** Union of each fragment with a kinetochore with one without a kinetochore, resulting in one chromosome with a duplication and another with a deficiency.

the inverted section: chromosomes with single kinetochores deficient for some genes and duplicated for others.

If breaks occur in two separate chromosomes, still more genetic newness may result (Fig. 73, A). If two homologous chromosomes break at different places, reunion of the four fragments may lead to formation of a chromosome with two kinetochores and a fragment without one (Fig. 73, B). Such recombination of broken chromosomes may result in a breakage-fusion-bridge similar to the one already discussed. Or reunion of the fragments may result in an exchange of nonidentical parts and the creation of two new stable chromosomes, one without a middle section, the other with this section duplicated (Fig. 73, C). A cell that contains both chromosomes retains the normal number of all alleles, but when the two chromosomes segregate into different gametes, during meiosis, these gametes possess either a *deficiency* or a *duplication* and will produce zygotes with new phenotypes.

Breaks in two nonhomologous chromosomes may result in reciprocal *translocations* (Fig. 74). Again, a cell containing both chromosomes retains the normal number of alleles. In meiosis, however, either the two translocation chromosomes go together to the opposite pole from the two normal homologues, or one translocation chromosome and one normal chromosome go to one pole and the other translocation chromosome and the other normal chromosome go to the other pole. In the first case, two gametes with full genic complements are formed; in the second case, both gametes have one duplicated section, different in the two gametes, and also lack a section.

With more than two breaks, the variety of possible recombinations increases, but no recombinations are essentially different from those already discussed. Basically, the behavior of broken chromosomes is controlled by two properties: (1) usually a broken chromosome has a tendency to unite with another broken chromosome, either a sister strand or part of a different chromosome, which may be either homologous or nonhomologous; and (2) a chromosome must have one, and not more than one, kinetochore if it and its daughter chromosomes are to be distributed normally in mitosis or meiosis.

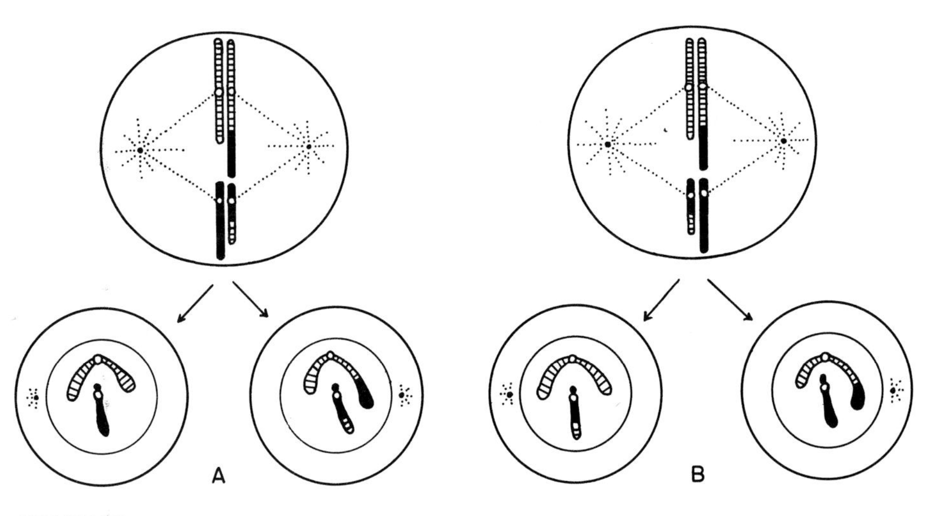

FIGURE 74
Chromosome breakage. Two breaks have occurred, in two nonhomologous chromosomes. Union of fragments has resulted in an exchange of end pieces (translocations). **A, B.** Meiosis in a translocation heterozygote. Two different meiotic arrangements of the two chromosomes involved in the translocations and of their nontranslocated homologues: **A.** The meiotic distribution results in the formation of balanced gametes. **B.** The meiotic distribution results in the formation of unbalanced gametes that lack one chromosomal section and duplicate another.

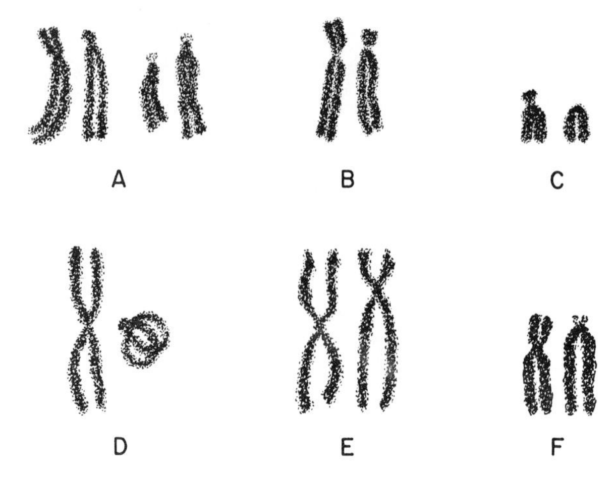

FIGURE 75
Abnormal human chromosomes. Each chromosome pair consists of, *left*, a normal and, *right*, an abnormal partner. **A.** Translocation of part of the short arm of chromosome 5 to the long arm of 13. **B.** Deficiency of about one-half of the short arm of chromosome 5. **C.** Deficiency of the short arm of chromosome 18. **D.** Ring chromosome 1. **E.** Pericentric inversion of chromosome 1. **F.** Pericentric inversion of a chromosome in the group C. (A, after Lejeune, Lafourcade, Berger, and Turpin, *Comptes Rendus Acad. Sci. Paris,* **258,** 1964; B, after Turpin and Lejeune, 1965; C, after Uchida, McRae, Wang, and Ray, *Amer. J. Hum. Genet.,* **17,** 1965; D, after Cooke and Gordon, *Ann. Hum. Genet.,* **29,** 1965; E, after Lele, Dent, and Delhanty, *Lancet,* **1,** 1965; F, after Jacobs et al., *Ann. Hum. Genet.,* **31,** 1968.)

The various types of abnormal chromosomes had first been observed in plants and experimental animals. Their analysis in man (Fig. 75) followed the basic discoveries in other organisms. At the Chicago Conference on Standardization in Human Cytogenetics (1966) an elaborate nomenclature has been proposed providing shorthand descriptions of the great variety of chromosomal abnormalities encountered in man. Details of this nomenclature will be found in the Report of the conference but a few examples may be informative. The notation 46, XY, 21 + designates an individual with 46 chromosomes, XY sex chromosomes, and an extra No. 21. The Report further proposed the use of the capital letters A to G for the seven morphologically distinguishable chromosome groups, the letters p and q for the short and long arm, respectively, of a chromosome, and the letter t for translocation. 45, XX, D–, G–, $t(D_qG_q)$ signifies 45 chromosomes, XX sex chromosomes, one chromosome missing from the D group and one from the G group, their long arms having united to form a DG translocation chromosome.

Chromosome Abnormalities and Spontaneous Abortions. It has been estimated that as many as 25-40 per cent or even more of all fertilized zygotes fail to develop up to the point at which there is either a normal or a premature birth but rather are aborted spontaneously during the earlier months of pregnancy. The causes of such losses are manifold, but it is now known that a large fraction, variously estimated as between 20 and 50 per cent, of abortuses expelled spontaneously are chromosomally abnormal. Thus, in a study by Carr, cultures of cells from 227 unselected abortuses expelled spontaneously yielded abnormal chromosome constitutions for 50 of the specimens, an incidence that is more than 50 times as high as that in live-born infants. In this and other studies, trisomics for autosomes of all groups were encountered but no monosomics, suggesting that the latter lead to death so early that they have remained undetected. The major sex-chromosomal abnormality in abortuses is the XO constitution, even though some zygotes having the abnormal XO sex-chromosome complement do come to term and survive into adulthood.

About 15 per cent of spontaneously aborted fetuses have 69 chromosomes and some have 92, being triploid and tetraploid, respectively. The mode of origin of polyploid embryos is not known with certainty. Triploids may possibly result from fusion of a normal haploid gamete with an abnormal diploid, the latter perhaps a meiotically reduced gamete derived from an abnormally tetraploid oogonial or spermatogonial cell. Tetraploid zygotes may possibly result from duplication of all chromosomes in a diploid fertilized egg. Another possible origin of polyploid zygotes is multiple fertilization. If a mature haploid egg were fertilized by two or three sperm and the multiple nuclear fusion product developed by typical mitosis, triploid and tetraploid embryos would be formed. This assumption that there can be chromosomally stable triploid or tetraploid zygotes derived from dispermic or trispermic fertilization runs counter to the well established fact that—in organisms as different as the threadworm *Ascaris* and various sea urchins—dispermic and trispermic zygotes during their first cleavage division form multipolar mitotic configurations that lead to widespread maldistribution of chromosomes in the daughter cells (Fig. 11, B, p. 24). It may, however, well be that polyspermic mammalian zygotes form a typical single spindle that controls normal separation of sister chromosomes. An analogous situation has been discovered in experiments in which fusion of two somatic mammalian cells in tissue culture may occur. When such fused cells undergo mitosis, a typical single spindle is formed instead of a tetrapolar spindle, which would be expected if each centrosome were to divide independently of the presence of the other. Whatever its origin, polyploidy in humans, as in other mammals, seems only very rarely to be compatible with later fetal survival, a surprising fact since the presence of three or four sets of genes of each kind in many plants and in insects provides a balance similar to that of the diploid and permits survival and rather normal development. No wholly polyploid human embryos have ever developed beyond a few

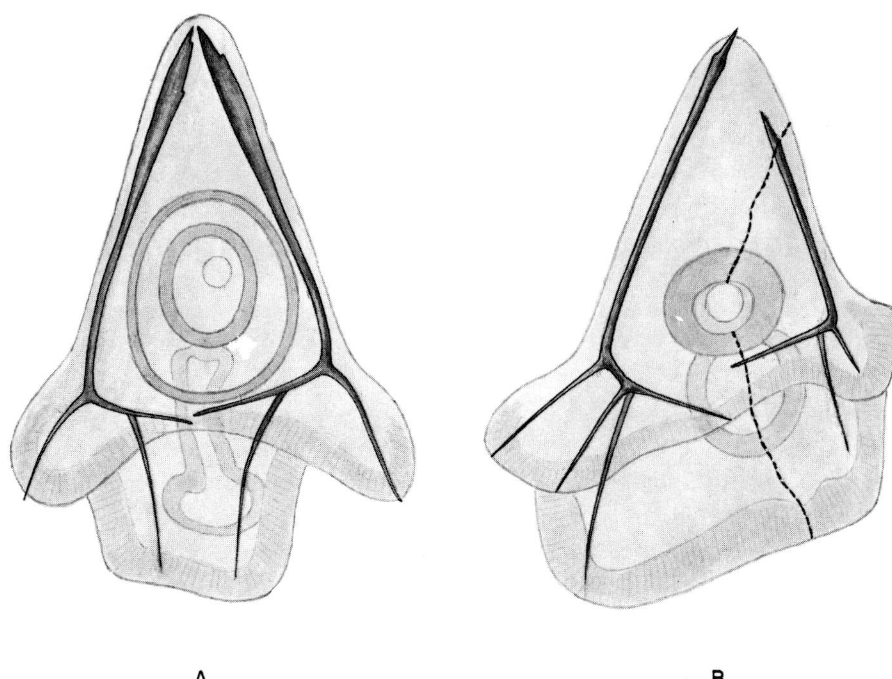

FIGURE 76
"Congenital malformations" in larvae of the sea urchin *Strongylocentrotus lividus*. Note the skeletal structures, shown darker than the rest of the larvae. The larvae were derived from eggs each fertilized by two sperm, resulting in a tripolar spindle during the first cleavage division of the egg. In **A**, the resulting distributions of chromosomes led to three blastomeres with approximately equal sized nuclei and normal, nearly symmetrical formation of skeletal structures. In **B**, one-third of the larva, shown to the right of the dotted line, which had derived from one of the initial three blastomeres, had abnormally small nuclei with a reduced number of chromosomes. The skeleton is abnormal and asymmetrical. (After Boveri, *Zellenstudien,* **6,** 1907. Fischer, Jena.)

days after birth, but several children, with serious defects, have been found to be mosaics of diploid and triploid cells.

A very significant relation exists between the anatomical development of an abortus expelled spontaneously and its chromosomal constitution. In Carr's study of 227 specimens only 84 were anatomically not obviously abnormal, the rest being either grossly abnormal ("blighted ova") or consisting of more-or-less empty embryonic sacs. Only six per cent of the "not obviously abnormal" abortuses had a chromosomal abnormality but approximately 50 per cent of the abnormal "blighted ova" had a chromosomal abnormality. It is obvious that chromosomal aberrations play a large role in the type of embryological maldevelopment that terminates in spontaneous abortions. The relation between fetal malformation and abnormal chromosomal constitution in man did not become known until the 1960's. Yet, evidence for this relation in larval stages of sea urchins (Fig. 76) had been presented 50 years earlier.

The Epidemiology of Chromosomal Aberrations. The cellular mishaps underlying nondisjunction and other abnormal chromosomal behavior are often spoken of as the results of chance. This is justified in the sense that any individual instance is unpredictable and likely to be the result of the coincidence of two or more rare and independent events. Given "normal" conditions the frequency of chromosomal aberrations in a population would be expected to be constant over extended periods of time, in different areas, and for different groups of people. In reality significant differences have been found in some surveys in the frequency of aberrations at different times, in different locations, and among different populations. Collmann and Stoller's study in the State of Victoria, Australia, may serve as an example. These investigators tabulated all cases of Down's syndrome among all births during the period 1942–1957. There was no change in the overall incidence of the syndrome but there were certain years in which the incidence was significantly higher than average and others in which it was significantly lower, the maxima and minima recurring in periods of 5–6 years. Some other data, from different parts of the globe, also suggest clustering of Down's syndrome in time although still others provide no evidence for more than random variations in its incidence. Clustering in space, in addition to that in time, was observed in the Australian population. Urban rates were higher than rural, and fluctuations in rural rates were smaller. Also, the rural maxima and minima lagged behind the urban variations by one year. In contrast to these variations in frequencies of Down's syndrome in Australia, no differences in the frequency of Down's syndrome was found among the births during the period December 1962–April 1966 at two hospitals, which we may refer to as A and B, in Denver, Colorado. This may not seem surprising but it gains importance from the fact that of 12 children with nondisjunctional sex-chromosomal abnormalities born during the same years in the same hospitals, all but one occurred at hospital B. Taking into account the slightly higher total number of children born at hospital B than at A, the percentage of infants born with sex-chromosomal errors is still 10.4 times higher in B than in A.

What is responsible for the observed clustering in time, area, and population of chromosomal maldistributions? In the Australian study, some correlation was noted between periods of increased incidence of Down's syndrome and epidemics of infectious hepatitis but other surveys did not show such relations. The Denver data indicate a relation between sex-chromosomal aberrations and socioeconomic status of the parents. Hospital A, where only a single infant with a sex-chromosomal abnormality was born, is a private facility, whose clientele comes mainly from upper and middle socioeconomic groups; hospital B where there were 11 such births, has a socioeconomically depressed clientele, including many mothers who are on welfare. Using a scale of five classes for socioeconomic status, with Class I for the highest and Class V for the lowest status, the one infant with a sex-chromosomal abnormality at hospital A was born to a Class II couple; three of the abnormal infants at hospital B were assignable to Class IV and eight to Class V parents. Although these data are few, they are nevertheless suggestive of a significant relation between socioeconomic status and the frequency of chromosomal errors in live-born. It remains unknown what specific factors are entailed in this relation be they nutritional in nature, or related to use or misuse of drugs, exposure to toxic environmental agents, frequency of sexual intercourse, or many other possibilities. And the fact that—in contrast to the relative incidence of the sex-chromosomal errors—the incidence of Down's syndrome at the two hospitals did not differ serves as a warning against making premature generalizations about frequencies of chromosomal aberrations and socioeconomic factors.

Human Cytogenetics and Genetic Counseling. Modern chromosomal discoveries in man have provided unexpected insights into the nature of a variety of physical and mental defects. Although the majority of inherited human ills are undoubtedly the result of submicroscopic variations in DNA molecules, the elucidation of some as being determined by gross chromosomal aberrations has had important practical consequences. Knowledge of the maternal age effect in Down's syndrome and of its similar, though weaker, effect in sex-chromosomal aberrations may lead to reduction of childbirth among older women and thereby to a decrease in the incidence of abnormal offspring.

From the point of view of human feeling, it is important that no longer need a couple who have had a child affected with Down's syndrome feel guilty about what they might have done to bring it on—about, for example an emotional upset to which the mother may have been subjected during her pregnancy. The nondisjunctional error concerning chromosome 21 that leads to the defective birth occurs close to the time of fertilization and does not depend on later events. Nor, under most circumstances, need a couple feel responsibility for a spontaneous abortion, most of which are due to early meiotic or mitotic mishaps: Spontaneous abortion is a benign way independent of human decision of preventing the survival of blighted offspring. Another sort of human advantage deriving from knowledge of chromosomal aberrations is demonstrated by the example that cytological study showing the presence of a translocation may yield a warning to parents that a defective child born to them may likely be followed by others of its kind, or, that cytological study showing normal chromosome sets in the parents may suggest a very good chance of normal future offspring. In addition, the method of amniocentesis (p. 808) permits the cytological study of the chromosomes of a fetus before birth and this may provide information highly relevant to counseling, as will be discussed in Chapter 30 (Aspects of Medical Genetics).

PROBLEMS

12. (a) How many different types of fertilized eggs that are trisomic for an autosome are possible in man?
    (b) How many different types are possible that are trisomic for two different autosomes?
13. Down's syndrome women and men are able to have offspring, though this is a rare occurrence. If matings of two affected persons produced several fertilized eggs, what chromosomal constitutions would be expected among them and in which ratios? What presumably would become of the zygotes?
14. A normal girl has 45 chromosomes. She has one normal brother and another brother who is affected with Down's syndrome. The number of chromosomes of both brothers is 46. What is a likely explanation of the constitutions of the three sibs?

REFERENCES

Apgar, Virginia (Ed.), 1970. Down's syndrome (mongolism). *Ann. N.Y. Acad. Sci.*, **171**:303–688.

Carr, D. H., 1971. Chromosome studies in selected spontaneous abortions: polyploidy in man. *J. Med. Genet.*, **8**:164–174.

Caspersson, T., Zech, L., and Johannson, C., 1970. Differential binding of alkylating fluorochromes in human chromosomes. *Exp. Cell Res.*, **60**:315–319

Chicago Conference, 1966. *Report of the Chicago Conference: Standardization in Human Cytogenetics.* 21 pp. National Foundation-March of Dimes, New York. (Birth Defects: Original Article Series, Vol. II, No. 2.)

Court-Brown, W. M., 1967. *Human Population Cytogenetics.* 107 pp. North Holland Publ. Co., Amsterdam.

Eberle, P., 1966. *Die Chromosomenstruktur des Menschen in Mitosis und Meiosis.* (Fortschritte der Evolutionsforschung, Band II). 261 pp. Fischer, Stuttgart.

German, J., 1970. Studying human chromosomes today. *Amer. Sci.*, **58**:182–201.

Hamerton, J. L., 1968. Robertsonian translocations in man: evidence for prezygotic selection. *Hum. Hered.*, **7**:260–276.

Hamerton, J. L., 1971. *Human Cytogenetics*, Vol. I, General Cytogenetics. 412 pp. Academic, New York.

Hamerton, J. L., 1971. *Human Cytogenetics,* Vol. II, Clinical Cytogenetics. 545 pp. Academic, New York.
Jacobs, Patricia A., Cruickshank, G., Faed, M. J. W., Frackiewicz, A., Robson, E. B., Harris, H., and Sutherland, I., 1967. Pericentric inversion of a group C autosome: a study of three families. *Ann. Hum. Genet.,* **31**:219–230.
Jacobs, Patricia A., Price, W. W., and Law, Pamela (Eds.), 1970. *Human Population Cytogenetics.* 325 pp. Edinburgh University Press, Edinburgh.
Lilienfeld, A. M., 1969. *Epidemiology of Mongolism.* 145 pp. Johns Hopkins University Press, Baltimore.
Mikkelsen, Margareta, and Stene, J., 1970. Genetic counselling in Down's syndrome. *Hum. Hered.,* **20**:457–464.
Montagu, M. F. A. (Ed.), 1961. *Genetic Mechanisms in Human Disease. Chromosomal Aberrations.* 592 pp. Thomas, Springfield, Ill.
Penrose, L. S., and Smith, G. F., 1966. *Down's Anomaly.* 218 pp. Little, Brown, Boston.
Swanson, C., 1967. *Cytogenetics.* 194 pp. Prentice-Hall, Englewood Cliffs, N.J.
Tokunaga, Chiyoko, 1969. The effects of low temperature and aging on nondisjunction in Drosophila. *Genetics,* **65**:75–94.
Turpin, R., and Lejeune, J., 1969. *Human Afflictions and Chromosomal Aberrations.* 392 pp. Pergamon, Oxford. (English translation of *Les Chromosomes Humaines,* 1965.)
Valentine, G. H. 1969. *The Chromosome Disorders; an Introduction for Clinicians.* 2d ed. 172 pp. Heinemann, London.

6

PROBABILITY

We have seen that an individual has a pair of alleles at each genic locus and that one allele of each pair is obtained from the mother and the other from the father. When the individual produces eggs or sperm, only one allele of each pair is transmitted to a gamete—one-half of the gametes receive the maternal allele, and the other half the paternal.

A child, therefore, receives only one allele from each of his mother's pairs of alleles and only one from each of his father's pairs. Whether any specific allele is the one which the parent received from the child's grandmother or from its grandfather is a question that cannot be answered with certainty. It is a matter of chance, depending on whether the fertilizing gametes happened to contain the maternal or paternal allele. Chance is not a vague concept but can be expressed quantitatively. Its evaluation in terms of probability plays a major role in all human activities. The probability that a healthy person will still be alive the next day determines much of his behavior. It also determines the behavior of all other persons with whom he has contact—although everyone is aware of the possibility of death. The probability that the sales volume of a store will not vary greatly for corresponding periods is the basis of ordering new merchandise. The probability that an automobile accident will occur is decisive in determining the rates of accident insurance. Most human actions are based on the probability that a certain event will or will not occur.

Probability Expressed as a Fraction. In tossing a coin, there is an even chance that heads or tails will turn up. The probability of throwing heads is therefore 1 out of 2, or 1/2. In throwing a die, the probability of obtaining a specific number—for example, three—is 1 out of 6, or 1/6. In general, *probability* is defined quantitatively as *the fraction obtained when the favorable event is divided by all possible events*—"favorable" denoting the event whose probability is under discussion. For the coin, the probability of turning up a head is the fraction formed by dividing this one favorable event by the sum of the possible events, head or tail, which is two. If the problem should be the probability of picking, blindfolded, a red marble out of a container in which 30 red, 50 white, and 20 black marbles are mixed, this probability is 30 out of 100, or 3/10.

Applying this to human genetics, we may ask: What is the probability that a child will inherit from its father the allele which the father received from his father? In other words: What is the probability that a child will inherit a certain allele from its paternal grandfather rather than from its paternal grandmother? The answer is 1/2, since the child could receive from its father either the "favorable" paternal allele or the "unfavorable" maternal allele.

The Probability of Independent Events Coinciding. A common problem is that of determining the probability that two independent events will both occur. What, for example, is the probability that a child will obtain from his mother the grandmother's allele (i.e., the mother's maternal allele) and from his father the grandfather's allele (i.e., the father's paternal allele)? Let us enumerate all possible cases and then obtain a fraction by dividing the favorable one by the total number of possible cases (Fig. 77):

1. A child may receive the maternal allele of each parent.
2. A child may receive the paternal allele of each parent.
3. A child may receive the maternal allele of the mother and the paternal allele of the father.
4. A child may receive the paternal allele of the mother and the maternal allele of the father.

Obviously, the four combinations are equally probable. The third is the favorable combination, whose probability we wish to determine. This combination takes place in 1 out of 4 cases, and the probability is thus 1/4.

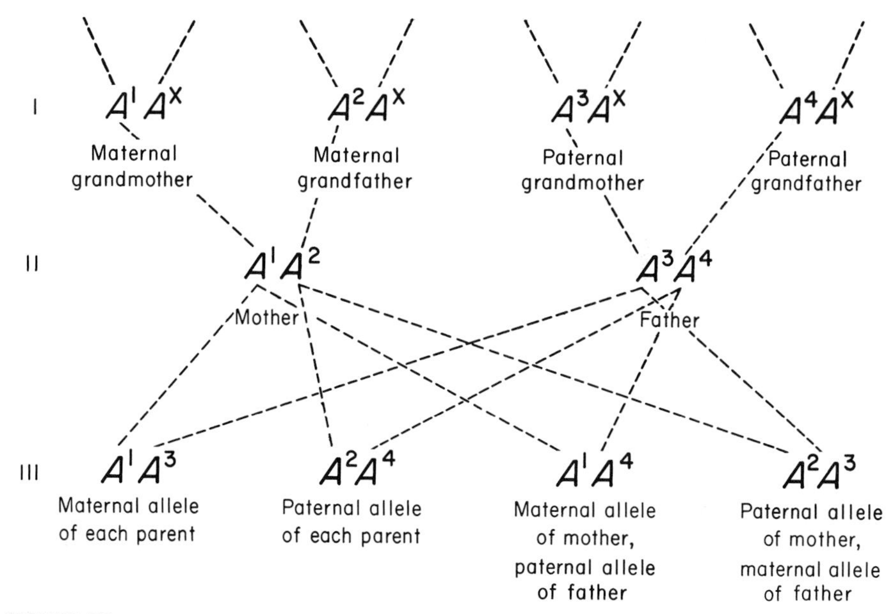

FIGURE 77
The distribution of maternal and paternal alleles through three generations. I. Generation of grandparents. II. Generation of parents. III. Present generation. A^X = alleles that were not transmitted to the parents in generation II.

We can derive this result in a different way. Since the probability of obtaining the paternal grandfather's allele is 1/2 and that of obtaining the maternal grandmother's allele is also 1/2, the first event is realized in only 1/2 of the cases in which the second occurs. This means that the chance of both events happening together is 1/2 of 1/2, or 1/4.

If, in a given population, half of all individuals are men and 1/10 of them belong to blood group B, then the probability of choosing at random an individual who is both a man and belongs to blood group B is 1/10 of 1/2, or 1/20. Generalizing, we obtain the theorem: *The probability of two independent events occurring together is the product of the two separate probabilities.* This theorem can be extended to any number of independent events; the probability that they will occur together is the product of the separate probabilities of each of the events.

The Probability of One or Another Event Occurring. What is the probability that *either* one *or* another of two mutually exclusive events will occur? (It is implied here that either alternative is "favorable.") For example, what is the probability that a child will inherit his two alleles either from the two grandfathers or from the two grandmothers, and not one allele from a grandfather and the other from a grandmother? Inheritance of the alleles from the two grandfathers has a probability of 1/4, and the probability of inheriting them from the two grandmothers is also 1/4. The over-all probability that a favorable event will occur is thus 1/4 *plus* 1/4, or 1/2. Generalized: *The probability that one or the other of any number of mutually exclusive events will occur is the sum of the separate probabilities.*

The limits of probability are 0 and 1. A probability of 0 signifies impossibility; that is, absence of a favorable possibility. For example, the chance that a child will inherit both alleles of a pair from its father and none from its mother is 0, since no such alternative is listed among the four possible cases on page 138. A probability of 1 means certainty; that is, *all* possible events are favorable. For example, the chance that a child will inherit either its father's paternal or maternal allele is 1, since these two alternatives are the only ones that exist.

Aspects of Probability in Human Genetics. It can be predicted with certainty that a child will receive from each parent one of the two alleles of all genic pairs, but it cannot be predicted which specific combination of either parent's maternal and paternal alleles will be received. In the gametes of an individual who is heterozygous for one pair of loci in each of the 23 chromosomes, 8,388,608 different combinations of alleles are possible, and equally probable. Therefore, the probability that a child will inherit any one specified combination is 1/8,388,608; there are 8,388,607 other combinations, any one of which the child is as likely to receive as the one specified. Thus, the chance of his receiving the specific combination is close to the lower limit of probability —zero.

Every person is a "half-breed" in the sense that he has inherited half of each pair of genes from one parent and the other half from the other parent. No one should be called "half-blood," since blood is no more transmitted directly than are limbs or eyes or gastric juices. The expression "quarter-blood," or any other designation that attempts to express the fraction of a genotype which a person owes to an ancestor more remote than his parents, is even more misleading. The genotype of each individual may well be considered to consist of two halves, namely, two complete single sets of genes; but neither of these single sets is composed of two equal parts, each contributed by one of the grandparents. The set of genes that the individual has received from his father may consist solely of the grandfather's alleles or solely of the grandmother's alleles, or any one of the numerous combinations of alleles of both grandparents. It is true that among the many combinations there are some in which one-half consist of alleles from the grandfather and the other half of alleles from the grandmother. In this sense, one could defend a statement like the following: It can happen that one-quarter of the genes of an individual may be derived from his father's father, one-quarter from his father's mother, and one-half from his mother—this latter half consisting of any one of the numerous combination of the alleles of his mother's parents.

This would be a biologically meaningless formulation. Each locus is a unique entity among a set of loci. To speak of one-quarter of all alleles of a set would be equivalent to enumerating all the different parts of a complex machine and then designating any randomly chosen quarter of these parts as "one-quarter of the machine."

Many times in this book, questions on human genetics and their answers will be expressed in terms of probability. One of the important things that can be learned from the study of human genetics is that many of the problems of inheritance are statistical ones.

Correlation. The height of individuals varies and so does the color of their hair. Should we seek to determine whether the color of hair is related to stature or whether these two variables are independent, one of at least three possibilities might be found to be true: (1) the taller the person, the darker the hair; (2) the taller the person, the lighter the hair; and (3) a short person is as likely to have light—or dark—hair as a tall one. If the last were true, we would say that there is no correlation between the two traits; if the first were true they would be "positively correlated"; if the second, "negatively correlated."

These terms are also applicable when a single trait is considered in different individuals. Thus, if we ask whether height is correlated in parents and offspring, we might expect to find that (1) the taller the parents, the taller the child; (2) the taller the parents, the smaller the child; or (3) the height of a child is unrelated to that of its parents.

The actual finding here is that there is a positive correlation between the heights of parents and offspring (Table 4). Such a correlation does not by itself indicate anything about its cause or causes. It could, for instance, signify that most tall parents grow up in homes in which nutrition and other environmental factors are favorable to the attainment of tallness and that they, in turn, provide the same favorable circumstances for their children. Or it could mean that tall parents come from a part of the country where conditions are favorable for growth and that their children share this geographic background. Or it could mean that tall parents have genes for tallness and that their children share some of these genes. A decision among these hypotheses would have to depend on independent evidence.

It is usually desirable to be able to express the strength of a given correlation. When, for example, an increase in the measurement of a given trait is accompanied by a proportional increase in the measurement of another, there is complete positive correlation. When an increase in the first is usually, but not always, accompanied by an increase in the second, or if the increase in the second is not always strictly proportional to the increase in the first, the

TABLE 4
Relation Between Heights of Parents and of Their Adult Children.

Mean Height of Parents (in Inches)	*Number of Children of Given Height (in Inches)*							
	60.7	62.7	64.7	66.7	68.7	70.7	72.7	74.7
64	2	7	10	14	4	–	–	–
66	1	15	19	56	41	11	1	–
68	1	15	56	130	148	69	11	–
70	1	2	21	48	83	66	22	8
72	–	–	1	7	11	17	20	6
74	–	–	–	–	–	–	4	–

Source: After data of Galton from Johannsen, *Erblichkeitslehre,* Jena, 1909.

correlation is positive, but incomplete (Table 4). When there is a tendency for the second to decrease when the first increases, the correlation is negative; it, too, may be complete or incomplete. The *correlation coefficient,* r, is a measure of the strength of correlation. No formula for r will be given here, but it should be noted that it is so defined that for complete positive correlation $r = +1$, and for complete negative correlation $r = -1$. In the absence of any correlation, $r = 0$, and the various degrees of incomplete positive and negative correlation are expressed by coefficients which lie between +1 and −1. An important application of the correlation coefficient to human genetics relates to the fractions of genes held in common by various groups of people. Except by chance, unrelated persons share no genes ($r = 0$), identical twins have all genes in common ($r = 1$), and various other related persons share intermediate fractions of genes—e.g., a parent and his child share one-half of their genes ($r = 0.5$).

Data for the calculation of correlation coefficients are often presented in the form of "scatter diagrams." Two such diagrams based on measurements of only a few individuals are shown in Figure 78. In part A, the absence of any trend indicates lack of correlation between length and width of the head in a sample of 12 adult males; in B, the increasing ratio of chest girth to height with decreasing height is evidence for incomplete negative correlation between chest girth/height and height. A tabular presentation of incomplete positive correlation—between heights of parents and offspring—has been given in Table 4. Numerical values of some correlation coefficients follow. In one study the correlation coefficient for the relation between the length of the first joint of the forefinger on the right hand and the length of that on the left hand of the same individuals was found to be +0.92; in other studies, that between stature of husband and wife, +0.28; between stature of fathers and sons,

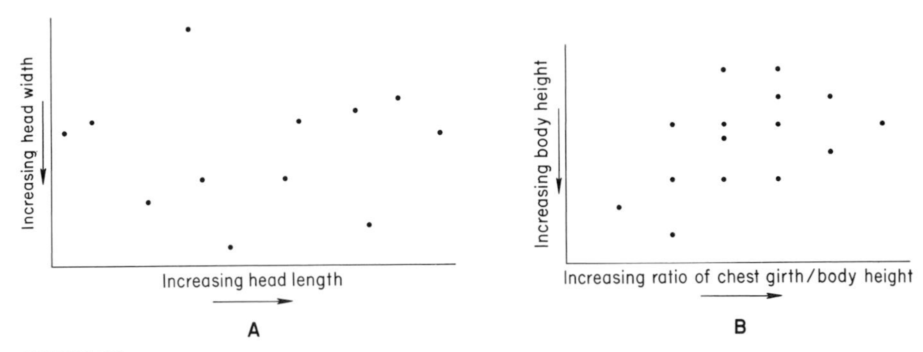

FIGURE 78
Scatter diagrams of the relation of two variable measurments in groups of individuals. **A.** Absence of correlation between head length and head width. **B.** Incomplete negative correlation between chest girth/body height and body height. (After Weber, *Variations-und Erblichkeits-Statistik,* Lehmann, Munich, 1935.)

+0.51; between height and body temperature, 0.00; and between adult age and body temperature, −0.15. In later chapters various coefficients of correlation will be given and their genetic significance discussed.

PROBLEMS

15. Assume the probabilities for a birth to occur are alike for all days of a year:
    (a) What is the probability of a girl having her birthday on Christmas?
    (b) A man was born on May 1. What is the probability of his marrying a woman born the same date? On June 1? On October 12?
    (c) In a random sampling what is the probability of finding a husband and wife who have their birthdays on May 1?
16. What is the probability that a child will inherit either the allele I^{O} or the allele I^{A} from his $I^{O}I^{A}$ mother?
17. What is the probability of a person inheriting one of the alleles of a given pair of his genes from his paternal grandmother and the other allele from his maternal grandmother?
18. A man's maternal grandfather belongs to blood group AB, but all other grandparents are O. What is the chance of the man being:
    (i) A? (ii) B? (iii) AB? (iv) O?
19. A man received the following genes, all located in different chromosomes: from his father *ABCDE* and from his mother *A'B'C'D'E'*. What is the probability of his giving to his child the combination (i) *AB'C'D'E?* (ii) Either *ABCDE* or *A'B'C'D'E'?*

20. If the genotypes of the parents are $I^A I^O$ and $I^A I^A$, what is the chance of their two children both being (i) $I^A I^O$? (ii) $I^A I^A$? (iii) $I^O I^O$?

21. (a) What fraction of his genes does a parent have in common with any one of his children?
    (b) What fraction have two sibs in common?

22. Assume that the probabilities of the birth of a boy or a girl are 1/2. What are the probabilities of the following sequences of births in a hospital:
    (a) A succession of 30 boys?
    (b) 15 of each sex born alternately?

23. If no crossing over occurs, what is the probability that a woman will inherit:
    (a) None of her 22 maternal autosomes from her grandmother?
    (b) All of her 22 maternal and all of her 22 paternal autosomes from her two grandfathers?

24. What is the probability that a man will transmit a specific allele to his great-granddaughter?

25. Assume that a certain desirable trait of an individual depends on the presence of the following rare alleles: $A'B'C'D'E'$ in heterozygous combination with their alleles $ABCDE$. What is the probability that a grandchild of this individual will receive the alleles $A'B'C'D'E'$ from his grandparent?

26. Would you expect positive, negative, or no correlation between the following variables:
    (a) Degree of schooling and longevity?
    (b) Degree of schooling and income?
    (c) Length and importance of a research report?
    (d) Age of adult and number of healthy teeth?

REFERENCES

See the books on statistics cited in references to Chapter 10.

7

SIMPLE AUTOSOMAL INHERITANCE

Genetic differences between two individuals may consist of differences between the alleles at a single locus or between those at more than one locus. In general, different individuals have different alleles at *many* loci. Mendel taught us not to attempt the study of the inheritance of all genetic differences at once, but to follow separately the inheritance of differences dependent upon single loci. This is reflected by the word "simple" in the title of the present chapter. In addition, this word implies that the discussion will be restricted to examples in which a given genotype is always expressed in the same manner in the phenotype. Later, we shall discuss inherited conditions for which this is not true. The word "autosomal," which is also part of this chapter's heading, signifies that discussion will be restricted to genes located in one or another of the 22 human autosomal pairs. Since the autosomes are present in equivalent pairs in both females and males, inheritance of autosomal genes follows a "symmetrical" mode in which it makes no difference in specified matings which one of the parents of a sibship carries one and which one the other genotype. This differs from sex-linked inheritance (Chapter 14), in which the XX-XY alternative entails an asymmetry of parental chromosomal and genic types.

Normal and Abnormal Traits. Phenotypic differences between two individuals may be such that one can be regarded as normal and the other abnormal, or both may be variants of either normal or abnormal types. The study of all kinds of differences is important. The difference between normal and abnormal is particularly significant to medical men and to psychologists who are interested in the inheritance of mental defects; the difference between normal variants is of interest to anthropologists and psychologists who investigate the normal variability of mankind. Studies have shown that many persons afflicted with striking abnormalities are genetically distinguished from normal persons by a difference at no more than a single locus, while the genetic differences between persons who have differing normal variants of a trait are often polygenic—that is, differences at many loci. That this is so is not surprising when we remember the concepts of genic action developed in Chapter 3. A trait depends on numerous genic loci, and differences between the alleles at any one of these loci may cause variations in the phenotype.

Many of the slight differences between normal individuals are, therefore, due to the sum of minor differences in the action of several loci. A rare, striking difference between a normal and an abnormal person, although it may be due to a combination of minor genic differences, is more likely to be brought about by the failure of a single locus to control a reaction necessary for normality. An analogy with an automobile might again be made. The "normal" variation between two different models often consists of a large number of minor differences in many parts; but the difference between a "normal" automobile, which starts, and an "abnormal" one, which doesn't, is likely to be traceable to a single cause.

These developmental considerations account for the fact that a genetic study of variations of such traits as facial features or body build within the range of normality usually is difficult although genetic study of abnormal variants is often easy. Indeed, we know more about the inheritance of striking defects in facial features or body build than of their normal variants. The distinction disappears, however, when traits are studied that, unlike those named, are not the distant results of the actions and interactions of many genes but are close to the primary action of single genes. The methods by means of which normal proteins and—in terms of disease causation—defective proteins can be distinguished are equally applicable to a differentiation between variants of normal proteins. The genetic study of such normal molecular variants is no more difficult than that of defect-causing variants.

DOMINANT INHERITANCE

We shall begin our discussion of transmission of hereditary traits by considering a dominant allele, *D*, whose presence or absence is made immediately apparent by the presence or absence of the corresponding phenotype. If the dominant allele is rare, then most individuals carrying it have it singly—that is, heterozygously with the recessive allele *d*. Homozygosity for *D* could only result from its presence in both parents—and this combination would be so much rarer than the postulated rarity of the gene that almost no doubt would exist that a person with the dominant phenotype was heterozygous. If, however, the dominant gene is a common one, then both homozygous (*DD*) and heterozygous (*Dd*) individuals should be common, and the dominant phenotype could no longer serve as a complete indication of the genotype.

An individual heterozygous for a dominant gene (*Dd*) produces equal numbers of gametes with the alleles *D* and *d*. Marriages between such a person and another without the dominant gene—that is, *dd*—result in children any one of whom has an equal chance of receiving either *D* or *d* from the *Dd* parent, and all of whom receive *d* from the other parent (Fig. 79). Consequently, half of the children of such marriages carry the trait dependent on *D* (i.e., are affected), and the other half lack it. Furthermore, it is obvious that only affected parents can have affected children—that all affected children have at least one affected parent. In simple dominant inheritance, a trait never skips a generation.

It should be emphasized that the statement that half of the children with *Dd* and *dd* parents carry the trait dependent on *D* does not mean that in any one family one-half of the sibs have it and the other half do not. The distribution of the trait is governed by the laws of probability (see Chap. 6), and any generalization concerning it is valid only for a large population. Thus, we often

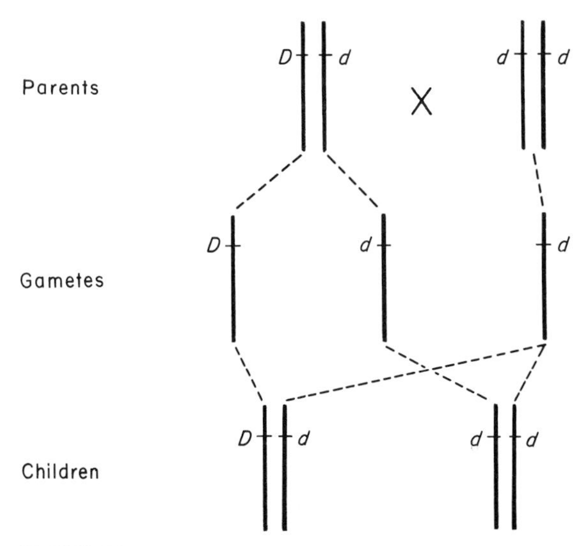

FIGURE 79
Marriage $Dd \times dd$. Transmission of alleles from parents to children.

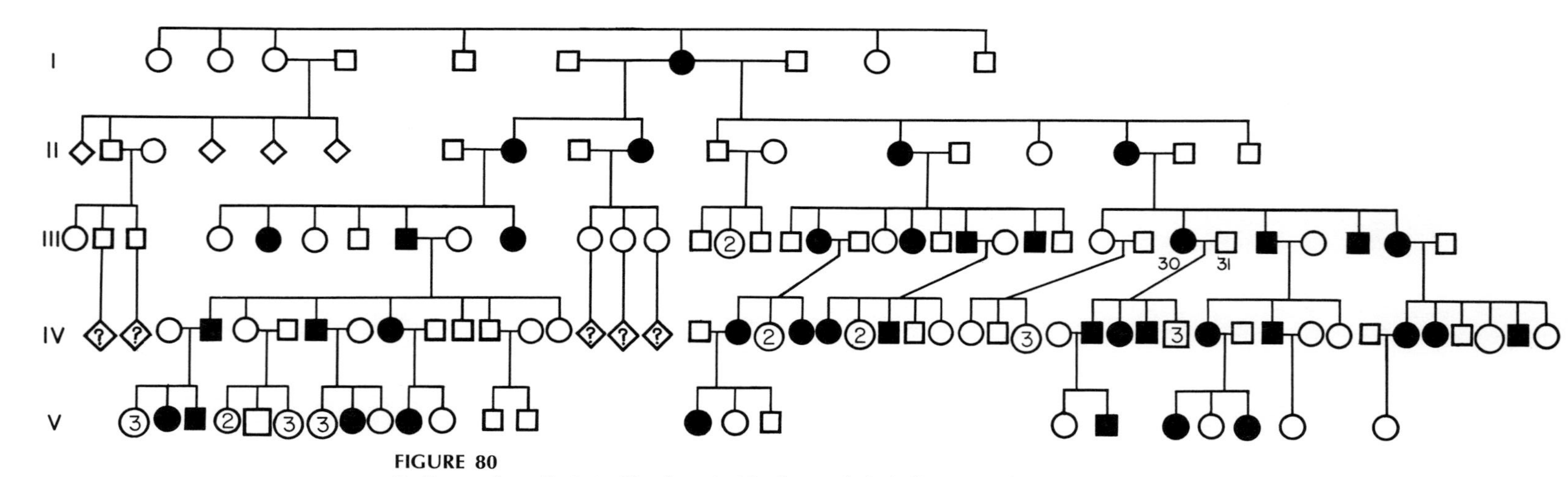

FIGURE 80
Pedigree of woolly hair. Numbers inside the symbols indicate numbers of sibs not listed separately. (After Mohr, *J. Hered.*, **23,** 1932.)

FIGURE 81
Segregation for woolly hair. Parents III-30 and III-31 (Fig. 80) and their children. (Mohr, *J. Hered.*, **23**, 1932.)

speak of the *ideal* 1:1 ratio. All the children of one pair of *Dd* and *dd* parents may have the trait; none of the children of another pair may have it. More will be said about this in Chapter 10, Genetic Ratios.

Woolly Hair. An example of simple dominant inheritance is seen in Figure 80. This extensive pedigree shows the inheritance of a rare type of unusually woolly hair (Fig. 81) in a Norwegian kindred. The curled or fuzzy texture of this woolly hair is similar to, but not identical with, that of Negroes' hair. Though of good growth, it breaks off at the tip and never gets very long. It has been propagated for at least five generations. Nothing is known about the first appearance of the trait, but it seems certain that it is not due to admixture of African alleles, since the possibility of early interbreeding of Norwegians and Africans is very remote. What appears to be the same condition has also been observed in Holland, in "Old American" stock, and elsewhere.

The pedigree demonstrates the principle of direct transmission from affected parent to half of the children. Summing all affected and nonaffected children from all 20 marriages in which one parent was affected, we find 38 individuals with woolly hair and 43 without it (omitting 3 persons, long deceased at the time the pedigree was made, for whom the hair type could not be ascertained).

Studies from various countries of all similar pedigrees recorded up to 1936, show 145 children with woolly hair and 130 with normal hair in sibships from marriages of one affected and one nonaffected parent. These numbers deviate from the ideal 1:1 ratio (137.5:137.5), but as no one expects, for example, when throwing a coin 276 times, to obtain exactly 138 heads and 138 tails, so the ratio of *Dd* to *dd* children cannot be expected to be exactly 1:1, since it is a matter of chance whether a gamete with *D* or one with *d* fertilizes a *d* gamete of the opposite sex when a child is conceived. We shall later discuss methods of judging whether the deviation between an observed and expected ratio may be considered as simply due to chance or whether it is so large as to be statistically significant.

More Rules of Dominant Inheritance. Other facts about the appearance of simple dominant traits may be learned from the study of pedigrees of inheritance of woolly hair:

1. There were 64 men and 66 women among the affected persons whose sex was given in the pedigrees. Thus, the trait does not appear more frequently in one sex than the other.
2. Altogether, woolly-haired fathers had 24 woolly-haired and 36 nonwoolly-haired children, and woolly-haired mothers had 34 woolly-haired and 26 nonwoolly-haired children. Thus, both affected men and women transmitted the characteristic hair type to their children. Furthermore, regardless of whether the father or the mother was affected, the proportion of woolly to nonwoolly-haired children deviated from the 1:1 ratio by no more than would be expected as a result of chance.

The facts derived from the genealogical charts are in agreement with expectations derived from the theoretical principles of transmission of a dominant gene in a chromosome and the chance fusion of sperm and egg carrying different alleles at a single locus. Whenever such facts are found in studying the inheritance of a rare trait, we may conclude that it is transmitted as a simple dominant. This conclusion is subject, however, to the provisions (1) that the pedigree is representative of the facts and is not a biased selection, and (2) that large enough numbers of individuals are covered to make the data truly representative.

It may be asked: What would result from the marriage of two individuals who both possessed woolly hair or some other simple dominant trait? Since each of the parents would normally possess one allele for woolly and one for nonwoolly hair, and since each of them would form the two kinds of germ cells, *D* and *d*, with equal frequency, chance fusion of egg and sperm would lead to the genotypes *DD*, *Dd*, and *dd* in the proportions 1:2:1.

Egg		Sperm		Child
D	+	D	=	DD
D	+	d	=	Dd
d	+	D	=	Dd
d	+	d	=	dd

If the dominance of *D* is complete, *DD* and *Dd* individuals would be indistinguishable and we would expect 3 woolly-haired children to 1 nonwoolly-haired child. Should *DD* produce a phenotype different from that produced by *Dd*, three types of offspring would result: the new *DD* type, woolly (*Dd*), and nonwoolly (*dd*)—in the Mendelian ratio 1:2:1. Since no marriages of two woolly-haired individuals have been recorded, the phenotype of a *DD* person is unknown.

Homozygotes for Dominant Alleles. Although the likelihood is great that any woolly-haired person is the offspring of one woolly-haired and one nonaffected parent, it is still possible that he might be the *DD* offspring of two affected parents. He would then endow all his germ cells with the *D* gene, so that if he married a nonaffected person, all his children would be *Dd* and, thus, woolly-haired. It would, however, be very difficult to establish his homozygosity for *D* unless he had an unusually large number of children. If only woolly-haired children are produced by one affected and one nonaffected parent, it may either be due to the *DD* homozygosity of one parent or be nothing but a chance deviation from the theoretical 1:1 ratio expected from a *Dd* parent.

In a kindred in which many persons are affected with the dominantly inherited disease of the nervous system, dystonia musculorum deformans (disordered tonicity of the muscles), a certain individual is, with high probability, homozygous for the abnormal allele. The disease expresses itself in involuntary and irregular contortions of the muscles of the trunk and extremities. Although the disease is rare when any large population is considered, more than a hundred affected individuals are known from a thinly populated and relatively isolated region in Sweden near the Lapland border. Most of the affected persons who married had normal spouses and produced both normal and diseased offspring in about equal proportions (Fig. 82). One affected woman, however, married a man who seems to have been affected himself (he had died several decades before the genetic study of the population). The marriage produced nine children of whom five were affected. The nine children of this marriage represent a ratio of five affected to four normal, which is compatible with the theoretical expectation of 3:1. Two of the affected sibs each had both types of offspring and thus were heterozygotes. A third affected person, the oldest son, married a normal woman and had nine children, everyone having "the disease" as the natives were used to refer to

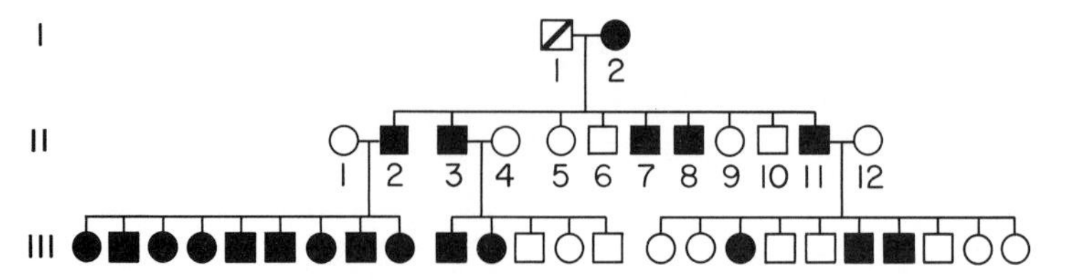

FIGURE 82
Pedigree of the neuromuscular disease dystonia musculorum deformans. I. Marriage of two affected individuals. (The symbol for I-1 signifies that the determination of his having been affected was only indirect). II-2 presumably was homozygous for the dominant gene responsible for the disease since all nine of his children were affected. (After Larsson and Sjögren, *Acta Neurol. Skand. Suppl.*, **17**, 1966.)

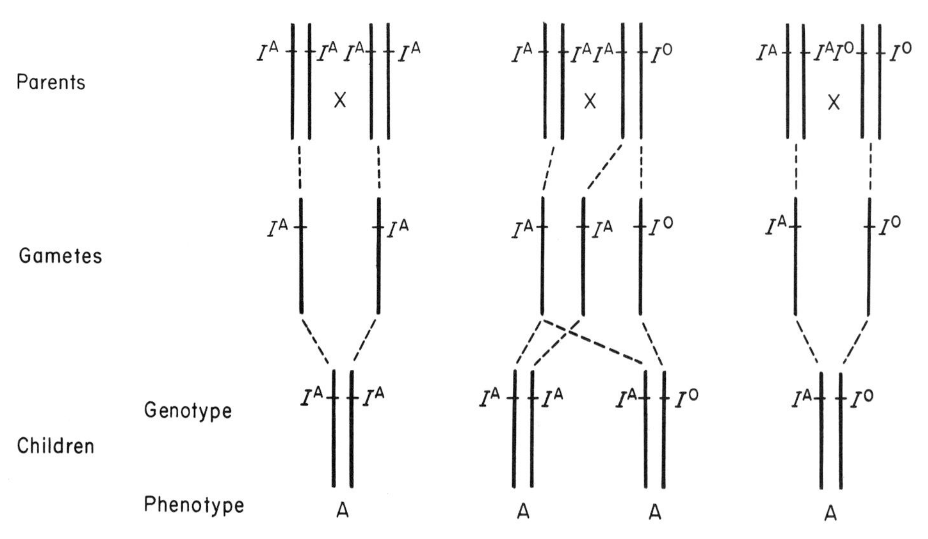

FIGURE 83
Marriages $I^AI^A \times I^AI^A$, $I^AI^A \times I^AI^O$, and $I^AI^A \times I^OI^O$ and the resulting offspring.

it. Most likely, he was homozygous for the dominant abnormal gene. The degree of his illness did not exceed that of many of the heterozygotes.

For many other traits, pedigrees including homozygous dominant individuals are well known. If a dominant allele is common enough in the population to permit numerous marriages between heterozygotes, homozygous dominant offspring will be encountered. Marriages of blood group A persons homozygous for I^AI^A to individuals who are either I^AI^A, I^AI^O, or I^OI^O all lead to group A children only (Fig. 83). Observational proof of this is not so simple as in the case of woolly hair, where the existence of an affected phenotype is practically an assurance of a heterozygous genotype. The phenotype of a person in the A blood group does not tell us whether he is heterozygous I^AI^O or homozygous I^AI^A. Even the phenotypes of his offspring may not give us the answer. Although the diagnosis of heterozygosity can be made whenever both parents belong to group A and have an O group child, no certain diagnosis is possible when only A group children result. An O group child is genetically I^OI^O, which signifies that he received an I^O allele from each parent and reveals the parents' heterozygosity, I^AI^O. The absence of I^OI^O children, however, does not prove that either parent is homozygous I^AI^A, since even if the parents were both I^AI^O, they might have no I^OI^O children; for, when the number of offspring is small, it is possible that fertilizations with only I^A gametes may occur.

To be certain that an A group phenotype is based on I^AI^A homozygosity, special pedigrees have to be selected. The best are those in which both parents are in the AB group, because of the codominant action of their genotype I^AI^B. Since such parents can produce only gametes with either I^A or I^B, and since presence of the I^B allele in the offspring would be phenotypically apparent, group A offspring from such a marriage can only be I^AI^A.

The discussion of the dominant inheritance of the blood-group property A shows the difficulties encountered in pedigree studies of a common trait. We shall see later that these difficulties can be overcome by applying statistical methods to pooled data from large samples of a population.

Further Examples of Dominant Inheritance. The examples of simple dominant autosomal inheritance of relatively rare traits that follow will not teach any new principles, but they will serve to indicate the wide range of traits so transmitted.

Figure 84 is a pedigree of a family in which a white forelock (Fig. 85) has been observed for five consecutive generations. It fulfills all requirements necessary to show dominant transmission of the trait. (This pedigree omits symbols for the spouses of affected parents, as is frequently done when a

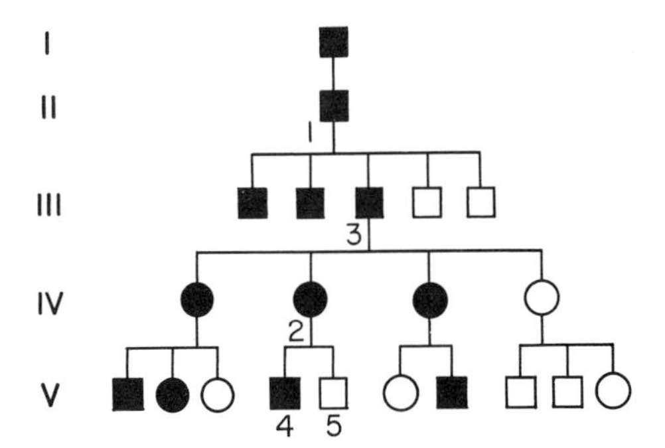

FIGURE 84
Pedigree of white forelock. Photographs of the individuals II-1, III-3, IV-2, V-4, and V-5 are shown in Figure 85. Note the uninterrupted line of transmission of the trait, its occurrence in both men and women, and its presence in both sexes of the offspring of an affected parent. (After Fitch, *J. Hered.*, **28**, 1937.)

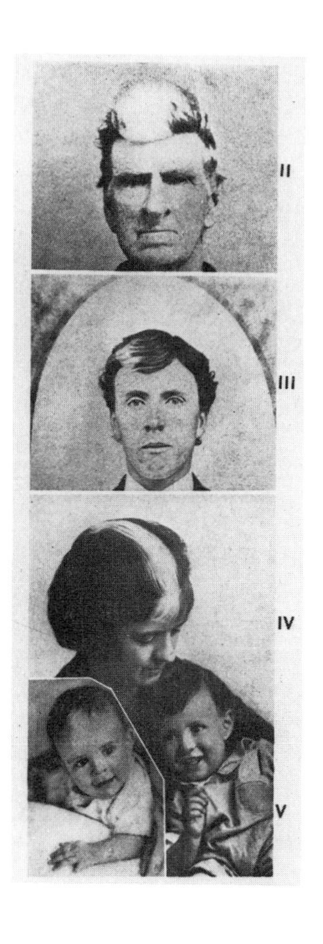

FIGURE 85
White forelock in four generations. The individuals shown are marked by numbers in the pedigree Figure 84. (Fitch, *J. Hered.*, **28,** 1937.)

trait is rare and it can be taken for granted that the spouses are not affected unless there is specific evidence that they are.)

The trait white forelock is akin to the piebaldness, which is quite common among the color patterns of animals (e.g., Holstein cattle, black and tan dogs). The old engraving reproduced here as Figure 86 correctly suggests the analogy between human and animal spotting.

Simple dominant inheritance can also be illustrated by various types of brachydactyly, or short-fingeredness (Fig. 87). One pedigree of this skeletal abnormality, from an American kindred, provided the first demonstration of dominant Mendelian inheritance in man (Fig. 88).

Another abnormality inherited as a dominant is congenital stationary night blindness, a type of defective twilight vision. One pedigree, famous for its extension over many generations, comes from France. Jean Nougaret, who was born about 1637, had, in the course of ten generations, 139 or more affected individuals among his descendants (Fig. 89).

A very thoroughly explored American pedigree of dominant dwarfism is reproduced in Figure 229 (p. 555). These dwarfs belong to a pseudo-achondro-

FIGURE 86
White spotting. Old engraving with inscription: "The spotted negro boy. George Alexander Opattan, the spotted boy died on the 3. Febr. 1813 aged 6 years, was buried at Great Marlow in Buckingham. . . . Painted from life by Dan. Orme and engraved under his Direction by his late pupil P. R. Cooper." (Goldschmidt, *Einführung in die Vererbungslehre*, 2nd ed., Engelmann, Leipzig, 1913.)

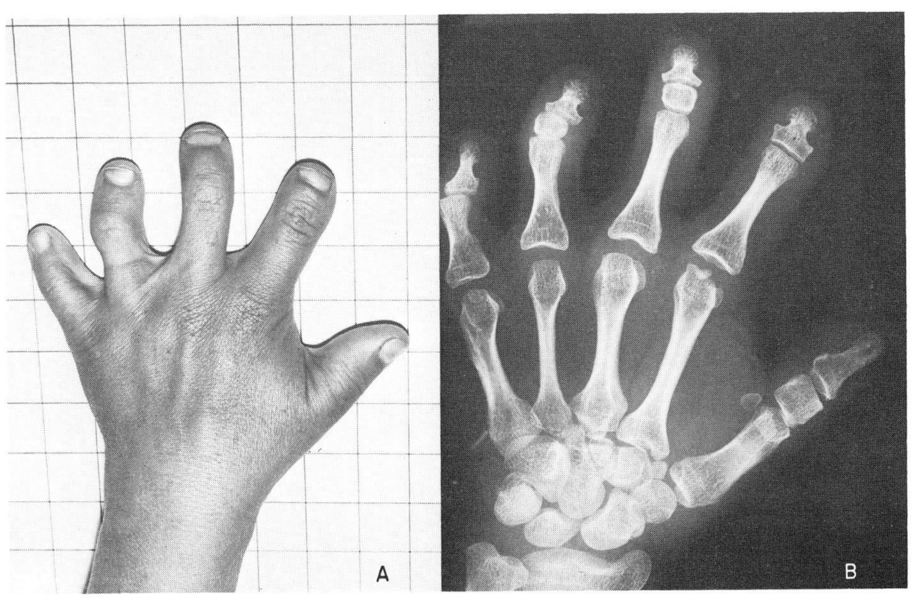

FIGURE 87
Brachydactylous hand. **A.** Photograph. **B.** X-ray photograph. Note, among the numerous skeletal abnormalities, the middle and terminal phalangeal bones that are abnormally short and partly fused. (Hoefnagel and Gerald, *Ann. Hum. Genet.*, **29**, 1966.)

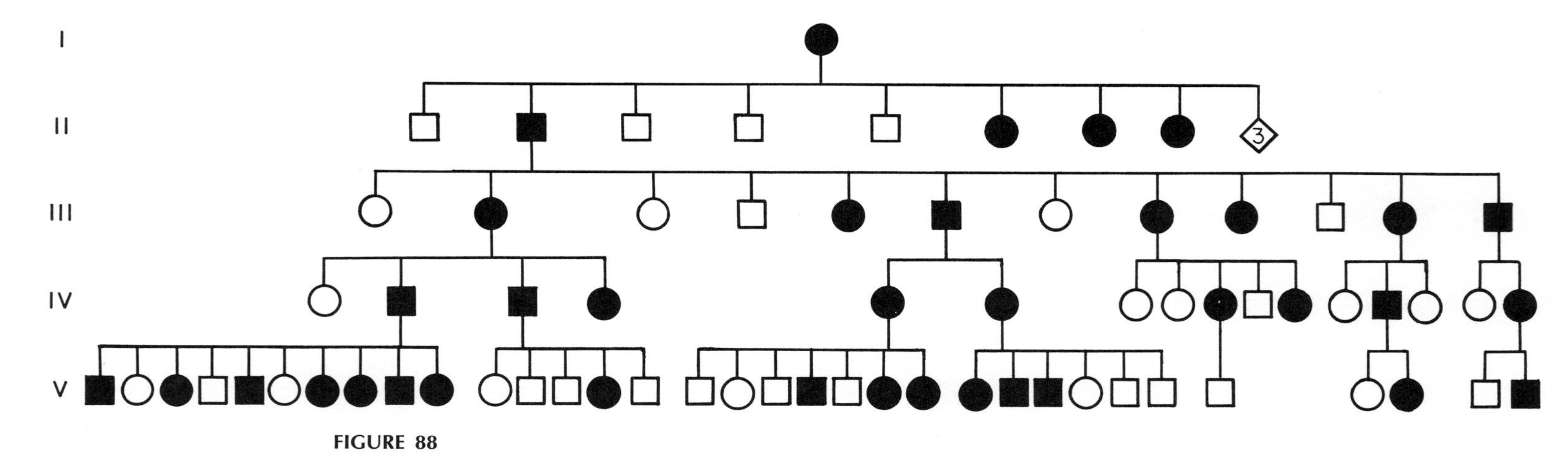

FIGURE 88
Pedigree of brachydactyly. The first demonstration of dominant Mendelian inheritance in man. (After Farabee, *Papers Peabody Mus.*, Harvard Univ., **3,** 1905.)

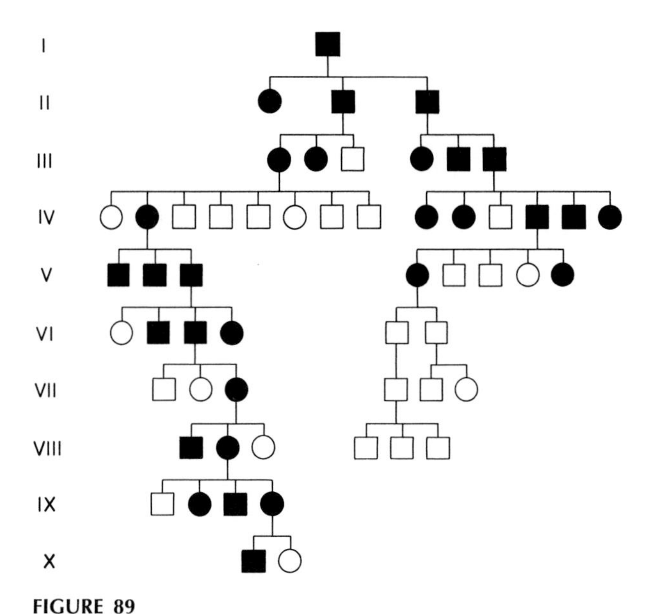

FIGURE 89
Congenital stationary night blindness. Part of the pedigree beginning with Jean Nougaret, born about 1637. (After Nettleship, *Ophth. Rev.*, **26,** 1907; Dejean and Gasseng, *Bull. Soc. Ophth. France,* **1,** 1949; and François, Verriest, De Rouck, and Dejean, *Ophthalmologica,* **132,** 1956. The pedigree sections in the last two publications do not fully agree with each other.)

plastic type, in which the size of the trunk and head is similar to that of unaffected people but the limbs are very short (Fig. 90).

Sometimes the term dominance is used when actually codominance of two alleles is exhibited in the heterozygotes. We may consider as an example a particular enzyme-producing locus. Most persons are homozygous for a normal allele at this locus and produce a normal indophenol oxidase, a tissue enzyme. But heterozygotes, who carry a rare variant allele that is transmitted as a dominant from parent to offspring (Fig. 91, A), produce both the normal enzyme and an abnormal form of it, as shown by electrophoresis (Fig. 91, B). Homozygotes for the variant allele are not known.

Our final example of a dominant phenotype is a trait found in various families, but particularly well documented from the Hapsburg dynasty. Many members of this house had a narrow, undershot lower jaw and protruding underlip, which resulted in a half-open mouth (Fig. 92). The trait can be traced back to the fourteenth century. It was present in, among others, Emperor Charles V (1500–1558), Maria Theresa of Austria (1717–1780), and Alfonso XIII of Spain (1886–1941).

FIGURE 90
Pseudo-achondroplastic dwarf (a South Indian) and a normal observer (J.J.). (Original photograph from Mr. Jacob John.)

It appears likely that many of the traits which characterize specific families are controlled by single dominant alleles. Family resemblances are often due to the recurrence of rather specific features—form of nose or mouth, position of eyes, shape of eyebrows, and so on. If such traits reappear in successive generations and are transmitted by affected parents only, a dominant allele may be the cause of such family-specific, relatively rare variants within the normal range of phenotypes.

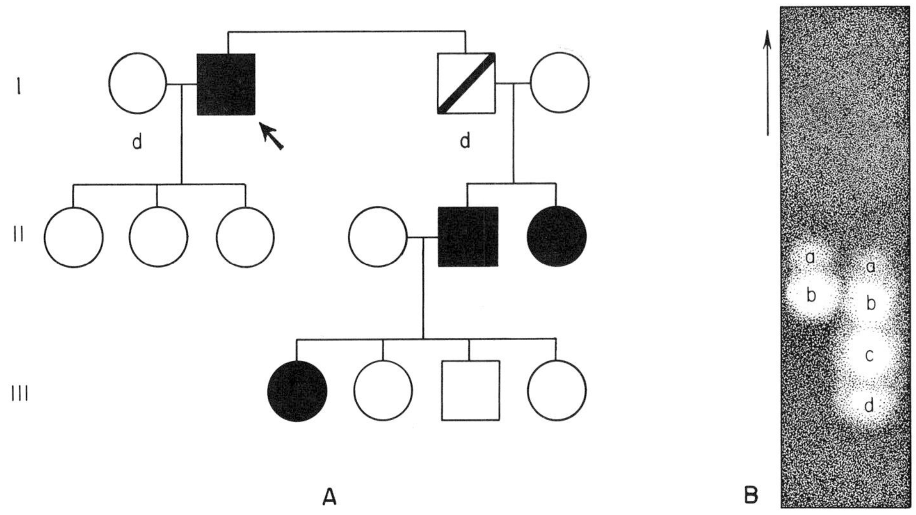

FIGURE 91
Codominance of a variant allele. See text. **A.** Pedigree of a kindred with a variant enzyme (an indophenol oxidase). d = died. The symbol for I-3 signifies that his affected phenotype was not directly observed but was deduced from the phenotypes of his descendants. **B.** Electrophoretic patterns: *left*, from a normal individual (having only bands a and b); *right*, from the heterozygous propositus (having bands c and d in addition to a and b). (After Brewer, *Amer. J. Hum. Genet.*, **19**, 1967.)

FIGURE 92
The Hapsburg lip through the centuries. **A.** Emperor Maximilian I (1459-1519). **B.** King Alfonso XIII of Spain (1886-1941). (A, Stromayer, *Nova Acta Leopoldina*, **5**, 1937.)

RECESSIVE INHERITANCE

By definition a single recessive allele will not affect the phenotype of a person; only when both of the alleles at a locus are recessive is there any effect. Therefore, each of the parents of a child with a recessive trait must carry at least one copy of the specific recessive allele. Three different types of marriages fulfill this condition (Table 5, where d denotes, as before, the recessive allele and D the dominant). The offspring can be expected to be exclusively dd in the first type of marriage, 50 per cent dd in the second, and 25 per cent dd in the third. Clearly, simple recessive inheritance is identical with simple dominant inheritance, but here the observer's emphasis is shifted from the dominant to the recessive allele.

TABLE 5
Marriages That Yield Homozygous Recessive Offspring (dd).

Parents	*Eggs*	*Sperm*	*Offspring*
$dd \times dd$	all d	all d	dd
$dd \times Dd$	all d	$\frac{1}{2}D$, $\frac{1}{2}d$	$\frac{1}{2}Dd$, $\frac{1}{2}dd$
$Dd \times Dd$	$\frac{1}{2}D$, $\frac{1}{2}d$	$\frac{1}{2}D$, $\frac{1}{2}d$	$\frac{1}{4}DD$, $\frac{1}{2}Dd$, $\frac{1}{4}dd$

All three types of marriages can be found when both the dominant and the recessive alleles, D and d, are common in an intermarrying population. This is true for the blood-group types A and O, which depend on the dominant allele I^{A} and the recessive I^{O}. The three types of marriages involving these two alleles from which homozygous $I^{O}I^{O}$ children can arise are $I^{O}I^{O} \times I^{O}I^{O}$, $I^{O}I^{O} \times I^{A}I^{O}$, and $I^{A}I^{O} \times I^{A}I^{O}$. They yield the different kinds of children in the relative proportions postulated by the theory, provided sufficient data are accumulated and analyzed.

Another recessive allele frequently found in man is one that causes inability to taste the substance phenylthiocarbamide (PTC; synonym, phenylthiourea) or related compounds. The majority of people are "tasters," a large minority "nontasters." Marriages of nontasters to nontasters yield, with few exceptions, nontaster children only. Marriages of tasters to tasters, or of tasters to nontasters, may bring forth children of either type. This suggests that nontasters are homozygous recessives, tt, and tasters homozygous or heterozygous dominants, TT or Tt. Support for this hypothesis cannot easily be supplied by a study of pedigrees alone but has been provided by the interesting method of allele frequency analysis, which is described and illustrated by the taster example in Chapter 10. On the other side, it must be admitted that detailed studies of the PTC traits show these to be more complex than the "simple" inheritance that is the subject of this chapter. If testing is done with a series

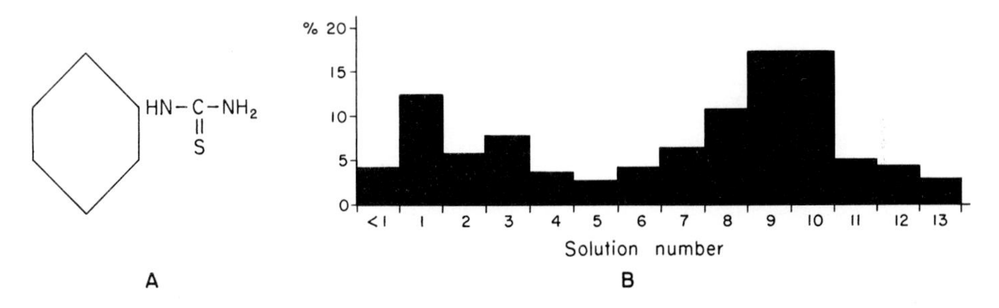

FIGURE 93
Tasting of phenylthiocarbamide. **A.** Chemical structure of the substance. **B.** Frequencies of individuals with various thresholds for tasting as determined by ability to taste a given solution. Each solution in the series 2-13 is one-half as concentrated as the preceding solution. (Harris and Kalmus, *Ann. Eugen.*, **15**, 1949.)

of solutions containing decreasing concentrations of PTC, it is found that nearly all persons are able to taste very strong solutions and that persons differ from one to another in the weakest PTC solution whose taste they are able to distinguish from that of plain water. The weakest solution recognized by a person as giving a PTC taste sensation is called his threshold solution. Figure 93 gives the distribution of taste thresholds for 142 British males. Instead of two clear-cut phenotypes, tasters and nontasters, the figure shows a continuous distribution with two main peaks. Solution 5 may be considered as that which separates the "nontasters" (left) from the "tasters" (right).

If the gene *d* that is responsible for a recessive condition is a rare one, *dd* persons will be very rare and even *Dd* persons uncommon. Only once in a great while will two affected people marry. Marriages between homozygous *dd* persons and normal-appearing but heterozygous *Dd* persons will also be rare. Most likely, *dd* individuals will marry *DD* persons, and children from such marriages will all be *Dd*. When they, in turn, marry normals of the genotype *DD*, half of their children will be *DD* and half *Dd*, all of normal phenotype. The heterozygotes will probably marry normal homozygotes, and the same will be the rule for later generations. Thus, the allele *d* may be carried "unseen" and unknown for centuries, even though the trait it produces in homozygotes is a striking one.

We shall see later that a *d* allele occasionally originates by the permanent transformation (mutation) of a *D* allele to *d* in one chromosome of a human cell. In a hypothetical population in which only the dominant *D* allele was originally present, a *d*-containing germ cell, newly created by the mutation of *D* to *d*, can meet only a *D* germ cell from the opposite sex in fertilization. The child originating from this fertilized egg will be *Dd*; subsequently, the *d* allele will be segregated to half of his germ cells, and thus potentially transmittable to his children. In the course of many generations, various *Dd* individuals, all going back in their pedigree to the first *d* germ cell, may be present in the population. A marriage of two such *Dd* persons may then, for the first time in the history of that population, produce a homozygous *dd* child. There may be only one such child—a "sporadic" case—or there may be more than one, since two of the parents' *d* germ cells will theoretically meet in one-fourth of all fertilizations. The appearance of several affected individuals in a sibship

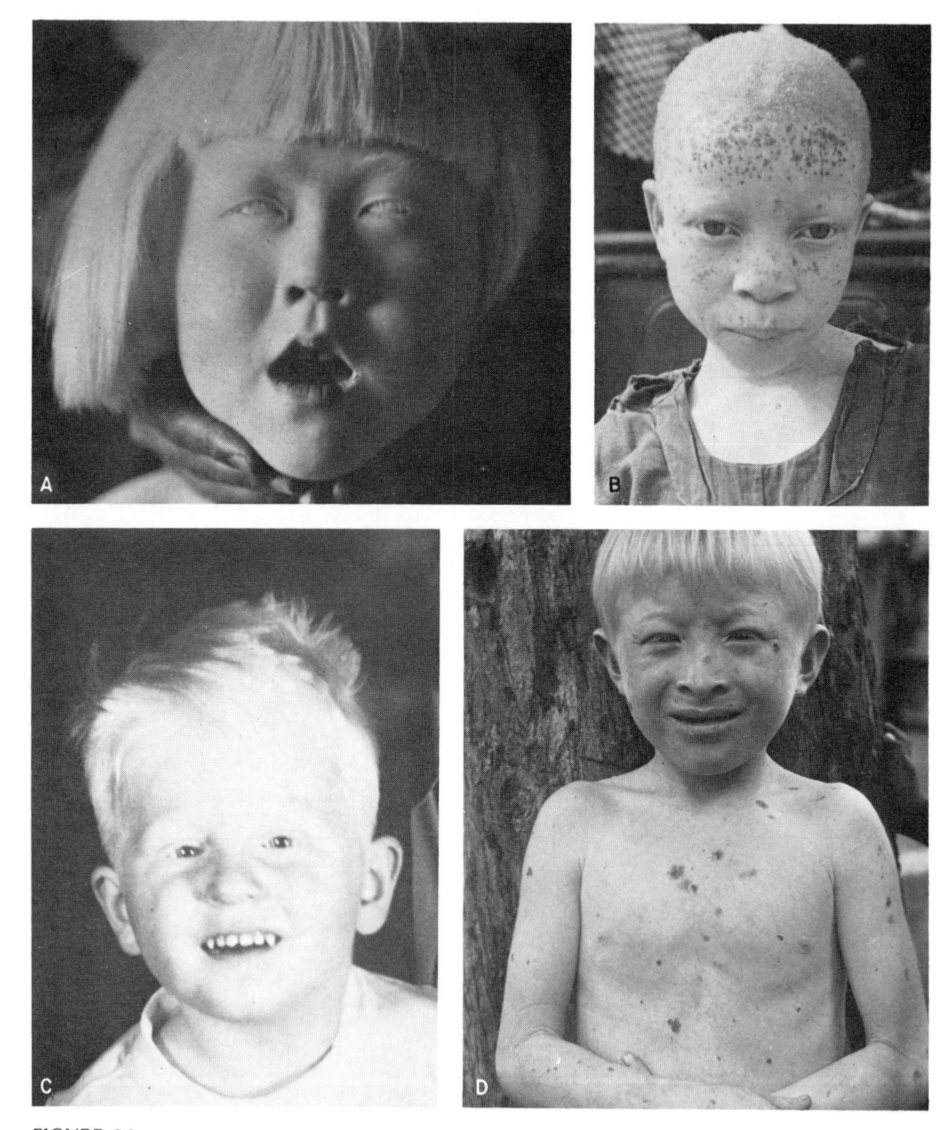

FIGURE 94
Albinism in four racial groups. **A.** Japanese, **B.** African from Nigeria, **C.** Caucasian, **D.** American Indian from San Blas, Panama. (A, original from Dr. M. Ito; B, Barnicot, *Ann. Eugen.*, **17**, 1952; C, original from Dr. J. V. Neel; D, original from Dr. C. E. Keeler.)

or in a larger family group has been called "familial" incidence of an abnormal trait.

All these facts are consonant with the general concepts of inheritance of alleles. They stress the point that inheritance is based on transmission of genes and not of traits. A trait may appear sporadically or familially; it may show up in consecutive generations or it may skip one or many. But the allele responsible will have been transmitted in an uninterrupted line to its carrier from a parent, a grandparent, and so on back to the ancestor in whom the allele originated.

Albinism. Albinism is a good example of inheritance of a trait dependent on a rare recessive gene (Fig. 94). It forms the topic of a monumental collection of pedigrees and data from all over the globe compiled by Pearson (1857–1936) and his collaborators. In individuals who are not albinos, numerous microscopically small granules containing melanin pigment are deposited in the skin, hair, and iris, giving these parts pigmentation. Albinos are nearly unable to transform tyrosine into melanin. They are, consequently, light-skinned (this is particularly striking in individuals belonging to "colored" races), have white or light hair, and their eyes are red because reflected light passes through the red blood vessels in the eyes, or green-blue if some pigment is indeed formed in the iris. Eye abnormalities of different kinds are usually part of the multiple expression of the albino gene.

Homozygous albinos are rare. It has been estimated that in various European countries about 1 in 20,000 individuals is an albino. If this low figure should also apply in the United States, there would still be about 10,000 albinotic individuals in this country. In certain other populations albinos are relatively more common. Thus, 5 albinos were found among 14,292 Negro children in Nigeria, giving a ratio of about 1 in 3,000; and among 20,000 San Blas Indians in Panama the ratio was 1 in 132. High frequencies of albinos have also been observed among the Hopi and Zuni Indians of the southwestern United States.

The pedigrees (Figs. 95–99) demonstrate the validity of the conclusions about the transmission of recessive traits which we have derived theoretically.

1. Marriages of albino with albino give albino children only (Fig. 95).
2. Marriages between an albino and a nonalbino either give no albino children (Fig. 96, II, IV), in which cases the normal mates were probably homozygous, or give both types of children (Fig. 97), obviously because of heterozygosity of the normal spouse.
3. Marriages between nonalbinos that result in albino offspring also yield normals (Fig. 98).

That both sexes must be equally involved in the transmission of the albino gene is obvious from its recessive nature, and that either sex may express the trait is seen from the pedigrees.

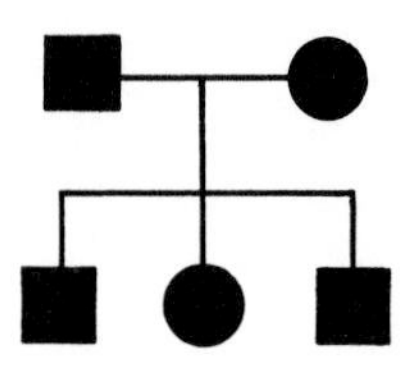

FIGURE 95
Pedigree of recessive albinism. Marriage of two affected individuals. (Snyder, *Principles of Genetics*, Heath, 1947.)

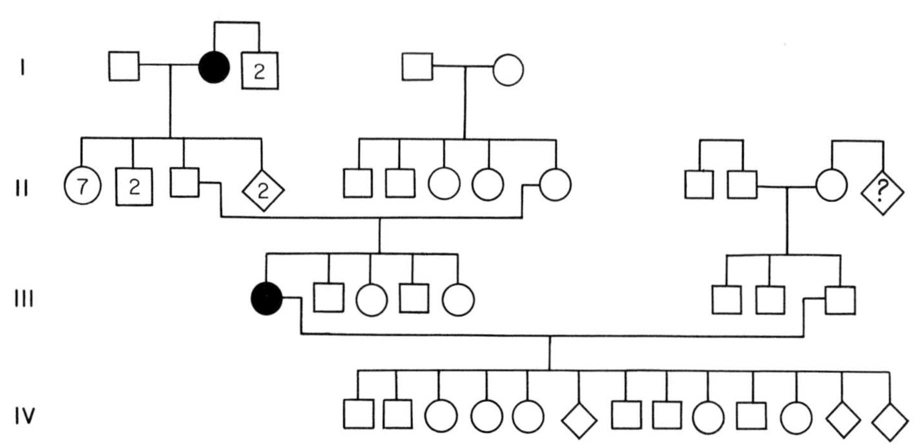

FIGURE 96
Pedigree of recessive albinism. Since I-2 was affected and therefore homozygous, all her children must be heterozygous. As shown by the affected state (homozygosity) of III-1, the youngest daughter of the second sibship was also heterozygous. (After Pearson, Nettleship, and Usher. Pedigree 98.)

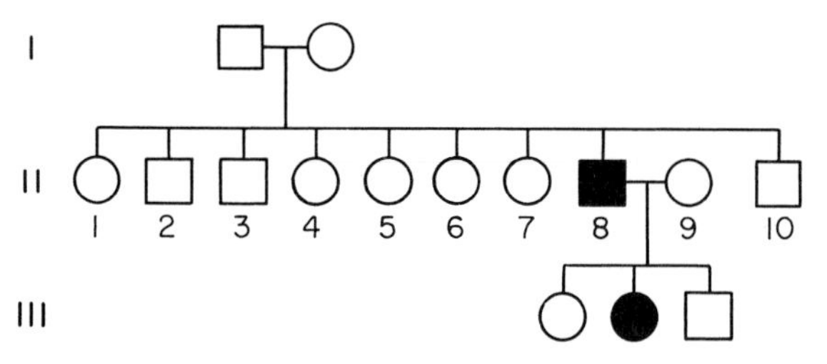

FIGURE 97
Pedigree of recessive albinism. (After Pearson, Nettleship, and Usher. Pedigree 54.)

Phenylketonuria. An example of familial occurrence of a rare recessive trait, phenylketonuria, is given by the pedigree reproduced in Figure 99. This condition consists of a metabolic abnormality which results in mental deficiency, as mentioned earlier. Affected individuals reproduce so very rarely that nearly all marriages from which the abnormality arises are between two normal-appearing persons who are both heterozygous for the recessive allele, usually without knowing it. Proof of simple single-factor inheritance in cases like these

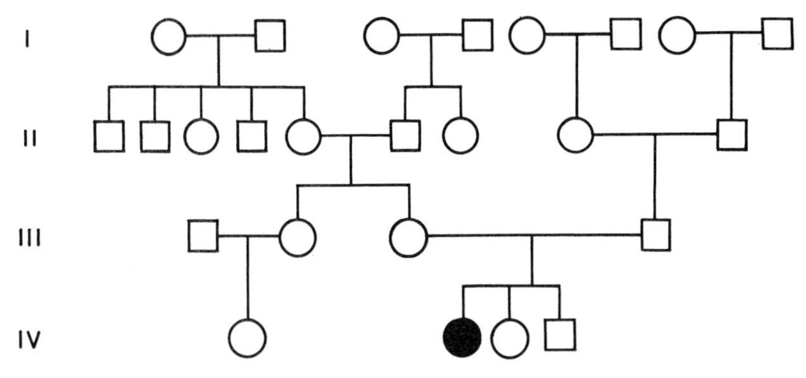

FIGURE 98
Pedigree of recessive albinism. (After Pearson, Nettleship, and Usher. Pedigree 113.)

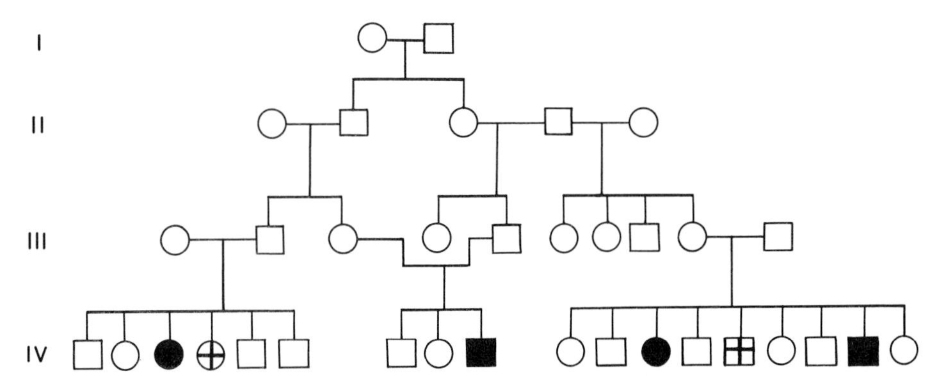

FIGURE 99
Pedigree of phenylketonuria and associated mental deficiency. The individuals marked with a cross were probably affected. They died young. These families lived in an isolated group of small islands in Norway. It may well be that certain apparently unrelated normal persons have some early ancestor in common from whom they inherited the recessive allele that made them heterozygous carriers. (After Følling, Mohr, and Ruud.)

requires special statistical techniques that make it possible to determine whether the theoretical expectation for recessive offspring from heterozygous parents is quantitatively fulfilled in that the homozygous affected children represent one-quarter of all sibs. The data on phenylketonuria agree with this theoretical expectation.

Total Color Blindness (Achromatopsia). Persons affected with this rare condition seem to see the world in shades of gray only. The cones of the retina that normally control color vision may be missing or defective. Thus, only the rods that are normally not color-sensitive are functional. When the light intensity is low the rods are superior in general vision to the cones. Totally

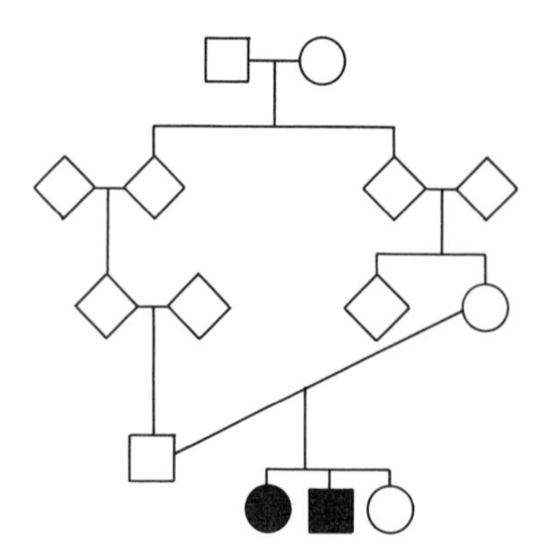

FIGURE 100
Pedigree of total color blindness. (After Waardenburg, *Bibliogr. Genet.*, **7**, 1932.)

color-blind individuals therefore see relatively better at night than at daytime and their condition is thus also known as "day blindness." The pedigree (Fig. 100) shows a sibship of three with an affected female and male. The parents were first cousins once removed and must have been heterozygous for the day-blindness recessive gene.

Genetic Control of Chromosomal Phenotypes. The phenotype of an individual includes the appearance and the behavior of his chromosomes. In most genotypes, whether they are normal or abnormal, chromosomes as seen under the light microscope are normal. Several inherited conditions, however, do express themselves in changed chromosomal phenotypes. One such condition consists of an incomplete mitotic coiling of a section of chromosome 1 (see Fig. 153, p. 367). It is transmitted as if a gene for incomplete coiling is located in a chromosome 1 at the site of incomplete coiling. A normal allele located in the homologous chromosome 1 leads to complete coiling of its chromosome section. This is an example of codominance at the chromosomal level. The affected heterozygous individuals are otherwise fully normal (homozygotes have not been observed).

A more general genetic influence on the phenotype of chromosomes is represented by some very rare chromosomal breakage syndromes including Bloom's syndrome. Individuals affected with this condition have characteristic skin abnormalities consisting of the presence of abnormally dilated blood vessels, abnormal facial traits, and stunted growth. Many of them die of leukemia or other malignant types of tumors before reaching age 20. Bloom's syndrome is inherited as an autosomal recessive. In addition to the features listed, the chromosomes of affected individuals frequently undergo breakage, abnormal reunion, and destruction by pulverization (Fig. 101). In different cells these phenomena are observed in a variety of the chromosomes. The cytological abnormalities have been studied in cell cultures only, and it is not

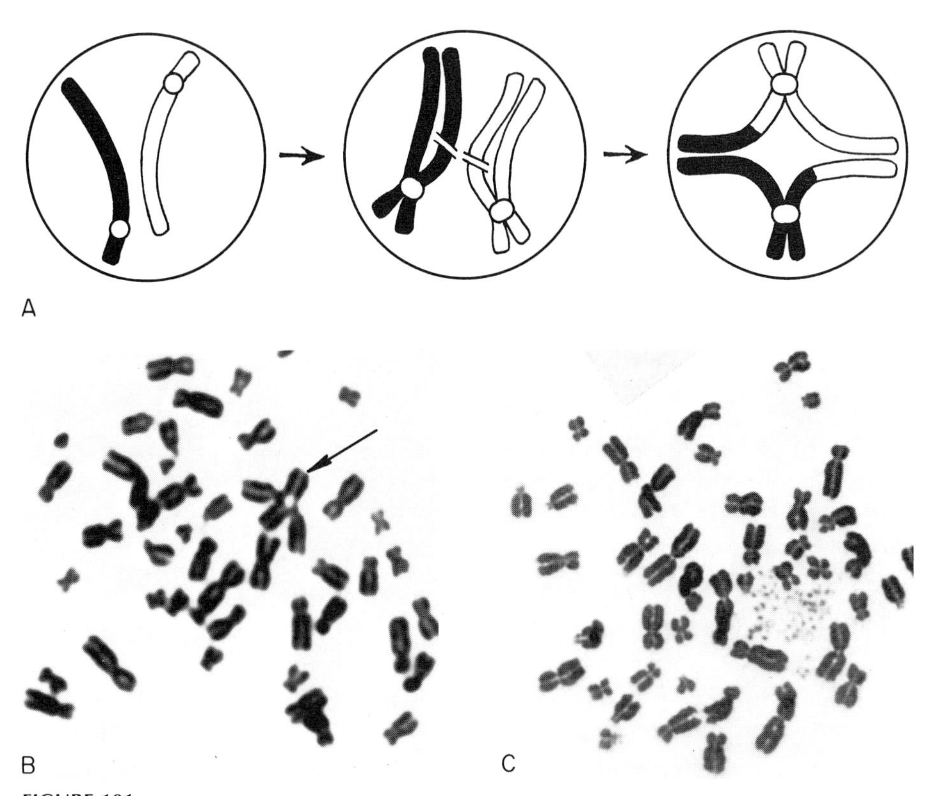

FIGURE 101
Abnormal chromosome behavior in Bloom's syndrome. **A.** Diagram illustrating somatic crossing over between two homologous chromosomes. **B.** Quadriradial configuration of chromosomes in a dividing cell in vitro, possibly the result of somatic crossing over. **C.** Pulverization of chromosomes. Detailed analysis shows that chromosome 2 and 16 and one chromosome of the G group are missing. (B, German, *Science,* **144,** 298, 1964, copyright © 1964 by the American Association for the Advancement of Science; C, German, *Birth Defects, Original Article Series,* **5,** 1969.)

known whether they also occur inside the body. If so, the limited growth of the bearers of the abnormal genotype may be a direct consequence of the loss of cells following chromosome breakage.

Further Criteria for Recessive Inheritance. Even with homozygous recessives of types that usually produce offspring—as do nontasters and albinos, in contrast to phenylketonurics—evidence for single factor inheritance can often be obtained only by special analyses of ratios of affected to nonaffected sibs (see Chap. 10), and by a special criterion, a higher than usual frequency of cousin marriages among the unions that produce affected children. Such marriages are likely to bring together in the offspring two recessive alleles derived from a single allele carried by one of the common parental grandparents (see Chap. 19). Cousin marriages, with their offspring, are recorded in Figures 99 and 100.

The reader should readily understand that the appearance of affected offspring from normal parents who, themselves, have only normal ancestry may not always be due to a recessive allele, but, for example, to special combinations of alleles at two or more genic loci, to a new dominant mutation, or even to externally caused developmental irregularities.

Different Genotypes for Albinism. It has been pointed out earlier that one and the same inherited phenotype, appearing in different pedigrees, may be the result of different genetic constitutions, all of which may bring about the same effect in development. Certain unusual pedigrees of albinism have been thought to be examples of this. One of these is that of two English albino parents whose four children were all nonalbinos. If the family history is correct—and blood tests for paternity are compatible with legitimacy—it would suggest that the two parents were albinotic for different genetic reasons, one having the usual albino genotype a_1a_1 and the other a genotype a_2a_2 at a different locus. In this case the genotypes of the parents must be expressed in terms of a more elaborate genetic formulation postulating two different genic loci whose recessive alleles, when homozygous, specify albinism and in which the first parent would be $a_1a_1A_2A_2$ and the second $A_1A_1a_2a_2$. This would imply that most normal persons are $A_1A_1A_2A_2$ and that replacement of either A_1A_1 by a_1a_1 or A_2A_2 by a_2a_2 blocks the formation of melanin. The children from marriages of $a_1a_1A_2A_2$ to $A_1A_1a_2a_2$ individuals would be $A_1a_1A_2a_2$ and, because of the dominance of A_1 over a_1 and A_2 over a_2, would be fully pigmented. The assumption of two different types of albinism in man is supported by the fact that two types are also known in another mammal (the mouse) as well as by biochemical tests. Unfixed hair bulbs from some human albinos incubated under specific conditions form melanin, while those of other human albinos fail to do so. Albinos from any one sibship are either all hair-test positive or all negative. Chemically, tissues from test-positive albinos contain the enzyme tyrosinase which, however, is inhibited from acting normally. In test-negative tissues tyrosinase seems to be absent or nearly so. The albino mother of the English family mentioned at the beginning of this paragraph was hair-test negative, her husband positive. Test-negative albinism, as in the mother, may be designated $a_1a_1A_2A_2$, test-positive albinism, as in the father, $A_1A_1a_2a_2$. The latter type of albinism seems to be the more frequent one. For purposes of discussion it is often sufficient to use the symbol A for normal and a for albinism.

Some unusual pedigrees of albinism have suggested dominant inheritance. Such type of inheritance might be suspected in the pedigree shown in Figure 97, in which the albino male of the second generation (II-8) married a nonalbino and had one albino and two nonalbino children. Although this fits the pattern of dominant inheritance, it fits that of recessive inheritance equally

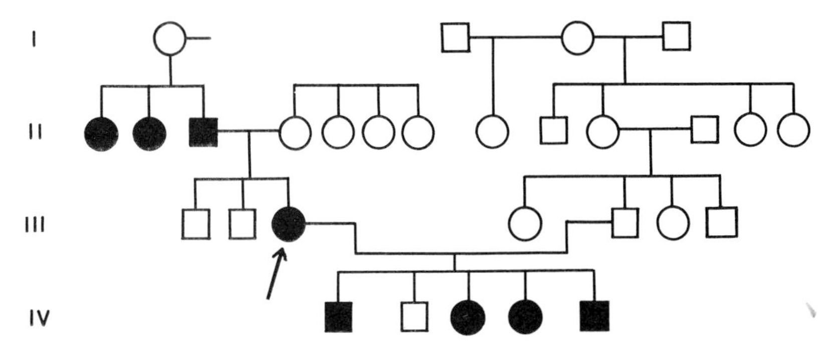

FIGURE 102
Pedigree of possibly dominant albinism in blacks. The proposita, i.e., the woman through whom the pedigree became known, is marked by an arrow. (After A. C. and S. B. Pipkin, *J. Hered.*, **33,** 1942.)

well, provided II-8 was homozygous *aa* and married a heterozygote *Aa*. The parents of II-8 were nonalbinos, but they may well have each carried one *a* allele. The hypothesis of a dominant gene for albinism would here require additional assumptions—for example, that II-8 had obtained a newly mutated dominant gene, or that a dominant gene was present in one of his parents but failed to produce its effect.

The hypothesis of a dominant gene for albinism is somewhat more reasonable when the trait appears not only in two but in three successive generations, as in the pedigree of an American black kindred shown in Figure 102. Here the father of the three albinos in generation II is unknown, but the albino II-3 had an albino daughter by a normally pigmented wife, and this daughter married a nonalbino man and had albinos among her children. Once more, however, the assumption of a dominant albino gene can be countered by assuming that the albinos were typical recessive *aa*'s and that the normal mates of the albinos in both generations II and III were *Aa*. Since *Aa* individuals are not too uncommon, it is reasonable to expect that, occasionally, successive generations of parents will include such heterozygotes. Moreover, in the pedigree under discussion, it is possible that the apparently unrelated individuals in the different generations actually shared a common ancestor in an earlier, unrecorded generation from whom the *a* allele was transmitted to all of them.

A similar situation is known for alkaptonuria. Most pedigrees fit recessive inheritance, but at least one, in which affected individuals descended from affected parents for three generations, has been frequently claimed to be evidence for dominant inheritance. Later genealogical studies show connections between members of this kindred with others in which the trait is un-

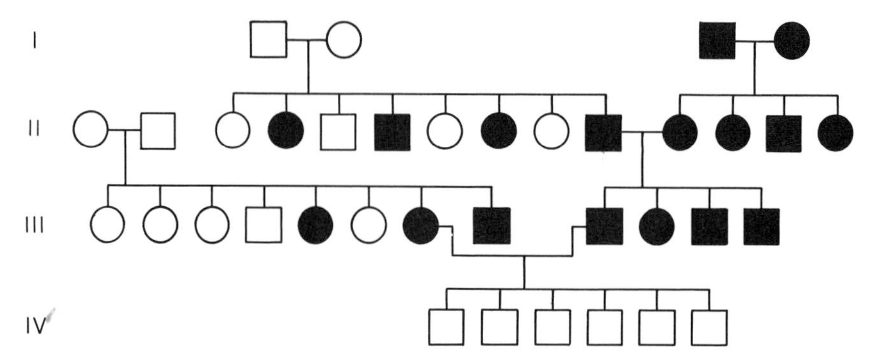

FIGURE 103
Pedigree of deaf-mutism in a kindred from northern Ireland. (After Stevenson and Cheeseman.)

doubtedly recessive. This suggests that the same recessive allele is responsible for alkaptonuria in all these kindreds.

Hereditary Deaf-mutism. Deafness that is present at birth and is later followed by muteness unless special training is provided can be caused either by external agents, such as infection of the fetus by the rubella virus in mothers who have an attack of German measles, or by genetic factors. Many genetic types of deafness seem to be caused by homozygosity for a pair of recessive alleles, since hearing parents who are both heterozygous carriers of such a gene usually have both hearing and deaf children in a ratio of 3:1; a heterozygous hearing person and a deaf-mute usually have equal numbers of both types of children; and two deaf-mutes usually have only deaf-mute children. Two of these relations are shown in generations I-III of Figure 103: the deviations from the theoretical ratios are explainable as chance phenomena.

The last generation (IV) consists of individuals who are not what we would expect them to be. Although both parents are deaf-mutes, presumably by heredity, their six children all hear normally. This suggests that III-7 and III-9 are deaf-mute for different reasons: III-7 may be considered to be affected because she is homozygous for a genotype *dd*; and III-9 to be homozygous for a genotype *ee*, which also causes the defect. Using a genetic scheme similar to that presented on page 168, III-7 would be *ddEE*; III-9, *DDee*; and their children, *DdEe*. This general genetic interpretation is supported by a detailed analysis of several hundred kindreds of hereditary deaf-mutism in Northern Ireland, made by Stevenson and Cheeseman. Moreover, it appears that there are considerably more than two recessive types of hereditary deafness: *ddEEFF, DDeeFF, DDEEff,* etc. In addition, many of the pedigrees of deaf-mutism from Northern Ireland can be accounted for only by assuming that they depend on dominant genes or interaction of genes at more than one locus (see Chap. 18). Still other, rarer pedigrees show the existence of an X-linked gene or genes for deaf-mutism.

TRAIT AND TYPE OF INHERITANCE

The example of deaf-mutism contains an important lesson: It is not possible to state without reservation that "deaf-mutism is a simple recessive." It is true that the majority of deaf-mutes owe their phenotype to a homozygous recessive genotype but, in any individual case, only a compilation of a pedigree and its study can result in a reasonably accurate statement about the type of inheritance involved. Even if all known pedigrees indicate clearly only one common type of inheritance for a trait, it cannot be assumed a priori that new cases may not belong to another class—to a different kind of recessive or dominant genotype.

Abnormalities that are based on dominant alleles in some pedigrees, and on recessive alleles in others, often differ from each other in their severity. This is not surprising, since different genes or different alleles may be expected to produce variations in effect. It is, however, remarkable that recessive genes usually cause greater deviations from normality than dominant genes that affect the same trait. It has been suggested that this phenomenon may be the result of natural selection. If alleles causing a severe abnormality arise in a population, and if some are dominant and others recessive, then, it is argued, the dominant alleles will be weeded out by selection but the recessive alleles will not. The dominant alleles will cause such serious defects that their carriers will probably die before having reproduced or even before having been born; therefore, the dominant alleles will not be retained. The recessive alleles, however, will be carried invisibly by heterozygotes and thus remain in the population. Genes causing slight abnormalities will be retained, whether they are dominant or recessive.

Although this selectionist explanation of the greater severity in effect of recessive genes may have some validity, a biochemical interpretation is probably of more general application. When normal genes control the production of specific enzymes, their abnormal alleles often fail to do so. Homozygotes for such abnormal alleles will lack the enzyme completely. This will be so in affected individuals who are recessive homozygotes. In heterozygosity, which is the common genotype for dominantly affected individuals, the one normal allele may well produce some of the enzyme. Obviously, absence of an enzyme would have a greater effect than its presence in a reduced amount. And though we have only limited evidence, it is likely that when abnormal dominant homozygotes occur, they are as severely affected as abnormal recessive homozygotes.

EXTRANUCLEAR INHERITANCE IN MAN?

Most inherited properties have been traced back to chromosomal genes. Some few specific phenotypes of various organisms depend on the transmission of self-replicating units that are located outside the nucleus, i.e., in the cytoplasm of the cell. The most generally occurring units of this kind, the mitochondria, are now known to contain DNA, which makes them very likely candidates for the bearers of extranuclear inheritance. Such inheritance is mostly recognized by exclusion. If a contrasting pair of phenotypes does not show Mendelian inheritance, i.e., lacks clear-cut segregation and corresponding ratios, then it does not conform to chromosomal behavior and suggests the possibility of nonchromosomal, extranuclear inheritance. Sometimes it is difficult to decide whether such extranuclear inheritance is a result of genetic variations in normal constituents of cells or whether it is really an infection in which the infectious agent is regularly transmitted in the cytoplasm (or perhaps even in the nucleus, but not as part of the chromosomes).

In man, there is no clear example of transmission of a trait that requires the assumption of extranuclear inheritance. In the simplest case—one encountered in other organisms—the extranuclear genic material might be transmitted by an affected mother through the cytoplasm of the egg to all her children, while an affected father would be unable to transmit his extranuclear genic material through the minute amount of cytoplasm in his sperm to any of his offspring. Possibly this interpretation can be applied to a very unusual case of a kindred in which 72 daughters and no sons were born (see p. 547).

The transmission of a type of degeneration of human optic nerves known as Leber's disease is not understood on simple assumptions of either nuclear or extranuclear inheritance. Among the pedigrees of this affliction, which may represent a number of different genetic types, there are many which show the following unusual features: much higher frequency of affected men than woman, a tendency for most men of a sibship to be affected and most women to be carriers, and, particularly remarkable, almost no transmission of the disease to any of the descendants of an affected man. Some of these features are shown in the pedigree in Figure 104. Primarily it is the lack of transmission by males that led to a hypothesis of extranuclear inheritance in Leber's disease, but this hypothesis requires additional assumptions in order to explain the different frequency of expression of the disease in males and females. Alternatively, Komai has constructed a hypothesis that assumes inheritance by means of an autosomal gene that is dominant in males and usually recessive in females. In order to account for the nontransmission of the gene by males this chromosomal (nuclear) hypothesis had to be complemented by making the unusual assumption of complete prezygotic selection against sperm carrying the defect-causing gene.

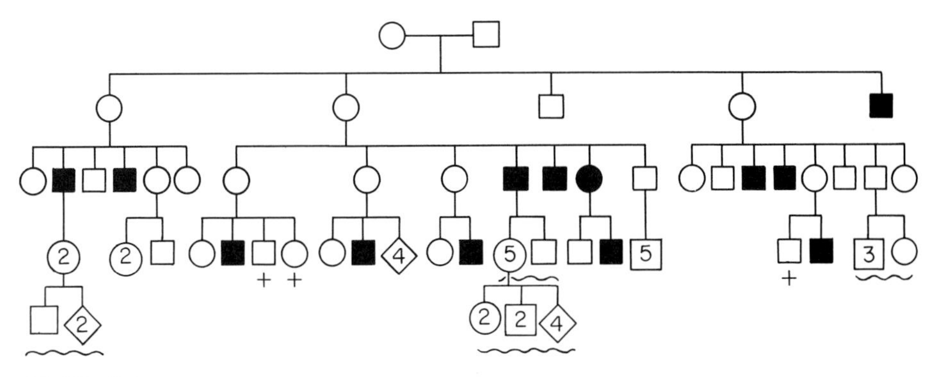

FIGURE 104
Part of a pedigree of hereditary optic atrophy (Leber's disease). A plus sign means died young. The wavy lines under two sibships indicate that the birth order was unknown to the recorder of the pedigree. (After Waardenburg from Bell, *Treas. Hum. Inher.*, **2,** 1931. Pedigree 708.)

EXCEPTIONS IN SIMPLE AUTOSOMAL INHERITANCE

Simple single-factor inheritance supplies the clearest insight into the working of two of the basic processes of human heredity: segregation, and chance combination of gametes. Nevertheless, occasional deviations from the simple expectations for single-factor inheritance should not cause consternation. No single gene pair that is involved in making a person affected or nonaffected works by itself. It takes part in an interaction among many or all genes of the genotype—an interaction whose course is in many ways determined by the environment. An unusual genotype at some locus other than that of the pair of alleles in question, or an unusual factor in the environment, may, for instance, result in the occasional production by *Aa* of the phenotype characteristic for the recessive *aa* or even of the phenotype characteristic of *AA* by *aa* or of *aa* by *AA*. Furthermore, occasional mishaps in the distribution of chromosomes, such as nondisjunction, result in deviations from regular transmission of their associated genes. If during meiosis, for instance, the two chromosomes that carry the alleles *A* and *a,* respectively, should move together to the same pole instead of going to opposite poles, a gamete would be formed which contains both *A* and *a,* and another which contains neither of them.

Another type of exception in simple autosomal inheritance consists in systematic occurrence of segregation ratios in heterozygotes other than 1:1. Thus, marriages of persons who are heterozygous for a dominant gene responsible for absence of the iris to normal spouses have produced offspring in the ratio 38 affected to 62 normal instead of the expected 50:50. The cause for this abnormal ratio is not known. Hypotheses to explain this disturbed ratio and some others include selection against eggs or sperm that carry the abnormal gene and abnormal meiotic segregation. In *Drosophila* the higher than 50 per cent frequency of certain segregants has been referred to as "meiotic drive."

PROBLEMS

In Problems 27–32 assume that a person affected with a rare dominant trait is heterozygous.

27. Construct a representative pedigree (i.e., one showing *all* essential features of dominant inheritance) of the three generations of a hypothetical family in which woolly hair occurs. Do not include more than a total of 15 persons. List the genotypes of all individuals.

28. A woman is woolly haired. What is the chance of woolly hair in:
    (a) Her first child?
    (b) Her first and second child?
    (c) All her six children?
    (d) In any one of her great-grandchildren?

29. A man of blood group O marries a woman of blood group A. The wife's father is of blood group O. What is the probability that their children will belong to blood group O?

30. A brachydactylous man marries a woolly-haired woman.
    (a) What are their genotypes?
    (b) Give all the possible genotypes and phenotypes of their prospective children.
    (c) What are the expected proportions of genotypes?

31. If two achondroplastic dwarfs marry, what genotypes, and in what proportions, would be theoretically expected among their children?

32. What type of inheritance pertains in the pedigree in Figure 111?
    (a) Determine the proportion of affected to nonaffected among the offspring of all relevant marriages.
    (b) What is the proportion of males to females among the affected?
    (c) What is the proportion of males to females among the nonaffected in sibships from an affected parent?
    (d) What is the proportion of males to females among the offspring of nonaffected parents?

In Problems 33–36 on recessive inheritance, assume that individuals unrelated to an affected person are homozygous dominant **unless evidence to the contrary is available.**

33. Construct a representative pedigree of three generations in which albinism is sporadic in the second generation. Let the first generation consist of two pairs

of parents, the next of two sibships of three children each, and the last of a single sibship of six. Give the most likely genotype or, if more than one genotype is likely, give all alternatives, of all individuals.

34. Two normal parents have an albino son and a normal daughter. The normal daughter marries a man from a family with all normal members. The third offspring (a daughter) of this marriage is an albino. What is the probability of the first child being an albino if the albino son mentioned above marries:
    (a) Someone normal in the general population?
    (b) A sister of his brother-in-law?

35. A man with normal pigmentation and woolly hair marries a woman with normal pigmentation and normal hair. Their first child is an albino with normal hair.
    (a) What other phenotypes may they expect in their children?
    (b) What will be the theoretical expectation of the proportions of all phenotypes?

36. In the pedigree in Figure 64, what are the genotypes of the following individuals: III–1, III–2, IV–3, IV–5, II–1, I–2? If more than one genotype is possible, give the alternatives.

37. From marriages of two hereditary deaf-mutes, the children are sometimes of two types, normal and deaf, in equal frequencies. How can this be explained?

38. A deaf-mute of the genotype *ddEE* is married to a person with normal hearing of the genotype *DdEe*. What are the possible genotypes and phenotypes of their children? What are their expected proportions?

REFERENCES

Blakeslee, A. F., and Fox, A. L., 1932. Our different taste worlds. *J. Hered.*, **23**:96–110.

Chung, C. S., Robinson, O. W., and Morton, N. E., 1959. A note on deaf mutism. *Ann. Hum. Genet.*, **23**:357–366.

German, J., 1969. Chromosomal breakage syndromes. *Birth Defects: Original Article Series*, **5**:117–132.

Haws, D. V., and McKusick, V. A., 1963. Farabee's brachydactylous kindred revisited. *Bull. Johns Hopkins Hospital*, **113**:20–30.

Imai, Y., and Moriwaki, D., 1936. A probable case of cytoplasmic inheritance in man: a critique of Leber's disease. *J. Genet.*, **33**:163–167.

Mendel, G., 1866. Experiments on plant hybrids. (English translation from *Versuche übér Pflanzen-Hybriden.* Pp 1–48 *in* Stern, C., and Sherwood, Eva R., (Eds.), 1966. *The Origin of Genetics*. W. H. Freeman and Company, San Francisco.)

Pearson, K., Nettleship, E., and Usher, C. H., 1911–1913. *A Monograph on Albinism in Man*. Dulau, London. (Draper's Company Research Memoirs. Biometr. ser., **6, 8, 9.**)

Stevenson, A. C., and Cheeseman, E. A., 1956. Heredity of deaf mutism with particular reference to Northern Ireland. *Ann. Hum. Genet.,* **20:**177–231.

Witkop, C. J., Jr., Nance, W. E., Rawls, Rachel F., and White, J. G., 1970. Autosomal recessive oculocutaneous albinism in man: evidence for genetic heterogeneity. *Amer. J. Hum. Genet.,* **22:**55–74.

Woolf, C. M., and Dukepoo, F. C., 1969. Hopi Indians, inbreeding, and albinism. *Science,* **164:**30–37.

8

LETHAL GENES

Different genes influence the ability of an organism to survive in varying degrees. The measurement of *viability,* as this ability is called, depends on the criteria used to define it. Some genotypes, for example, have greater ability to survive under adverse conditions than others, and viability of a given genotype may thus be defined in terms of frequency of survival relative to some standard genotype. A slightly different definition of viability may use for its criterion the average life span attained under normal conditions by individuals of a given genotype: this definition of viability will be used in the present chapter. From studies of other organisms, we know that different genes cover a whole spectrum of degrees of viability—from better than average, through average, to subnormal.

Average, Favorable, and Unfavorable Alleles. The normal variability of many human traits is due to the presence of somewhat different alleles in different individuals. The viability of these different, although normal, individuals is not affected strikingly by the slight differences in their alleles.

The discovery of unusually favorable alleles or genic combinations is made difficult because, in general, definitions of normality are not sharp, and all those phenotypes which are not clearly inferior are called normal. Consequently, superior traits are classified together with average traits instead of being classed separately, and the distinction is lost. It seems to be true that there are alleles that endow their bearers with unusually favorable characters, such as immunity to tuberculosis or other infectious diseases, or a lower than average tendency to the development of cancer. Similarly, the existence of special genotypes for longevity is probable, although it is not known whether they provide a person with a generally higher over-all vitality of tissues and organs, or whether they act primarily by way of single organs, such as the heart or some hormone-producing gland.

The discovery of *unfavorable* alleles is easier. Although there must be numerous genotypes which lower viability so slightly that an exact analysis in man is impossible, there are others which cause such decided physiological or morphological defects that their inheritance can be clearly traced.

Lethals. The lethals constitute a special class of alleles. *Lethal alleles* are defined as those that do not permit survival of the embryo or infant. No sharp division separates very early acting lethals from others that lead to death during childhood or, at the latest, before the reproductive age has been reached. The time of death may, moreover, vary considerably from one individual carrying a given lethal genotype to another. Lethals that lead to death relatively late after birth are sometimes called sublethals or semilethals.

Dominant Lethals. Lethal alleles may be dominant or recessive. A child who carries a dominant lethal allele is generally *not* the offspring of an affected parent, since, by definition, the bearer of this kind of dominant allele does not reproduce. Occasionally, however, a person with a dominant lethal allele may reach adulthood and reproduce, and afflicted children among his offspring demonstrate the dominant action of the allele. For example, most persons with the inherited condition known as epiloia (Fig. 105), in which abnormal growth of the skin occurs, accompanied by severe mental deficiency and epilepsy, as well as tumors in the heart, kidneys, and other parts, die young; a few, however, who are mildly affected, may reproduce (Fig. 106).

If we assume that typical lethal phenotypes that *never* reproduce are caused by dominant alleles, we must further assume that the appearance of such phenotypes is the consequence of a new mutation. One of the normal parents must have formed a gamete in which a dominant lethal replaced a normal allele, and thus an abnormal child developed. There is no intrinsic flaw in such a hypothesis, but it cannot be proved directly. Its support lies, rather, in the exclusion of alternative explanations, one of which might be that the

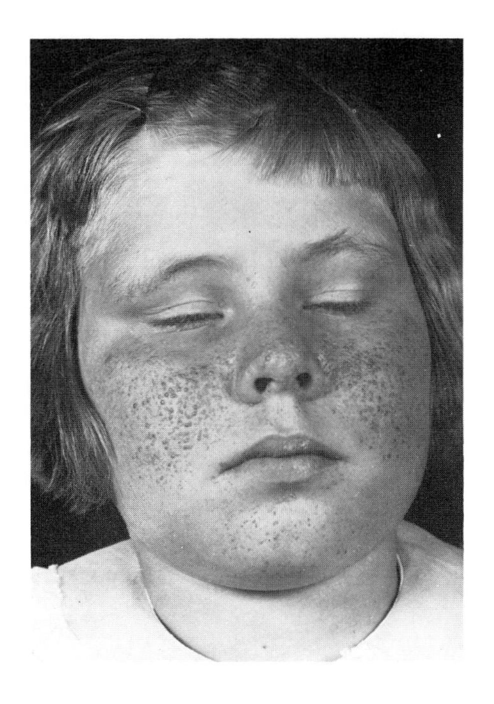

FIGURE 105
Epiloia, a sublethal syndrome. Note the "butterfly rash" which is part of the syndrome (adenoma sebaceum). (Original from Dr. V. McKusick.)

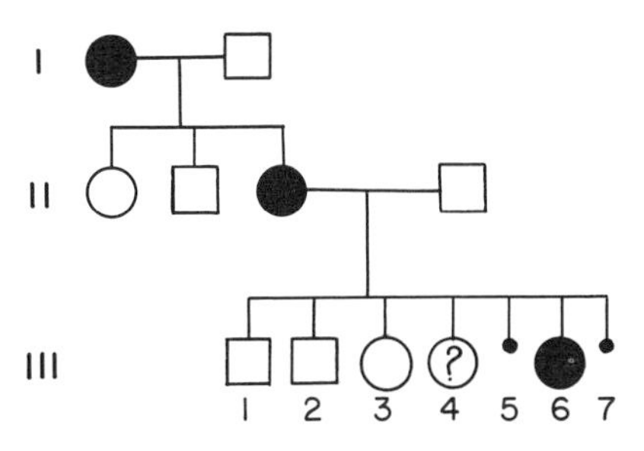

FIGURE 106
A pedigree of epiloia. Individuals III-5 and III-7 were miscarriages or stillbirths. The question mark in the symbol for III-4 indicates that information on the individual was incomplete. (After Gunther and Penrose, *J. Genet.*, **31,** 1935.)

lethality had been caused by a homozygous recessive genotype; another, that during embryonic life an unusual accident, purely environmental, had produced the abnormality of the fetus or child. The first alternative recessiveness, can often be excluded by showing that in most sibships in which the specific lethal phenotype thought to be newly arisen is found only a single individual is affected. If a recessive gene were responsible, more than one affected child would be expected in many sibships. The second alternative, the occurrence of a prenatal environmental accident, is more difficult to test. It can be excluded in the special class of dominant lethal conditions that are not due to abnormal genes but to excess or deficiency of normal genic material. The malformations caused by trisomy for any one of most autosomes are lethal in early fetal stages; 13- or 18-trisomics may survive until birth, but only to die within a few months or, at most, a few years. The lethality caused by monosomy or partial monosomy, as from the loss of part of a chromosome, also may be called dominant.

Recessive Lethals. Recessive lethal alleles are responsible for a considerable number of defects of the embryo or infant. Examples are abnormal leathery skin with deep, bleeding fissures (ichthyosis congenita, Fig. 107); a special type of dwarfism (Fig. 108); a paralytic condition called Werdnig-Hoffmann's disease, which commences in fetal life or within the first two years of life; and the infantile form of amaurotic idiocy (Tay-Sachs disease) in which complex fatty substances, called gangliosides, accumulate in the central nervous system (Fig. 109), leading to complete mental degeneration, blindness, and wasting-away of the body. The heterozygous parents of such homozygous affected children are fully normal but detectable by special biochemical tests. For reasons unknown, the allele for Tay-Sachs disease is much more frequent in Ashkenazy Jewish populations than in others.

It is likely that many very early spontaneous abortions are the result of embryonic malformations caused by lethal genotypes. This can be easily demonstrated in experimental animals such as mice: if adults known to be heterozygous for recessive lethals are mated, about a quarter of the resulting embryos will die in the uterus. An interesting phenomenon has been observed, particularly in *Drosophila* but likely also occurring in man. A gene exerts its lethal effect at a specific developmental stage; but, as usual, there is some variability in the genic action, so that a few homozygous individuals pass successfully through the critical period and, once having passed it, develop normally. Such survivors have been called "escapers from death" (Hadorn).

Some types of lethality are related to sex. Among aborted fetuses with certain kinds of severe malformations, males are more common than females; among those with other kinds, females are more common.

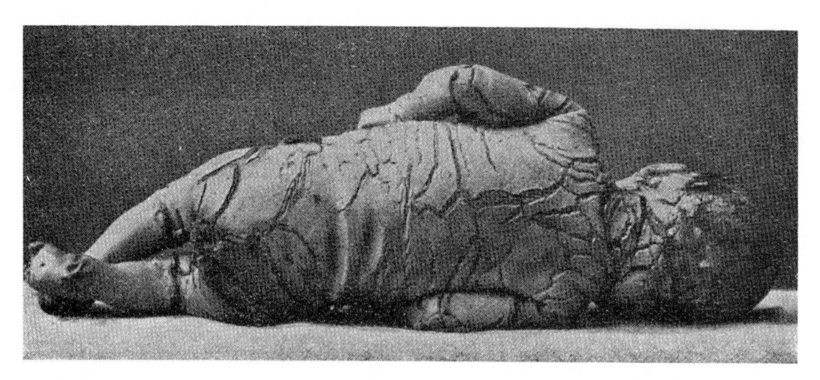

FIGURE 107
Ichthyosis congenita in a newborn infant, caused by homozygosity for a recessive lethal allele. (After Lesser, from Mohr, *Zeitschr. Abst. Vererb.*, **41.** 1926.)

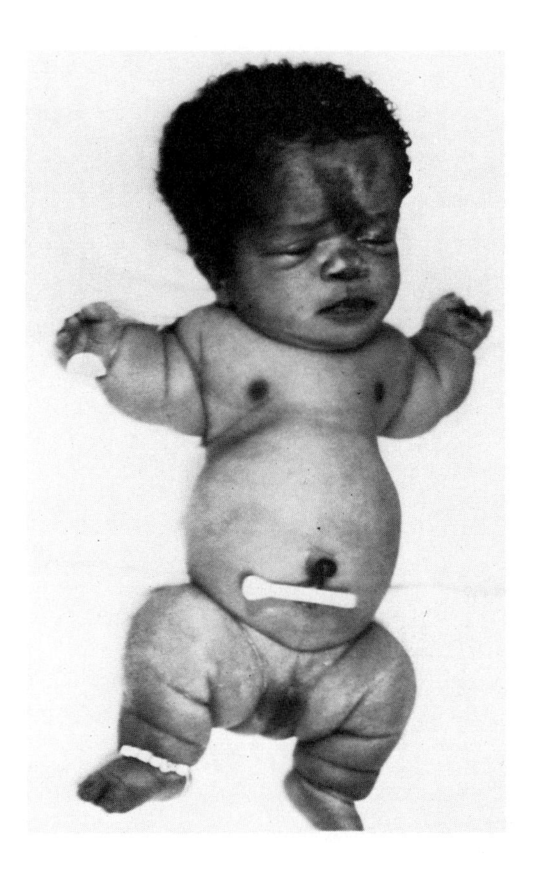

FIGURE 108
A recessive lethal type of dwarfism ("thanotophoric") in a newborn infant. Note the very small size of the chest. (Original from Dr. V. A. McKusick.)

Dominant Alleles with Recessive Lethal Effects. Some recessive lethals produce harmless phenotypic effects when present heterozygously. Homozygosity for a specific recessive allele is responsible for xeroderma pigmentosum, a sublethal condition in which skin abnormalities and, finally, multiple cancerous growths occur in those parts of the body that are exposed to light. Many carriers, although of normal health, are heavily freckled, obviously on account of the presence of the allele in heterozygous state (Fig. 245, p. 599).

Another presumptive example of a dominant allele with recessive lethal effects refers to achondroplastic dwarfism. The dominant gene that leads to the trait is usually present in heterozygous form. Two families are known, however, in which both parents were typical achondroplastic dwarfs; each family had one child with skeletal deformities similar to, but more severe than, those of the usual affected infant (Fig. 110, A, B). Both severely affected children died within a few months after birth. It is very likely that they were homozygotes for the achondroplasia gene.

A few more examples may show the diversity of dominant alleles with recessive lethal effects. In a Norwegian family studied by Mohr and Wriedt,

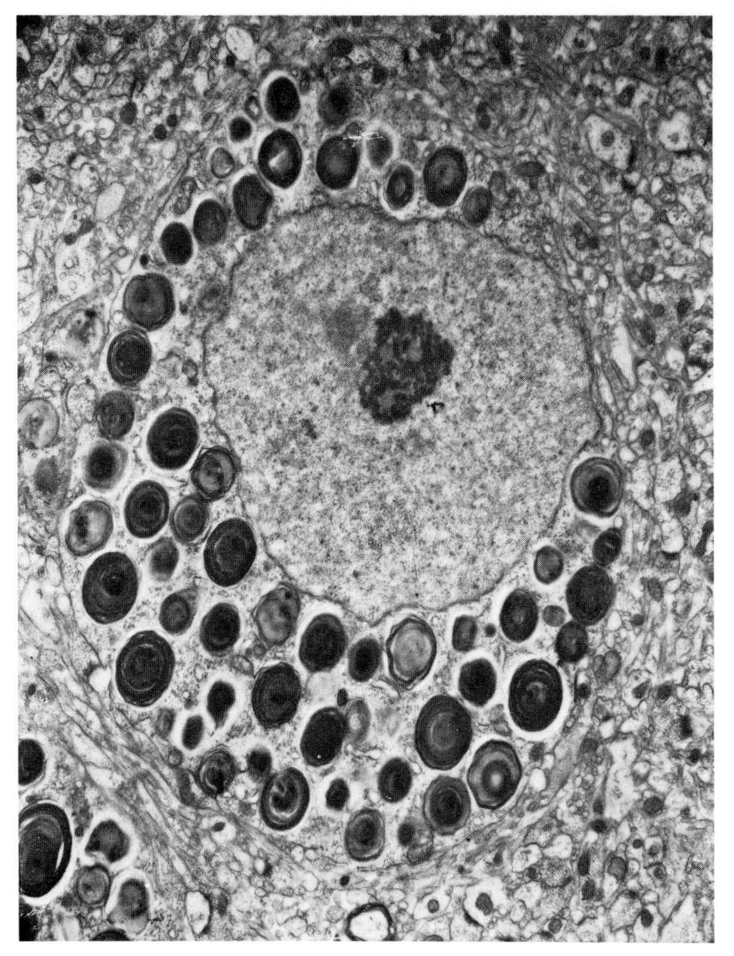

FIGURE 109
Tay-Sachs disease. Electron micrograph of a cerebral neuron of an affected child. The cytoplasm is crowded with laminated bodies, about one micron in diameter. Neurons of normal individuals do not contain these bodies. (Terry and Weiss, *J. Neuropathol. Exp. Neurol.*, **22,** 1963.)

a shortened or crooked forefinger and second toe occurred, a slight abnormality that did not interfere in the least with normal viability. In the pedigree (Fig. 168, p. 398), a marriage of two affected people is recorded (I-1 × I-2). Of the two children born to them, one had the typical short forefinger and was otherwise normal, but the other was a cripple—without fingers and toes and with severe disorders of the whole skeletal system. It is probable that this infant, who died at the age of one year, was homozygous for the allele responsible for short-forefingeredness. This allele, then, may be defined as dominant in regard to short-forefingeredness and as a recessive in its lethal

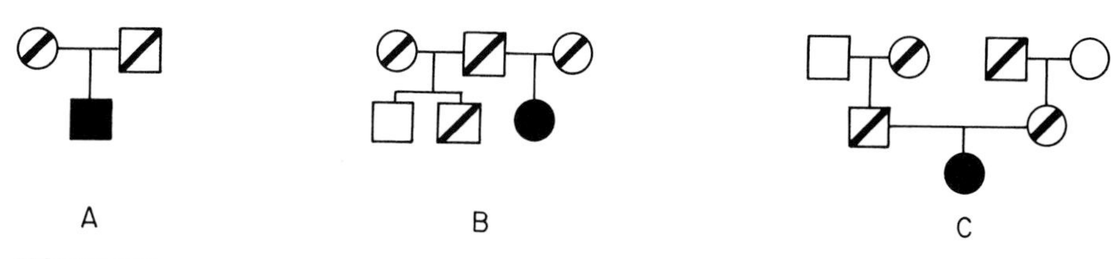

FIGURE 110
Presumed homozygotes for dominant alleles with recessive lethality. A diagonal bar indicates heterozygous effect, a filled-out symbol, homozygous effect. **A, B.** Families in which both parents were achondroplastic dwarfs ($A^1A \times A^1A$). **A.** Parents (P) and a severely affected child (F_1) who was presumably homozygous for dwarfism (A^1A^1). **B.** The first marriage of a typically affected man to a likewise affected woman produced a normal son (AA) and a typically affected son (A^1A). The same man's marriage to another affected wife produced a severely affected child. (After Hall, et al., *Birth Defects, Original Article Series,* **5,** 1969). **C.** Two generations of typical multiple telangiectasia and a severely affected sublethal child in the third generation who was probably homozygous for the dominant gene. (After Snyder and Doan, J. Lab. Clin. Med., 29, 1944.)

action. Another way of describing the genic action would be to say that heterozygotes with the allele have a skeletal system that is intermediate between those of normal and abnormal homozygotes.

In multiple telangiectasia a gene in heterozygous combination leads to enlargement of some of the finer blood vessels of the nose, tongue, lips, face, or fingers. This trait is usually inherited as a typical, rare dominant which is transmitted from one of the two parents to about half of the children. The pedigree (Fig. 110, C) shows the marriage of two affected individuals. Their newborn child possessed many abnormally dilated blood vessels; within a few weeks, other severe symptoms of multiple telangiectasia appeared; and after less than three months, breaking of numerous blood vessels resulted in death. The probability that this child was homozygous for the so-called dominant gene is high.

Hypercholesteremia, the presence of excessive quantities of cholesterol in the blood, is an inherited trait that frequently leads to disease of the coronary arteries in later life. A dominant gene often seems to be responsible for it. In a kindred with such dominant inheritance, two nonrelated, typically affected individuals had a lethally affected child, probably a homozygote.

Still another lethal condition is the hemoglobin disease thalassemia (Cooley's anemia). Homozygotes for the allele *c* die at an early age, but the health of *Cc* heterozygotes is very nearly as good as that of the normal *CC* individuals.

Many lethal effects that terminate early embryological development are caused by chromosome aberrations. These are either due to chromosomal mishaps occurring in chromosomally normal parents or in early stages of their conceptuses, or they are due to built-in aberrations such as translocations that may result in unbalanced genotypes. Many other early abortions may be the result of abnormal genic alleles. In man, such cases are not well known. We are, however, familiar with genetic conditions in many animals that lead to termination of development in the earliest stages of cleavage, or at various later steps in embryogeny. An example is provided by the yellow mouse, whose yellow fur is caused by a dominant allele for which it is heterozygous. Cleavage of a homozygous egg and formation of the hollow, ball-shaped blastocyst proceed normally, but disintegration occurs soon after the embryo has become embedded in the uterine wall. Thus, the allele for yellow fur acts as a recessive lethal. In man, such early failure of development would probably remain unnoticed, since the small embryo would be resorbed by the maternal tissue.

Homozygosity for another lethal allele in mice is known to end life at a later stage by causing incomplete development of the notochord, correlated with nonformation of the nervous system, of the tail, or even of legs. Embryos homozygous for still other lethal alleles never develop kidneys, form no (or inconspicuous) external genitalia, and have abnormalities of the anal region. Aborted human embryos exhibiting defects like those in mice are familiar to physicians, and many of the defects probably have similar genetic causes.

Gametic Lethals. In general a man who is heterozygous for a recessive lethal allele, mated to a woman who is homozygous for a nonlethal allele at the same locus produces two kinds of offspring in the ratio of one homozygous nonlethal to one heterozygote. From this it follows that the one-half of the man's sperm which is haploid for the lethal allele is as viable as the other half which is haploid for the nonlethal allele. In the same way, lethal-allele-carrying haploid eggs of a heterozygous woman are as viable as the eggs carrying the normal allele. The "nonfunctioning of the genes" in gametes—to use H. J. Muller's phrase—is particularly impressive when, as a result of nondisjunction or unusual segregation, whole chromosomes or at least long sections of chromosomes are absent in gametes. Such abnormal gametes usually are as viable as gametes with a full genic complement. It should, however, be kept in mind that the possibility exists that gametes may be subject to positive or negative selection on the basis of their genic content. The existence of prezygotic selection has indeed been assumed on occasion.

Infertility. The most obvious causes of infertility are disturbances of normal oogenesis or spermatogenesis that result in absence or insufficient presence of functional gametes. Genetic causes of such disturbances are clearly established in certain chromosome aberrations: XO females are essentially devoid of ovaries and in XXY as well as related types of males spermatogenesis is incomplete. Studies on testicular cells from subfertile or infertile men have shown 10 per cent or more of the men to have some type of abnormal chromosomes including a translocation, a ring chromosome, and inversions. It is not yet known how these chromosomal features are related to infertility. Absence or low frequencies of spermatozoa are also observed in men with normal chromosomal constitution. Some abnormalities in gametogenesis may be caused by nongenetic factors, either physiological, developmental, or infectious. Others are due to homozygosity for recessive genes causing infertility or to other more complex genotypes. In cattle, the mouse, and particularly, *Drosophila* a variety of genes for specific female or male types of sterility are known as well as genes that lead to sterility in both sexes. Among examples of infertility genes in man are the gene for testicular feminization, which transforms XY males into sterile female-like individuals (p. 516), and the gene or genes for pituitary dwarfism, a syndrome that usually includes infertility (p. 83). In other cases of sterility caused by genes, even when normal eggs or sperm are produced, fertilization may be prevented by abnormal form or function of the external or internal genital organs.

Infertility may also be encountered where there is normal fertilization. Failure of the embryo to become implanted or abnormal development may lead to abortion. Undoubtedly, in some of these events, especially when they occur repeatedly in a family, genetic factors are involved. The immunological incompatibility in certain combinations between the blood group of the mother and that of the fetus, particularly for the blood groups determined by the *Rh* alleles (Chap. 17) is an example. Homozygosity for a recessive lethal allele in the offspring of heterozygous parents may result in abortion in only one-quarter of the conceptuses. Nevertheless, chance may bring it about that several successive pregnancies terminate in this way, thus simulating infertility. The same may be true when abnormal chromosomal segregants from a parent heterozygous for a translocation form unbalanced genotypes in the offspring that result in early fetal death.

At one time, it would have been a contradiction in terms to speak of the inheritance of lethality or of sterility. Modern knowledge of the transmission of genes, as opposed to traits, has removed the apparent paradox.

PROBLEMS

39. List the genotypes of all individuals recorded in the pedigree in Figure 110, C. If additional children were present in the last generation, what phenotypes and genotypes, and in what proportions, would be expected?

40. A man is heterozygous for a recessive sublethal gene for ichthyosis. His wife is not a carrier for the ichthyosis gene, but is heterozygous for Tay-Sachs disease. What are the prospects of their future children being affected?

41. A man is known to have been heterozygous for a very rare recessive gene that, in the homozygous state, causes death during infancy. He married his half-sister (the daughter of his father and another mother) and had two children, one affected and one normal.
    (a) Draw a pedigree of the case and give the genotypes of all individuals.
    (b) With the genotypes of the parents as outlined above, what were the chances of both children being normal?

REFERENCES

Bedichek, S., and Haldane, J. B. S., 1938. A search for autosomal lethals in man. *Ann. Eugen.,* **8**:245–254.

Böök, J. A., and Rayner, S., 1950. A clinical and genetical study of anencephaly. *Amer. J. Hum. Genet.,* **2**:61–84.

Dunn, L. C., 1940. Heredity and development of early abnormalities in vertebrates. *Harvey Lectures,* **35**:135–165.

Eaton, G. J., and Green, M. M., 1963. Giant cell differentiation and lethality of homozygous *yellow* mouse embryos. *Genetica,* **34**:155–161.

Eaton, O. N., 1937. A summary of lethal characters in animals and man. *J. Hered.,* **28**:32–326

Gregory, P. W., Regan, W. M., and Mead, S. W., 1945. Evidence of genes for female sterility in dairy cows. *Genetics,* **30**:506–517.

Hadorn, E., 1961. *Developmental Genetics and Lethal Factors.* (English translation of *Letalfaktoren,* 1955.) 355 pp. Wiley, New York.

Hulten, M., Eliasson, R., and Tillinger, K. G., 1970. Low chiasma count and other meiotic irregularities in two infertile 46, XY men with spermatogenic arrest. *Hereditas,* **65**:285–290.

Kjessler, B., 1966. *Karyotype, Meiosis and Spermatogenesis in a Sample of Men Attending an Infertility Clinic.* 74 pp. Karger, Basel.

McIlree, Maureen, Price, W. H., Court-Brown, W. M., Tulloch, W. S., Newsom, J. E., and Maclean, N., 1966. Chromosome studies on testicular cells from 50 subfertile men. *Lancet,* **2**:69–71.

Nowakowski, H., and Lenz, W., 1961. Genetic aspects in male hypogonadism. *Rec. Progr. Hormone Res.,* **17**:53–95.

Pearson, P. L., Ellis, J. D., and Evans, H. J., 1970. A gross reduction in chiasma formation during meiotic prophase and a defective DNA repair mechanism associated with a case of human male infertility. *Cytogenetics,* **9**:460–467.

Penrose, L. S., 1957. Genetics of anencephaly. *J. Ment. Defic. Res.,* **1**:4–15.

Penrose, L. S., 1963. Genetical aspects of human infertility. *Proc. Roy. Soc. London, B,* **159**:93–106.

Robertson, G. G., 1942. An analysis of the development of homozygous yellow mouse embryos. *J. Exp. Zool.,* **89**:197–227.

Sloan, H. R., and Frederickson, D. S., 1972. G_{M2} gangliosidoses: Tay-Sachs disease. Ch. 29, pp. 615–638 *in* Stanbury, J. B., Wyngaarden, J. B., and Frederickson, D. S. (Eds.) *The Metabolic Basis of Inherited Disease,* 3rd ed. McGraw-Hill, New York.

9

GENETIC PROGNOSIS

The recognition of simple dominant or recessive inheritance of a given trait permits statements of practical value about the genotypes of individuals. Persons affected with unfavorable traits, and their close relatives, rightly wish to know whether their future children are likely to have the same traits. Genetic prognosis makes use of a wide range of information on human inheritance. In this chapter predictions made possible by the understanding of simple autosomal inheritance will be stressed, because only this type of inheritance has been covered in the preceding parts of this book. It may appear premature to discuss problems of genetic prognosis before more facts have been presented; yet even at this stage of the discussion, it seems worthwhile to show how knowledge of human genetics can be applied to problems of individuals and of society. In later chapters—for instance, those on multiple alleles, sex-linkage, and medical genetics—further material will be presented which may provide a basis for genetic counseling.

Dominant Genes. The answer to questions about the prospects of persons in families with affected individuals having affected children is simple when

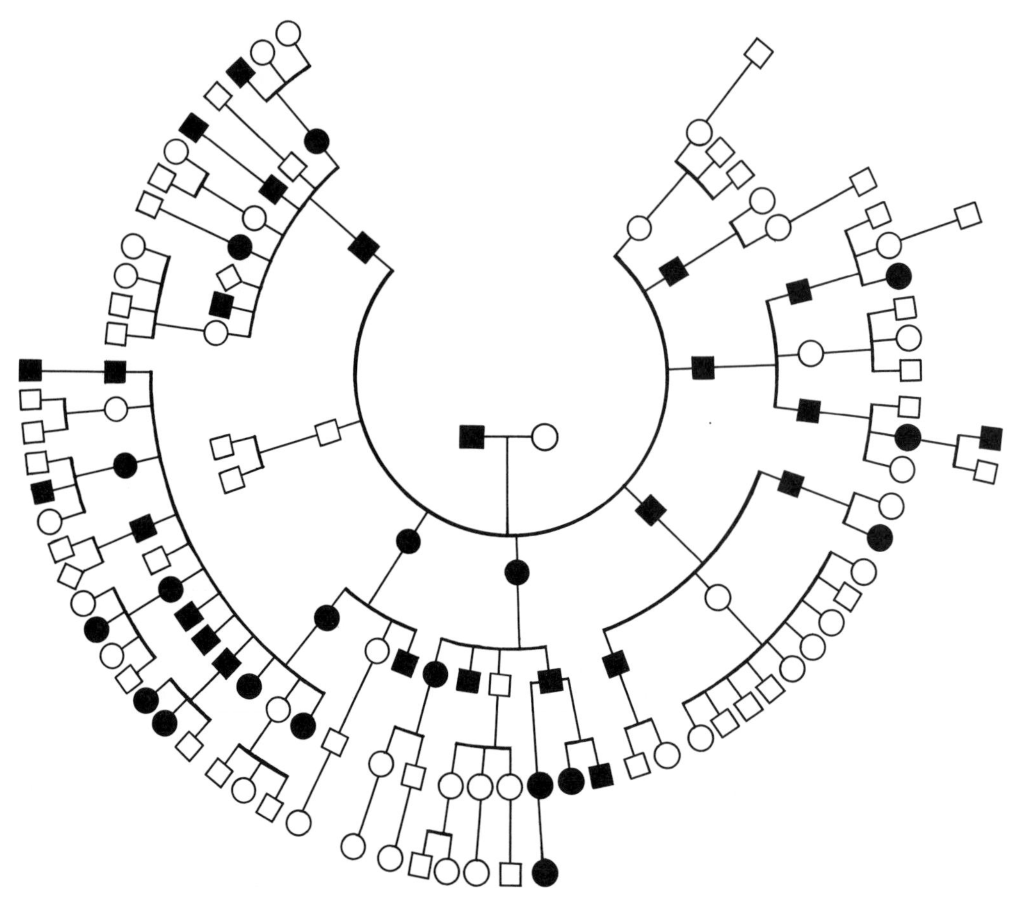

FIGURE 111
Pedigree of juvenile cataract. (After Lutman and Neel, *Arch. Ophth.*, **33**, 1945.)

the trait is dominant. Even the closest relatives of affected individuals, as long as they are normal themselves, will not transmit the trait, either to their immediate offspring or to later generations. Affected persons will, theoretically, transmit the dominant allele to one-half of their children, although there is always a possibility that none or all will receive it.

There are numerous more-or-less serious defects that are transmitted in simple dominant form. For example, a pedigree of cataract is shown in Figure 111. The condition, opacity of the lens of the eye, is present in many members of this family group from childhood on. Blindness can be avoided by an operation; but even after the operation and after providing the affected individuals with special glasses, they are greatly handicapped. The trait is clearly dominant. Before having been advised by a medical geneticist, the members of this kindred were aware that the affliction was hereditary, but they did not know the rules of its transmission. They knew only that it was uncommon for all children of an affected parent to develop cataracts and that all children of such a parent might be normal.

The apparent irregularity of appearance had kept the members of the family group from realizing that cataracts never appeared among the offspring of parents who were both normal. The physician who drew up the pedigree on the basis of information willingly supplied by the family was able to explain to them why cataract appeared in only some children of a sibship and why, in some small sibships, all the children were defective or all normal. Having supplied this enlightenment, he could further assure normal sibs of affected persons that they did not need to worry about transmitting cataract to their children. This example shows that much harm may be avoided and much worrying relieved if consultants in human genetics become widely available to the public.

Recessive Genes. While the genotypes of individuals can be given with certainty if simple dominant inheritance of a rare condition is under consideration, some genotypes of rare recessive conditions are less easy to define. An affected person, of course, is *dd,* but nonaffected relatives are either *DD* or *Dd.* If they are phenotypically normal sibs of a *dd* individual, and if their parents were normal, their genotypes could be either *DD* or *Dd,* since both parents must have been heterozygous. The two genotypes, *DD* and *Dd,* occur in a ratio of 1:2, which makes the probability that a normal sib does not have the recessive allele 1/3, and that he is a carrier 2/3.

These two probability fractions may seem to contradict the expectation that $Dd \times Dd$ marriages will yield 1/4 *DD* and 1/2 *Dd* children; but they do not. The probabilities 1/4 and 1/2 express the frequency with which any child from two heterozygous parents may be expected to be *DD* or *Dd,* considering also that he may be *dd.* The fractions 1/3 and 2/3, on the other hand, are derived from the number of individuals who are phenotypically normal, that is, those who are certainly not *dd.* This may be illustrated by an example: If two normal parents who have an affected child want to know the probability that their unborn second child will be homozygous normal, the answer would be 1/4, since he could be *dd, Dd,* or *DD.* If, however, the second child is normal, the answer to the same question would have to be 1/3, because his phenotype shows with certainty that he is not *dd.*

Determination of the probability that an affected individual or phenotypically normal relatives of such an individual will have affected homozygous recessive offspring has to take into account the various types of marriages possible, and also the frequency of the condition in the population at large. Marriages in which both partners are affected will, of course, result in affected children only. If one prospective parent is affected and the other normal and drawn from the general population, it is necessary to estimate the probability that a normal person chosen at random is heterozygous, i.e., a carrier.

A formula that permits such an estimate will be derived in Chapter 11, the Hardy-Weinberg Law. Here, it may be stressed that even genes for very rare conditions, which are due to a homozygous recessive constitution, are quite often carried unseen by heterozygotes.

Probabilities in Various Marriages of a First Child Being Affected. Knowledge of the frequency with which heterozygotes occur enables a genetic counselor to predict the chance of affected offspring in various types of marriages. What, for instance is the chance of a homozygous recessive *dd* child being born to two normal parents who have no affected relatives? Let us begin our considerations with firstborn children and consider albinism of one of the two types (see p. 168), assuming, somewhat arbitrarily, that the two types are of equal frequency. Since nonalbinos can have albino children only if both parents are heterozygotes, we have to know the frequency of such parental pairs. It has been estimated that about 1 in 100 persons is a heterozygote for one of the types of albinism, *Dd*. Therefore, the probability that two nonalbino heterozygotes will marry is $1/100 \times 1/100 = 1/10{,}000$. The probability of the child of two heterozygotes being an albino, *dd*, is 1/4 and thus the probability of an albino child of the type considered from two nonalbino parents is $1/4 \times 1/10{,}000$ or 1 in 40,000.

Another question may inquire about the probability of a firstborn child being an albino if one parent is normal and without affected relatives but the other an albino. This marriage type has the probability 1/100 (given the probability of *Dd* for the nonalbino parent) times 1 (given the fact of albinism in the other parent). The probability of the child of a $dd \times Dd$ marriage being an albino is 1/2, and thus the probability of an albino child from one albino and one nonalbino parent is $1/2 \times 1/100$ or 1 in 200.

A third question may concern the probability of a firstborn child being an albino if both parents are normal and unrelated but one has an albino sib. The probability of the normal sib of the albino being *Dd* is 2/3 and that of the spouse of the albino is 1/100. The probability of an albino child from the marriage is $1/4 \times 2/3 \times 1/100$ or 1 in 600.

If we compare the risks of albinism in the first child in the three types of marriages we see that the probability of a firstborn child being an albino from normal parents who have no reason to suspect heterozygosity is very low, namely 1 in 40,000; that it is almost sixty-seven times as great (1 in 600) if one of the parents has an albino sib and that it is two hundred times as great (1 in 200) if one of the parents himself is affected. Notwithstanding these relative figures, it may be important to realize that even 1 chance in 100 is a low risk.

Probabilities of Later Born Children Being Affected. Statements about the probability that an affected child will be born to parents one or both of whom are normal are subject to change when new information gives more specific knowledge of their genotypes. If a normal sib of an albino and his normal fiancee should ask how probable it is that their first child will be an albino, the best estimate possible, as we have just seen, is 1 in 600. Should the same couple, after having an albino child, ask the same question regarding their next, yet unborn, child, the answer would be 1 in 4 since it would then be certain that both parents are heterozygous. Similarly, if one parent is affected, the chance of a second child being *dd* if the first one actually is *dd* is 1/2.

Often the first child is normal. This may be brought about in two different ways, either by at least one of the parents being *DD* and therefore not able to have any *dd* children or by both parents carrying a *d* allele but the first conception not happening to include a *d* allele from each. In the following discussion of this situation, instead of using the probability 1/100 of being a heterozygote for albinism, we shall use the letter h for the probability of being a heterozygote for any given recessive allele. Then, in the case of two normal parents without affected relatives, the chance that both are *Dd* and their first child is not *dd* is $3/4 \times h^2$ (Table 6). The chance under these circumstances that the second child will be *dd* is $1/4 \times 3/4 \times h^2 = 3/16 \times h^2$ or only 3/4 of the probability for the first child. Similarly the chance of one normal and one affected parent not having a *dd* first child is $1/2 \times h$ and therefore their chance of having a *dd* second child, after a normal child has been born, is $1/2 \times 1/2 \times h = 1/4 \times h$ or only 1/2 of the probability for the first child. Finally, the chance of two normal parents, one of whom has an affected sib of not having a *dd* first child is $2/3 \times 3/4 \times h = 1/2 \times h$, and therefore their chance of having a *dd* second child, after a normal child has been born, is $1/4 \times 1/2 \times h = 1/8 \times h$,

TABLE 6
Some Probabilities of Affected Offspring for Recessive Genotypes.
(The probability that an individual in the general population is a carrier is h.
For a specific type of albinism h is about 1/100 and for many other rare recessives it lies somewhere between 1/40 and 1/200.)

Parents	*First Child Affected*	*Second Child*	
		Not Affected	Affected if First Is Affected
Both nonaffected; no affected relative	$\frac{1}{4}h^2$	$\frac{3}{16}h^2$	$\frac{1}{4}$
One affected	$\frac{1}{2}h$	$\frac{1}{4}h$	$\frac{1}{2}$
Both nonaffected; one has affected sib	$\frac{2}{3} \cdot \frac{1}{4} \cdot h$	$\frac{1}{8}h$	$\frac{1}{4}$
Both affected	1	–	1

or only 3/4 of the probability of the first child. In summary it is seen that the probability of having a *dd* child after the birth of a first non-*dd* child is only 3/4, 1/2, and 3/4 respectively of the chance for the first child being *dd*. Obviously, the more children born to given couples without any one of them being *dd*, the smaller is the likelihood that the parents both carry a *d* allele. (The derivations of the probabilities in this paragraph, while giving answers that are very close to being right, are not quite correct. The reader is referred to discussions of conditional probabilities in books dealing with probabilities in detailed fashion.)

Sometimes prospective parents are less interested in the probable genotypes of individual future children, considered from firstborn on as we have been doing, than in the more general prospect of how likely it is that any of their children will be homozygous recessives *dd*. To use a specific example, the probability that an albino will marry a normal pigmented unrelated person who is heterozygous for the same kind of allele is 1/100. If enough children come from such a marriage to make it reasonably certain that both genotypes, *Dd* and *dd*, will actually occur in the event that the normal parent is a carrier, the probability they will have some affected offspring approaches 1/100. Expressed in a different way: if prospective parents, one of whom is an albino, ask, "If our family is large, what chance is there of having albino children?" the answer is about 1/100. Should, however, the couple ask, "What chance is there that our first child will be an albino?" then the answer is $1/2 \times 1/100 = 1/200$, since the probability of the first child from $dd \times Dd$ parents being *dd* is 1/2.

Further Probabilities of Being a Heterozygous Carrier of a Recessive Gene. It is possible not only to determine the probability that individuals who are sibs of an affected person are heterozygotes, but also to calculate the probability of heterozygosity among individuals who are related to affected persons in other ways. These probabilities are consequences of the mechanism of genetic transmission, and many can be found by applying the theorem that the probability of two or more events occurring together is the product of the separate probabilities. Let us give a few examples: (1) The probability that the normal child of an affected parent is heterozygous is 1, or certainty. (2) The probability that a normal parent of an affected person is a carrier is 1, while that for the affected's uncles or aunts is usually 1/2. This follows from the fact that a heterozygous parent himself almost always comes from a marriage of a normal homozygote and a heterozygote and the chance that any offspring from such a marriage will be *Dd* is 1 out of 2. (3) The probability that the children of these aunts and uncles will be carriers is 1/2 that of their parents, or 1/2 of 1/2, i.e., 1/4. This last calculation does not take into consideration the rather rare chance that the spouse of the uncle or aunt may also be a carrier. All these probabilities are greatly increased when individuals from families with affected members marry close relatives (see Chap. 19).

The Independence of Probabilities from Preceding Events. It is important to stress that, if the genotypes of the parents are known, the probability prediction for any one child is not influenced by the type of offspring already born.

"Chance has no memory!" The truth of Le Châtelier's remark is obvious to the geneticist, who knows that the formation of any one combination of genes at the conception of a child depends solely on the union of gametes present at that time and not on any preceding, independent union of gametes. Thus, two parents heterozygous for a recessive allele that causes an abnormality have one chance in four that their first child will be affected. If the first child is affected, then the chance that the second child will also be affected is likewise one in four. However, the chance that two children will be affected is the product of the independent chances, or one in sixteen.

In a relevant case reported by Mohr, a couple had a child who developed a paralysis causing death before it was a year old. Since it was known that the condition is due to a rare recessive gene, the parents were marked as heterozygotes. They were informed that the risk was one in four that any other child would also be doomed. The couple greatly desired a healthy child and were willing to take the risk of another tragedy. They had a second child, and it too developed the deadly paralysis. When a third child was conceived, the dice of destiny determined finally that it was normal.

There are many traits for which a definite genetic interpretation is not available, and, for these, counseling cannot be based on theoretical ratios. However, the data that have been collected on a number of such traits show how often children or other relatives of affected individuals are similarly affected. Such *empiric risk figures* can be used in order to predict the probability of recurrence of the trait. Examples of risk figures are given in Tables 83 and 85 (pp. 672 and 692).

This chapter on genetic prognosis illustrates some principles that underlie genetic predictions. It should be emphasized that human genetics counseling involves more than the statement of probabilities and that the tasks of counselors must be based on many more insights than those provided here. Also, a great responsibility rests with the prospective parents. They may have to decide whether it is genetically desirable for them to have children and what the consequences will be, both for their personal lives and for society at large. These and other general problems will be discussed in Chapter 30, Medical Genetics.

PROBLEMS

42. In Figure 89:
    (a) What is the probability that the first child born to individual IX–2 is affected with night blindness?
    (b) What is the probability for the first child of her brother IX–1?

43. A normal woman has three sibs affected with juvenile cataract. Her father and paternal grandmother were also affected. What are the prospects of her future children being affected?

44. A couple, of normal ancestry, have two normal children and an infant affected with Tay-Sachs disease. The sister of the husband wishes to marry the brother of the wife. Assume that they may have many children. What is the chance of their having affected offspring?

In Problems 45–50, assume that all individuals are homozygous normal **unless there is evidence to the contrary.** *Assume that the different albinos all have the same albino genotype.*

45. In Figure 98, what is the probability that IV–3 is a carrier for albinism?

46. In Figure 98, what is the probability that IV–1 is a carrier for albinism?

47. List the genotypes of all individuals given in Figure 98. If more than one genotype is likely, list the alternatives and their probabilities.

48. If a man and his unrelated wife each have an albino sib, what is the probability:
    (a) That their firstborn will be an albino?
    (b) That albinism will occur if they have many children?
    (c) That if they will have three children, all will be albinos?
    (d) That if their first child is an albino, the next two will also be albinos?

49. (a) If the normal sister of an infant affected with Tay-Sachs disease wants to know if her hoped-for child will be affected, what would be your prognosis?
    (b) If it turned out that her husband had an affected cousin, what would your prognosis be then?

50. In Figure 99, individuals III–1 and III–10 were carriers for phenylketonuria. In addition, carriers must have been present among the individuals of generation II. Could the appearance of the affected individuals in generation IV be explained if the following individuals in generation II had been heterozygous: II–1 and II–2; or II–1 and II–4; or II–2 and II–3; or II–2 and II–4; or II–1 and II–5?

REFERENCES

For further discussion of genetic prognosis and genetic counseling see Chapter 30 and consult the books by Carter; Fuhrmann and Vogel; Hammons; Motulsky; Reed; Fraser Roberts; Scheinfeld; and Stevenson, Davison, and Oakes, which are listed at the end of Chapter 30.

10

GENETIC RATIOS

We have already seen that certain ratios of nonaffected to affected children are to be expected in various types of marriages. The two most important are the 1:1 ratio (in heterozygous *Dd* × homozygous *dd* marriages) and the 3:1 ratio (in *Dd* × *Dd* marriages). These ratios do not represent certainties in the sense that they can be predicted with accuracy for any one family, but are the result of chance occurrences that lead to predictable probabilities.

CHANCE PHENOMENA

It may be well to recount these chance occurrences. We know that in a *Dd* man (♂) spermatogenesis leads to the formation of equal numbers of *D* and *d* sperm and that in a *dd* woman (♀) oogenesis leads to *d* eggs exclusively. The role of chance in the production of *Dd* and *dd* children if such persons marry rests on at least three circumstances:

1. Some sperm cells, because of accidents in development, do not become mature sperm. Since there is no reason to believe that either the *D* or the *d* sperm is more subject to such accidents than the other, the cells will degenerate at random. Therefore, it is unlikely that exactly the same number of *D* and *d* cells will be eliminated, and slight inequalities in the two kinds of mature sperm will thus result from chance elimination.
2. Even if absolutely equal numbers of both kinds of sperm should be present in the female ducts, the endowment of the new child would depend on the fusion of either a *D* or a *d* sperm with the *d* egg. Again, there is no reason to believe that either kind of sperm is more likely to fertilize a *d* egg—at least, such "selective fertilization" has not been demonstrated. Thus, it seems a matter of chance whether a *D* or a *d* sperm fertilizes the egg. If the first pregnancy happens to be the result of fertilization by a *D* sperm, a later fertilization will not be influenced by this; the chance that the later fertilization will result in a *Dd* embryo is just as great as it was the first time, and equal to the chance that fertilization by a *d* sperm will occur and result in a *dd* embryo.
3. Even assuming absolute equality in the number of *Dd* and *dd* embryos in early pregnancy, birth of equal numbers of the two kinds of children will depend on normal completion of development. Since an appreciable number of pregnancies are not successfully brought to term, and since it may be assumed here that *Dd* and *dd* fetuses are equally viable, there would, once more, be an equal chance that a *Dd* or a *dd* child would not survive.

In the reciprocal marriage, *Dd* ♀ × *dd* ♂, we have only one kind of sperm, *d,* to consider. Two main chance processes determine the genotype of a child of this marriage. The first concerns the fate of the *Dd*-containing chromosome pair on the meiotic spindles of the egg cell. Two of the four chromosome strands of the pair possess a *D* allele and the other two a *d* allele, but only one of the four strands is included in the egg nucleus. Whether this strand contains a *D* or a *d* allele depends on the arrangement of the strands on the meiotic spindles—a matter of chance. Thus, there is equal probability that a *Dd* or a *dd* embryo will result from fertilization by the one kind of sperm produced by the *dd* man. The second chance process in the marriage *Dd* ♀ × *dd* ♂, is identical with the third in the marriage *dd* ♀ × *Dd* ♂. It is the equal probability that either a *Dd* or a *dd* embryo will die before birth.

In a *Dd* × *Dd* marriage, the various chance processes involved in the development of sperm cells into mature sperm, in the production of *D* or *d* eggs, in the fusion of *D* or *d* sperm with *D* or *d* eggs, and in the survival of *DD, Dd,* and *dd* embryos will all play a role.

RATIOS IN SMALL FAMILIES

Ideal Ratio 1:1. It is possible to express numerically the expected chance deviations from an ideal ratio. If, for instance, a *Dd* × *dd* marriage results in 2 children, the probability that either child will be *Dd* or *dd* is 1/2 each. Consequently, the probability that both the first and second child will be *Dd* is the product of the separate probabilities: $1/2 \times 1/2 = 1/4$. The probability of the first being *Dd* and the second *dd* is also 1/4; that of the first being *dd* and the second *Dd,* again 1/4; and that of the first and second being *dd,* once more 1/4. This means that among *Dd* × *dd* marriages with 2 children, chance will cause one-quarter of the families to have only *Dd* children, another quarter to have only *dd* children, and half to have one of each kind.

Similarly, we may calculate the chance deviations from the "ideal" expectations for any given number of children in a sibship. Before giving the general procedure, let us consider sibships with 3 children. The expected probability of either type of child remains, of course, 1/2. All 3 children will be *Dd* in $1/2 \times 1/2 \times 1/2$, or 1/8, of all sibships. Another 1/8 of all sibships will have 3 *dd* children. Two *Dd* children and 1 *dd* child are possible in three sequences: *Dd–Dd–dd, Dd–dd–Dd,* and *dd–Dd–Dd,* each sequence having the probability of $1/2 \times 1/2 \times 1/2 = 1/8$. Thus, the probability of 2 *Dd* children and 1 *dd* child is 3/8. Likewise, 3 out of 8 sibships will consist of 1 *Dd* child and 2 *dd* children.

It is evident that the procedure employed is the determination of all possible combinations of two events, the birth of a *Dd* or a *dd* child, in sets of three. A mathematical "recipe" for obtaining the results consists in expanding the binomial $(a + b)^s$. Applied to our example, a equals the probability of the occurrence of the genotype *Dd,* which is 1/2; b, the probability of the genotype *dd,* which is likewise 1/2; and s, the number of children in the sibship.

For sibships of 2, the binomial is $(1/2\ Dd + 1/2\ dd)^2$, which, expanded, becomes

$$1/4(Dd)^2 + 2/4(Dd)(dd) + 1/4(dd)^2.$$

The exponents assigned to the genotypes indicate how often they occur in any given sibship of 2 children: $(Dd)^2$ signifies 2 *Dd* children; $(Dd)(dd)$ signifies 1 *Dd* and 1 *dd* child; and $(dd)^2$ signifies 2 *dd* children. The fractions 1/4, 2/4, and 1/4 designate the probabilities of occurrence of the particular combinations of 2 children in a sibship.

For sibships of 3, the binomial is $(1/2\ Dd + 1/2\ dd)^3$, which, expanded, becomes

$$1/8(Dd)^3 + 3/8(Dd)^2\, dd + 3/8\, Dd\, (dd)^2 + 1/8(dd)^3.$$

The four terms of this series give the probabilities of the four kinds of sibships possible with 3 children: 1/8 that all 3 children will be *Dd*; 3/8 that there will be 2 *Dd* children and 1 *dd* child; and so on.

In the general case, where s indicates the number of children in a sibship, the expansion of the binomial $(1/2\ Dd + 1/2\ dd)^s$ yields the following series:

$$(1/2\ Dd)^s + s\ (1/2\ Dd)^{s-1}\ (1/2\ dd) + \frac{s(s-1)}{1 \times 2}\ (1/2\ Dd)^{s-2}\ (1/2\ dd)^2$$

$$+ \cdots + s\ (1/2\ Dd)\ (1/2\ dd)^{s-1} + (1/2\ dd)^s.$$

This may be written in the form

$$(1/2)^s\ (Dd)^s + s \times (1/2)^s \times [(Dd)^{s-1}\ dd] + \frac{s(s-1)}{1 \times 2} \times (1/2)^s \times [(Dd)^{s-2}\ (dd)^2]$$

$$+ \cdots + s \times (1/2)^s \times [(Dd)\ (dd)^{s-1}] + (1/2)^s\ (dd)^s.$$

Expressed in genetic terms, the series states, that $(1/2)^s$ of all sibships can be expected to have only *Dd* children, that $s \times (1/2)^s$ of all sibships have $(s-1)$ *Dd* children and 1 *dd* child, that

$$\frac{s(s-1)}{1 \times 2} \times (1/2)^s$$

of all sibships have $(s-2)$ *Dd* and 2 *dd* children, and so on. If we take as an example sibships with 8 children, then $(1/2)^8$ or 1/256 of all sibships have only *Dd* children, 8/256 have 7 *Dd* and 1 *dd*, 28/256 have 6 *Dd* and 2 *dd*, and so on.

In order to determine the probability that a sibship of 8 will consist of 5 *Dd* and 3 *dd* children, it is necessary to find the probability corresponding to $(Dd)^5\ (dd)^3$. This may be done by introducing one more general algebraic expression. If s is the total number of children per family, and x the number of *Dd* children, then $(s-x)$ is the number of *dd* children. The general binomial term that signifies the probability of a sibship with x *Dd* and $(s-x)$ *dd* children is

$$\frac{s!}{x!(s-x)!}\ (1/2\ Dd)^x\ (1/2\ dd)^{s-x},$$

or

$$\frac{s!}{x!(s-x)!}\ (1/2)^s\ [(Dd)^x\ (dd)^{s-x}.$$

An exclamation point following a letter or a combination of letters in parentheses, indicates that they are "factorials," that is, the product of all integers from 1 to the number signified; for example, if s = 5, then "s factorial" = s! = 1 × 2 × 3 × 4 × 5. Applying the binomial to the family of 8 children (s = 8), the probability that it consists of 5 *Dd* (x = 5) and 3 *dd* (s − x = 3) children is

$$\frac{8!}{5!3!} \times \frac{1}{2^8} = \frac{1 \times 2 \times 3 \times 4 \times 5 \times 6 \times 7 \times 8}{1 \times 2 \times 3 \times 4 \times 5 \times 1 \times 2 \times 3} \times \frac{1}{256} = \frac{56}{256} = \frac{14}{64}.$$

This calculation, then, shows that, in 14 out of 64 sibships of 8 children with *Dd* × *dd* parents, it is expected that 5 of the children will be *Dd* and that 3 will be *dd*.

Table 7 gives the probabilities of the occurrence of all possible combinations of heterozygous dominant (*Dd*) and homozygous recessive (*dd*) children in sibships of from 1 to 8 children with *Dd* × *dd* parents. This table reveals many interesting facts. It shows, for example, that the ideal 1:1 ratio in sibships of 4 is found in only 6 out of 16 families, i.e., 37.5 per cent, and that in sibships of 8, the 1:1 ratio is even less common, being found in 70 out of 256 families, or approximately 27 per cent. On the other hand, the larger the sibship, the more likely it is to be close to the 1:1 ratio. Thus, in sibships of 4, 10 out of 16, or 62.5 per cent, have ratios of 3:1, 1:3, or "worse"; while in sibships of

TABLE 7
Probabilities of All Possible Combinations of *Dd* and *dd* Children in Sibships of from 1 to 8 Children of *Dd* × *dd* Marriages.

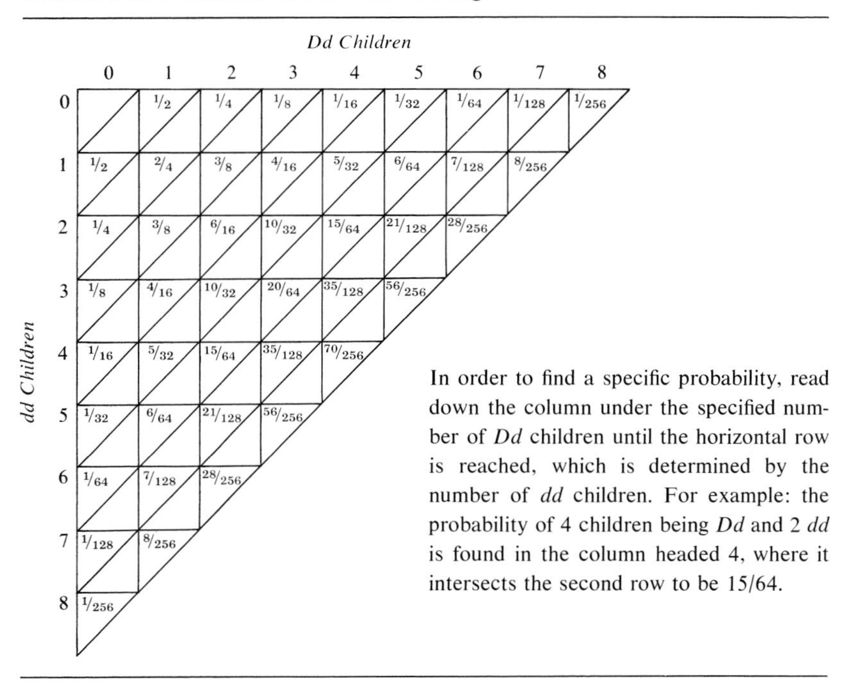

dd Children \ *Dd* Children	0	1	2	3	4	5	6	7	8
0		1/2	1/4	1/8	1/16	1/32	1/64	1/128	1/256
1	1/2	2/4	3/8	4/16	5/32	6/64	7/128	8/256	
2	1/4	3/8	6/16	10/32	15/64	21/128	28/256		
3	1/8	4/16	10/32	20/64	35/128	56/256			
4	1/16	5/32	15/64	35/128	70/256				
5	1/32	6/64	21/128	56/256					
6	1/64	7/128	28/256						
7	1/128	8/256							
8	1/256								

In order to find a specific probability, read down the column under the specified number of *Dd* children until the horizontal row is reached, which is determined by the number of *dd* children. For example: the probability of 4 children being *Dd* and 2 *dd* is found in the column headed 4, where it intersects the second row to be 15/64.

8, only 74 out of 256, or 28.9 per cent, show equally great deviations. Furthermore, in families with 4 children, a ratio of 3 *Dd*: 1 *dd* will be found in 4 out of 16, or 25 per cent, and a ratio of 1 *Dd*: 3 *dd* in another 25 per cent, although the ideal expectation is 1:1. Among sibships of all sizes, there are some in which only one genotype, *Dd* or *dd*, occurs.

Ideal Ratio 3:1. When both parents are heterozygous for a specific recessive allele, the ideal ratio is 3 dominant to 1 recessive offspring. Again, it is obvious that chance processes will not only permit, but necessarily cause, deviations from the ideal ratio.

Let us consider, in detail, families with 4 children.

Five different types of 4-child sibships with $Dd \times Dd$ parents are possible: 4, 3, 2, 1, or 0 *dd* children, or, stated differently, 0, 1, 2, 3, or 4 *D*– children (*D*– signifies either *DD* or *Dd*). In the second column of Table 8, all 16 possible combinations of *D*– and *dd* children in a set of 4 are enumerated. In the third column, the probabilities of the combinations are listed. These probabilities follow from the fact that (1) the occurrence of a *D*– child has a probability of 3/4; (2) the occurrence of a *dd* child has a probability of 1/4; and (3)

TABLE 8
Types and Probabilities of Sibships of 4 from $Dd \times Dd$ Marriages.

Type of Sibship	*Sequence of Children*				*Probability of Sibships*	*No. of D– and dd in 256 Families*	
	First	Second	Third	Fourth		*D*–	*dd*
4*D*–:0*dd*	*D*–	*D*–	*D*–	*D*–	$\frac{3}{4} \cdot \frac{3}{4} \cdot \frac{3}{4} \cdot \frac{3}{4} = \frac{81}{256}$	324	0
3*D*–:1*dd*	*D*– *D*– *D*– *dd*	*D*– *D*– *dd* *D*–	*D*– *dd* *D*– *D*–	*dd* *D*– *D*– *D*–	$4 \cdot \frac{3}{4} \cdot \frac{3}{4} \cdot \frac{3}{4} \cdot \frac{1}{4} = \frac{108}{256}$	324	108
2*D*–:2*dd*	*D*– *D*– *dd* *D*– *dd* *dd*	*D*– *dd* *dd* *dd* *D*– *D*–	*dd* *dd* *D*– *D*– *dd* *D*–	*dd* *D*– *D*– *dd* *D*– *dd*	$6 \cdot \frac{3}{4} \cdot \frac{3}{4} \cdot \frac{1}{4} \cdot \frac{1}{4} = \frac{54}{256}$	108	108
1*D*–:3*dd*	*D*– *dd* *dd* *dd*	*dd* *D*– *dd* *dd*	*dd* *dd* *D*– *dd*	*dd* *dd* *dd* *D*–	$4 \cdot \frac{3}{4} \cdot \frac{1}{4} \cdot \frac{1}{4} \cdot \frac{1}{4} = \frac{12}{256}$	12	36
0*D*–:4*dd*	*dd*	*dd*	*dd*	*dd*	$\frac{1}{4} \cdot \frac{1}{4} \cdot \frac{1}{4} \cdot \frac{1}{4} = \frac{1}{256}$	0	4
Grand total						768	256
Total, omitting the 4*D*–:0*dd* sibships						444	256

the probability of any specific combination of 4 children is the product of the four separate probabilities. The fourth column gives the total number of *D*– and *dd* children, calculated on the assumption that 256 families are being studied. As shown in the third column, 81 families out of 256 would have *D*– children only, yielding a total of $4 \times 81 = 324$ *D*– and no *dd* children. Similarly, 108 families have 3 *D*– and 1 *dd*, giving a total of $3 \times 108 = 324$ *D*– and 108 *dd*, and so on.

The ideal ratio of 3 *D*–:1 *dd* is realized in only 108 out of 256 sibships; that is, in less than 42 per cent of all sibships of 4 can the "expected" ratio (3:1) actually be expected. Instead, 54 sibships will have a 1:1 ratio; 12 sibships, a 1:3 ratio; 81 sibships, a 4:0 ratio; and 1 sibship, a 0:4 ratio!

Rather than laboriously enumerating all combinations, we may again use the binomial theorem to calculate the probabilities. In $Dd \times Dd$ marriages, the binomial takes the general form

$$(3/4\ D{-} + 1/4\ dd)^{s},$$

with any member of the series of terms equal to

$$\frac{s!}{x!(s-x)!} \times (3/4\ D{-})^{x} \times (1/4\ dd)^{s-x}$$

$$= \frac{s!}{x!(s-x)!} \times (3/4)^{x} \times (1/4)^{s-x} \times [D{-}^{x} \times dd^{s-x}].$$

An example may demonstrate the usefulness of this formula: The probability that a sibship of 8 children with two *Dd* parents, will consist of 7 *D*– children and 1 *dd* child is

$$\frac{8!}{7!} \times \left(\frac{3}{4}\right)^{7} \times \frac{1}{4} = 8 \times \frac{2{,}187}{65{,}536} = \frac{2{,}187}{8{,}192}$$

or approximately 1/4; the remaining 3/4 of such sibships will be made up of all other combinations of *D*– and *dd* children.

It would be instructive to work out a diagram similar to Table 7 that would list the probabilities of all combinations of *D*– and *dd* children from $Dd \times Dd$ marriages in sibships of from 1 to 8.

RATIOS IN POOLED DATA

Dominant Traits. It should now be clear that there are great deviations from expected "ideal" ratios in single families. An obvious way to correct the incomplete picture that study of single families conveys would seem to be to pool many sibships with identical parental genotypes and study them as a single unit. The chance deviations which make certain ratios too large in some, it might be reasoned, would be compensated for by the chance deviations which make the same ratios too small in others. This is fully true for a *dominant* trait, if we pool the offspring of all marriages in which one of the parents has the dominant phenotype. Here, sibships in which the ratio of affected to nonaffected children is greater than 1:1 will be compensated for by sibships in which it is smaller. Similarly, in extreme cases, sibships consisting only of affected children and those of nonaffected children only will balance each other.

Even in dominant inheritance, systematic deviations from expected ratios will be encountered if the genotypes of the parents are unknown and the persons studied come from sibships that are detected only because they contain affected individuals. This may be shown by an example dealing with the sex ratio. If we count a random sample of the *children* in a community, the ratio of girls to boys will be close to 1:1. If, however, we restrict the inquiry to those *families* that contain at least one girl—by asking, for example, the parents of all girls in the community to state the sex ratio among their children—a ratio other than 1:1 will be found. Among 2-child families, for example, the ratio of sibships with 2 girls (♀ ♀) to sibships with 1 girl and 1 boy (either ♀ ♂ or ♂ ♀) is 1:2. Consequently, the ratio obtained if these sibships are pooled is 4 girls to 2 boys, or 2:1. A still different ratio is obtained if we begin our inquiry by selecting at random *girls* of a given age and determine the sex ratio in the sibships of those from 2-child families. Here the chance that a 2-girl sibship will come to notice is twice as great as that a 2-child sibship with 1 girl only will be discovered. Therefore, there will be 2 ♀♀ sibships for each ♀♂ and ♂♀ sibship, and the resulting ratio will be 6 girls to 2 boys, or 3:1.

Ascertainment. This example, in which three different sex ratios were obtained, shows how important the *method of ascertainment*—that is, the method used in gathering data—is in human genetics. The 1:1 ratio of girls to boys was obtained when ascertainment was by "complete selection," that is, based on complete random sampling of all available data; the 2:1 ratio, when ascertainment was by "incomplete truncate selection," a method that made use of all families with at least one girl but excluded an entire group (here, all families with no girls); and the 3:1 ratio, when ascertainment was by "incomplete single selection" of individual girls. Still other ratios would be obtained if the inquiry is based neither on a truncate selection of family units that gives each relevant family, regardless of its number of children, an equal chance to be ascertained, nor on a single selection of sibships that makes the probability of ascertainment proportional to the number of sibs. Such intermediate situations are based on "incomplete multiple selection."

Recessive Traits. If two parents are heterozygous for a recessive gene, a 3:1 ratio of normal to affected children is to be expected among their offspring. Any individual sibship from such parents may show deviations from the expected ratio—from all normal to all affected—but a pooling of many sibships will yield the 3:1 ratio.

Usually, in recessive inheritance of uncommon traits, ascertainment of sibships is not through the parents, who are normal, but through the fact that one or more sibs are affected. Under these circumstances a certain compensatory effect of pooling data still holds, but this compensation is not complete: a systematic error remains. It is due to the sibships from *Dd* × *Dd* parents, in which no affected (*dd*) children occur in spite of the "ideal" expectation that they will. Sibships from two heterozygous parents, in which all children appear normal (*D*–), make up $(3/4)^s$ of all sibships, where s is the number of sibs. When s = 4, these all-normal sibships constitute 81/256 or nearly 1/3 of all sibships of 4 children (Table 8), and the proportions of all-normal sibships are still higher when there are fewer children.

Ascertainment of sibships through affected sibs misses the all-normal sibships from *Dd* × *Dd* parents (truncate selection). Thus, there is insufficient compensation, when data are pooled, for those sibships which happen to contain too many *dd* children. Consequently, the ratio of dominants to recessives in pooled data of sibships from heterozygous parents is shifted from the expected 3 *D*–:1 *dd* value toward a higher value for the *dd* group. This is shown for 4-child sibships in Table 8. As seen earlier, none of the sibships of 4 *D*– children—81 out of 256—will be recognized. The 108 sibships with 3 *D*– and 1 *dd* children yield 324 and 108; the 54 sibships with 2 *D*– and 2 *dd* contribute 108 of each; the 12 sibships with 1 *D*– and 3 *dd* add 12 and 36; and the 1 sibship with 4 *dd* adds 4. This makes a total of 444 *D*– and 256 *dd*, or a ratio of 1.734:1, instead of 3:1 (or, stated differently, the fraction of *dd* individuals is 1/2.734 = 0.366 instead of 1/4 = 0.25).

The size of the deviation from the 3:1 ratio in pooled data decreases as the number of children per sibship increases, since the proportion of "lost," all-normal sibships decreases. Three-fourths of all children from 1-child sibships are normal, and in 9/16 of 2-child sibships both children are normal; but in 8-child sibships only $(3/4)^8$ or 6,561/65,536, approximately 1/10, consist of all normal children, and in 12-child sibships only $(3/4)^{12}$, or about 1 in 32 sibships.

In Table 9, the expectation for pooled data from sibships of varying sizes is compared with data concerning two recessive traits, albinism and phenylketonuria. Normal and albino children were pooled from three groups of sibships, arranged according to size. In the group of smallest sibships (from 1 to 4 sibs) there are more albinos than normals, but with increasing number of children the proportions approach more and more closely the 3 *D*–:1 *dd*

TABLE 9
Observed and Expected Proportions from Heterozygous Parents of Normal ($D-$) to Affected (dd) Children in Sibships with at Least One Affected.

Trait	*Sibs per Sibship*	*D–*	*dd*	*Proportion D–:dd*	
				Observed	Expected
	1–4	–	–	0.9:1	–
Albinism	5–7	–	–	1.6:1	–
	8 or more	–	–	2.3:1	–
	1	–	6	0:1	0:1
	2	6	8	0.75:1	0.75:1
	3	8	10	0.80:1	1.31:1
	4	12	8	1.50:1	1.74:1
	5	22	13	1.69:1	2.05:1
Phenylketonuria	6	18	12	1.50:1	2.29:1
	7	10	4	2.50:1	2.46:1
	8	16	8	2.00:1	2.60:1
	9	7	2	3.50:1	2.70:1
	10	15	5	3.00:1	2.77:1
	11	9	2	4.50:1	2.83:1
	12	9	3	3.00:1	2.87:1
	13	9	4	2.25:1	2.90:1

Source: Data on albinism from Roberts, *An Introduction to Medical Genetics*. Oxford, 1959, 2nd ed; on phenylketonuria from Munro.

ratio. For phenylketonuria, Table 9 shows even more strikingly the rise in the proportion of normals with increasing size of sibship. The occasional rather large differences between observed and expected ratios are probably due to chance.

Methods for Correcting Ratios from Pooled Data. Since, theoretically, a 3:1 ratio is one of the fundamental attributes of simple single factor recessive inheritance, the deviation due to truncate selection stands in the way of recognition of this important type of heredity. However, there are methods to correct for the systematic bias that results from the failure to detect many sibships from heterozygous parents, so that the corrected data may be compared with ideal expectations. Several such methods, sometimes called segregation analysis, were proposed by the physician Weinberg (1862–1937) who first recognized the problem of ascertainment in Mendelian ratios. Various other students, among them Bernstein, Hogben, Haldane, Fisher, and Morton, have further clarified the issues.

A method based on an a priori expectation. This method presupposes a specific expected ratio, e.g., 3:1, and endeavors to determine whether the observed ratio, corrected for truncate selection, agrees with the a priori expectation. The procedure is to work out the ratio that should be expected in the observed incompletely selected sibship material on the assumption that data from complete selection would give the a priori ratio. This a priori expectation for offspring from *Dd* × *Dd* parents is $p = 1/4$ for *dd* children, and $q = 3/4$ for *D–* children. We wish to find the fraction p′ of *dd* children to be expected in the sibships that supply our data.

These sibships are only those in which at least one *dd* child occurs—i.e., they are all possible sibships (= 1) minus those in which no *dd* child occurs. As was shown on page 203, the families with no *dd* children have the frequency q^s, where q is the probability of *D–* children (i.e., 3/4) and s the number of sibs per family. Therefore the detectable sibships that supply our data have the frequency $1 - q^s$. Affected children occur in the proportion p (i.e., 1/4) of the children from all possible sibships. Their proportion p′ in the detectable sibships is

$$p' = \frac{p}{1 - q^s}.$$

It is seen that p′ is larger than p.

We may apply the formula to 4-child sibships (s = 4). The expected, observable fraction of *dd* children is

$$p' = \frac{1/4}{1 - (3/4)^4} = 0.366.$$

This value agrees with the value which was derived from our enumeration of all types of 4-child sibships (p. 203).

It is often useful to determine the expected average *number* of affected children per sibship instead of the expected observable *fraction*, p′. This number of affected children is obtained by multiplying p′ by s, the total number of children in the sibship. It is listed, for families with from 2 to 10 children, in Table 10. A comparison of the actual expectation with the "naïve," uncorrected expectation based on direct use of the probability $p = 1/4$ shows how

TABLE 10
Average Number, $p' \cdot s$, of Affected Children in Sibships with at Least One Affected Sib from *Dd* × *Dd* Marriages. (Total number of children in sibship, s, from 2 to 10. A priori probability p of $dd = 1/4$.)

Children in Sibship	s								
	2	3	4	5	6	7	8	9	10
Expectation $p' \cdot s$	1.143	1.297	1.463	1.640	1.825	2.020	2.222	2.433	2.649
Uncorrected expectation $p \cdot s (p = 1/4)$	0.500	0.750	1.000	1.250	1.500	1.750	2.000	2.250	2.500

Source: Bernstein.

TABLE 11
Application of the A Priori Method for Determining the Presence of Recessive Inheritance to Myoclonic Epilepsy.

Size of Sibship (s)	*No. of Sibships* (x)	p′ · s (*from Table 10*)	*Total Affected*	
			Expected p′ · s · x	Observed
1	1	1	1	1
4	1	1.463	1.463	2
5	1	1.640	1.640	2
6	3	1.825	5.475	7
8	1	2.222	2.222	1
9	2	2.433	4.866	4
Total: 54	9		16.666	17

Source: Data from Bernstein after Lundborg, simplified.

great the difference between the two expectations is in the small sibships, and how it decreases with increasing number of children per sibship.

We shall, finally, demonstrate the use of the a priori method on data collected from Swedish families in which one or more children from normal parents were afflicted with a special type of progressive epilepsy (myoclonic epilepsy). The genetic hypothesis is that the diseased children are the *dd* offspring of *Dd* parents. Affected children were found in 9 sibships, ranging in size from 1 to 9 children. They are listed in the first two columns of Table 11. The third column gives the average number of affected children per sibship expected according to the a priori ratio. The fourth column gives the number obtained when the number of affected children is multiplied by the number of sibships, and it should be compared with the observed number listed in the final column. When we consider the small amount of data, observation and expectation diverge little: in fact, the total numbers of expected and observed affected children, 16.666 and 17, agree remarkably well. It may be concluded that myoclonic epilepsy fulfills the condition imposed by the hypothesis that it is based on a simple single recessive gene.

The a priori method entails the silent assumption that all families have an equal probability of being included in the data. This was true in the study of myoclonic epilepsy but often is not the case. Other more generally applicable methods have therefore been designed to test for ratios from pooled data.

Weinberg's proband method. This method corrects for the loss of data on *D*– children, caused by nonascertainment of *Dd* × *Dd* families without *dd* children, by omitting from consideration an appropriate number of *dd* children so that their omission compensates for the loss of the *D*–. Specifically the *dd* propositi (or probands) of each sibship are discounted and the ratio of *D*– to *dd* sibs is determined among the sibs of the propositi. This ratio represents an unbiased estimate of *D–* : *dd*.

How this method works may be demonstrated by applying it first to an artificial example of a single sibship and then to a collection of observed sibships. Assume a sibship of 13 children, 4 of whom are homozygous *dd*, thus presenting a proportion of recessives of $4/13 = 0.31$. Assume that the sibship was ascertained by one of the *dd* sibs. To compensate for the bias in ascertainment this *dd* sib is discounted so that the corrected sibship has only 12 children, 3 of whom are *dd*, thus giving a proportion of 0.25, which fits the expectation for recessive inheritance. Let us now apply the proband method to a sample of 13 sibships that were ascertained because they included one or more individuals homozygous for cystic fibrosis of the pancreas, a recessive disease that is often lethal in childhood. A sum of the raw data (Table 12, columns 1, 2) gives a ratio of recessives to total number of children of $17/36 = 0.472$. To correct for the mode of ascertainment each of the sibships is to be treated like the artificial example just discussed, thus providing a ratio of "affected minus the propositus," $(r - 1)$, to "total children in the sibship minus the propositus," $(s - 1)$, equal to $(r - 1)/(s - 1)$. By pooling all sibships the ratio becomes [sum of all $(r - 1)$] / [sum of all $(s - 1)$]. There is one additional consideration. Sibships that were independently ascertained twice, as was the case in two sibships (number of propositi $a = 2$) are being given double weight, as has been done in the table, by

TABLE 12
Distribution by Sibships of 17 Individuals with Cystic Fibrosis. Data obtained from multiple selection; modes of ascertainments have been recorded.

No. in sibship s	*No. affected* r	*No. propositi* a	a(r − 1)	a(s − 1)
1	1	1	0	0
1	1	1	0	0
2	1	1	0	1
2	1	1	0	1
2	1	1	0	1
2	2	2	2	2
3	1	1	0	2
3	1	1	0	2
4	1	1	0	3
4	1	1	0	3
4	1	1	0	3
4	2	1	1	3
4	3	2	4	6
36	17	15	7	27

Source: From data of Dr. Charles Lobeck cited by Crow, 1965.

multiplying by a the respective (r − 1) and (s − 1) terms. This use of the proband method results in a corrected ratio of $7/27 = 0.259$. Given the small size of the chosen sample of families with cystic fibrosis the fraction 0.259 is close enough to 0.25 to regard the data as in accord with the hypothesis of simple recessive inheritance of the disease.

The proband method can be applied not only to data where selection is multiple and ascertainment has been recorded, but also to truncate selection with complete ascertainment, and to single selection with very incomplete ascertainment. In the first instance each affected individual is ascertained independently and $a = r$. In the second instance the likelihood of a family being ascertained more than once is close to zero, and $a = 1$.

Method of discarding the singles. More than fifty years after Weinberg first worked out methods that compensate for ascertainment bias, Li and Mantel devised an unusually simple further method. It is applicable to truncate selection with complete ascertainment of all families with affected children. In this method a fraction (f) is formed of the "sum of all affected children (r) minus those that occur as the only affected children in their sibships (= singles) (γ)" divided by the "sum of all children (t) minus those that occur as singles in their sibships (γ)":

$$f = \frac{r - \gamma}{t - \gamma}.$$

The determination of this fraction requires no more data than those obtained by counting three numbers—the number of affected, of all, and of affected children that have no affected sibs. If the trait under consideration is a simple recessive, then it will be shown that f is equal to 1/4. If it differs significantly from 1/4 and the data do indeed derive from complete ascertainment of affected individuals then it may be concluded that simple recessive inheritance is not involved.

Ideally with truncate selection and complete ascertainment of affected children, among x sibships with s children each, the sum of all *dd* children is

$$r = 1/4\ xs.$$

The sum of singles, being proportional to the second term in the expansion of the binomial $(3/4\ D{-} + 1/4\ dd)^s$ is

$$\gamma = xs\ (3/4)^{s-1}\ (1/4).$$

Finally, the total number of children is s times the number of sibships ascertained. This latter is equal to $x\ [1 - (3/4)^s]$ since $(3/4)^s$ is the frequency of the not-ascertained sibships with only *D*– children. Therefore the total number of children is

$$t = xs\,[1 - (3/4)^s].$$

Inserting the terms for r, γ, and t into the formula for f and cancelling out xs in each term, we obtain

$$f = \frac{1/4 - (3/4)^{s-1}\,(1/4)}{[1 - (3/4)^s] - (3/4)^{s-1}\,(1/4)}$$

which reduces to

$$\frac{1/4\,[1 - (3/4)^{s-1}]}{1 - (3/4)^{s-1}} = 1/4 = 0.25.$$

Since $f = 0.25$ holds for sibships of any size and since in most studies sibships of different sizes occur, we can pool the data from all sibships and the fraction of (sum of all *dd* − sum of singles)/(total no. of children − sum of singles) should again be 0.25.

Li and Mantel's method requires complete ascertainment. This is a restrictive condition. No such restriction is necessary in Weinberg's proband method, which is a more general one.

We have seen that various types of selection affect in different ways the ratios obtained from pooling of data. It is appropriate to mention a special type of selection not treated above. It occurs when data from pedigrees are pooled that were obtained independently by various authors. Many such pedigrees owe their publication to peculiarities in the number of affected individuals or to other unusual characteristics—they represent a conscious "selection for oddity-interest"—and since, for a general analysis of inheritance, ordinary cases are as valuable as "interesting" ones, their omission tends to give a false picture of a hereditary problem.

In the preceding pages it was shown that a 3:1 ratio is *not* to be expected in recessive inheritance if the pooled sibships have been ascertained because they include affected children. Before this was recognized, it was mistakenly believed that simple recessive inheritance was proved when pooled sibships gave a 3:1 ratio! Conversely, when the proportion of affected children was found to be higher than 1 in 4, it was—equally mistakenly—thought that simple recessive inheritance was excluded, although an excess of affected sibs was just what would have been expected.

GOODNESS OF FIT OF RATIOS

We have investigated the varieties of individual sibships expected as a result of chance. The frequencies of sibships with ratios different from any of the ones expected follow binomial distributions. We have seen how pooling sibships and making necessary corrections can result in a ratio equivalent to the expected. It is, however, clear that an expected ratio is a probability fraction and not an absolute value. Even in pooled and corrected data, we do not expect *exact* agreement with, for example, a 1:1 or a 3:1 ratio. Deviations from the expected ratio can be predicted, and their degree estimated quantitatively.

The Probability That a Specific Deviation from Expectation Is Likely to Be Due to Chance. When an observed ratio deviates from expectation, the question arises whether the deviation should be considered to be due to chance or whether some specific causes should be assumed to be responsible for it. In order to decide this question, we calculate the probability that a deviation of the given magnitude may occur purely as a result of chance. If that probability is high, then the observed ratio may be considered compatible with the expected one. If it is low, then the observed ratio is not considered compatible.

Two different questions may be raised regarding an observed ratio which differs from that expected: First, we may want to know the probability that the observed ratio will occur. Thus, particularly in small samples, one may be interested in determining the probability of finding a *specific* ratio. This kind of question has been dealt with earlier in this chapter, in examples such as that concerning the probability of finding a sibship of 2 affected and 6 nonaffected when the expectation is 1:1. Second, we may want to know the probability of finding a ratio that differs as much from expectation as the observed ratio does. This second way of looking at the deviation divides, so to speak, the whole range of possible deviations into two sections: one includes all possible ratios that are closer to the expected ratio than the observed ratio; the other includes the observed ratio and all possible ratios that are further from expectation than it.

It is the second question which is of interest if we wish to judge whether an observed ratio should be considered a chance deviation from the expected one. This will be apparent if we consider an example: Assume that a coin is tossed 100 times and that 45 heads and 55 tails are counted (Fig. 112). The probability of obtaining exactly this ratio is

$$\frac{100!}{45!55!} \times \frac{1}{2^{100}},$$

or about 1 in 20. This is a low probability. Should we, then, conclude that the observed deviation from 50:50 is not due to chance but that some specific cause is responsible for it—for example, an inherently biased coin? This seems to be unreasonable, since the probability of observing any specific ratio in a set of 100 tosses is small, even that of the ideal ratio 50:50 being only about

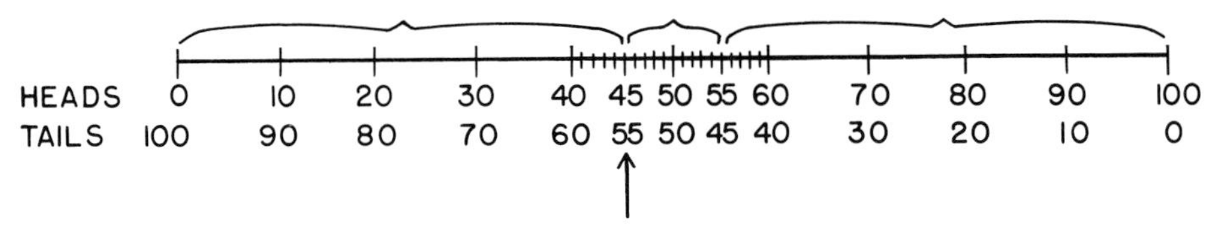

FIGURE 112
The range of possible results of flipping a coin one hundred times. The bracket in the center includes those ratios that deviate from 50:50 less than 45:55, or 55:45. The two other brackets include all ratios that deviate from 50:50 as much as, or more than, 45:55 or 55:45.

1 in 13. If one calculates (in a way to be shown), as an alternative, the probability of obtaining a sample which deviates as much from 50:50 as 45:55, or even more, it is found that such deviations occur about once in three series of 100 tosses. This high probability of a deviation as great as, or even greater than, 45:55 appears quite compatible with the hypothesis that the coin was not biased and that the observed ratio deviated from the expected as a result of chance.

How was the last probability calculated? One method would be to add up the probability of occurrence of the observed ratio 45:55 and of all probabilities of occurrence of larger deviations, namely, 44:56, 43:57, 42:58, . . . , 0:100. This sum of all probabilities would still have to be multiplied by 2, since the question "What is the probability of obtaining a deviation from 50:50 as great as, or greater than, the observed ratio 45:55?" refers equally to the ratios 45 heads:55 tails and 55 heads:45 tails; 44 heads:56 tails and 56 heads:44 tails; and so on. This elaborate method of calculation is not used in practice. A much simpler procedure, the "chi-square test," will be presented in the following pages.

The Significance of Various Probabilities. What meaning shall we attribute to various probabilities of deviations? It can be calculated that, in a group of sibships totaling 30 individuals, a deviation from a 1:1 ratio as great as, or greater than, 18:12 will be found in about 1 out of 4 cases. Obviously, such a deviation will not be regarded as unusual; therefore, the 18:12 ratio may be considered to be a chance deviation from the expected 1:1.

Had a ratio of 22:8 been found, calculation would show that a deviation as great as, or greater than, this would occur in only 1 out of about 100 cases. It would be a mistake to conclude from this that the 22:8 ratio *cannot* be a deviation from a 1:1 ratio. The possibility that even the greatest possible deviation is a result of chance, although small, is not excluded, but it must be regarded as unlikely. A more likely explanation of the 22:8 ratio would be to assume that it represents a less extreme and, therefore, more probable deviation from some other theoretical ratio, e.g., 3:1. An investigator confronted with the 22:8 ratio would, therefore, retrace the hereditary analysis of his material in order to find out whether the facts in the pedigrees are compatible with some other type of interpretation leading to a theoretical ratio other than 1:1 and closer to that actually observed.

Had a ratio of 20:10 been found, calculation would show that this or a greater deviation from 1:1 would have a probability of about 1 in 15. Obviously, no definite statement about its compatibility with the 1:1 expectation can be made. However, by common agreement, a probability higher than 1/20 that an observed ratio is a chance deviation from an expected one is regarded as signifying that chance is a reasonable cause for the deviation, and a probability lower than 1/20 is interpreted as signifying that the deviation is not the obvious result of chance.

Specifically, observations are said to deviate from expectation "nonsignificantly," in a statistical sense, if the probability that any deviation as great or greater will occur is higher than 1/20 (0.05). If a probability lies between 0.05 and 0.01 (1/100), we speak of statistically "doubtful significance"; and if the probability is lower than 0.01, of statistical "significance." Alternatively, we refer to a result as "significant at the 5 per cent (or 1 per cent) level." The term "significance" denotes that some specific reason or reasons, and not chance, probably underlies such a deviation.

The Chi-square Test of Goodness of Fit. Various tests of significance by means of which the "goodness of fit" of observed ratios can be measured have been devised. Only one of them, the χ^2 test (read: chi-square, Greek letter χ; of Karl Pearson (1857–1936), will be described here. In brief, it consists of (1) calculating the absolute difference between observed and expected numbers in each class of a given sample; (2) squaring each difference and dividing the square by the number of expected; and (3) adding the figures obtained in (2). This final sum is called χ^2. It is 0 if there are no differences between observed and expected numbers, and increases with increasing size of the differences. The probability, P, that any deviation of observed from expected ratios is due to chance has been calculated for many values of χ^2 and can be read directly from tables.

The chi-square test applied to a sample consisting of two classes. The procedure will be illustrated for the frequency of woolly and nonwoolly hair in sibships (p. 150). As shown in the upper table on p. 213, the calculation yields a χ^2 value of 0.818.

This value, 0.818, of χ^2 corresponds to a specific probability of obtaining by chance a set of observed results as far from the expected as 145 woolly and 130 nonwoolly, or further. This probability can be obtained by consulting Table 13. The first column of this table is headed "Degrees of Freedom."

	Class 1 (affected)	*Class 2* (nonaffected)	*Total*
Observed	145	130	275
Expected (1:1)	137.5	137.5	275
Absolute difference (observed–expected)	7.5	7.5	

$$\chi^2 = \frac{(\text{obs.}-\text{exp.})^2}{\text{exp.}} + \frac{(\text{obs.}-\text{exp.})^2}{\text{exp.}} = \frac{(7.5)^2}{137.5} + \frac{(7.5)^2}{137.5} = 0.818$$

This expression refers to the fact that if we know only that a given number of observations, say, 275, fall into two or more classes, the number in each class is still undecided ("free"). When, however, the number in Class 1 of a group with only two classes has been determined—in our example, 145—the number in Class 2 is fixed, not "free," since it is the difference between the total (275) and the number in Class 1 (145), that is, 130. The number of degrees of freedom in any group consisting of two classes is therefore 1, because only one number is "free," and the degrees of freedom in a group of more than two classes is usually 1 less than the number of classes. In the woolly-nonwoolly study there was only 1 degree of freedom; so the probability corresponding to the calculated χ^2, 0.818, is to be found in the top row in Table 13. It lies between two values, 0.45 and 1.07. The heading of the table shows that these χ^2 values correspond to the probabilities 0.50 and 0.30, respectively. Thus, the probability of obtaining a χ^2 value of 0.818 or larger lies between these two probabilities. The deviation of observation from expectation is, therefore, statistically nonsignificant, since in from 30 to 50

TABLE 13
Table of Chi-square. (Depending on the row in which it is entered, a listed χ^2 value corresponds to the probability given at the top of obtaining, by chance, results deviating from the expectation by as much as, or more than, the observed results.)

Degrees of Freedom	*Probability,* P .70	.50	.30	.05	.01
1	.15	.45	1.07	3.84	6.63
2	.71	1.39	2.41	5.99	9.21
3	1.42	2.37	3.66	7.81	11.34
4	2.19	3.36	4.88	9.49	13.28
5	3.00	4.35	6.06	11.07	15.09
6	3.83	5.35	7.23	12.59	16.82

Source: Abridged from Table III of Fisher, *Statistical Methods for Research Workers,* Edinburgh: Oliver and Boyd Ltd., and by permission of the author and publishers.

per cent of all samples of 275 individuals it would be pure chance that the observed ratio would deviate from the expected 1:1 ratio as much as 145:130, or even more. Accordingly, the statistical analysis shows agreement with the hypothesis of dominant inheritance, which is the basis of the 1:1 expectation.

For the sake of accuracy the chi-square test applied to two classes needs a correction (Yates' correction) when an expected class is smaller than 50, but the correction is usually unimportant unless *the expected class is 5 or smaller.* The correction consists in subtracting 0.5 from each of the two absolute differences between the observed and expected values.

The chi-square test applied to samples of more than two classes. Genetic data are sometimes subdivided into three or more classes. For example, progeny from marriages between two parents of the genotype $I^A I^B$ belong to three blood groups: A, AB, and B. They are expected in the proportions 1:2:1, and we may want to compare a specific observation with this expectation. The procedure, using the chi-square test, is practically identical with that applied to data which fall into two classes. The only difference is that the calculated value signifies a different probability if it is derived from three instead of two classes.

Three classes permit 2 degrees of freedom for any given number of observations: any of a variety of numbers of these observations may be expected in the first of the classes, and hence it is "free"; even after the number in the first class has been determined, the second class is also "free," since any one of many different numbers of remaining observations may be found in it; but after the numbers in two classes have been determined, the number in the third class is fixed, because it must be the difference between the total and the sum of the other two classes. An example of the chi-square method applied to three classes is presented in Table 14, where $\chi^2 = 0.920$. This value must be related to the probabilities listed in Table 13 under 2 degrees of freedom. It lies between two values listed there, 0.71 and 1.39, which correspond to probabilities of 0.70 and 0.50. There is, therefore, a probability greater than 0.5, but smaller than 0.7, that deviations as large as those observed, or larger, would occur by chance, and the deviations are statistically nonsignificant.

TABLE 14
Computation of χ^2 for Frequencies of the Three Blood Groups A, AB, and B among 151 children from AB × AB Marriages.

Frequencies	A	AB	B	*Total*
Observed	39	70	42	151
Expected (1:2:1)	37.75	75.5	37.75	151

$$\chi^2 = \frac{(1.25)^2}{37.75} + \frac{(5.5)^2}{75.5} + \frac{(4.25)^2}{37.75} = 0.920$$

TABLE 15
Computation of χ^2 for a Comparison of Two Sets of Observations on Offspring of Parents One of Whom is Woollyhaired.

	Parents		*Children*		
	Woolly	Non-woolly	Class 1 Woolly	Class 2 Nonwoolly	*Total*
Observation A.	Father	Mother	24	36	60
Observation B.	Mother	Father	34	26	60
Total			58	62	120
Expectation A.	Father	Mother	29	31	60
Expectation B.	Mother	Father	29	31	60
Total			58	62	120

$$\chi^2 = \frac{(24-29)^2}{29} + \frac{(34-29)^2}{29} + \frac{(36-31)^2}{31} + \frac{(26-31)^2}{31} = 3.34$$

When more than three classes are studied, the calculated χ^2 value must be related to the values listed in the row of Table 13 that is 1 less than the number of classes.

Comparison between two sets of observations. We have earlier noted (p. 150) that offspring of (A) woolly-haired fathers consisted of 24 woolly and 36 nonwoolly children, and of (B) woolly-haired mothers consisted of 34 woolly and 26 nonwoolly children. Do the two ratios, 24:36 and 34:26, deviate from each other by chance only? In order to answer this question, both observations A and B must be compared with a theoretical expectation. This expectation is not based on any a priori genetic ratio; the problem is, rather, whether the observed ratios deviate significantly from those that would be expected if both sets of observations were samples of the same general population of children with one parent with woolly hair, or, expressed in another way, whether the two observations do not differ from one another significantly ("null hypothesis"). The expected numbers are given in Table 15 and were calculated as follows: There was a total of 58 woolly-haired (Class 1) individuals. If both sets of observations, A and B, each of 60 individuals, came from the same general population, the 58 woolly-haired individuals should ideally have been divided into an equal number from both types of parents, namely, 29. Similarly, the expected division of the 62 nonwoolly-haired (Class 2) individuals from both types of parents would have been 31 and 31.

It is now possible to apply the chi-square test in the same manner as in former examples by calculating the squares of the differences between the four observed and the four expected numbers, dividing each square by the expected number and adding the quotients. The value of χ^2 is 3.34. This comparison has only 1 degree of freedom, since determination of the number in

TABLE 16
Computation of χ^2 for a Comparison of Two Sets of Observations on Feeblemindedness in Twins.

Twins	*Class 1 (Both Affected)*	*Class 2 (One Affected)*	*Total*
Observation A. Identical	115	11	126
Observation B. Nonidentical	42	51	93
Total	157	62	219
Expectation A. Identical	90	36	126
Expectation B. Nonidentical	67	26	93
Total	157	62	219

$$\chi^2 = \frac{(115-90)^2}{90} + \frac{(42-67)^2}{67} + \frac{(11-36)^2}{36} + \frac{(51-26)^2}{26} = 57.67$$

one group fixes the numbers in the three other groups. The 29 individuals of Class 1 expected in set A require that there be 29 from Class 1 in set B because 58 individuals of Class 1 have been observed, and the same 29 individuals of Class 1 in set A require that there be 31 individuals of Class 2 in set A, since that set contains 60 individuals. Finally, that there must be 31 individuals of Class 2 in set B follows rigidly from the observed totals and the single "free" expectation.

The probability for 1 degree of freedom which corresponds to $\chi^2 = 3.34$ lies between 0.30 and 0.05 (Table 13), rather close to the latter. The deviation in proportions of woolly to nonwoolly offspring from the two sets of parents could, therefore, be attributed to chance only.

The two sets of observations on the offspring of woolly-haired parents consisted of equal numbers of individuals. More often than not, we must compare two sets in which the number of observations is different. For an example we will compare the frequencies of feeble-mindedness in one or both (A) identical twins and (B) nonidentical twins. As shown in Table 16 for data compiled by various investigators, there were altogether 219 pairs of twins, 126 identical and 93 nonidentical. In order to test whether the difference between the two sets of data is statistically significant, the expected numbers are calculated by dividing the number of doubly affected pairs, 157, into two numbers that have the same proportion as the two whole samples A and B, namely, 126:93. This yields expectations of 90 and 67 for Class 1 in A and B, respectively, and by simple subtraction, 36 and 26 for Class 2. The calculated χ^2 value is very large, corresponding to an extremely low probability that observations would be so different from expectations if the two kinds of twins were basically alike in regard to feeble-mindedness. The two sets of observations are thus said to differ at a highly significant probability level.

Comparison between any number of sets of observations each of which may contain any number of classes can also be made. The procedure follows from the foregoing discussion. The degrees of freedom to which the obtained value of χ^2 must be related in Table 13 may usually be found by multiplying (number of sets of observations minus one) × (number of classes minus one): e.g., number of sets of observations is 3, number of classes is 4; therefore, number of degrees of freedom $= (3 - 1) \times (4 - 1) = 6$.

This chapter has dealt with some intricate considerations of ratios in small families, ratios in pooled data, and statistical tests of the variability inherent in many phenomena which are influenced by chance. For a more thorough treatment of these topics, the reader should study articles and books especially devoted to them, such as those listed under References.

As we look back over these discussions, we may well be impressed by the ingenuity of the human mind, which has been able to overcome some of the difficulties in understanding caused by the smallness of human families and the working of chance.

PROBLEMS

51. A man is brachydactylous, his wife normal. If they have ten children, what is the probability of their having:
    (a) The first, third, fifth, seventh and ninth child brachydactylous, the others normal?
    (b) Only the first five brachydactylous?
    (c) Five brachydactylous?
    (d) All brachydactylous?

52. Two taster parents have a nontaster child.
    (a) What is the probability of the second child being a nontaster?
    (b) What is the probability of a sibship of four children in which the first born is a nontaster, the second born a taster, the third born a nontaster, and the fourth born a taster?
    (c) What proportion of families with four children from heterozygous taster parents will have two tasters and two nontasters?

53. Owing to a recessive lethal condition, a couple's first child is stillborn. After this stillbirth, what is the probability of the couple having:
    (a) Five conceptions, all resulting in homozygous normal genotypes?
    (b) Five conceptions, all resulting in viable children?
    (c) Five conceptions, two of which result in homozygous normal and three in heterozygous children?

54. A man is brachydactylous, his wife is normal. Both are tasters, but both of their mothers were nontasters.

(a) Give the genotypes of the couple. If they have 8 children, what is the probability:
(b) Of no brachydactylous child among them?
(c) Of all being brachydactylous?
(d) Of all being nontasters?
(e) Of all being tasters?
(f) Of four being brachydactylous?
(g) Of four being tasters?
(h) Of the first, third, fifth, seventh being brachydactylous and nontasters, and the others having normal fingers and being tasters?

55. Assume that the probability of the birth of a boy or a girl is 1/2. What are the probabilities of the following proportions of births in a hospital:
(a) A total of 10 boys to 10 girls?
(b) A total of 15 of one sex to 5 of the other sex?
(c) At least 2 girls in a total of 30?

In Problems 56-58, assume that no crossing over occurs.

56. What is the probability of a child inheriting only one autosome from its paternal grandmother?

57. What is the probability of a man inheriting 2 of his paternal and all of his maternal autosomes from his grandmothers?

58. What is the probability of a child inheriting 11 of his 22 paternal autosomes from his grandmother?

59. Assume that uncorrected, pooled data of sibships with at least one albino child from normal parents gave a ratio of normal to albino of 1.8:1 in the preceding century, but of only 0.8:1 in the present one. Assume that the data are reliable and that the difference between the two ratios is statistically significant. What cause can be suggested for the occurrence of these ratios and their difference?

60. What proportion of sibships with two children will have no boys?

61. What proportion of boys from two-children sibships will have a brother? A sister?

62. What sex ratio do you expect from a random sample of 27 two-children sibships each of which contains at least one boy?

63. In a certain population, it is found that from normal parents there are 50 sibships containing at least one albino. Of these sibships, 6 consist of one child only, 7 of two children, and 37 of three children.
(a) How many of each type of sibship would be expected to have one, two, or three albinos?
(b) What ratio of normal to affected would be expected from pooling all data?

64. Pooling of sibships from several marriages of woolly-haired $\times$ normal parents yields 37 woolly-haired and 53 normal children. Determine χ^2 and P. Do you regard these data as conforming to expectation?

65. In a group of children, there are 15 girls and 25 boys.
(a) Is this a significant deviation from an expected 1:1 ratio?

(b) Had there been ten times as many children in the same proportions, what would your answer have been?

66. If you expect a 3:1 ratio in the offspring of parents both known to be heterozygous, but observe 14:1 in a sibship, would you regard this deviation as significant?

67. The frequency of tasters in a general population is 70 per cent. In a specific group of 150 people, 135 were tasters. Is the deviation significant?

68. Of the 15 grandchildren of a certain individual, 4 are college graduates. In the general population the proportion of college graduates in the socioeconomic class of this family is only 10 per cent. Does the family under discussion represent a significant deviation from general expectation?

69. The frequency of illegitimate children in the white population of Mississippi in 1957 was 12.8 per cent. If, in a certain town, 25 per cent of 20 children born were illegitimate, would you regard this as a significant increase?

70. Among 240 families with 3 children each, the children in 50 of them were either all girls or all boys. Compare this observation with the expectation, based on a sex ratio of 1:1, and judge the significance of the deviation.

71. In two samples each of 1,000 African Pygmies, two investigators found that 270 in one sample and 300 in the other belonged to the blood group O. Are these results consistent with each other?

72. Among 100 Chinese from Canton, 6 were AB; among 100 Chinese from Szechwan, only 3 were AB. Do these data indicate a significant difference?

73. In classes at the University of California in Berkeley, there were 178 whites who could taste PTC and 68 who could not. There were also 36 oriental tasters and 4 nontasters. What is the statistical significance of the difference in taster frequency between the two groups?

REFERENCES

Bailey, N. T. J., 1951. A classification of methods of ascertainment and analysis in estimating the frequencies of recessives in man. *Ann. Eugen.*, **16**:223–225.

Brownlee, K. A., 1960. *Statistical Theory and Methodology in Science and Engineering*. 570 pp. Wiley, New York.

Burdette, W. J. (Ed.), 1962. *Methodology in Human Genetics*. 436 pp. Holden-Day, San Francisco.

Crow, J. F., 1965. Problems of ascertainment in the analysis of family data. Pp. 23–44 *in* Neel, J. V., Shaw, Margery W., and Schull, W. J. (Eds.), *Genetics and the Epidemiology of Chronic Diseases*. Government Printing Office, Washington, D.C. (U.S. Public Health Service Publ. No. 1163.)

Elandt-Johnson, Regina C., 1971. *Probability Models and Statistical Methods in Genetics*. 592 pp. Wiley, New York.

Fisher, R. A., 1934. The effect of methods of ascertainment upon the estimation of frequencies. *Ann Eugen.,* **6:**13–25.

Gnedenko, B. V., and Khinchin, A. Ya, 1961. *An Elementary Introduction to the Theory of Probability.* 139 pp. W. H. Freeman and Company, San Francisco.

Hodges, J. L., Jr., and Lehmann, E. L., 1964. *Basic Concepts of Probability and Statistics.* 375 pp. Holden-Day, San Francisco.

Kempthorne, O., 1954. *Statistics and Mathematics in Biology.* 632 pp. Iowa State College Press, Ames.

Li, C. C., 1961. *Human Genetics.* 218 pp. McGraw-Hill, New York.

Li, C. C., and Mantel, N., 1968. A simple method of estimation of segregation ratio under complete ascertainment. *Amer. J. Hum. Genet.,* **20:**61–81.

Morton, N. E., 1958. Segregation analysis in human genetics. *Science,* **127:** 79–80.

Morton, N. E., 1962. Segregation and linkage. Pp. 17–52 *in* Burdette, W. J. (Ed.), *Methodology in Human Genetics.* Holden-Day, San Francisco.

Mosiman, J. E., 1968. *Elementary Probability for the Biological Sciences.* 255 pp. Appleton-Century-Crofts, New York.

Munro, T. A., 1947. Phenylketonuria: data on forty-seven British families. *Ann. Eugen.,* **14:**60–88.

Snedecor, G. W., 1956. *Statistical Methods,* 5th ed. 534 pp. Iowa State College Press, Ames.

Weinberg, W., 1912. Weitere Beiträge zur Theorie der Vererbung. 4. Uber Methode und Fehlerquellen der Untersuchung auf Mendelsche Zahlen beim Menschen. *Arch. Rass. u. Ges. Biol.,* **9:**165–174.

Woolf, C. M., 1968. *Principles of Biometry.* 359 pp. Van Nostrand, Princeton.

11

THE HARDY-WEINBERG LAW

It is obvious that individuals owe their genotypes to their parents. It is somewhat less obvious, but equally true, that the kinds of genotypes and their frequencies in any one generation of a particular group of persons—we call such a group a *population*—depend on the kinds of genotypes and their frequencies that are found in all parents in the preceding generation of the population. If, among these parents, the three genotypes AA, AA', and $A'A'$ occur, the frequency of any one genotype will depend on the frequency of the two alleles. Thus, for instance, if most of the alleles are A, and only a few A', we would expect more persons to be AA than AA' or $A'A'$. In the present chapter, we shall consider such questions as these: What are the proportions of different genotypes in a population? How will the proportions existing in one generation be related to those in the next generation?

Selective Forces. The answers to these questions will depend on a number of variables. Among these, selective agents may play a role. If a certain allele should lead to a serious defect in persons endowed with it, such persons may be more likely to die before they become parents than persons without it; or their chances of marrying may be decreased; or, if they marry, the number of children they produce may be below the average. As a result, there will be fewer persons in the next generation who have the allele with the unfavorable effect than there would have been had it not had the adverse effect.

An allele with a low survival property does not necessarily produce a phenotype which is defective in other respects. Thus, genetically caused sterility of an individual may be present side-by-side with outstanding physical or mental attributes.

Mutation. Mutations also may affect the genetic composition of the population. If heritable changes from A to A' occur in the germ cells of the parent generation, then the frequency of specific genotypes in the next generation will depend not only on their frequency in the preceding one but also on the mutation rate $A \rightarrow A'$. If mutations should occur not only in one direction but also in reverse ($A \leftarrow A'$), both rates would have to be taken into account in predicting the genotypes and their proportions in the next generation.

Systems of Mating. Still another factor, sometimes referred to as the system of mating, is significant. Given individuals of three genotypes and assuming that all three types of persons are equally likely to marry and have offspring, those of any one genotype may select their spouses without showing preference for any one of the three genotypes, or they may prefer spouses of a particular genotype. The former type of mating is called *random mating,* or *panmixis*; the latter *nonrandom,* or *assortative, mating.*

Assortative mating is not uncommon. One of the best-known examples refers to body size: tall people prefer to marry tall ones; short people, short ones. Since body size is dependent on specific genetic constitutions, like genotypes occur more often among couples than would be expected by chance. If we assume that assortative mating by body size is absolute, so that all marriages are between genotypically identical spouses, and if (in deliberate oversimplification) we write the genotypes of short, medium, and tall people as SS, Ss, and ss (Fig. 113), it is obvious that the marriages $SS \times SS$ and $ss \times ss$ will give rise to SS and ss homozygotes only, while $Ss \times Ss$ will lead to SS, Ss, and ss children in the ratio 1/4:2/4:1/4. The total number of heterozygotes (Ss) in the children's generation will then be equal to one-half of the children issuing from marriages between heterozygotes, and, if the sizes of the parental and filial generations are the same, there will be only one-half as many heterozygous children as parents in the population. Any system of mating other than this absolute assortative mating will lead to a higher proportion of heterozygotes, since, in addition to the $Ss \times Ss$ marriages, marriages between unlike spouses will also produce heterozygotes. The mating $SS \times ss$ will produce heterozygotes only, and in the matings $SS \times Ss$ and $ss \times Ss$ half of the

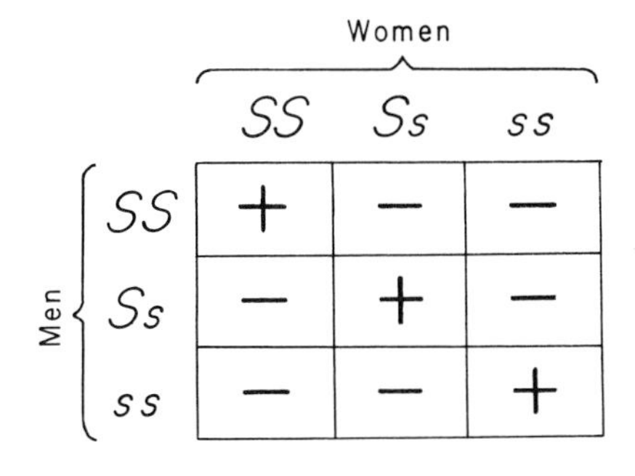

FIGURE 113
Possible combinations of short, medium-sized, and tall spouses. Plus sign indicates marriages between identical genotypes, minus sign between nonidentical.

offspring will be heterozygous. Thus, a union between an *Ss* and either of the two other genotypes will produce the same number of heterozygotes as a union between two *Ss,* and unions between *SS* and *ss* will lead to still more heterozygotes. Assortative mating, on the contrary, leads to fewer heterozygotes in each successive generation, since only one-half of the children from *Ss* × *Ss* unions are *Ss,* like their parents.

Assortative mating as measured by correlation coefficients, r, of traits in husbands and wives has been shown to exist for traits within the range of normality such as stature (r = 0.28), eye color (0.26), and performance in intelligence tests (0.55). Assortative mating is also known for abnormal traits such as deaf-mutism. The assortative mating demonstrated is never absolute and consists of a limited preference for similar partners. Of course, even in random mating, marriages of persons with the same specific traits are bound to occur, but the frequency of such marriages is no higher than expected by chance.

Besides preference for similar marriage partners (positive assortative mating, or *homogamy*), there may also be preference for opposite types (negative assortative mating) among persons with certain traits. Possibly this is true of red-headed persons, since marriages between two of them seem to be rarer than would be expected from random mating—but no careful studies have been conducted to verify this supposition.

In spite of the fact that people are, in general, highly aware of conscious choice in the selection of their marriage partners, such choice does not extend to numerous hereditary conditions. This is true when the differences between the phenotypic expression of the different genotypes either remain unknown or are of no importance to the people concerned. Examples of such differences are the blood groups, or the common situations in which one genotype, *A'A'*, is extremely rare, and the other two, *AA* and *AA'*, because of dominance, are indistinguishable, so that preferential mating in the *AA*-*AA'* population is not even possible.

Random mating is, therefore, of very wide occurrence. In this chapter its consequences for the composition of populations will be discussed, using for our example the simplest situation, namely, one in which selection for survival or reproductive ability plays no role, and in which mutations are absent. We shall see that, in a randomly mating population of this type, individuals of the three autosomal genotypes *AA, AA'*, and *A'A'* occur in predictable and constant proportions. This was first shown in 1908 by both the mathematician Hardy and the physician Weinberg, working independently of each other. The proportions may be expressed in terms of the frequencies – not of the three genotypes – but of the two alleles *A* and *A'* in a given population. If the frequency of *A* is designated as a fraction p and that of *A'* as a fraction q, where the two fractions add up to totality, $p + q = 1$, then the proportions of the three genotypes *AA, AA'*, and *A'A'* are $p^2:2pq:q^2$.

As long as the allele frequencies p and q remain unchanged, a random-mating population in which the three genotypes occur in the proportions $p^2:2pq:q^2$ is said to be *in equilibrium,* since even if a temporary intervention of assortative mating or some other disturbance causes a different proportion of the three genotypes, *AA, AA'*, and *A'A'*, to be produced, future generations of random mating will reestablish this proportion. The formula $p^2AA + 2pqAA' + q^2A'A'$ and its implications are known as the Hardy-Weinberg Law. We shall now proceed to prove this law, restricting the treatment to autosomal genes.

THE RESULTS OF RANDOM MATING

Mixture of Equal Numbers of AA and A'A'. Let us assume that a certain population consists of *AA* individuals only, and that another population is made up exclusively of *A'A'* persons. Let these two populations be of equal size, and let them mix and form a new population in which random mating occurs. What will be the genotypes and their proportions among the children of this panmictic group?

The children will come from four different types of marriages:

AA♀ × *AA*♂ *A'A'*♀ × *AA*♂
AA♀ × *A'A'*♂ *A'A'*♀ × *A'A'*♂

Since the two genotypes are assumed to be equally common, it follows that, with random mating, the four kinds of marriages are also equally common.

It does not make any difference in final results whether, in marriages of unlike genotypes, the female or the male has one or the other genetic constitution. We may, therefore, simplify our enumeration, letting a fraction indicate the proportion of all marriages which each specific type represents:

1/4 *AA* × *AA*
1/2 *AA* × *A'A'*
1/4 *A'A'* × *A'A'*

TABLE 17
Types and Frequencies of Marriages in a Random-mating Population Consisting of Equal Numbers of Two Homozygous Allelic Genotypes.

		Women	
		$\frac{1}{2}AA$	$\frac{1}{2}A'A'$
Men	$\frac{1}{2}AA$	$\frac{1}{4}AA \times AA$	$\frac{1}{4}A'A' \times AA$
	$\frac{1}{2}A'A'$	$\frac{1}{4}AA \times A'A'$	$\frac{1}{4}A'A' \times A'A'$

It is important to understand why marriages of unlike types are twice as common as marriages of either one of the two like types. One way of making this clear is provided by our enumeration of all four types of marriages, which shows two between unlike types and only one between each kind of like types. Another approach is: For each genotype, there is only one way in which both spouses are alike, but there are two ways in which they are unlike – the first spouse may be of type 1 and the second of type 2, or the first of type 2 and the second of type 1. Still another method of showing the relationship makes use of the checkerboard scheme as given in Table 17.

Having established the types of marriages and their relative frequencies, we can now find the genotypes of children and their proportions in the population. We assume, of course, that the average number of children from each marriage type is the same. Marriages $AA \times AA$, one-quarter of all marriages, yield AA children only; consequently, one-quarter of all children are of this type. Similarly, $A'A' \times A'A'$ marriages yield another quarter, all $A'A'$ children. The $AA \times A'A'$ marriages produce AA' offspring, and since these marriages constitute one-half of all marriages, one-half of all children are theirs. Adding all progeny, we obtain:

Marriages		*Offspring*		
Type	Frequency			
$AA \times AA$	$\frac{1}{2} \cdot \frac{1}{2}$	$\frac{1}{4}AA$		
$AA \times A'A'$	$2 \cdot \frac{1}{2} \cdot \frac{1}{2}$		$\frac{1}{2}AA'$	
$A'A' \times A'A'$	$\frac{1}{2} \cdot \frac{1}{2}$			$\frac{1}{4}A'A'$
Sum of all marriages	1	$\frac{1}{4}AA + \frac{1}{2}AA' + \frac{1}{4}A'A'$		

Our conclusion is that random mating leads to all three possible genotypes in the proportions 1:2:1.

What proportions of these genotypes will be found in the next generation, after further random mating? We may investigate this problem by enumerating again all types of marriages, their relative frequencies, and their offspring:

Marriages		*Offspring*
Type	Frequency	
$AA \times AA$	$\frac{1}{4} \cdot \frac{1}{4}$	$\frac{1}{16}AA$
$AA \times AA'$	$2 \cdot \frac{1}{4} \cdot \frac{1}{2}$	$\frac{1}{8}AA + \frac{1}{8}AA'$
$AA \times A'A'$	$2 \cdot \frac{1}{4} \cdot \frac{1}{4}$	$\frac{1}{8}AA'$
$AA' \times AA'$	$\frac{1}{2} \cdot \frac{1}{2}$	$\frac{1}{16}AA + \frac{1}{8}AA' + \frac{1}{16}A'A'$
$AA' \times A'A'$	$2 \cdot \frac{1}{2} \cdot \frac{1}{4}$	$\frac{1}{8}AA' + \frac{1}{8}A'A'$
$A'A' \times A'A'$	$\frac{1}{4} \cdot \frac{1}{4}$	$\frac{1}{16}A'A'$
Sum of all marriages	1	$\frac{1}{4}AA + \frac{1}{2}AA' + \frac{1}{4}A'A'$

The types and proportions of children in the second generation of random mating have remained unchanged! Consequently, these proportions of genotypes will remain the same in all successive generations, provided that no changes in the system of mating occur. (Here, of course, we disregard the fact that the frequency of each type of marriage may vary slightly from the expected figure and that the average number of children from each may not be exactly the same.)

Mixture of AA and A′A′ in Ratio 9:1. We have followed a population consisting originally of equal numbers of AA and $A'A'$ individuals. It may be thought that the final result, the consistent 1:2:1 proportions of the three genotypes, AA, AA', and $A'A'$, is due to this special "Mendelian" setup. This, however, is not true. Although the actual proportions of the three genotypes will depend on the proportions of types present at the beginning, the constancy in the composition of later generations is determined by a law that has general validity. Before deriving this general law, we shall consider one more special case. Let us assume that the original mixed population consists of 90 per cent (9/10) AA and 10 per cent (1/10) $A'A'$ persons. Under these circumstances the frequency of each type of marriage and the resulting children will be:

Marriages		*Offspring*
Type	Frequency	
$AA \times AA$	$\frac{9}{10} \cdot \frac{9}{10}$	$\frac{81}{100}AA$
$AA \times A'A'$	$2 \cdot \frac{9}{10} \cdot \frac{1}{10}$	$\frac{18}{100}AA'$
$A'A' \times A'A'$	$\frac{1}{10} \cdot \frac{1}{10}$	$\frac{1}{100}A'A'$
Sum of all marriages	1	$\frac{81}{100}AA + \frac{18}{100}AA' + \frac{1}{100}A'A'$

All three genotypes occur among the children, but not, this time, in any well-known proportions.

In the next generation, the following results are to be expected:

Marriages		*Offspring*
Type	Frequency	
$AA \times AA$	$^{81}/_{100} \cdot {}^{81}/_{100}$	$^{6561}/_{10000}\,AA$
$AA \times AA'$	$2 \cdot {}^{81}/_{100} \cdot {}^{18}/_{100}$	$^{1458}/_{10000}\,AA + {}^{1458}/_{10000}\,AA'$
$AA \times A'A'$	$2 \cdot {}^{81}/_{100} \cdot {}^{1}/_{100}$	$^{162}/_{10000}\,AA'$
$AA' \times AA'$	$^{18}/_{100} \cdot {}^{18}/_{100}$	$^{81}/_{10000}\,AA + {}^{162}/_{10000}\,AA' + {}^{81}/_{10000}\,A'A'$
$AA' \times A'A'$	$2 \cdot {}^{18}/_{100} \cdot {}^{1}/_{100}$	$^{18}/_{10000}\,AA' + {}^{18}/_{10000}\,A'A'$
$A'A' \times A'A'$	$^{1}/_{100} \cdot {}^{1}/_{100}$	$^{1}/_{10000}\,A'A'$
Sum of all marriages	1	$^{8100}/_{10000}\,AA + {}^{1800}/_{10000}\,AA' + {}^{100}/_{10000}\,A'A'$

We see that the proportions of the three genotypes have remained unchanged from one generation to the next. They are still 81 AA:18 AA':1 $A'A'$.

Mixture of AA and A′A′ in Ratio p:q. To establish the general law, we designate the proportion of AA individuals in the original population by p, and the proportion of $A'A'$ individuals by q. It is obvious that $p + q = 1$, since the two kinds of people together make up the whole group. We then obtain:

Marriages		*Offspring*		
Type	Frequency			
$AA \times AA$	p^2	p^2AA		
$AA \times A'A'$	$2pq$		$2pqAA'$	
$A'A' \times A'A'$	q^2			$q^2A'A'$
Sum of all marriages: $p^2 + 2pq + q^2 = (p + q)^2 = 1^2 = 1$		Sum of all offspring: $p^2AA + 2pqAA' + q^2A'A'$		

Thus, the first generation produced by random marriage of the two original populations contains the three genotypes AA, AA', and $A'A'$ in the proportions $p^2:2pq:q^2$.

Random mating among the individuals of the offspring generation leads to the marriages and their offspring listed in the table on the next page. By addition, the proportion of all AA offspring is found to be $p^4 + 2p^3q + p^2q^2$, which is equal to $p^2(p^2 + 2pq + q^2) = p^2(p + q)^2$. Since $p + q = 1$, this term reduces to p^2. Similarly, the proportion of AA' offspring is $2p^3q + 4p^2q^2 + 2pq^3 =$

Marriages		*Offspring*
Type	Frequency	
$AA \times AA$	$p^2 \cdot p^2$	p^4AA
$AA \times AA'$	$2 \cdot p^2 \cdot 2pq$	$2p^3qAA + 2p^3qAA'$
$AA \times A'A'$	$2 \cdot p^2 \cdot q^2$	$2p^2q^2AA'$
$AA' \times AA'$	$2pq \cdot 2pq$	$p^2q^2AA + 2p^2q^2AA' + p^2q^2A'A'$
$AA' \times A'A'$	$2 \cdot 2pq \cdot q^2$	$2pq^3AA' + 2pq^3A'A'$
$A'A' \times A'A'$	$q^2 \cdot q^2$	$q^4A'A'$

$2pq(p^2 + 2pq + q^2) = 2pq$, and the proportion of $A'A'$ offspring is $p^2q^2 + 2pq^3 + q^4 = q^2(p^2 + 2pq + q^2) = q^2$. Thus, the proportions of the three genotypes have remained the same as in the preceding generation, i.e., $p^2{:}2pq{:}q^2$.

Allele Frequencies and the Hardy-Weinberg Law. Having arrived at this conclusion by a consideration of the kinds and proportions of various marriages, we shall now investigate the same problem—that of the constitution of a random-mating population—by a less familiar but really much simpler method. The essential biological entities in the origin of a generation are not the individual parental pairs, but the gametes supplied by them—the eggs and sperm. Random mating primarily signifies random union of different types of germ cells in the process of fertilization. We can, therefore, determine the composition of a population simply by enumerating the kinds and relative frequencies of the gametes produced by the preceding generation and by finding out how frequently the different combinations of genotypes are formed in fertilization. In a population made up of individuals with A and A' in any possible combination, A and A' gametes will be produced. The relative frequencies of these gametes will be the same as the relative frequencies of the alleles A and A'. The allele frequencies in a population may lie anywhere between two extremes: in the first, $A = 1$ and $A' = 0$ (i.e., all individuals are AA); in the other, $A = 0$ and $A' = 1$ (i.e., all individuals are $A'A'$). From information already given for the three examples of mixed populations considered in the preceding pages, we may note the frequencies within them of the alleles:

1. There are equal numbers of AA and $A'A'$ individuals in the mixed population before random mating, representing equal frequencies of A and A' alleles. Stated differently, the AA individuals in the mixed population make up one-half of all individuals and the $A'A'$ individuals make up the other half. Correspondingly, the A alleles make up one-half of all alleles

and the A' alleles the other half. Expressed in decimal fractions, the frequency of A is 0.5, and A' is likewise 0.5.

2. There are 90 per cent AA and 10 per cent $A'A'$ in the mixed population before random mating. Thus, the proportion of A to A' alleles is 9:1, i.e., the frequency of A is 0.9; of A', 0.1.
3. There are p AA individuals in the mixed population before random mating and q $A'A'$ individuals. Therefore, the frequencies of A and A' alleles are p and q, respectively.

   (Note that in all three examples the *number* of alleles is twice that of individuals since each individual carries 2 alleles. In stating the relative *frequencies* of the alleles the factor 2 cancels out. Thus, if 80 AA individuals are mixed with 20 $A'A'$ giving frequencies of individuals of 0.8 and 0.2 respectively, the numbers of A and A' alleles are 160 and 40 but their relative frequencies are 0.8 and 0.2.).

We see that p and q characterize a population much more fundamentally than do the original proportions of homozygotes in a mixture of two groups of individuals before random mating (Fig. 114). Rather, p and q represent the relative frequencies of the two alleles present in the mixed population, regardless of how these alleles are distributed among the individuals in the population. The allele frequency of A or A' is 0.5, not only in a population in which half of all people are AA and the other half $A'A'$, but also in one in which, for example, three-sevenths are AA, one-seventh AA', and three-sevenths $A'A'$; in one in which one-eighth is AA, six-eighths AA', and one-eighth $A'A'$; or in one in which every individual is AA' (Fig. 115, generation 0). Each of these different populations, after one generation of random mating, will have the *same* composition: AA individuals will originate by fertilization of A eggs by A sperm in the frequency $1/2 \times 1/2 = 1/4$; AA' individuals, by fertilization of A eggs by A' sperm and of A' eggs by A sperm in the frequency $2 \times 1/2 \times 1/2 = 1/2$; and $A'A'$ individuals, by fertilization of A' eggs by A' sperm in the fre-

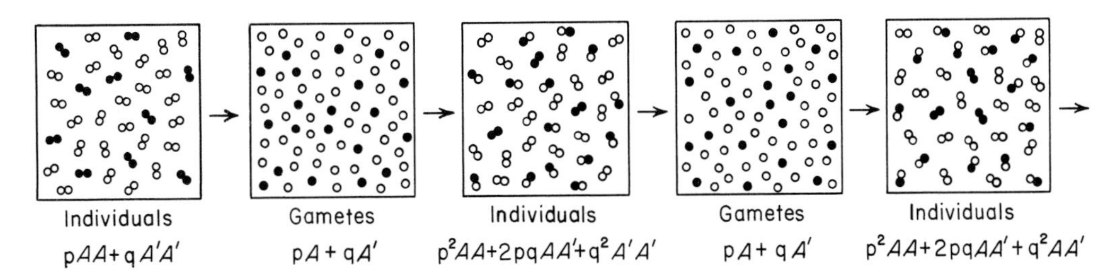

FIGURE 114
The Hardy-Weinberg equilibrium in successive generations of individuals and gametes of a population with allele frequenceis p for A and q for A'.

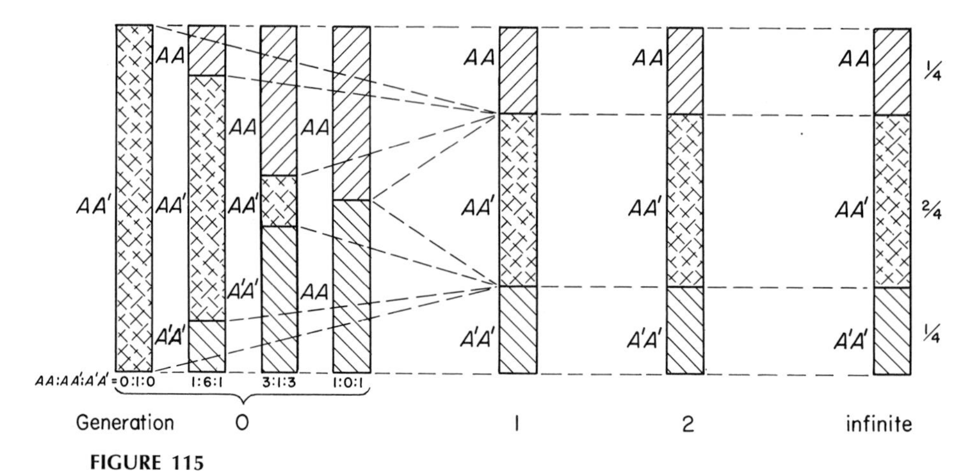

FIGURE 115
Examples of four different populations in generation 0 whose allele frequencies p_A and $q_{A'}$ are 1/2 each. Attainment of the Hardy-Weinberg proportions in one generation of panmixis.

quency $1/2 \times 1/2 = 1/4$ (Fig. 115, generation 1). An infinite number of populations containing various proportions of the three genotypes will fit the general allele frequencies p and q. Again, each of these different populations, after one generation of random mating, will have the same composition: *AA* individuals in the frequency p^2, *AA'* individuals in the frequency 2pq, and *A'A'* individuals in the frequency q^2.

It is now obvious, without further deductions, that the next and all future generations will repeat the $p^2 + 2pq + q^2$ composition of the first randomly produced generation. Since both preferential survival of one or the other allele and mutation were excluded in this discussion of random mating, the relative frequencies of the alleles will remain the same, namely, p and q. In any generation, therefore, p*A* and q*A'* eggs would be met by p*A* and q*A'* sperm, $(pA + qA') \times (pA + qA') = (pA + qA')^2$, and the composition of the resulting population would be $p^2AA + 2pqAA' + q^2A'A'$. The constancy, for later generations, of the specific proportions 1/4:2/4:1/4(1:2:1) is shown on the right side of Figure 115.

HUMAN POPULATIONS AND THE HARDY-WEINBERG LAW

The MN Blood Groups. The red blood cells of an individual carry on their surfaces a variety of chemical substances called *antigens*. Different individuals may possess either the same or different antigens. The next chapter will give

TABLE 18
Heredity of MN Blood Groups in 286 Families.

Parents		*No. of Families*	*Offspring**		
Phenotypes	Genotypes		M	MN	N
M × M	$L^{M}L^{M} \times L^{M}L^{M}$	24	98	–	–
N × N	$L^{N}L^{N} \times L^{N}L^{N}$	6	–	–	27
M × N	$L^{M}L^{M} \times L^{N}L^{N}$	30	–	43	–
M × MN	$L^{M}L^{M} \times L^{M}L^{N}$	86	183	196	–
N × MN	$L^{N}L^{N} \times L^{M}L^{N}$	71	–	156	167
MN × MN	$L^{M}L^{N} \times L^{M}L^{N}$	69	71	141	63

Source: After Wiener.
*Two exceptions, probably caused by illegitimacy, have been omitted.

a more detailed description of the different antigens and of the methods of recognizing them. Here, for an application of the Hardy-Weinberg Law to human populations, we shall select two specific antigens, M and N, and simply state that there are three kinds of persons: those who have M, others who have N, and still others who have both M and N.

Genetics of MN Blood Groups. Family data show that these blood groups are determined genetically. Inheritance is by the transmission of a single pair of alleles that will here be labeled L^{M} and L^{N}, though the letters M and N are often used not only for the antigens of the blood cells, but also for the alleles responsible for the antigens. Table 18 shows the results of the six possible kinds of marriages. If both parents are type M, all children are M, and type N parents have N children only. Marriages of M and N individuals lead to MN children exclusively. M and MN spouses have one-half N and one-half MN offspring; and, finally, MN and MN parents have children of all blood types in the proportions 1/4M:1/2MN:1/4N. Clearly, the phenotypes M and N are those of the homozygotes $L^{M}L^{M}$ and $L^{N}L^{N}$, respectively, and the heterozygote $L^{M}L^{N}$ has the codominant phenotype MN.

Proportions of MN Blood Groups. If the principle of random mating, as expressed in the Hardy-Weinberg Law, is valid for M and N blood groups, the proportions of M, MN, and N individuals in a population should conform to the ratio p^2:2pq:q^2. This expectation may be tested by comparison with the proportions of the three blood groups actually observed. These proportions in six different populations are listed in Table 19. It is seen that the frequencies vary from one population to another, and hence each population provides independent material with which to test whether expectation and observation agree.

TABLE 19
Frequencies of MN Blood Groups in Different Populations.

No. of Individuals	*Population*		*Percentages of Blood Groups*			*Allele Frequencies*	
			M	MN	N	L^M(p)	L^N(q)
6,129	Whites	obs.	29.16	49.58	21.26	0.540	0.460
	(U.S.)	exp.	29.16	49.68	21.16	–	–
278	Blacks	obs.	28.42	49.64	21.94	0.532	0.468
	(U.S.)	exp.	28.35	49.89	21.86	–	–
205	Indians	obs.	60.00	35.12	4.88	0.776	0.224
	(U.S.)	exp.	60.15	34.81	5.04	–	–
569	Eskimos (E.	obs.	83.48	15.64	0.88	0.913	0.087
	Greenland)	exp.	83.35	15.89	0.76	–	–
504	Ainus	obs.	17.86	50.20	31.94	0.430	0.570
		exp.	18.45	49.01	32.34	–	–
730	Australian	obs.	3.00	29.6	67.4	0.178	0.882
	aborigines	exp.	3.17	29.26	67.57	–	–

Source: Wiener.

The Frequencies of the Alleles L^M and L^M. In order to test this, it is necessary to find the frequencies of the alleles L^M (p) and L^N (q). They are easily derived from the frequencies of M, MN, and N individuals. Each M individual carries two L^M alleles; each N individual, two L^N alleles; and each MN individual, one L^M and one L^N allele. The total number of L^M alleles is, therefore, twice the number of M individuals plus the number of MN individuals. Correspondingly, the number of L^N alleles is twice the number of N individuals plus the number of MN's.

The total number of alleles in a population is twice the number of individuals. It then follows that the frequency p of the allele L^M, expressed as a fraction of all alleles, is

$$p = \frac{2\overline{M} + \overline{MN}}{2(\overline{M} + \overline{MN} + \overline{N})},$$

where $\overline{M}$, $\overline{MN}$, and $\overline{N}$ signify the relative frequencies of the three blood groups observed in the population. Expressed in percentages, as in Table 19,

$$\overline{M} + \overline{MN} + \overline{N} = 100,$$

so that

$$p = \frac{\overline{M} + 1/2\ \overline{MN}}{100}. \tag{1}$$

Similarly, the frequency of allele L^N is

$$q = \frac{\overline{N} + 1/2\ \overline{MN}}{100}. \qquad (2)$$

This frequency can also be derived from the formula $p + q = 1$, which yields $q = 1 - p$ (i.e., substracting the frequency of L^M from 1).

By applying formulas (1) and (2), the allele frequencies p and q have been found for each of the six populations listed in Table 19. Among the whites, for example, the frequency p of allele L^M is

$$\frac{29.16 + (\frac{1}{2} \times 49.58)}{100} = 0.54,$$

and the frequency q of allele L^N is $(1 - p) = 0.46$. These allele frequencies are given in the final column of the table.

The MN Blood Groups and the Hardy-Weinberg Law. The frequencies of the three blood groups M, MN, and N that are expected according to the Hardy-Weinberg Law are calculated from the values found for p and q. Thus, among the whites, $p^2 = (0.54)^2 = 0.2916$, $2pq = 2 \times 0.54 \times 0.46 = 0.4968$, and $q^2 = (0.46)^2 = 0.2116$. This expected distribution of the three blood groups agrees well with that observed, and similarly close agreement is shown for the other populations (Table 19). Though the gene frequencies p and q vary in different populations, the ratios $p^2{:}2pq{:}q^2$ are closely approximated within each population. This indicates that the MN groups in the populations are derived from random mating in the preceding generation.

The significance of this finding will be appreciated more clearly if we consider an artificially constructed population. Let us assume that an investigator has found a population of 1,000 individuals that consists of about 56 per cent M, 33 per cent MN, and 11 per cent N persons. In this population, the frequency of the allele L^M (p), is $0.56 + (1/2)(0.33) = 0.725$, and that of L^M (q) is 0.275. Under random mating, the allele frequencies should be reflected in the proportions of phenotypes of M (p^2), which is 52.6 per cent; of MN (2pq), which is 39.8 per cent; and of N (q^2), which is 7.6 per cent. In this population, the expectations do not conform closely to the observed frequencies (Table 20), and a chi-square test would indicate an extremely small probability of observing a deviation as large or larger than that found if it were solely due to chance. It may therefore be considered that the allele-frequency test has established that the population did not arise from random mating in the preceding generation. Indeed, the "artificial" population was obtained by combining data on equal numbers of American whites and Eskimos from East Greenland. Within each of these groups, the Hardy-Weinberg Law holds true, as shown in Table

TABLE 20
"Observed" and Expected Frequencies of the MN Blood Groups in an "Artificial" Population of 1,000 Individuals.

	Percentages of Blood Groups			*Allele Frequencies*	
	M	MN	N	L^M(p)	L^N(q)
obs.	56	33	11	0.725	0.275
exp.	52.6	39.8	7.6	–	–

19, but the artificial mixture is not a result of random mating and does not obey theory. It can be predicted, however, that after a single generation of random mating the offspring of such a mixture would have approximately the expected proportions listed in Table 20. This example shows the importance of the Hardy-Weinberg formula in finding out whether a population is interbreeding panmictically or whether it consists of subgroups that are more or less separated reproductively. Such subgroups are known as *isolates*.

The Proportions of Genotypes in Recessiveness

The MN blood groups supply a complete test of the Hardy-Weinberg formula, since all three genotypes are distinguishable by their phenotypes. Whenever the heterozygote of two alleles resembles one of the homozygotes, such direct comparison between expectation and observation is not possible. In these cases, the formula may be used to find the proportions of homozygotes and heterozygotes. Such procedure, of course, presupposes the existence of random mating in regard to the pair of alleles under discussion.

Albinism. One of the two types of albinism (see p. 168) may be used as an illustration. Since approximately 1 out of 20,000 individuals belongs to one or the other type of albinism, and assuming that each type is of equal frequency in a population (which may not be true), we take the frequency 1 out of 40,000 as the expression of the frequency of one albinotic type *dd*. Most of the remaining 39,999 individuals will be *DD*, and some *Dd*. It is the proportion of the latter, the carriers of the undesirable allele *d*, that is significant for both the population and the individual. The frequency of the albino allele, q, can be calculated from the frequency of homozygotes, q^2. Since $q^2 = 1/40,000$, $q = 1/200$, and p(frequency of the normal, nonalbino allele) $= 1 - q = 199/200$. Therefore, the frequency of persons who are *Dd* is

$$2pq = 2 \times \frac{1}{200} \times \frac{199}{200} = \text{(about) } \frac{1}{100}.$$

(Because the estimate of the frequency of albinos is only approximate, it is sufficiently accurate to consider the fraction 199/200 as equivalent to 1.)

The result of this calculation comes as a surprise to most persons, who are inclined to reason that the great rarity of albinism must signify a comparable rarity of heterozygous carriers. The quantitative estimate, however, shows that nearly one per cent of all persons are heterozygous for the albino allele: there are 400 times as many heterozygotes as affected individuals.

A number of different frequencies for homozygous recessive traits are listed in Table 21, together with the corresponding frequencies of heterozygotes and the ratio of these to affected persons. It is seen that the frequency of heterozygotes decreases with decreasing frequency of homozygotes. Since heterozygote frequency decreases more slowly, the ratio between the two increases. Thus, an extremely rare recessive condition like alkaptonuria (p. 48), which has been estimated to occur in approximately 1 out of 1,000,000 persons, is carried heterozygously by 1 out of 500 persons; i.e., it is carried by 2,000 times as many people as show the character. It should be apparent that these facts are significant if one wishes to propose public measures intended to eliminate, or at least reduce, the incidence of undesirable genetic traits in a population. An explicit discussion of the problem will be found in Chapter 29, Selection in Civilization.

The Frequencies of Different Types of Marriages. The principle of random mating provides a means for determining the relative frequencies of all different kinds of marriages between individuals of the three genotypes *DD, Dd,* and *dd*. Listing only those marriages which may produce affected (*dd*) offspring and designating by q the frequency of the recessive allele *d*, we obtain

Marriages		*Proportion of dd Offspring*	*dd Offspring*	*dd Offspring as Fraction of all dd* $(=q^2)$	*Percentage of dd Offspring if* $q^2 = 1/40{,}000$
Type	Frequency				
$Dd \times Dd$	$4p^2q^2$	1/4	p^2q^2	p^2	99.025
$Dd \times dd$	$4pq^3$	1/2	$2pq^3$	$2pq$	0.9950
$dd \times dd$	q^4	1	q^4	q^2	0.0025

In a rare recessive condition, where q is small, most affected children will come from marriages of two carriers. This will be illustrated by again using as an example one of the types of albinism which approximately occurs in $q^2 = 1$ out of 40,000 individuals. Here, since $q = 1/200$ and $p = 199/200$, p^2q^2, the frequency of *dd* offspring from $Dd \times Dd$ marriages, = 39,601 out of 1,600,000,000 individuals; $2pq^3$, the frequency of *dd* from $Dd \times dd$ marriages, = 398 out of 1,600,000,000; and q^4, the frequency of *dd* from $dd \times dd$, = 1 out

TABLE 21
Frequencies of Affected and Carrier Individuals under Random Mating in Various Cases of Simple Single Factor Recessive Inheritance.

Frequency of Affected (q^2)	*Frequency of Carriers* ($2pq$)	*Ratio of Carriers to Affected* ($2pq/q^2 = 2p/q$)
1 in 10	1 in 2.3	4.3:1
1 in 100	1 in 5.6	18:1
1 in 1,000	1 in 16	61:1
1 in 10,000	1 in 51	198:1
1 in 100,000	1 in 159	630:1
1 in 1,000,000	1 in 501	1,998:1

of 1,600,000,000. Thus, among 40,000 albinos, 39,601 (more than 99 per cent) come from two nonalbino parents, 398 (less than 1 per cent) from parents one of whom is an albino, and only 1 (0.0025 per cent) from parents both of whom are albinos. In other words, more than 99 per cent of all albinos are expected to be the offspring of parents who look normal.

Limitations of Random Mating. Although an analysis such as the one just given leads to considerable insight into the genetics of a human population, it must be realized that the underlying assumption of random mating is at best only an approximation. Slight deviations from random mating are common. Marriages of close relatives often are more frequent than would be expected by chance, and since relatives have a greater probability of sharing the same genes than two unrelated persons, such marriages represent a certain degree of assortative mating of like genotypes. However, the effect of marriages of close relatives on the genetic composition of a population is not very great (see Chap. 19).

It may also be true that marriages of two homozygous albinos are more common than expected by chance. Several marriages in which both spouses were albinos who were traveling in a circus troupe have been recorded. Considering the rarity of homozygous albinos, such preferential mating, if it really occurs, will have no appreciable effect on the composition of the population.

A more important exception to random mating, namely, significant negative assortative mating, has been reported from the San Blas tribe of Central American Indians. It has been known for nearly three centuries that this tribe contains an appreciable number of albinos. In a relatively recent survey of 20,000 San Blas Indians, 152 albinos were found. It is reported that the group prohibits marriages between albinos; consequently, in contrast to a population mating at random, there would be no albinos who are the offspring of albino × albino marriages. This slight deficit would, of course, be partly counteracted by the greater number of marriages of albinos to heterozygous or homozygous normals, which would increase the number of albinos and heterozygotes. Clearly, under this system of mating, the composition of a population would not agree with the distribution expected from the Hardy-Weinberg Law. It may be added that the San Blas Indians also take selective measures against albinism: not only are albino women prohibited from marrying albino men, but many of them do not marry at all. In the course of generations this introduces further deviations from the structure of the population to be expected from random mating.

In determining the relative frequency of individuals homozygous for a given recessive gene, the larger the sample, the more reliable the estimate. In countries with compulsory military training, for instance, this sample may comprise the total number of young men. Or, where persons with certain afflictions must be hospitalized or registered, data from the whole population may be available for use in determining the frequency of a trait. It is, however, generally known that even the most thorough census is likely to underestimate the number of affected individuals because of the tendency of such persons or their relatives to suppress unfavorable information.

Isolates. Even if the determined frequency is close to the actual one, there are difficulties in equating it with q^2 in applying the Hardy-Weinberg formula. Most populations are composed of subgroups which were originally relatively separate and have not yet intermingled enough to permit random mating in the strictest sense of the term: the existence of isolates within populations is the rule, not the exception. Such isolates may be, for example, whole villages or small towns, to which there has been little immigration for several generations, or various socioeconomic or racial strata in larger cities or in whole countries. In the United States, as elsewhere, positive assortative mating is prevalent in regard to presence or absence of African ancestry, in regard to phenotypically less obvious ethnic origins, in regard to religious affliation and many other attributes.

Whenever a group of individuals that is part of a larger population has had a tendency to intermarry for several generations and thus limit the exchange of alleles with the rest of the population, the distribution of alleles within the whole population will be uneven. Isolates, whether large or small, are actually relatively inbred associations, and many rare genes present in one such group may be absent in another. Consequently, the proportions of homozygous dominant, heterozygous dominant, and homozygous recessive affected persons may be quite different in different isolates, and also different in the population that is the sum of these isolates.

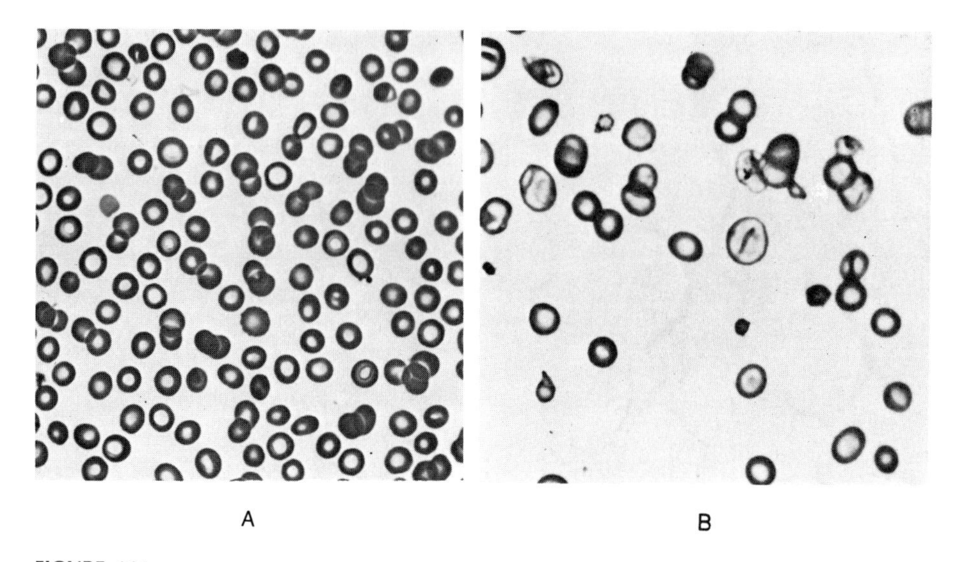

FIGURE 116
Blood picture in thalassemia. **A.** Normal blood. **B.** Anemic blood. The red blood cells contain little hemoglobin and the shape of many cells is irregular. (Originals from Dr. C. L. Conley.)

Cooley's Anemia, or Thalassemia. As an example approximating the situation just outlined, we will discuss Cooley's anemia (Fig. 116). This blood disease, whose primary cause seems to be a quantitative defect in synthesis of hemoglobin A, is still frequently fatal within the first years of life. It is caused by homozygosity for a recessive allele *c*. The trait is known in various parts of the world, and its frequency has been studied in a medium-sized American city. Hospital records showed that 11 individuals with the disease were born between 1928 and 1942. Since the total number of births during this fifteen-year span was approximately 100,000, it would seem that the frequency of heterozygous carriers, 2pq, could be derived from the frequency of those affected, $q^2 = 11/100{,}000$. On this basis, the calculation indicates that about 1 in 50 persons in the city was a heterozygous carrier of this rare disease.

Cooley's anemia is known not only by the name of the physician who first gave a clear description of it but also as Mediterranean anemia or thalassemia (from the Greek *thalassa* = sea; meaning anemia occurring near the sea), because it is largely confined to people living on or near the shores of the Mediterranean—Greeks, Italians, Syrians, Armenians—or to their descendants. It happens that the American city studied contained a rather large isolate of persons of Italian descent: all 11 cases of the disease on record came from within this group. The total number of births in the Italian isolate from 1928 to 1942 was estimated to have been about 26,000. Thus the frequency of thalassemia was 11 in 26,000, or slightly less than 1 in 2,400 in the Italian isolate. Since random mating is probably reasonably approximated in this group, an application of the Hardy-Weinberg formula, when $q^2 = 1/2{,}400$, gives the frequency of carriers (2pq) as about 1 in 25.

The over-all frequency of carriers in the city may be obtained if one assumes that the groups of non-Italian descent do not contain carriers. Since the total population consists of 25 per cent who are of Italian descent and 75 per cent who are not, the over-all frequency of carriers is (1 in 25) + (0 in 75), or 1 in 100 persons.

It is seen that the original estimate, based on failure to associate thalassemia with ancestral background, gave a figure between the zero frequency figure for the non-Italian group and the high value of 1 in 25 for the Italian group. As the Italian isolate is broken down by intermarriage with the population at large, shifts in the relative frequencies of the genotypes will occur.

The Breakdown of Isolates. Let us assume that a population is composed of two isolates, I and II, each consisting of 100 individuals, and that a deleterious recessive allele *a* has a frequency $q_I = 0.2$ in I, but is absent in II ($p_{II} = 1$, $q_{II} = 0$). Then the frequency of affected individuals (*aa*) in I is $q_I^2 = 0.04$, or 4 in the group. There are no affected persons in II. If we take a census of *aa* persons in the *total* population, which is the sum of I and II, we obtain a frequency of 4 in 200, or 2 per cent. What kind of an offspring population will be produced if the two isolates merge and mate panmictically?

The answer is provided by the Hardy-Weinberg Law (Fig. 117). Since the frequency of *a* alleles in isolate I is $q_I = 0.2$ and that in isolate II is $q_{II} = 0$, the allele frequency in the joint population (I + II) is $(q_I + q_{II})/2$ or $q_{I+II} = 0.1$. Therefore, after random mating, the frequency of affected individuals (*aa*) is $(q_{I+II})^2 = 0.01$, or 1 per cent: the frequency of affected individuals has decreased from 2 per cent to 1 per cent.

This result is of general significance and highly important, since it typifies the fate of all recessive alleles, which are present in frequencies that differ from one isolate to another, in isolates that merge with each other or with the general population. Breakdown of isolation followed by panmixis must result in a lowering of the incidence of affected individuals. The mean allele frequencies are, of course, not changed by panmixis, but any reduction of high allele frequencies in some isolates by "dilution" with the lower frequencies in others results in a proportionally greater reduction in the number of affected persons: it is greater because the frequency with which two recessive alleles form a homozygote is proportional to the square of the frequency of a single allele. This may be illustrated by one more example. Let isolate I consist of 100,000 people and have an allele frequency $q_I = 0.01$. Then the number of affected persons in I is $(0.01)^2 \times 100{,}000 = 10$. Let isolate II consist of 900,000 people and have an allele frequency $q_{II} = 0$, thus having no affected individuals.

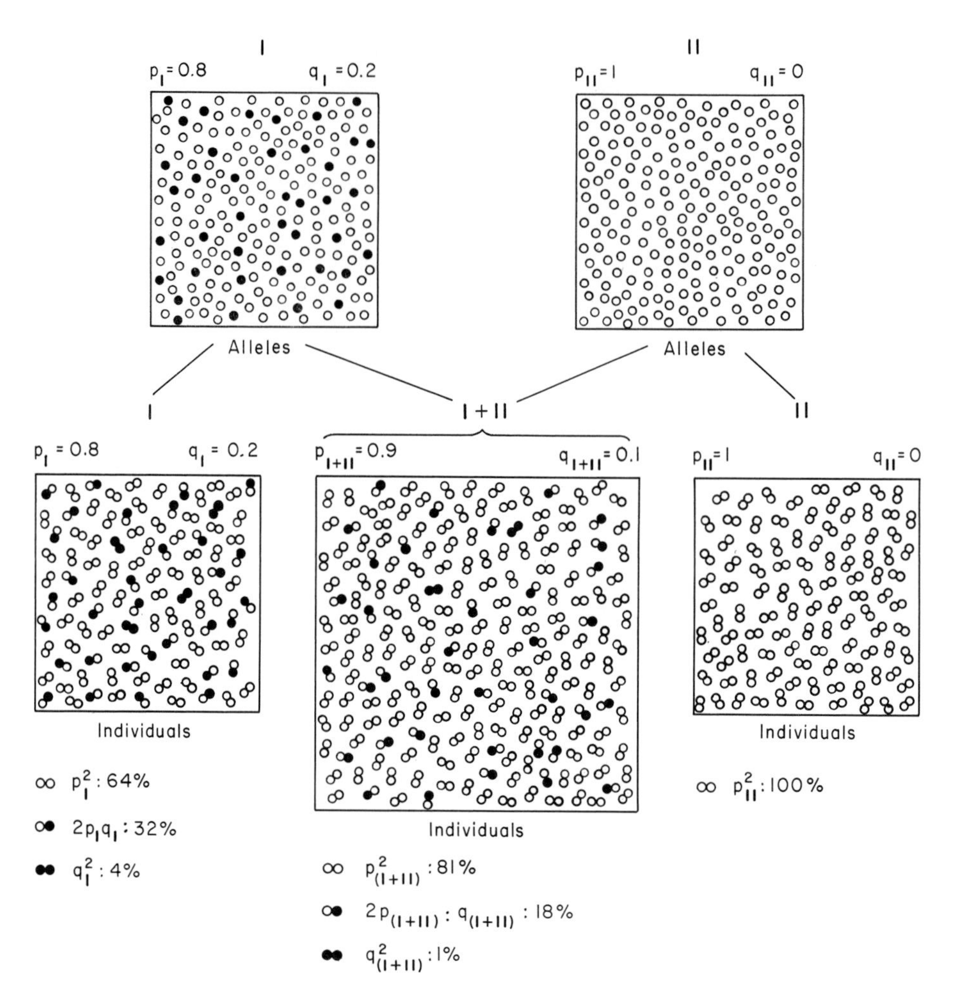

FIGURE 117
The consequences of breakdown of isolates. *Upper and lower, left and right:* Two isolates I and II with different allele frequencies, $q_I = 0.2$ and $q_{II} = 0$, produce two groups of individuals with 4 and 0 per cent homozygous recessive individuals, respectively. *Lower middle:* A panmictic population derived from I and II produces 1 per cent homozygous recessives.

Before panmixis the two populations together consist of 1,000,000 people, 10 of whom are affected. After breakdown of isolation only one-tenth of the new population (I + II) will be derived from isolate I so that the allele frequency $q_{(I+II)}$ will be only $1/10\ (q_I + q_{II}) = 0.001$. If the new population (I + II) mates at random and produces an offspring generation of 1,000,000 people, the number of affected among them will be $(0.001)^2 \times 1{,}000{,}000 = 1$. Thus the breakdown of the two isolates reduces the number of affected from 10 to 1.

The Use of the Hardy-Weinberg Law When Different Genes Cause Similar Phenotypes. We have seen in our discussion of genic action that phenotypic similarities may be due to agents other than identical genotypes. Different genes may produce similar, or identical, end effects and thus cause similar-appearing traits. Unrecognized cases of this type are another obstacle to the uncritical application of the Hardy-Weinberg formula. Let us assume the correctness of the hypothesis that albinism may be caused by recessive alleles at either of two different loci. In this event we would not be justified in equating the observed frequency of albinos, say 1 in 20,000, to a single q^2 value. Should, for instance, two different recessive alleles a_1 and a_2 in different loci each cause albinism when homozygous, then the number of observed albinos would be made up of the sum of the frequencies of a_1a_1 ($q^2_{a_1}$) and of a_2a_2 ($q^2_{a_2}$)—neglecting the very small number of individuals who are simultaneously homozygous for both a_1a_1 and a_2a_2. If q_{a_1} and q_{a_2} were equal, $q^2_{a_1}$ and $q^2_{a_2}$ would each be 1 in 40,000, so that q_{a_1} and q_{a_2} would each be 1/200. Therefore, the number of carriers for each albino allele would approximate 1 in 100.

Since, if two different loci are involved with equal frequency, 1 out of 100 persons is a carrier of the allele a_1 and 1 out of 100 of a_2, it may be thought that the frequency of carriers should be given as the sum of the two kinds of carriers, or 1 in 50. Genetically, however, each kind of carrier must be considered by itself, since only marriages between individuals who both transmit the same recessive allele can give rise to affected offspring. Therefore, a statement of frequency of carriers is more meaningful if made separately for each locus.

ALLELE-FREQUENCY TESTS OF GENETIC HYPOTHESES

When a character is present in two alternative forms in a population and the indications are that it is genetically determined, a first working hypothesis is that it is due to a specific pair of alleles: one dominant and the other recessive. If both phenotypes are relatively frequent and no good reasons exist for doubting the existence of random mating, the allele-frequency analysis can be used for a test of the working hypothesis.

"Taste-blindness." This will be shown for the so-called taste-blindness for the organic compound phenylthiocarbamide (PTC). It has been found that in an American white population, about 70 per cent experience a striking, bitter taste if this substance is applied to the tongue, and about 30 per cent find the substance tasteless. (The few individuals who have different taste experiences

TABLE 22
Data on the Inheritance of Ability to Taste Phenylthiocarbamide.

Parents	*No. of Families*	*Offspring*		*Fraction of Nontasters among Offspring*
		Tasters	Nontasters	
Taster × Taster	425	929	130	0.123
Taster × Nontaster	289	483	278	0.366
Nontaster × Nontaster	86	5	218	0.978

Source: After Synder.

will be disregarded.) The reaction of any one individual is more or less constant, and pedigree data show inheritance of the trait. By adding the offspring of numerous marriages of the three possible phenotypes, namely, taster × taster, taster × nontaster, and nontaster × nontaster, Snyder obtained the data given in Table 22.

If the ability or inability to taste phenylthiocarbamide is due to a pair of alleles T and t, persons with one of the two phenotypes must be homozygous recessive, and those with the other can be either heterozygous or homozygous dominant. The table shows that the tasters belong to the group which includes the heterozygotes, since parents who are both tasters have both types of children in appreciable numbers. Consequently, the nontasters are the homozygous recessives, and nontaster couples should be expected to have nontaster offspring exclusively. This expectation is closely approached but not completely fulfilled, since 5 tasters were found among 223 children. Such discrepancies do not need to be taken too seriously in genetic interpretation. In a large collection of human data, errors of misclassification are difficult to exclude, and there is always a possibility that some children in any group are illegitimate. Moreover, as discussed earlier, variations in the expression of the taster genotypes sometimes do not permit a specific assignment. Different individuals have different degrees of taste sensitivity: some can taste PTC in very low concentration; others only in higher concentrations.

If the taster parents actually consist of both TT and Tt genotypes, the allele-frequency relations, assuming random mating, give the proportion of these types as p^2 and $2pq$, and that of the nontasters (tt) as q^2. By the use of these frequencies, it is possible to predict the proportions of taster and nontaster children resulting from *groups* of marriages of two tasters, even though it is unknown whether an *individual* taster parent is homozygous or heterozygous.

Phenotypically, there are two kinds of relevant marriages: (A) that in which one spouse is a taster and the other a nontaster; and (B) that in which both are tasters.

A. Marriages between a taster and a nontaster are of two types. These two types, their frequencies, and their offspring are as follows:

Marriages		*Offspring*	
Type	Frequency	Tasters	Nontasters
$TT \times tt$	$2p^2q^2$	$2p^2q^2$	–
$Tt \times tt$	$4pq^3$	$2pq^3$	$2pq^3$

Adding all offspring, we find the total to be $2p^2q^2 + 4pq^3$. Among these there are $2pq^3$ nontasters; hence, the expected *fraction of nontasters* among all offspring of these marriages is

$$\frac{2pq^3}{2p^2q^2 + 4pq^3} = \frac{q}{p + 2q} = \frac{q}{1 - q + 2q} = \frac{q}{1 + q}. \tag{3}$$

B. Similarly, we can calculate the fraction of nontaster offspring from marriages of tasters to tasters. There are three types of such marriages.

Marriages		*Offspring*	
Type	Frequency	Tasters	Nontasters
$TT \times TT$	p^4	p^4	–
$TT \times Tt$	$4p^3q$	$4p^3q$	–
$Tt \times Tt$	$4p^2q^2$	$3p^2q^2$	p^2q^2

Here, the *fraction of nontasters* among all offspring is

$$\frac{p^2q^2}{p^4 + 4p^3q + 4p^2q^2} = \frac{q^2}{p^2 + 4pq + 4q^2} = \left(\frac{q}{p + 2q}\right)^2 = \left(\frac{q}{1 + q}\right)^2. \tag{4}$$

Formulas (3) and (4) enable us to test the hypothesis of single factor inheritance of taster ability, provided the frequency, q, of the *t* allele is known. It may be derived from the combined data on the population sample listed in Table 22. Of the 3,643 persons tested, 70.2 per cent were tasters and 29.8 per cent nontasters. Therefore, $q^2 = 0.298$, and $q = 0.545$. Given this value for q, the expected fraction of nontaster children from marriages of tasters to nontasters,

$$\frac{q}{1 + q},$$

is 0.353; and

$$\left(\frac{q}{1+q}\right)^2,$$

the fraction of nontaster children from marriages of tasters to tasters, is 0.124. The observations listed in Table 22 yielded 0.366 and 0.123, respectively, in close agreement with expectation. From this, it may be concluded that the assumption that a single pair of alleles *T* and *t* and random mating are responsible for the inheritance of taste reactions to phenylthiocarbamide fits the facts. (Nevertheless, the expectations do not fit the facts uniquely. It has been found that the two fractions given above also fit a trait which is caused by multiple homozygosity involving several loci.)

THE HARDY-WEINBERG RELATIONS FOR TWO PAIRS OF GENES

The Hardy-Weinberg Law states that, for a single pair of alleles, a panmictic population reaches equilibrium in a single generation. Such immediate attainment of equilibrium after panmixis is not characteristic for the combinations of alleles at more than one locus. Let us consider a population with the two alleles *A* and *A'*, at an *A* locus, and the two alleles *B* and *B'*, at a *B* locus; and let $p_A = 2/3$ and $q_{A'} = 1/3$, and $p_B = 3/4$ and $q_{B'} = 1/4$. At equilibrium, this population will contain nine different genotypes, namely, *AABB*, *AABB'*, and *AAB'B'*; *AA'BB*, *AA'BB'*, and *AA'B'B'*; and *A'A'BB*, *A'A'BB'*, and *A'A'B'B'*. Each one will have a frequency that is the product of the two independent genotype frequencies at the two loci. Thus, $AABB = (2/3)^2 \times (3/4)^2$; $AABB' = (2/3)^2 \times 2(3/4 \times 1/4)$; and so on. Generalizing for any values of the allele frequencies, the genotypic frequencies for *AABB* will be $p_A^2 \times p_B^2$; for *AABB'*, $p_A^2 \times 2p_B q_{B'}$; and so on.

Now let us consider a population which initially consisted exclusively of *AA'BB'* heterozygotes. The gametes produced will be of four kinds (*AB*, *AB'*, *A'B*, and *A'B'*) and will occur in equal frequencies (1/4:1/4:1/4:1/4). Although these can combine in 16 different ways, they can produce only nine different genotypes, as indicated in the upper half of Figure 118, where the frequencies are also noted. As shown in this figure, the first generation population, so different from the original one, is in equilibrium, just like the first generation in a population with only one pair of allelic differences.

Ordinarily, however, equilibrium for two allele pairs is not attained in a single generation. Indeed, only if the initial population consists solely of *AA'BB'* heterozygotes is the equilibrium attained so speedily. Let us consider another initial population: one that consists of equal numbers of *AABB* and

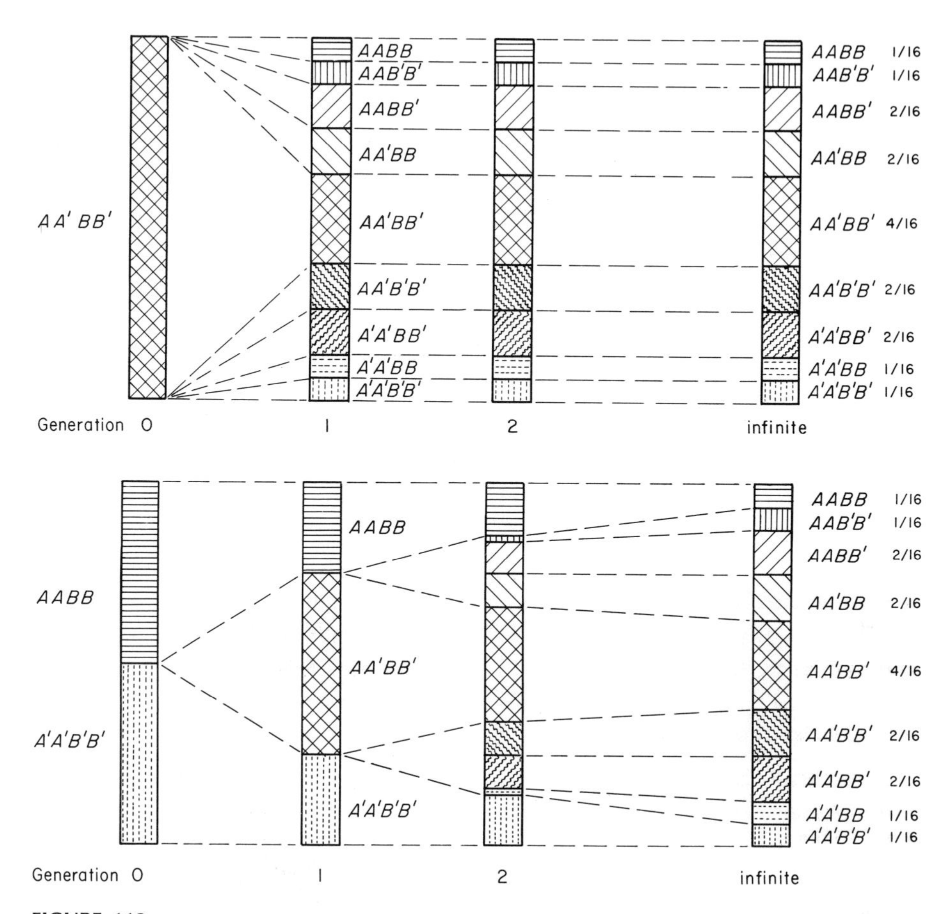

FIGURE 118
Above: A population consisting, in generation 0, of *AA'BB'* individuals only. Attainment of equilibrium in one generation. *Below:* A population consisting, in generation 0, of equal numbers of *AABB* and *A'A'B'B'* individuals. Stepwise attainment of equilibrium in successive generations.

A'A'B'B' homozygotes and undergoes panmixis (Fig. 118, lower half). Since only two types of gametes segregate in the population, namely, *AB* and *A'B'*, only three genotypes, *AABB*, *AA'BB'*, and *A'A'B'B'*, are present in the next generation in the ratio 1/4:2/4:1/4. All four possible kinds of gametes, *AB*, *AB'*, *A'B*, and *A'B'*, are produced by this first panmictic generation. The double heterozygote *AA'BB'* contributes equal numbers of these, but the *AABB* and *A'A'B'B'* homozygotes supply only *AB* and *A'B'*. The result is a second generation in which all nine possible genotypes occur but not at random frequencies. There are too many *AABB*, *AA'BB'*, and *A'A'B'B'* genotypes and too few of the other six genotypes, particularly *AAB'B'* and *A'A'BB*. Specifically, the first generation *AABB* individuals contribute one-quarter of all gametes, of the genotype AB, and the first generation $A'A'B'B'$ individuals another quarter, of the genotype $A'B'$. The $AA'BB'$ individuals contribute one-half of all gametes, namely one-eighth each of the four kinds AB, AB', $A'B$ and $A'B'$. In sum the gametic frequencies are 3/8 AB, 1/8 AB', 1/8 $A'B$, and 3/8 $A'B'$. Random fertilization results in a second generation of the following composition: 9/64 $AABB$, 6/64 $AABB'$, 1/64 $AAB'B'$; 6/64 $AA'BB$, 20/64 $AA'BB'$, 6/64 $AA'B'B'$; 1/64 $A'A'BB$, 6/64 $A'A'BB'$, and 9/64 $A'A'B'B'$. With later generations the excesses and deficiencies diminish until equilibrium is reached.

The gradual approach to equilibrium is characteristic for populations which segregate for more than one pair of alleles, no matter whether the frequencies of all alleles are equal or whether the values of p_A, $q_{A'}$, p_B, $q_{B'}$, and so on, vary greatly. It was Weinberg who soon after his discovery of the equilibrium formula for a single locus also recognized the essential difference between the attainment of equilibrium for single and multiple loci.

The approach to equilibrium in populations which segregate for two alleles at each of two loci is comparatively rapid, provided that the two loci recombine freely. If they are linked, the stronger the linkage, the slower the approach to equilibrium. However, given enough generations, equilibrium is reached even with close linkage. This important fact will be discussed in more detail in Chapter 15.

The Recognition of Recent Intermixture in Populations. If two populations containing different alleles at any given locus intermarry panmictically, equilibrium for the two alleles of any one locus will be reached in one generation.

Let us assume that two populations differ in two traits that depend on two separate pairs of genes. To be specific, let us assume that one pair of alleles governs the color of eyes and the other the color of hair, and that the heterozygotes are intermediate in both (in reality, differences in these traits are seldom controlled by single pairs of genes). Population I is to consist entirely of individuals who have light eyes and light hair; population II, of persons who have dark eyes and dark hair. Given panmixis between the two populations, the proportion of dark-eyed to light-eyed persons after a single generation, and forever after, will follow the Hardy-Weinberg Law, and so will the proportion of dark-haired to light-haired. Studying the proportions of either trait alone will not reveal whether the presence of dark and light individuals is the result of recent or long-past intermixture, or whether it arose independently by mutation.

But if both traits are studied together, the proportions of persons having the various possible combinations will change with each successive generation (see Fig. 118, below, for two populations of equal sizes; if sizes of the populations are unequal, the changes in genotypic composition in successive panmictic generations are similar to those with equal sizes but the proportions of the genotypes are different.) After a single generation of panmixis, only three types of persons will exist: those who, like population I, are light-eyed and light-haired; those who are dark-eyed and dark-haired, like population II; and those who are first-generation hybrids for both traits. There will be no light-eyed and dark-haired, or dark-eyed and light-haired individuals, and no individuals with other combinations. In the second generation all possible combinations will appear, but the phenotypes of the original populations and first-generation hybrids will still be more numerous than expected at equilibrium. With every new successive generation, each combination will approach its equilibrium frequency more closely. Thus, a study of the frequencies of combinations of traits that depend on different genes may throw light on the origin of a population. If the different combinations of genes do not occur at equilibrium frequencies, it is possible that the population is of relatively recent mixed origin. This, however, is only one of various other possibilities such as incomplete break-up of formerly isolated original populations or assortative mating.

POPULATION GENETICS

This chapter has provided an introduction to a particularly important part of the study of human genetics: the study of the genetic composition of whole populations in contrast to that of individual families or kindreds. Both population genetics and pedigree genetics are significant, and both rest on the Mendelian analysis of inheritance.

We have seen that the Hardy-Weinberg Law is a basic guide to the understanding of population genetics. In the following chapters, the Hardy-Weinberg Law and such concepts as random mating and allele frequency will often prove useful in our study of the genetics of human populations.

PROBLEMS

74. Assume people of the three genotypes *AA, Aa,* and *aa* to occur in the proportions of 1/4:2/4:1/4. If there is completely positive assortative mating between like genotypes:
    (a) What proportions of the whole would each of the three types represent in the next generation?
    (b) What proportions would result in the following generation?

75. In a given population, how would you determine the percentage frequency of the gene for:
    (a) Brachydactyly? (b) Nontasting?

76. Which of the following populations are in genetic equilibrium:
    (a) 100 per cent *AA*,
    (b) 100 per cent *aa*,
    (c) 100 per cent *Aa*,
    (d) 1 per cent *AA*, 98 per cent *Aa*, 1 per cent *aa*,
    (e) 32 per cent *AA*, 64 per cent *Aa*, 4 per cent *aa*,
    (f) 4 per cent *AA*, 32 per cent *Aa*, 64 per cent *aa*?

77. In a certain population, 16 per cent of all people belong to blood group N. Assuming panmixis, how many do you expect to be M and how many MN?

78. What can be concluded regarding a population of 1,000 people which has the following composition: M 33 per cent, MN 34 per cent, N 33 per cent?

79. Direct observations of a very large population have shown that three phenotypes occur in the ratios of 70:21:9. Are these frequencies compatible with the theory that this is a case of simple single gene-pair inheritance, the three phenotypes representing the genotypes *AA, Aa,* and *aa*?

80. Assume that an isolated population consists of people of blood group M only. At a certain time immigration occurs, the immigrants equalling the original population in number. Twenty-five per cent of the immigrants belong to group N. Since the immigrants were a thoroughly random-mated group, you can calculate the frequencies of the L^M and L^N genes among them. After the immigrants and the original population have married at random, what proportions of the blood groups will you expect in future generations?

81. Although panmictic population I contains, for gene *A*, only homozygous individuals, it contains, in regard to a different locus, *BB, Bb,* and *bb* individuals, the last in a frequency of 0.04. Another equally large and panmictic population II is homozygous for *BB*, but contains *AA, Aa,* and *aa* individuals, the last in a frequency of 0.01. If populations I and II intermarry completely and there is random mating in later generations, what frequencies, expressed in per cent, of *aa, bb,* and *aabb* people do you expect?

82. The frequency of the supposed recessive allele for inability to roll one's tongue lengthwise has been estimated as $q = 0.6$.
    (a) What are the frequencies of rollers and nonrollers?
    (b) In pooled data, what proportion of rollers to nonrollers are expected among the children from marriages of two roller parents?
    (c) What proportion are expected if one spouse is a nonroller and the other a roller?

83. Table 31 (p. 271) gives data on the offspring from marriages of Rh-positive and Rh-negative persons. Read the paragraph in which Table 31 is discussed and apply the allele-frequency tests to the simple genetic interpretation given there.

REFERENCES

Burdette, W. J. (Ed.), 1962. *Methodology in Human Genetics.* 436 pp. Holden-Day, San Francisco

Dahlberg, G., 1948. Genetics of Human populations. *Advan. Genet.,* **2**:67–98.

Fourth Princeton Conference on Population Genetics and Demography. Six articles on assortative mating, by various authors, 1968. *Eugen. Quart.,* **15**:71–143.

Garrison, R. J., Anderson, V. E., and Reed, S. C., 1968. Assortative marriage. *Eugen. Quart.,* **15**:113–127.

Hardy, G. H., 1908. Mendelian proportions in a mixed population. *Science,* **28**:49–50.

Li, C. C., 1961. *Human Genetics.* 218 pp. McGraw-Hill, New York.

Neel, J. V., and Valentine, W. N., 1947. Further studies on the genetics of thalassemia. *Genetics,* **32**:38–63.

Snyder, L. H., 1932. The inheritance of taste deficiency in man. *Ohio J. Sci.,* **32**:436–440.

Spuhler, J. N., 1968. Assortative mating with respect to physical characteristics. *Eugen. Quart.,* **15**:128–140.

Weinberg, W., 1908. Über den Nachweis der Vererbung beim Menschen. *Jahreshefte Verein f. vaterl. Naturk. in Württemberg,* **64**:368–382. (An important section of this paper is translated in Stern, C., 1943. The Hardy-Weinberg law. *Science,* **97**:137–138.)

Wiener, A. S., 1943. *Blood Groups and Transfusion,* 3rd ed. Chs. 13 and 14, pp. 219–244. Thomas, Springfield, Ill.

12

MULTIPLE ALLELES AND THE BLOOD GROUPS

Most of the examples of inheritance that have been given in the preceding chapters dealt with two alleles only, e.g., one for normal and the other for woolly hair, or one for pigmentation and the other for albinism. However, as discussed earlier, many genes have more than two allelic forms. Indeed, it is to be expected that the multitude of possible substitutions in the sequence of nucleotide pairs that constitute a gene would lead to the existence of many alleles in a population.

The best-known types of multiple allelism in man concern the blood groups. Here, the availability of very sensitive immunological tests has led to the recognition of numerous alleles at a variety of loci. This chapter, therefore, will be concerned particularly with the genetics of blood groups. Excellent summaries have been published by Race and Sanger, Wiener and Wexler, and Giblett, all of whom are in the forefront of workers in immunogenetics.

Since alleles are alternative varieties of a gene at a given locus, multiple alleles obey the same rules of transmission as alleles of which there are only two kinds. An individual may be homozygous for any one of the alleles or heterozygous for a combination of any two of them, and segregation, in meiosis, results in gametes that contain only a single allele.

If we wish to know how many alleles there are for a specific locus within a population, we must study many individuals since in a single person, each locus is only represented twice and no more than two different alleles can be found. If, on the other hand, 100 individuals are tested, their 200 homologous loci could theoretically be occupied by 200 different alleles. In reality, no such diversity can be expected within an interbreeding population in which many different individuals must have inherited the same alleles from common ancestors.

The Number of Possible Genotypes. When a gene has only two alleles, the total number of genotypes is three, namely, the two homozygotes and the heterozygote. The more alleles a gene possesses, the larger, of course, is the number of possible genotypes. Thus, with three alleles, A^1, A^2, and A^3, there are three homozygotes, A^1A^1, A^2A^2, A^3A^3; two heterozygotes of A^1 with the other alleles, A^1A^2 and A^1A^3; and one heterozygote, A^2A^3. With n alleles, $A^1, A^2, \ldots, A^n$, the following genotypes may occur:

n homozygotes	A^1A^1	A^2A^2	...	...	...	...	A^nA^n
(n − 1) heterozygotes with A^1	A^1A^2	A^1A^3	...	...	...	A^1A^n	
(n − 2) heterozygotes with A^2	A^2A^3	A^2A^4	...	...	A^2A^n		
(n − 3) heterozygotes with A^3	A^3A^4	A^3A^5	...	A^3A^n			
⋮	⋮	⋮	⋮				
2 heterozygotes with A^{n-2}	$A^{n-2}A^{n-1}$	$A^{n-2}A^n$					
1 heterozygote with A^{n-1}	$A^{n-1}A^n$						

The sum of all these genotypes is $n + (n - 1) + (n - 2) + \ldots + 1$, which is equal to $(1/2)[n \times (n + 1)]$. If there are five alleles, the number of genotypes is thus $5 + 4 + 3 + 2 + 1 = (1/2) \times (5 \times 6) = 15$; if $n = 20$, the number of genotypes is 210.

THE ABO BLOOD GROUPS

The Four ABO Blood Groups. It has been known, since Landsteiner's discovery in 1900, that the blood of an individual may belong to one of several different types, according to the reactions observed in mixtures of blood of different persons. Essentially, these are reactions between the red blood cells

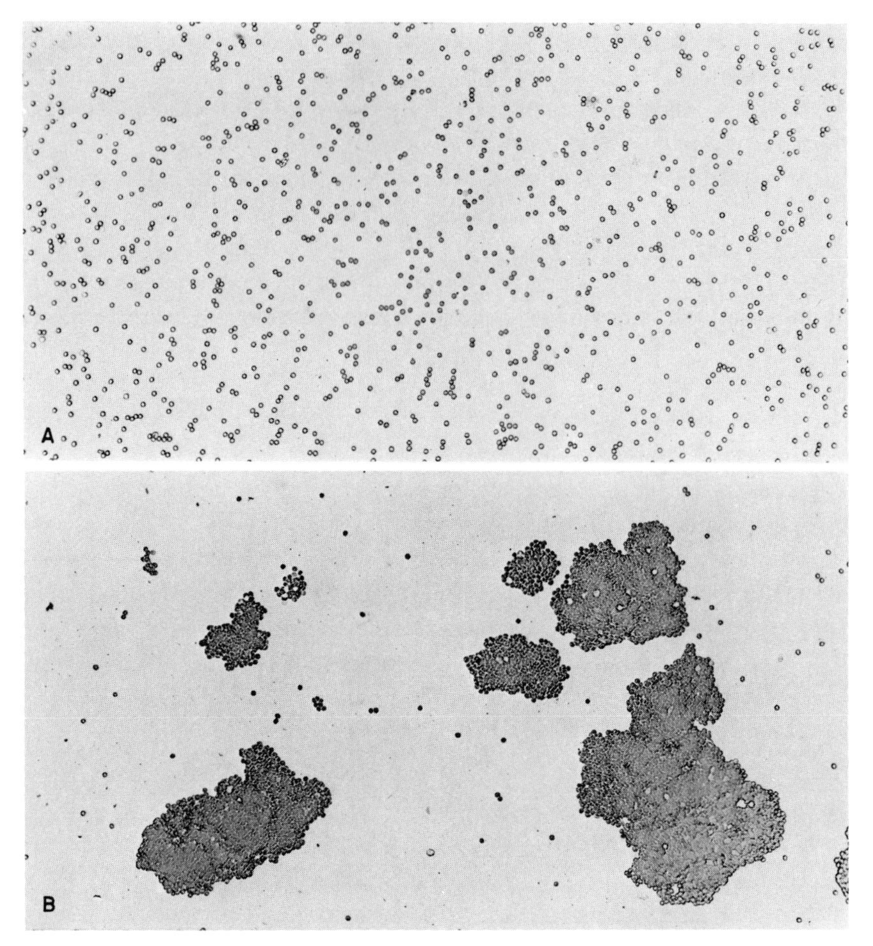

FIGURE 119
Photograph of (**A**) nonagglutinated and (**B**) agglutinated red blood cells. (Originals from Dr. C. L. Conley.)

(= red cells) of one individual and the fluid part, or *plasma,* of the blood of another. The fibrinogen, which enables blood to clot, can be removed from the plasma, and the remaining fluid is called the *serum.* By centifugation, red cells may be removed from whole blood. Then, when red cells and serum from the same individual are brought together again, the cells will become freely distributed in the fluid. If, however, red cells from one individual are mixed with serum from another, one of two different results will be observed. The cells may become freely distributed in the serum, or they may clump together (Fig. 119). These results are constant for any two individuals. Such observations divide mankind into at least two different groups. Further tests have established the existence of four main groups of human beings with respect

Group	Antigens in red blood cells	Antibody present in serum	Reaction to serum (listed to left) of red blood cells from group			
			O	A	B	AB
O	O	Anti-A Anti-B				
A	A	Anti-B				
B	B	Anti-A				
AB	AB	—				

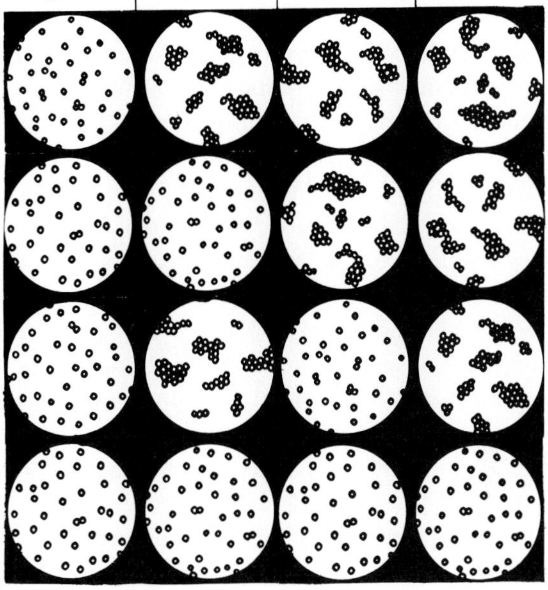

FIGURE 120
Reactions of red blood cells of O, A, B, and AB individuals to antibodies anti-A and anti-B. (After Hardin, *Biology, Its Human Implications.* W. H. Freeman and Company. Copyright © 1949.)

to ABO *isoagglutination.* The term agglutination refers to the clumping of the blood cells, and the prefix iso (from the Greek *isos,* meaning equal) signifies that agglutination is caused by sera from the same species, man.

These four groups of persons are distinguished from one another by the immunological properties of both their red cells and their serum. The red cells of an individual possess either one or the other, both, or neither of two substances (or groups of substances) called *antigens,* or *agglutinogens,* A and B; and his serum possesses either one or the other, neither, or both of two substances (or groups of substances) called *antibodies,* or *agglutinins,* anti-B and anti-A (Fig. 120). Red cells containing antigen A are agglutinated by anti-A, cells containing antigen B by anti-B. The four groups of persons are named after their antigens: A, B, AB, and O. Not only does every kind of human blood lack the antibodies which would agglutinate its own red cells —a necessary condition, since any appreciable clumping would be fatal—but also every kind of human blood contains those antibodies which are compatible with the antigens of its cells. It may be added that some human sera agglutinate the cells of O individuals and also, though less thoroughly, those of the other groups. Such sera are called anti-H, and the corresponding antigen, which is common to all groups, H. As discussed earlier, the H substance is regarded as a precursor of the A and B antigens (pp. 53-54 and Fig. 32).

The ABO Blood Groups in Transfusion. The importance of distinguishing the isoagglutinin blood groups for transfusion is obvious. Transfusion of whole blood is safe only between members of the same group. However, if necessary, it is possible to use certain other combinations of blood groups of donor and receiver. This can be done because the antibodies in the plasma of transfused blood are partly adsorbed by the tissues of an incompatible recipient before they can agglutinate appreciable numbers of his red cells, and the transfused blood is so diluted by the recipient's plasma that the concentration of the donor's antibodies may be below that causing a serious amount of agglutination. Such adsorption and dilution do not apply to antigens which are bound to the red cells of the donor. The properties of the cells are, therefore, of greater importance in determining the consequences of transfusions than are the properties of the sera. This fact has found expression in the terms "universal donor" and "universal recipient": the former applying to group O persons, whose red cells cannot be agglutinated by any recipient (barring a strong anti-H serum); the latter to group AB persons, whose serum cannot act on the cells of any donor (Fig. 120). Since, however, the effect of the donor's antibodies, if present, is not wholly negligible, the terms universal donor and recipient are not fully correct and the use of whole blood from any donor different in blood group from the recipient is avoided.

GENETICS OF THE ABO BLOOD GROUPS

These facts are significant in studies of human genetics because the blood groups are inherited. Studies of pedigrees have shown that children do not have the A antigen unless at least one of their parents has it. Similarly, the B antigen is found only among the offspring of a parent who has B. This indicates dominance for A and B. On the other hand, O persons occur not only in the progeny of O parents, but also among the children of A and of B parents, thus indicating recessive inheritance for O. A summary of numerous data on children from all ten possible combinations of persons of the four blood groups is given in Table 23. It is seen that some exceptions have been found to the rule that A and B antigens occur only among children of parents who carry A and B. These exceptions, as well as some others which are open to suspicion because of their rarity, will be discussed after we have dealt with the more common facts.

TABLE 23
Data on ABO Blood Groups Among Children from All Types of Marriages.

Marriage	*No. of Families*	No. of Children in Each Group* O	A	B	AB	*Total*
O × O	1,563	3,772	(14)	(9)	0	3,795
O × A	2,903	2,707	3,749	(10)	(1)	6,467
O × B	1,456	1,418	(7)	1,831	(1)	3,257
A × A	1,385	556	2,538	0	(2)	3,096
A × B	1,400	605	957	771	848	3,181
B × B	554	203	(1)	1,009	0	1,213
O × AB	530	(8)	633	646	(3)	1,290
A × AB	455	0	533	247	312	1,092
B × AB	323	(2)	183	406	232	823
AB × AB	59	0	28	36	65	129
Total	10,628	9,271	8,643	4,965	1,464	24,343

Source: Wiener, *Blood Groups and Transfusion,* 3rd ed., Thomas, 1943.
*Numbers in parentheses represent "exceptions" (see text).

It is obvious that the existence of four blood groups means that the gene that controls them must have more than two allelic forms, since two alleles may, at most, give rise to three different phenotypes. The now well-established explanation that three multiple alleles govern the inheritance of the main ABO blood groups was historically preceded by another hypothesis. Instead of passing over this hypothesis, we shall show why it became untenable. Such a treatment will be offered not for the sake of past history but to show the power of population-genetics methods.

The Earlier Hypothesis of Two Loci. The earlier hypothesis was based on the assumption that an individual's blood group was determined by two genes at independent loci in two pairs of chromosomes. The scheme may be represented by the symbols *A, a* and *B, b*; allele *A* would be responsible for antigen A, and allele *B* for antigen B, so that the existence of the following genotypes characterizing the four blood groups would be deduced:

aabb . . . . group O
AAbb, Aabb . . . . group A
aaBB, aaBb . . . . group B
AABB, AaBB, AABb, AaBb . . . . . group AB

Analysis of family data. As shown in Table 24 (middle column), this genetic scheme is able to account for the kinds of children resulting from the first six of the ten types of marriages, but not for those from the four types of

TABLE 24
Comparison of the Two Theories of the Inheritance of the ABO Blood Groups.

Groups of Parents	*Groups of Children Expected According to Theory*	
	Two Loci*	Multiple Alleles
O × O	O	O
O × A	O, A	O, A
O × B	O, B	O, B
A × A	O, A	O, A
A × B	O, A, B, AB	O, A, B, AB
B × B	O, B	O, B
O × AB	(O), A, B, (AB)	A, B
A × AB	(O), A, B, AB	A, B, AB
B × AB	(O), A, B, AB	A, B, AB
AB × AB	(O), A, B, AB	A, B, AB

Source: Wiener, *Blood Groups and Transfusion,* 3rd ed., Thomas, 1943.
*Parentheses indicate absence of the group in actual data.

marriages in which at least one of the parents belongs to group AB. Thus, parents O × AB, should have AB children exclusively if the AB parent was *AABB*; B and AB children if the AB parent was *AaBB*; A and AB children if the AB parent was *AABb*; and children of all types (A, B, AB, and O) if the AB parent was heterozygous for both loci, namely, *AaBb.* In reality, neither AB nor O children occur, if we omit the 11 exceptions out of 1,290 individuals (Table 23). Similarly, contrary to observations, marriages A × AB, B × AB, and AB × AB should, on the basis of the two-loci hypothesis, give rise not only to children of groups A, B, and AB, as they actually do, but also to O children.

These contradictions of the theory were not apparent for many years because of several circumstances. Among these is the fact that people of group AB constitute less than 5 per cent of most white populations, so that in the great majority of marriages neither spouse is AB, and thus the overwhelming proportion of marriages are of one of the six types that have children of the expected groups only. A second reason was that the techniques of blood-group determination were imperfect, so that some children assigned to group O really belonged to another group. Such erroneous determination meant that group O children were listed among the offspring of AB persons, where they were expected according to the two-locus theory, although more correct determinations have since shown that such parents have no O children. Finally, a certain amount of illegitimacy must be reckoned with, even in carefully selected data, and this can explain the occurrence of AB children from O ×

AB marriages (Table 23), where, again, they were expected according to the two-locus theory, but where they do not occur among legitimate offspring.

Allele-frequency analysis. The replacement of the hypothesis of two loci, each having two alleles, by the theory of multiple alleles was not based on an analysis of family data and of the discrepancies in groups of children expected and observed, but on a consideration, by the mathematician Felix Bernstein (1878–1956) of the relative frequencies of the four types of individuals in various populations. The two-locus hypothesis can be subjected to the allele-frequency analysis with which we became acquainted in the chapter on the Hardy-Weinberg Law. We shall assign the frequency p_A to the dominant allele *A* of the *A, a* pair, and the frequency p_B to the allele *B* of the *B, b* pair. The frequencies of the recessive alleles *a* and *b* are, therefore, $(1 - p_A)$ and $(1 - p_B)$, respectively. It is, then, a simple matter of multiplying the frequencies of genotypes *AA, aa,* and *Aa*, p_A^2, $(1 - p_A)^2$, and $2p_A\,(1 - p_A)$, by those of *BB, bb,* and *Bb*, p_B^2, $(1 - p_B)^2$ and $2p_B\,(1 - p_B)$, in order to obtain the frequencies of the different combinations of the genotypes at both loci (Table 25A). These calculations give us the expected frequencies of the four blood groups according to the two-locus hypothesis (last column of Table 25A).

By adding the frequencies of all A and AB persons, we obtain the following relation, in which $\overline{A}$ and $\overline{AB}$ indicate the frequency of A and of AB, respectively:

$$\overline{A} + \overline{AB} = 2p_A - p_A^2.$$

TABLE 25A
ABO Blood-group Composition of a Population According to the Two-loci Hypothesis.

Blood Group	*Genotype*	*Frequency*	
		Genotype	Blood Group (Sum of Appropriate Genotypes)
O	*aa bb*	$(1 - p_A)^2(1 - p_B)^2$	$(1 - p_A)^2(1 - p_B)^2$
A	*AA bb*	$p_A^2(1 - p_B)^2$	$(2p_A - p_A^2)(1 - p_B)^2$
	Aa bb	$2p_A(1 - p_A)(1 - p_B)^2$	
B	*aa BB*	$(1 - p_A)^2 \cdot p_B^2$	$(1 - p_A)^2(2p_B - p_B^2)$
	aa Bb	$(1 - p_A)^2 \cdot 2p_B(1 - p_B)$	
AB	*AA BB*	$p_A^2 \cdot p_B^2$	$(2p_A - p_A^2)(2p_B - p_B^2)$
	Aa BB	$2p_A(1 - p_A) \cdot p_B^2$	
	AA Bb	$p_A^2 \cdot 2p_B(1 - p_B)$	
	Aa Bb	$2p_A(1 - p_A) \cdot 2p_B(1 - p_B)$	

Similarly,

$$\overline{\text{B}} + \overline{\text{AB}} = 2p_B - p_B^2.$$

Since the frequency of AB persons is $(2p_A - p_A^2) \times (2p_B - p_B^2)$, the two pre- equations may be combined into the equation:

$$(\overline{\text{A}} + \overline{\text{AB}}) \times (\overline{\text{B}} + \overline{\text{AB}}) = \overline{\text{AB}}.$$

This equation and with it the two-locus hypothesis *can be tested.* All three terms, $\overline{\text{A}}$, $\overline{\text{B}}$, and $\overline{\text{AB}}$, indicate frequencies known from observation. If we add the percentage of A persons in any given population to the percentage of AB persons and multiply the sum by the sum of the percentages of B and AB persons, the resulting product should be the same as the percentage of AB persons.

This test of the hypothesis of two independent genes may be applied, not only to one population, but to a great number of populations, since it has been discovered that different racial groups show different frequencies of the four blood groups. Each racial group should fit the equation, since it should hold for any frequency of A, B, and AB.

However, the product $(\overline{\text{A}} + \overline{\text{AB}}) \times (\overline{\text{B}} + \overline{\text{AB}})$ is not equal to, but consistently larger than, $\overline{\text{AB}}$. This will be shown specifically for samples of two populations, American whites and Japanese. For the whites, the observed frequencies of the blood groups are

$$\overline{\text{O}} = 0.45,\ \overline{\text{A}} = 0.41,\ \overline{\text{B}} = 0.10,\ \overline{\text{AB}} = 0.04.$$

Entering these values in the equation, we obtain

$$(\overline{\text{A}} + \overline{\text{AB}}) \times (\overline{\text{B}} + \overline{\text{AB}}) = 0.45 \times 0.14 = 0.063,$$

which is more than 50 per cent greater than the observed frequency of AB, 0.04.

For the Japanese population, observation gave

$$\overline{\text{O}} = 0.294,\ \overline{\text{A}} = 0.422,\ \overline{\text{B}} = 0.206,\ \overline{\text{AB}} = 0.078.$$

Thus,

$$(\overline{\text{A}} + \overline{\text{AB}}) \times (\overline{\text{B}} + \overline{\text{AB}}) = 0.500 \times 0.284 = 0.142,$$

which is nearly twice as large as the observed frequency, 0.078. Similar discrepancies were found in many other populations. The disagreement between expected and observed frequencies means that either the underlying theory of two loci, each having two alleles, is false or that, of all different populations tested, none was the result of random mating. This latter possibility could

TABLE 25B
ABO Blood-group Composition of a Population According to the Theory of Three Multiple Alleles.

Blood Group	*Genotype*	*Frequency*	
		Genotype	Blood Group (Sum of Appropriate Genotypes)
O	$I^{O}I^{O}$	r_{O}^{2}	r_{O}^{2}
A	$I^{A}I^{A}$ $I^{A}I^{O}$	p_{A}^{2} $2p_{A}r_{O}$	$2p_{A}r_{O} + p_{A}^{2}$
B	$I^{B}I^{B}$ $I^{B}I^{O}$	q_{B}^{2} $2q_{B}r_{O}$	$2q_{B}r_{O} + q_{B}^{2}$
AB	$I^{A}I^{B}$	$2p_{A}q_{B}$	$2p_{A}q_{B}$

clearly be excluded, since family data show that there is certainly not enough, if any, assortative mating by blood groups to account for the observed deviation in frequencies. Therefore, the two-locus hypothesis had to be abandoned.

The Theory of Multiple Alleles. Having proved the inadequacy of the two-locus hypothesis, Bernstein assumed the existence of three multiple alleles, best called I^{A}, I^{B}, and I^{O} (some writers use such symbols as *A*, *B*, and *R*, or *A*, *B*, and *O*). It was further assumed that the alleles I^{A} and I^{B} are codominant if combined in the genotype $I^{A}I^{B}$, but that either allele is dominant in heterozygous combination with I^{O}.

From these assumptions, the frequencies of the six possible genotypes can be calculated as shown in the third column of Table 25B. Then, by adding the frequencies of those genotypes that have identical phenotypes, the frequencies of the four blood groups are obtained (final column).

Allele-frequency analysis. The frequencies p_{A}, q_{B}, and r_{O}, which were assigned to the three alleles, must add up to 1 ($p_{A} + q_{B} + r_{O} = 1$), just as the sum of the frequencies of two alleles, p and q, equals 1. Our goal is to compare the theoretical expectation that $p_{A} + q_{B} + r_{O} = 1$ (an expectation which holds only if the theory of multiple alleles is correct) with data derived from actual observation.

The formulas in the last column of Table 25B enable us to express the allele frequencies p, q, and r in terms of observed frequencies of blood groups. Since the frequency of O individuals is r_{O}^{2},

$$r_{O} = \sqrt{\overline{O}}. \qquad (1)$$

The frequency of O and A persons together is

$$\overline{O} + \overline{A} = r_O^2 + 2p_A r_O + p_A^2 = (r_O + p_A)^2,$$

or

$$p_A + r_O = \sqrt{\overline{O} + \overline{A}}. \qquad (2)$$

Similarly,

$$\overline{O} + \overline{B} = r_O^2 + 2q_B r_O + q_B^2 = (r_O + q_B)^2,$$

or

$$q_B + r_O = \sqrt{\overline{O} + \overline{B}}. \qquad (3)$$

By substitution, according to (1), of $\sqrt{\overline{O}}$ for r_O in (2), we obtain

$$p_A = \sqrt{\overline{O} + \overline{A}} - \sqrt{\overline{O}}, \qquad (4)$$

and by the same substitution in (3),

$$q_B = \sqrt{\overline{O} + \overline{B}} - \sqrt{\overline{O}}. \qquad (5)$$

Adding the three allele frequencies as given by the three equations (4), (5), and (1), we find

$$p_A + q_B + r_O = \sqrt{\overline{O} + \overline{A}} - \sqrt{\overline{O}} + \sqrt{\overline{O} + \overline{B}} - \sqrt{\overline{O}} + \sqrt{\overline{O}},$$

which, since $p_A + q_B + r_O = 1$, may be expressed finally as

$$\sqrt{\overline{O} + \overline{A}} + \sqrt{\overline{O} + \overline{B}} - \sqrt{\overline{O}} = 1. \qquad (6)$$

This equation, (6), and with it the theory of multiple alleles *can be tested* by determining the frequencies of O, B, and A persons in different populations and entering these frequencies in the left side of the equation. If the theory of multiple alleles is correct, the left side of the equation will always equal 1.

When we apply this procedure to the same two populations, American whites and Japanese, for which the two-locus hypothesis failed to give a satisfactory explanation, we find excellent agreement between observation and expectation, the difference between 1 and the calculated values being only about 1 per cent. This is shown in Table 26. This test for the theory of multiple alleles has been carried out for hundreds of different populations, always with close agreement between observed and expected values.

Instead of using the abstract terms listed in the headings of Table 26, it is customary to calculate, by means of formulas (4), (5), and (1), the actual fre-

TABLE 26
ABO Blood Groups. Test of the Theory of Multiple Alleles by Means of Equation (6).

Population	$\sqrt{\bar{O}+\bar{A}}$	$\sqrt{\bar{O}+\bar{B}}$	$\sqrt{\bar{O}}$	$\sqrt{\bar{O}+\bar{A}}+\sqrt{\bar{O}+\bar{B}}-\sqrt{\bar{O}}$
Whites	$\sqrt{0.45+0.41}$ $=0.927$	$\sqrt{0.45+0.10}$ $=0.742$	$\sqrt{0.45}$ $=0.671$	0.998
Japanese	$\sqrt{0.294+0.422}$ $=0.846$	$\sqrt{0.294+0.206}$ $=0.707$	$\sqrt{0.294}$ $=0.542$	1.011

quencies p_A, q_B, and r_O of the three alleles I^A, I^B, and I^O. Addition of the three values for p_A, q_B, and r_O should, and does, give a value close to 1. In our two examples, p_A, q_B, and r_O are 0.256, 0.071, and 0.671 for whites, and 0.304, 0.165, and 0.542 for Japanese, which add up, respectively, to 0.998 and 1.011. The results of some other calculations of p_A, q_B, and r_O from various populations are given in Table 27. (It may be noted that the derivation of the allele frequencies p, q, and r is not "fully efficient" in the statistical sense. Fisher and others have devised more efficient but also more complex formulas.)

Analysis of family data. The theory of multiple alleles not only gives a satisfactory explanation for the relative proportions of the four blood groups in different populations but also, in contrast to the two-locus hypothesis, fits the results of all possible matings (Table 24, last column). It is interesting to note that when the theory of multiple alleles was first proposed, full agreement between the theory and the available family data was not obtained. Many discrepancies, such as those listed as "exceptions" in Table 23, were on record. Indeed, this table summarizes only more recent data, and "exceptions" were considerably more numerous in the earlier determinations. Thus, among 2,270

TABLE 27
Allele Frequencies of I^A, I^B, and I^O in Six Different Populations.

Population	p_A	q_B	r_O	$p_A+q_B+r_O$
English	0.268	0.052	0.681	1.001
French	0.262	0.074	0.657	0.993
Bulgarians	0.271	0.108	0.624	1.003
Arabs	0.209	0.129	0.660	0.998
Senegal Negroes	0.149	0.189	0.657	0.995
Hindus	0.149	0.291	0.560	1.000

Source: Wiener, *Blood Groups and Transfusion,* 3rd ed., Thomas, 1943.

children from 973 families tested before 1924, the year of Bernstein's first publication, 1.32 per cent represented exceptions, but among 12,614 children from 5,559 families tested between 1927 and 1937, only 0.2 per cent were exceptions.

That many seeming exceptions were really faulty determinations of blood groups has been shown in an ingenious way. Some "exceptions" are the result not of faulty determination but of illegitimacy. Such exceptions can be excluded if data on blood groups of children and mothers only are selected, leaving the fathers out of the picture. Suitable for such a study are AB mothers, who should not have O children; and O mothers, who should not have AB children. Among 946 children of 675 AB mothers, only 3 O children were found; and among 5,454 children from 4,370 O mothers, 5 AB children. These data came from fifteen different reports, and the 8 exceptions were not evenly distributed through these reports but came from only three of them. Since these three studies contain less than 5 per cent of the total number of cases examined and were all made without knowledge of the multiple-allele theory—and hence were not rechecked—the exceptions may be regarded as due to errors of observation and not to inadequacy of the theory.

The Possible Occurrence of True Exceptions from Expected Genotypes. It is, of course, possible that true exceptions to the rules of inheritance of the ABO or other blood groups, or any other character, may occur. Extremely rare genes have been discovered which suppress the appearance of the ABO antigens even though the I genes are present (see p. 401). Or, although very unlikely, there is always the chance that a mutation will cause a child to have an allele that neither parent possessed. Furthermore, irregular chromosome distribution, as in nondisjunction, could account for true exceptions. If, in an AB individual, the two chromosomes which carry the I^A and I^B alleles do not segregate from each other in meiosis, gametes may be formed which carry both I^A and I^B alleles, or which lack both. If the mate of the AB individual belonged to group O, then fertilization with the abnormal gametes would result in AB and O children, both of which would be true exceptions.

It is now known that such chromosomally abnormal zygotes either die early during embryogenesis or, if able to come to term, exhibit severely abnormal phenotypes. The fact that one of the very few carefully checked exceptions in a study of blood groups, an O child from an AB mother, was severly pathological—mentally deficient and nearly blind—suggests that a chromosomal abnormality was involved.

There are, however, some cases on record in which normal individuals represent true exceptions from the ABO constitutions expected on the basis of their ancestry and offspring. In one relevant family an O mother had an AB daughter. The daughter married an O man and had two AB children. Presumably the daughter received the AB combination as a single I^{AB} allele from her father and transmitted the combination to her two children. Later in this chapter we shall discuss the possible origin of the I^{AB} combination. Here it might be appropriate to stress that the existence of the AB exceptions does not detract from the use of the ABO blood groups in questions of paternity (see Chapter 12). This is true since the exceptional AB individuals differ recognizably from typical individuals in whom I^A and I^B are located in two homologous chromosomes. The B antigen of persons in whom A and B are on the same chromosome appears weaker than normal and the serum contains some kind of anti-B which, however, does not react with their weak B antigen.

The Subgroups of A. Refined immunological tests have shown that blood group A consists of persons belonging to one or another of three different subgroups, called A_1, A_2, and A_3, and that group AB contains persons who are classed as A_1B, A_2B, and A_3B. Groups A_3 and A_3B are very small, constituting only a fraction of a per cent of all A or AB individuals. A_1 and A_1B, and A_2 and A_2B, make up about three-quarters and one-quarter, respectively, of all A or AB American whites. On the whole, pedigree studies indicate that the subgroups A_1, A_2, and A_3 depend on three alleles at the *I* locus, which might be called I^{A_1}, I^{A_2}, and I^{A_3}. The allele I^{A_1} is dominant over both I^{A_2} and I^{A_3}, and I^{A_2} is dominant over I^{A_3}.

The recognition of the four important alleles I^{A_1}, I^{A_2}, I^B, and I^O leads to the expectation of ten different genotypes and of six recognizable phenotypes. If one includes I^{A_3} and a fifth allele, I^{A_4}, which is even rarer than I^{A_3}, many more different genotypes and phenotypes are to be expected. Moreover still other rare kinds of I^A have been found as well as some very rare subgroups of I^B type alleles.

The foregoing account of alleles at the ABO locus is incomplete. Even more so are the following accounts of other blood-group genes, which do justice neither to the intricacies of the blood groups nor to the ingenuity and diligence of workers in this field. For more detailed presentations the reader is referred to the admirable most recent summary *Blood Groups in Man* by Race and Sanger, to somewhat older reviews by Wiener, alone or with collaborators, and finally to the vast original literature.

SECRETORS AND NONSECRETORS

The ABO antigens occur in an alcohol-soluble form, not only on the red cells but also in many other tissues. The presence of ABO antigens in different tissues is perhaps not particularly remarkable, since the *I* genes of an individual are presumably present in all his cells. A more significant fact is that many persons have water-soluble forms of the ABO antigens in their *secretions*, particularly the saliva, while the secretions of other persons do not contain any ABO antigens. The "secretors" and "nonsecretors" are differentiated by a pair of alleles: the former are *SeSe* and *Sese*; and the latter, *sese*. The ABO blood group of a secretor can be determined either by testing his blood or his saliva, but the saliva of a nonsecretor gives no clue. Secretion is restricted to the ABO antigens and to the so-called Lewis antigens, which will be discussed later: none of the antigens in the other blood groups are secreted.

THE MN BLOOD GROUPS

Landsteiner and Levine, in 1927, found two new human antigens, which they called M and N. One or the other is found in almost all human blood: the red cells of some persons possess M; of others, N; and of still others, both M and N. Antibodies against M and N are not usually found in humans, but they can be obtained by injecting either M or N blood into rabbits or other mammals, which induces the animal to produce the appropriate antibody. The serum of untreated rabbits does not agglutinate human red cells, but that of a rabbit injected with M blood clumps both M and MN cells, and that of a rabbit injected with N blood clumps both N and MN cells.

The inheritance of the M and N antigens through a codominant pair of alleles L^M and L^N has been described in the preceding chapter. That the *L* alleles, named *L* in honor of Landsteiner (1868–1943), are inherited independently of the *I* alleles can be shown by family data. Thus, marriages of persons having the phenotype OM to ABMN spouses produce four types of children, AM, BM, AMN, and BMN, in equal numbers. This shows that two separate randomly recombining loci are involved, with the genotypes of the parents being $I^OI^OL^ML^M$ and $I^AI^BL^ML^N$. The same information can be obtained from a survey of a population. According to the Hardy-Weinberg Law, as applied simultaneously to two different loci, the frequencies of M, MN, and N individuals should be the same in each of the groups O, A, B, and AB. This is true, within chance limits, as shown in Table 28.

MN Subgroups. A number of subgroups, similar to the A_1, A_2, and other subgroups of the ABO system, have been found in the MN system. Among these is one consisting of persons who have the rare, weak antigen N_2, which depends on the presence of the allele L^{N_2}; another, of those who have the antigen M^g, transmitted by the very rare allele designated as L^{M^g}. The antigen M^g is remarkable in that it does not react with either antibody anti-M or anti-N. Up to about 1965, among more than one hundred thousand Caucasians tested in Boston and England no persons with L^{M^g} had been found, but in a sample

TABLE 28
Percentage Frequencies of the M, MN, and N Blood Groups among the O, A, B, and AB Groups.

Group	*Percentage*		
	M	MN	N
O	32.5	46.5	21.0
A	32.6	48.0	19.4
B	31.8	47.7	20.5
AB	30.6	50.5	18.9

Source: Wellisch, *Z. Rassenphys.*, **10**, 1938.

of 6,530 Swiss, 10 were heterozygous for L^{Mg}. Two of the heterozygotes were married to each other. One of their two children was homozygous $L^{Mg}L^{Mg}$. She was neither M nor N nor MN! A still more striking, very rare *L* allele is that called L^{Mk}. It does not result in either M, N or Mg antigens nor in the S, s antigens to be described in the next section. Only heterozygotes for L^{Mk} are known. They form solely the antigens controlled by the specific "normal" *L* allele, which accompanies L^{Mk} in the heterozygote.

Presence of such alleles as L^{Mg} and L^{Mk} may result in apparent exclusions of paternity or maternity when actually no exclusion is justified. If, for instance, a woman is $L^{N}L^{Mk}$, she has the phenotype N. If she marries an $L^{M}L^{M}$ man, the expectation is that the apparent N × M union would produce only MN children. Actually, one-half of the children would be $L^{M}L^{Mk}$—i.e., phenotypically M. Only refined immunological details, pedigree data, and other evidence would prevent the exclusion of maternity in these very rare families.

The S and s Antigens. In 1947, twenty years after the discovery of the antigens M and N, a new antigen, S, was found. It differs serologically from M and N, occurs among individuals of all three MN blood types, and has an intimate genetic relation to them. Family studies have made it clear that individuals who have S are homozygous or heterozygous for one allele and that individuals who do not have S are homozygous for another allele. An antibody which agglutinates the red cells of the non-S homozygotes as well as those of the heterozygotes was soon discovered, thus indicating that there are two antigens, S and s, and three blood types, S, Ss, and s. (The nomenclature for antigens has grown up haphazardly. There is no good reason why one pair of antigens is designated by two different capital letters, namely M and N, and another pair by the capital and lower-case forms of a single letter, namely, S and s. Since heterozygotes have both antigens of a pair, it would perhaps be best to have a single letter for each pair and a subscript for each partner, e.g., M_1 and M_2, and S_1 and S_2. Since, however, MN and Ss are in general use, we shall continue to use these symbols.)

The Relation between MN and Ss. If S and s were determined by genes different from those determining M and N, then the frequencies of M, MN, and N should be the same among both S and s individuals. As shown in Table 29, this is not true. There are appreciably more M individuals among those with S, and more N individuals among those with s. Family data also show an association between M and S, and N and s. Thus, the offspring of MNSs × Ms parents consists of only two types of children: in some families these are MSs and MNs; in others they are Ms and MNSs. Clearly, in the first group all the children received either M and S or N and s from their heterozygous parent; in the second group all the children received either M and s or N and S. Thus, sibships consist of no more than two, complementary phenotypes, either MS and Ns, or Ms and NS. This phenomenon has a simple explanation. Expressed in the nomenclature for multiple alleles there are in any population four different alleles: L^{MS}, L^{Ms}, L^{NS}, and L^{Ns}. Each allele is responsible for the appearance of those two antigens noted in its superscript. MNSs persons are of two different genotypes: $L^{MS}L^{Ns}$ and $L^{Ms}L^{NS}$.

The frequencies of the four alleles are not alike. Among whites the percentage of alleles coding for M is 53; for N, 47; for S, 33; and for s, 67. Random association between the M, N and S, s determinants, calculated by multiplying with each other the appropriate frequencies (e.g., 0.53 × 0.33 for MS) would yield, in percentages, MS, 17; Ms, 36; NS, 16; Ns, 31. Observation, however, yields MS, 25; Ms, 28; NS, 8; Ns, 39. Relative to random expectations, M and S occur more often together than M and s, and N and s more often than N and S.

There is a kind of blood, called u, which is exceedingly rare in whites but relatively frequent in blacks, that does not react with either anti-S or anti-s serum—just as the rare antigen M^g does not react with either anti-M or anti-N. Genetic evidence indicates that u depends on one or the other of two alleles at the *L* locus, L^{Mu} and L^{Nu}. Disregarding for the moment the rare antigens N^2,

TABLE 29
Percentage Frequencies of the M, MN, and N Groups among 221 *S*– and 173 *ss* Individuals.

Individuals	M	MN	N
S–	22.3	26.9	7.0
ss	9.4	22.0	12.4

Source: After Wiener and Wexler.

M^g, and others, we have thus six *L* alleles, namely, L^{MS}, L^{Ms}, L^{NS}, L^{Ns}, L^{Mu}, and L^{Nu}. There is a serum, called anti-U, which agglutinates the blood cells of all individuals who have at least one of the first four alleles, and thus indicates the presence of a very common antigen, U. Absence of agglutination, which indicates that the blood is type u, is characteristic for the genotypes $L^{Mu}L^{Mu}$, $L^{Nu}L^{Nu}$, and $L^{Mu}L^{Nu}$, which signifies that the alleles with u as a superscript are recessive, as far as u is concerned.

The Hunter (Hu) and Henshaw (He) Antigens. Two more antigens among still others that are controlled by *L* alleles will be mentioned. They were named after the two blacks, one from New York and the other from Nigeria, in whom they were discovered. Both antigens are rare and the majority of persons tested do not have them. Family as well as population data show that heterozygotes for either Hu or He invariably transmit the ability to form these antigens together with a specific combination of M or N with S or s. The anti-Hu and anti-He sera have made it possible to distinguish further subtypes among the already known *L* alleles: L^{Ms} becomes L^{MsHu} and $L^{Ms\ not-Hu}$; and other alleles are L^{NSHu}, $L^{NS\ not-Hu}$, L^{NsHu}, $L^{Ns\ not-Hu}$, L^{NSHe}, $L^{NS\ not-He}$, L^{NsHe}, and $L^{Ns\ not-He}$. This nomenclature of the *L* alleles, while informative, is unwieldy. We shall return to it.

GENES, ANTIGENS, AND ANTIBODIES

The existence of the many diverse subgroups of *L* antigens poses questions concerning the organization of the various *L* genes responsible for the antigens. Is the *L* gene a single unit of function, a cistron of polynucleotides, or is it a "compound locus" consisting of two or more separate cistrons: one, for instance, that controls the M-N specificity; another, the S-s specificity of the antigens? The genetic evidence shows that in heterozygotes the two kinds of specificities contributed by each parent always segregate together, leading to two kinds of gametes in a ratio of 1:1. If the *L* gene is regarded as representing a compound locus, it is therefore necessary to assume that its components are so close together in the chromosome as to be "absolutely linked."

A decision between the alternatives (single versus compound locus) could be made if the chemical constitution of the polypeptide products of the *L* locus were known. With a single cistron only one type of polypeptide chain would exist; with two or more cistrons, two or more polypeptide chains would be discernible. These criteria cannot be applied at present since the polypeptide chains have not yet been isolated and analyzed. Judging from the extensive evidence on the molecular action of alleles of the *I* gene and from the more limited evidence from other blood-group genes, the polypeptide chains, alone or as parts of larger molecules, may be enzymes that control the addition of specific monosaccharides to polysaccharide side chains of glycoprotein and glycolipid precursor molecules (see p. 54).

The completed molecules are antigenic in nature. They may be responsible for the production of more than a single antigen since different regions (= sites) of the molecules may act independently as antigens. Thus instead of a gene controlling the appearance of a single antigen it may be responsible for a whole cluster of antigens.

Immunologically, specific antigens are recognized by the antibodies that they cause to be produced when they are injected into an experimental animal (or, as will be seen later, into a human individual who does not contain the antigen). Here again there is no simple relation between antigen and antibody. Landsteiner, in pioneering experiments, prepared chemicals with rather simple molecular structures and combined them with proteins. The ensuing substances acted as antigens that, after injection into rabbits, induced the production of antibodies. One of Landsteiner's remarkable findings was that an artificial antigen causes the production of a whole spectrum of antibodies. It seems justified to extrapolate from these experiments and to postulate: One gene leads to a variety of antigens, each of which may cause the production of a variety of antibodies.

If one gene leads to a variety of antigens, how can the roles of its different alleles be defined? Again the details are not known, but it may be postulated that different alleles lead to the formation of different enzymes, which, in turn, act on precursor molecules by adding specific side chains to them or otherwise endowing them with specific atomic configurations. These atomic configurations then express themselves as specific antigens and become observable due to their different antigenic properties.

Apart from these immunogenetic considerations, a study of a gene with multiple effects in *Drosophila* by Carlson may be mentioned. The effects of the gene "dumpy" have been divided into the following categories: (o) on wing size and shape; (v) on growth of the thorax surface; and (l) lethality. Seven main types of alleles have been established, each allele being characterized by its specific effect. By means of conventional recombination mapping it was deduced, with some uncertainties left, that there are several sites for each specific effect. The (o) effect can be produced by at least four different sites, the (v) effect likewise by at least four sites, and the (l) effect by at least three sites. Using the term "sublocus" for a given section of the dumpy locus Carlson summarizes a relevant part of his findings by stating: "a given type of effect is not confined to a given sublocus and a given sublocus is not confined to a given type of effect. Nevertheless, the region acts for the most part as functionally single gene."

No human blood-group gene has yet been mapped in detail by the use of recombination data. In cattle, however, such an analysis has been possible to a considerable degree. The genetic determinants of the "B blood-group system," which controls some 30 antigenic specificities, have been assigned—tentatively—to at least 16 sites along a linear gene map. The more than one hundred known combinations of specificities controlled by these sites represent more than one hundred different B alleles.

The foregoing general discussion of the relation between genes and their products, particularly the antigens, will be illustrated by a specific example relating to the ABO blood groups. It concerns the different antigenic effects of the two alleles I^A and I^B. Compared with the structure of the I^O allele, are there two different nucleotide sites on the DNA molecule, change at one of which results in an I^A allele and change at the other which results in an I^B allele? Or do I^A and I^B differ, in different ways, at the same site? Or do they differ in more complex ways from I^O? We have no direct information to answer these questions. It may be tempting, however, to make use of the existence of the exceptional I^{AB} alleles mentioned earlier which singly lead to both antigens A and B. Could I^{AB} be interpreted as an allele which combines specific properties at two separate sites, namely that normally related to an A antigen and another normally related to a B antigen? This hypothesis could account for the existence of I^{AB} either as a result of crossing over between sites A and B in a $I^A I^B$ heterozygote or as a result of a mutation at the B site in an I^A allele or at the A site in an I^B allele. Such crossing over would be an intragenic event that, although rare, is well known to occur in various experimental organisms. Nevertheless the "two specific sites" interpretation of the control of the A and B antigens is not binding. It is noteworthy that the B antigen that is dependent on an I^{AB} allele differs from the typical antigen that is dependent on an I^B allele: the heterozygote $I^{AB}\,I^O$ has somewhat different, "weaker" B effects than $I^A I^B$. This shows that the effects of the I^{AB} allele are not simply the sum of the effects at two autonomous sites. Either there is some specific interaction between the sites, dependent on whether they are in the same chromosome ($= I^{AB}$, *cis* arrangement) or in opposite chromosomes ($= I^A I^B$, *trans* arrangement), or the I^{AB} allele does not represent a combination of two separate A and B sites but has a unique molecular configuration at a site different from those determining either the I^A or I^B allele.

The two views according to which different antigenic specificities inherited as a unit are either controlled by different subunits within a complex blood-group gene, or are the derived products of a gene that cannot at present be resolved into subunits defined by different antigens have both had proponents, particularly as applied to the *Rh* locus, which will be discussed in the next section. Some, like R. A. Fisher, Race, and Sanger, were in favor of the "compound hypothesis;" others, like Wiener, insisted that no molecular subdivision of blood-group genes according to their antigenic effects was justified. The two different views found their expression in different nomenclatures: labeling

TABLE 30
Six of the L Alleles As Defined by the Reactions of Antigenic Properties Controlled by Them.

Alleles	*Reactions with These Antisera*							
	M	N	M^g	S	s	U	Hu	He
L^1	+	−	−	+	−	+		
L^2	+	−	−	−	+	+	+	
L^3	+	−	−	−	+	+	−	
L^4	−	+	−	+	−	+		+
L^5	−	−	+	+	+			
L^6	−	+	−	−	−	−		

NOTE: Alternative designations of the alleles: $L^1 = L^{MSU}$; $L^2 = L^{MsUHu}$; $L^3 = L^{MsU\ not\text{-}Hu}$; and so on. Absence of a + or − sign indicates that no test with the respective antiserum has been made.

genes by a series of different letters that stand for the series of antigens controlled by a given allele as against a single letter for all alleles, with different, often arbitrary, superscripts. The presentation in the rest of this chapter of various blood-group systems other than ABO and MN will provide further examples of complexity and of the nomenclatures used to describe it. It would perhaps be wisest to label different allelic units of transmission by simple superscripts, e.g., L^1, L^2, L^3, and to list in a table their antigenic properties as known by reactions to antibodies (Table 30).

Regardless of the symbols used, genetic transmission is simple: any individual has pairs of the segregating units and transmits members of the pairs in Mendelian fashion.

THE RH BLOOD GROUPS

Rh Positives and Rh Negatives. At the time of their discovery, the M and N groups seemed uncomplicated, but later studies showed the existence of a long series of finer divisions. The same to an even greater degree applies to Rh groups. In 1940, Landsteiner and Wiener injected blood from the Rhesus monkey into rabbits and guinea pigs. Serum containing the resulting antibodies agglutinated not only Rhesus red cells, but also those of about 85 per cent of a population of white New Yorkers; the cells of the other 15 per cent of the population did not react with the "antiserum." The majority of persons tested were termed Rh positives, the minority Rh negatives.

It was soon demonstrated that the presence or absence of the antigen responsible for the positive reaction is heritable (Table 31). Two Rh-negative parents yield only Rh-negative children, but marriages of two Rh positives or of an Rh

TABLE 31
Inheritance of Rh Blood Groups.

Type of Marriage	*No. of Families*	*Children*		*Percentage Rh Negative*
		Rh Positive	Rh Negative	
Rh positive × Rh positive	73	248	16	6.1
Rh positive × Rh negative	20	54	23	29.9
Rh negative × Rh negative	7	–	34	100

Source: Wiener, *Blood Groups and Transfusion*, 3rd ed., Thomas, 1943.

positive and an Rh negative may produce children of both types. The proportions of the two types of offspring in the two latter marriages fit a simple single factor interpretation, according to which Rh positives are either homo- or heterozygous for a dominant allele (*R*) and Rh negatives are homozygous for a recessive allele (*r*). The reader can easily verify this statement by applying equations (3) and (4) on page 243 to the data in Table 31. The frequency of the recessive allele in a large white population is obtained from the frequency of the recessive homozygote, which is $q_r^2 = 0.15$; thus, $q_r = 0.39$.

Since the discovery of the Rh groups, many different Rh antisera have been found, and these permit subdivision of what seemed at first to be simply two types of human blood. Most of these antisera have not been obtained from rabbits injected with monkey blood, but from humans who have produced antibodies against Rh antigens of other humans, either as a result of blood transfusions, the experimental injection of volunteers with the blood of others, or the passage of antigens from a fetus into the blood of the mother. It is this last process and its consequences for the health of the fetus that has made the Rh blood groups of unusual medical importance. The Rh interactions between mother and fetus will be dealt with in detail in Chapter 17.

Rh Nomenclature. Two different styles of nomenclature are used to designate the phenotypes and genotypes of the Rh blood groups. Fisher, Race, Sanger, and others represent the data by postulating a series of very closely linked genes, or sites within a single gene, which they call *C*, *D*, *E*, etc., and assigning to each of these genes, or sites, two or more allelic forms, e.g., *C*, *c*, and C^W; *D* and *d*; *E* and *e*. Correspondingly, they speak of C antigens and of anti-C antibodies, and so on. Any given gamete will contain a combination of alleles at the several loci or sites, such as *CDe* . . . , *cDe* . . . , *Cde* . . . , *cde* . . . , etc. Actually, certain data led to the suggestion that the proposed sequence of locations of the arbitrarily chosen letters C, D, and E is D-C-E.

Wiener and others interpret the data on the assumption that a single locus has any one of numerous different alleles. These are given the base letters *R* and r and distinguishing superscripts, e.g., R^1, R^0, R', and r^2. The antigens are called Rh, rh, and hr, with super- or subscripts; the antibodies, anti-Rh, anti-rh, and anti-hr, again with specifying super- or subscripts. For reasons indicated in the preceding general discussion of the relations between genes, antigens, and antibodies, in this book preference will be given to the use of R, as opposed to CDE, but the latter will often be given also to facilitate comparisons.

The Eight Rhesus Alleles Distinguishable by Three Specific Antisera. Three important and readily available antisera are anti-Rh_0, anti-rh′, and anti-rh″ (anti-D, anti-C, and anti-E). They permit recognition of eight different alleles (Table 32; at present consider the first three columns of reactions only). The allele r does not control antigens that react with any one of the three sera; the alleles r', r'', and R^0 control only one antigen each; the alleles r^y, R^1, and R^2, two antigens; and R^z, all three. Each antiserum, it is seen from the table, divides the blood tested into two groups, one that reacts positively and one that reacts negatively, and the combinations of the two alternatives in sets of three indicate the existence of eight different alleles. Their frequencies vary greatly, from 41 per cent to only 0.01 per cent.

Actually, blood tests do not provide a direct answer to the question: Which antigens are controlled by a specific allele? What such tests really determine are diploid genotypes, not single alleles. Genetic data show that the antigens produced by any one Rh heterozygote are the codominant sum of the antigens controlled by their two alleles or, viewed from a different angle, that the presence of an antigen reacting with any one of the three antisera is dominant over its absence. The homozygote rr is uniquely defined by its triple negative reaction, − − −, but more than one genotype can be assigned to all red cells that

TABLE 32
Eight of the Rh Alleles As Tested with the Three Antisera, Anti-Rh_0, Anti-rh′, and Anti-rh″. (Last two columns refer to additional tests with anti-hr′ and anti-hr″.)

Alleles	*Allele Frequency in Whites (%)*	*Reactions with Antisera*				
		Anti-Rh_0(D)	Anti-rh′(C)	Anti-rh″(E)	Anti-hr′(c)	Anti-hr″(e)
r	38	−	−	−	+	+
r'	0.6	−	+	−	−	+
r''	0.5	−	−	+	+	−
r^y	0.01	−	+	+	−	−
R^0	2.7	+	−	−	+	+
R^1	41	+	+	−	−	+
R^2	15	+	−	+	+	−
R^z	0.2	+	+	+	−	−

give at least one positive reaction. Thus, red cells which give the reaction $-+-$ are genetically either $r'r'$ or $r'r$, and individuals whose cells react $+++$ could belong to any one of the following genotypes: R^zR^z, R^zR^2, R^zR^1, R^zR^0, R^zr^y, R^zr'', R^zr', R^zr, R^2R^1, R^2r^y, R^2r', R^1r^y, R^1r'', R^0r^y. Some of these genotypes are very rare among whites, e.g., R^zr^y, and statistically, the most likely genotype for the $+++$ reaction is R^1R^2. Pedigree data, of course, often make possible a clear decision about a specific genotype or narrow down the range of alternatives. Note that $+++$ would be called CDE in the corresponding nomenclature, a designation that while suggestively simple obscures the multitude of genotypes which result in the three positive reactions.

Some further attributes of the eight Rh alleles. Two antisera, besides the three named, are the readily available anti-hr′ and the rare anti-hr″ (or anti-c and anti-e). No new alleles can be distinguished by the use of these sera, but they make it possible to distinguish various homozygotes from heterozygotes. Anti-hr′ gives positive reactions when anti-rh′ fails to do so, and anti-hr″ gives positive reactions when anti-rh″ does not (see Table 32, last columns). When anti-rh′ or anti-rh″ give negative reactions then anti-hr′ or, respectively, anti-hr″ give positive ones. In other words, rh′ and hr′, like M and N, are a pair of alternative antigens, and so are rh″ and hr″. Since heterozygotes have the antigens controlled by each of the two alleles, such genotypes as $r'r'$ and $r'r$ give different reactions to the battery of five antisera. For $r'r'$ they are $-+--+$; and for $r'r$, $-+-++$.

Much effort has been expended to find an antiserum to which the cells of Rh_0-negative bloods would react, but no anti-Hr_0 (anti-d) serum has been discovered. Although the symbols c (hr′) and e (hr″) stand for specific antigenic properties, the symbol d does not stand for a recognizable antigen, but only for the absence of an antigen.

Additional Rhesus Alleles. The reader may have noticed that in Table 32 the sum of the allele frequencies of the eight alleles adds up to less than 100 per cent. The reason for this is that there are still other alleles—more than 40 being known by 1968. The most common of these is called R^{1w}; it has an allele frequency of nearly 2 per cent and is similar in its action to R^1. The two alleles can be distinguished from each other by a special anti-rh^w (anti-C^W) serum, which reacts with an antigenic product of R^{1w} but not with R^1. Both R^1 and R^{1w} control positive reactions with anti-rh′ (anti-C). The anti-rh^w serum has also led to the discovery of an exceedingly rare r'^w allele, which is similar to r' but different from it in controlling a positive reaction with anti-rh^w.

Only two more examples of additional R alleles will be cited. One comprises a whole group of alleles which have been called "variants of R^0" (cD^ue) and which differ from the typical R^0 by controlling antigens that give only weak reactions with anti-Rh_0, or reactions with anti-Rh_0 sera from some but not all bloods. These variant alleles are, in a way, intermediate between typical R^0, which controls strong reactions with all anti-Rh_0 sera, and typical r, which leads to no reaction with any anti-Rh_0 sera. Intermediate reactions to anti-Rh_0 sera are also found with variants of R^1 and other R alleles. Still other variant alleles of a corresponding intermediate type are known for reactions to anti-rh′ and other antisera.

A remarkable R allele is $\overline{R^0}$ (–D–). Blood of the very rare persons who are homozygous for this allele reacts positively with anti-Rh_0 but not with anti-rh′, anti-hr′, anti-rh″, or anti-hr″. Thus, the $\overline{R^0}$ allele is one of a group of exceptions to the otherwise nearly universal rule that an R allele controls one or the other of the two antigens rh′ or hr′ (C or c), and also either rh″ or hr″ (E or e). The allele $\overline{R^0}$ is analogous in its expression to the L^{Mg} and L^{Mk} alleles of the MNSs blood groups.

The Heterozygote R^1r'. In heterozygotes the different R alleles usually act as codominants, leading to the production of all antigens controlled by each of the two alleles. According to Ceppellini, the R^1r' heterozygote is exceptional in that its anti-Rh_0 reaction is not the reaction typical for R^1 in other genetic combinations. Rather, it reacts with anti-Rh_0 sera in an "intermediate" manner, like some of the variants of R^1 (CD^ue). Pedigree studies make it clear that the intermediate reactions of R^1r' are not caused by an actual R^1 variant, but only by the combination of typical R^1 with r'. A certain R^1r' man with the atypical intermediate reaction who was married to an rr woman produced an R^1r daughter with the typical reaction to anti-Rh_0; and a typically reacting R^1r woman married to an $r'r$ man had, among her children, a typically reacting R^1r son and an atypically reacting R^1r' daughter.

The reason for the exception to typical dominance in the heterozygote R^1r' is not known. The C–D–E notation has suggested an explanation for this phenomenon. According to it, the genotype R^1r' is written as CDe/Cde; and R^1r, as CDe/cde. If we consider reactions to anti-Rh_0 as expressions of D and d, we must then ascribe the normal reaction of R^1r to the fact that d is next to c, and the intermediate reaction of R^1r' to d being next to C. Such *position effects* are known in other organisms, in which they have been tested by comparison between the two double heterozygous combinations of linked loci A and B, namely, Ab/aB and AB/ab (the oblique line separates two genes in one chromosome from their alleles in the homologous chromosome). In these tests it was certain that the A, a and the B, b alleles in the two heterozygotes were the same, since AB and ab had been synthesized by crossing over from Ab and aB. In the R alleles no crossover syntheses have been possible, so that it remains unknown whether the supposed d site in the Cde combination is identical with the supposed d site in the cde combination. It is therefore not clear why R^1r' shows intermediate reactions against anti-Rh_0, even on the C–D–E interpretation.

"Compound" Antigens. An antigen is known that seemed to be independent of the CDE series and was therefore given the designation f. It was later recognized that f was present in persons who had an allele leading to the c and e specificities, *ce*, but was not present in heterozygotes in which the two specificities were controlled by the separate alleles *c* and *e*. The antigen f may be a compound controlled by the cis combination *ce*—an interpretation that may reflect the CDE nomenclature more than the nature of the differential allelic reactions.

Differences in Allele Frequencies in Different Racial Groups. Most references to the frequencies of the Rh alleles are qualified by the statement that they relate either to whites or to blacks. This is necessary since there are striking differences in allele frequencies between different populations. We have mentioned this earlier in regard to the PTC-taster alleles, and also in regard to the *I* alleles for the ABO blood groups, and the same applies to alleles at other blood-group loci discussed in this chapter. A general treatment of the anthropological significance of variations in allele frequencies and detailed examples will be found in Chapter 31.

A Brief Summary of the Genetics of Rh. The analysis of genic loci may be compared with that of atomic "species." For a long time it was believed that there was a single type of atom for each element—carbon, phosphorus, uranium, etc.—but we now know that each comprises a family of isotopes that have similar or nearly identical chemical properties and yet differ in atomic weight, nuclear composition, and other respects. Physicists and chemists are not expected to know by heart all isotopes and their properties. Instead, tables are provided in which they can look up these facts. Similarly, any one of the numerous Rh alleles that are defined by their reactions with a series of antisera will be known to most persons concerned only after consultation of appropriate tables.

For some general purposes it is still adequate to distinguish Rh-positive and Rh-negative individuals by the reaction of their blood to a single serum containing anti-Rh_0. Use of two additional sera, anti-rh′ and anti-rh″, permits recognition of a total of eight Rh alleles (Table 32). Two other sera define the same eight alleles in more detail and permit the recognition of distinctions between some homo- and heterozygous genotypes which otherwise appear phenotypically alike.

OTHER BLOOD-GROUP SYSTEMS

By 1970 more than 12 other blood systems were known that seem to be inherited independently of each other and of the ABO, MN, and Rh systems. Some of these are listed in Table 33. Independence from all others has not yet been

TABLE 33
Some Other Blood-group Systems and Their Alleles.

Blood System	*Designations of Alleles**	*Blood System*	*Designations of Alleles**
P (= Q of Furuhata)	P^1, P^2, p	Diego	Di^a, Di^b
Kell	K, k, k^P	Yt	Yt^a, Yt^b
Lutheran	Lu^a, Lu^b	Dombrock	Do^a, Do
Duffy	Fy^a, Fy^b		
Kidd	Jk^a, Jk^b	Auberger	Au^a, Au
Lewis	Le, le	Stoltzfus	Sf^a, Sf

*After either Race and Sanger, or Wiener and Wexler.

proved for some of the most recently discovered systems. This is due to the fact that simultaneous data are needed on the population frequencies of the two systems whose independence is being tested, or simultaneous family data. When even one of the alleles of one of the systems is very rare it is difficult to obtain suitable data; if there is a very rare allele in both systems it is even more difficult.

The various alleles of each group either determine the presence or absence of an antigen or control more complex situations such as those described for the MN and Rh systems. In every blood-group system, alleles that lead to the appearance of specific antigenic properties are always dominant over alleles that do not produce them. If two alleles each lead to different antigens, they are codominant in the heterozygotes.

The basic genetics of most blood groups is well understood even though many of the precise and theoretically intriguing details are still obscure. For some time the Lewis groups were the outstanding exceptions, but even the basic features of their genetics now seem to be understood. The difficulties to understanding were connected with a peculiar interaction between the alleles *Le* and *le* in the control of the antigen Le^a and the secretor alleles *Se* and *se* (see p. 263). Accordingly, it is necessary to determine the presence or absence of Le^a separately for the saliva and the red cells. In the saliva the genotypes *LeLe* and *Lele* cause the appearance of Le^a regardless of the secretor genotype—be it *SeSe*, *Sese*, or *sese*—but on the red cells Le^a is present only when the dominant allele *Le*, homozygously or heterozygously, is accompanied by homozygous *sese*. No antigen is known which would be dependent on the recessive allele *le*. A special antigen Le^b which appears on the red cells of *Se–Le*–persons is believed to be an interaction product of the *H* and *Le* genes. It is present on the secreted glycoprotein molecules and is only secondarily absorbed onto the red cell surface (Table 34).

TABLE 34
The ABH Secretor and Lewis Systems Affecting the Saliva, According to the Theory of Grubb and Ceppellini.

Genotypes	*Antigen* of Saliva ABH	*Antigen* of Saliva Le^a	*Antigen* of Red Cells Le^a
SeSe or *Sese*, *LeLe* or *Lele*	+	+	−
sese, *LeLe*, or *sese*, *Lele*	−	+	+
SeSe or *Sese*, *lele*	+	−	−
sese, *lele*	−	−	−

Source: After Race and Sanger.

The typical Mendelian mode of inheritance of the blood groups assigns the location of all but one of their genes to the autosomes. The exception is the blood-group gene *Xg* whose inheritance is X-linked and which therefore is known to be located on the X-chromosome (see Chapter 14). It has been of special significance for the mapping of genes in this chromosome. The autosomal blood-group genes are similarily important in testing for autosomal linkages between them and the genes for other traits. One of the genes connected with the Lewis group—namely, the secretor gene—is located on the same chromosome as the Lutheran gene. This is the only known linkage of two blood-group loci.

"Private" and "Public" Blood Antigens. A number of antigens have been found only in single kindreds; others are found in nearly all persons. It is difficult to investigate whether specific private, or family, antigens belong to any one of the well-established blood-group systems since they are so rare, but some have been found to be controlled by very rare alleles of a known system. The public antigens are hard to recognize because their existence becomes known only if the rare individuals in whom they are lacking are discovered.

In the past, some antigens that were originally believed to be private have later been found to be common in racial groups other than that in which they were discovered. A striking example is the Diego-antigen, Di^a, which is inherited as a dominant. It is very uncommon in whites, but abundant in American Indians, Chinese, and Japanese.

The Number of Blood-group Combinations. Since there are at least fifteen blood-group systems, each represented by at least two blood groups and some by many, it is obvious that genetic recombination produces a great variety of

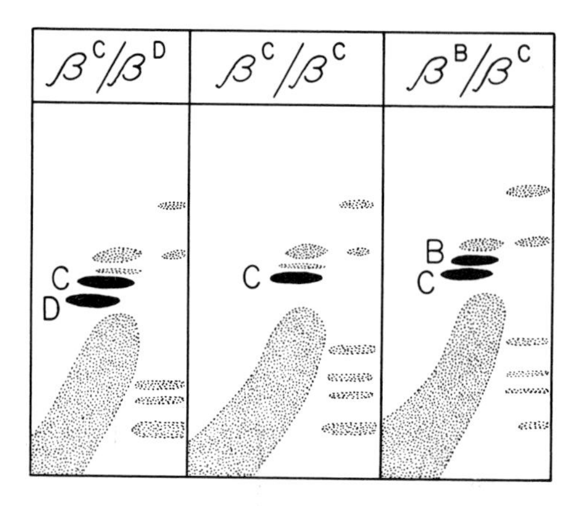

FIGURE 121
The results of electrophoresis of β-globulins from the sera of three individuals. The β-globulins were initially placed in the left-hand corner of each area and the electric fields applied once in the upward direction and once from left to right. *Left:* From a $\beta^C\beta^D$ heterozygote who possesses β-globulins C and D. *Center:* From a $\beta^C\beta^C$ homozygote who possesses C and not D. *Right:* From a $\beta^B\beta^C$ heterozygote who possesses B and C. The stippled areas represent other β-globulin components, among them the small amounts of B found in individuals who do not carry the β^B allele. (After Smithies, *Nature,* **181,** 1958.)

combinations of blood types in different persons. Some of these combinations are rare and others common, but even relatively common combinations occur in only one out of several hundred persons. Race and Sanger point out that the commonest combination of groups in England—O, M, N, s, Rh_1, rh, P_1, Lu^a negative, K negative, Fy^a and Fy^b positive, Jk^a and Jk^b positive, and Le^a negative and Le^b positive—is found in only 1 out of 270 persons. They have also encountered a person with the combination B, M, N, S, Rh_{1w}, Rh_1, P_2, Lu^a negative, KK, Fy^a negative, Jk^a positive, and Le^a positive, which should have a calculated frequency of only 1.4 in 100,000,000.

The great diversity of human blood types may be illustrated by the results of tests on 132 staff members of the Lister Institute in London. Using sera defining nine blood-group systems, 129 persons were shown to belong to different combinations, of which 126 occurred only once and 3 occurred only twice. Additional tests will make it possible to identify a person unambiguously from his blood alone. Such tests may include genetically controlled serum protein variants, e.g., the α2-globulins or haptoglobins (p. 55–58), β-globulins (Fig. 121), and γ-globulins; as well as the antigen groups of the white blood cells, and a number of enzyme variants found in serum.

THE FATE OF TISSUE GRAFTS: HISTO-INCOMPATIBILITY GENES

When pieces of skin taken from one person are grafted to another, as in treatment of severe burns, the transplant does not "take" permanently but sloughs off after a few weeks. Such *histo-incompatibility* can also be observed in transplants of other tissues or organs, e.g., kidneys. The destruction of the foreign tissue by the host is the result of immunological reactions. The transplant possesses antigens not present in the host, these antigens cause the production of antibodies by the host, and the interaction between the antibodies and antigens leads to the death of the transplant's cells.

In mice, detailed evidence of a genetic basis of histo-incompatibility has been obtained. At least 14 different histo-incompatibility loci are known, some with multiple alleles. These loci are not all of equal strength in their incompatibility reactions. The locus with the major effect is called *H*-2. Transplants between two strains homozygous for different alleles of a histo-incompatibility locus, H^1H^1 and H^2H^2, are not successful, nor are transplants from H^1H^2 hybrids to either parental type. However, the reciprocal transplants from either parent to the hybrid are permanent.

Why is this true? Apparently, the alleles H^1 and H^2 control the production of different antigens. The presence of an antigen within tissue transplanted to an animal lacking this antigen causes the production of specific antibodies, but an animal whose tissues already carry the antigen will not produce the antibodies and will therefore tolerate the transplant. Most experiments can be explained on the basis of the assumption that in a heterozygote the two different alleles codominantly lead to the production of different antigens. The outcome of transplantation experiments in certain other hybrids requires the assumption that the heterozygote does not possess either of the antigens that are present separately in its parents, but that a new antigenic "hybrid substance" is formed under the influence of the heterozygous genotype.

It is a remarkable and obviously necessary phenomenon that an animal or man does not ordinarily form antibodies against his own antigens. This tolerance is established at an early embryonic stage. Indeed, tolerance even against foreign antigens can be attained if the foreign tissue is grafted on the embryo early enough. It has been known for some time that, in cattle twins of different genotypes, blood-forming cells can pass from one embryo to the other, and that this "foreign" tissue becomes permanently established in the host. Human blood mosaics—defined as producers of a mixture of genetically different cells—have also been discovered in which a similar embryonic "transplantation" had led to permanent tolerance.

The pioneer analyses of histo-incompatibility in the mouse by Snell and Gorer have in recent years led to important basic and applied studies in man. In humans as in mice, survival of tissue and organ grafts depend on gene-controlled histo-incompatibility antigens (and, in addition on compatibility for the ABO blood groups). It is necessary, therefore, to use donors whose tissues are compatible with those of the recipients. A method for determining compatibility prior to grafting is the "mixed-leucocyte culture test." Leucocytes (white blood cells) of the prospective recipient and those of a series of potential donors, preferably from the recipient's family, are mixed and the reactions observed. If the cells in a mixture are alike or similar in regard to incompatibility genes, no or little stimulation of cell enlargement and mitosis is seen, contrary to the mixtures in which there are differences in incompatibility genes, in which, depending on the genes involved, there may be strong stimulation. A significant correlation has been found between degree of matching of leucocytes in these tests and survival of skin and kidney grafts. Genetic analyses of the leucocyte interactions of different individuals have led to the recognition of a major autosomal locus, *HL-A*, with more than twenty multiple alleles.

GENETICS OF ANTIBODIES

Antibodies are gamma-globulins that consist of four polypeptide chains: an identical pair of longer ("heavy") chains and an identical pair of shorter ("light") chains. Both types of chains have two different regions, one that is the same in its amino acid sequence in different antibodies of an individual and one that is greatly variable from one antibody to the next. The origin of this variability is a matter of much study in recent years. If we applied the one gene–one polypeptide chain concept to the polypeptides of antibodies, we would have to conclude that a very great number of different genes are responsible for the very great number of different antibody sequences. It seems unlikely that the genes of the fertilized egg include thousands that would be required to be coded for antibodies. Rather, it seems more probable that only a few genes in the zygote are destined to code for antibodies and that the diversity of antibody-controlling genes originated during development in the somatic cells that become the producers of antibodies. How this unusual type of differentiation is brought about is the subject of various theories. One of these suggests that the DNA sequence that codes for the variable regions of the immunoglobulin chains is distinguished by a very high mutation rate resulting in a variety of cells differing in immunoglobulin genotypes. Another theory, first proposed by Smithies, assumes that the variable DNA sequence frequently undergoes somatic recombination processes that result in new kinds of sequences. Somatic crossing over is known to occur in yeast, fungi, *Drosophila*, and other organisms, including some suggestive instances in man but in general, like mutation, it is a rare process. Various models have been proposed to account for new polynucleotide sequences produced by crossing over. One such model is outlined in Figure 122. The model is not intended to represent the actual situation, which must be much more complex, but to illustrate the kind of thinking that is being applied to the problem of antibody diversity as possibly generated by somatic recombination.

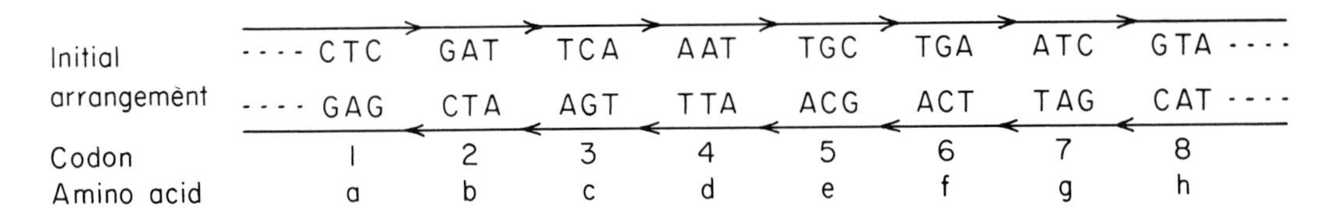

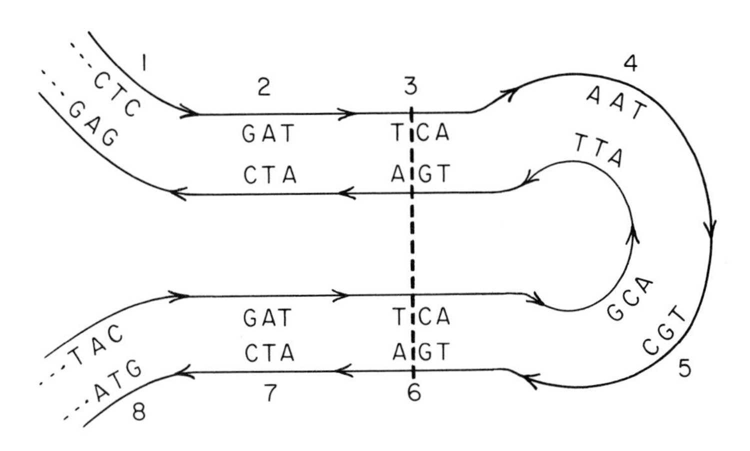

Arrangement after recombination	···· CTC	GAT	TCA	GCA	ATT	TGA	ATC	GTA ····
	···· GAG	CTA	AGT	CGT	TAA	ACT	TAG	CAT ····
Codon	1	2	3	4'	5'	6	7	8
Amino acid	a	b	c	i	j	f	g	h

FIGURE 122
Diagram of a section of a DNA gene consisting of 8 codons (a codon = a triplet coding for an amino acid). *Top:* The triplets 2 and 3 are duplicated, in inverted sequence, as triplets 6 and 7. *Middle:* The duplicated triplets undergo pairing. As a result the section bends back on itself. Recombination occurs within triplets 3 and 6 between complementary strands. *Bottom:* The section of 8 codons after intragenic recombination. Triplets 1, 2, 3, 6, 7, and 8 are unchanged, but the original triplets 4 and 5 are replaced by two new triplets 4′ and 5′ as a result of inversion of the stretch between the duplications. (After Smithies, *Nature,* **199,** 1963.)

Multiple Isoalleles. In general, different alleles are recognized by clearly different phenotypic effects. If two or more alleles are very similar in their effects, it may often be difficult to discriminate between them—even to be aware of their separate existence. If, for instance, two alleles of a hemophilia gene control blood-coagulation times that differ only slightly in length and if there is a variability in this time from one individual to another within each of the two genetically different groups, it may not easily become known that two alleles are actually present in the population. In such cases, one speaks of abnormal, or mutant, isoalleles.

Perhaps of more fundamental importance are normal isoalleles; that is, alleles that each control a phenotype within the range of normality but different in that range. Much normal variation may have its basis in the presence of many normal isoalleles at many loci.

It is seldom possible to find specific evidence for normal isoalleles. One way of demonstrating their existence becomes available when heterozygotes for a dominant abnormal allele show somewhat different phenotypes if the allele is combined with different normal isoalleles. Cases suggestive of this have been found in pedigrees in which there is great similarity between the phenotypes of affected sibs but little similarity between phenotypes of affected parents and their affected children. In pedigrees of the type of muscular dystrophy, which is due to a dominant autosomal gene, there is a much lower correlation between the ages of onset of the affliction of parents and children ($r = 0.32$) than between the ages of onset of paired sibs ($r = 0.66$). Similarly, variabilities in expression of a dominant autosomal gene for congenital abnormalities of the nails and of the patellae (kneecaps) show no significant parent-child correlation but very significant sib-sib correlation (0.46 for nails; 0.58 for kneecaps). According to Penrose, a genetic explanation for this phenomenon might be as follows: In certain families the affected parent is heterozygous for the mutant allele D and a normal isoallele d^1, and this combination (Dd^1) causes early onset or strong expression of the trait. The nonaffected parent is homozygous for a different isoallele (d^2d^2). The affected children are all Dd^2, a genotype that leads to late onset or weak expression of the trait. The affected sibs therefore will be similar among themselves (positive correlation), but different from the affected parent (negative correlation). In other families, the affected parent may be Dd^2, resulting in late onset or weak expression of the trait, and the nonaffected parent d^1d^1. The affected children will be Dd^1, with early onset or strong expression (again positive sib-sib and negative parent-sib correlation). Still other families will have other isoallelic combinations—$Dd^1 \times d^1d^1$, $Dd^1 \times d^1d^2$, etc.—some of which give positive correlations between parents and children as well as among sibs. Given high frequencies of both d^1 and d^2, or of additional isoalleles, the combined data will add up to absence of parent-child correlation but will retain positive sib-sib correlation.

This chapter has given various examples of multiple alleles in man. They are important in both family and population genetics. Their recognition has provided the genetic interpretation of the striking diversities of human blood types and of other human differences. In addition they play a role in the theoretical discussions of genes, gene structure, and gene action.

PROBLEMS

84. Six alleles (A^1, A^2, A^3, A^4, A^5, and A^6) occur at a certain gene locus. Enumerate, in two columns, all homozygous and all heterozygous combinations.

85. Give all possible genotypes for each of the following phenotypes relating to the ABO blood groups:

    (i) O (iii) A_2 (v) A_1B
    (ii) A (iv) B (vi) A_2B.

86. Given the alleles I^{A_1}, I^{A_2}, I^{A_3}, I^B, and I^O and the alleles L^{MS}, L^{Ms}, L^{NS}, and L^{Ns}, how many different genotypes are possible? How many phenotypes?

87. Assume the existence of 4 alleles of the *I* locus, 4 of *L*, 8 of *R*, 3 of *P*, 3 of *K*, and 2 each of *Lu*, *Fy*, *Jk*, *Le*, and *Se*. How many homozygous genotypes can be formed? How many genotypes, either homozygous or heterozygous?

88. From the allele frequencies given in Table 27, determine the frequencies of the four ABO blood groups among:
    (a) Bulgarians, (b) Arabs, (c) Hindus.

89. The frequencies of the ABO blood group alleles among Arabs and Hindus are given in Table 27.
    (a) If equal numbers of these two peoples intermarry at random, what will be the allele frequencies of the resulting population?
    (b) What will be the frequencies of the four blood groups?

90. In marriages of two I^AI^O L^ML^N people, what is the probability:
    (a) Of a child being A, M?
    (b) Of having one A, M child in a sibship of four?
    (c) Of having at least one (i.e., one or more) A, M child in a sibship of four?

91. With the use of Table 32, calculate the expected frequencies of the following genotypes in a population: rr, rr^y, r^yr^y, R^1r, R^1R^1.

92. The blood of a man gives positive reactions with all five antisera listed in Table 32.
    (a) What are his possible genotypes?
    (b) If it is found that his son has blood which reacts as $- - - + +$, what is the man's genotype?

93. List all possible types of transplantations, and their ultimate success or failure, between animals of the following histo-incompatibility genotypes: H^1H^1, H^2H^2, H^3H^3, H^1H^2.

REFERENCES

Bach, F. H., 1969. Histocompatibility in man—genetic and practical considerations. *Progr. Med. Genet.* **6**:201–240.

Bach, F. H., 1970. Transplantation: Pairing of donor and recipient. *Science*, **168**:1170–1179

Bach, F. H., and Amos, D. B., 1967. Hu-1, major histocompatibility locus in man. *Science,* **156:**1506–1508.

Bernstein, F., 1925. Zusammenfassende Betrachtungen über die erblichen Blutstrukturen des Menschen. *Zeitschr. Abst. Vererb.,* **37:**237–270.

Carlson, E. A., 1958. The bearing of a complex-locus in Drosophila on the interpretation of the Rh series. *Amer. J. Hum. Genet.,* **10:**465–473.

Cold Spring Harbor Laboratory of Quantitative Biology, 1967. *Antibodies.* 619 pp. Cold Spring Harbor Laboratory, Cold Spring Harbor, New York. (Cold Spring Harbor Symposium on Quantitative Biology, **32**.)

Edwards, J. H., 1968. The Rhesus locus. *Vox Sang.*, **15:**392–395.

Giblett, Eloise R., 1969. *Genetic Markers in Human Blood.* 629 pp. Blackwell. Oxford.

Grubb, R., 1970. *The Genetic Markers of Human Immunoglobulins.* 152 pp. Springer, New York.

Hildemann, W. H., 1970. *Immunogenetics.* 262 pp. Holden-Day, San Francisco

Kabat, E. A., 1968. *Structural Concepts in Immunology and Immunochemistry.* 310 pp. Holt, Rinehart, and Winston, New York.

Morgan, W. T. J., and Watkins, Winifred M., 1969. Genetic and biochemical aspects of human blood-group A-, B-, H-, Le^a-, and Le^b specificity. *Brit. Med. Bull.*, **25:**30–34.

Penrose, L. S., 1948. The problem of anticipation in pedigrees of dystrophia myotonica. *Ann. Eugen.*, **14:**125–132.

Prokop, O., and Uhlenbruck, G., 1969. *Human Blood and Serum Groups.* 891 pp. (English translation of revision of 2nd German edition.) Wiley, New York.

Race, R. R., and Sanger, Ruth, 1968. *Blood Groups in Man*, 5th ed. 599 pp. Blackwell, Oxford.

Renwick, J. H., 1956. Nail-patella syndrome: evidence for modification by alleles at the main locus. *Ann Hum. Genet.,* **21:**159–169.

Shreffler, D. C., and Klein, J., 1970. Genetic organization and gene action of mouse H-2 region. *Transpl. Proc.*, **2:**5–14.

Wiener, A. S., 1943. *Blood Groups and Transfusion*, 3rd ed. 438 pp. Thomas, Springfield, Ill.

Wiener, A. S., 1961. *Advances in Blood Grouping.* 549 pp. Grune and Stratton, New York.

Wiener, A. S., with a section by Shapiro, M., 1965. *Advances in Blood Grouping II.* 452 pp. Grune and Stratton, New York.

13

MEDICOLEGAL APPLICATIONS OF GENETICS

The genetic parent, or parents, of a given individual are not always known with certainty. It has happened, though very rarely, that newly born babies in a hospital have been assigned to the wrong mothers. A person may claim to be the long-lost child of a couple (usually a wealthy one!). War and social upheaval may separate children and parents for years, so that it is difficult to establish their relationship. Occasionally, the claims of individuals born outside the United States that they are citizens by American parentage have been doubted. By far the most common cause of uncertain relationship is illegitimacy. In such cases it is the paternity of the child that is often in doubt. Nearly 100 of every 1,000 live births in the United States are registered as illegitimate; specifically, in 1968, there were approximately 339,000 illegitimate births out of a total of 3,502,000. Many thousands of cases of disputed paternity and a few of disputed maternity come before the courts each year, mainly to judge whether a "suspected father" should support his supposed offspring. Hereditary characters are useful in helping to decide whether or not a person could be the parent of a specific child. Traits that depend on simple single factor inheritance, including multiple alleles, are particularly suitable for medico-legal analysis, since the relation between presence of the trait and presence of its determining allele is clear-cut.

EXCLUSION OF PATERNITY

There is no question that genetic evidence often shows conclusively that a specific man can *not* be the father of a given child: he is clearly not the father if the child carries a gene which is not carried either by him or the mother. However, genetic evidence that is compatible with the possibility that a specific man is the father of a given child does not constitute proof of his actual paternity. Almost always, other men have genotypes that are also compatible with their possible parenthood of the child. In general, exclusion of paternity is decisive; nonexclusion is not. In some situations, however, exclusion of one individual may be sufficient for positive designation of another. There may be nongenetic, social evidence that one of a known group of individuals must be the parent of a child. For example, a married woman may be suspected of adultery with a specific man, and it is desired to determine whether the husband or the suspect is the father of a child. If genetic evidence excludes the suspect, the husband will be judged to be the father, and vice versa. Such a judgment depends on showing that one of the two men could not be the father.

In many countries or states where genetic evidence is acceptable in court procedures, only decisions based on exclusion of paternity (or maternity) are considered valid. In some countries positive evidence of paternity that is based on genetic facts is also accepted. The justification for positive assignment that is not based on exclusion of another person will be discussed at the end of this chapter. First, genetic methods of exclusion will be presented. These methods depend on the analysis of the genotypes of certain traits of mother, child, and putative father. The most widely used traits are the blood groups.

Evidence Using the ABO Blood Groups. If evidence of paternity is derived from a study of the ABO blood groups of the individuals concerned, it is obvious that in many instances no decision can be reached. This is true not only when all three individuals belong to the same blood group, but even when the child and the putative father both have the same *I* allele and that allele is not present in the mother. Thus, an A child from an O mother can have an A father, but no specific A man can be so designated in the absence of other evidence; if the putative father is O or B, however, there is no possibility that the child is his.

When it can be assumed that only one or the other of two men is the father of a child, a first prerequisite to a genetic determination of parenthood is that the genotypes of the two men be different. Both of them, therefore, must not belong to groups O or AB. If both belong to either group A or group B, it will seldom be known whether they are homozygous or heterozygous for the I^A or I^B allele, but knowledge of the blood groups of one or both parents of the two men may permit recognition of their genotypes. Even if the genotypes of the two men are known to be different, an assignment of genetic paternity will depend also on knowing the genotype of the mother as well as that of the child. Assuming that one of the men is I^OI^O and the other I^AI^B and that the child and its mother both belong to group A, then no decision can be made. As shown in Table 35, all four genotypically different kinds of matings can give rise to a group A child. On the contrary, a clear conclusion as to paternity can be reached if the mother belongs to group O. In this case the father of an A child would be the AB man; and children derived from the O man could only be O.

Genetic information is more frequently used to exclude the possibility that a suspected man is the father of the child in question. Clearly, regardless of the group of mother, an O man cannot be the father of an AB child, or an AB man of an O child. If the man belongs to group A or B and it is not known whether he is homozygous, or heterozygous for I^O, an exclusion of paternity depends on the genotype of the mother; if neither she nor the man carry a gene present in the child, paternity of the suspect is excluded. The reader will find it instructive to compile a table showing the 16 combinations of the four ABO blood groups of alleged father and known mother and listing both the possible and the impossible blood groups of children from these unions. Additional exclusions can be arrived at if the subgroups dependent on I^{A_1} and I^{A_2} are taken into account.

Evidence Using the MN Blood Groups. The ABO blood groups were the first genetic traits employed on a large scale in cases of doubtful parenthood. Since the discovery of the existence and kind of inheritance of the MN groups, extensive use has also been made of these.

TABLE 35
An Example in Which It Is Equally Possible Genetically That Either One of Two Men May Be the Father of a Certain Child.

Individual	*Group*	*Possible Genotypes*	*Possible Matings*			*Genotype of a Child*
			Mother		Father (I or II)	
Mother	A	I^AI^A, I^AI^O	I^AI^A	×	I^OI^O	I^AI^O
Child	A	I^AI^A, I^AI^O	I^AI^A	×	I^AI^B	I^AI^A
Possible Father I	O	I^OI^O	I^AI^O	×	I^OI^O	I^AI^O
Possible Father II	AB	I^AI^B	I^AI^O	×	I^AI^B	I^AI^A, I^AI^O

Use of the antibodies for M and N make exclusion of paternity, if possible at all, particularly easy, since the three genotypes $L^M L^M$, $L^M L^N$, and $L^N L^N$ are all phenotypically recognizable. A child who possesses an M or N antigen not present in the putative parent cannot be his offspring. However, this test is not able to exclude paternity for a putative father who is MN, since such a man could be the father of any of the three types of children, M, MN, or N. Tests employing the antibodies against the S and s antigens of the *L* alleles allow still further exclusions.

Evidence Using the Rh Blood Groups. The numerous Rh alleles provide additional possibilities for exclusion of parentage. Because of the rarity of some of the sera containing specific antibodies, medicolegal tests are usually restricted to the use of the more common of them—namely, anti-Rh_0 (anti-D), anti-rh′ (anti-C), anti-rh″ (anti-E), and perhaps anti-hr′ (anti-c).

Other Genetic Evidence. Any of the other genetically well-understood systems of blood groups can be employed to exclude paternity. Nor is there a necessary restriction to blood groups. Any set of genotypes that are regularly expressed as specific phenotypes may serve the same purpose. Preference in tests for characteristics of the blood is primarily due to their easy recognition. Future paternity tests will, if necessary, almost certainly involve not only various blood groups but also hemoglobin and serum protein traits. Apart from blood, determination of the secretion of ABO antigens is useful. Less satisfactory, if taken by themselves, are traits in which there is variability in expression. Thus, the presence of hair on the middle digit of one or more of the four fingers is indicative of the presence of specific alleles which are dominant over alleles causing absence of such hair, and a man who lacks this hair may therfore be presumed not to be the father of a child with mid-digital hair from a mother without it. Although the genetic hypothesis accounting for this trait is not definitive, its use may, in otherwise undecided cases, add to the probability of exclusion or nonexclusion.

Cytological Evidence. Morphologically variant chromosomes if present in a child and his putative father may furnish evidence for paternity. Size variations of the Y-chromosome are particularly important, although, of course, only for cases in which the children are males. All sons of a man inherit his Y-chromosome. Presence of an unusually long or unusually short Y-chromosome in only one of an alleged father and son pair would be proof of nonpaternity, and presence in both individuals would be near-proof of paternity, how near depending on the frequency of the particular Y-chromosome variant in the population. Other cytological evidence for paternity has been furnished by a translocation involving an E and a G chromosome, which was found in both a putative father and his alleged illegitimate son.

The Probabilities of Exclusion. The usefulness in paternity cases of any test based on a pair or group of alleles depends on the frequencies of the alleles in the population. In most aboriginal American Indian populations the frequency of the I^O allele is 1: all individuals belong to blood group O. Obviously, in such populations the ABO blood groups supply no information on paternity. Formulas which give the probability that a man who is not the father can be excluded, based on allele frequencies, have been worked out by Wiener and Boyd.

For the ABO groups the maximum theoretical probability of exclusion of a wrongly accused man is about 20 per cent, but in most populations the allele frequencies permit only a lower percentage of exclusions. For the MN groups exclusion has a maximum theoretical probability of 18.75 per cent, this maximum probably being actually approached in many populations. For the Rh groups the chance of exclusion in Caucasoid populations is approximately 25 per cent.

Evidence Using More than One Blood-group System. In many cases in which knowledge of the ABO phenotypes of the individuals under investigation does not permit exclusion, analysis of the MN blood groups may do so. Conversely, the ABO groups may supply an answer where the MN groups fail to do so. An example would be the case of an O, MN woman and an AB, N man who is alleged to be the father of one or more of her four children—the children being O, MN; A, M; O, M; and A, MN (Table 36). The man cannot be the father of the first child because of the lack of genetic relation between his AB group and the child's O group, though there is genetic com-

TABLE 36
An Example in Which Exclusion of Paternity Is Based on Both ABO and MN Blood Groups.

Possible Mating		*Children*			
Mother	Putative Father	First	Second	Third	Fourth
O, MN	AB, N	O, MN	A, M	O, M	A, MN
Parentship of putative father excluded on basis of child being		O	M	O, M	?

patibility of the MN constitutions. For the second child, no decision is possible on the basis of the ABO relation, but the N man cannot be the father because of the lack of N antigen in the child. The third child cannot be his on the double score of genetic incompatibility between their ABO and between their MN blood groups. Finally, no decision is possible about the fourth child, since neither his A nor MN phenotypes contradict the constitution of the putative father. In general, it is obviously desirable, in medicolegal cases, to base a verdict on the results of tests for a whole battery of antigens since the probability of exclusion—if justified—increases with each additional test. If, in the example presented in Table 36, both the mother and the putative father had been Rh negative but the last two children Rh positive, then the man would also be excluded as the father of the fourth child. The third child, who had been shown not to be the man's child by the ABO and MN groupings, would now be excluded in triplicate by his ABO, MN, and Rh constitution.

Although the probabilities of exclusion of wrongly accused men increase with a multiple use of blood groups, the increase becomes smaller with each additional test. If two blood groups each provide a 20 per cent probability of exclusion, then using the second after the first has been tried will lead to exclusion of 20 per cent of the 80 per cent for whom the first was not decisive; i.e., it will add only 16 per cent. Multiple use of the ABO, MN, and Rh groups in a white population may provide about 50 per cent exclusion, and additional evidence based on other known groups will raise the chances to slightly more than 70 per cent.

Among a sample of several thousand court cases the frequency of actual exclusions fell far short of the theoretical expectation, and the discrepancy suggests that perhaps 2 out of 3 men imputed to be the fathers of illegitimate children were indeed correctly designated.

POSITIVE ASSIGNMENT OF PATERNITY

Mohr reports a case in Norway in which the normal mother of a brachyphalangic child designated a specific man as the father. The man denied his involvement; but when requested by the court to show his hands, he was seen to be short-fingered. Consequently, he was adjudged to be the father and ordered to support the child. This decision was made because brachyphalangy is caused by a dominant gene and so rare that the probability of another individual with the same genotype being the father was most remote. The brachyphalangic man must have been heterozygous for the rare dominant gene and hence have had a 50 per cent chance of begetting a normal child. The absence of the abnormality in the child would not have absolved him from possible paternity, but would have left the case undecided unless other evidence had been forthcoming.

A similar basis for assignment of paternity exists in other cases, where the putative father and the child both carry a very rare allele that is not present in the mother. Thus, a child heterozygous for an r^y allele of the Rh blood-group system born to a mother who does not possess r^y can only have been sired by an r^y– man. The frequency of such men in a white population is considerably less than 1 in 1,000, so that if a man accused of being the father of an r^y– child proves to be r^y– himself, the likelihood that he is the father is very high. It is clear, however, that in a large population there are many other men who have the r^y allele. A positive assignment of paternity could not be based solely on the "suitable" genotype of the man in question, but would depend also on the strength of the nongenetic evidence that led to his being named as the possible parent.

If a child carries very rare alleles, at more than one locus, that he must have inherited from his male parent and if a designated man is shown to carry all of them, then, of course, it is highly probable on purely genetic grounds that he is the father. Even a combination of very rare, with relatively common, alleles may yield decisive information. Thus, a man of the blood types B, NS, $R^{I(w,\ u)}$ r, Lu(a+), K+ has a frequency among whites of only 1 in 10 million. If he were the alleged father of a child having the same types and if the mother had only common types, paternity could be assigned with near certainty, "provided"—as pointed out by Race and Sanger—"that the brothers of the accused had alibis."

A more generally applicable method than that of basing positive assignment of paternity on rare alleles is to base the assignment on combinations of many genes, none of which by itself is particularly rare. Consider five loci and assume that a child has the genotype A'–, B'–, C'–, D'–, E'– (where the dashes stand for the alleles received from the mother, which are assumed to be different from the A', B', C', D', E' alleles received from the father). If the allele frequencies (p) of the paternal genes are, for instance, 0.3, 0.2, 0.6, 0.05, and 0.4, respectively, then the probability that a man will have a genotype that enables one of his sperm to be $A'B'C'D'E'$ can be calculated as follows. In order to be able to transmit A', he must be either homo- or heterozygous for this allele. The probability that he does not have A' is $q^2 = 0.7^2$, and that of his having it is $(1 - 0.7^2)$. Similarly the probability that he carries B' is $(1 - 0.8^2)$, and that of his carrying $A'B'C'D'E'$ together is the product of the separate probabilities $(1 - 0.7^2) \times (1 - 0.8^2) \cdot\ .\ .\ . = 0.0096$. The smaller the probability of finding a given combination of genes, the greater the probability that a man who carries it is the father of a child who has it too.

It must be emphasized again that positive assignment of paternity is based on probability and thus can never be entirely certain, no matter how great the probability; whereas genetically justified exclusions are certain. In many countries, therefore, the law does not recognize positive assignment of paternity. In Germany—one of the countries where positive assignments are acceptable—expert testimony is couched in terms of probability. Designations of the degree of the possibility of paternity that are used in the German courts are (1) not determinable, (2) more probable than not, (3) probable, (4) very probable, (5) probable to a degree bordering on certainty, (6) more improbable than not, (7) improbable, (8) very improbable, and (9) improbable to a degree bordering on certainty.

These conclusions are normally based not on a determination of genetic constitutions but on a comparison of physical similarities. Facial appearance and other traits of the mother, child, and putative father are compared in great detail, and the degree of similarity between the father and child is one of the bases for the probability designations. Even though such procedures cannot fail to involve subjective evaluation, they may, in expert hands, reach a fair level of validity. This has been tested on a sample of 100 children of known paternity and their parents. In no case was a man mistakenly declared not to be the father of his child: none of the conclusions was of improbability of the degrees (6) to (9) above. For 7 father-child pairs the similarity evaluation led to the verdict "not determinable," but for the remaining 93 children different degrees of positive probability assignments were possible.

Formulas have been worked out which would permit an exact numerical evaluation of the probability of correctness of a paternity judgment if exact figures on the relative frequencies with which individual traits of children are found in their true fathers and in men of the general population were known. But such figures are only rarely known.

Superfecundation. The use of genes as markers of paternity has provided an answer to the question of whether it is possible for twins to have different fathers. A case is on record in which an O, M mother bore as twins a boy who was B, M and a girl who was A, MN. Two men were considered, on nongenetic evidence, to be the only possible fathers of the children. One of the men was A, MN; the other, B, M. The first man was thus excluded from being the father of the boy and the second man from being the father of the girl. If it is granted that the twins were not the offspring of a third man who was AB, MN, this case may be regarded as proof of "superfecundation."

Telegony. An elementary knowledge of genetics shows that it is not possible for one male to influence the genotype of the child of another male. Among breeders of dogs, there was a widespread superstition that mating of a purebred bitch with a dog of another breed not only leads to an immediate litter of mongrels (which is true), but also gives rise to mongrel traits in puppies that are later sired by males of the same breed as the bitch. This erroneous belief in *telegony* (influence of one sire on the "distant offspring" of a later sire) has also been held by some to be true of man: the procreation of a child by a man and woman who belong to different anthropological groups has been thought to have an effect on the appearance of a child from a later union between the woman and a man who belongs to the same group as she. Some opponents of racial intermixture have spread falsehoods of this kind deliberately.

PROBLEMS

94. For a given year, find the following data, if available, in *Vital Statistics of the United States, 1, Natality*. U.S. Dept. of Health, Educ., and Welfare, Public Health Service, Washington, D.C.
    (a) Total number of births for the United States Birth Registration Area.
    (b) Total birth rate per 1,000 population.
    (c) Total number of illegitimate births for the United States.
    (d) Total illegitimate birth rate per 1,000 live births.
    (e) Illegitimate birth rate per 1,000 live births in Louisiana, New York, Ohio, Vermont.

95. Construct a table listing all combinations of putative father, mother, and child which would allow a falsely designated man to establish nonpaternity by MN types.

96. Four babies were born in a hospital on the same night and their blood groups were later found to be O, A, B, and AB. The four pairs of parents were: (i) O and O, (ii) AB and O, (iii) A and B, (iv) B and B. All four babies can be definitely assigned to their parents. How?

97. For each of the following six mother-child combinations, list:
    (a) The phenotypes of men who could be the father of the child.
    (b) The phenotypes of men who could not be the father.

Combination	1	2	3	4	5	6
Mother	O, M	O, N	O, MN	AB, MN	A, N	B, MN
Child	O, M	A, MN	O, MN	AB, MN	AB, MN	O, N

98. Assume a case of disputed paternity involving a woman X, two men Y and Z, and four children, 1, 2, 3, and 4. On the basis of the blood properties, assign each child to his father, whenever possible. Let the constitutions be:

Adults		*Children*		
X	A MN	1	AB	N
Y	B MN	2	B	MN
Z	AB N	3	O	N
		4	AB	M

99. A woman of the blood properties A, M has a child B, MN. Her husband, who is A, N, accuses a certain man of being the father of the child. If the accused man turns out to be B, N, how would you judge the case?

100. Analyze the possibility of exclusion of paternity in the following cases:

Case	*Putative father*	*Mother*	*Child*
(a)	I^OI^O, L^ML^N, R^1r	I^OI^O, L^ML^N, R^1R^2	I^OI^O, L^ML^N, R^1r
(b)	$I^{A_2}I^{A_2}$ or $I^{A_2}I^O$, L^ML^M, rr	$I^{A_1}I^{A_1}$ or $I^{A_1}I^O$, L^NL^N, R^1R^1	$I^{A_1}I^{A_1}$ or $I^{A_1}I^O$, L^ML^N, R^1R^2
(c)	$I^{A_2}I^{A_2}$ or $I^{A_2}I^O$, $L^NL^NR^1R^1$	I^OI^O, L^NL^N, $r''r$	I^BI^B or I^BI^O, L^ML^N, rr

REFERENCES

Brownlie, A. R., 1965. Blood and the blood groups. A developing field for expert evidence. *J. Forens. Sci. Soc.*, **5**:124–174.

Geyer, E., 1940. Ein Zwillingspärchen mit zwei Vätern. *Arch. Rass. w. Ges. Biol.*, **34**:226–236

Koch, G., Gröpl, V., Meyer-Robish, M. and Schwanitz, G., 1968. E/G–Translokation bei Vater und Sohn. Pp. 176-183 in *Anthropologie u. Humangenetik.* Fischer, Stuttgart.

Race, R. R., and Sanger, Ruth, 1968. Ch. 22: Blood groups and problems of parentage and identity, pp. 447–466, in *Blood Groups in Man*, 5th ed. 599 pp. Blackwell, Oxford.

Schatkin, S. B., 1953. *Disputed Paternity Proceedings*, 3rd ed. 823 pp. Bender, New York.

Schwidetzky, I., 1954. Forensic anthropology in Germany. *Hum. Biol.*, **26**:1–20.

Wiener, A. S., 1943. Ch. 21: Medicolegal applications of blood tests in disputed parentage, in *Blood Groups and Blood Transfusion*, 3rd ed. 380–398. Thomas, Springfield, Ill.

Wiener, A. S., Owen, R. D., Stormont, C., and Wexler, I. B., 1957. Medicolegal applications of blood grouping tests. *J. Amer. Med. Ass.*, **164**:2036–2044.

14

SEX LINKAGE

The examples of simple dominant and recessive inheritance discussed in the preceding parts of this book do not show any association with sex. Individuals of one sex are no more frequently affected than those of the other, and there is only a random relation between the sex of the affected parent and the sex of affected offspring. This is to be expected if the gene responsible for a given trait is carried in any one of the 22 pairs of autosomes. Since men and women alike possess these autosomes, the hereditary roles of the two sexes should be equivalent.

The ordinary rules of inheritance, however, are not valid for all human traits. For some, the frequencies of affected individuals are very different in the two sexes, and the reappearance of the trait in children depends on whether it is handed down by the father or the mother. Some of these traits are now recognized as being due to ordinary autosomal alleles that have different phenotypic expressions in the two sexes; examples of these will be discussed in Chapter 16, Variations in the Expression of Genes. This chapter deals with traits that are due to alleles whose transmission is specifically related to sex.

Our understanding of these has been made possible only by the discovery of the sex chromosomes (see p. 98). The twenty-third pair of chromosomes is not the same in both sexes, for the female has two X-chromosomes in her nuclei and the male one X- and one Y-chromosome. Therefore, it can be expected that the two sexes are not equivalent in respect to genes located in the sex chromosomes. Traits based on such genes are called sex linked; their mode of transmission, sex-linked inheritance.

Obviously, two main types of sex-linked inheritance are possible, depending upon whether the sex-linked genes are located in the X-chromosome or the Y-chromosome. Many X-linked genes are known, but, except for sex determiners, there is only limited evidence for Y-linked genes.

In addition to genes restricted to either the X- or the Y-chromosome, i.e., absolutely or completely X- or Y-linked, it has been suspected that there are others which, by crossing over in the human male, are able to change their localization in successive generations: to go back and forth from the X- to the Y-chromosome and from the Y- to the X-chromosome. Such genes have been called partially or incompletely sex linked. They are known to exist in various animal species, such as the fish *Lebistes reticulatus* (guppy), the mosquito *Culex molestus*, and other insects. A search for partially sex-linked genes in man, initiated by J. B. S. Haldane (1892–1964) and continued by other investigators, at first seemed successful. Later work, however, has produced no evidence that partially sex-linked genes exist in man. The terms X-linkage and Y-linkage will, therefore, designate complete linkage of the respective types only.

A female, with her two X-chromosomes, has two alleles of all X-linked genes and may be either homozygous or heterozygous for them. A male, who has only one X-chromosome, is called *hemizygous* for the single allele of his X-linked genes, since the terms homozygous and heterozygous imply the presence of two alleles.

Y-LINKAGE

The inheritance of a gene which is permanently located in the Y-chromosome is exceedingly simple: (1) Only men possess a Y-chromosome, and the gene should, therefore, be found in the male sex only; (2) since all sons, and no daughters, receive a Y-chromosome from their father (Fig. 123), only the sons receive the gene, which they in turn transmit to all of their sons but to none of their daughters (*holandric inheritance*). If the gene expresses itself phenotypically whenever it is present in an individual, the inherited trait will be strictly limited to the male sex.

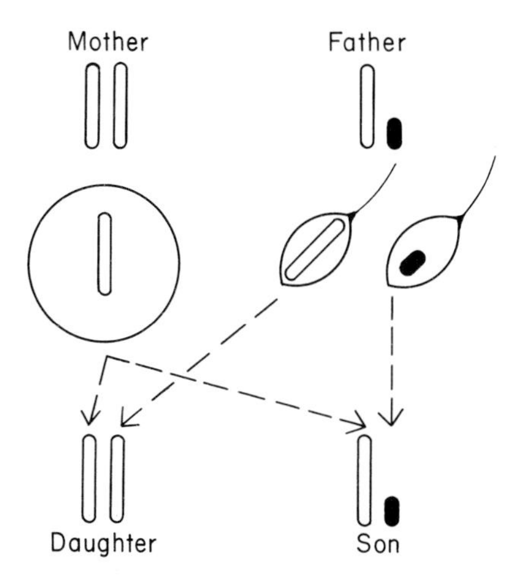

FIGURE 123
Transmission of the Y-chromosome from one generation to the next.

Restriction of a trait to males is not, however, sufficient evidence for the presence of a Y-linked gene. Some traits that occur only in the male sex, such as tenor, baritone, and bass voices, do not depend on Y-linked alleles; rather, they depend on autosomal alleles that are present in both men and women, but that express themselves differently in the two sexes. Male sex-limited autosomal and male sex-limited Y-linked inheritance are distinguished by the fact that the autosomal traits depend equally on alleles transmitted by both parents, while Y-linked traits neither appear in women nor are transmitted by them.

"Porcupine Men." Only a few of the many hundreds of inherited characteristics in man have been reported to be transmitted in Y-linked fashion. The most remarkable case of such transmission was that of the so-called porcupine men, who lived in England during the eighteenth and nineteenth centuries. In 1716, a normal-appearing baby, Edward Lambert, was born to two normal parents. They had many other children, all of whom remained normal throughout life. Edwards's skin, however, began to yellow when he was seven or eight weeks old. It gradually became black, and then began to thicken until his whole body—except palms, soles, head, and face—was covered with rough, bristly scales and cylindrical bristle-like outgrowths nearly an inch long.

Edward Lambert was reported to have had 6 sons, all afflicted with the same condition. The trait was said to have appeared in four later generations and was said to have been present in every son of an affected father, absent in every daughter, and never transmitted by any daughter. A pedigree, reprinted in numerous publications, including the original edition of this book, showed 12 affected males and 7 unaffected female sibs in six generations (Fig. 124, A).

It is hardly possible for an author to verify each statment that he finds in earlier, presumably careful publications. The pedigree of the porcupine men, however, seemed to deserve some special inquiry. By going back to the comtemporary accounts of the affected Lamberts, who were exhibited for money in Great Britain, Germany, France, Italy, and possibly Russia, and by consulting parish registers of births, baptisms, marriages, and burials, it was ascertained that important parts of the supposed family tree do not correspond to the facts. The most reliable evidence available indicates that there were only three instead of six generations of affected males, and that 4 or 5 instead of 12 persons were affected. The 6 children of the first affected male were not all sons but included 2 daughters. If their father, who had reported correctly the presence of both sexes, was also right in his statement that all 6 children were affected, then the existence of 2 affected daughters immediately rules out Y-linkage (Fig. 124, B). If we disbelieve his report of

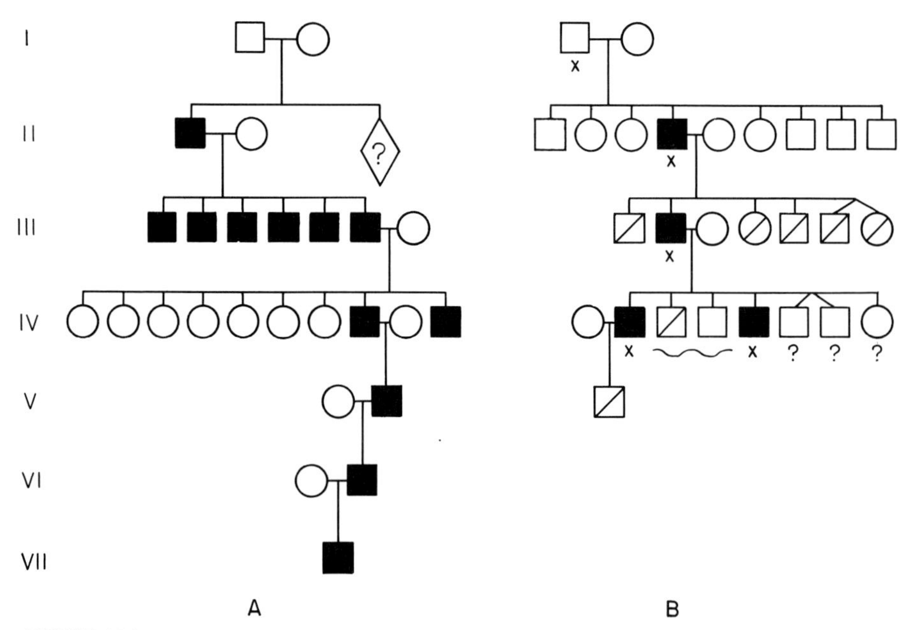

FIGURE 124
Pedigree of the "porcupine men." **A.** Formerly conventional pedigree. **B.** Revised pedigree. Symbols with an oblique line indicate that these individuals were reported to have been affected but have not been seen by independent witnesses. The wavy line in IV indicates that it is not known whether IV-3 or IV-4 is the older. The zygosity of the twins in IV is unknown. The question mark in A signifies absence of information concerning the number of sibs of II-1. The question marks in B signify absence of information concerning the trait. (Stern, *Amer. J. Hum. Genet.*, **9**, 1957.)

the presence of the skin disease in all his children, then we must disregard all but the one son who was carefully described by several contemporary observers. Similar considerations apply to the next generation (IV), the last in which affected individuals are definitely known to have appeared. We now are certain that there were at least 6 sons and probably 1 daughter (not 7) in this generation, and it is nearly certain that only 3 of these sons were affected (Fig. 124, B). Altogether, this reappraisal leads to the conclusion that the porcupine men cannot be used as an example of Y-linkage. Presumably they owed their abnormality to an extremely rare, dominant autosomal gene. This interpretation is not invalidated by the fact that all individuals definitely known to be affected were males. If there were really only 5 affected persons, then chance could easily account for a distribution of 5 males and no females instead of the expected near equality. It is more likely, however, that there actually may have been affected females, as the statement of the original porcupine man seems to indicate.

Webbed Toes. Another record often quoted to substantiate Y-linked inheritance concerns a web-like connection between the second and third toes of the 14 male members of a Schofield family in the United States. None, it was stated, lacked the trait, and none of the 11 daughters of the affected men were said to possess it. At the time the pedigree was published, only 3 of the daughters had married, and their 7 children were all normal.

Before accepting the sex association of webbed toes in this pedigree as proof for Y-linkage, it must be stated that there are numerous pedigrees for webbing between toes that do not show this type of inheritance, since both men and women possess and transmit webbed toes. Most of these pedigrees, however, show a higher incidence of the trait in males than in females, which makes them somewhat similar to the Schofield pedigree. The relatively high frequency of inherited webbed toes raises the question of "selection for curiosity's sake" in singling out the Schofield pedigree, and the observed preponderance of affected males over females in other pedigrees must be taken into account in any consideration of the expected probability of the appearance of the trait.

The higher incidence of webbed toes in males is probably due to variable expression of a dominant autosomal allele during the development of the two sexes. In order to calculate the chance of finding a pedigree like that of the Schofield family if inheritance were of the dominant autosomal type, the probabilities of phenotypic expression in all 13 male descendants of affected men and of phenotypic absence in all 11 female descendants must be determined. If it is assumed, for example, that the webbed-toe allele is always expressed visibly in the male but in only 1/10 of all females, then the probability that a son of an affected father will be affected is 1/2; however, the probability of a daughter being normal is not only 1/2 (the chance that she does not inherit the allele) but in addition, $1/2 \times 9/10$ (the chance that she does inherit the allele but that it does not show in her phenotype). The total chance that a woman would not have webbed toes would then be

$$1/2 + (1/2 \times 9/10) = 19/20.$$

Under these various assumptions, and omitting from consideration the marriages of 3 normal women who had 7 normal children, the probability of a deviation in the observed direction from an expected sex distribution of affected and nonaffected individuals as large as that found in the Schofield family is

$$(1/2)^{13}(\text{♂})^{13} \times (19/20)^{11}(\text{♀})^{11},$$

which is equal to about one chance in 14,000. Such a small probability makes it seem very unlikely that the trait can be attributed to an autosomal dominant gene. On the other hand, some questions remain unanswered. Webbing of toes is a variable trait—clearly discernible in some persons, less discernible in others. How reliable was the classification in this family? Was there an unconscious inclination to regard borderline cases which barely suggested webbing as positive in males, but as negative in females? Only a restudy of the still-living individuals in the original Schofield family and of their descendants could resolve these doubts. In the meantime this pedigree of webbed toes cannot be considered evidence of Y-linkage.

Hairy Ear Rims. A few other kindreds have been described in which some trait has been transmitted in a fashion that seems to indicate Y-linkage. In most of these the evidence for Y-linked inheritance is incomplete. The best candidate for Y-linkage in man is the genetic determinant for presence of long, stiff hairs on the rim of the ears, as found particularly in Indian populations, but also in Caucasians and Australian aborigines, and more rarely in Japanese and Nigerians (Fig. 125). The trait is restricted to males. Dronamraju has published a very large pedigree of his own family, in which hairy ear rims are present in every male who is descended in the male line from a common affected ancestor. Y-linked inheritance in this kindred seems well established, but other pedigrees frequently reveal males who carry the same Y-chromosome but do not show the trait. In order to explain how the trait can be absent when the Y-chromosome with the putative genetic determinant is present, we have to assume that presence of the gene often does not lead to phenotypic expression. This assumption is necessary, not only because there are nonaffected relatives with the same Y-chromosome as the affected ones, but also because population surveys according to age classes

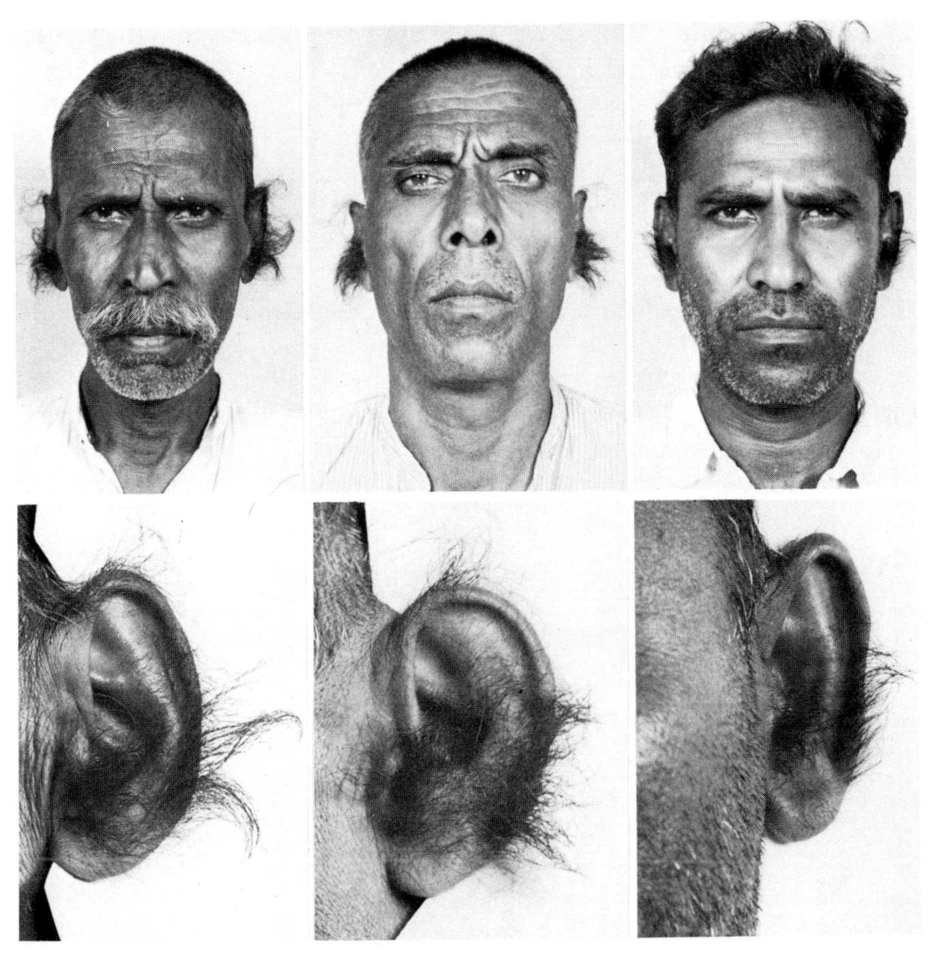

FIGURE 125
Three Muslim brothers from South India, aged ±60, 50-55, and 45-50, respectively, showing striking degrees of hairy pinnae. (Photographs by Mr. S. D. Sigamoni, Photography Dept., Christian Medical College Hospital, Vellore. From Stern, Centerwall, and Sarkar, *Amer. J. Hum. Genet.*, **16,** 1964.)

have shown invariably that the trait becomes increasingly more frequent the older the men; that is, the evidence shows nonexpression of the trait in a considerable fraction not only of young men but also of middle-aged and even older men.

The assumption of variability of gene expression (including nonexpression) is by no means unlikely per se. Such variability is frequently seen for other traits, and will be discussed in detail in Chapter 16. Nevertheless, the observed variability makes delineation of the type of transmission more difficult than it is for genes with uniform expression. For hairy ear rims, the variability of the trait from male to male extends in a continous distribution from complete absence, to presence of one, two, or more hairs, up to heavy bushiness. It seems that a variety of genotypes are involved in the genetics of hairy ear rims. In Dronamraju's family, the Y-chromosome seems to carry a gene that always leads to expression in postpubertal males of any age. In other families, we would have to assume the presence of "weaker" alleles for hairy ear rims or more complex genetic determination involving the autosomes. There is no reason why the human Y-chromosome should not carry typical genes, but comparison with other organisms does not necessarily lead to such an expectation. Some fish have numerous Y-linked genes, but the Y-chromosome in *Drosophila* is nearly devoid of genes. In mice a Y-linked gene controls an antigen important in histoincompatibility. In man, apart from "hairy ear rims," the only known role of the Y-chromosome is in sex determination. In both men and mice, the Y-chromosome controls the development of maleness during embryogeny.

X-LINKAGE

Color Blindness. If red, green, and blue lights are projected on a screen so that they overlap, the triple mixture of these primaries will make a white light. A person with normal vision will require particular intensity ratios of red:green:blue to "see" white—he is a *normal trichromate* (requiring three primaries to match white). Some persons, however, need more of one of the primary colors and less of the other two—these are *anomalous trichromates.* Normal color vision is based on the presence of three kinds of color-vision pigment in the retina: red, green, and blue. Anomalous trichromates also have three such pigments but one of them is reduced in amount and displaced in spectrum. Some persons need only two primaries to see white—these are dichromates (requiring two primaries). The dichromates are of three kinds: *protanopes* who lack the red pigment, *deuteranopes* who lack the green, and *tritanopes* who lack the blue pigment. A very few people are totally color blind. They can match a white light with just one primary color (any one); these monochromates (or achromates) perceive no hues and see only white or gray. Monochromates lack cones or have defective cones.

Most of these conditions have a genetic basis. They are all called "color blindness," but have different modes of transmission. Total color blindness (achromatopasia) is a very rare autosomal recessive condition (Fig. 100), and the rare tritanopes, though not fully understood genetically, seem also to be affected through the agency of autosomal genes. The remaining types of partial color blindness or color anomalies are often popularly referred to jointly as red-green color blindness. Something will be said about each of them, but, for the present, they will be treated as if they constituted a single genetic entity called color blindness.

During the late eighteenth and the nineteenth century, it became known that color blindness is inherited. Observations showed a number of regularities. Thus, if a color-blind man marries a normal woman, all of their children are usually normal. If a color-blind woman marries a normal man, inheritance follows a peculiar "crisscross" pattern: all sons are color blind, like the mother; and all daughters are normal, like the father. The phenotypically normal daughters of a color-blind parent are, however, able to transmit the defect to their sons. In spite of other observations on the transmission of color blindness which added to the already complicated picture, there seemed to be certain simple rules, which were given the name "Horner's Law."

A true understanding of the empirical findings was obtained only after similar rules had been formulated for the inheritance of quite different traits (e.g., pigmentation) in birds, moths, and flies, and after geneticists and cytologists had applied to these the knowledge gained from microscopic studies of chromosomes. It became clear, about 1910, that the peculiar features of the inheritance of color blindness could be understood if it were assumed that (1) the genes concerned are located in the X-chromosomes, and that (2) normal color vision is dominant over color blindness. Assuming that the Y-chromosome plays no role in the determination of color vision, men are hemizygous for either the normal or the color-blindness allele (located in their one X-chromosome), while women (who have two X-chromosomes) may be of any of three types: homozygous normal, homozygous color blind, or heterozygous.

Every man receives his X-chromosome from his mother and does not transmit it to his sons (Fig. 126). Every woman receives an X-chromosome

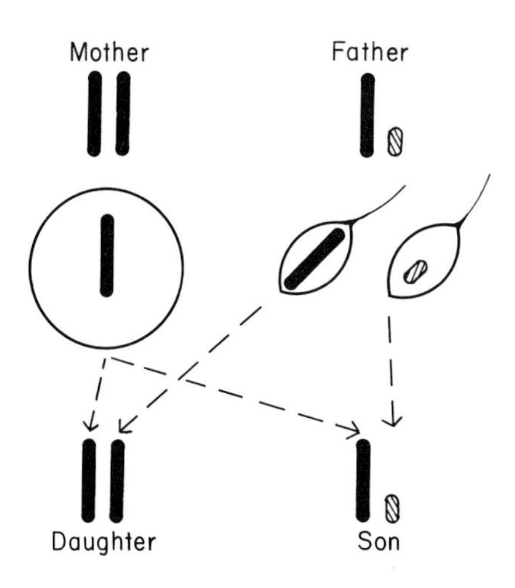

FIGURE 126
Transmission of the X-chromosomes from one generation to the next.

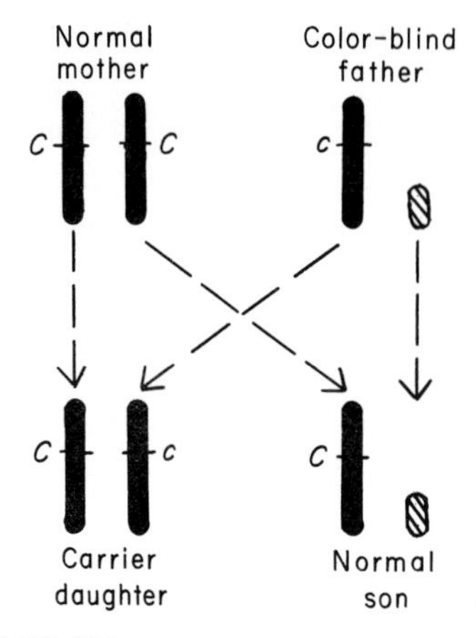

FIGURE 127
Transmission of color blindness. Normal woman × color-blind man.

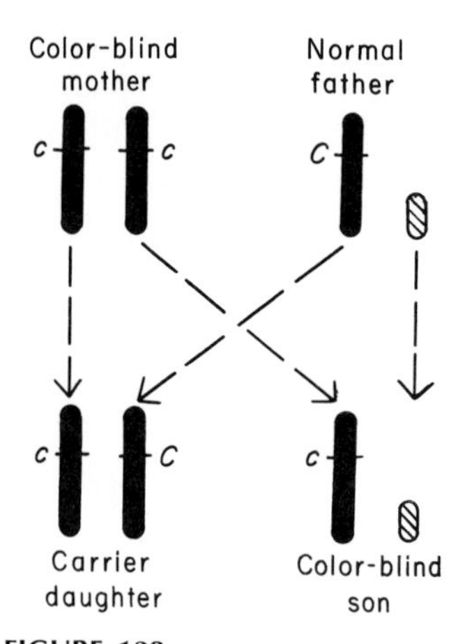

FIGURE 128
Transmission of color blindness. Color-blind woman × normal man.

from each of her parents. Her sons inherit either one or the other of her X-chromosomes, as do her daughters, who, in addition, receive a second X-chromosome from their father. From these facts, it follows that the sons of a color-blind man do not inherit their father's defect (Fig. 127). All daughters, however, are heterozygous carriers who, on the average, will produce normal and color-blind sons regardless of the color-vision status of their husbands (Fig. 129). Should the husband be color blind, then half of the daughters will be carriers and half will be color-blind homozygotes. Finally, all the sons of a normal man and a color-blind woman will be color blind, like the mother, and all the daughters will be heterozygotes, though phenotypically

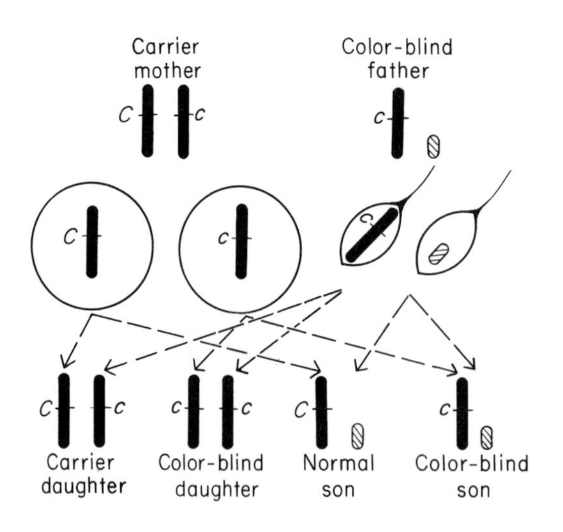

FIGURE 129
Transmission of color blindness. Carrier woman × color-blind man.

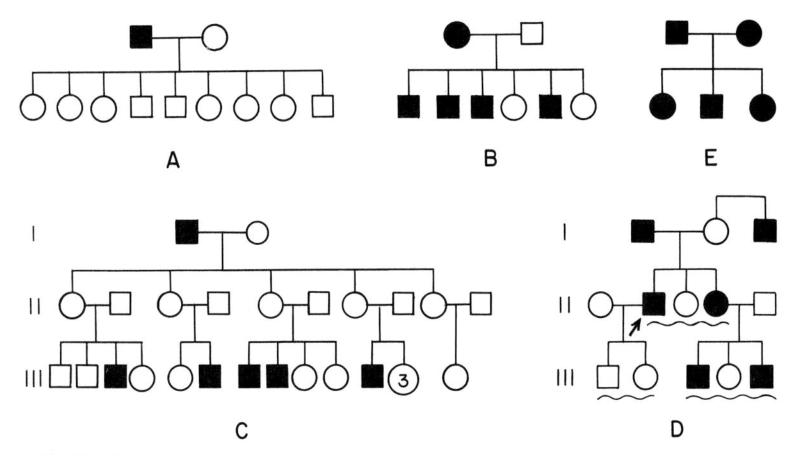

FIGURE 130
Pedigrees of color blindness. **A.** Part of Pedigree No. 406, Nettleship. **B.** Pedigree No. 584. **C.** Part of Horner's pedigree. **D.** Pedigree, Whisson, 1778. This is the first known pedigree of color blindness over more than one generation. One year earlier color blindness had been recognized for the first time in history. It occurred in three brothers of an English sibship of five boys and two girls. Both parents had normal color vision. **E.** Pedigree, Vogt. (A, B after Bell, *Treas. Hum. Inher.*, **II, 2,** 1926; C, after Gates, *Human Genetics,* Macmillan; E, after Baur, Fischer, Lenz.)

normal, like the father (Fig. 128). Pedigrees showing various types of marriages are reproduced in Figure 130. In general, genes which are transmitted in the fashion outlined for color blindness are thereby recognized as X-linked. Since most sex-linked genes are located on the X-chromosome (and not on the Y-chromosome), the terms X-linked and sex-linked are often used as synonyms.

Frequencies of Affected Men and Women. The male is more frequently affected by X-linked recessive traits than is the female. This is readily understood both from a consideration of individual marriages and from population genetics. All couples in which the female is heterozygous or homozygous for the X-linked abnormal allele may have affected sons, but only in the rare cases where the husband is affected and the wife is also affected or is a carrier can there be affected daughters. If q is the frequency of the abnormal X-linked allele, and p ($= 1 - q$) is that of the normal allele, then a man in the general population will have the probability q of having an X-chromosome with the abnormal allele and the probability p of having the normal allele. The frequency of the two kinds of men is, therefore, identical with the allele frequencies. For women, the proportions of the homozygous normal, heterozygous, and homozygous affected individuals are like those for autosomal genes. There is a probability of p^2 that both X-chromosomes will carry the normal allele, of 2pq that one X-chromosome will have the normal and the other will have the abnormal allele, and of q^2 that both X-chromosomes will carry the abnormal allele.

These considerations show an interesting relation between the relative frequencies of men and women affected with X-linked recessive traits. The frequency of the affected women, q^2, should be the square of the frequency of affected men, q. In different white populations, from less than 5 to 9 per cent of all men are color blind; i.e., q varies from less than 0.05 to 0.09. If there are 0.05 color-blind men, there should be only 0.0025 color-blind women; or, if 0.09 color-blind men, 0.0081 color-blind women. Thus, in a population in which 9 per cent of the men are color blind, color-blind women should be one-eleventh as frequent as color-blind men. If 5 per cent of the men are color blind, affected women should be one-twentieth as frequent, theoretically.

Actual figures lie close to the expected ones, but the agreement is not perfect, since, for example, an occasional heterozygous woman is color blind because of imperfect recessiveness of the gene concerned. Moreover, certain rare abnormalities of sex determination introduce slight deviations from expectation (see pp. 318–320; Chap. 20). Most important, not all kinds of color blindness are actually caused by the same genotype, as we have assumed in our discussion. As will be shown below, this results in an expectation of color-blind women that is significantly lower than it would be if a single genotype were responsible for the defect of color vision. Homozygous color-blind women are rare, but heterozygous carriers are relatively frequent. In a population in which the incidence of color blindness among men is 0.08 (q) that of carrier women theoretically is 0.15 (2pq).

The Chief Kinds of Partial Color Blindness. The X-linked partial color blindnesses fall into two different groups, called protans and deuterans (or, for short, prot and deuter). Physiologically, these two groups seem to be fundamentally distinct. In each group, there are three subtypes that may be regarded as variations of the same kind of defect. The two most common subtypes in the protan group are the anomalous trichromatic (ordinary *protanomaly*, red weakness) and the dichromatic (*protanopia*, red blindness). The two most common subtypes in the deuteran group are the anomalous trichromatic (ordinary *deuteranomaly*, green weakness) and the dichromatic (*deuteranopia*, green blindness); in the pronunciation of these four terms the accent is on the syllable "no."

A very extensive study of the incidence of these four conditions that was published by Waaler in 1927 was based on tests given to secondary-school children in Oslo, Norway. In all, 9,049 boys and 9,072 girls were available, and parents and other relatives were studied whenever this seemed important. Waaler found that about 5 per cent of the males were deuteranomals, and that each of the other three defects was present in about 1 per cent of the males. Since all four of these abnormalities are transmitted in an X-linked fashion, the percentages of affected males indicate corresponding allele frequencies of 0.05 and 0.01. Females homozygous for the different genes are expected at frequencies equal to the squares of 0.05 and 0.01, or 0.0025 and 0.0001, respectively. Thus, homozygous deuteranomaly should be more common in women than any other homozygous type of color blindness, and Waaler's observations confirmed this.

The Number of Loci for X-linked Color Blindnesses. The question arises: Do four different genes at two or more different loci in the X-chromosome control the common types of color blindness, or are they controlled by different alleles at a single locus? This question could most easily be answered if there were numerous women (with sons available for study) known to carry one of the four genes in one X-chromosome and another of the four in the other X-chromosome. Six different combinations of two X-chromosomes carrying different genes are possible. They would usually originate from marriages of men affected by any one type of color blindness to women heterozygous for normal color vision and some type of color blindness different from that of their husbands.

Women who possess two X-chromosomes with different genes for color blindness can be ascertained through close relatives. Thus, if a woman has two or more sons and if two types of color blindness occur among them, it may be concluded that she is heterozygous for two different color-blindness genes. The same conclusion applies if her father exhibits one type of color blindness, and a son another. Such women have been discovered. When heterozygous for protanomaly and protanopia, they are phenotypically protanomalous. This suggests that protanomaly and protanopia are allelic to each other, with the former (the "lesser" deviation from normality) dominant. Women who are heterozygous for deuteranomaly and deuteranopia are phenotypically deuteranomalous. This suggests that these two conditions are also allelic to each other, and that the minor deviation from normality is again dominant over the major one. Thus, there are two sets of three alleles, the first with increasing dominance in the order protanopia-protanomaly-normality, and the second in the order deuteranopia-deuteranomaly-normality. Although it has been suggested that each of the two sets has at least one more allele, which is intermediate in effect between the two abnormal alleles, this need not concern us here.

A rigorous demonstration of the allelic nature of protanomaly and protanopia or of deuteranomaly and deuteranopia would have to come from evidence of segregation, based on a study of numerous sons of women heterozygous for both prot or for both deuter genes. If the two prot genes are alleles, protanomaly and protanopia should segregate in oogenesis, and all sons should be affected with either one or the other of the prot conditions. Similarly, if the two deuter genes are alleles, deuteranomaly and deuteranopia should segregate, and all sons should be affected with either one or the other of the deuter conditions. In the very few suitable pedigrees known, this has been found to be true.

If the evidence points strongly toward allelism within the prot group as well as within the deuter group, the question arises whether the alleles of the two groups belong to two separate loci or whether they all belong to one single allelic group at one locus. One method of approach is to study women who have one X-chromosome with a prot defect and another with a deuter defect. All four possible combinations—(1) protanomalous/deuteranomalous; (2) protanopic/deuteranopic; (3) protanopic/deuteranomalous; and (4) protanomalous/deuteranopic—have been identified with reasonable certainty. The identification was, in two cases, based on the direct knowledge that one parent of the "compound" women was protanopic and the other parent deuteranopic (Fig. 131, A, B) and, in other cases, on the fact that the sons were of two different color-defective types (Fig. 131, C, D). In the fourth pedigree reproduced in Figure 131, the phenotypes of other members of the family add to the evidence (from her sons) that the woman II-2 was a protanopic/deuteranomalous heterozygote.

A remarkable fact about prot-deuter compound women (Fig. 131) is that they have *normal* color vision (a few do not, but their abnormal vision seems not to be connected with their compound genotype). A simple calculation lends support to the belief that women heterozygous for both a prot and a dueter allele should have normal color vision. If the frequencies of the four atypical genes are designated r, s, t, and u, then the expected frequencies of the mixed prot-deuter heterozygotes are 2rt, 2ru, 2st, and 2su, respectively,

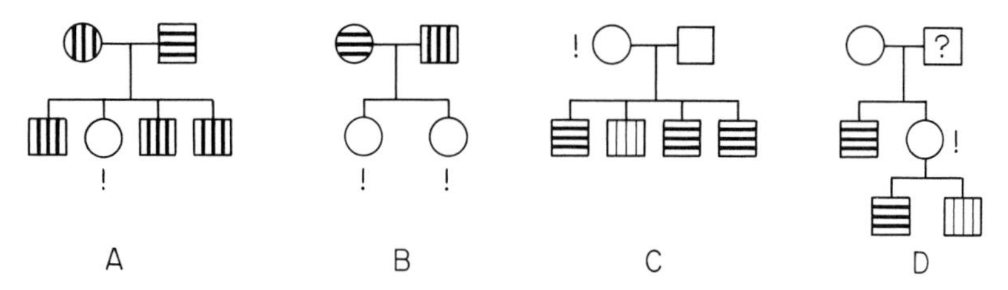

FIGURE 131
Pedigrees including women (marked "!") whose genotypes are compounds of two different X-linked alleles for defective color vision. Heavy horizontal lines = protanopia; heavy vertical lines = deuteranopia; light vertical lines = deuteranomaly. (A, B, after Franceschetti, *Bull. Acad. Suisse Sc. Médic.*, **5**, 1949, and Franceschetti and Klein. *Acta Genet.*, **7**, 1957; C, after Komai, 1947; D, after Waaler.)

provided that all four genes are allelic to one another, so that no single X-chromosome can carry more than one of them. (If the color-blindness genes are on two separate loci, the expected frequencies are different, but not strikingly so.) Since r, s, t, and u are known to be approximatly 0.01 (protanomaly), 0.01 (protanopia), 0.05 (deuteranomaly), and 0.01 (deuteranopia), respectively, the frequencies of the expected mixed heterozygotes are around 0.001 each for protanomalous/deuteranomalous and protanopic/deuteranomalous, and around 0.0002 each for protanomalous/deuteranopic and protanopic/deuteranopic. The sum of these four values is 0.0024, or 24 per 10,000. Among the more than 9,000 Oslo schoolgirls, about 20 should have shown one or another of the four compound conditions. Actually, however, none of the 40 color-blind girls belonged to any one of these types, but some were homozygous for a single kind of defect, and others heterozygous for protanomaly/protanopia or for deuteranomaly/deuteranopia (being, in either of the latter two cases, phenotypically anomalously trichromatic). The frequencies were in good agreement with expectations calculated on the basis of the allele frequencies. The fact that no girls exhibited any sort of prot-deuter combination of color blindnesses is interpreted to mean that typical prot-deuter heterozygotes have normal color vision. This is also the main reason for the earlier-mentioned discrepancy between the expected and observed frequencies of color-blind women. Since prot/deuter compounds are not color blind, the observed frequencies of color-blind women is lower than the expected.

The normal color vision of women who are heterozygous for one abnormal prot allele and one abnormal deuter allele, in contrast to the defective color vision of compound genotypes of two abnormal prot alleles or of two abnormal deuter alleles (Fig. 132), could be easily understood if the prot and deuter genes occupy two separate loci (Fig. 133, A). In general, if two nonallelic recessive genes are both present heterozygously, they permit the development of the double dominant phenotype; i.e., they complement each other. Thus, a person who is heterozygous for phenylketonuria as well as alkaptonuria would be neither a phenylketonuric nor an alkaptonuric, since the dominant normal allele at the first locus permits the transformation of phenylalanine to tyrosine and the dominant normal allele at the second locus that of homogentisic acid to acetoacetic acid (see Fig. 31, p.54).

It does not follow from this that only nonallelic genes complement each other in a compound genotype and thus lead to the dominant phenotype. An organism that possesses two different recessive alleles at the same locus may, nevertheless, exhibit the dominant phenotype, although this is not common. In *Drosophila*, for example, the X-linked gene for normal formation of certain bristle organs leads to development of these organs on both head and thorax. The abnormal recessive allele "scute-5" causes absence of the

	Normal	Protanom.	Protanop.	Deuteranom.	Deuteranop.
Normal	normal 1	normal 2	normal 3	normal 4	normal 5
Protanomalous		protanom. 6	protanom. 7	normal 12	probably normal 13
Protanopic			protanop. 8	normal 14	normal 15
Deuteranomalous				deuteranom. 9	deuteranom. 10
Deuteranopic					deuteranop. 11

FIGURE 132
The phenotypes of women possessing various combinations of X-chromosomes with a genotype for normal vision and with any one of the four chief genotypes for partial color blindness.

organs on the thorax and the abnormal recessive allele "scute-6" causes their absence on the head. The compound "scute-5/scute-6" is normal in regard to the presence of bristle organs on both head and thorax. The normal effects of the two complementing abnormal alleles are dominant in the compound genotype.

Returning to color blindness, it is clear that the normal phenotype of the combinations of prot and deuter genes is compatible with—but of course does not validate—the assumption that the four genes that cause defects are allelic to one another (see Fig. 133, B) and occupy the same locus as a fifth allele for normal color vision.

It should be possible to decide between the one-locus and two-locus hypothesis. A clear-cut decision could be reached by studying the sons of many mothers who have compound genotypes of prot/deuter alleles. If the two alleles are at one locus, rather than at two separate loci, then only color-blind sons should appear. The occurrence of any normal sons or of hemizygous prot-deuter combinations in sons would show that there was crossing over, and would thus establish the existence of two recombinable loci or sites controlling the four common types of color blindness. Both types of sons of prot/deuter mothers have indeed been observed. In a Belgian family studied by Vanderdonck and Verriest, a woman who apparently has a protanomalous/deuteranope compound genotype had five sons: two deuteranop, one protanomalous and two *normal.* In a Sardinian family studied by Siniscalco and his associates, a woman who has a protanop/deuteranop compound genotype had three sons: one deuteranop, one protanop, and one who on exten-

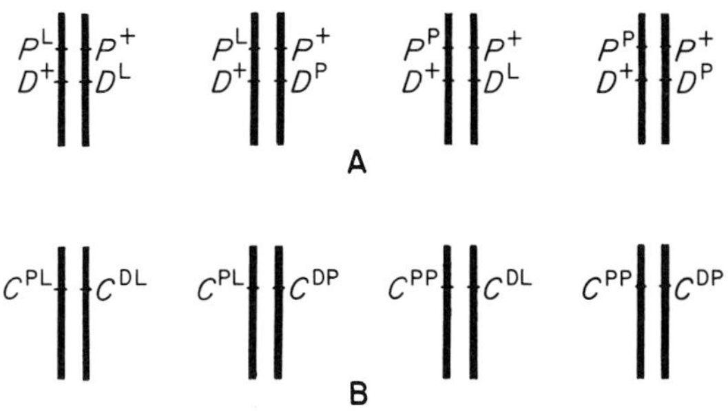

FIGURE 133
Heterozygotes for protanomaly/deuteranomaly, protanomaly/deuteranopia, protanopia/deuteranomaly, and protanopia/deuteranopia. **A.** Interpretation in terms of two loci with three alleles each: P^L (protanomaly), P^P (protanopia), and P^+ (normal); and D^L (deuteranomaly), D^P (deuteranopia), and D^+ (normal). **B.** Interpretation in terms of one locus with four abnormal alleles: C^{PL} (protanomaly), C^{DL} (deuteranomaly), C^{DP} (deuteranopia), and C^{PP} (protanopia).

sive multiple-color vision tests always "made mistakes typical for *protanopia and*/or for *deuteranopia*." Siniscalco concludes from these observations that the unusual son, as a result of recombination between two separate loci, prot and deuter, is carrying the genes for both prot and deuter color blindness. Very likely the interpretation of the findings in the two pedigrees as evidence for two separate loci is correct, but some doubts remain. In the Belgian family the protanomalous/deuteranop mother does not have normal color vision but is protanomalous. This heterozygous expression of protanomaly is unexpected. In addition, one of the two daughters of the mother who is married to a normal man, is again protanomalous, although only heterozygous for the condition. Could the protanomaly in mother and daughter be of a peculiar type constituting a new mutation, which might invalidate the interpretation of the normal sons as recombinants? And is it just chance that the Belgian family includes not just one but two brothers who appear to be cross-overs when the over-all frequency of such males in six other relevant pedigrees, including the Sardinian one, is only 1 in 21? In the Sardinian family, the interpretation of the unusual son as carrying both prot and deuter genes on his X-chromosome is highly suggestive, but is it the only possible one? Could a new mutation be involved? Such doubts may be overcritical. In any case, more analyses of such rare families are desirable. At present, the difference between the one-locus and two-locus hypotheses are of little practical importance; but the facts established regarding the effects of the different gene combinations for color blindness are of considerable interest.

An independent argument for the two-locus theory may be derived from molecular studies. It is known that the human red- and green-sensitive pigments differ in their protein component. Wald regards this as strong presumptive evidence for the operation of two different genes. If true, new questions arise. How different are these genes? How far away from each other are they located?

Other X-linked Traits. More than a hundred abnormal traits are due to abnormal X-linked alleles, and the corresponding normal traits are thus known to be controlled by normal alleles in the X-chromosomes. Among the abnormal traits may be mentioned certain types of congenital night blindness; atrophy of the optic nerve (X-linked type of Leber's disease); X-linked ichthyosis, a rough, scaly condition of the skin; hypogammaglobulinemia, the strange inability of the body to produce sufficient gamma globulin in the blood, which is thus unable adequately to produce antibodies against bacterial infections; brown teeth; absence of the enzyme hypoxanthine-guanine-phosphoribosyl transferase (HGPRT), which is active in purine metabolism, and associated excess production of uric acid leading to a rare neurological disorder of young boys manifested in severe mental and behavioral defects; rickets due to vitamin-D resistance; X-linked (Duchenne) type of muscular dystrophy; X-linked hemophilias; two different types of diabetes insipidus, a condition in which the patient may void as many as 10 quarts of urine daily and require a similarly large intake of fluid—one type of the trait being caused by deficiency of a pituitary hormone (vasopressin), the other by abnormal properties of the kidney; and ocular albinism, the nearly complete absence of pigmentation in the eyes of affected males, a trait with slight but recognizable expression in heterozygous females. One more X-linked condition is anhidrotic ectodermal dysplasia, a rare abnormality found in various populations and made famous by Charles Darwin in his book *Variation of Animals and Plants Under Domestication* (2nd ed., 1875, p. 319). He wrote as follows:

> I may give an analogous case . . . of a Hindoo family in Scinde, in which ten men, in the course of four generations, were furnished, in both jaws taken together, with only four small and weak incisor teeth and with eight posterior molars. The men thus affected have very little hair on the body, and become bald early in life. They also suffer much during hot weather from excessive dryness of the skin. It is remarkable that no instance has occurred of a daughter being affected. . . . Though the daughters in the above family are never affected, they transmit the tendency to their sons and no case has occurred of a son transmitting it to his sons. The affection thus appears only in alternate generations, or after long intervals.

In general, the inheritance of these traits is basically like that of color blindness, but a few of them will be used to demonstrate some special phenomena.

X-linked Dominants. Defective enamel leading to brown teeth is inherited in many families as an autosomal dominant. Some pedigrees exist, however, in which the defect seems to be due to an X-linked gene (Fig. 134). Unlike color blindness and most of the other abnormal X-linked traits, it behaves as a dominant. Consequently, all the daughters of an affected man, to whom he transmits his X-chromosome, will be affected, but his sons will not. This fits generation IV of the pedigree with its 8 affected daughters and 4 unaffected sons. The outcome of the reciprocal marriage, that of an affected woman and a normal man, will not betray the possible X-linked nature of the defect, since all affected women known are heterozygous for the rare dominant allele. Consequently, half of their sons will be hemizygous for the abnormal allele and half normal. Similarly, half of the daughters will receive from the mother the X-chromosome with the dominant atypical allele and the other half the normal allele. In other words, an affected woman will transmit the trait to half of her offspring, regardless of sex, exactly as in autosomal dominant inheritance.

Thus, the presumption that the dominant gene for brown enamel in the kindred depicted in Figure 134 is X-linked rests solely on the 12 children of generation IV. If the gene were an autosomal one, there would only be $(1/2)^{12}$, or 1 chance in 4,096, of obtaining the specific sex distribution of the trait. This speaks in favor of its X-linked nature, which is supported by other pedigrees that have been described more recently.

A still stronger case can be made for X-linked transmission of a dominant gene that leads to deficiency of inorganic phosphorus in the blood and often results in rickets resistant to normal doses of Vitamin D (see p. 379 and Fig. 157). In an extensive kindred, 7 affected males married to normal females had 11 daughters, all affected, and 10 sons, all normal. The reciprocal marriages of 9 affected females and normal males produced affected and normal

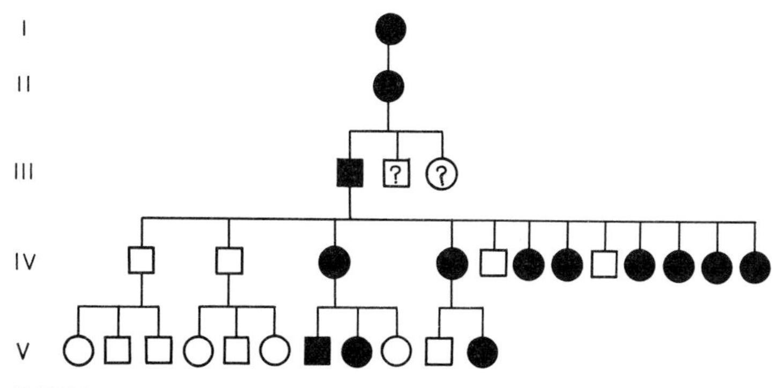

FIGURE 134
Pedigree of brown enamel of teeth. (After Haldane, *J. Hered.*, **28**, 1937.)

children of both sexes in a ratio compatible with a 1:1:1:1 expectation. Although the latter ratio from affected mothers could be due to transmission of either an autosomal or an X-linked dominant gene, the ratio of 11:0:0:10 from affected fathers rules out autosomal inheritance and fits X-linkage perfectly.

Glucose-6-phosphate Dehydrogenase (G-6-PD). The distinction between a "normal" and an "abnormal" trait is not always clear-cut. A striking illustration of this is provided by an X-linked trait characterized by an enzymatic defect in the red cells. Somewhat less than 10 per cent of all American blacks show a greatly reduced activity of glucose-6-phosphate dehydrogenase, an enzyme that participates in carbohydrate metabolism. The deficiency is rare in whites, but has been encountered particularly in individuals of Mediterranean stock. Persons with the enzyme deficiency are completely normal, and so are most properties of their red cells. The low activity of the enzyme is usually apparent only in laboratory tests, but under special circumstances persons with the defect suffer from sudden destruction of many of their red cells and resulting severe hemolytic anemia. The "special circumstances" are the inhalation of the pollen of broad beans (*Vicia faba*) or the ingestion of the raw bean, or the ingestion of certain drugs such as sulfanilamide, naphthalene (used in moth balls), and primaquine, an antimalarial agent. The illness brought about by the pollen or the eating of the bean is known as favism. An afflicted person recovers when the offending substances are eliminated.

Many more males than females have the deficiency. Pedigree studies show that an X-linked gene is responsible for the trait. The hemizygous males and the homozygous females are strongly affected. The enzyme activity in heterozygous females is usually more-or-less intermediate, but actually varies from as low as in homozygotes to as high as normal.

There are many different alleles at the G-6-PD locus, as shown by variant enzymes. The normal enzyme may belong to one or the other of two subtypes called A and B that can be distinguished by different speed of motility in electrophoresis. Subtype A moves faster than B.

X-linked Muscular Dystrophy. Wasting of muscles may be caused by an abnormality of the nervous system, or the degeneration may be independent of the nervous system. The term muscular dystrophy is usually applied to the latter type of defect. A number of different kinds of muscular dystrophy can be distinguished; partly on the basis of their different phenotypes, including time of onset, and partly by their different modes of inheritance—namely, recessive X-linked or dominant or recessive autosomal.

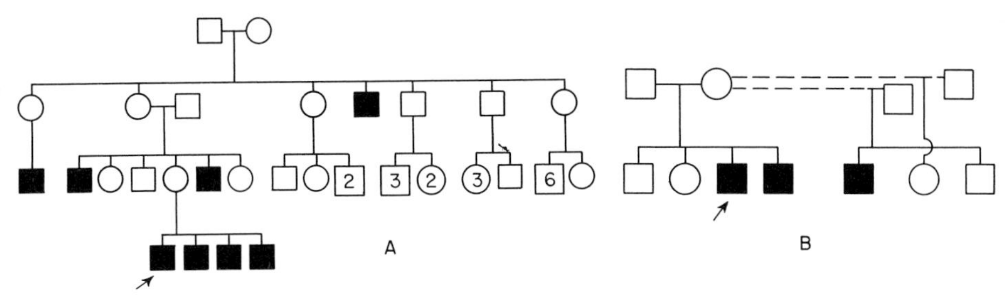

FIGURE 135
Two pedigrees showing inheritance of Duchenne-type muscular dystrophy. **A.** Three successive generations in which the disease occurred. **B.** "A complicated family" showing the disease in three males who have a common mother but two different fathers (see Problem 110, p. 331). (A, after Stevenson, *Ann. Eugen.* **18,** 1953; B, after Walton, *Ann. Human Genet.*, **21,** 1956.)

The X-linked type of muscular dystrophy, which most patients who have been classified as Duchenne-type have, usually shows its first symptoms in early childhood and progressively leads to invalidism and death, the afflicted persons seldom surviving to the age of twenty. Since affected males are almost never able to have offspring, the allele responsible for the disease belongs to the class of sublethals. It is a recessive, since women who are heterozygous for it are normal (Fig. 135). Homozygous affected women are not to be expected, since they would typically be the daughters of carrier women and affected men; but the pattern of invalidism and death of the defect almost entirely excludes such marriages. Nevertheless, there are a few records of affected females, but with normal fathers. Some of these females were probably misclassified as Duchenne-type muscular dystrophics, but at least one girl, who had 9 affected male relatives (grand-uncle, uncles, cousins), seems to have been affected with the Duchenne type of the disease.

There are several ways to account for the occurrence of the affected females. It is possible, for instance, that as a result of mutation a sperm of a normal father actually carries an abnormal allele. If this sperm fertilizes an egg which has received an abnormal allele from a heterozygous mother, a homozygous affected daughter would be produced. Another possibility is provided by abnormal distribution of the X-chromosomes in cases of nondisjunction (p. 318), which may result in an egg that after fertilization has a single X- and no Y-chromosome, and develops into a somewhat aberrant female. If the allele for X-linked muscular dystrophy were present on the single X-chromosome, such a female would be affected. Mutation and nondisjunction are rare events; but, fundamentally, this does not speak against them. Although one must attempt to explain all observations on the most likely assumptions, it is obvious that even the most unlikely of all possible events may be encountered when, as in human genetics, very large numbers of events are recorded.

X-linked Hemophilia. One of the most famous inherited abnormalities in man is the bleeder disease, hemophilia. It is one of several genetic conditions that may result in excessive bleeding at points of injury or stress, usually because deficiency of certain components of the blood delays clotting. Different abnormal genes, some autosomal and others X-linked, are responsible for the deficiency of different components. In classical X-linked hemophilia, hemophilia A, "anti-hemophilic globulin" is lacking or present only in small quantities. In another more recently discovered and rarer, X-linked bleeder disease, called hemophilia B, or Christmas disease (after a kindred which contained several affected persons), there is a deficiency for another substance, the "plasma thromboplastin component." Hemophilia is well-known, even among laymen, because of its seriousness and of the mystical feelings which are still commonly associated with true or imagined properties of blood, and because, during the past hundred years, it has occurred in many members of the royal families of Europe (Fig. 231, p. 556).

The inheritance of hemophilia has been partly known, though, of course, not understood, for a long time. The Talmud, in the second century A.D., contained rules regarding circumcision of boys in families in which death had resulted from excessive bleeding following the operation. Not only was it stated that the later-born sons of a woman who lost two boys because of bleeding should not be circumcised, but even the sons of her sisters were to be exempted from the ritual. Half-brothers of the dead children, by the same father and a different mother, were treated as normal individuals. These regulations imply that the Jews knew that the tendency was inherited from the mother, that the sisters of a known carrier could also be carriers, and that the occurrence of the disease among the sons is independent of the father.

In 1820, a more definite rule for the inheritance of the trait was given by Nasse, which, somewhat freely translated from the German, reads: "Women whose fathers were bleeders transmit the trait to their children, even if married to normal men. In these women themselves, and, in general, in all females, the trait is never expressed." A final formulation of the transmission of hemophilia, based on the discovery of sex chromosomes and their recognition as carriers of sex-linked genes, became possible only in the twentieith century (Fig. 136).

It had often been wondered whether female bleeders actually exist, even though some women of this type had been listed in the rare autosomal pedigrees of bleeder diseases. Until 1951, no X-linked hemophilic female had been diagnosed, a fact which led to the suggestion that *hh* females either die before birth or that they live but are phenotypically normal nonbleeders. Geneticists were not inclined to accept either hypothesis, but pointed out that *hh* females would only occur among the offspring of heterozygous (*Hh*) women and affected *h* men and that the very small number of such marriages

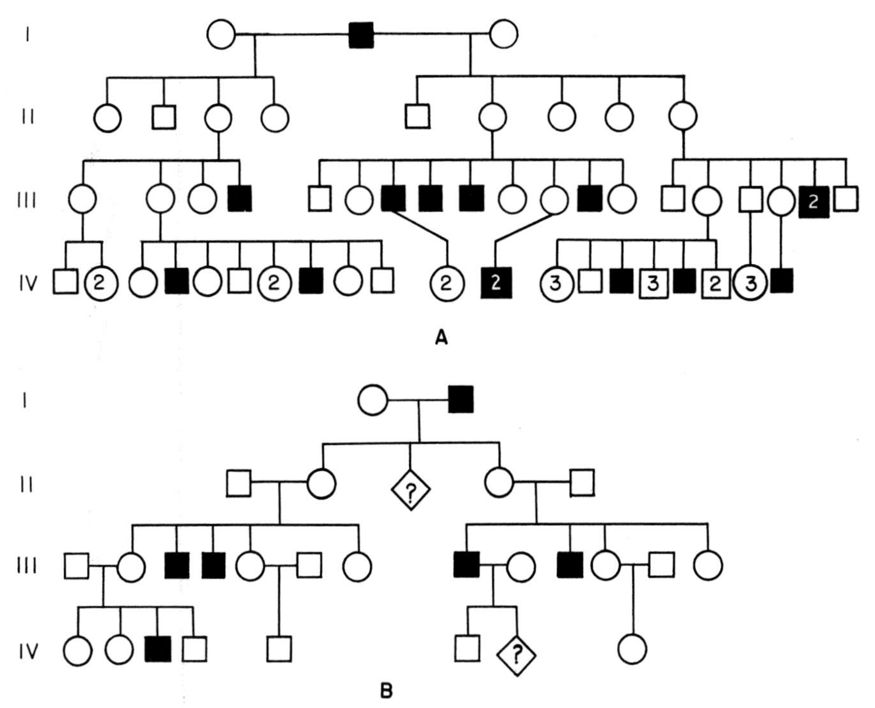

FIGURE 136
Pedigrees of hemophilia. **A.** Pedigree from Scotland. **B.** Pedigree from Germany. (A, after Bulloch and Fildes, *Treas. Hum. Inher.*, **1**, 1911; pedigree No. 490; B, part of the Schloessmann pedigree, after Baur, Fischer, Lenz.)

had, by chance, not produced *hh* daughters. This view was strengthened when typical X-linked hemophilia was discovered in a kindred of dogs and when matings between the heterozygous female and affected male dogs produced hemophilic females as well as males. Studies of the blood of these dogs contributed greatly to the understanding of hemophilia in man.

Soon after the hemophilic dogs had become known, human females were described who were *hh* homozygous and exhibited, according to careful tests, all symptoms of true X-linked hemophilia. One such woman had a history of easy bruising and free bleeding, but had successfully given birth to a child. She probably escaped serious bleeding at the time of delivery because, as is generally thought, the control of hemorrhage from the placental site is by muscular contraction of the uterine wall and not by coagulation of blood. This hemophilic woman's father had died of hemophilia, as had one of her mother's brothers (Fig. 137). Her mother may therefore be classified as a carrier, and the genotypes of the parents as *Hh* and *h*.

Oral-facial-digital Syndrome. A striking pleiotropic abnormality of mouth, face, fingers, toes, and still other parts of the body constitutes this syndrome.

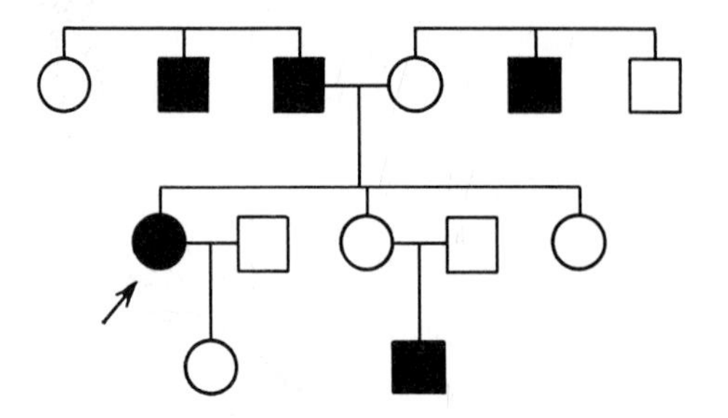

FIGURE 137
Pedigree of hemophilia containing a hemophilic woman. (After Israels, Lempert, and Gilbertson, *Lancet,* **260,** 1951.)

The great majority of cases occur in females. Affected women transmit the syndrome to about half of their daughters, and the sex ratio of their children tends to be two females to one male. These findings suggest that the syndrome is caused by a dominant allele that is lethal in males. Strong support for this interpretation is provided by the chromosomal analysis of one of the very rare affected male children, by Wahrman and his associates. They found that the boy had 47 chromosomes, the sex-chromosomal constitution being XXY. Apparently the usual lethality of the syndrome in XY males is not due to the sexual constitution of the individuals, but to the effect of a hemizygously lethal X-linked allele. The XXY child, in spite of his maleness, survived because he carried a normal X-chromosome in addition to the abnormal one. A similar genetic basis may be involved in a rare condition called incontinentia pigmenti. This syndrome is found almost exclusively in females, although its symptoms, which include abnormal pigmentation areas of skin, have no obvious relation to any sex difference. As in the oral-facial-digital syndrome, mothers affected with incontinentia pigmenti have approximately twice as many daughters as sons—half of the daughters being affected.

Nondisjunction and X-linkage. In the extensive investigations made by experimental geneticists and cytologists, many cases have been uncovered in which apparent exceptions from normal hereditary behavior can be explained by exceptional behavior of chromosomes. A well-known kind of "abnormal" transmission is that involving nondisjunction of the X-chromosomes in *Drosophila.*

A like mishap in the normal distribution of the X-chromosomes is now known to occur occasionally in man. Nondisjunction of the X-chromosome during either one of the two meiotic divisions of an egg leads to one or another kind of chromosomal constitution of the unferilized egg nucleus: XX when the two not-segregated X-chromosomes remain in the egg, or O when they jointly pass into a polar body (Fig. 138, upper rows). After fertilization by either an X- or a Y-carrying sperm, four types of zygotes are produced: XXX, XXY, XO, and YO. The first three types may develop into individuals which, in part, can be recognized by their characteristic phenotypes as well as by

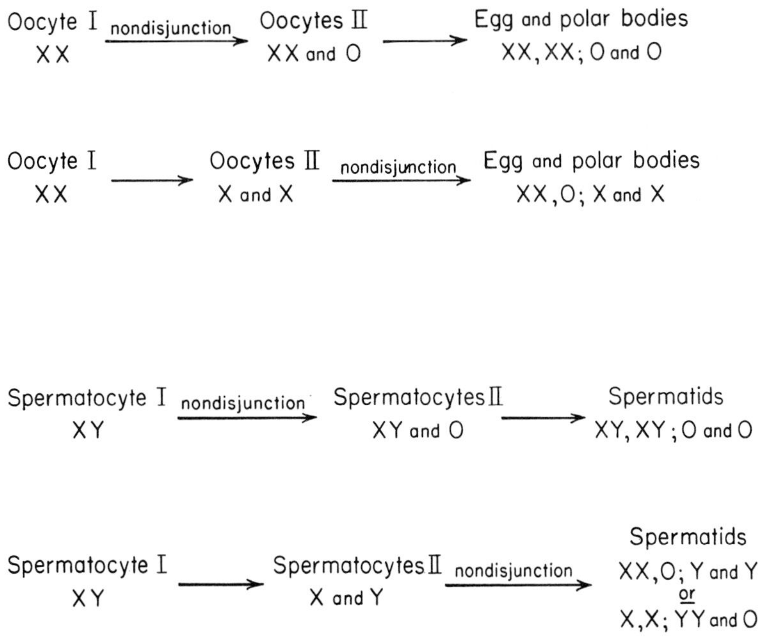

Oocyte I XX —nondisjunction→ Oocytes II XX and O → Egg and polar bodies XX, XX; O and O

Oocyte I XX → Oocytes II X and X —nondisjunction→ Egg and polar bodies XX, O; X and X

Spermatocyte I XY —nondisjunction→ Spermatocytes II XY and O → Spermatids XY, XY; O and O

Spermatocyte I XY → Spermatocytes II X and Y —nondisjunction→ Spermatids XX, O; Y and Y or X, X; YY and O

FIGURE 138
Diagrams of nondisjunction during meiotic divisions of oocytes and spermatocytes and the resulting chromosomal constitutions of eggs, polar bodies, and spermatids.

counting and studying their chromosomes; the YO type is unable to develop far.

The consequences of meiotic nondisjunction of the sex chromosomes during spermatogenesis will vary, depending on whether the abnormal distribution of chromosomes occurs at the first or second meiotic division of the spermatocytes (Fig. 138, lower rows). Normally, the first division segregates the X- and Y-chromosomes. If the division is nondisjunctional, the resulting secondary spermatocytes will by XY and O, respectively, and the second divisions will result in two XY and two O spermatids. If the first division is normal but the second division of one or the other secondary spermatocyte is nondisjunctional, then an XX or a YY sperm will be formed. Altogether, nondisjunction at the meiotic divisions will result in XY, O, XX, and YY sperm. Fertilization of regular X eggs with the four types of nondisjunctional sperm will lead to zygotes of the constitutions XXY, XO, XXX, and XYY. All of these are viable. The foregoing considerations imply that nondisjunction occurs only once in a meiotic sequence. This is probably true in the great majority of cases. Nevertheless, it is expected that occasionally more than one nondisjunctional process may occur during the meiotic divisions of an oocyte or spermatocyte. There is, indeed, genetic evidence, for a special individual, according to which he possibly originated from the fertilization of a normal egg by an XYY sperm. Such sperm would have to have undergone nondisjunction of X and Y at the first meiotic division, followed by nondisjunction of the sister chromosomes of the Y in the second division. Considering the exceptional nature of nondisjunction, it will be extremely rare that both egg and sperm will be abnormal in any given fertilization. Thus XX and O eggs will usually fuse with X or Y sperm, and the various nondisjunctional types of sperm will usually fuse with X eggs.

Since typical X-linked inheritance of genes is based on normal transmission of the sex chromosomes, unusual types of transmission of these chromosomes should lead to exceptions from typical X-linked inheritance. It has been shown that in man XXY zygotes develop into males, and XO zygotes into females (pp. 514-515). If sperm of a color-blind man that carries a Y-chromosome should fertilize a nondisjunctional XX egg of a woman normal in color vision, the resulting XXY son would be as normal in color vision as an XY son. If, however, sperm of a color-blind man carrying an X-chromosome should fertilize a nondisjunctional O egg, then the resulting XO daughter would be color blind, like her father, contrary to regular expectations. Furthermore, if a nondisjunctional XY sperm should fertilize an X egg, the resulting XXY son may constitute an exception to regular X-linked inheritance. If, for instance, the parents were a color-blind female and a normal male, the XXY son would not be color blind, contrary to usual expectation. Conversely, fertilization of an X egg by a nondisjunctional O sperm may result in an XO daughter who, given a color-blind mother and a normal father, would be color blind, again contrary to usual expectation.

Pedigrees in which the X-chromosomes were suitably marked with X-linked genes have provided evidence for nondisjunction in both females and males (Fig. 139, A-D). Approximately three-quarters of XO persons receive their X-chromosome from their mother, the rest from their father, and approximately two-thirds of XXY persons receive their "extra" X-chromosome from the father and the rest from their mother. (As has been discussed in Chapter 5, the origin of the XO and XXY constitutions is not necessarily by nondisjunctional events in meiosis, but may be, at least in part, by mitotic events in gonial cells or in the fertilized egg.)

Dosage Compensation

Normal development of an organism depends on a harmonious coordination of the effects of numerous genes. Autosomal genes are always present in pairs, and it is known from experimental organisms that the loss of one of

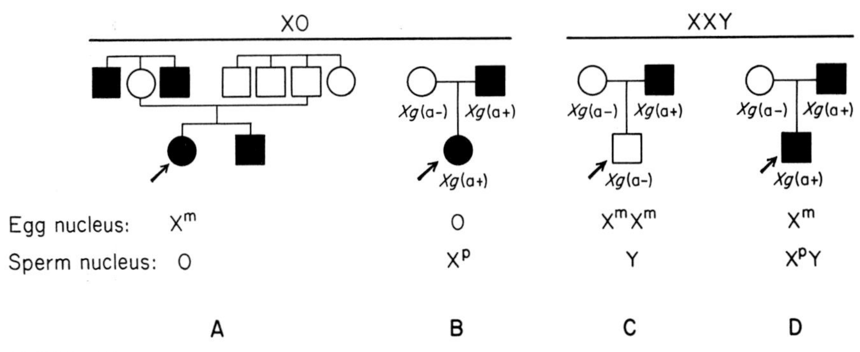

FIGURE 139
Pedigrees indicating the occurrence of two kinds each of nondisjunctional egg and sperm nuclei. (Marker genes: A, deuteranopia; B–D, Xg antigen. X^m, maternal X-chromosome; X^p, paternal X-chromosome. **A.** A deuteranopic XO female lacking a paternal sex chromosome. **B.** An XO female lacking a maternal X-chromosome. **C.** An XXY male possessing two maternal X-chromosomes. **D.** An XXY male having paternal X- and Y-chromosomes. (A, after Lenz; B–D, from data in Race and Sanger.)

a pair often has unfavorable effects on development. The balance of normal development obviously depends on a double dose of each autosomal gene, and a single dose of one in the presence of a double dose of the others produces an imbalance of processes. This, however, does not hold for X-linked genes. They are present as pairs in females but singly in males, and yet their effects are similar if not identical. The color blindness of a hemizygous male is like that of a homozygous female, and the same is true of other traits transmitted by X-linked genes, whether recessive or dominant, mutant or normal.

Since there are different effects from one as against two doses of autosomal genes, some developmental mechanism must exist that compensates for the dosage differences of X-linked genes. This dosage compensation of X-linked genes applies not only to X-chromosomally normal males and females but also to X-chromosomally aberrant types. Grumbach, Marks, and Morishima found the activity of glucose-6-phosphate dehydrogenase to be essentially alike in XY and XX individuals, as well as in XO, XXY, XXX, and XXXX individuals (apart from one exception of probably secondary causation).

Dosage compensation was discovered in *Drosophila,* in which the mechanism consists of greater metabolic activity of the single X-chromosome in a male than of each of the two X-chromosomes in a female. In a way, the *Drosophila* male works hard to lift itself to the level of the female. In man, dosage compensation has been recognized only relatively recently, and its mechanism is the opposite from that in the fly. In man, the human female reduces the activity of her two X-chromosomes so as to equal that of the single X of the male. Specifically, the reduction of activity of the two X-chromosomes in the female consists in the inactivation of one of them: the female descends to the level of the male.

The Single-active-X Hypothesis. The hypothesis of the "single-active-X chromosome" as proposed by Mary Lyon, Liane B. Russell, Beutler and others initially rested on two sets of observations. One of these was based on a cytological discovery concerning the nucleus of nondividing nerve cells in cats. In 1949 Barr and Bertram noticed that many cell nuclei of some cats possess a special, small, stainable body that is not present in most nuclei of other cats. The cats with the "Barr body" were females, those without it males. It was soon established that Barr bodies occur also in human females, in cells from many tissues including among others the epidermis, the oral mucosa, and the amniotic fluid (Fig. 211, p. 511). In human males, no Barr bodies are found. Later cytological evidence showed that the Barr body constitutes one of the two X-chromosomes of female cells. This chromosome replicates later than its homologue. Instead of uncoiling during the interphase stage between consecutive mitoses, the late-replicating X-chromosome remains condensed and appears as the Barr body. Since genic action seems to depend on an extended state of the genic thread, the condensed state of the Barr body would therefore correspond to inactivity of the genes of one X-chromosome in female cells.

In agreement with this interpretation, the nuclei of XXX females have two Barr bodies and those of XXXX females three, so that a single active X-chromosome is present alike in the nuclei of XX, XXX, and XXXX individuals (Fig. 211). This cytological finding is a counterpart of the finding reported above that there is equal content of G-6-PD in women who have the three different X-chromosomal constitutions.

The interpretation of the Barr body as an inactive X-chromosome fitted, with some additional assumptions, the second set of observations basic to the single-active-X, or "Lyon," hypothesis. These observations related to the pigmentation of the fur of mice. Most fur colorations that depend on autosomal genes either are uniform over the mice or follow a more-or-less regular morphological pattern, but X-linked genes for fur colors express themselves differently. In heterozygotes for these genes, random patterns of phenotypically different areas are observed, some exhibiting the coloration due to the allele in one X-chromosome and others exhibiting the coloration due to the allele in the other X-chromosome.

It was the mosaic nature of pigmentation controlled by X-linked genes which required the assumption that the inactivated X-chromosome, which cytologically appears as the Barr body, is not one and the same in all cells. Rather, in some cells it is the maternal X-chromosome and in the other cells it is the paternal. When the maternal X is inactivated, the genes on the paternal X are expressed phenotypically, and when the paternal X is inactivated, the maternal X-linked genes are expressed. It was postulated that the time in development at which the X-chromosomes become inactivated lies early in the life of the embryo. This was deduced from the facts that Barr bodies are seen first as early as the preimplantation period of the blastocyst and that the fur pattern of mosaic mice was a coarse one consisting of large coherent areas of each of the two alternative colors. Late inactivation, it was reasoned, should produce a finer mosaic of the two cell types. Coarse mosaicism as observed required the further assumption that descendants of an inactivated X-chromosome do not revert to activity. If such back-and-forth switching took place, a pepper-and-salt type of phenotypic mosaic would have been expected.

Evidence bearing on the single-active-X hypothesis in man comes from a variety of findings. One of these concerns the X-linked condition, anhidrotic ectodermal displasia, which affected the "toothless men of Sind" (= Scinde). Heterozygous women exhibit a mosaic of jaw areas with and without teeth and a mosaic of their skin with and without sweat glands (Fig. 140). There is considerable variation from heterozygote to heterozygote of the relative extent and location of the two tissue types, as would be expected from a more-or-less random inactivation of one or the other X-chromosome early in development. Such random inactivation also can account for the fact that heterozygotes for X-linked traits that are measured without reference to visible mosaicism often show unexpectedly large variability. In hemophilia, for instance, heterozygotes may have less than 20 per cent of the clotting factor of a normal person, or may have 100 per cent, or may have any value in between. The single-active-X hypothesis accounts for this variability as the result of random variability of alternative inactivation. Assume, for the sake of the argument, that at the time of embryonic inactivation there are only four cells whose descendants will form the clotting factors; chance will then lead to inactivation of the normal X-chromosome in all four cells in $(1/2)^4 = 6.25$ per cent of all heterozygotes, to inactivation of the hemophilia-carrying X-chromosomes in all four cells of another $(1/2)^4$ of heterozygotes, and to inactivation of equal numbers of the two kinds of X-chromosomes in $6 \times (1/2)^4 = 37.5$ per cent of heterozygotes, and there will be 3:1 and 1:3 ratios of inactivated cells in the remaining heterozygotes, each contributing 25 per cent of them.

The strongest support for the alternative inactivation hypothesis comes from studies of the enzymatic constitution or metabolic activities of individual cells and their descendants. In these studies the X-linked enzyme glucose-6-phosphate dehydrogenase, discussed in an earlier section of this chapter, has played a major role. It was first shown that women heterozygous for a normal allele and an abnormal one that leads to deficiency of the enzyme have two

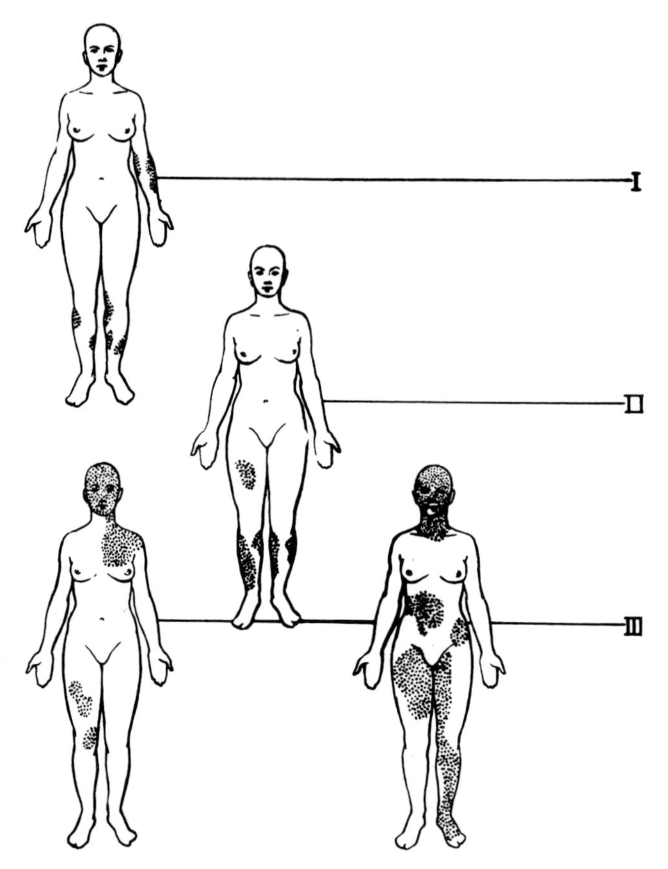

FIGURE 140
Three generations of women heterozygous for presumably X-linked anhidrotic dysplasia. Stippled areas indicate regions of increased skin resistance to an electric current due to absence or low frequency of sweat glands. The two females of generation III are identical twins. Note the striking differences in the distribution of the high-resistance areas. (Kline, Sidbury, and Richter, *J. Pediat.*, **55**, 1959, The C.V. Mosby Company, St. Louis.)

populations of red cells, one with normal G-6-PD activity and one with deficient activity. This is in agreement with the inactivation hypothesis, but could also be accounted for if some mechanism existed which led each individual blood cell to fall into one or the other activity type according to a threshold superimposed on some inherent variability of enzyme production. This possibility was excluded when Davidson, Nitowsky, and Childs obtained tissue cultures from women heterozygous for the deficiency allele and from women heterozygous for the two electrophoretic variant nondeficient alleles Gd^{A} and Gd^{B}. Initially tissue cultures of skin cells from the women heterozygous

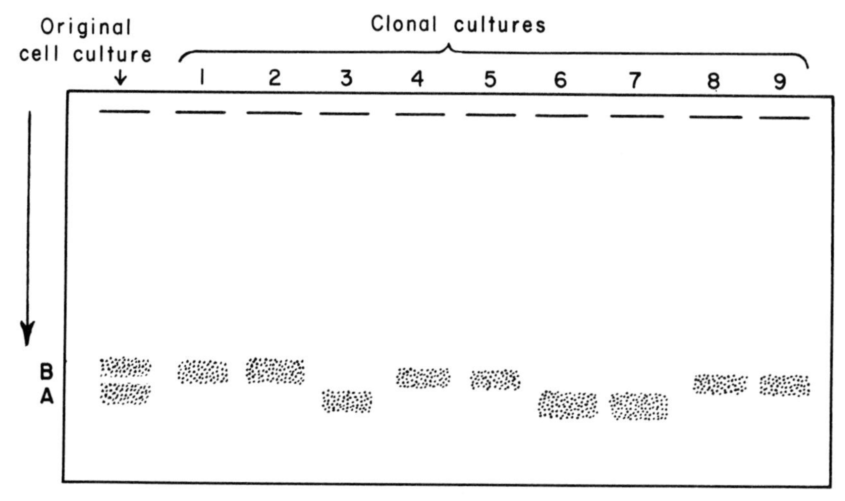

FIGURE 141
Electrophoretic pattern of G-6-PD properties of cell cultures from a woman who possesses the AB phenotype. Samples were started at the top of the figure. *Left edge,* from a culture before cloning showing two bands. **B.** Slow band. **A.** Fast band. Clonal cultures 1-9: six cultures showing the B, three showing the A band only. (After Davidson, Nitowsky, and Childs.)

for the deficiency allele showed an enzyme activity between that of non-deficient and deficient homo- or hemizygotes. Later, when cultures were assayed that originated from single cells ("cloned" cultures), they were seen to fall into two separate groups, either with fully normal or with typically deficient activity. Similarly, when initial tissue cultures of skin cells from the Gd^{A}/Gd^{B} heterozygous women were tested, they showed the presence of both A and B variants. When, later, derived cloned cultures were tested, they were either exclusively A or exclusively B (Fig. 141). Specifically, out of 54 cloned cultures from 6 heterozygous women, 24 produced the A variety of the enzyme and 30 the B variety. To these findings may be added the results of "natural" experiments. Linder and Gartler tested for the presence of A and B variants in certain benign uterine tumors in heterozygous Gd^{A}/Gd^{B} women. All but one of 29 tumors were pure for either A or B, the single exception probably due to admixture from nontumor cells. If, as is likely, the tumors start from single cells, the fact that they produce only one or the other of the two G-6-PD variants is consistent with the single-active-X hypothesis.

Another X-linked gene that has furnished evidence for the single-active-X hypothesis is that responsible for Hunter's syndrome (Fig. 142, A). The main clinical features of this syndrome are dwarfism, deafness, mental defect, and grotesque skeletal deformities that have given it the name gargoylism. Biochemically, the syndrome is characterized by an inborn error of mucopolysaccharide metabolism that results in an accumulation of the substance in

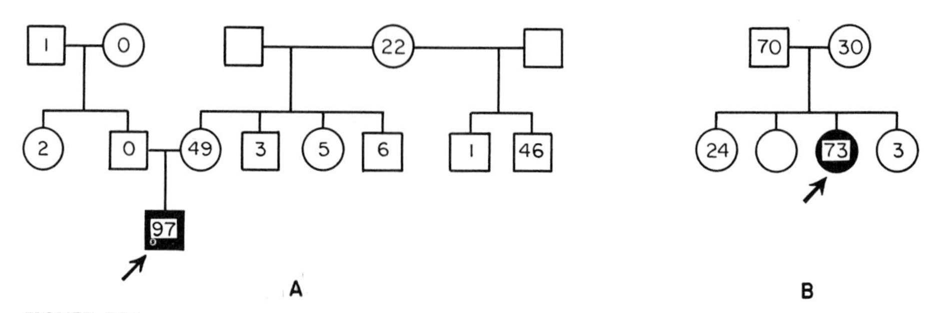

FIGURE 142
Pedigrees of Hunter's and Hurler's syndrome. **A.** An X-linked pedigree. (Hunter) **B.** An autosomal pedigree. (Hurler) Numbers inside symbols indicate the per cent of fibroblasts in culture that contained increased amounts of intracellular mucopolysaccharides. Absence of symbols indicates that no cell cultures were studied. Note that in the autosomal kindred the percentage of affected fibroblasts in the normal father is similar to that in the defective daughter. (After Danes and Bearn, *J. Exp. Med.*, **123**, 1966.)

various tissues of the body. A very similar syndrome is caused by the autosomal recessive gene responsible for Hurler's syndrome (Fig. 142, B). Thus, a comparison is possible between the autosomal condition that in heterozygotes should not result in two different cell populations, and the X-linked condition that in heterozygotes should provide for two alternative cell types. Tests for gene expression in tissue cultures depended on staining procedures specific for intracellular mucopolysaccharides. Uncloned tissue cultures of *autosomal* heterozygotes showed many of the cells to be positive for increased amounts of mucopolysaccharides (and increased amount of uronic acid). Even more important, cloned cultures were all of one type, again showing increased amounts. In contrast to this, several hundred clones from heterozygotes for the *X-linked* gene fell into two sharply distinct groups, with or without cells having increased amounts of mucopolysaccharides and uronic acid. Clearly, these data fit the single-active-X hypothesis.

There are, however, some facts that do not conform with the single-active-X hypothesis as originally proposed. The most striking of these is the abnormal phenotype of XO and XXY individuals. Although the former are females and the latter males, they are physically distinguishable from normal XX females or XY males, respectively. If one X-chromosome is inactivated in XX women, they would be effectively XO, but they *are* normal females. And if one X-chromosome is inactivated in XXY men, who have Barr bodies like XX women, they would be effectively XY, but they are *not* normal males. It is possible to reconcile these facts with the hypothesis if it is assumed that sexual differentiation is influenced at a very early embryonic stage prior to Barr-body-directed inactivation. In that case XO and XX, and XXY and XY are not equivalent in developmental control. Another assumption is possibly closer to the truth. If inactivation does not affect a whole X-chromosome but only part of it, and if the not-inactivated section carries genes that determine the phenotypic difference between XO and XX, and between XXY and XY, then XX is not equivalent to XO, and XXY not to XY.

If such a not-inactivated section does exist, we might expect some other X-linked genes likewise to be located in it. Such genes in heterozygous women should not lead to mosaic expression. Color blindness of the protanopia type, it has been found, does not show mosaicism in heterozygotes, neither from one eye to another nor within the retina of a single eye. Does this uniformity suggest that the mosaic pattern for protanopia heterozygotes is so fine as to have escaped detection? Or would it suggest location of the protanopia gene in a not-activated segment of the X-chromosome? Originally an equivalent hypothesis seemed to be needed also for explaining the fact that the red cells of women heterozygous for the Xg^{a} blood-group allele could not be separated into two populations, Xg(a+) and Xg(a−). Later, successful separation of two cell populations from heterozygous women was reported. Still later, no success in separation was achieved in several studies. In 1971, the balance of opinion was inclined to regard the evidence concerning Xg as "perhaps stronger against than for inactivation."

A searching critique of the whole inactivation hypothesis of X-linked genes has been offered by Grüneberg, who stresses the relative paucity of fully convincing cases, and the not-infrequent mosaic expression of autosomal genes which is explained, not by inactivation of alternate chromosomes, but by variation of gene expression at later stages in the pedigree of causes (see p. 74). On the whole, the single-active-X hypothesis has withstood the impact of Grüneberg's analysis. However, questions remain, among them that regarding the time of inactivation and its supposed irreversibility. In this connection, it would have been expected that there would be coarse mosaicism of tissues in women heterozygous for G-6-PD alleles and for alleles for other cellular materials. However, areas of skin as small as one square millimeter have been proven to contain cells of the two alternative varieties.

X-linked and Autosomal Genes and Differences in Sex Incidence of Traits. Many inherited human traits occur more frequently among males than among females. This phenomenon is, of course, fully understandable when an X-linked locus is responsible; for example, in partial color blindness. When, however, autosomal inheritance is operating, a higher frequency of affected males than females requires a special explanation. Two alternative explanations have been proposed: one, developmental; the other, genetic. The former assumes that a higher incidence of some inherited conditions among males is a result of developmental reactions induced by the specific genotypes' taking a different course in males than they do in females. The genetic explanation assumes that a given trait becomes apparent only when, in addition to an autosomal genotype, a recessive X-linked allele is present homozygously in the female or hemizygously in the male. Thus, if an abnormal autosomal genotype is recessive (*aa*) and the X-linked allele is designated by *s*, then, only *s aa* males and *ss aa* females will be affected; but *S aa* males and *Ss aa* or *SS aa* females will be normal. (All *AA* or *Aa* would also be normal regardless of whether they are *S* or *s* males, or *SS*, *Ss*, or *ss* females.) Since the ratio of *s* males to *ss* females is $q:q^2$ (where q equals the frequency of the allele *s*) *s aa* males and *ss aa* females should occur in the proportion $q:q^2$. Thus the genetic hypothesis fulfills the basic requirement; namely, it explains the surplus of affected males.

The hypothesis entails further genetic considerations, however, and these have been tested for four kinds of inherited abnormalities that have a higher male than female incidence: the Laurence-Moon-Biedl syndrome, harelip and cleft palate, allergy, and mental deficiency. The method of testing, suggested by R. A. Fisher and applied by Csik and Mather, requires determination of the mode of transmission of these traits. If an X-linked gene participates in causing them, X-linked inheritance should be involved; but if no evidence for X-linked inheritance is found, then the genetic hypothesis is untenable. Specifically, the expectations that would accord with the genetic theory were derived in the following way. The presence of the autosomal allele *a* in the parents, either homozygously or heterozygously, and homozygously in the children, is a prerequisite for the possible appearance of the trait. Since transmission of *a* has no relation to sex, it does not need to be considered further. Therefore, only the hypothetical X-linked allele pair *S* and *s* must be studied.

1. If a father had a normal X-linked allele *S*, then all his daughters would be normal, but his sons could be affected if the mother carried *s* homozygously or heterozygously.
2. If, on the contrary, the father transmitted *s*, then both his daughters and his sons could show the abnormal trait if the mother was *Ss* or *ss*. The frequency of affected daughters should be the same as that of affected sons, since a homozygous *ss* woman and a hemizygous *s* man will have *ss* daughters and *s* sons only; and a heterozygous *Ss* woman and hemizygous *s* man will have *Ss* and *ss* daughters and *S* and *s* sons in equal proportions.

Thus, two kinds of sibships would be expected: one having affected boys only, and the other having equal numbers of affected girls and boys.

A test for this simple expectation must take into account the variability of genetic ratios in small families: sibships of the second type will not always contain equal numbers of affected girls and boys. Moreover, ascertainment by means of affected sibs requires correction of the data before they may be compared with expectation (see pp. 202–209). These secondary difficulties were overcome in the tests by determining the proportions of affected girls to boys among the sibs of affected individuals. If a trait is sex-linked, the expectation is an equal number of affected girls and boys among the sibs of affected girls, but more affected boys than girls among the sibs of affected boys.

An examination of data from numerous pedigrees shows that a surplus of affected males occurs among the sibs of *both* affected girls and affected boys. Consequently, the genetic hypothesis, which assumes an X-linked component, is disproved, and it may be concluded that the developmental interpretation is correct: i.e., the higher frequency of affected males is due to a more pronounced action of certain autosomal genotypes in the male sex. This conclusion applies also to traits other than those tested by Csik and Mather, including baldness.

X-linkage and Counseling. The recognition of X-linkage of human traits provides an important tool for the genetic adviser. Predictions can be made with certainty about phenotypes and genotypes of the sons and daughters of men affected with an X-linked trait, provided their wives can be assumed to be homozygously normal. Absolute predictions can also be made about the types of offspring of affected women who are homozygous for a recessive allele and married to normal men. The progeny of a heterozygous woman married to a normal man can be foretold in terms of probabilities only, her sons having an even chance of being normal or affected, and her daughters an even chance of being homozygous normal or carriers.

The presence of a recessive X-linked allele in a heterozygous mother may be cause for reasonable concern; whereas presence of a recessive autosomal allele may not. A woman heterozygous for an uncommon autosomal gene will very rarely happen to marry a man who is also heterozygous for it—unless the spouses are closely related to each other—and only if both parents transmit the atypical gene to the same zygote will an affected child be born. But if the gene is X-linked, half of the sons of a heterozygous mother are likely to be affected, regardless of the genotype of the father.

A woman whose father was affected by a recessive sex-linked allele or who has an affected son is a carrier. The probability that daughters of such a woman will be carriers is 1/2. That probability is often a sufficient cause for abstaining from childbearing if the abnormality is severe. If methods were available to diagnose, by medical tests, the presence or absence of heterozygosity, advice on the desirability of having children could be based not on probability but on determinate evidence. In hemophilia, such a diagnosis, which makes use of the slightly prolonged clotting time of the blood of some carrier women, is sometimes possible. Another approach to more specific counseling in connection with X-linked genes will be described in Chapter 15. It makes use of information concerning chromosomal linkage between a deleterious gene, like that for hemophilia, and a closely linked other gene that may serve as a marker. Still another approach to problems of counseling is based on chromosomal and biochemical studies of the cells of a fetus, present in the amniotic fluid (see the discussion of amniocentesis in Chap. 30).

The Relative Frequency of Sex-linked Traits. The phenomenon of sex linkage has some dramatic features. Dosage compensation for X-linked genes and its explanation in terms of the single-active-X theory constitute phenomena that offer some understanding of regulation of gene activity. The usual "disappearance" of sex-linked traits in the children of affected men and the reappearance among some of their daughters' sons, the rarity of affected women as compared to men or even their virtual absence, are causes for wonder to the uninitiated. Knowledge of the chromosomal basis of inheritance provides clear insight into such facts, and also explains why there are relatively so many more known examples of sex-linked than of autosomal inheritance. Statistically, the sex chromosome pair should not be expected to carry more than about 1/23 of all genes, since it is only one of 23 pairs of chromosomes, yet the proportion of known sex-linked genes is much higher that 1 in 23. This is a result of greater ease of detection: a rare, recessive X-linked gene with an allele frequency q will show its phenotype in a fraction q of all males, while an equally rare autosomal recessive will become apparent in only q^2 of all individuals.

PROBLEMS

101. For a Y-linked trait, what phenotypes do you expect in the following descendants of an affected man?
    (a) His sons.
    (b) His daughters.
    (c) The sons of his sons.
    (d) The daughters of his daughters.
    (e) The daughters of his sons.
    (f) The sons of his daughters.
102. List the genotypes of color blindness for all individuals in the pedigrees in Figure 130. If more than one genotype is possible, list the alternatives and state their probabilities.
103. A woman has normal parents and a color-blind brother. What is the probability that her first son will be color blind?

104. A normal woman whose father was color blind marries a man with normal vision.
(a) What proportion of her sons is expected to be color blind?
(b) If her husband had been color blind, what would have been the expectation for the sons?

105. List the genotypes of hemophilia for all individuals in the pedigrees in Figure 136. If more than one genotype is possible, list the alternatives and state their probabilities.

106. (a) Determine the genotypes of all individuals shown in the pedigree in Figure 137.
(b) If the proposita had had a son, what would have been the probability of his being normal? If her daughter marries a normal man, what kinds of children can they expect?

107. In a certain population, the frequency of women affected with a harmless X-linked recessive trait is 1 in 10,000. What is the frequency of affected men?

108. In a considerable number of pedigrees, color-blind fathers have color-blind sons. Is this evidence for genetic transmission of color blindness?

109. In the first known pedigree of red-green color blindness in several generations, illustrated in Figure 130, D, J. Scott (II-2) was affected and had an affected father and a normal mother. What, in the general population, is the probability of a color-blind man having an affected father and a not-affected mother?

110. In several sibships, such as that shown in Figure 135, B, two or more sons affected with Duchenne-type muscular dystrophy have the same mother but different fathers. The parents are unrelated to each other. Why are these sibships strong indications for X-linked recessive, and against autosomal recessive, inheritance of the disease?

111. In population I, the frequency of a sex-linked recessive gene is 20 per cent. In population II, the frequency of the same sex-linked recessive is 4 per cent. After panmixis, established as a result of several generations of random mating of equal numbers of the two initial populations, what will be the frequency of all possible genotypes?

112. A man is affected with a dominant X-linked trait. His wife is not affected. Give the possible genotypes of their sons and daughters and of their sons' and daughters' children.

113. A large number of females are heterozygous for the two X-linked alleles at the G-6-PD locus: Gd^A and Gd^B. Assume that inactivation of one or the other of the X-chromosomes occurs at random at the embryonic stage where four cells are present whose descendants form all the future red cells of the individual.
(a) What fraction of females is expected to have only Gd^A active cells?
(b) What fraction is expected to have half Gd^A and half Gd^B active cells?

114. Can the pedigree of Leber's disease, shown in Figure 104 be explained on the basis of X-linked inheritance?

REFERENCES

Barr, M. L., 1965. The sex chromosomes of man. *Amer. J. Obst. Gynecol.,* **93**:608–616.

Bell, J., 1926. Colour blindness. *Treas. Hum. Inher.,* II, **2**:125–267.

Beutler, E., Yeh, M., and Fairbanks, V. F., 1962. The normal human female as a mosaic of X-chromosome activity: Studies using the gene for G-6-PD-deficiency as a marker. *Proc. Nat. Acad. Sci.,* **48**:9–16.

Chakravartti, M. R., 1968. Hairy Pinnae in Indian Populations. *Acta Genet.,* **18**:511–520.

Court-Brown, W. M., Harnden, D. G., Jacobs, Patricia A., Maclean, N., and Mantle, D. J., 1964. *Abnormalities of the Sex Chromosome Complement in Man.* 239 pp. Her Majesty's Stationery Office, London.

Csik, L., and Mather, K., 1938. The sex incidence of certain hereditary traits in man. *Ann. Eugen.,* **8**:126–145.

Curth, Helen O., 1966. Incontinentia pigmenti. Pp. 813–825 *in* Gottron, H. A., and Schnyder, U. W., (Eds.), *Hdbch. Haut-und Geschlechtskrankh,* vol. 7, *Vererbung von Hautkrankheiten,* Springer, Berlin.

Davidson, R. G., Nitowsky, H. M., and Childs, B., 1963. Demonstration of two populations of cells in the human female heterozygous for glucose-6-phosphate dehydrogenase variants. *Proc. Nat. Acad. Sci.,* **50**:481–485.

Dronamraju, K. R., 1965. The function of the Y-chromosome in man, animals, and plants. *Advan. Genet.,* **13**:227–310.

Fialkow, P. J., 1970. X-chromosome inactivity and the *Xg* locus. *Amer. J. Hum. Genet.,* **22**:460–463.

Forssmann, H., 1955. Two different mutations of the X-chromosome causing diabetes insipidus. *Amer. J. Hum. Genet.,* **7**:21–27.

Fujimoto, W. Y., and Seegmiller, J. E., 1970. Hypoxanthine-guanine phosphoribosyltransferase deficiency: activity in normal, mutant and heterozygote-cultured human skin fibroblasts. *Proc. Nat. Acad. Sci.,* **65**:577–584.

Grumbach, M. M., Marks, P. A., and Morishima, A., 1962. Erythrocyte glucose-6-phosphate dehydrogenase activity and X-chromosome polysomy. *Lancet,* **1**:1330–1332.

Grüneberg, H., 1967. Sex-linked genes in man and the Lyon hypothesis. *Ann. Hum. Genet.,* **30**:239–257.

Kalmus, H., 1965. *Diagnosis and Genetics of Defective Colour Vision,* 114 pp. Pergamon, Oxford.

Linder, D., and Gartler, S. M., 1965. Glucose-6-phosphate dehydrogenase mosaicism: utilization as a cell marker in the study of leiomyomas. *Science,* **150**:67–69.

Lyon, Mary F., 1961. Gene action in the X-chromosome of the mouse (Mus musculus L.). *Nature,* **190**:372–373.

Lyon, Mary F., 1968. Chromosomal and subchromosomal inactivation. *Ann. Rev. Genet.,* **2**:31–52.

McKusick, V. A., 1962. On the X-chromosome of man. *Quart. Rev. Biol.,* **37**:69–175.

Mittwoch, Ursula, 1967. *Sex Chromosomes.* 306 pp. Academic, New York.

Ohno, S., 1967. *Sex Chromosomes and Sex-linked Genes.* 192 pp. Springer, Berlin.

Penrose, L. S., and Stern, C., 1958. Reconsideration of the Lambert pedigree (Ichthyosis hystrix gravior). *Ann. Hum. Genet.,* **22**:258–283.

Pickford, R. W., 1965. *The genetics of colour blindness.* Pp. 228–248 *in* de-Reuck, A. V. S., and Knight, Julie, (Eds.), Ciba Found. Symposium on Colour Vision. Little, Brown, Boston.

Stern, C., 1960. Dosage compensation—development of a concept and new facts. *Can. J. Genet. Cytol.,* **2**:105–118. Reprinted (pp. 524–544) *in* Montagu, M. F. A. (Ed.), *Genetic Mechanisms in Human Disease: Chromosomal Aberrations.* Thomas, Springfield, Ill.

Stern, C., Centerwall, W. R., and Sarkar, S., 1964. New data on the problem of Y-linkage of hairy pinnae. *Amer. J. Hum. Genet.,* **16**:455–471.

Waaler, G. H. M., 1927. Über die Erblichkeitsverhältnisse der verschiedenen Arten von angeborener Rotgrünblindheit. *Zeitschr. Abst. Vererb.,* **45**: 279–333.

Waaler, G. H. M., 1967. Heredity of two types of normal colour vision. *Nature,* **215**:406.

Waardenburg, P. J., Franceschetti, A., and Klein, D., 1961–1963. *Genetics and Opthalmology,* 2 vols., 1914 pp. Thomas, Springfield, Ill.

Wahrman, J., Berant, M., Jacobs, J., Aviad, I., and Ben-Hur, N., 1966. The oral-facial-digital syndrome: a male lethal condition in a boy with 47/XXY chromosomes. *Pediatrics,* **37**:812–821.

Wald, G., 1966. Defective color vision and its inheritance. *Proc. Nat. Acad. Sci.,* **55**:1347–1363.

Weatherall, D. J., Pembrey, M. E., Hall, E. G., Sanger, Ruth, Tippett, Patricia, and Gavin, June, 1970. Familial sideroblastic anemia: problem of *Xg* and X chromosome inactivation. *Lancet,* **2**:744–748.

15

LINKAGE, CROSSING OVER, AND CHROMOSOME MAPPING

In earlier chapters on the biological basis of inheritance and on meiosis (Chaps. 2 and 4), we discussed the fact that two different nonallelic genes in man may either be located in two different chromosomes or at different loci in the same chromosome. If in different chromosomes, they are transmitted independently of each other. An individual who receives the alleles A^1 and B^1 in two different chromosomes from his mother and A^2 and B^2 in the homologous chromosomes from his father forms four different kinds of gametes in equal numbers: A^1B^1, A^1B^2, A^2B^1, and A^2B^2. Thus, among the gametes, the "new" combinations, A^1B^2 and A^2B^1, are as common as the "old" combinations, A^1B^1 and A^2B^2. We have seen that independent recombination of the maternal and paternal alleles at the two loci results from the independent arrangement of different chromosome pairs on the meiotic spindles (Fig. 48, p. 89).

If the genes A and B are in the same chromosome, segregation in an individual who receives A^1B^1 from his mother and A^2B^2 from his father might or might not result in equal numbers of gametes with old and new combinations. The old combinations, A^1B^1 and A^2B^2, are formed by meiotic chromosome strands which do not cross over in the section between A and B; the new combinations, A^1B^2 and A^2B^1, are formed by crossover chromosome strands (see Fig. 49, p. 91; the less common multiple crossovers will be disregarded). Since the frequency of crossing over in the section between two loci is positively correlated with the distance between them, new, crossover combinations of two closely linked loci are much rarer than the old, noncrossover combinations; but crossover combinations of two distantly linked loci may be as common as the noncrossover combinations.

It follows that chromosomal linkage of two genes can be easily recognized if an individual heterozygous for two loci forms gametes in unequal numbers, that is, if the two noncrossover gametes are more common than the two crossovers. If, however, the four kinds of gametes are equally common, it is not obvious whether the two gene pairs are located in different chromosome pairs, or whether they are in the same pair, but distant from each other.

Crossover Frequency and Recombination Frequency. The frequency of crossing over in the section between two loci in a pair of homologous chromosomes is not always identical with the frequency of new combinations at the loci. When the two loci are so close to each other that only single crossovers occur, then each crossover results in new genic combinations (e.g., A^1B^2 and $A^2\mathrm{B}^1$ from A^1B^1 and A^2B^2). When, however, the distance between the two loci is great enough to permit double (or higher multiple) crossing over in the section between them, the frequency of new combinations is lower than the frequency of crossovers. This is so because a double crossover may lead to the exchange of a section between the loci of A and B without separating A^1 from B^1 or A^2 from B^2.

In genetic studies we cannot determine the actual frequency of crossing over but only the frequency of crossovers that result in recombination of genes. In other words, the frequency of new combinations gives us the *recombination frequency* but not the *crossover frequency*. A low frequency of recombination is a close measure of the crossover frequency, but high recombination frequencies indicate only that crossover frequencies are still higher.

Recombination frequencies seldom rise above 50 per cent, even if the two loci considered lie very far apart. Two genes at widely separated loci will therefore be present as frequently in old as in new combinations, just as if the loci were in different chromosome pairs.

In experimental organisms it is often possible to decide whether genes are in different chromosome or in the same chromosome but far apart. In man it is at present rarely possible to decide between these alternatives.

X-linked and Autosomal Genes. Such a decision is simple if one of two genes that recombine freely with each other is sex linked but the other is transmitted

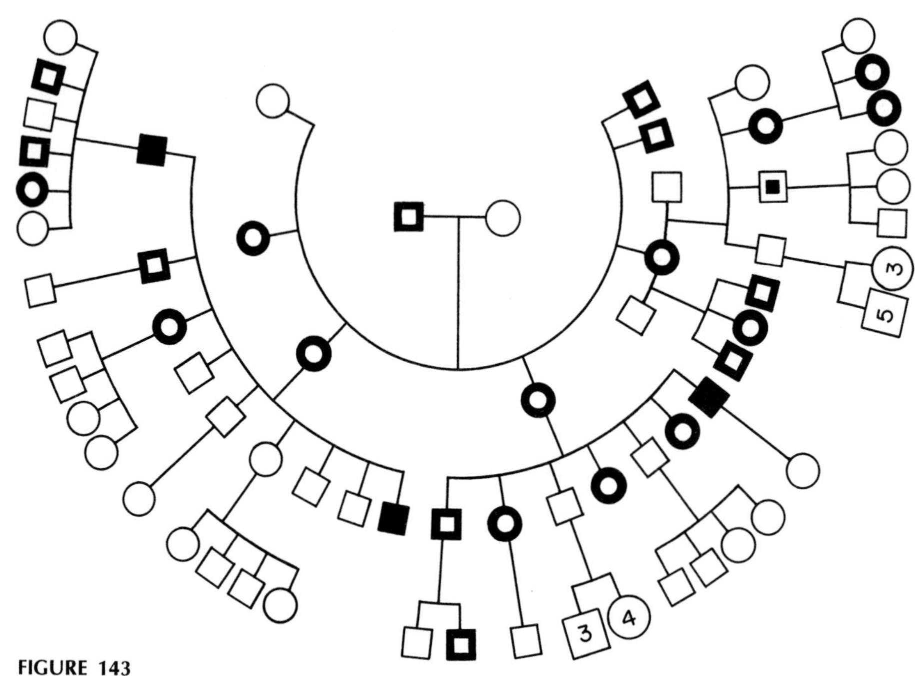

FIGURE 143
A pedigree in which both blue sclerotics and X-linked color blindness are transmitted. Symbols with heavy border and light center = blue sclerotics; with black center and light border = color blind; solid symbols = blue sclerotics and color blind. (After Riddell, *Ann. Eugen.*, **10**, 1940.)

autosomally. This marks their location in different chromosomes. Such is the case with X-linked color blindness and the autosomal dominant brittle-bone defect blue sclerotics (pp. 377–378), both of which occur together in the pedigree in Figure 143.

Autosomal Genes. If both genes are autosomal, the problem of possible chromosomal linkage between two freely combining genes *A* and *B* must be attacked in a different way. If each of these genes can be shown to be linked with a third intermediately located gene, *C*, then it is obvious that *A* and *B* are in the same chromosome. However, failure to find a third gene linked with both *A* and *B* does not prove that these genes are in different chromosomes: the negative result may be due to a lack of known, suitably located genes, or it may even be that the distance between the loci *A* and *B* is so great that no intermediate locus is close enough to both to be linked with each of them. If the latter is true, it may require two intermediately located genes, *C* and *D*, which show (1) genetic linkage with each other, (2) linkage of *C* with *A*, and (3) linkage of *D* with *B*, to demonstrate that *A* and *B* lie in the same chromosome.

In summary, genetic linkage, if once established, definitely assigns linked genes to the same chromosome; but free combination of different genes does not establish that they are located in different chromosomes. New information will often show that such freely recombining genes are very loosely linked. Even two groups of linked genes, one group segregating independently of the other, may become recognized as widely separated members of one linkage group (that is, genes in the same chromosome) if a gene is found to be linked with both groups. In time, it may be expected that twenty-three human linkage groups will be known.

Linkage of X-linked Genes. Linkage and crossing over may be studied most directly in pedigrees in which two X-linked genes can be followed simultaneously. If gene A is shown by means of typical X-linked inheritance to be in the X-chromosome, and if the same can be shown for gene B, then, obviously, A and B must be located in the same chromosome. If a female's two X-chromosomes are A^1B^1 and A^2B^2, respectively, so that she may be symbolized as A^1B^1/A^2B^2 (the diagonal line separates the genotypes of the two homologous chromosomes) her hemizygous sons will provide direct phenotypic information on the behavior of her X-chromosomes: the phenotypes $\mathrm{A^1B^1}$ and $\mathrm{A^2B^2}$ of her sons are the effects of noncrossovers and $\mathrm{A^1B^2}$ and $\mathrm{A^2B^1}$ are those of crossovers between the loci of A and B.

The simplest method of determining the frequency of crossing over between two loci, i.e., their distance from each other, consists in analyses of pedigrees in which both loci are heterozygous. When one of the alleles at each locus is rare in the population, pedigrees that segregate for both of them are extremely rare. Therefore linkage studies usually require at least one locus with relatively high frequencies for both of its alleles.

Among X-linked loci those for the antigen Xg and for partial color blindnesses are relatively frequent. They have been used extensively in attempts to determine map distances between them and loci with only rare not-normal alleles. Selection of pedigrees for these studies begins with those which segregate for the locus with the rare allele and proceeds with those pedigrees which simultaneously segregate for alleles at the *Xg* and the color blindnesses loci. For autosomal linkages the various autosomal blood-group genes have been tested among one another. These blood-group genes have also served to test for linkage between them and loci with rare autosomal alleles not concerned with blood-group antigens.

Color blindness and hemophilia. An example of linkage data on two X-linked genes will make these general discussions more concrete. Many pedigrees are now available in which both color blindness and hemophilia occur. Unfortunately, neither the type of red-green color blindness nor that of hemophilia is known for most of these pedigrees. The absence of information on the type of color blindness is perhaps not too serious, since it seems that if two loci are involved their distance from each other is not great. In respect to hemophilia A and B it seems that the respective genes are located far apart. Since, however, hemophilia A is several times more frequent than B, it is probably true that the majority of pedigrees with hemophilia refers to the classical, A type. In more recent studies involving the color blindnesses and hemophilia, specific information is obtained as to the types of color blindness and of hemophilia. In much of the following discussions we shall, reluctantly, follow the example of the earlier students of X-linkage in man, particularly Haldane, and treat X-linked color blindness and X-linked hemophilia as though they were each controlled by genes at a single locus.

Figure 144, A gives a pedigree in which both color blindness and hemophilia are transmitted. The earliest known male ancestor, I–1, was affected with both abnormalities, which were therefore both controlled by genes (*c* and *h*) in his one X-chromosome. His daughter, II–1, who was normal, must have received this double-recessive X-chromosome and a double-dominant one from her mother. She became the mother of 5 children. Two of her sons, III–3 and III–5, received both recessive alleles, *c* and *h*, and one, III–6, the normal alleles *C* and *H*. In other words, the three gametes of this woman that gave rise to sons retained the original linkages, either *c* and *h* (2 sons) or *C* and *H* (1 son). Her first daughter, III–2, of normal phenotype, had 2 sons, IV–1 and IV–2, the first being *C* and *H* and the other *c* and *h*. She had received a noncrossover X-chromosome, containing *c* and *h* from her mother and a normal X-chromosome, with *C* and *H*, from her father. Her 2 sons had been given noncrossover X-chromosomes. The constitution of six gametes out of seven from two double-heterozygous women can thus be deduced from this

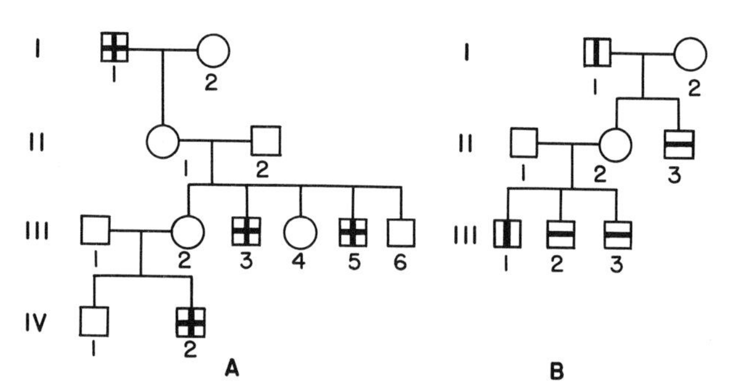

FIGURE 144

A: Part of a pedigree showing transmission of color blindness and hemophilia together. Symbols with cross inside = color blind and hemophilic. **B:** A pedigree showing transmission of color blindness and hemophilia alternatively. Vertical bar inside symbol = hemophilic; horizontal bar = color blind. (A, after Madlener, *Arch. Rassenbiol.*, **20,** 1928; B, after Birch, *Illinois Medic. Dental Monogr.*, **4,** 1937.)

pedigree. The two women were *CH/ch* and the gametes were of two kinds only, namely, four *ch* and two *CH*. Therefore, the two loci are so closely linked that no crossovers occurred among the six gametes.

The phenotypes (of three males) from another family (Fig. 144, B) were also the result of noncrossover gametes. Unlike the first pedigree, where *c* and *h* were found together in the X-chromosome of a male ancestor, the second pedigree shows the descendants of a male, I–1, who was hemophilic but not color blind: his normal-appearing wife must have brought in the allele for color blindness. The daughter, II–2, was, therefore, *Ch/cH*, and her 3 sons, III–1, III–2, and III–3, received noncrossover X-chromosomes, *Ch* and *cH*, respectively. The original pedigree contained other individuals who were not tested for color vision and therefore have been omitted here.

It may be noted that the male in generation I in each of the two pedigrees just discussed provided important evidence for the chromosomal genotype of his normal daughter, whose X-chromosomes were transmitted to her offspring. In the pedigree reproduced as Figure 144, A the male in generation I had the recessive alleles for color blindness and hemophilia in the same X-chromosome, and his wife transmitted the dominant alleles for normal color perception and nonbleeder status to their daughter. The daughter was therefore *ch/CH*. In the pedigree reproduced as Figure 144, B the male in generation I had the alleles for normal color perception and hemophilia in his X-chromosome, and his wife transmitted a chromosome with the alleles for color blindness and not-bleeding to their daughter who therefore was *Ch/cH*.

The knowledge that the daughter in pedigree Fig. 144,A had *c* and *h* in the "*coupling*" or "*cis*" phase, and that the daughter in pedigree Fig. 144,B had *c* and *h* in the "*repulsion*" or "*trans*" phase enabled us to recognize the genotypes of their offspring as resulting from noncrossovers. When it is only known that a woman is heterozygous for *c* and *h* but not known whether these alleles are in cis or trans phase, assignment of noncrossovers and crossovers becomes less certain. This will be seen in the two pedigrees of Figures 145. The pedigree Figure 145,A, with examples of both noncrossover and crossover gametes, was historically the first of its kind. A normal-appearing woman, II–1, whose father was hemophilic, had four sons, among whom both hemophilia and color blindness occurred. Her own ancestors had not been tested for color vision, so that it is not known whether her hemophilic father also carried *c* in his X-chromosome or whether she received this allele from her mother. If the first alternative was true, her constitution was *ch/CH*; if the second, *cH/Ch*. The 4 sons were all different: one was both hemophilic and color blind; the next, only hemophilic; the third, only color blind; and the last, normal in both respects. Thus, two of the four X-chromosomes which the woman transmitted to these sons were noncrossovers between the loci of *C* and *H*, and two were crossovers. Which of the two were noncrossovers

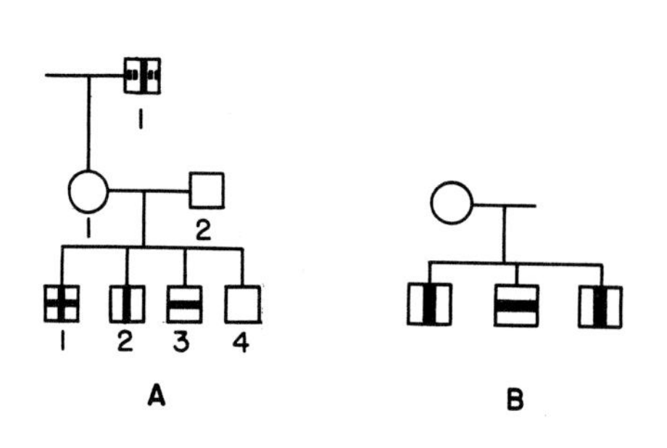

FIGURE 145
More pedigrees showing color blindness and hemophilia with uncertain assignment of noncrossovers and crossovers. The bars inside symbols have the same meaning as in Fig. 144. **A.** All four possible combinations. The broken bar in I-1 denotes uncertainty about this person's color vision. The high frequency of crossovers – 2 out of 4 – suggests that the type of hemophilia in this family was hemophilia B, whose locus, in contrast to that of hemophilia A, is very distant from that of color blindness. **B.** Two combinations. See text for explanation. (A, after Rath, *Arch. Rassenbiol.,* 1938; B, after Hoogvliet, *Genetica,* **23,** 1942.)

and which crossovers cannot be stated, since the decision depends on the unknown item in the constitution of the mother, II–1. If she was *ch*/*CH*, then her first and last sons were the noncrossovers, and the second and third the crossovers. If she was *cH*/*Ch*, the relations would be reversed. The 2 color-blind males in this classical pedigree were reinvestigated in 1952 by Jaeger, who found that III–1 was protanopic and III–3 protanomalous. Since the mother (II–1) was clearly not color defective, it has been surmised that, in contrast to other well-established findings, the two kinds of color defect in this family are alternative expressions of a single allele that depend on the presence or absence of a specific allele at another locus.

One more pedigree, consisting of only a normal-appearing mother and her 3 sons, presents us with another uncertainty (Fig. 145,B). Segregation among the sons shows that the mother was heterozygous for both color blindness and hemophilia but, lacking knowledge of her ancestry, it cannot be decided whether she was *ch*/*CH* or *Ch*/*cH*. If she was the former, then all 3 sons came from crossover gametes; but if she was the latter, all 3 sons came from non-crossover gametes.

The frequency of recombination of the genes for hemophilia and color blindness was zero among nine gametes in the first two pedigrees, 2 out of 4 in the third, and either 3 or zero out of 3 in the last pedigree. Pooling of all information from these and other relevant pedigrees by the method of likelihood ratio described on pages 354–357 gave Haldane and Smith (1947) an estimate of 9.8 per cent as the most likely frequency for recombination of the two loci.

Color blindness and myopic night blindness. A less close linkage has been found between deuteranopic color blindness and another X-linked gene, that for myopic night blindness. They appeared together in a kindred that

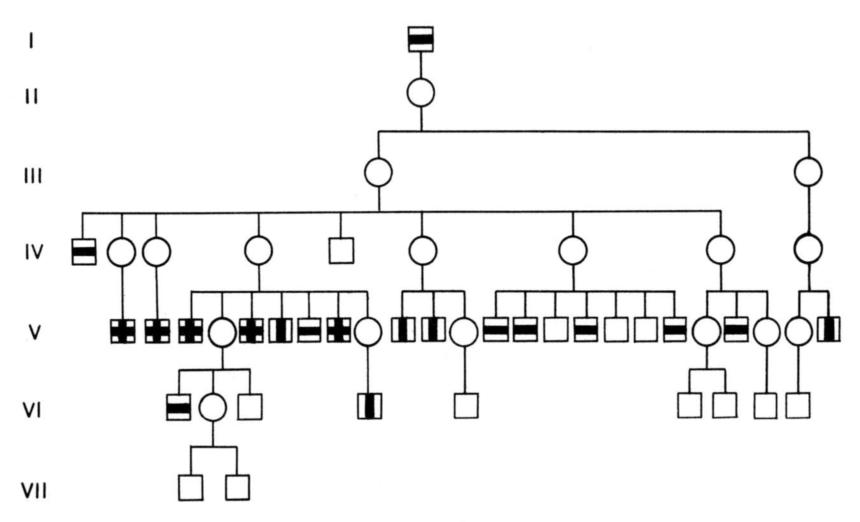

FIGURE 146
A pedigree in which color blindness (deuteranopia) and myopic night blindness occur in all possible combinations and without numerical indications of linkage between the two X-linked traits. Horizontal bar inside symbols = night blind, vertical bar = color blind. (After White, *J. Genet.*, **40,** 1940.)

was, subsequently, well investigated for linkage and crossing over (Fig. 146). An analysis of the data shows that the genes for color blindness and myopic night blindness recombine so freely that the frequency of crossovers is as high as that of noncrossovers. The verdict of linkage is based only on the knowledge that both are completely X-linked.

THE RANDOM OCCURRENCE OF DIFFERENT ALLELIC ASSOCIATIONS IN LINKAGE

Two different constitutions of females heterozygous for both color blindness and hemophilia were seen to exist, namely, *ch/CH* and *Ch/cH*. Their presence presupposes that all four possible kinds of chromosomes, namely, *ch*, *CH*, *Ch*, and *cH*, occur in the population. What are their frequencies? This question has often been answered incorrectly. The correct statement "color blindness and hemophilia are linked," was mistakenly interpreted to mean that the combinations *ch* (color blind and hemophilic) and *CH* (not color blind, not hemophilic) are both more common than the combinations *Ch* (not color blind, hemophilic) and *cH* (color blind, not hemophilic). That this is not so may be seen from the following considerations: Assume a population in which the X-linked alleles *C* and *c* are equally common ($p_1 = q_1 = 0.5$) and in which the X-linked alleles *H* and *h* are equally common ($p_2 = q_2 = 0.5$). If this population, at one time, contained only two types of chromosomes, *CH* and *ch*, in equal numbers, then, as a result of random mating, there would be three types of females, *CH*/*CH*, *ch*/*ch*, and *CH*/*ch*, and two types of males, *CH* and *ch*. This population would not be in equilibrium but would change from generation to generation. The homozygous females and the males would form gametes containing the original *CH* and *ch* chromosomes only. The heterozygous *CH*/*ch* females, however, would form gametes with not only the original chromosomes, *CH* and *ch*, but also new, crossover chromosomes, *cH* and *Ch*.

Some females in the next generation would be *cH*/*CH* and *Ch*/*CH*, and some males *cH* and *Ch*. Mating between these would lead, one generation later, to some *cH*/*Ch* females. Thus, the population would then contain both *CH*/*ch* and *cH*/*Ch* heterozygotes, with the majority of women being *CH*/*ch*, derived from noncrossover gametes, within whom *cH* and *Ch* crossover chromosomes would continue to segregate during meiosis. The minority would be *cH*/*Ch* women, who would propagate mainly their two kinds of chromosomes, *cH* and *Ch*. As long as the *CH*/*ch* combination is more common that is *cH*/*Ch* counterpart, it diminishes, and the latter increases, until an equilibrium is reached at which there are equal numbers of all four chromosome types: *ch*, *cH*, *Ch*, and *CH*.

It is noteworthy that this equality of all combinations is independent of the degree of linkage between the two loci. If the linkage is so close that relatively few crossovers take place between the two loci, it will require more generations until equilibrium is reached than if linkage is loose. Equilibrium is reached only when the two heterozygotes *CH*/*ch* and *cH*/*Ch* are equally

TABLE 37
Types of Association Between Two Inherited Traits.

Genetic Basis	*Correlation in*	
	General Population	Segregating Sibships
(1) Multiple effects of a gene	+	+
(2) Group differences in gene frequencies		
(a) Genes not linked	+	0
(b) Genes linked	+	+ or −
(3) Panmictic population		
(a) Genes not linked	0	0
(b) Genes linked	0	+ or −

common, so that one recreates, by crossing over, those combinations that the other loses by crossing over.

The initial assumption that the alleles *C* and *c*, and *H* and *h* are equally common was introduced in order to simplify the discussion. In reality, the allele frequencies are very unlike and, perhaps, roughly: p_1 (of *C*) = 0.9, q_1 (of *c*) = 0.1, p_2 (of *H*) = 0.9999, and q_2 (of *h*) = 0.0001. In such a population, equilibrium is reached when the four combinations occur at random frequencies; that is, chromosomes which carry *C* and *H* and those which carry *c* and *H* occur in proportion to the frequencies of *C* and *c*, and chromosomes which carry *C* and *h* and those which carry *c* and *h* also occur in proportion to the frequencies of *C* and *c*. Using the allelic frequencies given above, we obtain the following chromosomal frequencies:

$$CH = p_1 \times p_2 = 0.89991$$
$$ch = q_1 \times q_2 = 0.00001$$
$$cH = q_1 \times p_2 = 0.09999$$
$$Ch = p_1 \times q_2 = 0.00009$$

It is seen that the combination of the alleles for color blindness and hemophilia (*ch*) is *not* more common than those of the alleles for hemophilia and normal vision, and nonhemophilia and color blindness; but, on the contrary, is by far the least common chromosomal constitution.

That the random association of the *C* and *c* with the *H* and *h* alleles does indeed represent equilibrium may be seen if any other frequencies of association are considered. Let us assume, for instance, that at a given time, more *ch* chromosomes are present than in a population at equilibrium. They would cross over with *CH* chromosomes in *CH*/*ch* women and give rise to new *Hc* and *hC* chromosomes, so that, after some generations, the number of these chromosomes will have increased, and the number of the *ch* chromosomes decreased, until *cH*/*Ch* women produce as many *CH* and *ch* chromosomes as are lost by crossing over in *CH*/*ch* women. This condition is fulfilled when the population contains equal numbers of *CH*/*ch* and *cH*/*Ch* heterozygotes, the former having the equilibrium frequency $2 \times (p_1 \times p_2) \times (q_1 \times q_2)$, the latter $2 \times (p_1 \times q_2) \times (q_1 \times p_2)$, the two products being identical.

The statement that the four chromosomal constitutions of two pairs of linked genes occur at random frequencies is valid not only for X-linked genes but for any pair of linked genes. But it is also evident that even if the two genes were not linked, but located in different chromosome pairs, the same random frequencies would apply to the four genic combinations. Because of this, the existence of linkage cannot be deduced from a consideration of the frequencies of combinations of traits in a long-established panmictic population (Table 37:3, a, b, middle column). This may be shown for the distribution of two traits: the ABO blood groups, and an abnormality of the

TABLE 38
Random Association of Traits Based on Two Linked Loci, ABO Blood Group and Nail-patella Syndrome.

Nail-patella Syndrome	*0*		*Not 0*	
	Obs.	Exp.	Obs.	Exp.
Affected	35	(32.9)	50	(52.1)
Nonaffected	39	(41.1)	67	(64.9)

$\chi^2 = 0.37$ P ~ 0.55

Source: After Renwick and Lawler.
Note: The table compares the associations in blood group O individuals with those in not O (=A + B + AB) individuals. A more detailed tabulation shows random association of the syndrome with any one of the four blood groups individually.

nails and the patella. The genes for the two traits—blood group and nail-patella—are rather closely linked; yet the frequency with which any of the blood groups is associated with the nail-patella abnormality is no different than the frequency of the blood group in the general population (Table 38).

Bernstein has offered an amusing simile of the principle of random association between the alleles of two linked genes:

> Imagine a number of ladies and gentlemen in a dance hall, who possess exactly equal attraction toward each other. Assume further the peculiarity, which corresponds to the linkage hypothesis, that a gentleman while he is dancing with a certain lady has a greater inclination to dance the next dance with her than with another one. Also be it the property of the ladies, to make their partner forget with whom he had danced earlier so that while he favors his momentary partner, he faces all other ladies indiscriminately. In this case, even if the attractive force is very great, but not infinitely so, then—provided the party lasts long enough—each gentleman would get together with each lady with equal frequency. This then would be the same result as if the momentary partner had not offered a special attraction. The only difference would be in the time required.

To the naïve observer, it seems plausible that traits based on linked genes should more often be associated than separate. Often enough, two characters which show a frequent or even absolute association have been described as new examples of linkage. In reality, however, such a finding is, in itself, likely to be evidence against linkage, since in a population mating at random chance governs the association of genes responsible for two traits: if the traits are associated more frequently than may be expected by chance, the reason must be sought in causes other than chromosomal linkage. To these we will now turn.

Causes of Nonrandom Association of Different Traits

Multiple Effects of a Gene. A high frequency of association between different traits is often simply due to multiple effects of one and the same gene (Table 37:1). This is clearly so when the association is absolute, that is, when the two traits are always found together as, for instance, excretion of phenylpyruvic acid and absence of phenylalanine oxidase. If the two traits were caused by two separate closely linked genes, crossing over in former generations should have led to their separation, so that excretion of phenylpyruvic acid unaccompanied by the enzyme deficiency, and the enzyme deficiency unaccompanied by abnormal excretion; would occur.

The known multiple effects of a gene may not always be seen in the same person. In Marfan's syndrome, for example, long-fingeredness, misplaced eye lens, and heart defects are frequently, but not invariably, found together in the same individual. Each of these traits, separately, is so rare that a chance association of two or three of them would be most exceptional. A study of pedigrees shows that a long-fingered person who does not have a displaced lens may have the latter defect appear in his long-fingered progeny and that a long-fingered person with a displaced lens may have offspring with only the finger anomaly. This shows that the same dominant gene is responsible for both traits in this syndrome, but that, developmentally, sometimes only one trait is expressed. The incomplete association of the different traits is therefore not caused by linked genes whose abnormal allelic varieties may or may not be present together in the same chromosome, but by a single gene with irregularly expressed multiple effects (a partly expressed, "frustrated" syndrome is called a *forme fruste*).

Some pedigrees contain individuals with misplaced lenses, none of whom is arachnodactylous. In these families, misplaced lenses are inherited in strictly dominant fashion, in contrast to the irregular appearance of the trait when it is associated with long-fingeredness. The different type of inheritance is a strong indication that misplaced lenses without arachnodactyly do not depend on the same locus, or at least not on the same allele, that controls arachnodactyly. This eliminates the possibility that misplaced lenses without long-fingeredness are the result of crossing over between two linked genes. There is, however, a possible objection to this argument: it could be true that two separate genes, one for misplaced lenses and another for long-fingeredness, so interact with each other in development that the effect of the lens gene is irregular when they are present in the same individual but regularly dominant if they are not. The main, decisive argument against interpreting frequent association of traits as evidence of linkage is the statistical one given above.

It is possible to look at the problem in still a different way. If we assume the existence of a separate atypical gene for each trait of a syndrome, it is also necessary to assume that, originally, mutations from normal to abnormal took place simultaneously at the different linked loci. It is most unlikely, from what we know about mutations, that simultaneous mutations of separate genes occur with sufficient frequency—if at all—to account for the origin of syndromes.

Heterogeneous Populations. We have seen that, in a population which mates at random and is at equilibrium, associations of traits due to separate linked genes occur at random. If a population does not fulfill these conditions, such traits may not be associated at random frequencies. In the population of the United States there is a high association between straight hair and light skin, and between kinky hair and dark skin. This, of course, is because whites have genes for both straight hair and light skin in their chromosomal set, and blacks have genes for kinky hair and dark skin. That these traits are controlled by separate genes and not by multiple effects of one gene is shown by the segregation of hair and skin traits in the immediate and subsequent progeny of hybrids. Whether such segregation is the result of free recombination of genes in different chromosome pairs or of crossovers between loci of linked genes is not known. Investigation may someday show which of these alternatives is correct, since the speed of obtaining random association after generations of intermarriage would depend on whether recombination took place freely or was restricted by linkage (Table 37:2, a, b). Rife has pointed out that nonrandom association of two genetic traits and the presence of random association of either with a third, within a hybrid population of relatively recent origin, may under certain conditions be indicative of linkage. In a hybrid population, presence of some random associations of traits that are nonrandomly associated in the unmixed groups might indicate that enough generations had elapsed to have resulted in full recombination of independently inherited or loosely linked genes; nonrandom association of other traits might be the expression of linkages, which had not yet had time to become randomized. In Sudanese and American blacks, who are both of mixed African-Caucasian descent, Rife found nonrandom association between degree of pigmentation and specific patterns of lines on the palms of hands. Both of these traits showed random association with ability to taste PTC. There is thus a suggestion that the genes involved in pigmentation and palm pattern are linked, but because, among other difficulties, the genetics of both traits is complex, it is no more than a suggestion.

THE DETECTION OF LINKAGE

It may seem an easy matter to decide whether two genes recombine freely or with the limited frequency of crossovers. This is indeed true in most organisms but only rarely true in man, largely because of the relatively small size of human families. If the number of children in a sibship is less than 4, it is impossible for all four phenotypes, *A-B-*, *A-bb*, *aaB-*, and *aabb*, to appear. If, in a larger sibship, all four phenotypes do appear, the number of each may well be statistically compatible with the assumption of either independent recombination or moderate, if not close, linkage.

Some of these difficulties are similar to those pertaining to expected ratios in single factor recessive inheritance. There, many marriages of two heterozygotes do not come to the attention of the investigator because many sibships with such parents do not include recessive children, and the mark by which marriages of two heterozygotes are recognized is the production of at least one homozygous recessive child. In single factor inheritance, these difficulties can be overcome by adding together data from many sibships with at least one homozygous recessive child and by applying appropriate statistical corrections to these pooled data. In determining possible linkage, pooling of data by simple addition of sibships is of no help. If a parent is heterozygous for each of two linked pairs of genes, *Aa* and *Bb*, he may have either of two different chromosomal constitutions: *AB/ab* or *Ab/aB*. The proportions of the four combinations *AB*, *Ab*, *aB*, and *ab* in the gametes of the two kinds of parents are quite different. An *AB/ab* parent will form more *AB* and *ab* noncrossover gametes and fewer *Ab* and *aB* crossover gametes, resulting in positive correlation of *A* and *B* and of *a* and *b*. On the contrary, in the gametes of an *Ab/aB* parent, the more common noncrossover combinations will be *Ab* and *aB*, and the less common crossover combinations *AB* and *ab*, resulting in a negative correlation of *A* and *B* and of *a* and *b*. If x is the frequency of recombination of *A* and *B*, then the proportions of gametes produced by the two kinds of parents are as listed in the first two lines of Table 39.

As we have seen, in a population at equilibrium the two kinds of parents are equally common. Consequently, by adding the gametes from many parents (more accurately, by adding the bearers of phenotypes of sibships derived from the gametes), we obtain the result, given in the last line of Table 39: equality of all kinds of gametes, no matter what the strength of linkage!

TABLE 39
Gametes Produced by Parents Heterozygous for Two Linked Gene Pairs *A*, *a* and *B*, *b*. (The frequency of recombination is x.)

Parents	*Gametes*			
	AB	*Ab*	*aB*	*ab*
AB/ab	$\frac{1}{2}(1-x)$	$\frac{1}{2}x$	$\frac{1}{2}x$	$\frac{1}{2}(1-x)$
Ab/aB	$\frac{1}{2}x$	$\frac{1}{2}(1-x)$	$\frac{1}{2}(1-x)$	$\frac{1}{2}x$
Sum of gametes of both parents	$\frac{1}{2}$	$\frac{1}{2}$	$\frac{1}{2}$	$\frac{1}{2}$

There are, however, several different ways in which the difficulties inherent in the exploration of human linkage can be met. A simple direct study is possible if the genotypic phase of a double heterozygous parent is known. Indirect approaches make use of statistical phenomena that are consequences of linkage, and can be found even in collections of pedigrees where the genotypes of parents are incompletely known.

Direct Analysis for Linkage. When the parental genotypes for two pairs of alleles are fully known the detection of linkage is simply based on the proportions of nonrecombinant and recombinant genotypes among the children. For X-chromosomal linkage it is only necessary to know the genotype of the mother in order to determine recombination as shown by her sons since the father does not contribute an X-chromosome to them. Moreover, the mother's genotype for X-linked alleles may be relatively easy to determine if that of her father is known. Examples of pedigrees permitting this sort of analysis were shown in Figure 144.

In autosomal linkage matters are often less transparent and the genotypes of the parents of sibships remain doubtful. Only in some cases can they be deduced with certainty from (1) genetic information on the sib's grandparents, (2) investigation of the kinds of children in large sibships, and (3) inspection of the parental phenotypes. Sometimes all these sources have to be tapped and the results combined. To illustrate this by an example, let us consider a group of parents, each of normal pigmentation and belonging to blood group A and all having children with the following phenotypes: normally pigmented, group A; normally pigmented, group O; albino, group A; and albino, group O. From this, we can derive that the parents were all Aa for pigmentation and $I^{\mathrm{A}}I^{\mathrm{O}}$ for blood group. If the genes for pigmentation and blood group were linked, these pairs of parents could be of four different kinds, namely,

Females		Males
$AI^{\mathrm{A}}/aI^{\mathrm{O}}$	$\times$	$AI^{\mathrm{A}}/aI^{\mathrm{O}}$
$AI^{\mathrm{A}}/aI^{\mathrm{O}}$	$\times$	$AI^{\mathrm{O}}/aI^{\mathrm{A}}$
$AI^{\mathrm{O}}/aI^{\mathrm{A}}$	$\times$	$AI^{\mathrm{A}}/aI^{\mathrm{O}}$
$AI^{\mathrm{O}}/aI^{\mathrm{A}}$	$\times$	$AI^{\mathrm{O}}/aI^{\mathrm{A}}$

Unless it is known to which of these kinds a specific pair belongs, it is not clear which children are noncrossovers and which crossovers. Knowledge of the grandparents can sometimes provide the necessary information. If, for instance, the maternal grandmother were an albino and group O, then her normally pigmented daughter would receive from her the combination aI^{O} and would, therefore, be $AI^{\mathrm{A}}/aI^{\mathrm{O}}$. If the paternal grandfather belonged to group O and, although he was normally pigmented, was known to be a carrier of albinism, his son would most likely owe his genes aI^{O} to this parent and would, therefore, be AI^{A}/aI^{O} like his wife, so that the mating would thus be $AI^{A}/aI^{O} \times AI^{A}/aI^{O}$. On the basis of this information, assuming linkage was involved, the albino, group O children (aI^{O}/aI^{O}) from this couple could be diagnosed as noncrossovers, and the albino, group A children as crossovers. The nonalbino children could be of various genotypes, some representing noncrossovers and others crossover gametes.

Circumstances are not always as unfavorable as in the example given, where both parents were heterozygous for both allelic pairs. Sometimes, one parent may be known to be homozygous double recessive (ab/ab), or at least a homozygous recessive for one of the two gene pairs (aB/ab or Ab/ab). Usually, however, there will be even less information than in the example of albinism and blood groups. If, for instance, the pigmentation and the blood group of the maternal grandmother were unknown, or if she was normally pigmented and belonged to group A, and if the paternal grandfather's genotype was less clearly defined, then it would be impossible to proceed by the direct method.

Indirect Analysis for Linkage. A number of different methods of obtaining information on linkage from incompletely known genotypes have been devised. It is not the purpose of the following pages to describe these methods in detail. Rather, an attempt will be made to give some insight into the ingenious thinking which has been applied to the problems of human linkage, and to open the reader's mind to novel approaches to genetic problems.

The y statistics. The earliest method used data from two generations. Restricting ourselves to one of the simplest cases, we consider again families with one parent heterozygous for two different recessive genes and the other homozygous for them. In a random sample of such families the two types of marriages $AB/ab \times ab/ab$ and $Ab/aB \times ab/ab$ can be expected to be equally common, but given typically small human families the frequencies of the different types of offspring do not tell us the type of any particular parental pair. It is therefore impossible to separate the sibships with an AB/ab parent from those with an Ab/aB parent, and without such separation no information on linkage seems to be obtainable (see Table 39).

It was this situation that led Bernstein to a "new thought." In the absence of linkage—that is, in independent assortment—the four types of gametes AB, Ab, aB, and ab are equally common, but with linkage either AB and ab are more common than Ab and aB, or the reverse is true, depending on whether a parent was AB/ab or Ab/aB. The recognition that, from either of the two linkage phases, two more-common and less-common types of gametes were to be expected, opened a way to distinguish between independent assortment and linkage. If we obtain for each sibship the product y of the sum of gametes AB and ab and the sum of gametes Ab and aB, so that

$$y = (AB + ab) \times (Ab + aB),$$

the value of y, which depends on the frequency of recombination, is the same whether the parent was AB/ab or Ab/aB. This may be shown by a very simple example. Consider families of 4 sibs. If there is free recombination (recombination value $x = 0.5$), the most probable sibship from either type of parent would consist of one individual of each gametic type, so that

$$y = (1 + 1) \times (1 + 1) = 4.$$

If, on the other hand, linkage is present and if, say, $x = 0.25$, then the most probable sibships from an AB/ab parent would contain 3 nonrecombinant individuals from gametes AB or ab, and 1 recombinant individual of type Ab or aB, making

$$y = (3) \times (1) = 3.$$

Conversely, from an Ab/aB parent, the 3 nonrecombinants would be Ab or aB and the 1 recombinant AB or ab. This would give the same result, namely

$$y = (1) \times (3) = 3.$$

Not all 4-child sibships from gametes of independently assorted A and B pairs will have $y = 4$, since chance may lead to many sibships less probable than the 1:1:1:1 type. Likewise, when $x = 0.25$, many sibships of 4 will not yield $y = 3$. On the average, however, the value of y will be higher for $x = 0.5$ than for $x = 0.25$.

One can construct a table in which the mean values of y are given for an array of recombination values in the whole range from $x = 0.0$ (complete linkage) to $x = 0.5$ (independence) adjusted for any number of sibs in each sibship (Table 40). Such a table can then be consulted for the detection and estimation of linkage. Assume, for instance, that among a total of 16 sibships the sum of y for 6 sibships of 2 children is 2, for 7 sibships of 3 is 8, and for 3 sibships of 4 is 6. The observed mean values of y are thus 0.33, 1.14, and 2.0 for the three types of sibships. Table 40 shows that for $x = 0.5$ (absence of linkage) the expected mean value of y for sibships containing 2 children is 0.5, 3 children 1.5, and 4 children 3.0—values from which the observed mean values deviate considerably. The disagreement becomes even more obvious when one compares the sum of the observed y values $2 + 8 + 6 = 16$ with the sum of the expected ones, $(6 \times 0.5) + (7 \times 1.5) + (3 \times 3.0) = 22.5$.

TABLE 40
Product Method for Detection of Linkage (y statistics). Mean Values of the Product $y = (AB + ab) \times (Ab + aB)$ for Sibships of from 2 to 4 Individuals from Parents of the Genotypes $AB/ab \times ab/ab$ and $Ab/aB \times ab/ab$.

Recombination Value (x)	*Mean Values of* y *for Different Numbers of Sibs*		
	2	3	4
0.1	0.18	0.54	1.08
0.2	0.32	0.96	1.92
0.5	0.5	1.5	3.0

Source: Bernstein, 1931.

Since the y values for our families do not fit those expected for independently transmitted genes, they should be compared with y values characteristic of linkage; and the table shows that they fit rather well those for a recombination frequency of x = 0.2. For this frequency the table gives y values of 0.32 for sibships of 2, 0.96 for those of 3, and 1.92 for those of 4; and these give a sum of 14.4 for all sibships, close to the observed sum of 16. It is concluded, therefore, that the families under discussion are compatible with the hypothesis that the two traits are controlled by linked loci. The recombination value seems to be about 20 per cent, but no definite statement about linkage or nonlinkage can be made without studying more families.

The mating of a double heterozygote with a double recessive homozygote is, of course, only one of many different possible unions. Each type of mating requires a separate treatment and separate tables. Several of these were supplied in Bernstein's original paper. Independently of Bernstein, another method for the detection of linkage was proposed by Wiener, and important extensions of Bernstein's basic principle have been made by Fisher, Finney, and others.

The Sib-pair method. This method is based on the relative frequencies of pairs of sibs being alike or unlike for two traits whose linkage is under study. As it is not always easy to evaluate the possibility that a higher than chance frequency of like sib pairs is not due to factors other than linkage, the sib-pair method is likely to give false positives—that is, "evidence" for linkage where it does not really exist.

The sib-pair method is based on Penrose and Burks' striking realization that evidence of human linkage can be obtained from data on a single generation. Let us again consider matings $AB/ab \times ab/ab$ and $Ab/aB \times ab/ab$. In 2-child sibships, the following ten types of sib pairs would be found (the

TABLE 41
Tabulating the Results of Paired Sibship Analysis.
Similarity within Pairs of Sibs.

		Phenotype A or a	
		Sibs Like	Sibs Unlike
Phenotype B *or* b	Sibs Like	(1)	(3)
	Sibs Unlike	(2)	(4)

capital letters A and B stand for the phenotypes of *AA* or *Aa*, and *BB* or *Bb*; and the lower-case letters a and b for *aa* and *bb*, respectively):

Type 1	AB, AB	Type 6	AB, aB
Type 2	Ab, Ab	Type 7	AB, ab
Type 3	aB, aB	Type 8	Ab, aB
Type 4	ab, ab	Type 9	Ab, ab
Type 5	AB, Ab	Type 10	aB, ab

These ten types can be easily classified in four groups:

Group 1. Sibs alike in both traits (types 1, 2, 3, and 4—both sibs being either A and B, A and b, a and B, or a and b).

Group 2. Sibs alike in the first but unlike in the second trait (types 5 and 10—both sibs A or both a, but one B and the other b).

Group 3. Sibs unlike in the first but alike in the second trait (types 6 and 9—one sib A and the other a, but both B or both b).

Group 4. Sibs unlike in both traits (types 7 and 8—one sib AB and the other ab, or one Ab and the other aB).

If the genes *A* and *B* are not linked, then the four groups are equally common, since the pairs formed by unlike partners, types 5 to 10, are each alike in frequency but twice as frequent as the pairs formed by like partners, types 1 to 4. An entry of the numbers of sib pairs of the four groups in a "fourfold" table (see Table 41) should therefore result in equality of its four cells. If, however, the genes *A* and *B* are linked, some combinations will be more common than others regardless of the linkage phase in the heterozygous parent. Consider the extreme case of a linkage so close that, in a sample of families, only the original, noncrossover combinations of the parent appear among the sibs. The sib pairs will be

AB/ab Parents		Ab/aB Parents	
Type 1	AB, AB	Type 2	Ab, Ab
Type 4	ab, ab	Type 3	aB, aB
Type 7	AB, ab	Type 8	Ab, aB

In the fourfold table, only the upper-left and lower-right cell will be filled, the former with types 1 to 4, the latter with 7 and 8.

If the genes *A* and *B* are less closely linked, so that all possible types of offspring are produced, types 5, 6, 9, and 10, which consist of one noncrossover and one crossover sib, will also occur. These sib pairs will provide entries for the upper-right and lower-left cells of the fourfold table, but, since crossovers are less frequent than noncrossovers, only a minority of all sib pairs will go into these cells.

It is seen that the sib-pair method provides a means of distinguishing between independent recombination and linkage of different genes without knowing whether the parental genotype is *AB*/*ab* or *Ab*/*aB*. The entries of paired sibs in a fourfold table will show a random distribution when there is no linkage but, when linkage exists, an excess of sib pairs in those cells where the two sibs are alike in both traits or unlike in both.

There are, of course, in any population, many more genotypes relating to the *A* and *a* and *B* and *b* loci than the three (*AB*/*ab*, *Ab*/*aB* and *ab*/*ab*) which formed the basis of our discussion of the sib-pair method. Some types of parental pairs, for example, *AB*/*AB* × *ab*/*ab*, will not give any evidence on linkage or its absence, and other types, such as *AB*/*ab* × *Ab*/*aB*, will provide only partial evidence. But it can be shown that, taking all types of parents together, those sib pairs which do not contribute decisive information fall into the four cells in a random manner in contrast to those which do contribute such information and which fall into the doubly like or doubly unlike classes. The net result is that the sib-pair method, although essentially valid, is not very efficient, since, by disregarding the parents, it fills the fourfold table with much useless information. Whenever possible it is therefore preferable to use other methods for the analysis of linkage.

Two actual examples of the use of the sib-pair method are given in Tables 42 and 43, where linkage between the genes for the nail-patella syndrome and those for either the MNS or the ABO blood groups is tested. It is seen that the entries in the fourfold table for the syndrome and MNS do not deviate significantly from random distribution. There is, therefore, no evidence of linkage. But in the table concerned with the syndrome and ABO groups, the entries for pairs alike in possessing or lacking the syndrome and alike in ABO blood group and for pairs unlike for both traits are significantly greater than would accord with random expectation. The genes for these two traits are

TABLE 42
The Sib-pair Method Applied to the Loci for the Nail-patella Syndrome and the MNS Blood Groups.

Type of Pair		*Syndrome* Sibs Like		Sibs Unlike	
		Obs.	Random Exp.	Obs.	Random Exp.
MNS	Sibs Like	29	(30.5)	76	(74.5)
	Sibs Unlike	35	(33.5)	80	(81.5)
	$\chi^2 = 0.2$	P = 0.9	No linkage!		

Source: Renwick and Lawler.

TABLE 43
The Sib-pair Method Applied to the Loci for the Nail-patella Syndrome and the ABO Blood Groups.

Type of Pair		*Syndrome* Sibs Like		Sibs Unlike	
		Obs.	Random Exp.	Obs.	Random Exp.
ABO	Sibs Like	78	(55.8)	44	(66.2)
	Sibs Unlike	30	(52.2)	84	(61.8)
	$\chi^2 = 33.7$	P < 0.0001	Linkage!		

Source: Renwick and Lawler.

obviously linked. Special methods permit estimation of the frequency of recombination from the amount of deviation from random expectation. Estimates of recombination between the nail-patella and the ABO group loci suggest a frequency of about 10 per cent. This estimate, which will be discussed below, is based on the method of likelihood ratio to which we will now turn.

Odds: the method of likelihood ratio. A third method for the detection and estimation of the strength of linkage is conceptually the simplest. Devised by J. B. S. Haldane and C. A. B. Smith, it consists of determining the amount of information available in a collection of data on two loci and comparing the probability of obtaining such data if the two loci are linked with the probability if they are not. The ratio of these two probabilities gives the odds for or against linkage. If, for example, the probability of obtaining the observed distribution of two traits in a pedigree is much higher if the genes are linked than if they are not, then the odds are obviously in favor of linkage. Conversely, if the probability of obtaining the information given by a pedigree is much higher if the genes are not linked than if they are, then the odds are against linkage. This method of likelihood ratio, is now the method of general choice in linkage studies. It is also known as the *lod* (log odds) method, since the necessary calculations make use of tables of logarithms of odds ("z scores," Morton).

Linkage itself may be strong or weak: strong if the recombination value x is low (0.0 indicates complete linkage); weak if it is high (0.5 indicates no linkage). The probability of obtaining the particular distribution of traits shown by a given pedigree or collection of pedigrees is therefore expressed as a function of x, different degrees of probability corresponding to different degrees of linkage. The recombination value that gives the highest probability that the observed distribution would be obtained is also the best estimate of the degree of linkage between the two loci.

The nail-patella syndrome and the ABO blood groups may once more serve as a simplified example, illustrating the method of likelihood ratio. Figure 147,A, gives a small part of the extensive pedigree of one of the kindreds which were classified for the syndrome and the blood groups. The father of the family shown carried in the heterozygous state the syndrome-producing allele, N, and also heterozygously the I^{A_1} allele, as seen from the fact that the two children indicate segregation for both loci. He was therefore either NI^{A_1}/nI^{O} or NI^{O}/nI^{A_1}. The first son obviously received an NI^{O} sperm from his father, and the second son an nI^{A_1} sperm. What is the probability of having two such sons if N and I are linked and the recombination value is x? If the father is NI^{A_1}/nI^{O}, then the probability of producing an NI^{O} crossover gamete is $\frac{1}{2}$ x and that of forming an nI^{A_1} crossover gamete is also $\frac{1}{2}$ x (the sum of the two probabilities, $\frac{1}{2}x + \frac{1}{2}x = x$, represents the total probability of a recombinant gamete). The probability of obtaining the particular sibship of two sons would therefore be $(\frac{1}{2}x)^2$. On the other hand, if the father was NI^{O}/nI^{A_1}, then the probability of producing an NI^{O} or an nI^{A_1} noncrossover gamete was $\frac{1}{2}(1-x)$ each (the sum of the two probabilities, $\frac{1}{2}(1-x)+\frac{1}{2}(1-x)=1-x$, represents the total probability of a noncrossover gamete). The probability of obtaining the sibship would thus be $[\frac{1}{2}(1-x)]^2$. The probability of obtaining the sibship from either an NI^{A_1}/nI^{O} or an NI^{O}/nI^{A_1} father is the sum of the two

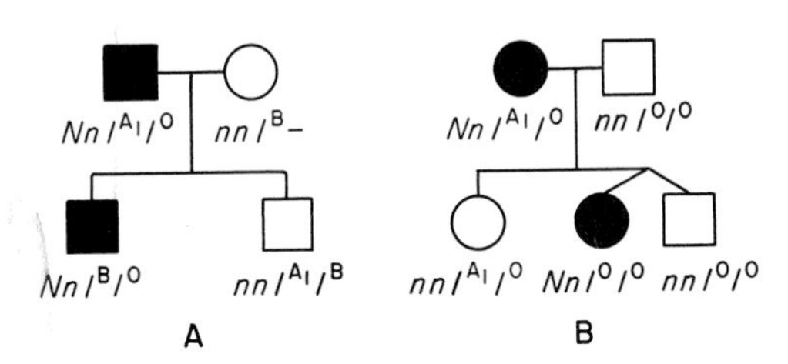

FIGURE 147
Two families in which one of the parents is heterozygous at both the nail-patella locus and the ABO blood-group locus. (After Renwick and Lawler.)

probabilities, each of which multiplied by $\frac{1}{2}$ to account for the two alternative genotypes of the father.

$$P = \tfrac{1}{2}(\tfrac{1}{2}\,x)^2 + \tfrac{1}{2}[\tfrac{1}{2}(1-x)]^2 = \tfrac{1}{8}(1 - 2x + 2x^2).$$

In order to determine the numerical probability of obtaining the sibship if there were no linkage, we make $x = 0.5$, and thus obtain $P = 1/16$. If now we make $x = 0.0$, we can determine the probability of obtaining the sibship if there were complete linkage: $P = 2/16$. The odds in favor of complete linkage as against no linkage are therefore two to one. In a similar manner the odds can be determined for any intermediate value of x. The odds for six values of x are listed under sibship 1 in Table 44. It can be seen that it is most likely that the sibship resulted from the behavior of two completely linked loci, but it is also apparent that the lesser likelihood that the sibship resulted from looser linkage, or even no linkage, cannot be excluded on the basis of this small sibship.

Let us therefore consider another family with data on the same syndrome and the same blood groups (Fig. 147,B). This time it was the mother who was heterozygous for both loci, her genotype being either NI^{O}/nI^{A_1} (Case 1) or NI^{A_1}/nI^{O} (Case 2). The probabilities that she would produce eggs resulting in the 3 children listed were

Case 1: $\frac{1}{2}(1-x)$, $\frac{1}{2}(1-x)$, and $\frac{1}{2}x$
Case 2: $\frac{1}{2}x$, $\frac{1}{2}x$, and $\frac{1}{2}(1-x)$.

The joint probabilities of the sibships from either phase were

$$P = \tfrac{1}{2}[\tfrac{1}{2}(1-x)]^2 \times \tfrac{1}{2}\,x + \tfrac{1}{2}(\tfrac{1}{2}\,x)^2 \times \tfrac{1}{2}(1-x) = \tfrac{1}{16}(x - x^2).$$

TABLE 44
Linkage Between the Loci for the Nail-patella Syndrome and the ABO Blood Groups. (The table lists the odds in favor of linkage, as compared to free assortment, for various recombination values.)

Recombination Value x	*Odds*			
	Sibship 1	Sibship 2	Sibship 1 + 2	All Data
0	2.00	0	0	0
0.1	1.64	0.36	0.59	4,000,000,000
0.2	1.36	0.64	0.87	200,000,000
0.3	1.16	0.84	0.97	600,000
0.4	1.04	0.96	0.998	500
0.5	1.00	1.00	1.00	1

Source: After Renwick and Lawler.

Making $x = 0.5$, to find the probability of no linkage, $P = 1/64$ whereas if $x = 0.0$ (complete linkage), $P = 0$. The odds on complete linkage are thus zero; that is, complete linkage of the two genes is excluded. The odds for intermediate values of x entered under sibship 2 in Table 44 show that it is most likely that in the second family the two loci are independent, although the lesser likelihood of any degree of partial linkage cannot be excluded.

To obtain a joint estimate of the situation, it is only necessary to find the product of the odds for each of the two families for each value of x. The results, given in the fourth column of the table, show that there is a maximum likelihood of obtaining the two sibships if the two loci are not linked, but the odds against linkage of even close degree are not high.

Obviously, with so few data, no reliable estimate of the value of x can be made. However, the totality of the data of which those on our two families constitute only a small fraction yields undeniable evidence for rather close linkage. By combining the individual odds for all families, we find that the over-all odds in favor of various degrees of linkage as against no linkage have the values listed in the final column of Table 44 and shown graphically in Figure 148. This final column shows once more that there is no possibility of explaining the data by assuming that the syndrome and the ABO blood groups are completely linked. The distribution of odds shows a very high peak in the neighborhood of the recombination value 0.1; in fact, the odds are four billion to one in favor of such linkage as against independence ($x = 0.5$), and the odds in favor of more frequent recombination ($x = 0.2$, 0.3, or 0.4) are also high. The most likely recombination value, that which gives the highest odds against absence of linkage, has been estimated as 0.104, or 10.4 per cent. This estimate may still be subject to considerable revision when new pedigrees become available. In order to show how carefully such an estimate should be expressed, we mention that the authors of this linkage study imply that, although the most likely value is close to 10 per cent, there may be a chance of 5 in 100 that the true value is less than 5 per cent or more than 14 per cent.

The methods for the detection of linkage that have been described here entail the analysis of given sets of pooled observations. Morton has adapted the "sequential analysis method" to the study of problems of linkage—a statistical method whose importance for problems of human genetics was first pointed out by Madge Macklin (1893–1962). This method makes use of the fact that information on possible linkage is accumulated as a succession of samples, each of which is quite small relative to the amount of data required to detect even moderately close linkage. It consists of determining, at successive stages during the collection of data, whether enough information has been collected to make it possible to decide that there is no linkage (H_0), that there is linkage of a specified degree (H_1), or that such a decision cannot be made one way or the other. The decision itself is given in terms of probability and makes use of lods, the logarithms of likelihood ratios.

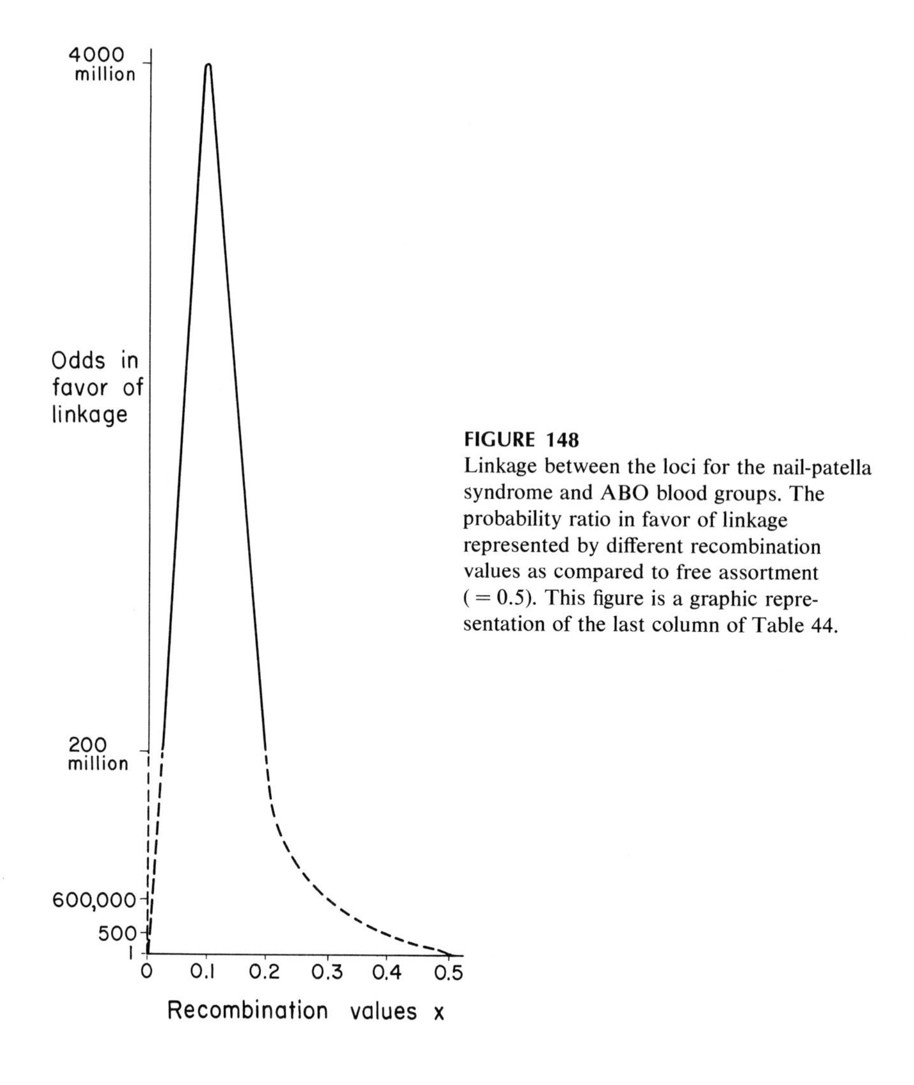

FIGURE 148
Linkage between the loci for the nail-patella syndrome and ABO blood groups. The probability ratio in favor of linkage represented by different recombination values as compared to free assortment ($= 0.5$). This figure is a graphic representation of the last column of Table 44.

It should not be surprising that only a few autosomal linkages have been established in man. Whenever two loci are tested for linkage, the a priori probability is small that they are in the same chromosome; it is much more likely that if one locus is in a certain autosome, the other will be in any one of the other 21. In addition, two loci in the same chromosome may be so far apart that they assort freely and thus appear unlinked. Conversely, apparent cases of linkage must be treated very critically, since such extraneous circumstances as interaction between different alleles of the two loci that reduce the viability of their bearers may simulate linkage ratios (for instance, if A^1B^1 and A^2B^2 confer on their bearers higher viabilities than A^1B^2 and A^2B^1). Acceptance of supposed linkages in man must be based on a very high level of confidence. In this connection it should be reiterated that information about three or more generations in a pedigree may often establish the exact genotypes of relevant parents and thus lead to clearly significant findings.

The discoveries of autosomal linkages are the result of large-scale surveys. When, for instance, linkage involving the nail-patella syndrome was to be studied, separate tests for linkage between the syndrome and a whole array of blood-group systems were made: ABO, Rh, MNS, P, Le, K, Lu, Fy, and Se. Each one of these showed independent recombination with the syndrome except ABO. Tests for linkage between various autosomal loci now include even more blood groups as well as loci for various serum proteins and other biochemical groups. Renwick estimates that in 1966 there had been published "perhaps 2,000 pedigrees with segregation data for 2 to 25 loci" and that a complete evaluation of the likelihoods of these data for ten values of the recombination fraction would mean approximately one-fifth of a million calculations, where one calculation concerns one set of ten values for two loci segregating in one pedigree. It is easily seen that the use of computers is necessary to deal with this embarrassment of riches.

Crossing Over in Males and Females. In *Drosophila,* crossing over occurs only in the female, and in the silkworm *Bombyx*, it is restricted to the male; but in man, as in mice, rats, and other mammals, crossing over has been found in both sexes. All known examples of linkage and crossing over between two pairs of X-linked genes furnish, of course, evidence of crossing over between the two X-chromosomes in human females. Crossing over of autosomal genes has been recorded for both human females and males.

It is likely that the strength of linkage is different in the two sexes. Thus, the recombination value for the nail-patella and ABO genes is 14.6 per cent in females but only 8.4 per cent in males. Similarly, the recombination values for the *Lu* and *Se* loci in females and males are as 1.7:1. These sex differences are, however, not yet significant statistically. Even if they were, it is possible that other linkages might not show differences or might show higher frequencies in the male. Also, within the same sex the frequency with which two loci cross over may not be constant. In many experimental organisms, such variables as temperature, nutrition, and age are known to influence it, and the relation between the age of an animal and the number of crossover gametes it produces may be a complex one, falling and rising several times as the organism grows older. Nothing is known about such variations in man, though it is likely that they occur. A time may come in which it may seem desirable to transmit unchanged a certain linked arrangement of alleles present in a parent to his or her offspring, or, on the contrary, to have such an association broken by crossing over. Knowledge of controllable factors that influence the frequency of crossovers would then become a practical tool. Such a speculation may appear farfetched, but even more unexpected developments have arisen from small beginnings.

Heterogeneity in Linkage Data. One of the well-established autosomal linkages is that between the loci that determine the Rh blood groups and either the usual round or the rare oval shape of red blood cells (Fig. 149). Elliptocytosis or ovalocytosis is caused by a dominant gene *El*. By combining data from numerous pedigrees, it has been shown that the recombination value is approximately 20 per cent. A study of the individual pedigrees indicates that they fall into at least two different groups: one of kindreds with linkage much closer than 20 per cent, namely 3 per cent, and one of kindreds with apparently free assortment. Statistical tests confirm the impression that the recombination values in the different pedigrees differ from one another more than chance alone can account for—in other words, that the data are heterogeneous. On the other hand, no significant heterogeneity for linkage has been encountered in the fourteen known nail-patella syndrome kindreds or in the Lutheran-secretor kindreds.

Indications of heterogeneity will probably be found in various future studies. There are two principal, but not mutually exclusive, explanations of such—at first—disconcerting findings. One is based on the knowledge that similar or seemingly identical phenotypes may be caused by action of entirely unrelated genes. If there were two different loci, El_1 and El_2, for elliptocytosis, and if one of these was in the same chromosome as the *R* locus while the other was in a different chromosome, then clearly some kindreds would show linkage and others free assortment. The finding of heterogeneity in linkage studies may thus serve as an incentive to search for biochemical, physiological, or morphologic differences between apparently identical phenotypes from different pedigrees.

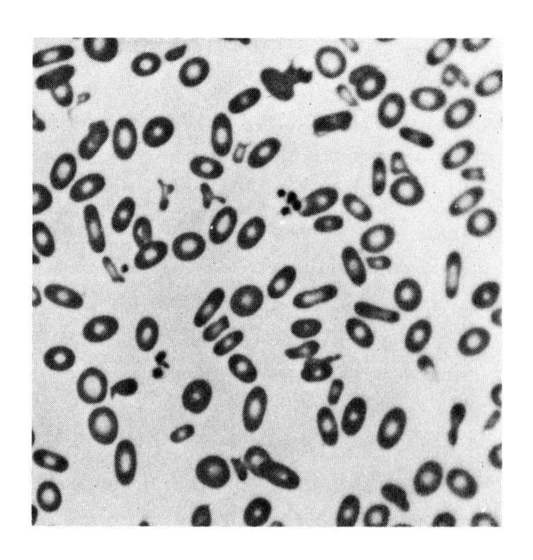

FIGURE 149
Hereditary elliptocytosis. (Original from Dr. C. L. Conley.)

Another explanation of heterogeneity in linkage data assumes that in all families one and the same locus controls a given phenotype, but considers the possibility of variations in the chromosomal constitutions of different persons. It is well known that more than one kind of arrangement of genetic material is found in different populations of many animals and plants. Thus, one group of individuals may have a chromosome pair containing the two linked loci, A and B, and another pair in which two other loci, C and D, are linked (Fig. 150, A). In another group, A and C may be together in one chromosome pair, and B and D in another (Fig. 150, B). Still a third group may have an "arrangement-heterozygote" consisting of four different chromosomes (Fig. 150, C). Such situations are the result of an exchange of segments between nonhomologous chromosomes, a rare process which may lead to a translocation. Once having been produced, translocation chromosomes reproduce their own kind. It is obvious that individuals of the first type would show linkage between A and B and between C and D, but that individuals of the second type would show independence of the loci for A and B and of the loci for C and D, but linkage between A and C and between B and D.

Translocation between different pairs of chromosomes is only one kind of the different known causes of chromosomal heterogeneity in plants and animals. Another common type of chromosomal variation is *inversion*. Some individuals may have in a certain pair of chromosomes a linear sequence of loci M–N–O–P–Q–R; others may be homozygous for the partly inverted sequence M–Q–P–O–N–R; and still others may be inversion heterozygotes with one chromosome of each kind (Fig. 150, D, E, F). If the loci M and Q in individuals with a regular sequence of loci are located very far from each other, they may assort so freely that they appear not to be linked. In those with an inverted sequence, the same two loci may show rather close linkage. Finally, in inversion heterozygotes, crossing over at the four-strand stage

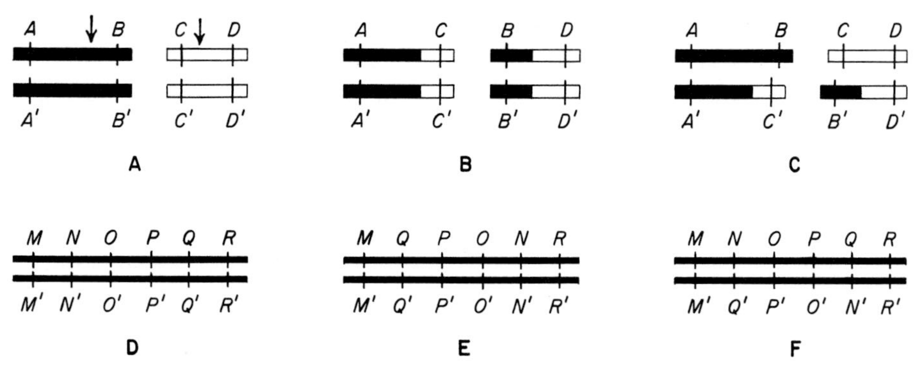

FIGURE 150
Chromosomal translocation (**A–C**) and inversion (**D–F**). See text.

(see pp. 90–93) will frequently result in two chromosomes which have undergone no crossing over and two crossover chromosomes. The crossover chromosomes are usually so abnormal that if they are transmitted by a gamete, they do not lead to a viable zygote (Fig. 72, p. 128); hence, the loci M and Q would seem to be almost absolutely linked. We see, then, that according to the sequences of loci in chromosomes which are inverted in relation to each other and the homo- or heterozygosity for such sequences, two gene pairs may assort freely, be linked to some degree, or seem to be absolutely linked.

MAPS OF HUMAN CHROMOSOMES

The strengths of various linkages are measured by the frequency of recombination—a frequency that depends on the physical distance between the linked loci in the chromosomes. This makes it possible to construct chromosome maps in which the loci of known genes are represented by points on a line whose distances from one another are proportional to their recombination percentages. Detailed chromosome maps have been made for microorganisms and many higher plants, insects, chickens, mice, rats, and a few other mammals, but the mapping of man's chromosomes in terms of genic loci has only begun.

Autosomal Linkages. The first clear case of autosomal linkage in man was found by J. Mohr in 1954. It established that the Lutheran and Secretor loci are linked. Fifteen years later ten different autosomal linkages had become known, with recombination frequencies ranging from essentially zero to 18 per cent (Fig. 151). Most of the maps consist of two loci only but two of them, (3) and (8), are "3-loci" maps (though in both cases since two of the loci have not undergone any recombination with each other in the limited pedigrees available, they actually are only "2-point" maps). With future discoveries of linkages between additional loci more 2-point maps will become 3-point, 4-point and many-point maps. Some of the additional loci will undoubtedly lie between those now located and others on either side of the chromosome region now delimited by them.

At present, useful linkage maps can be constructed only for well-defined pairs of single loci. Suggestions for the existence of further linkages can also come from studies of traits that may be less well defined genetically. We have referred to data on linkage between palm print and skin pigmentation in northern Sudanese and American black populations. In another investigation 300 whites from New England have been tested, by the sib-pair method, for linkage between body build and 9 other phenotypes: the data point to an autosomal linkage of loci involved in body build and in freckling. In a some-

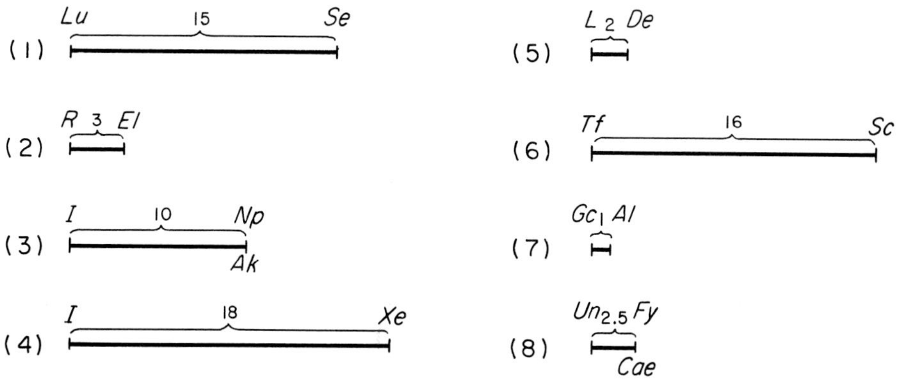

(1) Lu 15 Se

(2) R 3 El

(3) I 10 Np, Ak

(4) I 18 Xe

(5) L 2 De

(6) Tf 16 Sc

(7) Gc_1 Al

(8) Un 2.5 Fy, Cae

FIGURE 151
Eight autosomal linkages in man. Gene symbols: *Lu*, Lutheran blood group; *Se*, secretor; *R*, Rhesus blood group; *El*, elliptocytosis; *I*, ABO blood group; *Np*, nail patella syndrome; *Xe*, xeroderma pigmentosum; *L*, blood group; *De*, dermatologic syndrome; *Tf*, transferrin; *Sc*, serum cholinesterase; *Gc*, serum group; *Al*, albumen variant; *Un*, uncoiler chromosome locus; *Fg*, Duffy blood group; *Cae*, type of congenital cataract. All map distances are the most likely ones within wide ranges of also possible values.

what similar study of 75 pairs of white brothers, 37 physical traits were tested in 666 paired comparisons, one of which was highly suggestive of linkage between genes for "nose-tip thickness" and "Darwin's point" of the external ear. One more autosomal linkage is suggested by one of the earliest applications of the sib-pair method: between loci concerned with a tooth deficiency and hair color.

The sum of the map distances of all autosomal linkages recorded in Figure 151 is less than 70. This is a small fraction of the theoretical length of all autosomal maps added together. The theoretical length is determined very approximately as follows: If a pair of homologous chromosomes shows one and only one crossover configuration in all meiotic cells then, keeping in mind that crossing over occurs between two of the four strands only, the resulting four chromatids will consist of two crossovers and two noncrossovers. Thus, the map length of the chromosome is 2 crossovers out of a total of 4 strands, equal to 50 per cent. If all 22 autosomal pairs underwent one meiotic exchange the combined map length would be 1100 units. Observations of meiotic stages by Ford, Hamerton, and Kjessler show a total number of exchanges in the whole group of autosomes of about 50, i.e., on the average somewhat more than 2 exchanges per pair. Since each of the 50 exchanges may be equivalent to a map length of 50 per cent, all 50 would suggest a total autosomal map length of about 2500. The sum total of 70 autosomal map units known from 2-point linkages is less than 3 per cent of 2500; 97 per cent of the maps are still unknown.

Mapping the X-chromosome. Autosomal loci that freely recombine with one another are either in different chromosomes or far apart in the same chromosome. X-linked genes that freely recombine are known to be in the same chromosome, the X, but their free recombination is compatible with a map distance of 50 units, or any longer distance. Tests for linkage of a variety of two X-linked loci often yield 50 per cent recombination, thus, the two loci are said not be "within measurable distance" (Fig. 152,B). Some tests, however, have clearly shown closer linkages (Fig. 152,A). Among these are the 2-point maps for color blindnesses and hemophilia A, for color blindnesses and G-6-PD, for the Hunter syndrome and the Xm serum group and

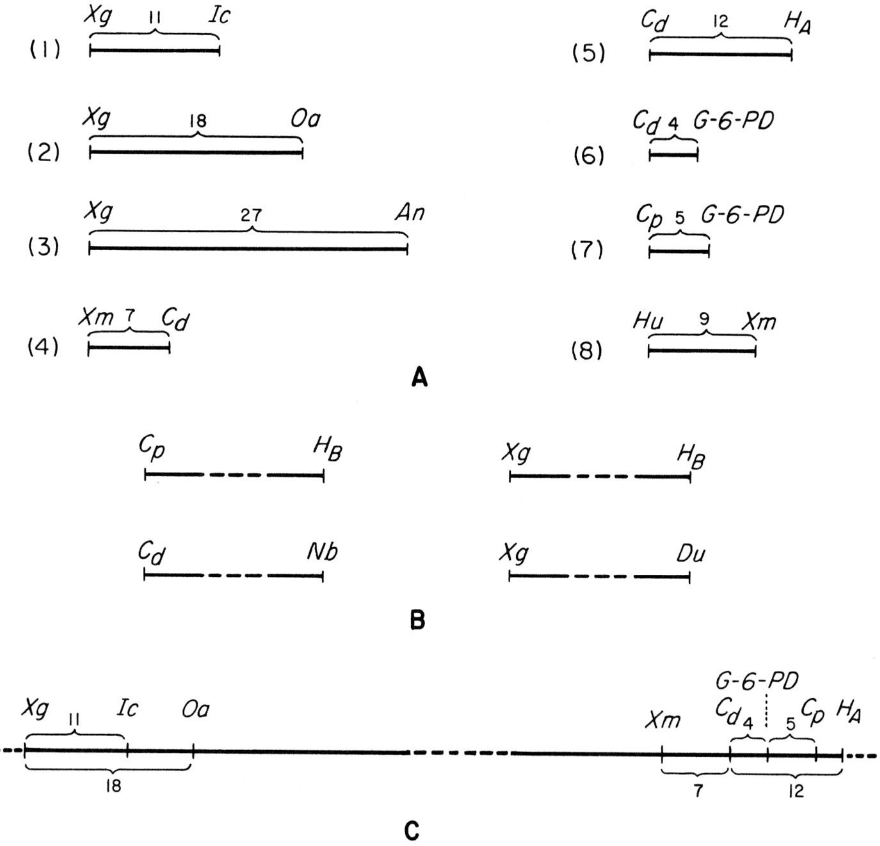

FIGURE 152
Some X-chromosomal linkages in man. Gene symbols: *Xg*, blood group; *Ic*, ichthyosis; *Oa*, ocular albinism; *An*, angiokeratoma (Fabry's disease, a lipid syndrome); *Xm*, serum property; C_d, deutan color blindness; H_A, hemophilia A; *G-6-PD*, serum enzyme; C_p, protan color blindness; *Hu*, Hunter's syndrome; H_B, hemophilia B; *Nb*, night blindness; *Du*, Duchenne muscular dystrophy. **A.** Linkages within measurable distances. **B.** Linkages not within measurable distances. **C.** A partial map of the X-chromosome. The sequence of loci is only one of many that are compatible with available information. (C, after Race and Sanger.)

for *Xg* and a variety of other loci. It would seem that it should be easy to combine the various 2-point maps to make one single map representing much of the length of the X-chromosome (Fig. 152,C). Actually, the data are not yet sufficient for such a task. It is true that we can place most of the X-linked genes studied so far in one or the other of two regions of the map, which are separated by a not-measurable distance: *Xg*, ichthyosis, and ocular albinism are in one "cluster" and *Xm*, color blindnesses, G-6-PD, and hemophilia A in another. This leaves out of consideration the unknown location of the two-point, 27 unit stretch from the *Xg* locus to that of angiokeratoma. Moreover, the sequence of loci within each cluster is uncertain. Take, for instance the three loci *Xg*, ichthyosis (*Ic*) and ocular albinism (*Oa*). Starting with the left end of the over-all map the two 2-point maps are compatible with four different sequences. (1) *Xg-Ic-Oa*; (2) *Oa-Ic-Xg*; (3) *Ic-Xg-Oa*; (4) *Oa-Xg-Ic*. In addition, the sequences (5) *Xg-Oa-Ic* and (6) *Ic-Oa-Xg* are only excluded if the most likely recombination value for *Xg-Ic*, 11 per cent, is truly significantly smaller than that for *Xg-Oa*, 18 per cent; and that is not the case.

Assignment of Genes to Specific Autosomes. Independently of detailed knowledge of human linkage maps the question arises which specific individual or linked loci belong to specific autosomes. Several paths are available to approach this question, some of which have already led to specific assignments. All these depend on simultaneous studies of the transmission or frequencies of genes and of microscopically visible chromosomal abnormalities.

The earliest method for chromosomal assignment of genes makes use of data from trisomic individuals, particularly those affected with Down's syndrome. Genes located in chromosome 21 should result in genotypic frequencies among 21-trisomics that are different from those in normal disomics. Making certain assumptions concerning the abnormal meiotic processes leading to 21-trisomy and assuming for illustration that the ABO locus is in chromosome 21, it would follow that the frequency of blood group O among Down's individuals ($I^OI^OI^O$) is the cube of the frequency of the I^O allele in contrast to being the square of the allele frequency among normal individuals (I^OI^O). Similarly the frequencies of A, B, and AB genotypes should vary between tri- and disomic individuals. Tests of these expectations for a variety of genic loci indicate strongly that none of them including the ABO locus belongs to chromosome 21. This result, although negative, is useful since it narrows down the possible assignment of these loci to one less autosome than the whole set of 22.

A second method that makes use of 21-trisomic individuals employs comparative measurements of the activities of a variety of enzymes in trisomics and disomics. If there is a gene dosage effect, those enzymes whose genic determiners are in chromosome 21 might have a higher concentration in 21-trisomics than in disomics. This deduction is valid only if there are no regulating processes in the organism that lead to variations in enzyme concentrations independent of gene dosage. Most, if not all, enzymes tested so far seem not to be suitable for assignment or nonassignment by this method of their respective genes to chromosome 21. Nevertheless, the study of dosage effects in cases of trisomy of whole chromosomes or in partial trisomy (i.e., duplications of chromosome segments), as well as in partial monosomy (i.e., deficiencies of chromosome segments) may in the future lead to chromosome assignments for specific genes.

Independent of dosage effects, such an assignment has been possible for the haptoglobin locus. In a certain pedigree a translocation between chromosome 2 and 16 was transmitted, and a simultaneous study of the transmission of alleles for the α chain of the haptoglobins showed close linkage between the haptoglobin locus (α-*Hp*) and the break points of the translocation (no recombination among 11 gametes). This indicated that *Hp* lies either in chromosome 2 or 16. Other pedigrees showing abnormalities of chromosome 2 indicated that there was no linkage between *Hp* and chromosome 2. This leaves chromosome 16 as the bearer of the *Hp* locus, a conclusion confirmed by study of a pedigree in which one chromosome 16 shows frequent breaks at a "fragile site" and where only 3 recombinants between the site and the α-*Hp* locus were observed out of 33 opportunities for recombination.

Another assignment of a specific locus, that for the Duffy blood group, to a specific chromosome concerns chromosome 1. In several phenotypically normal families, in the United States and in the Soviet Union, this chromosome deviates from a normal chromosome 1 by having a part of its long arm relatively uncoiled (Fig. 153). The variant chromosome is transmitted in a regular way. This suggests that the uncoiled section carries an "uncoiler" gene *Un* causing the unusual appearance. A large number of loci for red-cell antigens and serum proteins were studied in these families. All of these loci but one, *Fy* (= Duffy), proved to recombine freely with *Un*. The linkage of *Un* with *Fy* is close, the most likely recombination frequency being 2.5 per cent (within wide limits). Independent family studies have yielded no recombinations between *Fy* and a gene *Cae* for a congenital cataract, thus assigning *Cae* likewise to chromosome 1.

A most unexpected method of assignment of a human gene to a specific chromosome comes from experiments by Mary Weiss and H. Green with somatic man-mouse hybrid cells. When human and mouse cells are grown together, a human cell and a mouse cell may fuse: cytoplasm with cytoplasm and nucleus with nucleus. The result is a cell that possesses all 46 chromosomes of man plus all 40 of the mouse. In subsequent cultures these hybrid cells lose chromosomes, resulting in cells with a variety of chromosome mixtures and numbers. From these, cell cultures have been obtained that

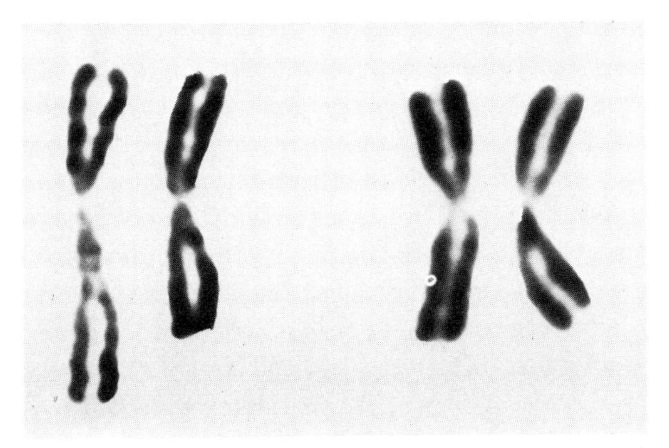

FIGURE 153
Two pairs of chromosomes 1. *Left:* From a person heterozygous for the variant "uncoiler" chromosome. *Right:* From a related person who does not carry the variant. (Donahue, Bias, Renwick, and McKusick, *Proc. Nat. Acad. Sci.,* **61,** 1968.)

possess all mouse chromosomes but only one or a few human chromosomes. These cells originated in a medium in which all cells die that cannot produce the enzyme thymidine kinase. Since the mouse cells that had been used in the hybridization lacked the enzyme while the human cells possessed it, only those hybrid cells could survive that contained the human chromosome with the locus of thymidine kinase. This chromosome, common to all survivors is 17.

In addition to the assignment of specific genes to specific autosomes attempts have been made to relate the loss or gain of specific regions of autosomes to the appearance of clinically abnormal phenotypes. Deletions for part of the short or long arm of chromosome 18 are responsible for a variety of traits: mental retardation, abnormal eye folds, short fingers, abnormal palm folds, syndactyly and others. By comparing the different clinical abnormalities that occur in individuals with different chromosomal deletions, including some that give rise to ring chromosomes, de Grouchy has constructed a tentative map of segments of chromosome 18 whose absence is responsible for specific abnormalities. It is not yet possible to judge fully the validity of this "topological approach."

LINKAGE AND GENETIC COUNSELING

Linkage Involving Two Loci. When extensive linkage maps of all human chromosomes are finally available, more than just an intimate knowledge of man's biological make-up will have been achieved. The maps will also help determine the genotypes of individuals and thus remove some of the great difficulties confronting genetic advisers. These are the inability to distinguish carriers of recessive atypical genes from homozygous normal individuals; to recognize, at an early stage, atypical dominant or homozygous recessive genotypes for genetically conditioned abnormal traits that do not become apparent before a certain age; or to differentiate between genetically normal and abnormal individuals in those cases in which atypical genes may not be expressed phenotypically. It is of relatively little help to tell a healthy young man or woman with a parent who carried the dominant gene for Huntington's chorea that he has 1 chance in 2 of having inherited the gene and of becoming afflicted in later life. If it were known, however, that this gene, *Ht*, is located in the same chromosome as some other gene and is close to it, and if the presence or absence of the other gene could be recognized more easily than that of *Ht*, then the diagnosis of the desirable genotype *ht/ht* or the ominous genotype *Ht/ht* could be made with greater certainty.

We may clarify this statement by a hypothetical example. Assume that a deleterious dominant allele V', whose effect does not appear until middle age, is linked to the blood-group gene I and that V and I yield crossovers 5 per cent of the time. A man without the allele V' and belonging to blood group O marries a woman belonging to blood group A who later develops the defect. Their young son belongs to group A. What are his chances of being a carrier of V'? The accuracy of the answer will depend upon knowledge of the blood-group genotype of the mother. If it were known that her father belonged to blood group O and had the abnormal trait caused by V', then she must be $V'I^{\text{O}}/VI^{\text{A}}$. Her son, who received his I^{A} allele from his mother, has therefore a chance of 95 per cent of not having the allele V', since this is the frequency with which V is linked with I^{A} in the mother's gametes. Differently expressed, there is only 1 chance in 20 that he is a carrier.

This example assumed linkage between the V locus and I locus whose alleles I^{O} and I^{A} are both common. To serve as a useful marker for the presence or absence of an allele at another locus, an allele must occur frequently in heterozygous combinations. Had the mother of the young son been homozygous for I^{A}, I^{B}, or I^{O}, no marker for the V allele would have been available.

There are already a few cases on record in which a knowledge of linkage has been employed in counseling. Hoogvliet, in his studies of hemophilia and color blindness, concluded that a certain mother was heterozygous for genes for these two traits and that her genotype almost certainly was of the cis phase, *CH/ch* (the *c* gene was one of the deuter alleles). Her husband was a deuteranope, and the mating was thus $CH/ch \times cH$. One of the daughters was color blind, but four others appeared to have normal color vision. Accepting the recombination value for the linkage of *c* and *h* as about 10 per cent, we can conclude that the color-blind daughter had a 90 per cent chance of having received from her mother a nonrecombinant X-chromosome, *ch*, making her *ch/cH*; if so, the probability that any son she might have would be hemophilic would be 50 per cent. Against this distressing report to the color-blind daughter may be placed the more reassuring one to her normal-visioned sisters. They have a 90 per cent chance of having received from their mother the other nonrecombinant X-chromosome, *CH*, and thus of being *CH*/*cH*; hence there is only a slight chance that they would have any hemophilic offspring. Similar counsel was given families in which the mother was also doubly heterozygous for hemophilia and color blindness but, probably, in the trans phase, *cH*/*Ch*. More recently, the closer linkage between G-6-PD and hemophilia, 4 per cent, than between color blindness and hemophilia has been used in counseling. In suitable cases this increases the probability of correct counseling to 96 per cent.

Linkage Involving Three Loci. Further improvement of the method of determining the genotype of an indivdual by means of linked marker genes will result from future detailed human chromosome maps. In the example of *V* and *I*, it was assumed that *V* and *I* would be crossovers 5 per cent of the time; so the uncertainty of predicting the presence or absence of *V* on the basis of the *I* marker was still 5 per cent. If *two* marker genes were known, located not too far to the right and left of the *V* locus, the accuracy of prediction would approach certainty. This follows from the knowledge that double crossovers are so rare within short map distances that they can practically be ignored. Consequently, if the marker alleles retain their original linkage relations, there is a very high probability that the test gene has remained between them.

The utilization of marker genes on both sides of the test gene has its limitations. It depends, of course, on heterozygosity for all loci concerned, since recombination is discoverable only in heterozygotes. Moreover, if a crossover has occurred between the two marker genes, then there may be no clear indication whether the test gene is present or absent in the recombined chromosome. These limitations could be overcome if enough loci in each chromosome were known, so that it was possible to select those that gave maximum information. Although we are far from having the requisite knowledge, the search for new marker genes will continue to provide new loci for future chromosome maps. When enough marker genes are known to make the method feasible for general use, information on several generations preceding the individuals under consideration will be desirable. We may foresee that, at some future date, genealogical records containing the greatest possible amount of data on every common, easily determinable genetic property will be available.

Linkage and the Direct Evidence on Genotypes. It is, of course, possible that other methods of recognizing hidden genotypes will develop faster than the linkage method. It is likely that most, if not all, heterozygotes for recessive genes will be found to be different from the dominant homozygote in some slight, detectable way, and that every genotype will someday become phenotypically distinguishable from every other. If this should be true, direct methods of ascertaining the presence or absence of any given allele would be available, and no recourse to marker genes would be necessary. Whatever methods may finally be employed, it should not be forgotten that they will apply not only to unfavorable alleles but also to the less obvious favorable ones.

The value of linkage studies in man is not limited to the construction of maps of human chromosomes and to their application in counseling. Data on linkage can throw light on such questions as whether there is only one locus, or several loci, able to produce a given phenotype, or whether sex linkage or autosomal linkage exists in inherited traits that include nonreproduction of affected males and absence of affected females (see testicular feminization, p. 516). An answer to the first question would have particular significance. If the frequency with which a given trait arises by mutation is determined, it is obviously important to know whether mutation has occurred at the same single locus each time or whether a variety of loci—one at a time—are involved, so that the frequency of mutation per locus is only a fraction of the total rate of mutation for the trait. Another value of linkage studies, already mentioned in connection with elliptocytosis, is their potentiality of revealing genetic heterogeneity of seemingly uniform traits. Such findings may stimulate biochemical and developmental exploration to differentiate between the different genetic pathways which lead to the trait and then to design appropriate methods of control.

The Genetic Organization of Human Chromosomes. The assignment of specific genes to specific chromosomes and the construction of chromosome maps by means of linkage data is an example of the present status of age-old efforts to gain an understanding of the structure of organisms. Even before recorded history, man presumably endeavored to gain knowledge of his own body. The science of anatomy gradually grew up from incidental and isolated observations until the locations and interrelations of the various organs, the bones and tendons, the muscles and nerves, became ever better known. When the skill of the anatomist became inadequate for the elucidation of the finer structures of the body, the microscope, the electron microscope, methods of differential staining, and microchemical methods led to further advances. Outrunning these delicate probes, the subtle tools of the mind now lay bare the topography of molecular entities within the chromosomes—the genes whose allelic forms so differentiate the individual's development that he may turn out to be dark- or light-skinned, endowed with normal limbs or misshapen ones, able to see the beauty and ugliness of the physical world or assigned to lifelong blindness.

PROBLEMS

115. (a) If a woman has three sons, one hemophilic and two color blind, what, most likely, is her genetic constitution?
(b) If another woman also has three sons, two normal and one both hemophilic and color blind, what, most likely, is her genetic constitution?

116. A woman's father is color blind. Two of her brothers, as well as an uncle on her mother's side, are hemophilic.
(a) What are the possible genotypes of all individuals mentioned?
(b) If the woman has a hemophilic son, what must be her genotype?
(c) What is the probability that a sister of the hemophilic son is a carrier for hemophilia?
(d) If the woman's husband is color blind, what is the probability that a color-blind sister of her hemophilic son is a carrier for hemophilia?
(e) What is the probability that a sister whose color vision is normal does not carry hemophilia?

117. Among whites the frequency of the dominant secretor allele *Se* is 0.5; that of the Lutheran blood-group allele Lu^a, 0.35. The two genes are linked with a recombination frequency of 10 per cent.
(a) What is the frequency of $Se\ Lu^a/se\ Lu^b$ people in a population at equilibrium?
(b) Of $Se\ Lu^b/se\ Lu^a$ people?
(c) What would these frequencies be if the recombination frequency were (i) only 1 per cent, or (ii) 50 per cent?

118. People with blue sclerae frequently have brittle bones. Is this due to linkage?

119. Why can even a large number of one-child sibships give no information on linkage when it is unknown whether the parental genotypes are in the coupling or repulsion phase? Why can two-child sibships provide such information?

120. Assume that 2 autosomal genes are linked with a recombination value of 20 per cent. Among 100 sibships of 2 sibs each, from a random sample of 100 marriages of the genotypes $AaBb \times aabb$, what are the expected numbers of all possible types of sibships? (Proceed in the following sequence: First, list all possible combinations of two sibs. Second, determine which of these combinations consist of
(a) identical crossovers or noncrossovers,
(b) complementary crossovers or noncrossovers,
(c) crossovers and noncrossovers.
Third, determine the frequencies of (a), (b), and (c). Fourth, assign to each combination of two sibs its frequency among 100 sibships.)

121. (a) By means of the method of likelihood ratio, calculate the odds in favor of linkage between hemophilia and color blindness for recombination values $x = 0, 0.1, 0.2, 0.3, 0.4$, and 0.5 for each of the pedigrees shown in Figure 145, A and B.
(b) Combine the odds for the two pedigrees. If these two pedigrees were the only basis for your decision, which one of the x values in your table is closest to the most likely recombination value?

122. A woman is heterozygous for the three X-linked loci *Xg*, protan color blindness, and hemophilia B: Xg^a Xg; C c; H h. The sequence of these loci is uncertain. Assume it is *Xg-C-H*.
(a) List the four possible "options" according to which the three allele pairs may be distributed over the two X-chromosomes.
(b) Analyze for each option which of the woman's five sons listed below contained a noncrossover X-chromosome (0), or a crossover in region *Xg-C* (1), or a crossover in region *C-H* (2), or a double crossover, one in each region (1, 2). The genotypes of the sons were (i) $Xg\ C\ h$; (ii) $Xg\ c\ h$; (iii) $Xg\ c\ H$; (iv) $Xg^a\ c\ h$; and (v) $Xg^a\ c\ H$. (This is an actual family described by Graham, Tarleton, Race, and Sanger, *Nature*, **195,** 1962.)

123. Assume recombination frequencies as follows: *Xg* and X-linked ichthyosis, 11 per cent; *Xg* and ocular albinism, 18 per cent.
(a) What are the possible sequences of the three loci?
(b) What further information is needed to establish the true sequence?

124. Assume: that a gene F has two alleles F^1 and F^2 which control the presence of antigens F_1 and F_2; that a gene G has two alleles G^1 and G^2 which control the presence of antigens G_1 and G_2; that the loci of F and G are in the X-chromosome and close to each other; and that y, a recessive X-linked gene causing a serious affliction, lies between F and G. A woman with the antigenic properties F_1, G_1, and G_2 has several brothers afflicted with the y trait. The following genotypes of her ancestors are known:
(i) maternal grandmother F^1yG^2/F^1YG^2;
(ii) maternal grandfather F^2YG^1;
(iii) father F^1YG^1.
(a) Draw a pedigree of the kindred.
(b) What is the chance that the woman is a carrier for the atypical allele y which would lead to the production of affected sons? (Assume absence of double crossovers.)
(c) What would have been the answer to (b) if the woman had been of the phenotype F_1, F_2, G_1?

REFERENCES

Bailey, N. T. J., 1961. *Introduction to the Mathematical Theory of Genetic Linkage*. 298 pp. Clarendon, Oxford.

Bateman, A. J., 1960. Blood-group distribution to be expected in persons trisomic for the ABO gene. *Lancet* **1:**1293–1294.

Bernstein, F., 1931. Zur Grundlegung der Chromosomentheorie der Vererbung beim Menschen. *Zeitschr. Abst. Vererb.*, **57:**113–138 (also **63:**181–184, 1932).

Bostian, C. H., Whittinghill, M., Pollitzer, W. S., and Muro, L., 1969. Evidence of iris-earlobe linkage in a mosaic man. *J. Hered.*, **60:**3–9

Cook, P. J. L., 1965. The Lutheran-secretor recombination fraction in man: a possible sex difference. *Ann. Hum. Genet.*, **28:**393–397

Crandall, Barbara F., and Sparkes R. S., 1971. Thymidine kinase activity and human chromosome No. 18. *Biochem. Genet.*, **5**:451–456.

de Grouchy, J., 1965. Chromosome 18: a topologic approach. *J. Pediat.*, **66**: 414–431.

Donahue, R. P., Donahue, R. R., Bias, Wilma B., Renwick, J. H., and McKusick, V. A., 1968. Probable assignment of the Duffy blood group locus to chromosome 1 in man. *Proc. Nat. Acad. Sci.*, **61**:949–955.

Ephrussi, B., and Weiss, Mary L., 1969. Hybrid somatic cells. *Sci. Amer.*, **220**: 26–35.

Fisher, R. A., 1935. The detection of linkage. *Ann. Eugen.*, **6**:187–201;339–357.

Gedde-Dahl, T., Jr., Grimstad, A. L., Gunderson, S., and Vogt, E., 1967. A probably crossing over or mutation in the MNSs blood group system. *Acta Genet.*, **17**:193–210

Haldane, J. B. S., 1936. A search for incomplete sex-linkage in man. *Ann. Eugen.*, **7**:28–57.

Haldane, J. B. S., and Smith, C. A. B., 1947. A new estimate of the linkage between the genes for colour blindness and hemophilia in man. *Ann Eugen.*, **14**:10–31.

Hoogvliet, B., 1942. Genetische en klinische beschouwing naar aanleiding van bloederziekte en kleurenblindheit in dezelfde familie. *Genetica*, **23**:93–220.

Howells, W. W., and Slowey, A. P., 1956. "Linkage studies" in morphological traits. *Amer. J. Hum. Genet.*, **8**:154–161.

Kjessler, B., 1966. *Karyotype, Meiosis and Spermatogenesis in a Sample of Men Attending an Infertility Clinic.* 74 pp. Karger, Basel.

Kloepfer, H. W., 1946. An investigation of 171 possible linkage relationships in man. *Ann. Eugen.*, **13**:35–71.

Lawler, S. D., and Sandler, M., 1954. Data on linkage in man: elliptocytosis and blood groups. IV. Families 5, 6 and 7. *Ann. Eugen.*, **18**:328–334.

Macklin, M. T., 1954. The use of sequential analysis method in problems in human genetics. *Amer. J. Hum. Genet.*, **6**:346–353.

Matsuya, Y., Green, H., and Basilico, C., 1968. Properties and uses of human-mouse hybrid cell lines. *Nature*, **220**:1199–1202.

Miller, O. J., and Siniscalco, M., 1971. Mitotic separation of two human X-linked genes in man-mouse somatic cell hybrids. *Proc. Nat. Acad. Sci.*, **68**: 116–120.

Mishu, Mona, and Nance, W. E., 1969. Further evidence for close linkage of the Hb^{β} and Hb^{δ} loci in man. *J. Med. Genet.*, **6**:190–192.

Mohr, J., 1954. *A Study of Linkage in Man.* 119 pp. Munksgaard, Copenhagen

Mohr, J., 1966. Genetics of fourteen marker systems: associations and linkage relations. *Acta Genet.*, **16**:1–58.

Morton, N. E., 1957. Further scoring types in sequential linkage tests, with a critical review of autosomal and partial sex linkage in man. *Amer. J. Hum. Genet.*, **9**:55–75.

Morton, N. E., 1962. Segregation and linkage. Pp. 17–52 *in* Burdette, W. J. (Ed.), *Methodology in Human Genetics.* Holden-Day, San Francisco.

Nyhan, W. L., Lesch, J., Sweetman, L., Carpenter, D. G., and Carter, C. H., 1967. Genetics of an X-linked disorder of uric acid metabolism and cerebral function. *Pediat. Res.*, **1**:5–13.

Pearce, W. G., Sanger, Ruth, and Race, R. R., 1968. Ocular albinism and Xg. *Lancet*, **1:**1282–1283.

Penrose, L. S., 1935. The detection of autosomal linkage in data which consist of pairs of brothers and sisters of unspecified parentage. *Ann. Eugen.*, **6:**133–138.

Perkoff, G. T., 1967. Hereditary renal diseases. *New Eng. J. Med.*, **277:**79–85, 129–138 (See p. 81).

Perkoff, G. T., Nugent, C. A., Dolowitz, D. A., Stephens, F. E., Carnes, W. H., and Tyler, F. H., 1958. A follow-up study of hereditary chronic nephritis. *Arch. Intern. Med.*, **102:**733–746.

Reitalu, J., 1970. Observations on the behavior pattern of the sex chromosome complex during spermatogenesis in man. *Hereditas*, **64:**283–290.

Renwick, J. H., 1969. Progress in mapping human autosomes. *Brit. Med. Bull.*, **25:**65–73.

Renwick, J. H., 1971. The mapping of human chromosomes. Ann. Rev. Genet., **5:**81–120.

Renwick, J. H., and Lawler, S. D., 1955. Genetical linkage between the ABO and nail-patella loci. *Ann. Hum. Genet.*, **19:**312–331.

Rife, D. C., 1954. Populations of hybrid origin as source material for the detection of linkage. *Am. J. Hum. Genet.*, **6:**26–33.

Shaw, Margery W., and Gershowitz, H., 1962. A search for autosomal linkage in a trisomic population: blood group frequencies in Mongols. *Amer. J. Hum. Genet.*, **14:**317–334. (Additional data in Shaw and Gershowitz, 1963, Blood group frequencies in Mongols, *Amer. J. Hum. Genet.*, **15:** 495–496.)

Shaw, R. F., and Glover, R. A., 1961. Abnormal segregation in hereditary renal disease with deafness. *Amer. J. Hum. Genet.*, **13:**89–97.

Smith, C. A. B., 1953. The detection of linkage in human genetics. *J. Roy. Stat. Soc., B*, **15:**153–192.

Yunchen, Catherine, 1968. Meiosis in the human female. *Cytogenetics,* **7:**234–238.

16

VARIATIONS IN THE EXPRESSION OF GENES

The study of inheritance would be simpler than it is if a given genotype always expressed itself in exactly the same way in all individuals. Although there are many genes whose effect on development seems to be constant under all known circumstances, there are many others of which this is not true. Examples of unequivocal expression like albinism or woolly hair best demonstrate the regularities of gene transmission, but it is obviously equally important to consider genes whose effects are variable.

There is no inherent difficulty in understanding variable gene effects if we remember the concept of a network of developmental reactions. Whenever gene and effect are not related to each other directly, but through numerous interconnecting steps, it is easy to realize that a change anywhere along the network of interconnections may lead to a change in the expression of the gene.

It may be wondered whether variable expression of a gene may not be caused by variability of the gene itself—that is, mutation—rather than by variablility in the network of reactions. This is not impossible, but the probability that it is true is very small. Stability or instability of genes has been ascertained by study of their transmission through germ cells from generation to generation. Such studies have established that genes at any given locus are transmitted unchanged in the overwhelming majority of all gametes, an observable mutation occurring in usually less than one of 50,000 cells. We have little justification for believing that the genes in body cells are less stable than those in germ cells. It is therefore not reasonable to explain the high variability of certain phenotypic effects by a low stability of the genes that control them.

The concept of interacting developmental processes, on the other hand, gives us an interpretation of variable phenotypes that accords with our knowledge of essentially invariable genes. We regard it as a sound premise that *differences in genic expression are variable consequences of constant genes.*

INCOMPLETE PENETRANCE OF GENES

Stiff Little Finger. We have defined a dominant gene as one whose effect is recognizable in the heterozygous condition. A pedigree for a dominant gene (D) that causes a permanently bent and stiff little finger (camptodactyly) is reproduced in Figure 154. The condition is due not to an abnormal joint but to defects in the attachments of some of the muscles controlling this joint. Obviously, it is not a serious aberration, although it may cause slight inconvenience. Most of the individuals in this pedigree have a bent finger on one hand only, and the trait seems to appear equally on the right and the left hand. The appearance of the trait on one hand only is the result of a variable effect of the Dd genotype on different parts of the same person. As these individuals developed from the fertilized Dd eggs, a stiff little finger was formed on one hand of the embryo and a normal finger on the other, because the action of

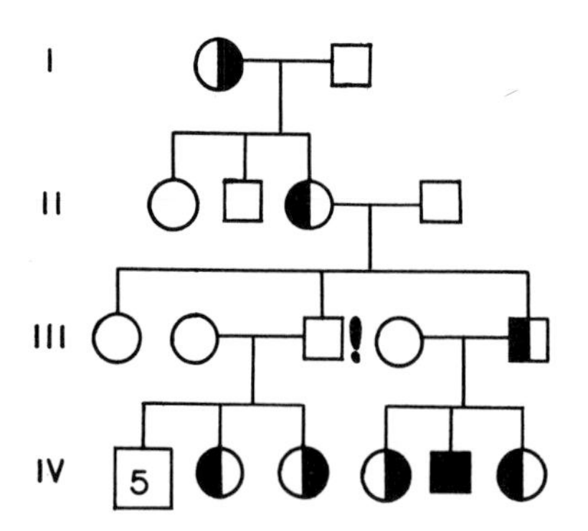

FIGURE 154
Pedigree of stiff little finger (camptodactyly). Solid left or right half of symbol = left or right hand affected only; ! = no penetrance. (After Moore and Messina, *J. Hered.*, **27**, 1936.)

the gene *D* failed to "penetrate" the developmental network on the normal side and to create the abnormal phenotype. A genotype which may or may not produce a given trait is said to be *incompletely penetrant* or to show *incomplete manifestation.*

The reasons for unilateral penetrance of the gene *D* for camptodactyly presumably lie in some delicately balanced embryological events that may tip the scale toward abnormal expression in one hand but toward normality in the other. It is to be expected that in some individuals the balance will be such as to lead to the formation of stiff little fingers on both hands, and in others to normal fingers on both. The pedigree contains 1 bilaterally affected male in generation IV and 1 who is not affected in either hand in generation III. The latter, although normal himself, has 2 affected children. We conclude that he had the same *Dd* constitution as his mother and his 2 affected children, but that the gene *D* failed to express itself during the development of his hands.

The penetrance of such a gene may be quantitatively expressed in terms of the percentage of all those who have the gene who show the trait. A dominant gene with phenotypic expression in all individuals who carry it has 100 per cent penetrance; one expressed in only half the individuals, 50 per cent. To determine the penetrance of a specific gene, we must know the number of carriers who do not show its effect as well as the number who do. A method for such determination consists of counting the numbers of affected and unaffected parental pairs in whose progeny the incompletely penetrant gene expresses itself. In Figure 154, there are four sibships with affected individuals. Three of the four parental pairs contained an affected member, but both of one pair (III–2 and III–3) were normal. The normal father, III–3, was a carrier, as proven by his affected offspring. Therefore, the penetrance of stiff little finger was 3 out of 4, or 75 per cent. The statistical uncertainty of this value is, of course, high, since it is based on a sample of four parental pairs only.

Blue Sclerotics. A slightly different kind of example in which an abnormality is due to a dominant gene with incomplete penetrance is supplied by blue sclerotics (van der Hoeve's syndrome). Persons with this trait have an unusually thin, bluish outer wall of the eye instead of the white sclerotic (or sclera) of normals. The eye condition itself is harmless but is commonly associated with serious defects in other parts of the body. Among these are otosclerosis which leads to deafness, and excessive fragility of the bones (osteogenesis imperfecta), causing frequent fractures (Fig. 155).

The medical literature contains many descriptions of inherited brittle bones, among them the following: "In Oslo, a young heterozygous man, turning around suddenly in the street in order to look at the legs of a pretty girl,

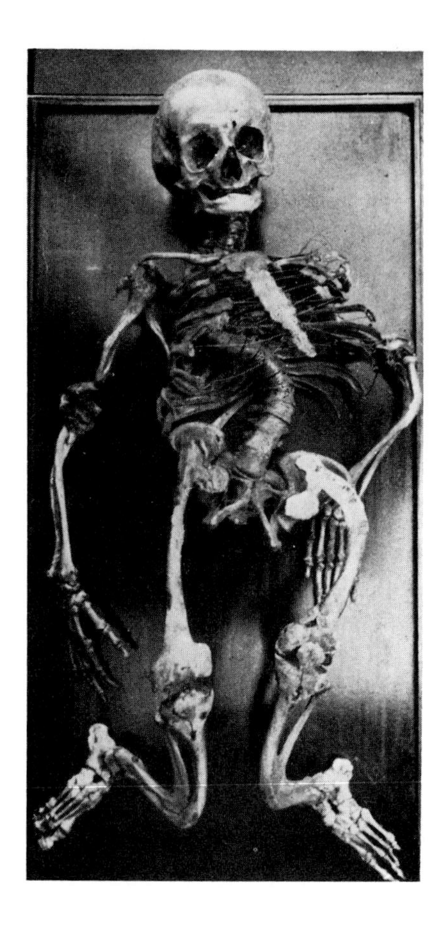

FIGURE 155
Skeleton of a man who was affected with blue sclerotics and brittle bones. (Bell, *Treas. Hum. Inher.*, II, **3**, 1928.)

fractured his leg bone. Another time, when he took his fiancee on his lap, his thigh-bone broke" (Mohr).

It is not clear how the three abnormalities are developmentally related to one another, whether by some general metabolic condition that involves the mineral content of the blood or by control of fundamentally different processes in different parts of the body by the gene for blue scleras. The gene effect on the sclera itself has a very high, but not complete, penetrance, but fractures due to bone fragility were met with in only 63 per cent of a sample of 400 persons carrying the gene (see Fig. 156). A similarly lower degree of penetrance of 60 per cent seems to apply to the deafness caused by the gene. The three defects were found together in 44 per cent of all individuals known to carry the gene. Actually, these percentages vary in different groups. Thus, in two large Japanese kindreds the frequency of fragile bones among persons with blue scleras was 60 per cent in one kindred, 29 per cent in the other; of deafness, 36 and 26 per cent; and simultaneous presence of all these traits, 28 and 7 per cent.

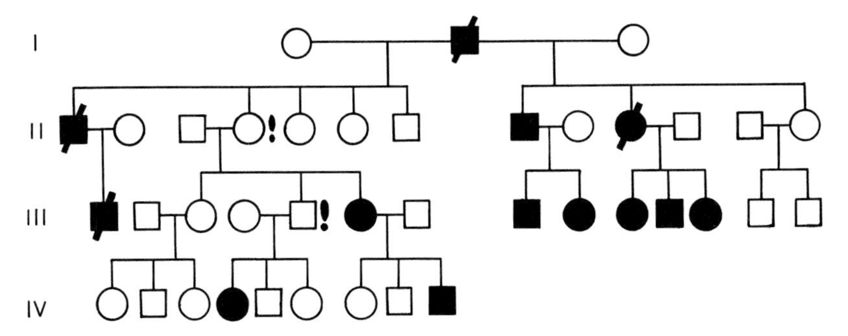

FIGURE 156
Pedigree of blue sclerotics and brittle bones. Solid symbol = blue sclerotics; diagonal bar = brittle bones; ! = no penetrance. (Bell, *Treas. Hum. Inher., II,* 3, 1928; pedigree No. 621.)

Vitamin-D-resistant Rickets. A type of rickets that was very common before the need of vitamin D for bone development was recognized has now practically disappeared in many parts of the world. Some children, however, develop rickets even though provided with a normally adequate diet containing from 400 to 1,000 international units of vitamin D daily, and do not improve when given several times the normal amount of vitamin D. Only when very large doses of the vitamin, up to 150,000 units per day, are administered do the rickets disappear.

In a large pedigree from North Carolina, with at least 28 affected individuals in five generations, a pattern of inheritance is discernible which suggests presence of a dominant gene with incomplete penetrance (Fig. 157, A: note nonpenetrance in the normal woman II–2 as evidenced by her affected son).

A careful clinical study of this kindred brought out the fact that all individuals with bone defects had an abnormally low concentration of inorganic phosphorus in their blood serum (hypophosphatemia)—a condition which is in part responsible for rickets. The low concentration of phosphorus was also found in some individuals with normal bone development, but all these "normal" individuals had both an affected parent and affected offspring. Part B of Figure 157 consists of the same pedigree as part A, but in it the affected persons are those who have a low phosphorus concentration. Clearly, the gene transmits hypophosphatemia in a fully penetrant dominant fashion.

In this pedigree the gene has incomplete penetrance so far as rickets is concerned, and this is obviously due to a variation in some secondary process in the network of development. The gene invariably causes a low phosphorus concentration; but the processes of bone formation, which are also ultimately dependent on the gene, may sometimes lead to normality, sometimes to defectiveness.

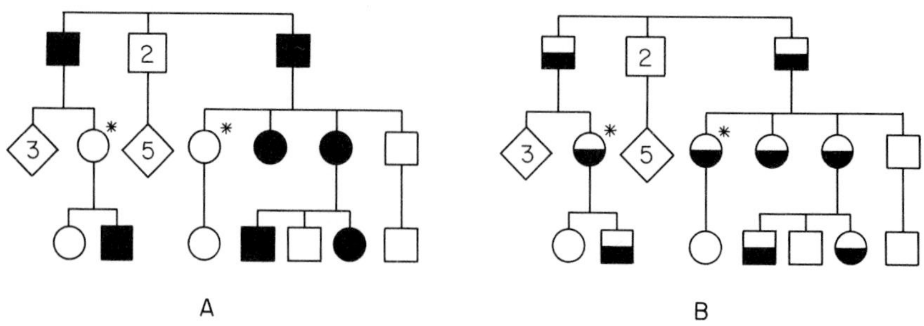

FIGURE 157
Part of a pedigree of vitamin-D-resistant rickets. **A.** The individuals indicated as affected show active rickets or deformities. **B.** The same part of the pedigree, showing as affected individuals those having an abnormally low concentration of inorganic phosphorus in their serum. The low phosphorus concentration in the two individuals marked with an asterisk signifies that they carry the gene responsible for it although it did not lead to bone defects. (After Winters, Graham, Williams, McFalls, and Burnett, *Medicine,* **37,** 1958.)

Variable Age of Onset of Genetically Caused Diseases

Most of the phenotypes we have discussed were determined during prenatal life or were clearly present within a few days after birth. The fragility of the bones of blue-sclerotic individuals is a partial exception. The bones of normal babies are poorly ossified but highly cartilaginous. Gradually, during childhood, the mineral content of the bones increases, the minerals replacing the soft cartilage. In many blue-sclerotic individuals, therefore, the bone fragility becomes apparent only when the process of ossification nears completion. Since the time it takes bones to harden varies somewhat from child to child, the fragility of bones in affected individuals is not apparent at the same age in all. This variation in age of onset of the disease is so great that, in some families, the bones are defective before birth and the condition may lead to stillbirth or death in early infancy.

Many other inherited conditions become apparent only in an adult, particularly certain nervous diseases, most kinds of cancer, diseases of the heart, some eye abnormalities, and, obviously, all specific afflictions of old age. The speed of aging itself is variable, and it is not to be wondered that the onset of the expression of these characters varies over a wide range of years in different persons.

Huntington's Chorea. Huntington's chorea, a disease characterized by involuntary jerking movements of body and limbs, well illustrates differences in age of onset of a genetically caused affliction. The disease consists of a

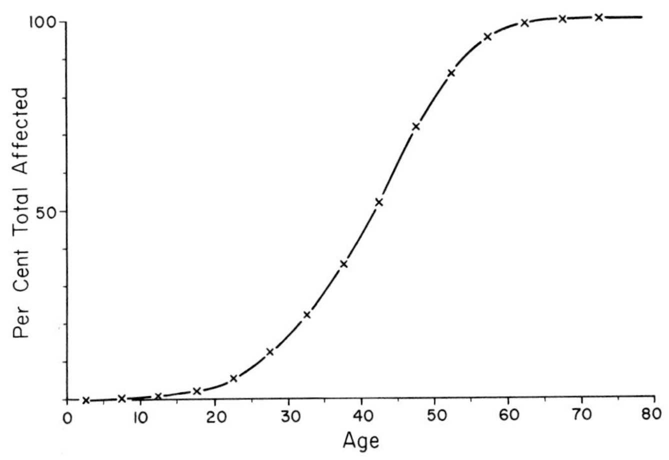

FIGURE 158
Huntington's chorea. Distribution of age of onset in 762 patients. (After Landzettel, Unterreiner, and Wendt, *Acta Genet. Stat. Med.*, **9**, 1959.)

progressive degeneration of the nervous system, and leads to gradual physical and mental impairment of affected individuals and eventually to death. The mean age of onset of this affliction lies between forty and forty-five, but Figure 158, based on a survey of age of onset of many cases, shows that the disease sometimes appears in the first years of life and sometimes after sixty. The trait is clearly hereditary, occurring only in certain families and usually being transmitted from an affected parent to half of the children. This simple picture of a dominant trait is disturbed by the variable age of onset. Choreic offspring may be born to parents who appear normal. In such cases one of the parents usually has an affected parent himself. Apparently he inherited the abnormal allele and transmitted it to his offspring. If he should die before having shown choreic symptoms, and any of the children develop the disease, an apparent exception to typical dominance may be recorded. The older the parents at death, the less likely it is that that a carrier of the abnormal allele will have remained unafflicted.

If the inheritance of Huntington's chorea is to be studied in detail, data such as those given in Figure 158 are useful, since they permit adjustment of the observations to account for seemingly normal persons who may be carriers although they do not yet show it. Obviously, in determining the type of inheritance of traits with a variable age of onset, it is necessary to know the age of persons in a pedigree at the time it is recorded and the age of death of persons who appeared normal.

Relatively few individuals who have an affliction with a rather early mean age of onset—like Huntington's chorea—die of natural causes before they show signs of their affliction. But those hereditary diseases that usually do not appear until later—such as certain defects of the heart or the general circulatory system, and particularly many afflictions of old age—will remain undiscovered in a large number of persons, since these persons will die of other causes before symptoms of the disease in question can be noticed. In order to investigate the inheritance of such a condition, several corrections must be applied to raw data on its frequency in families with affected members. These corrections are based on tables which give the probabilities of survival of an individual in the general population, and also the percentages of individuals that can be identified as affected in each age class.

Huntington's chorea is an example of a hereditary condition that is not present at birth, i.e., not "congenital." Albinism is both hereditary and congenital. The defects produced by infection of the embryo with the rubella virus (German measles) are not hereditary but congenital and injuries sustained after birth are neither hereditary nor congenital.

Further Aspects of Penetrance. The term penetrance is applicable not only to heterozygously dominant genes but also to either dominant or recessive homozygous genotypes. This use expresses the fact that not only the phenotypes of heterozygotes but also those of homozygotes may vary. This is well established in experimental organisms such as *Drosophila* in which analysis of phenotypes can be based on strains with clearly defined genotypes. In man, standardization of genotypes for groups of individuals is usually not possible, and, as a result, unequivocal examples of incomplete penetrance of homozygotes are rare. Perhaps the variability of the ability to taste phenylthiocarbamide (PTC) or related chemicals is relevant here. This trait was discussed earlier as an example of "simple" inheritance. Careful tests show that persons who can taste PTC differ widely in their sensitivity to it: some can taste highly diluted solutions of the substance; others can recognize it only in strong solutions. Plots of the frequency distribution of tasters and their response to various concentrations may be either continuous, with two peaks, or discontinuous (Fig. 159). The two ends of such distributions may be regarded as corresponding to the genotypes *tt* (left) and *TT* and *Tt* (right), but it is not possible to decide exactly where to place the dividing line. The components of the variability in the taster reaction are manifold: there are racial differences; the same person will give different responses in different tests; responses differ in different age groups; and women are more sensitive tasters than men. In other words, the penetrance of one or all three *T* and *t* genotypes is variable, but details are still to be worked out.

Because of low penetrance, some abnormal genotypes may result in abnormal phenotypes so rarely that proof of their inherited nature may be impossible. Many abnormalities that occur singly in human pedigrees are

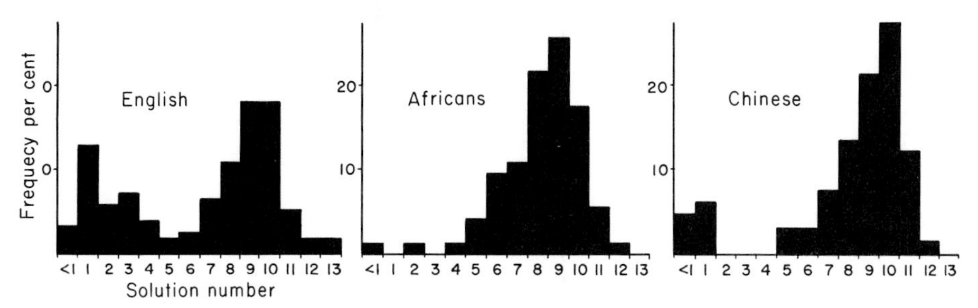

FIGURE 159
Distributions of taste thresholds for phenylthiocarbamide in 155 English males, 74 African Negroes, and 66 Chinese. The strongest solution (1) had a concentration of 0.13 per cent in water, the next (2) half this strength, and so on through 13. (Barnicot, *Ann. Eugen.*, **15,** 1950.)

probably genetically conditioned, although it seems strange to apply the term "inherited" to traits that were absent in all known ancestors and, in all likelihood, will not reappear in future generations.

On many occasions, students of human genetics have attempted to interpret the inheritance of certain traits in terms of single dominant or recessive genes even when the frequencies of affected individuals in successive generations or within sibships were much lower than demanded by the genetic hypothesis. In order to obtain agreement between observation and expectation, incomplete penetrance was invoked and its degree so chosen as to fit the data. Often this procedure proved valid, as in the traits discussed in the early pages of this chapter: stiff little finger, blue sclerotics, Huntington's chorea, and others. Also often, however, detailed analyses of the frequency of affected individuals among various kinds of relatives led to contradictions to the assumed degree of penetrance. It was then realized that deviations from single gene expectations could be caused not only by incomplete penetrance but also by a basically different type of inheritance called "polygenic." As will be shown in Chapter 18, many hereditary traits are the results not of single gene effects but of the cumulative effects of whole groups of genes. The appearance and disappearance of polygenic traits in successive generations often resembles the appearance and disappearance in pedigrees of dominant genes with incomplete penetrance.

Returning to single-gene inheritance with incomplete penetrance, there is a simple consideration which helps to decide whether dominance or recessiveness is the mode of inheritance of specific uncommon traits. In dominance, most affected children, *Dd,* will come from *Dd* × *dd* marriages, and the sibs will be *Dd* and *dd* in a ratio of 1:1. It follows that whatever the degree of penetrance, the frequency of affected sibs should be the same as that of affected parents in *Dd* × *dd* marriages. In recessiveness, most affected propositi *aa* will come from $Aa \times Aa$ marriages and the sibs will be AA, Aa, and aa. This means that the frequency of affected sibs should be higher than that of affected parents.

EXPRESSIVITY OF GENE EFFECTS

In a very large Norwegian pedigree compiled by Mohr and Wriedt, a short index finger (minor brachydactyly) was inherited in simple dominant fashion. However, there were two degrees of shortness of the second phalange: some individuals had a very short bone; but others, only a slightly shortened one (Fig. 160). It was shown that the short-fingeredness, regardless of the strength of its expression, was caused by the same dominant allele. Such production of different abnormal phenotypes by the same gene is called *variable expressivity*. The different degrees of expressivity may often form a continuous series grading from extreme expression to "no penetrance." *Penetrance* thus refers to the expression—presence or absence—of a gene, regardless of degree of expression; *expressivity* applies to the variability of the expression. The two terms were coined by the neurologist and biologist O. Vogt.

Variable expressivity is a very common attribute of genes, as demonstrated by innumerable examples among experimental animals and plants. However, it should not be assumed that similar phenotypes, different only in degree, are always due to variable expressivity of the same gene. Different genes at a variety of loci may result in similar phenotypes that are different in degree of expression. Similarly, multiple alleles of a single gene may differ in their degree of expression. Only when the expression of a rare trait varies in the

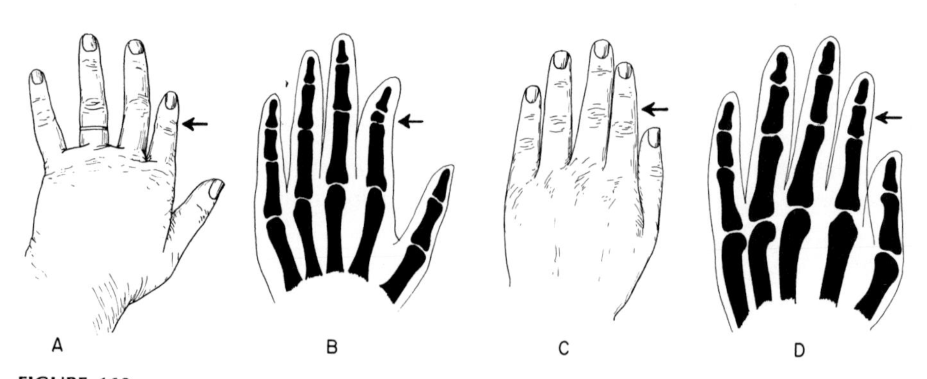

FIGURE 160
Minor brachydactyly of the index finger. **A, B.** Strong expression **C, D.** Weak expression. (After Mohr and Wriedt.)

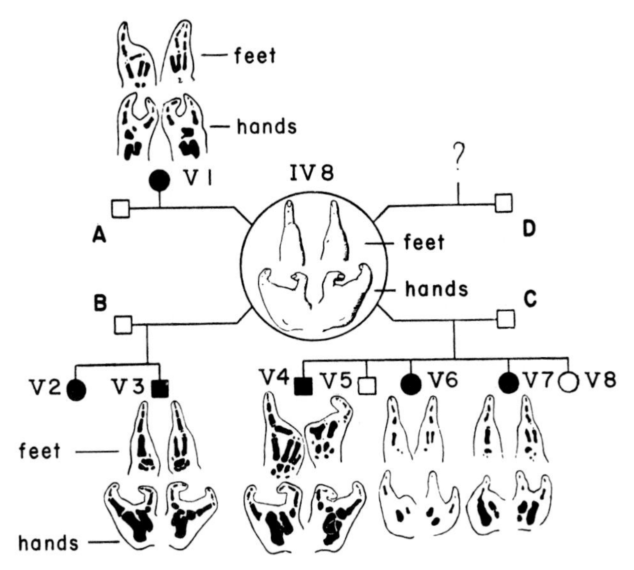

FIGURE 161
Pedigree of cleft hand including an affected woman and her offspring by three different fathers. The feet and hands, and associated skeletal structures, of affected individuals are shown. The feet have from one to about four metatarsals and one or two rows of phalanges. The large bones in the hands were probably formed by fusion of carpals and metacarpals. The skeletal structure of each extremity is unique. (Müller, after Ströer.)

same pedigree can we be reasonably certain that we are dealing with variable expressivity of a single gene.

A trait, cleft hand or lobster claw, which fulfills this requirement is illustrated in Figure 161. It is inherited as a dominant. The hands and feet of affected persons show severe defects of the skeletal system and abnormal external form. The six affected individuals about whom detailed information is available—i.e., the mother, IV–8, and five of her affected children from three different husbands, A, B, and C—all differ in degree of malformation.

Both incomplete penetrance and variable expressivity may be illustrated in a pedigree of polydactyly (Figs. 162, 163). Polydactyly occurs in many forms: sometimes an extra digit grows next to the thumb, sometimes one next to the little finger. The mode of inheritance varies from family to family. It includes families like the one shown in Fig. 163, in which a dominant gene (D) controls the numbers of bony rays formed in the embryonic buds of hands and feet. In normal dd genotypes five rays of metacarpals or metatarsals and phalangeal bones lead to the formation of the normal number of digits. The

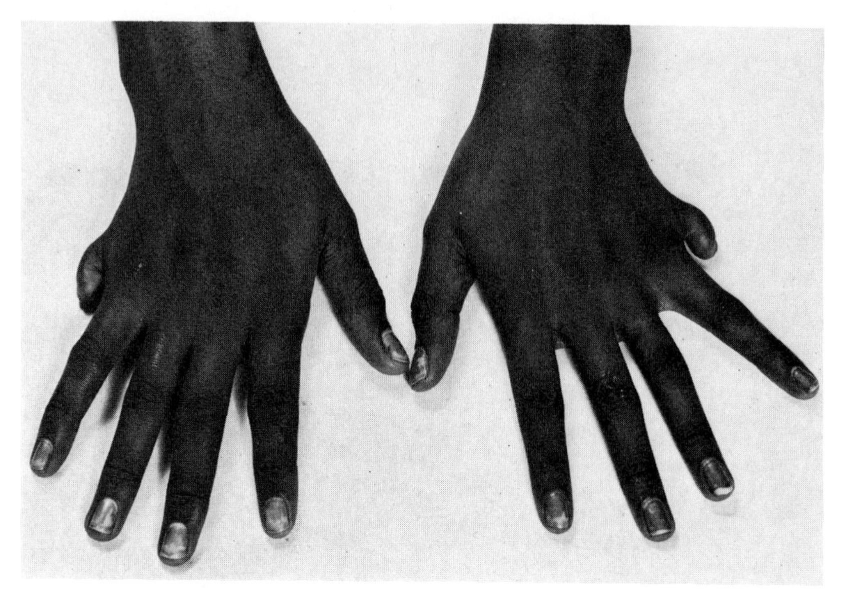

FIGURE 162
The hands of a polydactylous girl. This individual does not belong to the polydactylous kindred presented in Figure 163. (Original from Dr. V. McKusick.)

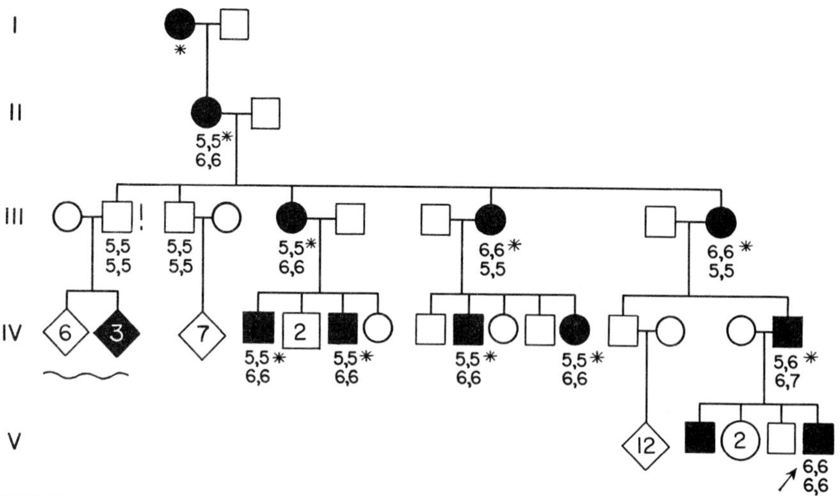

FIGURE 163
Selected individuals in a pedigree of polydactyly. In each group of four numbers, the upper two represent the numbers of fingers on the left and right hand, respectively, and the lower two the numbers of toes. An asterisk indicates that the type of polydactyly was not stated unequivocally in the original report. (After Lucas, *Guy's Hosp. Rpts.* 3rd ser. **25,** 1881.)

Dd genotype varies in its penetrance and expressivity in different extremities and in different individuals. Thus, in III–2 there was no penetrance at all, and in the last individual of generation IV no penetrance for one hand but penetrance with different degrees of expressivity in the other extremities: six fingers on the hand and six and seven toes, respectively, on the feet. If all four extremities are considered the pedigree shows four different degrees of expressivity: 5,5–6,6; 6,6–5,5; 5,6–6,7; and 6,6–6,6. Many more degrees would probably have been discerned had the kindred been described after, rather than before, the discovery of X-rays.

CAUSAL FACTORS IN VARIATIONS OF GENIC EXPRESSION

The type of expression of a gene is dependent on its interaction with the other genes of the two genic sets with which it is associated in the nucleus of the cell of a developing and aging individual, as well as on the cytoplasm and on the environment in which the development and aging proceed. The term environment is here used in its widest meaning to designate every factor, except the genic content of the individual, that may influence the phenotype. Environment thus consists of not only the external physical world which surrounds a human being from birth to death, but also the maternal body which encloses the developing embryo. Through the placental connection, the physiologic state of the mother and the hormones, minerals, and other substances circulating in her blood may influence the expression of the embryo's genes. Even the position of the embryo within the uterus, or the presence of a twin, is an aspect of the embryonic environment. Finally, environment includes the intellectual, emotional, and cultural atmosphere that is provided by family, school, church, nation, social class, and the historically changing "climate of opinion."

Penetrance and expressivity are thus not intrinsic properties of a given gene but results of its interaction with other genes and of nongenetic factors. In order to avoid misconceptions, some students of human genetics do not use the term penetrance and prefer to speak of "probability or rate of manifestation" of a gene or genotype.

Penetrance and expressivity as consequences of developmental events in gene action are interrelated. Many genes with a low degree of penetrance express themselves weakly when they are penetrant, and high penetrance and strong expressivity often go together. Harelip and cleft palate will be cited later as examples of these interrelations.

Internal Environmental Factors

Internal Environment. In many cases, it is clear that extragenic influences—i.e., environmental factors—are responsible for the variability of gene effects, even though it may not be possible to define these factors specifically. Examples of this can be seen in the cases of the stiff little finger, the cleft hand and feet, and the polydactyly already discussed in this chapter. One peculiarity of these conditions is the inequality of expression on the two hands or feet of an affected individual. Since we assume that all cells of the same individual contain the same genes, his two hands should be genetically identical. Therefore, we consider that differences in their phenotypes are caused by environment, although we cannot yet state what kind of environmental factors are at work. The determination of the characters of hands and feet occurs very early during development, and the influencing factors are probably intrinsic to the embryo itself. The individual, so to speak, provides an "internal environment" for the genes acting within him.

Multiple Effects of Blebs in the Mouse. Embryological studies of an inherited syndrome of abnormalities in the mouse by Bagg, Kristine Bonnevie (1872–1945), and others have shed considerable light on some aspects of penetrance and expressivity. In the syndrome, a recessive gene, *mb,* causes abnormal development of the eyes, ears, limbs, and coat, as well as a variety of other defects.

The variable effects on eyes and feet only will be considered. The eye defects may be completely absent, present in one eye only, or present bilaterally. If present, they may vary from slight atrophy of the eyelids to missing eyelids, and from slight to marked atrophy of the whole eye. The feet also vary from normal to such diverse phenotypes as syndactylism (joined toes), hypodactylism (absence of toes), and polydactylism (presence of extra toes) (Fig. 164). As with the eye effects, penetrance and expressivity may vary on the two sides of the body.

The bewildering multitude of phenotypes, their incomplete penetrance and variable expressivity, may be largely explained by a single developmental peculiarity of *mb mb* embryos. At a stage of development corresponding approximately to that of five-week-old human embryos, blisters (or blebs) filled with fluid appear under the epidermis. Sometime later, blood clots may form beneath the blebs and blood may escape into many of the blebs. The maldevelopment of eyes and feet is caused by the presence of blebs and blood clots (Fig. 165).

The essential point, so far as penetrance and expressivity are concerned, is that the location, size, and time of appearance (and later disappearance) of the blebs are variable from animal to animal and from one side to the other

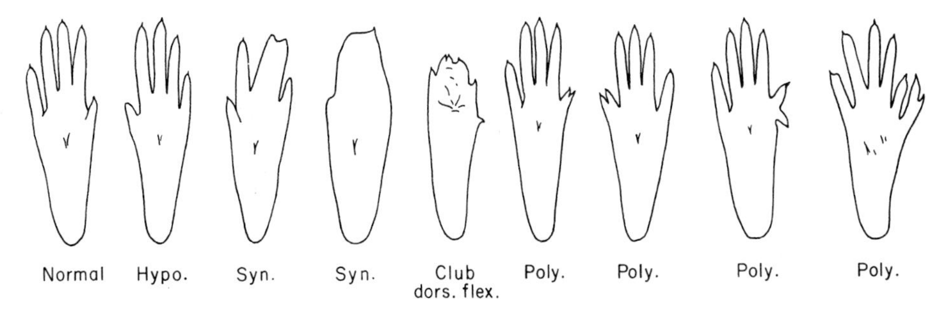

FIGURE 164
Myelencephalic blebs in mice. Normal hind foot (plantar view) and eight different defective types. (From Bagg, *Amer. J. Anat.*, **43,** 1929.)

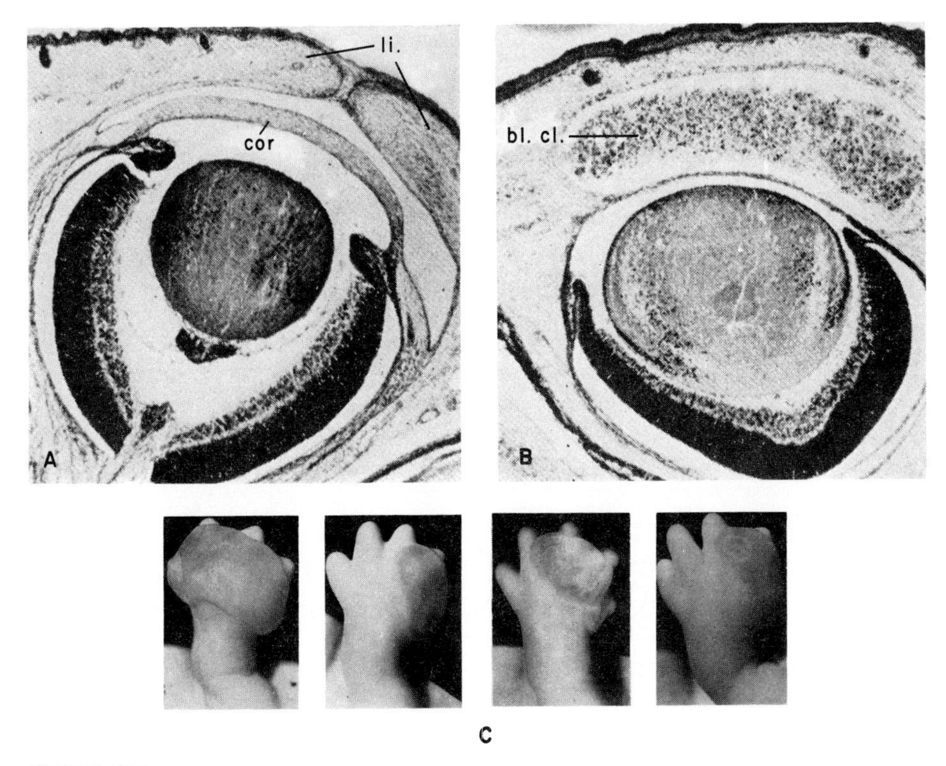

FIGURE 165
Myelencephalic blebs in mice. **A.** Normal eye in a 15-mm embryo; li = lid-forming tissue; cor = cornea. **B.** Abnormal eye in a 15-mm embryo. The eye is covered with a blood clot (bl. cl.). **C.** Abnormalities of hind feet in four embryos. Note the swollen appearance caused by large blebs. (Bonnevie, 1934.)

of a single animal. The points at which fluid accumulates under the skin seem to be largely determined by chance.

The variability of the blebs is the direct cause of variability in the later phenotype. If the blister happens to be near a developing eye or limb bud, the effect of *mb mb* will be penetrant and its degree of expression will depend on the size of the bleb, its nearness to the eye or limb area, and the developmental stage of the embryo at the time the bleb forms. If no bleb forms near an eye or limb area, normal development—i.e., no penetrance of *mb mb*—results.

The symbol *mb*, it may be remarked, was originally chosen to fit the theory that the blebs (b) were derived from cerebrospinal fluid that was extruded by a part of the brain, the myelencephalon (m), and migrated to the regions of the eyes, limbs, and other parts. This theory was abandoned when it was found that the blebs seem to originate directly under the skin in the positions where they are observed.

In Figure 166 a human pedigree is shown in which the variety of expressivity of a dominant gene resembles that of the *mb* gene in the mouse. The human gene produces abnormalities of the anterior limbs that include atrophy of one or both arms, absence of thumbs, and presence of two small thumbs on each hand or it may be nonpenetrant. The developmental mechanism of gene action is not known in this case but it is clear that the gene is not one "for absent thumbs" or "for double thumbs" but one "for disturbance of limb formation, including the thumbs."

Finger Pattern. A normal character that shows variable expression on homologous parts of the same individual is the pattern of finger ridges. Although each individual has a unique pattern, it is possible to group the many dif-

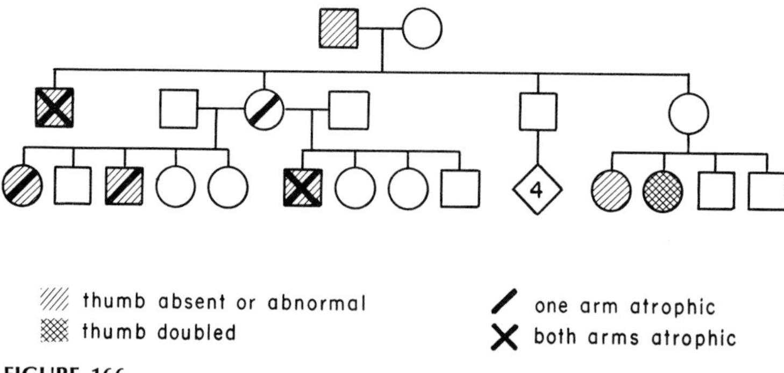

FIGURE 166
Abbreviated pedigree of a kindred in which diverse malformations of the arms and thumbs are transmitted by an incompletely penetrant, dominant, autosomal gene with variable expressivity. (After original data from Dr. Leonard Roth.)

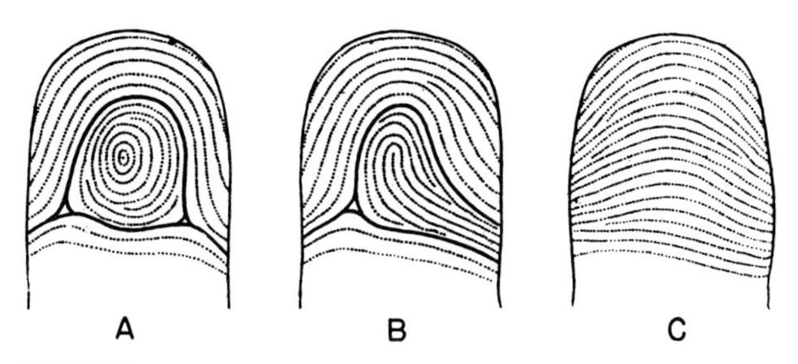

FIGURE 167
The three main types of finger pattern: **A.** Whorl. **B.** Loop. **C.** Arch. (Cummins and Midlo, *Finger Prints, Palms and Soles.* Copyright 1943 by the Blakiston Co. Used by permission of McGraw-Hill Book Co.)

ferent patterns in a few classes, three of which are shown in Figure 167. Individuals belonging to these different classes are genetically different, but the exact genotypes are not known. Superimposed on the genetic variability is a developmental, environmental one. In no person are the patterns of like fingers on the left and right hands identical in detail, even though they usually belong to the same general class. We have only very uncertain indications of the causes of this nongenetic variability. Some features of the finger pattern are possibly determined early in embryonic life by the arrangement of nerve fibers that grow from the spinal cord toward the surface of the finger rudiments. Although the growth and branching of these nerves, on the whole, is genetically determined, the details are sufficiently flexible to be subject to nongenetic influences. The growth of these nerves may be compared to the formation of the course of a small stream caused by a sudden downpour in the desert. The general outline of the course is determined by the arrangement and shape of mountains and valleys, but it is also dependent on small "accidental" conditions, such as the position of a rock or a tree, or previous weathering of the surface structure of the landscape.

Changing Phenotypes. Many phenotypes are fixed for life once embryonic processes have taken their course. The expression of other genetically determined characters is conditioned by the specific internal environment that the individual provides at a later stage. One of these characters is hair color. Although many very dark-haired persons are born with fully developed pigmentation, it is well known that others start with light hair that becomes darker as they grow older. The specific genotype of such individuals does not express itself by a static phenotype but, rather, by a phenotype dependent on time: or, stated differently, the genes for pigmentation express themselves differently in the bodily environment of a young person than in that of an older one.

Change of Dominance. This dependence of the phenotype on the age of the individual may give rise to the phenomenon called "change of dominance." Often, a child of a dark- and a light-haired parental pair will first show dominance of the genes derived from the light-haired parent. However, as he grows older, the genes from his dark-haired parent become dominant and his hair darkens. Similar changes in dominance may be observed in the phenotypes of other features, for example, shape of face or nose. Such traits are not fixed at birth but change all through life. Even if one assumes that genic action does not change, the action takes place in a different internal environment at different developmental stages, thus resulting in different phenotypes.

The picture of continuous, unchanging genic action in a changing environment is, of course, far too simple. A gene may be turned on or off at different times and it may be an agent which itself causes changes in its own environment. Any effect at a late stage of development may be due to an accumulation of the products of genic action in the preceding stages. In many cases, it may be impossible to define this relation with certainty. Is the onset of Huntington's chorea at the age of forty due to genic action that has not changed since birth, the difference between earlier health and later affliction resulting from changes in the nervous system, independent of the genotype *Ht ht*, that have made the nerve cells susceptible to damage? Or has the action of *Ht ht* itself caused continuous, slight damage, which, after forty years, is great enough to result in manifestation of the disease? Or does a regulator mechanism repress the action of the *Ht* allele in younger years but derepress the allele at later ages?

Changes at Puberty. In some instances the changes in internal environment that cause changes in the expression of a genotype can be specified. One of these is the change in the male voice that occurs at puberty. The voices of boys before puberty seem to be genetically determined. At puberty, each type of voice becomes lower. The changes as such are independent of genotype but dependent on the internal production of male sex hormones at this time. This is known from the fact that castrated boys, in whom no male hormones are secreted, retain their child-like voices, and from the fact that women who produce abnormally high amounts of male sex hormones—as those with diseases of the adrenal glands may do—develop deep, male voices.

External Environmental Factors

External Environment. If we pass to external environmental factors—i.e., those not located within the individual himself—that influence penetrance or expressivity, we enter the large field of the general interrelation of heredity and environment. We shall consider these problems at length later (Chaps. 25–27). Here only a few examples will be given.

Diabetes mellitus, the metabolic disease that leads to excretion of sugar in the urine, has a genetic basis with incomplete penetrance. A pair of genetically identical adult male twins shows how the same genotype may express itself differently according to different external circumstances. One of the twins was slightly diabetic; the other was healthy. However, a clinical test showed that the healthy brother reacted abnormally to a high sugar diet. The reason for the presence of diabetes in one twin and its near absence in the other was easily explained by their different modes of life. One had been a restaurant owner who had subjected his diabetically inclined body to an additional strain by heavy beer drinking, and thus made his genotype penetrant. The other twin, who had lived a more temperate life, had been spared the illness.

Differences in the frequency of diabetes have been observed in different populations. Some of these populations are from racially different groups, others from socioeconomic subgroups of racially similar people. The differences in the frequencies of the disease may be partly due to different frequencies of the responsible genotypes, but it is certain that they are also partly due to differences in penetrance controlled by factors in the external environment—poverty or wealth, type of occupation, quantity and quality of diet.

Differences in human stature have a genetic basis but also depend on the external environment. It is well known that the average height of many populations in the United States, Europe, and Japan has increased steadily during the past one hundred or more years. Such an increase does not seem to be due primarily to changes in genic composition (see p. 457): it must be largely due to alterations in the external environment. Undoubtedly, these external influences consist mainly of the nutritional advantages which large groups of mankind have enjoyed more recently. The same genotypes which gave certain heights during an earlier historical period now result in greater stature.

Right- and Left-sidedness. The striking separation of mankind into right-handed and left-handed individuals has been variously interpreted in terms of primarily genetic determination, of exclusively environmental determination, as well as in terms of interactions between genotypes and environment. Genetic hypotheses were based primarily on observations showing that the frequency of left-handed children is correlated with the handedness of their parents, two right-handed parents having the least number of left-handed offspring, two left-handed parents having the most, and parental couples comprising one of each type having an intermediate number. These data clearly point to transmission of handedness from parents to children but not necessarily to transmission by means of genes. Indeed, it has recently been shown by Collins that a genic interpretation is unable to account for a number of important facts. They refer to the proportions of handedness (1) in pairs of individuals with identical genotypes (i.e., identical twins), (2) in pairs of nonidentical twins and (3) in random pairs of sibs. It turns out that the proportions of pairs both partners of whom are right-handed, R-R, to those with one left-handed, R-L, to those with both left-handed, L-L, are the same in the three groups (1), (2), and (3) and follow completely chance expectation based on the proportion p:q of R:L individuals in each population of pairs. In other words, the identity in genotypes of identical twin partners and the nonidentity of nonidentical twins and of other sibs play no role in the determination of the handedness phenotypes.

There are some findings which seem to contradict the conclusion that genic differences are not involved in type of handedness. Rife has summarized and contributed studies which indicate that the frequencies of certain ridge patterns on the palms of right-handed individuals differ slightly, but significantly from those of left-handed individuals and that comparisons of the finger and palm prints of right handers with those of left handers reveal slight but significant trends towards greater bilateral symmetry among left handers. Since finger and palm patterns are in part genetically determined and are fixed from the beginning of their appearance in early fetal development, it would seem that handedness has a genetic component. Similarly, Strangmann-Koehler and Ludwig found a relation between handedness and the number and types of branches in the network of veins in the hands, the hand with preferred use having a more complex network value than the other hand of the person. Since the network of veins is said to be fixed prenatally, a relation between its complexity and handedness has been postulated. None of the examples of correlation between morphological traits and handedness seem fully substantiated yet.

Any theory of the determination of handedness must explain why left-handed parents have more left-handed children than right-handed ones. This is easily done if we adopt a primarily genic interpretation but requires different hypotheses if such an interpretation is ruled out. Two alternatives are the assumption of extrachromosomal biological influences and, the more likely, of parental behavioral influences. It is interesting that the frequency of left-handed offspring is higher when the mother is left-handed and the father right-handed than in the reciprocal mating, but this fact is compatible with either biological or cultural causation.

THE GENETIC BACKGROUND

In the preceding pages examples were given of environmental circumstances that result in different expressions of a given genotype. We have sometimes used the term genotype to refer to a genetic situation controlled by a single pair of alleles, as in cleft hands, and at other times to refer to a more complex case involving a group of loci, not known in detail, as exemplified by finger pattern or stature. In either case, the concept of a network of genetically controlled reactions leads to the conclusion that the rest of the genotype, that overwhelming majority other than the one or few genes under primary consideration, may have an effect on the expression of the trait. It is customary to speak of the specific gene or genes that are primarily responsible for the appearance of a trait as the *main* gene or genes, and of the rest as their *genetic background.* Since the genetic background of the main gene differs from individual to individual, we can expect that individuals alike for a given assembly of main genes may often show differences in the manifestation of the trait that they control.

Whenever gene effects are variable, it may be assumed that both environment and genetic background have important influences on penetrance and expressivity. In experimental animals or plants it is not too difficult to separate these genetic and environmental factors. The simplest experimental procedures are (1) to stabilize the genotype—main gene or genes, plus genetic background—and to investigate the influence of different environments, and (2) to keep the environment constant and vary the genetic background. Stabilization of the genotype is achieved by establishing a strain of experimental organisms as alike as possible in their total genetic constitution.

Isogenicity. A group of individuals that are identical genetically is called *isogenic.* In general, the ideal isogenic strain or population is one in which all individuals are homozygous for the same alleles at all loci. The term isogenic is, however, applicable also to a population of heterozygotes in which all individuals are genetically alike.

One of the most important ways of creating an isogenic strain is *close inbreeding*, repeated for numerous generations. Individuals of common ancestry share many alleles. Matings between sibs, therefore, provide an opportunity for the production of individuals homozygous for the same alleles, at many loci. After many generations of inbreeding, the probability of isogeneity at any one locus becomes very high. It is true, however, that there remains an appreciable probability that some loci may not have become isogenic and that complete stabilization of the genotype of a strain is only approached asymptotically.

The success of inbreeding for isogeneity can be judged in various ways. One method consists of evaluating phenotypic variability. When the establishment of an isogenic strain is begun the heterozygosity of many loci is expressed in a wide range of differences among sibs and between parents and offspring. The more isogenic the strain becomes, the more all individuals resemble each other.

Another test of high isogeneity is the ability of tissues transplanted from one animal to another to survive. In mammals, as we have seen earlier, transplants generally do not persist in the host if the donor is genetically different. In an isogenic strain, the tissues of one animal are genetically like those of any other, and no donor-host incompatibilities are encountered.

Much of our knowledge of human inheritance has to be based on the more accurately determinable facts derived from studies of laboratory animals. This is why long-inbred strains of mice, rats, guinea pigs, and various other organisms have been established in certain laboratories with considerable cost and made available to other investigators. In order that their results may be comparable, students in the United States, in England, on the Continent, and elsewhere do not work with any available nonstandardized mouse (formerly often referred to as "the" mouse), but with such animals as the "dbA" mouse from the Bar Harbor Laboratory in Maine or the inbred "Wistar" rat from the Wistar Institute in Philadelphia. The importance of such standardization may be illustrated by a report that showed that treatment of four different strains of mice with cancer-producing chemicals gave the following widely divergent percentages of animals showing effects of the treatment during a given period: strain I, 88 per cent; strain II, 48 per cent; strain III, 34 per cent; and strain IV, 15 per cent.

In man, isogenic individuals exist only in pairs or in small multiples. Identical twins, triplets, etc., are derived from single eggs, and the partners in such multiple births are therefore isogenic among themselves. The isogeneity, of course, is for genotypes which are homo- or heterozygous at many loci. The five Dionne sisters were the largest isogenic group of humans ever to survive earliest infancy.

Genetic Modifiers

In the study of the genetics of experimental organisms, so-called modifying genes have been encountered. In *Drosophila*, the typical wild-type variety has a group of four bristles on the back of its thorax, while the variety "Dichaete" has fewer. The name of this variety suggests that only two (di-) out of the four bristles (chaetae) are present, but this was true only for the original Dichaete fly from which the strain was derived. A detailed analysis has shown that the condition is due to a dominant gene (*D*) present heterozygously (it is lethal when homozygous). The specific expression of the heterozygous genotype *Dd* may vary from apparent normality—that is, presence of four bristles—to absence of all four.

This variability is to some degree due to controllable environmental conditions. Dichaetes which develop at high temperatures or under crowded conditions have fewer bristles than those which develop at low temperatures or with less larval crowding. Part of the variability is uncontrollable and resembles the differences in expression of polydactyly on the hands of the same person. Even in highly constant environments, left and right halves of the same fly will often have different bristle patterns; there may, for instance, be two bristles on the left side and one on the right.

Finally, different isogenic strains of heterozygous Dichaetes have shown differences dependent on the genetic background. The average bristle number of *Dd* flies varies significantly from strain to strain even if the environment has been standardized for all of them. Genetic analysis has shown that in these strains different genes, which do not affect the four bristles of the normal type, are powerful determinants of the penetrance and expressivity of the *D* gene. The genetic background of two strains may differ only in a single modifying gene—so they may have different alleles at only one modifying locus—or they may differ in a few or many loci. Such different loci may have similar or different effects—some increasing penetrance or expressivity, others decreasing them. It is often difficult, if not impossible, to disentangle, by genetic analysis, all the numerous "plus and minus modifiers" which can shift the expression of a main gene toward or away from an extreme phenotype.

The full nature of the interaction between *D*, the main gene, and the other modifying genes is not known. It has been speculated that the formation of bristles may require the presence of a substance at some decisive stage in development and that the normal *d* allele assures the production of a large amount of this substance, but that *D* so affects its production that the amount is usually insufficient for formation of four bristles. Within this hypothesis, modifying genes are considered to cause a slight increase or decrease in the amount of the critical substance. This slight variability is of no phenotypic consequence in normal, *dd* individuals because of the large amount of the substance they form, but it may have striking phenotypic expression if the presence of *D* lowers the amount of the substance to a level close to the minimum for bristle formation. Presence of plus modifiers may assure enough of the substance to form bristles, and presence of minus modifiers may tend to keep the substance below the minimum level. Alternatively, or in addition to this hypothesis, plus modifiers may lower the threshold level that permits a minimum of bristle formation and minus modifiers may raise the level. It is immaterial in our context how true these specific hypotheses of the effect of modifying factors are. In any case, they provide a useful picture of their action.

Minor Brachydactyly. An example of a modifying allele in man is provided by minor brachydactyly in a Norwegian kindred (Figs. 160, 168). As described earlier, this character has two clearly different forms of expression: one in which the second finger is very short; and another in which only careful inspection of the hands reveals the slightly expressed brachydactyly.

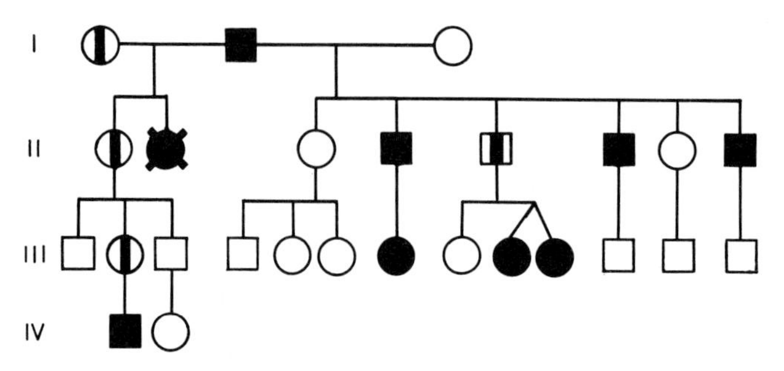

FIGURE 168
Pedigree of minor brachydactyly of the index finger. Solid symbols = strongly affected; vertical bar in symbols = weakly affected; diagonally crossed symbol of II-2 = sublethal crippled child, presumably homozygous for the allele *B* for brachydactyly. (After Mohr and Wriedt.)

Disregarding for a moment the variability in expression, the pedigree shows that the trait follows simple, dominant inheritance. Since the trait is rare, most affected individuals are heterozygous for its gene, *Bb*. It is possible to account for the trait's two sharply different phenotypes by assuming the presence of a single modifying locus at which there may be either of two common alleles, one dominant over the other in the heterozygote. If, for instance, the dominant allele *M* is necessary for full expression of brachydactyly, then all strongly affected individuals have the genotype *Bb MM* or *Bb Mm*, and the slightly affected ones have *Bb mm*. On this basis, the affected persons in the pedigree may be assigned their appropriate genetic constitutions.

The normal individuals also belong to different genotypes, namely, *bb MM*, *bb Mm*, and *bb mm*. They cannot be distinguished phenotypically, but the affected offspring of normal parents sometimes enable the geneticist to judge the parents' constitutions. The normal husband of II–1, for instance, could not have been *bb MM*, since if he had, there would have been no offspring with weak expression of *B*, such as III–2. The data are not sufficient, however, to enable us to distinguish between the possible genotypes *bb Mm* and *bb mm* for this husband.

An examination of the pedigree will show that its peculiarities may be explained equally well by the assumption of a homozygous recessive allele *n*, a modifier responsible for extreme expression, and a dominant allele *N* as the agent of slight expression. If this hypothesis were true, then the strongly brachydactylous individuals would be *Bb nn*, and the slightly affected ones would be either *Bb Nn* or *Bb NN*. Normal persons, again, would comprise three indistinguishable types, namely, *bb NN*, *bb Nn*, and *bb nn*.

It is very difficult to determine the modifiers that constitute the genetic background in man. Therefore, recognizing the presence of modifiers is of no positive help in predicting the specific type of a trait that will appear in individuals of future generations. Rather, it is important to be aware of the complications that can be caused by possible modifiers, both as a deterrent to rushing to conclusions and as a concept that helps us understand variability in inheritance.

Favism. We have earlier seen that the X-linked deficiency for glucose-6-phosphate dehydrogenase often leads to a hemolytic disease when the deficient individuals are exposed to fava beans or to certain drugs (p. 314). Not all deficient persons, however, become ill. Affected individuals show a familial aggregation that seems not to be due to biases in the collection of data or to intrafamilial similarities in exposure but suggests a genetic predisposition to favism in some deficient persons. The data fit the hypothesis that an autosomal gene enhances the susceptibility to favism of G-6-PD deficient individuals. The susceptibility is often transmitted from one generation to the next. This can be explained either by assuming that the autosomal enhancer is a dominant or, if a recessive, that many homozygous recessive individuals have heterozygous mates.

Cataract. In Chapter 9, reference was made to an extensive pedigree of early cataract (Fig. 111). It was shown that the trait followed the rules of simple dominant inheritance. It may now be added that the expressivity of the condition was extremely variable. Examinations of the eye lenses of different affected individuals showed three general types of cataract, each of which varied and changed with age. The two most important categories of cataract were that in which opacities developed mainly in the anterior region of the center of the lens and that in which the opacities were in the interior of the lens. Both eyes of the same person were always alike. The different modes of expressivity are significant for the affected individuals, since the interior type causes more interference with vision, and at an earlier age, than the anterior type.

There is evidence that the type of expressivity is inherited. The anterior type of cataract was present in one woman, her 2 affected sons, and 1 affected grandchild. The interior type of cataract was found in a niece of this woman, 5 out of 6 of the niece's affected children, and all 5 affected grandchildren. (One of the 6 affected children had a feather-like lens opacity unlike either of the two categories described.)

It thus seems that the dominant gene responsible for the development of the juvenile cataracts determines the occurrence of the affliction without specifying the particular type of opacity. The variations in type are probably due to simple modifiers.

Cleft Hand. We may return here to the striking abnormality cleft hand, which was used earlier as an illustration of differences in expressivity in the hands and feet of the same person. Such differences were clearly environmental in origin and uncontrollable. It was also pointed out that the trait showed different degrees of expressivity in different affected individuals. In the remarkable pedigree (Fig. 161) that lists an affected woman who had normal and affected children by three different normal husbands, a greater variation is found among the different children than in each child. The daughter V−6 has the fewest bones in hands and feet; her sister V−7 has the next greatest deviation from normality; their brother V−4 has the largest number of bones; and their half-sibs V−1 and V−3 are somewhere in between. This greater variability among sibs as compared with the lesser variability of hands or feet in a single person may have been brought about partly because the differences in the environments of embryos that develop at different times are greater than those on the two sides of the same embryo. It is probably also true that at least part of the variability among sibs is due to differences in their genetic backgrounds. Even though some have the same parents and all have one parent in common, normal heterozygosity will produce a great variety of different genotypes among them.

Different Types of Genic Interaction. Genetic modifiers are usually thought of as producing phenotypic effects not by themselves but as modifying traits dependent on a specific allele of a main gene. Although this view is useful as a first approximation, it is more likely that a modifying genotype, *MM* or *Mm*, will have effects different from those of *mm* not only when there is a specific allele at the main locus, but in general. Thus, it is possible that the *M* and *m* modifiers of the minor brachydactyly genotype *Bb* influence finger size in normal *bb* individuals, although the differences may be too small to be noticed. Or, even if *M* and *m* do not affect the size of normal fingers, a biochemical study of *MM*, *Mm*, and *mm* persons might reveal differences.

Modification of genic effects by other genes may be of many kinds. In *Drosophila*, as we have seen, the gene *D* may reduce the number of certain bristles on the back of the thorax, but only of these bristles. Another dominant gene, *H*, reduces the number of certain other bristles. A fly which carries both *D* and *H* lacks not only the bristles affected by the two different genes but also other bristles not affected by either *D* or *H* alone. In man, an analogous example is that of the interaction of the separate genes for thalassemia and sickle-cell anemia. Heterozygotes with either one or the other abnormal gene are fundamentally healthy, though tests of their blood cells demonstrate the effect of the gene. In an individual heterozygous for both genes they may so interact as to produce a disease difficult to distinguish from that caused by homozygosity for the sickle-cell gene alone.

Suppressor Genes. The interactions of human genes at different loci are rarely understood in detail but in some cases the biochemical analysis has progressed sufficiently to provide specific insights. This is particularly true for so-called suppressor genes—i.e., genes that inhibit the reactions controlled by other genes. The oldest example in man concerns the secretor alleles *Se* and *se* (p. 263). In the presence of *Se* in homo- or heterozygous state, the blood-group genes *H*, I^A and I^B do not only result in the occurrence of the respective antigens on the red cells but the antigens are also secreted into the body fluids including the saliva. When *Se* is absent—the genotype being *se se*—secretion does not occur. It is believed that *Se* controls the expression of the *H* gene in the biochemical pathway which leads from precursor substances to H, A, and B substances in the saliva (Fig. 32 p. 55). As described earlier the gene *Se* also interacts in a specific way with the gene *Le* of the Lewis blood system (pp. 276–277).

Another striking suppressor gene is responsible for the very rare "Bombay" phenotype discovered by Bendhe and associates. It is characterized by absence of the A, B, and H antigens on red cells and in saliva. Formerly given the symbol *x*, the gene is now recognized as a recessive allele *h* at the *H* locus. The great majority of people are *HH* or *Hh* and thus able to convert the ABO precursor substance into the H substance, which, in turn, may be converted into A or B. In contrast *hh* persons cannot form H and, consequently, neither A nor B. Thus, *hh* individuals who, lacking A and B, appear to belong to blood group O may actually carry I^A and I^B alleles and transmit them to their *Hh* offspring. There they become expressed phenotypically.

It may be instructive to describe a specific family in which the *hh* genotype was present. An O woman married to an A_1 man gave birth to an A_1B child. The unexpected appearance of both A and B antigens in the child of an O mother cannot be explained by illegitimacy since no regular genotype of a father could cause this. Detailed blood tests showed that the mother possessed not only the anti-A and anti-B antibodies normally present in O individuals but also the anti-H antibody, which is not found in typical O persons. She had older identical twin sisters who had the same unusual blood properties. The appearance of the B antigen in the A_1B child of the woman with antibodies against A, B, and H led to the following interpretation. The woman herself must have carried the I^B allele in spite of lacking the B antigen. Her father was a typical O, her mother B. Both these parents, who were cousins, carried heterozygously a recessive gene *h* which became homozygous in the I^BI^O woman. The *hh* genotype suppressed B antigen formation, but not that of anti-H antibody: it was also shown to suppress the action of the dominant secretor gene. Phenotypically, both parents of the A_1B child were nonsecretors and would therefore normally have been *se se*. The A_1B child was a nonsecretor, as expected, but a younger sister, of true O type, was a secretor. The genotypes of the two children and their parents thus were as follows:

Parents (P) $I^{B}I^{O}$; *hh*; *Se se* ♀ × $I^{A_1}I^{O}$; *HH*; *se se* ♂

Offspring (F_1) (1) $I^{A_1}I^{B}$; *Hh*; *se se* (2) $I^{O}I^{O}$; *Hh*; *Se se*

The genotype *hh* is a very rare one and its failure to suppress the formation of anti-H serves to distinguish *hh* individuals from true O individuals.

Still another very rare recessive suppressor gene *y* causes in red cells a block of the metabolic step from H to A so that persons of the genotype *yy* I^{A}– lack the A antigen in their red cells (though not in saliva). The actions of genes I^{B} and *H* are not suppressed by *yy*.

It is clear that suppressor genes can produce genetic situations which may seem contrary to established facts. In reality, suppressor genes have contributed greatly to an understanding of gene-controlled metabolic pathways, and therewith of their suppressor actions. In other organisms than man, types of suppressor genes are known that are characterized by changes in the polynucleotide sequence of the suppressed genes themselves or that affect the basic processes of transcription of DNA to RNA and the latter's translation into polypeptide chains.

Interaction Between Alleles. The genetic background of a main gene or genes is usually defined in terms of genes at other loci. In its widest sense the background may also be regarded as including, in heterozygotes, the allele with which a specific gene is paired. Possible examples of this type of interaction have been discussed earlier. Dominant genes for autosomal muscular dystrophy and for the nail-patella syndrome seem to be differently expressed in heterozygotes with different normal isoalleles (p. 282). One other example of allelic interaction was also described earlier (p. 274). The R^1 allele, homozygously or paired with *r*, causes a typically strong reaction with anti-Rh_0; but when it is paired with *r′*, only a weak reaction takes place. The expressivity of the R^1 allele is thus dependent on the type of allele with which it is combined in a heterozygote.

Sex-limited Traits

A very important part of the genetic background is the sexual constitution. Although penetrance and expressivity of many genes are alike in males and females, there is a great difference in the action of others in the two sexes, even though the genes in question are not themselves concerned with the determination of sex. Some such genes are *sex limited* in their effect; that is, they are expressed phenotypically in one sex only. In strict sex limitation, we may formulate the situation as follows: If S and s are two alleles with sex-limited expression, the three genotypes, SS, Ss, and ss, are indistinguishable in one sex but give rise to two or three different phenotypes in the other sex, depending on whether the heterozygote is like one of the homozygotes or different from both. Knowledge of sex limitation is of great practical importance in breeding dairy cattle. It is known that the yield and quality of milk are under the control of many genes, and that the genes are, on the whole, contributed equally by both parents. Since these genes are located in the autosomes, bulls and cows may have identical milk-yield genes. Obviously, the phenotype "milk yield and quality" is limited in expression to females, although, experimentally, it could be made to appear in males. No such experiments have been attempted in cattle, but male guinea pigs have secreted milk after they had been deprived of their testes, which produce the male hormones, and had been treated with a female sex hormone. The type and quantity of this milk would undoubtedly have shown indications of hereditary differences.

In man, sex-limited expression of genes occurs in uterine and prostate cancer. From general considerations it also seems certain that yield and quantity of milk of the human female are influenced by genes present in both sexes. Very likely, other genes contributed by both parents control various anatomical and physiological properties of the female sex, such as width of pelvis or age of onset of menstruation. Similarly, sex-limited male characters, such as type of beard growth or amount and distribution of body hair, probably depend on genes common to both sexes. This is indicated by the results of interracial mating. Caucasoid males, on the average, are more hairy than Mongoloids, but the sons from marriages between members of the two races are often intermediate in hairiness—indicating that the mother has contributed genes which find their expression in the male sex only.

It must be emphasized that sex limitation is not the same as sex linkage. The latter term refers to the localization of genes in a sex chromosome; the former, to the developmental expression of the genes in only one of the two sexes. Most sex-limited genes are autosomal, but a few are known which are X-linked: both females and males may carry these genes, and simple inspection can show their sex-limited effect. It would be more difficult to judge the situation if the existence of Y-linked genes were established. Such genes would be linked with the determiner of maleness, which is itself located in the Y-chromosome. Normally, therefore, it would not be possible to decide whether a Y-linked gene is limited in expression to the male because it is Y-linked, or whether it is developmentally sex limited regardless of its chromosomal location.

Sex-controlled Traits

Sex limitation is only the extreme example of control of the expression of certain genotypes by sex. When a genotype is expressed in both sexes but in a different manner in each, we speak of *sex-controlled*, or *sex-modified*, genic expression. The earliest study of a sex-controlled character in man is that of Bernstein on the inheritance of singing voices in adult Europeans. A single pair of alleles was regarded as responsible for the six different singing voices: bass, baritone, tenor, soprano, mezzo-soprano, and alto. Strangely enough, the low bass voice in males and the high soprano in females seemed to be determined by the same genotype, A^1A^1, and the high tenor in males and the low alto in females by A^2A^2, the heterozygote A^1A^2 leading to baritone and mezzo-soprano. However, Bernstein's specific interpretation has not withstood the test of time. While studies of twins, pedigrees, and populations point to genetic factors in the determination of singing voices, many independently variable properties are involved in type of voice. Some of these are indeed controlled by the sexual constitution of the individual: the development of the voice box in the divergent male and female direction of the adult takes place at puberty under the influence of the sex hormones. However, a single pair of alleles with simple expression in the two sexes is not sufficient to explain the facts, which still await detailed analysis.

A somewhat different type of sex control affects the penetrance and expressivity of many genotypes. Harelip and cleft palate are developmental abnormalities with a genetic basis. Penetrance is incomplete (Fig. 169), and expressivity varies from very slight external clefts to very severe clefts of the soft, and even the hard, palate. Sex control is apparent in the fact that penetrance is higher in males than in females (about 60 per cent of affected individuals are males), and that severe types of expression occur more frequently in males.

Even more extreme is the sex control of penetrance of gout. This condition is based on an excess of uric acid in the blood (hyperuricemia). Its genetic basis had been described as that of a dominant autosomal gene and its penetrance been estimated as more than 80 per cent in males but less than 12 per cent in females, resulting in the many more observed affected men than women. If the single-gene hypothesis should not be valid and hyperuricemia should be the result of polygenic determination, the different frequencies of affected men and women would have to be reinterpreted in terms of sex-controlled threshold phenomena as will be discussed in Chapter 18.

A greater frequency of affected males than females is typical for many abnormal traits other than gout which are under the control of autosomal genes. As shown earlier (pp. 327–329), a study of four such traits proved that the higher incidence of affected males is not due to X-linked modifying genes,

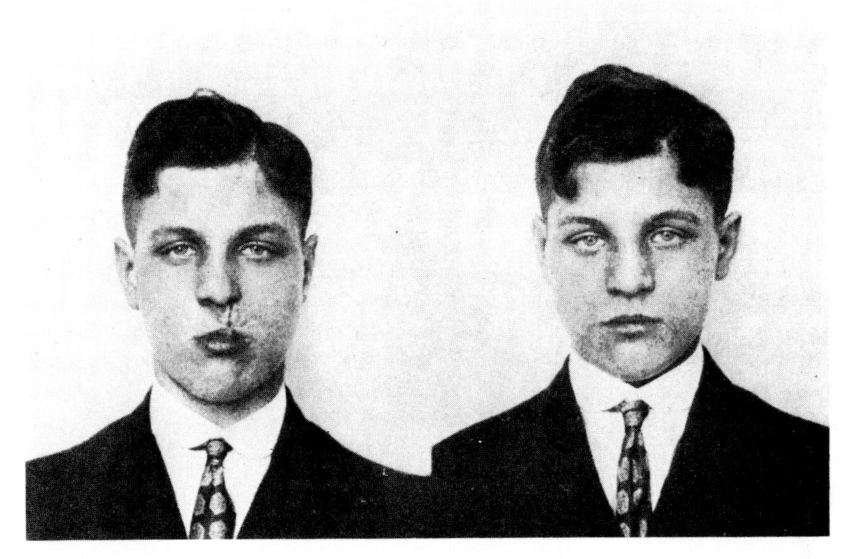

FIGURE 169
A pair of identical twins. Penetrance of harelip at left, lack of penetrance at right. (Claussen, *Zeitschr. Abst. Vererb.*, **76,** 1939.)

but to developmental control of phenotypic expression. This is probably true of the majority of traits that are more common in one sex than the other.

Sex control may also lead to higher frequencies of affected females. Some congenital malformations such as anencephaly (absence of the brain) or spina bifida (cleft vertebral column) are more often found in female than male zygotes. Females affected with congenital dislocation of the hip are six times as frequent as males. Apparently the penetrance of the genotype for dislocation depends on the normal difference in the shape of the pelvis of the two sexes.

Certain seeming examples of sex control do not actually belong in this category. These are the cases in which an X-linked recessive leads to death or infertility of hemizygous males but heterozygous females are able to survive and reproduce. The survival or fertility of the females may simply be due to their having one normal allele in addition to the abnormal ones; the affected male has only the abnormal allele. A fair comparison would match affected males with homozygous abnormal females: but, since affected males are unable to reproduce, such females do not exist. Special examples of such "pseudo sex control" have been discussed under X-linkage (oral-facial-digital syndrome; incontinentia pigmenti).

Sex control of the expression of alternative genotypes sometimes affects the dominance relation of two alleles. This has been thoroughly analyzed in crosses between breeds of sheep. In some breeds, both sexes have horns, but in others both males and females are hornless. A pair of autosomal alleles, H^1 and H^2, is responsible for this difference, the horned breeds being homozygous H^1H^1, and the hornless H^2H^2. All hybrid males are horned, but all hybrid females hornless. Since their genotypes are alike, H^1H^2, it follows that the allele for horns H^1, is dominant in males, but recessive in females: or, if we view it differently, that the allele for absence of horns, H^2, is recessive in males but dominant in females. In this case the male and female hormones in the growing animals provide two different internal environments, which interact with the H^1H^2 genotype and control its expression.

Baldness. Sex-controlled dominance has been suggested as an explanation of the pattern of inheritance of baldness in man. The investigation of this, as of many other human characters, is made difficult by the fact that the phenotype "baldness" varies greatly. Baldness may be slight or extreme, occur first on the crown or on the forehead, and appear early or late in life; moreover, some baldness seems to be due to specific abnormalities in thyroid metabolism or to infectious diseases. But most types of typical "pattern baldness" occur in healthy individuals, and many pedigrees exhibit a succession of affected individuals or show numerous affected sibs and other relatives (Fig. 170). A hereditary basis for baldness is therefore probable.

Both sexes may be affected, but the high relative frequency of affected males is notable. Pedigree studies show that there is no sex linkage. As an alternative, it has been suggested that an allele B^1 in homozygous state, B^1B^1, permits normal adult hair growth in both men and women; that the homozygotes B^2B^2 are bald regardless of sex; and that male B^1B^2 heterozygotes are bald, but females are not.

Like hornedness in sheep, hair growth in man is controlled by sex hormones. Few eunuchs become bald, but women with a tumor of the adrenal cortex, which results in a high production of male-type sex hormones, not only may develop such typical male-limited traits as beards and mustaches but also

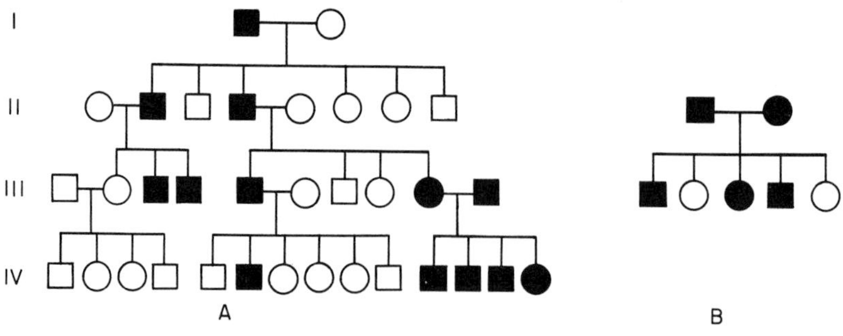

FIGURE 170
Pedigrees of baldness. (After Snyder and Yingling, *Hum. Biol.*, 7, 1935.)

may become bald. Upon removal of the tumor, both the sex-limited and sex-controlled traits disappear. According to Conway Zirkle, the sex control of baldness was known to Hippocrates and Aristotle. "Eunuchs neither get gout nor grow bald," noted the former; and "No boy ever gets bald, no woman, and no castrated man," wrote the latter.

Snyder and Yingling tested the hypothesis that baldness is governed by a single pair of alleles whose expression in heterozygotes is sex controlled, by means of the allele-frequency method. The theory designates men who are not bald as homozygous B^1B^1. Consequently, as with any autosomal gene in a random-mating population, the gene frequency of B^1 is obtained by extracting the square root of the frequency of B^1B^1 individuals:

$$p_{B_1} = \sqrt{\text{nonbald men.}} \qquad (1)$$

Conversely, the theory designates bald women as homozygous B^2B^2. It follows, therefore that

$$q_{B_2} = \sqrt{\text{bald women.}} \qquad (2)$$

Since the sum of the two frequencies p and q is unity, the theory predicts

$$\sqrt{\text{nonbald men}} + \sqrt{\text{bald women}} = 1. \qquad (3)$$

How do the observed data fit this expectation? Approximately 40 per cent of all men thirty-five years old and older in the general population were found to be bald. The criterion by which a person was judged to be bald must have included very slight degrees of thinness of hair. Given the percentage of bald men, the theory should predict the percentage of bald women. In order to obtain data on the frequency of baldness in women, it was necessary to determine the presence or absence of the trait in a fairly large sample of the population. This was difficult to accomplish in a general population so that data on the inmates of state mental hospitals in Ohio and Illinois were used. The frequency of bald women, based on a total of 1,883, was found to be 7.75 per cent (0.0775). This is regarded as a typical frequency, not in any way correlated with the mental condition of the women, since it was found that the frequency of bald men who were inmates of the same hospitals was not significantly different from that of mentally normal men in the general population. The specific percentage for the institutionalized men was 42.96 per cent, based on a total of nearly 4,000. The frequency of nonbald men, therefore, was 0.5704. Using this figure for nonbald men and 0.0775 for bald women, we obtain, by means of equations (1) and (2),

$$p_{B_1} = 0.755$$
$$q_{B_2} = 0.278$$

and testing equation (3), we find that

$$\sqrt{\text{nonbald men}} + \sqrt{\text{bald women}} = 1.033.$$

This is very close to the value of 1.000 required by the theory.

In spite of this agreement between expectation and observed data, some doubts regarding the correctness of the theory remain. The proof would be more convincing if different populations with different frequencies of baldness could be shown to conform to the theoretical requirements given in equation (3). Such multiple tests of the relevant theories were provided for various blood groups. It should likewise be possible to extend the test of the theory for baldness to various ethnic groups in which the incidence of the trait is known to be lower than in Snyder and Yingling's material.

There are recorded exceptions to the theory, such as a bald woman whose constitution, as judged by her parent's phenotype, must have been B^1B^2 and not B^2B^2, and men and women with very thin hair in a pedigree containing typical baldness who may have carried B^2 but did not show its typical effect.

The most serious objections to the theory are that the phenotypic class, baldness, seems to include several differently inherited types of lack of hair and that very little is known about baldness in women. Harris collected data that he divided into two groups: one of persons in whom baldness sets in before the age of thirty and is strong before forty; and another of persons in whom it begins later. Genetic examination shows that the two groups are hereditarily different. There is a suggestion that premature baldness may be due to an allele which is dominant in heterozygous men but not in heterozygous women, and is so rare that homozygous women are not known. It has also been suggested that no hormonally typical women are bald and that baldness is not really sex controlled but sex limited. According to this view, baldness in women would be a result of the interaction of a genotype normally unexpressed in females with a male-like hormonal condition. Further investigations are needed to clarify the genetics of baldness.

ANTICIPATION

We have seen that the age of onset of Huntington's chorea varies from one affected individual to another. Such variability is typical of many inherited diseases whose symptoms appear late in life. An opinion widely held among medical men—and some statistics seem to support it—is that the age of onset of these diseases becomes earlier and earlier in successive generations (Table 45). The phenomenon is call *anticipation*, or *antedating*. Furthermore, it is held that diseases whose severity varies increase in severity from

TABLE 45
Age of Onset of Various Hereditary Diseases.

Disease	*No. of Parent–Child Pairs*	*Age of Onset, Mean Values (in Years)*		
		Parent	Child	Difference
Peroneal atrophy (dominant)	86	24.30	19.36	4.94
Muscular dystrophy (dominant)	90	27.44	21.00	6.44
Hereditary glaucoma	113	42.08	30.66	11.42
Huntington's chorea	153	40.80	31.98	8.82
Diabetes mellitus	216	60.29	43.06	17.22
Mental illness (all diagnoses)	1,728	50.50	34.20	16.30
Dystrophia myotonica	51	38.48	15.24	23.24

Source: Penrose.

one generation to the next. This presumed phenomenon might exist independently of anticipation or could be a consequence of it, since many of these diseases entail progressive degeneration of the affected organs, and a disease that begins early in life will have more time to run its course than a disease with later onset.

The concept of anticipation does not readily fit in the system of genetic facts and interpretations that has proven so fruitful in the study of man and of a variety of experimental organisms. Geneticists have therefore carefully analyzed the data which suggest anticipation. That there are pedigrees which show an earlier and more severe onset of a disease in a younger generation than in a preceding one is not surprising. If the age of onset varies, one should expect such pedigrees, just as one should expect others in which onset is earlier in the older generation. These two types of pedigrees should be equally frequent, and it must therefore be explained why, for some diseases, pedigrees with earlier onset in older generations seem to be rarer than those with earlier onset in younger ones.

In some pedigrees this phenomenon is perhaps due to unknown environmental conditions which have changed with the times, and bring about earlier onset of the diseases in more recent generations. Or it may well be that the published data are not random samples of all cases, but represent a selected group. Probably the most common cause of selection stems from the fact that a serious disease that affects a person early in life greatly reduces his chances of leaving offspring. Therefore, only the individuals in the older generation with late onset will have children and thus become available for the records. If the average age of onset in their children were the same as that in an *unselected* preceding generation, it would be earlier than that in the *selected* group, who became known only because the disease began late.

Another source of selection in pedigrees showing anticipation may be that the early and more severe cases of the generation living at the time of the study have a greater chance of becoming known than the late and less severe ones. Hence, the average age of onset in the ascertained sample of the living generation may be earlier than if the whole unselected population were studied. If the age of onset in the affected parents of such propositi is determined it will tend to show a normal unselected distribution, and the average age of onset will be greater than that of their offspring.

Julia Bell (b. 1879), who collected extensive data on the age of onset of Huntington's chorea and other progressive diseases, has come to the conclusion that no anticipation can be demonstrated if allowances are made for the bias introduced into the data by methods of ascertainment. Indeed, anticipation seems a statistical rather than a biological phenomenon.

Correlation Between Ages of Onset among Relatives. Ages of onset in parents and children and in sibs often show a high degree of correlation. This suggests that within families, similar agents determine the age of onset—agents that may be either genetic or environmental. If the similarity in age of onset within one family was genetically determined, the differences found in different pedigrees could be due to the existence of different genes, each of which was responsible for the disease; or of different alleles of the same gene; or of modifying genes in the genetic background against which some single main gene acts. If, on the other hand, the similarity in age of onset within one family was environmentally conditioned, then the differences between different families would signify that the external agents affecting the age of outbreak of the hereditary disease are those that are similar within families but not from one family to another.

Haldane has found evidence that in some diseases—for example, peroneal atrophy (wasting of the calf muscles)—age of onset in different kindreds is determined by different main genes. In other diseases, for instance, Huntington's chorea, the main gene seems to be the same in all pedigrees, and the difference among individuals appears to be due to modifying genes.

An interesting situation has been discovered in dominant autosomal muscular distrophy. There is much less similarity between the ages of onset in affected parents and children than between pairs of sibs. As described in Chapter 12 (p. 282), Penrose suggested that the age of onset is partly dependent on the type of normal isoallele, d^1, d^2, etc., with which the defect-carrying allele D is combined.

There is still another way in which striking differences in age of onset of a disease may be genetically controlled. Figure 171 shows a pedigree of a type of spastic paraplegia (a degenerative condition of the nervous system). The individual I–3 remained healthy until the age of sixty-five, when his gait became uncertain. At the age of eighty-six, the failure of muscular coordi-

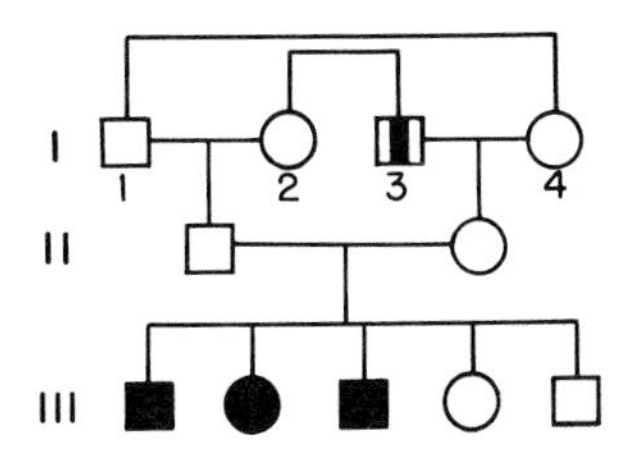

FIGURE 171
Pedigree of spastic paraplegia. Vertical bar in symbol = late onset; solid symbol = early onset. (After Haldane, *J. Genet.*, **41**, 1941.)

nation was marked; nevertheless, he lived to be ninety-one. His daughter (II–2) and her husband, who were double first cousins, were not affected, but 3 of their 5 children developed paraplegia at the age of ten. Haldane suggests, plausibly, that the affected grandfather of the children and his sister (I–2) or her husband (I–1) were both heterozygous for an abnormal allele of low penetrance which either causes no defective phenotype or causes one only late in life. The normal parents II–1 and II–2, according to this interpretation, were also heterozygotes, but their first 3 children were homozygous for the abnormal allele. Thus, homozygosity for the allele is believed to be responsible for the very early age of onset in the youngest generation.

When an abnormal condition is caused by a dominant allele, a heterozygote may be less severely affected than a homozygote. In several Swedish pedigrees a dominantly inherited muscular degeneration (distal myopathy) has a greatly variable age of onset, from thirty-four to eighty-two. In one family both spouses were affected. They had 16 children, of whom 7 were affected, 2 much more severely than the others. If these 2 were homozygous for the defect-causing gene, their children should all be carriers of the dominant gene. There are 8 of them, but most are still too young to show the effect of their genotype.

This chapter has provided a basis for understanding how the interplay of different components of the genotype affects phenotypic expression. It has also given examples that show the great significance of the various environmental factors that influence the appearance of genetic traits.

PROBLEMS

125. Construct a typical pedigree of blue scleras comprising, altogether, fifteen individuals in three generations.

126. Give the genotypes of all individuals in the pedigree on blue scleras in Figure 156. If more than one genotype is possible, give alternatives.

127. From the data in Figure 158, what is the chance that an individual heterozygous for Huntington's chorea will remain unaffected at
(a) 27.5 years of age? (b) At 47.5? (c) At 67.5?

128. The 42.5-year-old daughter of a man who is affected with Huntington's chorea is not affected. What is the probability of her carrying the allele for Huntington's chorea?

129. Assume that M is a dominant modifier for strong expression of minor brachydactyly. Give the genotypes of all individuals in the pedigree in Figure 168. When more than one genotype is possible, give alternatives.

130. If the difference in expressivity of minor brachydactyly were due to a dominant modifier N for weak expression, what would be the genotypes of all individuals in the pedigree in Figure 168? When more than one genotype is possible, give alternatives.

131. Two parents were $I^{B}I^{O}$ and $I^{A}I^{B}$, respectively. They had three $I^{B}I^{O}$ or $I^{B}I^{B}$ children, one $I^{A}I^{O}$, two $I^{A}I^{B}$, and one more $I^{A}I^{B}$. The genotype of the last child was determined by means of the properties of her saliva, into which the antigens A and B had been secreted. Her red cells, however, carried antigen B only—that is, they lacked A. It has been assumed that the parents were heterozygous for genotype Yy, and that yy suppresses the expression of the A antigen in the red cells. What are the possible Y, y genotypes of the children?

In Problems 132 and 133, assume the correctness of Bernstein's theory of inheritance of singing voices.

132. List all possible marriages in relation to types of singing voices. What types of offspring and in what proportions are expected in each type of marriage?

133. In a population, 25 per cent of all women are sopranos. What are the frequencies of all other types of female or male voices?

In Problems 134–137, assume the correctness of Snyder's theory of inheritance of baldness.

134. Give the genotypes of all individuals in the two pedigrees in Figure 170. Whenever appropriate, list alternative genotypes.

135. If a nonaffected woman in the sibship of Figure 170, B, married the individual IV–6 of Figure 170, A, what types of offspring, and in what proportions, could they expect?

136. In a certain population, 1 per cent of all women are bald.
   (a) How many women are heterozygous?
   (b) How many men are bald?

137. In a certain population, 51 per cent of all men are bald.
   (a) How many women are bald?
   (b) What is the frequency of marriages between bald men and nonbald women?

REFERENCES

Bell, Julia, 1928. Blue sclerotics and fragility of bone. *Treas. Hum. Inher.*, **II, 3**:269–324.

Bell, Julia, 1934. Huntington's chorea. *Treas. Hum. Inher.*, **IV, 1**:1–67.

Bell, Julia, 1942. On the age of onset and age of death in muscular dystrophy. *Ann. Eugen.*, **11**:272–289.

Bonnevie, K., 1934. Embryological analysis of gene manifestation in Little and Bagg's abnormal mouse tribe. *J. Exp. Zool.*, **67**:443–520.

Cavalli-Sforza, L. L., and Bodmer, W. F., 1971. *The Genetics of Human Populations.* Ch. 9, pp. 586–587. W. H. Freeman and Company, San Francisco.

Collins, R. L., 1970. The sound of one paw clapping: an inquiry into the origin of left-handedness. Pp 115–136 *in* Lindzey, G., and Thiessen, D. D. (Eds.), *Contributions to Behavior-Genetic Analysis: The Mouse as a Prototype.* Appleton-Century-Crofts, New York.

Falek, A., 1959. Handedness: a family study. **11**:52–62. *Amer. J. Hum. Genet.*,

Fogh-Andersen, P., 1943. Inheritance of harelip and cleft palate. *Opera Copenhagen Univers.*, **4**:266 pp.

Fraser, F. C., 1970. Review article: the genetics of cleft lip and cleft palate. *Amer. J. Hum. Genet.*, **22**:336–352.

Hamilton, J. B., 1958. Age, sex and genetic factors in the regulation of hair growth in man: a comparison of Caucasian and Japanese populations. Pp. 399–433. *in* Montagna, W., and Ellis, R. A., (Eds.), *The Biology of Hair Growth.* Academic, New York.

Holt, Sarah, 1968. *The Genetics of Dermal Ridges.* 195 pp. Thomas, Springfield, Ill.

Hoyme, L. E., 1955. Genetics, physiology and phenylthiocarbamide. *J. Hered.*, **46**:167–175.

Komai, T., Kunii, H., and Ozaki, Y., 1956. A note on the genetics of van der Hoeve's syndrome. *Amer. J. Hum. Genet.*, **8**:110–119.

Levine, P., Robinson, E., Celano, M., Briggs, O., and Falkinburg, L., 1955. Gene interaction resulting in suppression of blood group substance B. *Blood*, **10**:1100–1108.

Levit, S. G., 1936. The problem of dominance in man. *J. Genet.*, **33**:411–434.

Mohr, O. L., and Wriedt, Chr., 1919. *A New Type of Hereditary Brachyphalangy.* 65 pp. Carnegie Institution, Washington, D.C. (Carnegie Inst. Wash. Publ. 295.)

Myrianthopoulos, N. C., 1966. Huntington's chorea. *J. Med. Genet.*, **3**:298–314.

Neel, J. V., Fajans, S. S., Conn, J. W., and Davidson, Ruth T., 1965. Diabetes mellitus. Pp. 105–132 *in* Neel, J. V., Shaw, Margery W., and Schull, W. J., (Eds.), *Genetics and the Epidemiology of Chronic Diseases.* Government Printing Office, Washington, D.C. (U.S. Public Health Service Publ. No. 1163.)

Penrose, L. S., 1948. The problem of anticipation in pedigrees of dystrophia myotonica. *Ann. Eugen.*, **14**:125–132.

Pincus, G., and White, P., 1933. On the inheritance of diabetes mellitus. *Amer. J. Med. Sci.*, **186**:1–14.

Rife, D. C., 1955. Hand prints and handedness. *Amer. J. Hum. Genet*, **7**:170–179.

Sang, J. H., 1963. Penetrance, expressivity, and threshholds. *J. Hered.*, **54**: 143–151.

Smars, G., 1961. *Osteogenesis Imperfecta in Sweden.* 240 pp. Svenska Bokförlaget, Stockholm

Stamatoyannopolous, G., 1966. On the familial predisposition to favism. *Amer. J. Hum. Genet.*, **18**:253–263.

Ullrich, O., 1949. Turner's syndrome and status Bonnevie-Ullrich; a synthesis of animal phenogenetics and clinical observations on a typical complex of developmental anomalies. *Amer. J. Hum. Genet.*, **1**:179–202.
Zirkle, C., 1945. The discovery of sex-influenced, sex-limited and sex-linked heredity. Pp. 169–194 *in* Montagu, M.F.A, (Ed.), *Studies and Essays in the History of Science and Learning in Honor of George Sarton.* Schuman, New York.

17

CONGENITAL MALFORMATIONS AND PRENATAL INTERACTIONS

In general, the genotypes of children are independent of any variation in the physiology of their parents caused by such factors as age, number of preceding births, or health. Thus, a person heterozygous for a pair of alleles forms equal numbers of the two possible kinds of gametes regardless of his physiological condition. Likewise, the independent assortment of genes in the different chromosome pairs is a basic process of normal meiosis and is not influenced by parental condition. However, two genetically significant processes, mutation and crossing over, are known to be affected by the physiology of an individual. For mutation, this will be shown in Chapter 23; for crossing over, a brief discussion will be given here.

In experimental organisms, including *Drosophila* and mice, the frequency of crossing over has been shown to be dependent, in complex ways, on age as well as on various metabolic conditions. In man, nothing is yet known about variability in the frequency of crossing over, but it may well be found that gametes from parents of certain ages will show higher crossover frequencies than gametes from parents of other ages. Should this be true, the gametes produced by individuals during the period when the frequency of crossing over is low would more often contain blocks of alleles still linked in the same chromosomes in which they were received from the parents, than the gametes produced by the same individuals during the period when the frequency of crossing over is high. As a consequence of this still hypothetical situation, children produced during the period when there was a low rate of crossing over in a parent would be likely to possess associations of linked characters present in one of the grandparents, whose chromosomes they would receive with relatively little change, and would lack the alternative blocks of characters of the other grandparent, whose chromosomes they would not receive. On the contrary, children produced when the rate of crossing over was high would be likely to present a finer mosaic of grandparental characters, since many of the chromosomes the children would receive would comprise segments derived from both grandparents.

In pooled data from many sibships, no correlation would be expected between characters and the high or low crossover periods of the parents. This follows from considerations similar to those presented in Chapter 15 (pp. 341–344) regarding the lack of correlation in a population between different traits that are controlled by linked genes. Thus, in no population could segregation and recombination or linkage and crossing over lead one to expect correlations between the ages of parents and the genotypes of offspring.

MATERNAL PHYSIOLOGY AND PHENOTYPES OF CHILDREN

In spite of the general expectation, the occurrence of a number of inherited congenital characteristics of offspring has been found to be correlated with physiological variables of the mothers, particularly with her age.

It is not surprising that no clearly established example of a paternal influence of this type is known. The father's contribution, the sperm, consists almost entirely of the genes that are contained in its head and hence of little that could exert extragenic influence. The mother, on the contrary, not only supplies the egg, which contains a large volume of extragenic substance, primarily the cytoplasm; but in addition provides, in her uterus, the environment in which the embryo develops. It would seem reasonable that the physiological condition of a woman should affect the cytoplasmic content of the egg cells ripening in her ovaries, and that differing uterine environments might have variable effects on the phenotype of the developing child.

In snails and insects, maternal influences which affect the phenotype of the offspring by way of the egg's cytoplasm have been discovered. In a few laboratory mammals, too, the physiological condition of the mother has been shown to affect the expression of the offspring's genotype. Thus, in a strain

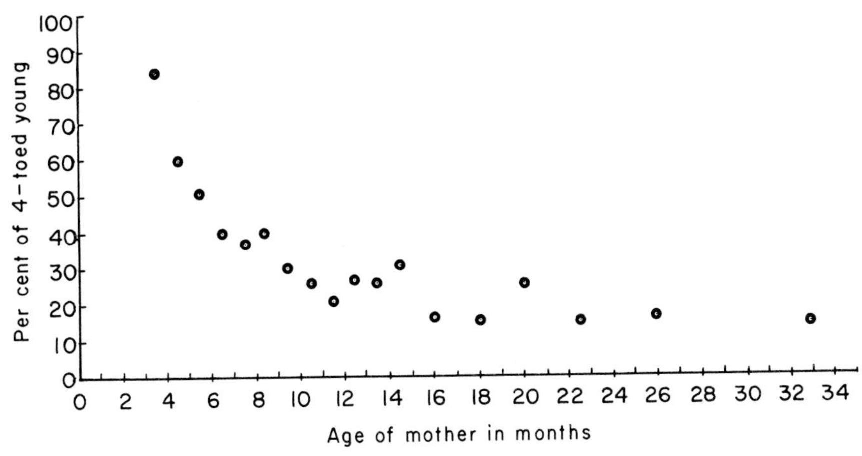

FIGURE 172
Influence of age of mother on number of extra toes in 416 polydactylous guinea pigs. (After Wright, *Amer. Natur.*, **60**, 1926.)

of polydactylous guinea pigs, Wright found that penetrance and expressivity are partly controlled by the age of the mother. The older the mother, the more young with the normal number of toes (Fig. 172). The same tendency has been found for a variety of skeletal abnormalities in certain strains of mice: polydactyly, absence of a third molar tooth, unusual position of a vertebral spine, and of an opening in a vertebral arch. Prenatal influence is also manifested in the size of white areas on the coat of the young in a piebald strain of guinea pigs, the areas becoming larger with increasing age of the mother.

Vertebrae in Mice. Mice normally have either twenty-six or twenty-seven presacral—that is, thoracic and lumbar—vertebrae. Significantly more mice from young mothers have twenty-seven vertebrae than do mice from older mothers. W. L. Russell has carried out experiments which, although not concerned directly with the age of the mother, show that differences in number of vertebrae can be the result of differences in uterine environment, which does, of course, vary with age. These experiments with mice will be described in detail in order to make clear what the problems are in understanding how the physiological condition of a human mother can influence the phenotype of her child.

Russell worked with two strains of mice in which the percentage of animals having five rather than six lumbar vertebrae varied greatly. In strain C57-black only 1.4 per cent, but in strain C3H 96.9 per cent, had five lumbar vertebrae. In crosses between these strains, the frequency of those with five

TABLE 46
Percentage Frequency, in Two Strains, of Mice with Five Lumbar Vertebrae, and in the Offspring of Reciprocal Crosses.

Type of Mice	*Per Cent with 5 Lumbar Vertebrae*		
	Female	Male and Female	Male
C57 black		1.4	
C3H		96.9	
C57 black ♀ × C3H ♂	24.1		33.4
C3H ♀ × C57 black ♂	45.8		68.1

Source: After Russell and Green, *Genetics,* **28,** 1943.

was only about half as high when the mother was C57-black and the father C3H as when the mother was C3H and the father C57-black (Table 46).

Such a difference in reciprocal crosses, in which the genetic constitution of the hybrids should be alike, could be due to any one of four factors. It is possible, first, that the cytoplasm in each strain contains different self-reproducing properties—similar in this respect to the genes in the nucleus—which lead to different kinds of cytoplasm in the eggs. Second, it could be that the cytoplasm of the eggs is different, not because it was derived from different kinds of maternal cytoplasm, but, rather, because it was built up in the ovaries of the mothers under the influence of different genes in the cells of the ovaries—those in the C57-black or those in the C3H strain (maternal prefertilization influence). A third possibility is that genes of the mother in cells outside the ovaries could influence the cytoplasm of the eggs while they grow in the ovaries (another type of maternal prefertilization influence). And the fourth possibility is that the difference in number of vertebrae is the result of the uterine environment in which the hybrid embryos develop—an environment which may vary in the two types of mothers (maternal post-fertilization influence).

A test eliminated the hypothesis of self-reproducing cytoplasmic differences. If they existed, hybrid daughters of C57-black mothers should form eggs with the cytoplasm characteristic of the C57-black strain, and hybrid daughters of C3H mothers should form eggs with the cytoplasm characteristic of the C3H strain. Therefore, the offspring from the two types of hybrid daughters should again differ in mean number of vertebrae. This, however, proved not to be the case: The daughters from reciprocal crosses had progeny with the same mean number of vertebrae.

In order to determine whether genes in the mother's ovarian cells affected the egg cytoplasm, ovaries from a strain, "129," with high mean numbers of lumbar vertebrae were transplanted into females of a strain, "L," with low numbers whose own ovaries had been removed.

The "L" females with "129" ovaries were then mated to "129" males, so that young with the "129" genotype developed in the uteri of "L" mothers. In another experimental group, young with the "129" genotype were raised in "129" mothers, either as a result of mating "129" males to unoperated "129" females, or to "129" females whose ovaries had been removed and replaced by ovaries from other "129" females. (This latter procedure served as a control to safeguard against the possibility that transplantation itself might be responsible for differences in number of vertebrae; it turned out to have no influence.)

Comparison of young who had developed in "L" females with those who had developed in "129" females showed a significantly lower average of vertebrae in the former group.

Thus it was concluded that the skeletal effect observed in the experiments is caused by a difference either in the influence of the mother's extraovarian genes on the egg cytoplasm or in the uterine environments as furnished by the different mothers.

To decide between these last two alternatives, one more experiment was carried out. Eggs from matings of C3H females (low mean number of lumbar vertebrae) with C57-black males (high number) were transplanted soon after fertilization into the uteri of C57-black "mothers." The young were carried to term and the number of their vertebrae determined. As shown in Table 47, the vertebral frequencies were characteristic of the C57-black strain and not of the C3H strain. This, then, leads to the conclusion that in this instance it is the uterine environment and not gene-dependent properties of the egg cytoplasm that causes the variation. In a similar transplantation experiment but with different strains, E. L. and M. C. Green have however, demonstrated that differences in numbers of vertebrae were not dependent on uter-

TABLE 47
Effect of the Uterine Environment on Skeletal Morphology in the Mouse.

Cross			*No. of Lumbar Vertebrae*		
Females		Males	5	5 one side, 6 other side	6
C57 black	×	C3H	16	11	34
C3H	×	C57 black	30	11	8
C3H	×	C57 black*	7	4	19

Source: After McLaren and Michie, *Nature*, **181**, 1958.
*Fertilized eggs transferred to C57 black uteri.

ine influences but could solely be explained in terms of cytoplasmatic differences between the eggs of the two strains. In man, similar differences in the number of vertebrae of specific regions of different individuals are also common. There is evidence for genetic differences causing variability of the vertebral column, as well as of variable manifestation of the genotypes involved. It may well be that part of this nongenetic variability is due to prenatal influences. Obviously, such influences, which vary with the condition of the mother, are not due to inherited self-reproducing properties of the cytoplasm.

Birth Weight. In crosses between Aberdeen-Angus and Herefordshire cattle, the weights of hybrid calves are similar regardless of which of the two breeds the mother comes from—birth weight depends mainly on the genotype of the fetus. In contrast to this, crosses between the large Shire horses and Shetland ponies yield hybrids whose size is greatly influenced by the breed of the mother. In man, both the mother and the fetal genotype contribute to variability in birth weight. The correlation between the birth weights of identical twins is high, namely, 0.67, and this seems to suggest that it is the identical genotype of the two fetuses that causes them to be so similar in weight. However, since the correlation between weights of genetically unlike nonidentical twins is 0.59, not much lower than that of identicals, it is clear that a large part of the similarity in both types of twins is due to their simultaneous development in the same mother.

Support for this conclusion comes from the observation that infant weight is correlated more highly with height of mothers than of fathers. This difference may persist for some time, but at maturity the correlations between weight of offspring and stature of mothers and fathers has become the same, the genotype of the offspring having overcome the prenatal maternal influence.

The correlation between the birth weights of separately born sibs, 0.50, is lower than that of nonidentical twins in spite of the fact that the genotypes of nonidentical twins are no more similar than those of separate sibs. A comparison of the two correlations suggests that the uterine environment provided the growing fetus by a mother is variable from one pregnancy to another. In order to discover whether this maternal environment also has a constant genetic component, comparisons have been made between the birth weights of first cousins. There was a small, but significant, positive correlation, 0.13, when the mothers of cousins were sisters, but a nonsignificant correlation of only 0.02 when the cousins were offspring of brothers or of brothers and sisters. Thus, the genetic similarity of sisters expressed itself in similarity of prenatal influences on their children—influences that outweighed by far the effect of similarity of the genotypes of the cousins themselves, which is the same regardless which of the parents are sibs.

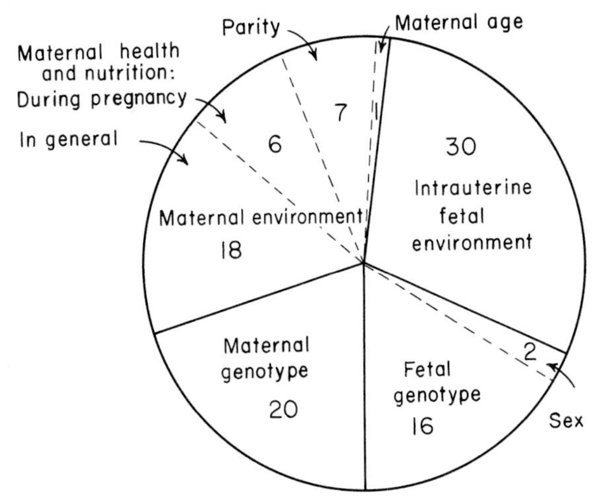

FIGURE 173
Birth weight of single born: partition of causes of variation. Numbers indicate partition in per cent. (After Penrose, *Sci. Prog.*, **169,** 1955; and *Proc. 9th Int. Congr. Genet.*, 1954.)

On the basis of these findings Penrose has attempted a quantitative partition of the causes of the variability of birth weights (Fig. 173). Four kinds of influences can be distinguished: fetal genotype, maternal genotype, maternal environment including socioeconomic factors, and an unanalyzed residue, "intrauterine fetal environment." Specific items within these four groups are indicated in the figure. The relative importance assigned to the causes of variation in birth weight must be regarded as tentative.

Congenital Defects. It is not surprising that the frequency of congenital malformations is in part related to the physiological state of a mother during pregnancy. Infection of the mother with German measles (rubella) during the first three months of pregnancy may cause destruction or severe defects of the embryo. The rubella virus is capable of crossing the placental barrier between mother and child, and the damage to the embryo is the result of prenatal infection. Similarly, maternal infection with the rare protozoan parasite toxoplasma can cause serious congenital defects of the fetus. Irradiation of the pelvis of pregnant women with X-rays may impinge on the embryo and lead to malformation. Sugar diabetes in the mother seems to increase the chance for developmental abnormalities of her offspring. Excess of the amniotic fluid surrounding the embryo may also lead to defects, but it is not clear whether the excess is caused by the state of the mother or that of the embryo itself.

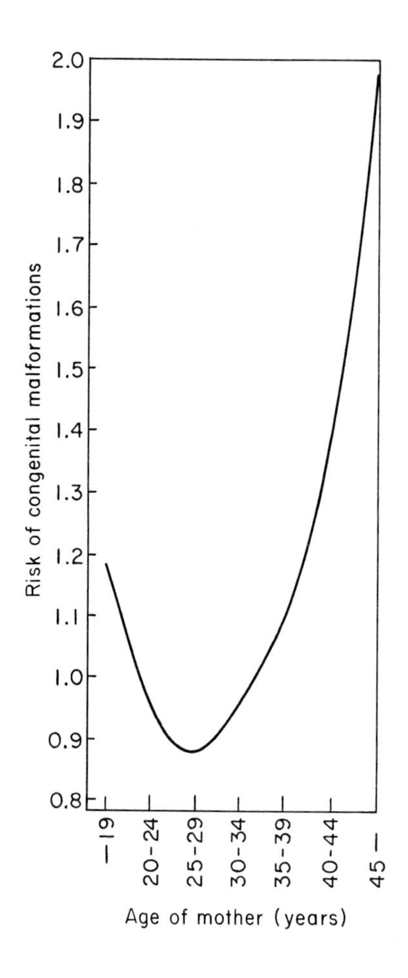

FIGURE 174
Frequencies of congenital malformations as related to age of mother expressed in proportion to average frequency. (Parkes, *Sci. J.*, **6**, 1970. Copyright 1970 by the American Association for the Advancement of Science.)

Age of the mother is correlated with the frequency of congenital defects, but the degree of increase is not very high and the relation is not linear (Fig. 174). Mothers of the youngest age group have slightly more defective children than those of mothers in the middle age groups, and only the oldest child-bearing groups have strikingly increased proportions of defective offspring. Other studies have given divergent results. Thus, among more than 100,000 births in the city of Hamburg, Germany, about 1 per cent were affected with obvious defects observable at birth. Among the mothers of ages 40 and older the incidence was 1.5 per cent. In a still larger study in Ohio, mothers younger than 40 years had 0.72 per cent abnormal children, as compared to those of all ages which was 0.74 per cent. In a way these studies give a false impression of the general frequency of congenital defects. They are found to be more frequent when attention is paid not only to major, easily recognizable defects but also to minor external defects and internal metabolic abnormalities. For a Japanese sample, Neel found 1.37 per cent of newborn with major congenital defect, a figure which increased to 3.12 per cent when the infants were re-examined at 9 months of age. In other studies still higher frequencies—up to higher than 7 per cent—were recorded.

Instead of determining the over-all rate of congenital defects, many studies are concerned with specific defects or groups of defects. The most striking case is that of Down's syndrome whose incidence increases with increasing age of the mother. As we have seen, this correlation is not caused by a prenatal interaction but is due to the larger proportion of fertilized eggs which, by means of nondisjunction of chromosome 21, have become trisomic. Congenital malformations of the circulatory system are more frequent in the children of older than of younger women, and the same is true for cleft palate as well as cleft lip and cleft palate, but not for cleft lip alone. Some abnormalities are more common in the children of younger mothers. This probably holds for clubfoot; for polydactyly; for hypospadia in the male, in which the opening of the urethra is shifted from its normal position; and for pyloric stenosis, an obstruction of the opening of the stomach into the intestine.

Anencephaly and Spina Bifida. Among the variety of congenital malformations anencephaly has been subjected to very detailed studies. This condition consists in a failure of part of the skull and brain to develop and is not compatible with survival. A developmentally related malformation is spina bifida, a defect of the vertebral column and the spinal cord. Although most babies born with spina bifida do not survive, a few of the more mildly afflicted have. Anencephaly and spina bifida seem to be different expressions of the same developmental abnormality. The combined frequency of the two defects varies in different populations from 0.1 in 1,000 (e.g., Ljubljana, Yugoslavia) to nearly 8 in 1,000 (e.g., South Wales and Northern Ireland). For anencephaly alone, Figure 175 shows that there are varying frequencies in different regions of England, Scotland, and Wales. There is a striking negative correlation, –0.84, between the frequencies of mortality due to spina bifida and the longitude of location of the different states in the United States, the highest frequencies occurring in the East, the lowest in the West. A similar east-west gradient for spina bifida exists in Canada.

In New England during the period 1946–1965, the incidence of both anencephaly and spina bifida decreased strikingly. The incidence was greatest among firstborn infants and was higher than average among the offspring of very young as well as of relatively old mothers, but the data from various sources are contradictory. There is also a seasonal variation in incidence, the conditions being most frequent among children conceived in early summer. In some populations no relation was found with social status of the parents but in Ireland the incidence among the children of unskilled workers was five times that among the children of professional and managerial classes.

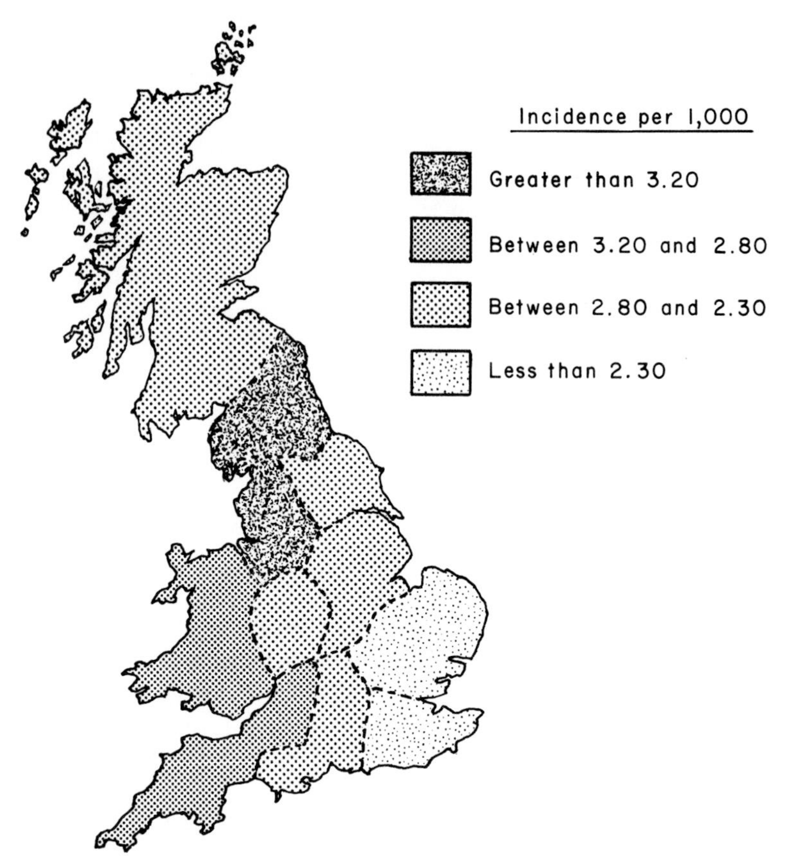

FIGURE 175
Estimated relative incidence of 491 cases of anencephaly by region in England, Scotland, and Wales, March-May, 1958. (After Fedrick, *Ann. Hum. Genet.*, **34**, 1970.)

A higher incidence was observed in urban than in rural populations. According to data from Caucasian as well as Japanese populations, the risk of recurrence of one or the other of anencephaly or spina bifida is of the order of 10 times that in the general population. When two affected children have been born in the same family, the risk of recurrence seems further increased, a rate of 10 per cent being suggested by limited data.

How can these manifold observations be understood? Clearly a purely genetic explanation is ruled out since it cannot account for effects of maternal age, effects of season of conception, recent decreases in incidence, as well as for the additional fact that there have been almost no cases in which both of two genetically identical twins have been affected. The epidemiologist MacMahon to whom we owe important publications on anencephaly and spina bifida is inclined to doubt any genetic interpretation as indicated by the question mark in the title of a recent paper by him and his associate Stella Yen, "Genetics of Anencephaly and Spina Bifida?" In order to interpret ethnic differences in incidence and increased recurrence risks MacMahon assumes the existence of persistent or recurring environmental factors to be as likely as that of genetic predispositions. If a genetic predisposition is involved, then very low penetrance must be invoked, a penetrance that is subject to a variety of environmental influences. Apart from penetrance, single-gene as against polygenic determination has been discussed, the latter model favored by Carter and Roberts. In Hawaii evidence has been obtained of a paternal influence on the frequency of spina bifida. This points toward a genetic component. Clearly, more study is demanded, and it should be emphasized that congenital defects other than the two abnormalities of the central nervous system discussed here pose similar questions.

Maternal Age and Correlated Properties. A correlation of frequency of congenital abnormality with age of mother does not necessarily indicate a causal connection between the two phenomena. Age of mother itself is highly correlated with other variables, such as weight of mother, age of father, occurrence of a first birth (primogeniture), and number of preceding births (birth rank, parity). It is possible, by means of statistical techniques, to unravel the various connections between these variables. Thus, if mothers are divided into groups of similar ages, the relation between the age of the fathers and the frequency of an abnormality in the offspring can be studied. Most such studies do not demonstrate a significant correlation between the age of the father and the incidence of a specific abnormality. An exception concerns cleft lip with or without cleft palate, which, according to C. M. Woolf, shows a higher frequency with increasing age of the father. Otherwise it is known from census data that the age of the father, independent of that of the mother, is positively associated with the rate of stillbirths and neonatal mortality (Yerushalmy; Sonneborn has found a similar relation for early fetal deaths and age of father). Penrose has suggested that fresh mutations accumulating with age of fathers may cause mortality of their offspring in early life and thus account for the paternal age effect. There is indeed evidence for higher mutation rate from normal to achondroplasia, as well as some other defects in older fathers (pp. 569–570), but it is not clear how much these findings can be generalized.

Various studies have indicated that parity, and not age alone, plays some role in the origin of malformations. The incidence of pyloric stenosis per thousand live births in Birmingham, England, at different birth ranks was found to be: first birth, 4.3; second, 2.8; third, 2.5; and fourth and higher 1.4. Birth rank is also a factor in the frequency of stillbirths. There is a steady increase of stillbirths from rank 2 to higher ranks, with rank 1 being exceptional in having more stillbirths than those at several of the ranks following it. Many stillbirths are caused by fetal defects.

Cancer of the chorionic membrane is more frequent in first than later pregnancies, and leukemia in childhood, though not strictly congenital, also has its highest incidence among firstborn. In many of these diseases or malformations it is apparent that maternal age itself, after separation from its correlated variables, is an important, and sometimes the only important, causal factor responsible for the statistical correlation between age of mother and frequency of the abnormality.

What specific physiological conditions change with the age of the mother or with other variables and are responsible for changes in the incidence of abnormalities is still to be discovered. One malformation whose incidence depends on maternal variables is an anatomical abnormality of the pregnant uterus. Instead of the embryo being implanted in the posterior part of the uterus, as is normal, the placenta may become attached near the orifice that opens into the vagina (Fig. 176). Such a placenta praevia often results in severe hemorrhage before childbirth. Penrose and Kalmus have shown that the frequency of the so-called central type of placenta praevia is correlated with maternal age and, apart from maternal age, also with parity. In this connection it may be mentioned that the frequency of "spontaneous" cleft lip in a certain strain of mice also depends on the position of the embryo in the uterus.

In some invertebrates, in rainbow trout, and in frogs, aging of eggs before they are fertilized results in defective development of embryos. As described in Chapter 5 aging of human ovulated eggs before fertilization may be a factor that contributes to the maternal age effect on the frequency of Down's syndrome. Whether overripeness of sperm may likewise affect the frequency of congenital defect is unknown.

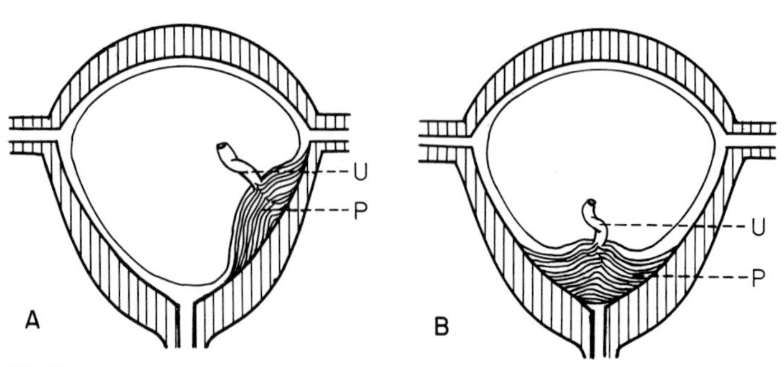

FIGURE 176
Diagrammatic longitudinal section through uterus. **A.** Normally located placenta. **B.** Placenta praevia. P = placenta; U = umbilical cord.

Independence of Traits from Maternal Physiology. Many traits are *not* affected by the age of the mother or correlated variables. Mental defect is one of these. In Table 48, the numbers of feebleminded children are given separately for the first and second halves of a group of affected sibships. No significant difference between birth order and frequency of the condition is shown.

It must be kept in mind, however, that the interrelations between maternal condition and penetrance or expression of traits are complex and that various maternal influences may partly cancel each other. Absence of an obvious correlation between the phenotypes of the offspring and known maternal variables cannot, therefore, prove absence of correlation between phenotypes of offspring and the physiological condition of the mother. The unexpected fact that, in mice, maternal influences are responsible for the number of lumbar vertebrae in the offspring suggests that prenatal influences of the uterine environment are, perhaps, of greater significance in the expression of traits, within their normal range of variability, than hitherto suspected.

A much discussed possibility of prenatal effects is an influence of season of conception on intelligence test performance of children. In some studies significant correlations between the two variables were observed showing higher scores for children conceived in winter than in summer, but in other studies no correlations were found. Final judgment must be postponed. If and when a correlation exists in certain populations, several different interpretations would be available. One of these would assume that in families whose children have high scores a higher proportion of conceptions occurs in winter than in families whose children have low scores. This interpretation implies that intelligence influences season of conception. Another interpretation would ascribe the seasonal effect to environmental influences exerted on the child after birth. This interpretation implies that season influences intelligence. Still another interpretation would assume that the environmental influence of the season on the pregnant mother affected the fetus. Such an

TABLE 48
Feeblemindedness
(excluding Down's syndrome).

Half of Family	*Affected**	*Normal**
First half	44.5	135.5
Second half	55.5	124.5

Source: M. Murphy, 1936.
*The fraction in the observed results arises from the division of cases occurring in the mid-position of families containing an odd number of children.

assumption could point to the fact that the number of abnormal lumbar vertebrae of mice depends on the temperature at which the animals are bred. More abnormal vertebrae were found among the young of mothers kept at –3° C than at 21° C.

The Causes of Prenatal Effects. Most of the abnormalities discussed run in families. Although the incidence of more than one case of a given abnormality in a sibship is low, it is considerably higher than expected by chance, and the appearance of a trait or some symptoms of it in relatives other than sibs is also relatively high.

Increased incidence of congenital malformations within certain families is, by itself, no proof of genetic predisposition, since it is also possible that environmental circumstances prevalent in such families or acquired physiologic conditions of the mother may be responsible. In experimental animals such as rats, mice, guinea pigs, rabbits, and pigs, severe "insults" to the pregnant females may cause various kinds of defects in the embryos. Warkany, a pioneer in teratology, the study of spontaneous and experimentally induced abnormal development, first showed that a critical deficiency of vitamin A in pregnant female rats may lead to congenital abnormalities of the eyes, lungs, and other organs in their offspring. Subsequently, he and other investigators were able to produce specific skeletal defects, such as cleft palate, or specific defects in the central nervous system, such as spina bifida or hydrocephaly, by a variety of agents, including vitamin deficiencies, reduced atmospheric pressure with an attendant decrease in oxygen, and injections of hormones and of various chemical substances. Many of the defects produced by experimental interference with embryonic development are also known to be caused under normal conditions by abnormal genes. The artificially induced malformations are thus phenocopies.

In most of these animal experiments the treatments applied are so drastic that they will rarely or never be encountered by pregnant women. Nevertheless, it is possible that less extreme deficiencies in vitamins or oxygen, or other disturbances that affect the maternal physiology, which may take place in times of famine or war or in other unfavorable social circumstances, may interfere with normal fetal development.

Extreme and chronic nutritional deficiencies of pregnant women are indeed responsible for a deleterious effect on the fetus. Newborns with this intrauterine malnutrition syndrome are smaller than normal and many suffer serious neurological consequences, including cerebral palsy and mental defect. Data of this kind often are based on joint prenatal and postnatal malnutrition and studies on rats suggest that prenatal deprivation may make the brain more vulnerable to a subsequent postnatal insult. Experiments with rats who were given before and during pregnancy a diet normal in calories but containing only one-third of their protein requirement produced young with 10 per cent less DNA in their brain than in controls—presumably due to fewer brain cells. In addition the protein content of the brain was reduced by 20 per cent.

Maternal Stress. The opinion has been expressed that some congenital malformations can be produced by severe emotional disturbance of the mother, particularly during the earlier months of pregnancy. Stress, as well as physical agents, is likely to result in increased secretion of the hormone cortisone, and experiments with pregnant mice have shown that injection of heavy doses of this substance will produce cleft palate in the embryos. A study of human mothers of children who were born with cleft lip with or without cleft palate, or with cleft palate alone, has shown that such mothers recalled a higher frequency of emotional upsets during the pregnancy that produced the child with the abnormality than during other pregnancies which resulted in normal siblings. However, mothers of children with genetically determined disease such as hemophilia or albinism that are fully determined at conception and not ascribable to prenatal stress also recalled a higher frequency of emotional upsets. The greater recalled frequency of such upsets during pregnancies resulting in cleft lip and cleft palate apparently represents "maternal memory bias" rather than reality. That stress of pregnant women usually does not lead to congenital malformations is also suggested by the fact that no increase in the frequencies of malformations occurred in the German cities of Berlin and Hamburg during the years of World War II in spite of the frequent bombing attacks.

Prenatal Effects on Behavior. Prenatal influences may result in changed behavior of the offspring after birth. This is primarily known from animal experiments. Rats were X-irradiated in early pregnancy; their young later showed greater "emotionality," as measured by frequency of defecation, and greater "intelligence" than nonirradiated controls. After the same radiation was applied to rats in later stages of pregnancy, more "docile" and less "intelligent" young were born. In man, information is available on fetuses exposed in utero to irradiation from the atomic bombs in Japan. Many of the children born have shown severe brain damage.

Perhaps radiation should not be included in prenatal interaction effects since it impinges on the fetuses directly, in addition to its being absorbed by the mother's body. Experiments in which alcohol or adrenaline was injected into pregnant rats or mice differ from the X-ray experiments by affecting the mother primarily and the embryos only secondarily. Again, however, effects on behavior of the young were produced. Interestingly, the effect of adrenaline varied with the strain of mice used, as demonstrated by differences in the characteristic amount of physical activity. Injected mothers of a low-activity strain yielded young with decreased activity. Anxiety of the mother, according to Thompson, also plays a role in changed behavior of the young. This investigator induced a conditioned avoidance response in nonpregnant female rats by sounding a buzzer while giving an electric shock to the animals. After mating, the buzzer stimulus was continued but no further shocks administered. The influence of the anxiety called forth in the pregnant females by the sound of the buzzer showed in the offspring when tested at ages 30–40 days and 130–140 days: they were changed emotionally as measured by defecation frequency, decreased activity, and high latency. (As the investigator himself pointed out, this result might possibly not have been due to a prenatal effect of the mother on her offspring but to a direct effect of the buzzer sound on the fetuses. This, however, seems unlikely.)

Effects of extreme prenatal malnutrition on the chemistry of the brain in rats have already been mentioned. Lesser dietary deficiencies may result in minor degrees of lowered mental performance. In a study by Harrell, Woodyard, and Gates, vitamin supplements were added to the diet of pregnant and lactating women from low-income families. Their children had intelligence quotients several points higher, on the average, than those in a control group whose mothers received no dietary supplement. Further inquiries of this kind are highly desirable.

Damage to the developing brain may be caused not only by dietary deficiencies but also by excess of a nutrient brought about by a genetically caused metabolic block. Some few women homozygous for the recessive gene causing phenylketonuria are mentally within the normal range even though they possess abnormally high amounts of phenylalanine in their blood. Married to homozygous normal men their heterozygous children should be unaffected by the PKU syndrome. The same would have been expected for the rare heterozygous children whose mothers exhibit the whole syndrome. Actually, the children are mentally deficient even though metabolically not affected. It appears that the high phenylalanine level in the PKU mother's blood damages the developing brain of the fetus.

There is a "natural experiment" that leads to a prenatal effect on later performance in intelligence tests. In general, two twins have different birth weights and score differently in intelligence tests. A study showed that there is no association between difference of score and of birth weight among genetically nonidentical twins but among genetically identical twins, the twin with heavier birth weight, on the average, scored several points higher. The association was not significant for the total study population of identical twins but was so for those twin pairs with large birth weight differences (300 grams or more). It seems likely that unequal intrauterine relations, perhaps of a circulatory type, are responsible for the later I.Q. differences.

Additional Prenatal Interrelations. The genotypes of both mother and fetus influence the degree to which external prenatal influences lead to the occurrence of abnormal phenotypes. This has been shown by the results of some animal experiments. Ingalls and his collaborators subjected mice of five different strains to reduced atmospheric pressure equivalent to that at an altitude of 29,000 feet—the height of Mount Everest. Among the malformations studied were those of the sternum (Fig. 177). From 0 to 29 per cent of offspring born to unexposed controls of the five strains showed deviations from a standard type of sternum. Reduced atmospheric pressure for five hours on the ninth day of pregnancy caused from more than 20 to more than 70 per cent deviations from the norm in offspring of the pregnant subjects. These figures show that the five strains differed in the spontaneous incidence of deviations as well as in the frequency of induced variants. Moreover, strain DBA, which had the highest frequency of induced variants, had a lower frequency of spontaneous ones than did some of the other strains.

Deeper insight into the prenatal interaction of mother and child is provided by another series of experiments on mice, in which a heavy dose of cortisone was injected into female mice of two different genotypes, pregnant with young of a variety of genotypes (Table 49). The same hormone treatment caused 100 per cent of the embryos of strain A to develop with cleft

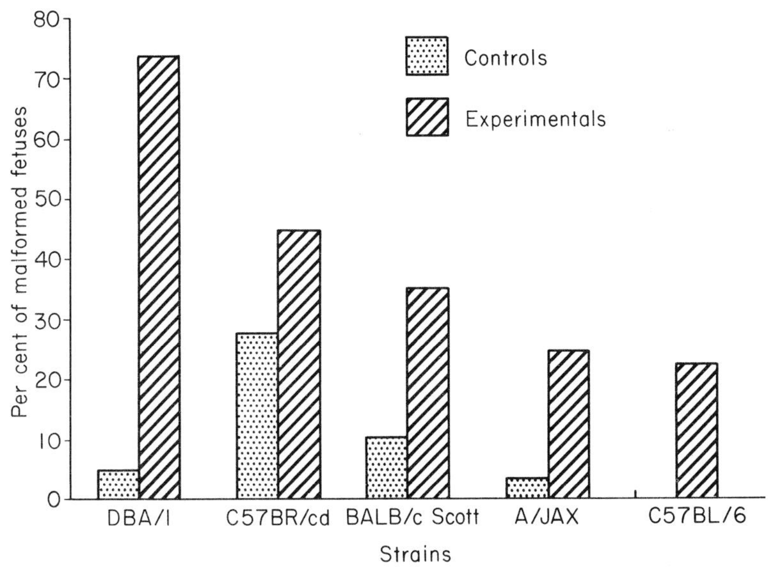

FIGURE 177
Frequencies of minor malformations of the sternum in five genetically different strains of mice. The experiments consisted of exposure to low atmospheric pressure. (After Ingalls, Avis, Curley, and Temin, *J. Hered.*, **44**, 1953.)

TABLE 49
Inheritance of Susceptibility to Induction of Cleft Palate in the Young of Pregnant Mice Treated with Cortisone. (A = strain A/Jax., B = strain C57 black/Jax.)

Mother	*Young*	*Treatment*	*Cleft Palate (%)*
A	A	−	0
A	A	+	100
B	B	−	0
B	B	+	17
A	A × B	+	43
B	A × B	+	4

Source: After Fraser, Kalter, and others. See Fraser.

palate but only 17 per cent of the embryos of strain B. When A mothers pregnant with A × B hybrid young were injected with cortisone, 43 per cent of the offspring had cleft palate while only 4 per cent of the same A × B young in B mothers were affected. The much greater frequency of the defect in A young from A mothers than in A × B young from A mothers shows the influence of the embryo's genotype; the greater frequency of the defect in A × B young from A mothers than in young of the same genotype from B mothers shows the influence of the mother's genotype.

Two final examples of prenatal interaction resulting in congenital malformations are not concerned with specific external treatment, but with the interaction of the genotypes of mother and embryos in producing abnormalities during an undisturbed pregnancy. A certain highly inbred strain of mice contains two types of individuals, one normal (*fu fu*) and the other heterozygous for an incompletely penetrant dominant gene, *Fu*, which causes fusion abnormalities of the skeleton of the tail and other parts of the vertebral column. Embryos heterozygous for *Fu* carried by two different types of mothers have been studied, and the frequencies of fused vertebrae were determined as follows:

in *fu fu* mothers, 65 per cent
in *Fu fu* mothers, 34 per cent

Clearly, the genotypes of both mother and child are involved in the production of the defect. Without the *Fu* allele the offspring is normal, but the penetrance of the *Fu* allele when present depends on the genotype of the mother. If she carries *Fu* she produces fewer affected offspring than if she herself is genotypically normal. Results like these suggest that even external influences so slight as to escape notice may lead to malformed offspring if particularly sensitive genotypes are involved.

It seems peculiar that the abnormal *Fu fu* constitution of mothers tends to act beneficially rather than harmfully on the developmental processes of her young. A somewhat related interaction between mother and fetus has been found by Hollander and Gowen. It concerns the recessive allele *hl* (hair-loss) in the mouse. Nothing unusual happens in matings of homozygous recessive hair-loss mice inter se, nor in matings of homozygous *hl hl* males with normal females. But in matings of *hl hl* females with heterozygous normal males, the expected 1:1 segregation is changed by differential mortality to roughly 2 hair-loss : 1 normal. Deaths occur principally at birth and during the first two weeks of age, the non-hair-loss young showing variably inferior growth with fragility of bones. When *hl hl* females are mated with homozygous normal males, all the young are affected. The physiologic nature of the antagonism between hair-loss mothers and normal-hair offspring may be related to their calcium metabolism. Apparently this antagonism is not immunological in nature (in contrast to such antagonisms described in the following section). It is interesting that the normal-hair progeny and not the abnormal progeny are at a disadvantage.

In experimental work it is often easier to cause the appearance of abnormalities than to hinder their development. Undoubtedly, however, it will be possible to work out methods by which animals and, ultimately, humans with genotypes usually highly penetrant in leading to congenital malformations can be treated in order to suppress defective development. A demonstration of such treatment has been furnished in mice. A certain mutant strain is characterized by lack of coordination of muscles, lack of equilibrium, and retraction of the head. A phenocopy of these abnormalities can be produced during prenatal development in genetically normal mice by feeding the gestating females a manganese-deficient diet. Both mutant and phenocopy effects are caused by defective development of the otoliths in the inner ear. When the diet of pregnant mutant mice was supplemented with manganese, the young, after birth, showed normal behavior!

The successful application of such prenatal therapy to human mothers of children with possibly unfavorable genotypes will prevent a great deal of human suffering. In addition, general public-health measures—better nutrition, lowered incidence of certain diseases, and others—may help to reduce the frequency of congenital malformations due to prenatal influences of environmental agents on the manifestation of those ill-defined genotypes that have a tendency to produce defects.

IMMUNOGENETIC PRENATAL INTERACTIONS

In general, the details of prenatal interactions between mother and fetus are obscure. There is, however, an important group of such interactions that is well understood. It results from genetically determined immunological incompatibility between mother and child and may lead to disease or death of the offspring. Fortunately, the knowledge of the genetic basis of these incompatibilities has led to methods to reduce greatly the incidence of the disease and nearly eliminate it.

Rh Incompatibility. Soon after the discovery of the Rh factor, Levine recognized that it played a part in hemolytic disease of the newborn, or erythroblastosis fetalis, which consists of an anemia due to hemolysis (breakdown of the blood) in the child and consequently results in jaundice. One aspect of the disease, which has given it its name, is the presence of immature red cells, the erythroblasts, in the blood stream. Normally, these immature cells are found only in the bone marrow and other organs, not in the circulating blood. In the past, fetal erythroblastosis frequently led to stillbirth or neonatal death. Most of the affected children who did survive were completely healthy following recovery.

The disease was known to have a familial occurrence, but in spite of various attempts to formulate a genetic interpretation, the hereditary mechanism remained unknown until the Rh antigen was found and its mode of inheritance determined. Levine then noticed that more than 90 per cent of erythroblastotic children were Rh positive—that is, had blood containing the antigen—and had Rh-negative mothers. He concluded that the blood cells of the Rh-positive newborn were hemolyzed by an antibody provided by the Rh-negative mother, and was able to show that such an antibody existed in the serum of Rh-negative mothers who were, or had been, pregnant with Rh-positive offspring. It was lacking in most Rh-negative women who had never been pregnant or had only nonaffected children: the few such women who had the antibody had received repeated blood transfusions, some of them many years earlier.

This led to the following explanation of the origin of erythroblastosis (Fig. 178). Normally, human blood does not contain specific antibodies against the

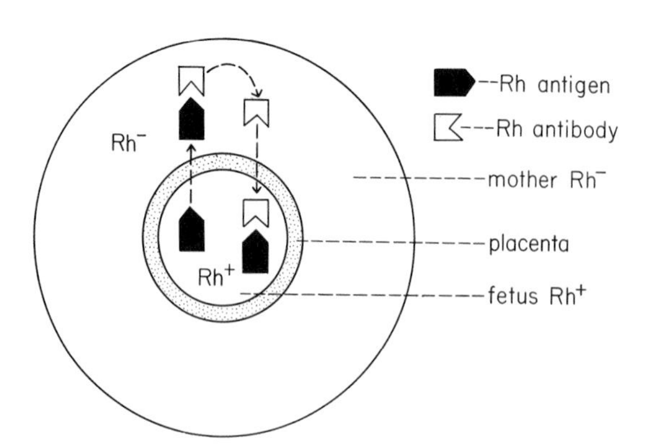

FIGURE 178
Diagram of the course of events leading to erythroblastosis fetalis of an Rh-positive embryo in an Rh-negative mother.

Rh antigen, but the antibodies may be produced in Rh-negative persons by repeated transfusions with Rh-positive blood. This iso-immunization, the production of antibodies in the same species (here, man) from which the antigen comes, is comparable to the experimental production of antibodies in rabbits or other animals by injection of blood from different species. When an Rh-negative woman carries an Rh-positive fetus the Rh antigen from the latter may find its way into the blood of the mother and cause the production of an antibody. This antibody does not harm the blood cells of the Rh-negative mother, since they lack the antigen. When, however, the antibody finds its way through the placenta to the Rh-positive fetus, an immunological reaction between the antibody and the red cells of the fetus takes place, resulting in erythroblastosis fetalis.

In the inverse case—that of an Rh-positive mother carrying an Rh-negative child—iso-immunization does not occur. The antibody-forming mechanism of a child does not mature until about six months after birth, and the fetus is therefore unable to produce antibodies against the antigens in its mother's blood.

Because Rh-positive erythroblastotic children from Rh-negative mothers trace their *R* allele to their fathers, the fathers of affected children are Rh positive. The frequency of erythroblastosis fetalis among all pregnancies is determined by the frequency of the *R* allele in the population. The frequency of the *R* allele (p) and that of the *r* allele (q) can be determined by the use of the Hardy-Weinberg Law. Since Rh-negative individuals are *rr*, their frequency is q^2, and the allele frequency of *r* is equal to the square root of the frequency of Rh-negative individuals. In a white population, this frequency is close to 16 per cent (0.16), so that $q = 0.4$ and $p = 0.6$. From this, it follows that the frequency of homozygous *RR* individuals is 36 per cent (p^2) and that of heterozygotes *Rr* is 48 per cent ($2 \times p \times q$).

Rh-positive pregnancies in Rh-negative women result from two kinds of marriages: *rr* (females) × *RR* (males), and *rr* (females) × *Rr* (males). The frequencies of these marriages are $p^2 \times q^2 = 5.76$ per cent, and $q^2 \times 2pq = 7.68$ per cent. In the first, all pregnancies result in production of an Rh-positive fetus, but only one-half of the children in the latter type of marriage are Rh positive. Thus, the frequency of all potentially unfavorable pregnancies in the population is $5.76 + 3.84 = 9.60$ per cent, or about one-tenth of all pregnancies.

Even before methods were devised to reduce greatly the incidence of the disease, it was very much lower than the incidence of potentially unfavorable pregnancies. It occurred about once in 200–500 pregnancies in white populations: that is, only in from 1 out of 20 to 1 out of 50 of all potential cases. The reasons for this low occurrence will be discussed beginning on page 438.

If the disease has occurred during one pregnancy, it recurs, often more severely, in every succeeding pregnancy in which the genotpye of the fetus is the unfavorable *Rr*. This means that if the father is homozygous *RR*, all succeeding pregnancies lead to illness of the child. If, on the other hand, he is heterozygous *Rr*, there is a 1:1 chance that the child will be healthy.

The practical consequences of the immunological and genetic understanding of erythroblastosis are great. Immunologically, it is clear that Rh-negative females should not be given transfusions of Rh-positive blood before the end of their reproductive period, since they may build up antibodies which may react with an Rh-positive fetus in a subsequent pregnancy, and that newly born children with serious cases of erythroblastosis should be given transfusions primarily to replace the affected red cells. In extreme cases, in order not only to replace the affected cells but also to remove the antibodies derived from the mother, all of the child's blood should be replaced with blood from Rh-negative donors.

It is now a routine procedure to test a pregnant woman for the presence or absence of the Rh antigen. If she is Rh positive, no special precautions are required. If she is Rh negative, the husband is tested. If he is also Rh negative there is no problem; but even if he is Rh positive, the chances that the child will be healthy are very high, since erythroblastosis, as we have seen, occurs in only a small fraction of Rh-positive pregnancies in Rh-negative mothers. Moreover, even the occurrence of this small fraction can now be prevented, as will be described at the end of this chapter.

It was shown in an earlier chapter that many different *R* alleles exist in human populations (see pp. 270–275). Erythroblastosis fetalis may occur whenever a fetus carries any allele at the *R* locus which produces antigens not present in the mother. The terms Rh positive and Rh negative do not take into account the great diversity of antigenic constitutions. Although the majority of children with erythroblastosis are the result of presence of the antigen Rh_0 (D) in the child and its absence in the mother, the disease may also be produced by presence of any one of the other Rh antigens in the child—rh′ (C), rh″ (E), or even hr′ (c) and hr″ (e)—and its absence in the mother. Thus, a family has been reported in which an Rh-positive mother had an affected child. Examination showed the presence of antigens Rh_0, rh′, and rh″ in the child and of Rh_0 and rh′ in the mother. Accordingly, it was suspected that the rh″ antigen was responsible for the disease, and this hypothesis was proved correct by finding antibodies against rh″ in the mother's blood.

ABO Incompatibility. Occasionally, erythroblastosis fetalis results from incompatibility between mother and child for blood-group antigens other than the Rh antigens. The disease has occurred in Kell combinations as well as in $L^{M,N,S,s}$ and some of the "private" family blood groups. Of particular interest are the results of incompatibilities involving the ABO blood groups. Although rare, erythroblastosis fetalis has been observed in A and B children from O mothers, and in other combinations in which the child forms an A or B antigen that is not present in the mother herself. Such mothers normally carry anti-A or anti-B isoantibodies in their blood. These antibodies may pass the placental barrier and occasionally lead to a mild form of hemolytic disease. Very rarely, the concentration of these antibodies becomes so high that enough diffuses into the child's blood to cause serious damage.

Much more important that the very uncommon erythroblastosis of the late fetus or newborn from ABO incompatibility is an embryologically earlier ABO interaction between mother and child. From an antigenic point of view, we can distinguish between "compatible" and "incompatible" marriages. The former are those in which the husband cannot transmit an I^A or I^B allele to his children which they could not also get from the wife; the latter are those in which the husband can transmit an I^A or I^B allele which is not present in the wife. Examples of compatible marriages are A♀× O♂, A♀× A♂, and AB♀× O or A or B or AB♂; of incompatible marriages, O♀× A or B or AB♂, A♀× B♂, and B♀× AB♂. In compatible marriages the mother's blood will contain only those anti-A or anti-B antibodies for which her children lack the corresponding antigens; in incompatible marriages children may be produced whose antigens can be attacked by the mother's antibodies. Data on the actual outcome of ABO-incompatible pregnancies are contradictory. In some surveys a deficiency of children with incompatible genotypes was observed, leading to the hypothesis that prenatal lethality accounted for the loss of such children. For instance, in Japanese families, Matsunaga found that in the compatible matings of A women to O men the proportion of A and O children was as expected from the allele frequencies but that in the reciprocal incompatible matings of O women to A men, there was a deficiency of A children (Table 50). This difference in the outcome of the reciprocal matings seemed to be the result of frequent loss of the incompatible A fetuses in O women. Since the total stillbirth rates of children in the two reciprocal matings was about the same, it was postulated that the loss of incompatible A fetuses occurred in the form of early spontaneous abortions. Indeed the frequencies of abortions in the matings A × O was less than that in O × A . Certain Caucasian family data support

TABLE 50
Number of O and A Children from Compatible and Incompatible O × A Matings.

Mother	*Father*	*No. of O Children*		*No. of A Children*	
		Obs.	Exp.	Obs.	Exp.
A	O	282	284.5	430	427.5
O	A	320	275.3	369	413.7

Source: Matsunaga, *Amer. J. Hum. Genet.*, 7, 1955.

the finding of increased fetal loss in ABO incompatibility but later Japanese studies by Matsunaga and associates have given no evidence for extensive fetal loss in ABO incompatibility. At the time of writing the issue remains unresolved.

Rh and ABO Incompatibilities. ABO incompatibility between mother and child does not always have undesirable effects: on the contrary, the effects are generally beneficial. This surprising state of affairs arises from an interaction of Rh and ABO incompatibility. We have seen earlier that only a fraction of the potentially incompatible Rh combinations of mother and child results in erythroblastosis. One of the reasons for this low incidence is clearly connected with the fact that it takes at least one Rh-positive pregnancy of an Rh-negative mother to build up enough antibodies to lead to hemolytic disease in the next Rh-positive child. The main reason for the low incidence of Rh-dependent erythroblastosis, however, is that ABO incompatibility between an Rh-negative mother and an Rh-positive child greatly diminishes the probability of Rh hemolytic disease. Levine noted this early in his studies, and others confirmed his findings. To take two extreme examples: Rh hemolytic disease occurs very rarely in the offspring of the ABO-incompatible mating of an O mother and an AB father; but in the ABO-compatible mating of an AB mother and an O father the second child may be affected. According to Levine, the explanation for this protection of Rh-incompatible fetuses by ABO incompatibility lies in the fate of the child's red cells if and when they reach the mother's blood stream. Rh-positive cells of the fetus that are ABO-incompatible with the mother will be destroyed by the normally present anti-A or anti-B antibodies of the mother before they are able to stimulate the production of anti-Rh antibodies in her blood. If, however, the Rh-positive blood cells that pass into the mother are ABO compatible, they will survive and stimulate the formation of anti-Rh antibodies. A statistical analysis based on considerations of allelic frequencies has shown that this explanation accounts well for the detailed frequencies of Rh hemolytic disease in the various types of ABO marriage combinations. In addition to the protection furnished by ABO against Rh fetal death, there seems to be reciprocal protection by Rh against ABO fetal death (Bernice Cohen). The rationale for this phenomenon is not known.

For some years the recognition of the protective nature of ABO incompatibility against Rh incompatibility remained as an intellectually satisfying but practically unimportant item. Gradually C. A. Clarke and his colleagues in Liverpool recognized that man can experimentally provide protection against Rh incompatibility when the ABO groups fail to do so. For months they, including Mrs. Clarke, had puzzled over the problem. Then, Clarke relates, one night he was awakened by his wife who said, "Give them anti-Rh." Irritated by being disturbed in deep sleep, he replied: "It is anti-Rh we are trying to *prevent* them from making," and went to sleep again. The next morning, however, the idea made sense. If, at the right time, anti-Rh is given to a pregnant Rh-negative woman, it will agglutinate and destroy any Rh-positive blood cells that she might acquire across the placenta from an Rh-positive fetus. The right time, it became apparent, is a short period around the birth of the child. During most of the pregnancy period not enough red cells leak from the child to the mother to stimulate the production of anti-Rh antibodies. Most of the leakage occurs shortly before or during birth. Injection of anti-Rh within up to 72 hours after birth will lead to destruction of leaked Rh-positive cells before they have had a chance to lead to anti-Rh production (in addition the presence of injected anti-Rh may have a depressing effect on the mother's potential anti-Rh producing immunological mechanism). It is now clear why without treatment first Rh-incompatible children are not affected. It is because anti-Rh is usually not made until after the birth of the first child. Anti-Rh injection shortly after this birth thus protects the next child, and so on. By 1971 in combined data from England and Baltimore only one out of 173 women who had born an ABO-compatible, Rh-incompatible first child and then had received treatment as just described had built up anti-Rh, while 38 out of 176 not-treated controls had a titer of anti-Rh in their blood. Only two out of 86 treated women showed anti-Rh immunization during subsequent pregnancies as compared to 20 out of 65 subsequent pregnancies in untreated controls. Altogether, since 1967 innumerable women at risk have been treated successfully.

There are many intriguing details that have to be left out of this account, including the work of a New York team who independently of that in Liverpool conceived the use of anti-Rh. Let it be mentioned only that most of the anti-Rh material injected into women is obtained from volunteer Rh-negative males who produce the antibody in response to injection of Rh-positive red cells.

PROBLEMS

138. Among 228 children affected with cleft palate or harelip, 40 per cent were first-born, and 14 per cent later-born.
   (a) What do these figures indicate concerning the relationship between birth order and defect?
   (b) What bearing on the preceding question has the information that, in a comparable sample of 15,000 children born in the general population, the distribution according to birth order was 38.1, 32.6, 15.1, and 14.2 per cent, respectively?

In Problems 139–143, assume the existence of two alleles only, R and r.

139. A baby girl affected with erythroblastosis fetalis recovers and reaches adulthood. Her fiancé was the last child born healthy in a family in which several later children died from erythroblastosis. Discuss the prospects for healthy or affected children in the proposed marriage.

140. Of a couple's three children, the third was erythroblastotic at birth. The couple asks for advice regarding the prognosis for a future child. Questioning discloses that a younger sister of the husband died from erythroblastosis as a baby. Which advice is the sounder:
    (a) No more healthy children to be expected?
    (b) A chance of 50 per cent for healthy children?

141. In a certain population, 4 per cent of all people are Rh negative. What is the frequency of marriages in which erythroblastosis may occur among the offspring?

142. All Chinese, but only about 91 per cent of blacks, are Rh positive.
    (a) What frequency of erythroblastosis fetalis is to be expected in each group separately if k, the fraction of actually affected among the potentially affected, is 0.05?
    (b) After intermarriage of equal numbers of Chinese and blacks and many generations of random mating, what proportion of Rh-positive and Rh-negative individuals are to be expected?
    (c) What frequency of erythroblastosis is to be expected in the mixed population ($k = 0.05$)?

143. If statistics were available from the preceding century on the ratio of erythroblastotic children to normal children in sibships in which this disease occurred, and regarding the ratio of albino to normal children in sibships in which albinism occurred, it would probably be found that the proportion of the erythroblastotic children has decreased in this century, while that of albinism has increased? Why would this be true?

REFERENCES

Banerjee, Papia, 1969. Birth weight of the Bengali new born: effect of the economic position of the mother. *Ann. Hum. Genet.,* **33**:99–108.

Böök, J. A., Fraccaro, M., Hagert, C. G., and Lindsten, J., 1958. Congenital malformations in children of mothers aged 42 and over. *Nature*, **181**:1545–1546.

Carter, C. O., and Roberts, J. A. F., 1967. The risk of recurrence after two children with central-nervous-system malformations. **Lancet**, **1**:306–308.

Clarke, C. A., 1968. The prevention of "Rhesus" babies. *Sci. Amer.*, **219**:46–52.

Conference on Parental Age and Characteristics of the Offspring. 1954. *Ann. N.Y. Acad. Sci.*, **57**:451–614.

Farley, F. H., 1969. Comment on low intelligence and month of birth. *Hum. Hered.*, **19**:307–311.

Fishbein, L., Flamm, W. G., and Falk, H. L., 1970. *Chemical Mutagens: Environmental Effects on Biological Systems.* 364 pp. Academic, New York.

Fraser, F. C., 1955. Thoughts on the etiology of clefts of the palate and lip. *Acta Genet. Stat. Med.*, **5**:358–369.

Green, E. L, and Green, Margaret C., 1959. Transplantation of ova in mice. *J. Hered.*, **50**:109–114.

Harrell, R. F., Woodyard, E., and Gates, A. I., 1955. The effect of mothers' diets on the intelligence of offspring. 71 pp. Bureau of Publ. Teachers College, Columbia University, New York.

Hewitt, D., 1965. Regional variations in the incidence of spina bifida. Pp. 295–303 *in* Neel, J. V., Shaw, Margery W., and Schull, W. J., (Eds.) *Genetics and the Epidemiology of Chronic Diseases.* Government Printing Office, Washington, D.C. (U.S. Public Health Service Publ. No. 1163.)

Hollander, W. F., and Gowen J. W., 1959. A single-gene antagonism between mother and fetus in the mouse. *Proc. Soc. Exp. Biol. Med.*, **101**:425–428.

James, W. H., 1969. Central nervous system malformation stillbirths, maternal age, and birth order. *Ann. Hum. Genet.*, **32**:223–236.

Joffee, J. M., 1969. *Prenatal Determinants of Behavior.* 366 pp. Pergamon, Oxford.

Kaelber, C. T., and Pugh, T. F., 1969. Influence of intrauterine relations on the intelligence of twins. *New Eng. J. Med.*, **280**:1030–1034.

Kalmus, H., 1947. The incidence of placenta praevia and antepartum haemorrhage according to maternal age and parity. *Ann. Eugen.*, **13**:283–290.

Kalter, H., and Warkany, J., 1959. Experimental production of congenital malformations in mammals by metabolic procedure. *Physiol. Rev.*, **39**: 69–115.

Lenz, W., 1959. Der Einfluss des Alters der Eltern und der Geburtennummer auf angeborene pathologische Zustände beim Kind. *Acta Genet. Stat. Med.*, **9**:169–201.

Levine, P. 1958. The influence of the ABO system on Rh hemolytic disease. *Hum. Biol.*, **30**:14–28.

Levine, P., Burnham, L., Katzin, E. M., and Vogel, P., 1941. The role of isoimmunization in the pathogenesis of erythroblastosis fetalis. *Amer. J. Obst. Gynecol.*, **42**:925–937.

McKeown, T., MacMahon, B., and Record, R. G., 1951. The incidence of congenital pyloric stenosis related to birth rank and maternal age. *Ann. Eugen.*, **16**:249–259.

Montagu, M. F. A., 1962. *Prenatal Influences.* 614 pp. Thomas, Springfield, Ill.

Murakami, U., 1955. Experimental embryologic and pathologic study on malformation of the central nervous system. *Acta Pathol. Jap.*, **5** Suppl.: 495–513.

Neel, J. V., 1958. A study of major congenital defects in Japanese infants. *Amer. J. Hum. Genet.*, **10**:398–445.

Newcombe, H. B., 1965. Environmental versus genetic interpretations of birth-order effects. *Eugen. Quart.*, **12**:90–101.

Newcombe, H. B., and Tavendale, O. G., 1964. Maternal age and birth order correlations. Problems of distinguishing mutational from environmental components. *Mut. Res.*, **1**:446–467.

Penrose, L. S., 1939. Maternal age, order of birth and developmental abnormalities. *J. Ment. Sci.*, **85:**1141–1150.

Russell, W. L., 1948. Maternal influence on number of lumbar vertebrae in mice raised from transplanted ovaries. *Genetics*, **33:**627–628.

Scrimshaw, N. S., and Gordon, E. (Eds.), 1968. *Malnutrition, Learning, and Behavior.* 566 pp. M.I.T. Press, Cambridge, Mass.

Thompson, W. R., Watson, J., and Charlesworth, W. R., 1962. *The Effects of Prenatal Maternal Stress on Offspring Behavior in Rats.* 26 pp. American Psychological Association, Washington, D.C. (Psychological Monograph 76.)

Yen, Stella, and MacMahon, B., 1968. Genetics of anencephaly and spina bifida? *Lancet*, **2:**623–626.

18

POLYGENIC INHERITANCE

Mendel's discovery of segregating alleles accounted immediately for the sharply defined segregation of phenotypes in single factor inheritance. The occurrence, in a population and in individual sibships, of albinos and pigmented persons or of brachydactylous and normal individuals, was readily explained by the existence of two different alleles. However, a large number of traits that are known to be inherited do not manifest themselves as sharply defined pairs of phenotypes. Normal body height, for instance, covers a wide range. Measurements of height in a population show a continuous gradation, and even sibs vary in size from their parents, as well as among themselves. Other examples of such "quantitative" characters are found among traits that can be classified according to a numerical scale: longevity, degree of resistance to diseases, age of onset of disease, score in mental tests, degree of hair pigmentation, amount of skin pigment in black-white hybrids, basal metabolism, rate of heart beat, level of vascular tension, and dimension of any particular bodily structure, such as length of finger or weight of thyroid gland.

Undoubtedly, some of the variability of these characters is due to a response of the genotypes to differences in the external or internal environment. If, for instance, there were three genotypes for body height, A^1A^1, A^1A^2, and A^2A^2, which in identical environments determined the development of three phenotypes, short, medium, and tall, and if each of these genotypes, in different environmental conditions, expressed itself in different phenotypes, then a continuous series of body heights might well be produced. There is, indeed, evidence that the environment influences the expression of body height and many other quantitative characters, although it is not the only agent responsible for the wide array of intergrading phenotypes. Thus, studies of genetically identical twins show that, in general, environmental differences produce differences in body height and other traits that are smaller than differences normally found between genetically nonidentical twins.

Multiple Allelism. It may be suggested that the genetic component of the continuous variablity of quantitative characters is multiple allelism at a single locus responsible for the variability. But although it may be true that part of the hereditary variability of quantitative traits in a population is due to the existence of genotypes A^1A^1, A^1A^2, A^1A^3, A^2A^3, etc., all of which determine a different degree of phenotypic expression, multiple allelism cannot explain the wide range of variation. This may be shown by considering the variability of a trait within a single sibship. Two parents can supply, for a single locus, no more than four different alleles, as in the marriage $A^1A^2 \times A^3A^4$. Among the children of such parents, the maximum number of genotypes based on multiple alleles at one locus that may appear is thus four—namely, A^1A^3, A^1A^4, A^2A^3, and A^2A^4. Observations of such traits as height or hair color in large families show that more than four different phenotypes often occur.

Polygenic Inheritance. A satisfactory genetic interpretation of the inheritance of quantitatively graded characters was first suggested by Mendel. In addition to his famous experiments with peas, he reported on a cross between white and purple-red flowering beans. The hybrids had flowers with less intense coloration than the purple-red parent. In the second generation, Mendel obtained a whole series of different colors—from white through pale violet to purple-red—instead of the expected two types in the simple 3:1 ratio discovered in other experiments. His tentative explanation was that *more than one pair* of genes operated in specifying color of flower. This hypothesis of *multifactor*, or *polygenic, inheritance* was later proven to be correct by Nilsson-Ehle (1873–1949) in an analysis of a graded series of seed pigmentation in wheat crosses and by Weinberg and R. A. Fisher, who provided a detailed mathematical model partly based on pioneering work by Karl Pearson (1857–1936).

DIFFERENCES IN PIGMENTATION BETWEEN BLACKS AND WHITES

We shall describe the theory of polygenic inheritance as it was first applied to a case of human genetics by Gertrude and Charles B. Davenport (1866-1944) in 1910 and 1913. These investigators were interested in the inheritance of differences in skin color and, since the normal differences between members of the Caucasian race are relatively slight, they collected data on black-white crosses.

There was considerable variation in skin pigmentation both within white groups without black ancestry and black groups without white ancestry, and a very great variation among those of mixed ancestry. The Davenports devised an objective classification of the different degrees of pigmentation by means of a "color top," which is a rotating circular disc on which different sectors bear different colors. In classifying the array of pigmentation found in the population investigated as expressed by the percentage of black on a color top, the investigators were able to distinguish five different classes of skin color.

Genetic Interpretation. The first-generation hybrids, the typical mulattoes, showed considerable variability in pigmentation, with the majority falling into Class 2 (Table 51). The two individuals in Class 1 were almost dark enough and the five individuals in Class 3 almost light enough to be in Class 2. This variability was probably due to the minor genetic factors which cause the observed variations within the original races—for instance, the light skin of many Northern Europeans in contrast to the dark skin of many Mediterranean whites, or the yellow-brown of Africans from the Sudan in contrast to the black of those from Nigeria.

A simple single-locus interpretation of the skin color of the first generation would assign the genotype A^1A^1 to the white parents, A^2A^2 to the blacks, and A^1A^2 to the hybrids. If this were correct, marriages between hybrids $A^1A^2 \times A^1A^2$ should segregate into 1/4 A^1A^1 (white), 1/2 A^1A^2 (mulatto), and 1/4 A^2A^2 (black).

TABLE 51
Pigmentation of White-Black Hybrids.

Black Area (%)	Color Class	Designation of Skin Color	First-generation Hybrids	Offspring of Two First-generation Hybrids
0–11	0	White	–	3
12–25	1	Light Mulatto	2	10
26–40	2	Mulatto	22	13
41–55	3	Dark Mulatto	5	5
56–78	4	Black	–	1

Source: Davenport.

The actual distribution was different. Marriages between first-generation hybrids produced offspring who were distributed over the whole range of pigmentation (Table 51). In the small number of sibships available for study, a few children were as light as typical whites (Class 0), one as dark as a typical black (Class 4), many were "light mulattoes" (Class 1), many "dark mulattoes" (Class 3), and the largest group typical mulattoes (Class 2). It is clear that a single-locus interpretation does not explain the manifold color types.

A more satisfactory hypothesis is based on the assumption that skin color is controlled by two independent loci; for each locus there are two alleles, A^1 and A^2 for one and B^1 and B^2 for the other. According to this hypothesis, whites are $A^1A^1B^1B^1$, and blacks $A^2A^2B^2B^2$. The darker pigmentation of blacks is considered to be due to *additive* action of the A^2 and B^2 alleles, and it is assumed that substitution of A^2 or B^2 for A^1 or B^1 leads to equal increases in pigmentation. The amount of pigment produced by A^2 in A^1A^2 heterozygotes is assumed to be intermediate between that in A^1A^1 and A^2A^2 homozygotes, and the same holds true for B^2.

First-generation mulattoes are $A^1A^2B^1B^2$. Their intermediate phenotype is the result of the presence of one of each of the A^2 and B^2 alleles, which makes the individuals darker than whites, who have neither of these alleles, and lighter than blacks, who have two of each. When first-generation mulattoes produce gametes, independent assortment will lead to the formation of four different kinds—A^1B^1, A^2B^1, A^1B^2, and A^2B^2—containing either none, one, or two alleles for increased pigmentation. The children of two mulattoes will have from zero to four such alleles and thus fall into five different pigmentation classes:

		$A^1A^1B^2B^2$		
	$A^1A^1B^1B^2$	$A^1A^2B^1B^2$	$A^1A^2B^2B^2$	
$A^1A^1B^1B^1$	$A^1A^2B^1B^1$	$A^2A^2B^1B^1$	$A^2A^2B^1B^2$	$A^2A^2B^2B^2$
Class 0	Class 1	Class 2	Class 3	Class 4

The two-locus hypothesis thus accounts, qualitatively, for the appearance of all five classes in the offspring of two mulattoes. It implies, further, that typical mulattoes may have any one of three different genotypes, and both light and dark mulattoes either of two different genotypes. Consequently, marriages of light, typical, or dark mulattoes may produce various numbers and proportions of pigmentation types in the offspring. For example, a marriage of two typical mulattoes, both of whom are $A^2A^2B^1B^1$, will result in children who are phenotypically and genotypically like their parents; a marriage between two typical mulattoes, one of whom is $A^2A^2B^1B^1$, and the other $A^1A^2B^1B^2$ with A^2B^1 gametes produced by one parent and A^1B^1, A^2B^1, A^1B^2, and A^2B^2 by the other, leads to three types of offspring—light mulatto ($A^1A^2B^1B^1$), typical mulatto ($A^2A^2B^1B^1$, $A^1A^2B^1B^2$), and dark mulatto ($A^2A^2B^1B^2$)—in the proportion 1:2:1.

Davenport concluded that his data on the proportions of pigmentation types of offspring from various types of parents showed that observation and expectation according to his theory agreed reasonably well. Later workers have pointed out discrepancies which require expansion of the simple two-locus theory to one involving a greater number of loci. For example, the original division of color types into five classes was artificial; actually, there is a continuous distribution. This is clear not only from the color-top studies but also from modern reflectance spectrophotometry, which measures the amount of melanin pigment in the skin by the amount of light reflected from it at different wavelengths (Fig. 179). Harrison and Owen, who have applied this method to the skin of a group of white Englishwomen, their West African black husbands, their hybrid children, and their few (at the time of study) grandchildren, found that the skin color of the Africans varied much more than that of the whites. Very likely, the genes that modify pigmentation have a better chance of phenotypic expression if a large amount of melanin is present than if only a little melanin is. The color of first-generation mulattoes varied considerably even in the same sibship. This is evidence for segregation of modifying genes—particularly, perhaps, those from the white parents. It may be added that, on

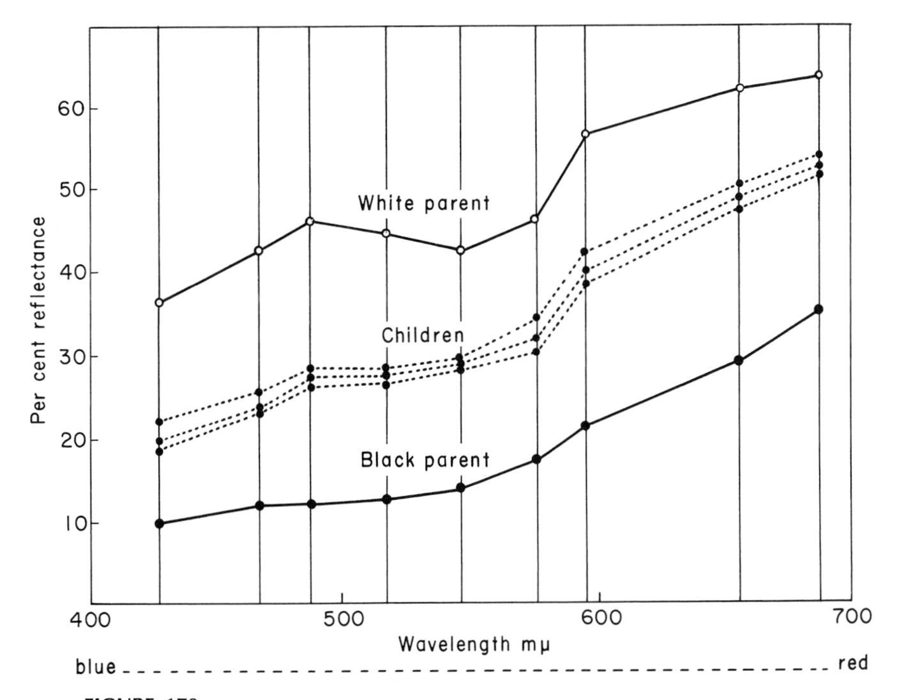

FIGURE 179
The reflectance curves, which measure skin color, for light of different wavelengths between 400 and 700 millimicrons for a white ♀ × black ♂ family. (After Harrison, *Eugen. Rev.*, **49,** 1957.)

the average, the pigmentation of the mulattoes was somewhat closer to that of the white than to that of the black parents.

Even refined determinations of the relative amounts of melanin in the skin of different individuals and groups do not provide data from which the exact type of inheritance of skin color can be deduced. However, hypotheses can be formulated on the number of gene pairs that are operating if some more-or-less arbitrary assumptions are made about whether the phenotypes of the heterozygotes for each pair are intermediate between those of the homozygotes or whether they are closer to one or the other homozygous phenotype, and whether the effects of the different genes on the amount of melanin formed are simply additive or more complex. Expectations from any specific hypothesis can then be compared with observations.

The result of such analyses have led to the conclusion that Davenport's two-gene-pair model does not fit the more extensive data that later became available. Models based on three to six pairs of additively acting genes seem to be closer to the facts than the original two-pair hypothesis, or hypotheses based on 10 or 20 pairs.

With a model of five equal and additive pairs, the color genotypes of Caucasians and Africans would be $A^1A^1B^1B^1C^1C^1D^1D^1E^1E^1$ and $A^2A^2B^2B^2C^2C^2D^2D^2E^2E^2$, respectively, and the American black population would be composed of $3^5 = 243$ different color genotypes, containing from zero to ten "dark" alleles and thus forming eleven pigmentation phenotypes.

Pigmentation of offspring from segregants. According to the two-locus hypothesis, white-skinned segregants in later generations of a black-white union are genetically $A^1A^1B^1B^1$, just like a Caucasian of unmixed ancestry. In fact, on any hypothesis not assuming dominance, white-skinned segregants should be of the same color genotype as their white forebears. Consequently, neither a marriage between a white segregant and a "pure" white or between two white segregants should give rise to darkly pigmented children. This expectation is fulfilled as far as our knowledge goes in spite of the popular belief that "black" (= very dark) children may be born to "white" parents one or both of whom had segregated out in the offspring of a past black-white cross. The belief may, perhaps, be based on marriages between two light mulattoes who may be nearly white-skinned and yet produce children darker than themselves. In marriages of a "near-white" person of black-white ancestry to a white, no child should be darker than the near-white parent, apart perhaps, from some effect of minor modifying alleles. But, since the genetic basis of skin-color differences is not established beyond doubt, it would be unscientific to deny the possibility that, in marriages of whites and near-whites, children somewhat darker than their near-white parent could be produced.

It is most improbable, however, that "black" or even very dark children can issue from such unions, and it can be stated with certainty that no well-established case of this nature has been reported. On the contrary, the few alleged instances which it has been possible to investigate either turned out to be based on hearsay and not on fact, or were, apparently, due to illegitimacy involving a darkly pigmented parent.

The genetic basis of other physical traits that differentiate the whites and blacks is independent from that of pigmentation, and it is possible that some genotypes which underlie phenotypes can be reconstituted among the offspring of two black-white segregants in whom these traits do not appear. If this should occur together with pigmentation that is somewhat increased but still within the normal range of variation, it may serve to make an observer aware of an individual's black ancestry when skin color alone would not have done so. By themselves, traits other than color are seldom used in social discrimination, for custom has selected as an index the rather meaningless but easily noticed property, "amount of skin pigment."

Pigmentation after Random Mating. It is illuminating to consider what kind of a population would result from complete intermarriage of whites and blacks after genetic equilibrium had been established. It is known that the American-black group is a hybrid population which has derived approximately 80 per cent of its genes from African and approximately 20 per cent from Caucasian ancestors. (See p. 830ff.) If the frequency of the African alleles in the white group is close to zero and in the black group 0.8, and if the blacks make up one-tenth of the whole population, then the frequency of African alleles in the population is $q_b = 0.08$ and that of the Caucasian alleles $p_w = 0.92$. If we assume the validity of the five-locus model, the frequencies at equilibrium of the eleven color types with from zero to ten African alleles are obtained from the binomial $(0.92w + 0.08b)^{10}$. The calculations show that 43.4 per cent of the population will have only white genes, 37.8 per cent nine white genes, 14.8 per cent eight, and 3.4 per cent seven. These light types total 99.4 per cent, the remaining 0.6 per cent being composed of seven types with from four to ten genes derived from Africans (Table 52). A very tentative diagram (Fig. 180) of the approximate distribution of color types in the large white and the small black segment

TABLE 52
Frequencies, in Per Cent, of the Eleven Types of Color Segregants Expected from a Model of Pairs of Alleles at Five Equally Additive Loci in a Panmictic Population at Equilibrium in Which $p_w = 0.92$ (Caucasian Alleles) and $q_b = 0.08$ (African Alleles).

Number of African Alleles										
0	1	2	3	4	5	6	7	8	9	10
43.4	37.8	14.8	3.4	5×10^{-1}	5×10^{-2}	4×10^{-3}	2×10^{-4}	6×10^{-6}	1×10^{-7}	1×10^{-9}

Source: Stern, *Acta Genet. Stat. Med.*, **4**, 1953.

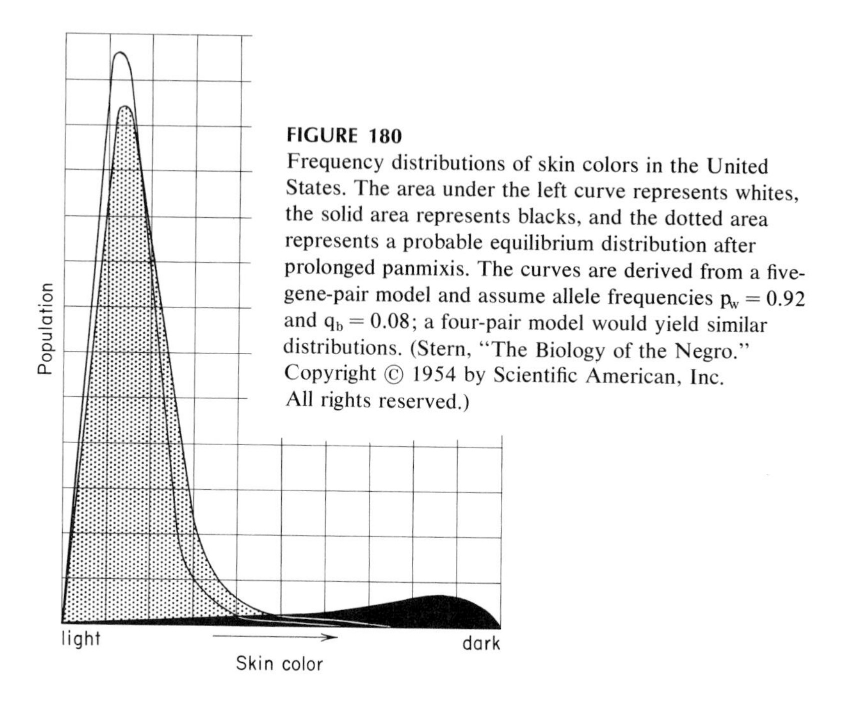

FIGURE 180
Frequency distributions of skin colors in the United States. The area under the left curve represents whites, the solid area represents blacks, and the dotted area represents a probable equilibrium distribution after prolonged panmixis. The curves are derived from a five-gene-pair model and assume allele frequencies $p_w = 0.92$ and $q_b = 0.08$; a four-pair model would yield similar distributions. (Stern, "The Biology of the Negro." Copyright © 1954 by Scientific American, Inc. All rights reserved.)

of the present American population and the distribution that would exist at equilibrium after panmixis shows that the average skin color would be only slightly darker and that very few deeply pigmented individuals would appear in each generation. This result may be surprising at first but it will appear less so if it is realized that at equilibrium the assumed ten alleles for dark pigmentation will be "diluted" by the ten-times more numerous alleles for lightness.

Frequency Distribution in Polygenic Inheritance. Not only do quantitative characters extend over a scale of measurable values, but many have a characteristic frequency distribution in which the largest group of individuals lies close to the mean measure of the range; and the greater the deviation from the mean, the smaller the groups become (Fig. 181, A). This distribution is represented graphically by the so-called normal curve. It will be shown that although the polygenic theory can account for this distribution, the same distribution may also be produced by action of the environment, or by combinations of genetic and environmental agents.

Let us assume that a trait is enhanced if the allele A^2 is substituted for the allele A^1, or if the alleles B^2, C^2, etc., are substituted for B^1, C^1, etc., at given loci. It will be assumed, further, that the alleles are equal in effect, cumulative in action, and all of equal frequency in the population ($p_{A^1,B^1,C^1,\ldots} = q_{A^2,B^2,C^2,\ldots} = 0.5$). An individual may have none, one, two, or any possible higher

1	0	0	1	5	7	7	22	25	26	27	17	11	17	4	4	1
4:10	4:11	5:0	5:1	5:2	5:3	5:4	5:5	5:6	5:7	5:8	5:9	5:10	5:11	6:0	6:1	6:2

A

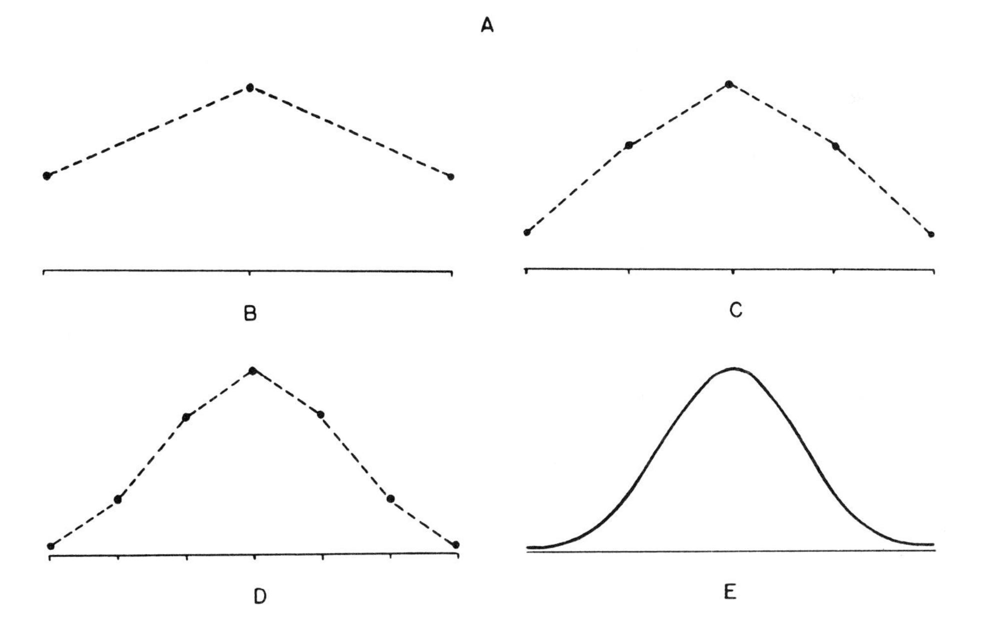

FIGURE 181
Frequency distributions of phenotypes in polygenic inheritance. **A.** Company of 175 soldiers arranged in groups according to height, from 4 feet 10 inches to 6 feet 2 inches. The lower row of numbers indicates the height of the groups, the upper row the number of men in the groups. **B.** One pair of alleles: distribution over three phenotypes. **C.** Two pairs of alleles: distribution over five phenotypes. **D.** Three pairs of alleles: distribution over seven phenotypes. **E.** An infinite number of pairs of alleles: distribution over a continuous array of phenotypes. For details of underlying assumptions, see text. (A, Blakeslee, *J. Hered.*, **5**, 1914.)

number of "enhancing" alleles; each number results in a different phenotype as measured on a scale for the trait.

If only one locus is considered, an individual may have none, one, or two enhancing (E) and two, one, or none nonenhancing (e) alleles, respectively. The frequencies of the three combinations, derived from the binomial $(1/2E + 1/2e)^2$, are in the ratios 1:2:1 (Fig. 181, B).

If two loci are considered, the probabilities for zero, one, two, three, or four enhancing alleles are derived from $(1/2E + 1/2e)^4$, which gives the ratios 1:4:6:4:1 (Fig. 181, C). With three loci, the formula is $(1/2E + 1/2e)^6$ and seven classes are expected, in the ratios 1:6:15:20:15:6:1 (Fig. 181, D).

All these distributions have one common feature: the largest number of individuals is found in the mean class, with a symmetrical decrease in frequency toward plus or minus deviation from the mean. With each increase in the number of loci considered, the number of classes increases, and the phenotypic differences between them decrease, so that each successive series consists of phenotypes that grade more and more closely into each other. The frequencies of the numerous classes follow a mathematical extension of the binomial ratios listed above to an extension which, for an infinite number of loci, would give a perfect "normal" distribution (Fig. 181, E).

Frequency Distributions in Environmental Variability. If a trait were controlled by the same genotype in all individuals, all variations in a quantitative measure of the trait in a population would be caused by variations in the environment. If the environment varied in only two alternative ways—for example, in the presence or absence of a specific factor—two phenotypic classes would exist: an enhanced phenotype if the factor were present; a nonenhanced phenotype if it were not. If presence and absence of the factor during the development of an individual were of equal likelihood, the two phenotypic classes would be equally common. With two such factors, and assuming identical and cumulative enhancing action, three phenotypic classes would be found, according to whether neither, one or the other, or both factors were present, and the frequencies of these classes would follow the ratios 1:2:1. Thus, if one of the factors in a plant's environment were fertilizer and the other good sunlight, both of which could be either present or absent, then one-quarter of the plants would not have either favorable agent, one-half lack one or the other, and one-quarter would have both.

With four different factors, five phenotypic classes, produced by the impact of zero, one, two, three, or four enhancing actions, would exist in the ratios 1:4:6:4:1. The greater the number of environmental factors concerned in phenotypic variability, the more nearly would its distribution approach "normal." It is thus clear that the same frequency distribution of graded phenotypes may be produced by genetic factors and by environmental agents, therefore also by combinations of both.

Fingerprint patterns. An example, carefully studied by Sarah Holt, of a characteristic whose determination is primarily genetic is that of the numbers of fingerprint ridges (Fig. 182). Adding the ridge counts of all 10 fingers of any person gives a value between 0 and about 300; the distribution of such ridge-count values in a population is close to that of a normal curve. The counts in different individuals are correlated with those of their relatives, the correlation coefficients between parent and child having been found to be +0.50, those between sibs +0.49, between nonidentical twins +0.49 and between identical twins +0.95. In contrast to these values the correlation between husband and wife is essentially 0, indicating absence of assortative mating for ridge count. It can be shown that the proportion of genes in common between parent and child, between sibs and between nonidentical twins is 1/2 and between identical twins is 1. It can further be shown that in the presence of only additive polygenic factors the correlation coefficients should be +0.5 for all the types of comparisons just mentioned except that between identical twins for which the coefficient should be 1. It is seen that the observed values closely approach the expected ones and thus fit the hypothesis of additive polygenes. There is, however, a slight effect of nongenetic factors as demonstrated by the fact that the observed correlation between genetically identical twins, 0.95, falls somewhat short of unity. It may be added that the finger ridges are fully differentiated by the end of the fourth month of intrauterine life. Any nongenetic influences on the formation of ridges must therefore be due to intrauterine variations of the external or internal environment of the fingers of the fetus.

FIGURE 182
The method of ridge-counting of fingerprint patterns as applied to a loop. The ridge count of this specific pattern, from the triradius (left) to the core (center), is 14. (From Holt, Sarah, 1968. Courtesy of Charles C Thomas, Publisher, Springfield, Ill.)

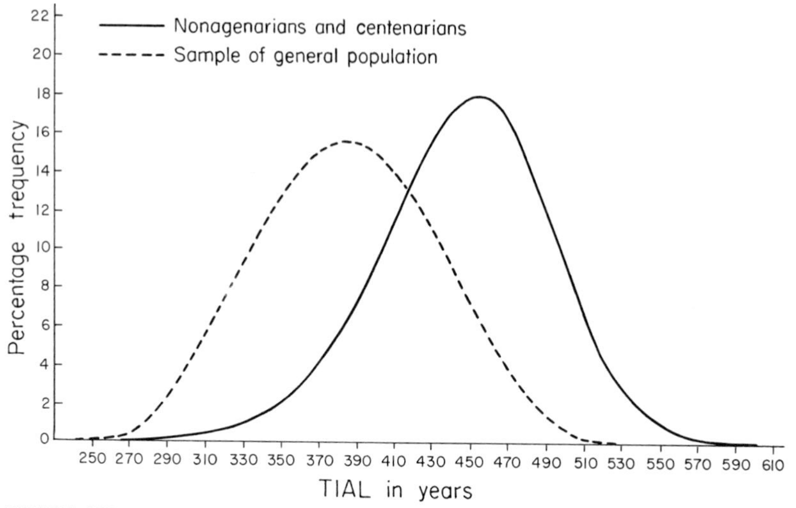

FIGURE 183
Fitted curves showing the distribution of variation in TIAL for a long-lived group and a comparison group. (After Pearl and Pearl, 1934.)

Length of life. Another example of continuously graded attributes is shown in Figure 183, which illustrates the distribution of length of life in the direct ancestors of nonagenarians and centenarians as compared to the distribution in the general population. As a measure of ancestral length of life, Raymond Pearl (1879–1940) and Ruth Pearl used the sums of the ages at death of the parents and grandparents of the propositi, a sum to which they gave the acronym TIAL (= Total Immediate Ancestral Longevity). The two curves show that the TIAL for the long-lived propositi is much greater than that for the comparison group. Specifically, the latter group's mean TIAL is 61 years less than that of the former. It is not possible in this case to separate clearly genetic and environmental factors in the determination of life span. Both kinds of factors are involved.

DOMINANCE IN POLYGENIC INHERITANCE

If the effects of alleles at different loci are not equal and cumulative, skewed frequency distributions, instead of the symmetrical normal distribution, will usually result. The same is true if dominance is involved in the expression of alleles at one or more loci.

In agricultural genetics, much attention has been paid to the analysis of quantitative variation of characters, such as the length of ears of corn or the body weight of fowls, and special statistical methods have been worked out to derive information from first- and second-generation data and those from crosses of hybrids to the parental types. The complexity of the interaction of many loci, each with at least two alleles, and of environmental variability, raises difficulties even in controlled experimental organisms. It is not surprising that our insight into the interactions that affect man is still more limited. Only a few theoretical situations in which dominance in polygenic inheritance has some interesting consequences will be outlined here.

When two individuals marry or two populations interbreed, and each is homozygous and isogenic for a different genotype controlled by two loci, and there is also complete dominance at each locus, the constitutions of the contrasting individuals or groups are either *AABB* and *aabb,* or *AAbb* and *aaBB.* (We shall assume that each of the dominant genes, *A* and *B*, has an enhancing, additive action, so that *A*, present either homozygously or heterozygously, makes the measurement of the trait 10 units greater than in *aa*, and that *B* acts similarly in relation to *bb*. With an index of measurement of 100 for *aabb*, the index is then 110 for *AAbb* and *aaBB,* and 120 for *AABB*.)

The cross *AABB* × *aabb* yields *AaBb* offspring whose index, 120, is like that of the "higher" parent, and, in later generations, segregation will produce all three possible phenotypes: 100, 110, 120. The cross *AAbb* × *aaBB*, in which the parents have equal indexes, namely 110, produces *AaBb* children, whose index, 120, is greater than that of their parents! Again, in later generations, segregation leads to all three phenotypes: 100 (that is, less than either original ancestor), 110, and 120. To use a slightly different example, let both *A* and *B* again be dominant and additive, but of different action, so that *A* leads to enhancement by 10 units and *B* by 5 units. In this case, the parental phenotypes of the cross *AAbb* × *aaBB* have indexes of 110 and 105, respectively, and the offspring *AaBb* an index of 115, again higher than either parent. This phenomenon—the stronger expression of a trait in certain polygenic heterozygotes than in their parental polygenic homozygotes—is at least partly the basis for heterosis.

Heterosis. Hybridization between different species or breeds of animals or plants often results in increased size, productiveness, and resistance to diseases or other unfavorable conditions of the environment. This phenomenon has been referred to as *the stimulating effects of hybridity, hybrid vigor,* or *heterosis*. The causes of heterosis may be diverse in different crosses, or different causes may be jointly effective in a single cross. Indeed, full clarification of the phenomenon of heterosis has not been obtained, but two main theories seem to contain important elements of truth, as demonstrated by experimental tests.

One theory, in simplified form, is that vigor results from the collaboration of many loci; that different species or breeds are not likely to carry all the favorable alleles at the various loci concerned; and that, therefore, hybrids may combine in their genotype favorable alleles from both parents. If these favorable alleles are dominant, then their effect in the first-generation hybrids will be apparent as increased vigor. Theoretical examples of genotypes fitting this explanation have been given in a preceding paragraph, the simplest case being a pair of *AAbb* and *aaBB* parents and their *AaBb* hybrid. The other important theory concerning causes of heterosis assumes that heterozygosity of single loci may result in increased vigor as compared to the vigor of the constituent homozygotes; perhaps on account of the production of two different proteins, one depending on an A^1 allele, the other on an A^2 allele, which might endow the heterozygote with greater functional versatility than either of the two homozygotes. Heterosis due to heterozygosity at a single locus is often called "overdominance."

The first of the two theories depends on the existence of dominant genes for vigor. Such dominance has been well established in specific cases—for example, by Mendel, in peas, for an allele for tall, dominant over dwarf, size. Similarly, in mice, normal size is dominant over pituitary dwarfism. It is possible that different normal strains of various organisms may differ in both dominant and recessive genes for vigor. If two isogenic breeds are, respectively, *AABB* and *aabb* where *A* is a dominant and *b* a recessive allele for increased vigor, then the hybrids, *AaBb*, would show no change in vigor if the *A* allele for increased and the *B* allele for decreased vigor balance each other (Fig. 184,I). If dominant genes for vigor are in excess over recessives, heterosis will result (Fig. 184,II); and if the recessive genes for vigor are in excess (or, to say it differently, if dominant genes for decreased vigor are in excess of those for increased vigor),

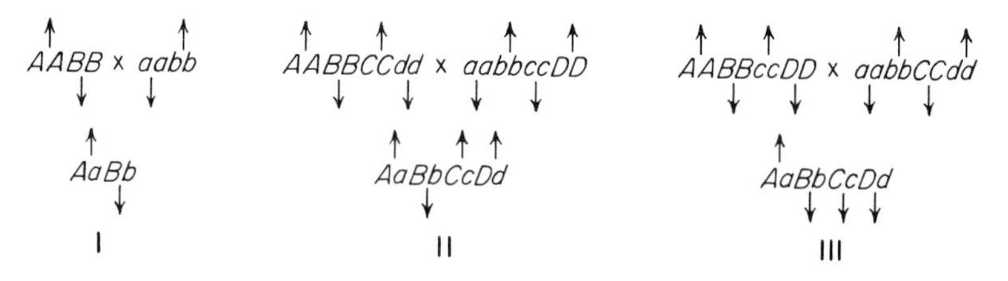

FIGURE 184
Three crosses between strains that differ at several loci concerned with increased or decreased vigor. Because of compensating effects of dominant and recessive genotypes at these loci, it is assumed that all strains are equally vigorous. Upward arrow = increased vigor; downward arrow = decreased vigor. I. If *A* increases and *B* decreases vigor, then the vigor of the hybrid is like that of the parents. II. If *A*, *C*, and *D* increase, and *B* decreases vigor, then the vigor of the hybrid is greater than that of parents. III. If *A* increases, and *B*, *C*, and *D* decrease vigor, then the vigor of the hybrid is less than that of the parents.

then the offspring of hybridization will be less vigorous than either parent (Fig. 184, III). The breakdown of human isolates within a given race, as well as hybridization between major racial groups, may well result in no change of some traits, improvement of others, and deterioration of still others.

The same may be true if the overdominance theory of heterosis holds. In addition to increased vigor of heterozygotes for a single locus, mutual interference in the action of two alleles may result in decreased phenotypic effectiveness. Very often, heterozygosity for single loci has no perceptible effect.

Heterosis and human stature. It is well known that the average body size has increased during the past century in various European countries, the United States, and Japan. Part of this increase undoubtedly is the result of improved external conditions. In addition, it has been suggested by Dahlberg (1893-1956) that the breakdown of isolates has furnished a genetic basis for heterosis. An ingenious study concerning possible heterosis in man is that by Hulse on various physical measurements of the descendants of villagers from the Swiss district of Ticino. Because the villages in this mountain district are rather isolated from one another, a very high percentage of marriages has been between individuals within the same village. Hulse has compared the stature of the offspring of parents from the same village (endogamous marriages) with that of the offspring of parents from two villages (exogamous marriages). Each of the two groups of offspring could be further subdivided into (1) those who remained in their native area, (2) those who had emigrated to California, and (3) those who were born in California of immigrant parents. The data show a decided influence of the environment, in that the men born in Switzerland were about 4 cm shorter than those born in California (Table 53). In addition, the mean stature of the exogamous groups was approximately 2 cm greater than that of the endogamous one. These findings have been interpreted as evidence for human heterosis. More recently, however, this interpretation has been challenged by Morton, Chung, and Mi, who found no indication of heterosis for metrical traits in the interracial crosses in Hawaii and whose statistical analysis of the Swiss data also lead them to the conclusion that they do not conform

TABLE 53
Mean Stature, in Centimeters, of Adult Male Offspring of Exogamous and Endogamous Marriages of Swiss Natives or Their Descendants.

Subgroup	*Exogamous*	*Endogamous*
(a) Swiss	168.51	166.21
(b) Emigrants	168.71	166.90
(c) Californians	172.33	170.50

Source: Hulse.

to expectation from models of heterosis. Notwithstanding these criticisms it seems possible that a minor component of heterosis is included in Hulse's data. Some other studies, such as those on inbreeding effects on the stature of Japanese and French children and the relation in the Polish city of Szczecin between growth of offspring and distance between parents' birthplaces, also seem to reveal indications of heterosis. There is, however, also some evidence for "negative heterosis" in man, i.e., a decrease in viability with increase of heterogeneity. Bresler has found for two American white populations that fetal loss in matings of a parental generation increases cumulatively with each additional country of birth in the great-grandparental generation.

POLYGENIC INHERITANCE OF ALTERNATIVE TRAITS

Although polygenic inheritance was first discovered in continuously variable traits, it is not confined to them. There are stages in the development of many organs and structures where growth and differentiation may take one of several alternative paths. Either a limb bud forms the normal number of digits, or too many, or too few; either the lateral parts of the embryonic face grow together and form a typical upper lip and palate, or the fusion is incomplete and results in harelip and cleft palate. For some traits, the choice of alternatives may be determined by the presence of a specific allele at a single locus; but for others, a polygenic system governs the developmental decision.

This was first demonstrated by Sewall Wright, who crossbred two strains of guinea pigs, one of which was normal, with three toes on the hind feet, the other polydactylous, with four toes. Nearly all first-generation hybrids were three-toed, and the second generation segregated into three-toed and four-toed individuals in a ratio of 3:1. This would seem to show simple single-factor inheritance in which the allele for three toes was dominant over that for four toes. But this interpretation was not supported by results of tests of the second-generation segregants. It became clear that approximately 4 pairs of additively acting genes differed in the two original strains, so that each strain had 8 alleles that controlled the number of toes. A minimum number of "four-toe" alleles—about 5, Wright deduced—was necessary to direct development of polydactyly (Fig. 185). The first-generation hybrids were nearly all three-toed, since they obtained only 4 "four-toe" alleles. In the second generation, segregation should lead to a series of nine types having from 0 to 8 "four-toe" alleles. Most with less than 5 of the alleles and some with 5 or more (variable penetrance could account for these) would be three-toed, but the great majority of those with 5, 6, 7, or 8 "four-toe" alleles would develop four toes.

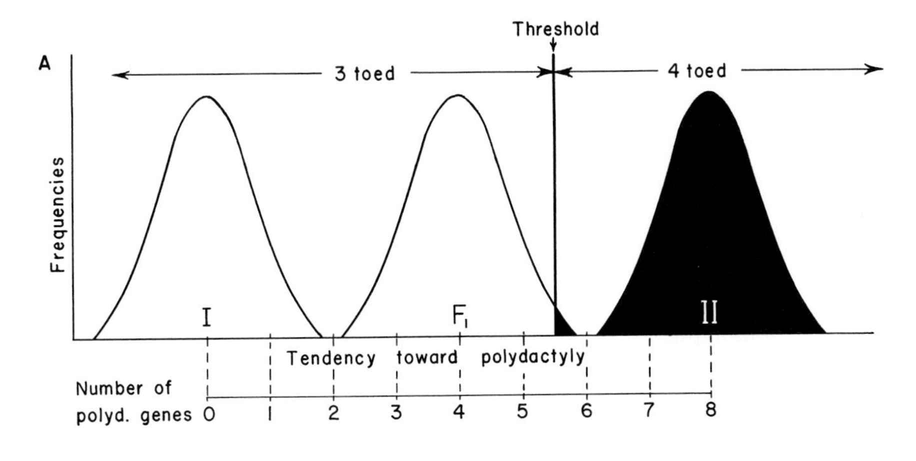

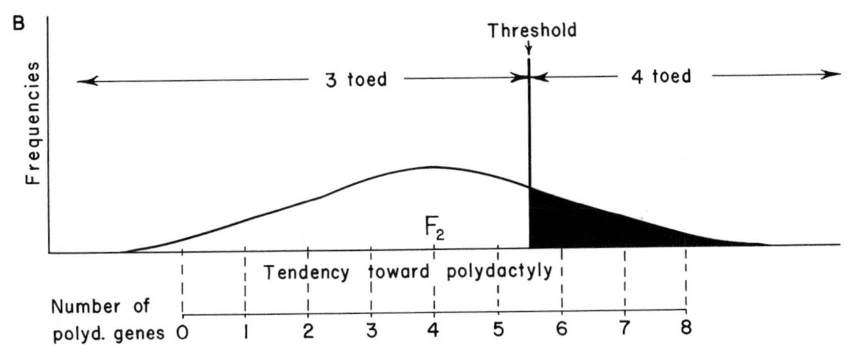

FIGURE 185
Genetic and developmental aspects of polygenic inheritance of polydactyly in guinea pigs. **A.** Strain I possesses no allele for polydactyly, Strain II possesses eight alleles (four pairs). The F_1 individuals have four such alleles. Animals with any one of the three genotypes vary in their tendency toward polydactylism (see curves), but only those with tendencies beyond the threshold form extra toes. It is assumed that a small fraction of F_1 animals would cross the threshold. **B.** The overlap of the nine segregated genotypes in the F_2 generation gives a continuous distribution of the tendency toward polydactylism. Approximately one-fourth cross the threshold and form extra toes. (After Wright, 1934.)

Quasi-continuous Traits. The alternative appearance or nonappearance of a trait may be less sharp than assumed above. Thus, the fourth toe of a polydactylous guinea pig may be either a well-developed digit or a "poor" toe, the latter being characteristic of animals who have only the critical number of 5, or perhaps 4 or 6, "four-toe" alleles. Grüneberg has studied inheritance of variations in the size of the third molar tooth in mice. Variability is continuous over a range from normal to small, and some mice have no third molars. A polygenic system determines both size and, at a certain threshold, presence or absence of the tooth. If only the alternatives presence or absence of a tooth are registered, an obvious discontinuity of phenotypes is obtained. This discontinuity, however, as Grüneberg expresses it is "not genetic but physiological in nature" since it is based on a continuous scale of polygenic determination. For characters of this kind he proposed the name "quasi-continuous variations."

Two types of quasi-continuous variation may be distinguished. The first includes traits that actually vary along a continuous scale but for which the observer chooses a point below which he calls the individuals nonaffected and above which he calls them affected. An example is provided by early studies of high blood pressure in man, in many of which this condition was attributed to a single dominant gene that produced "essential hypertension" (high pressure without a specific known cause); normal blood pressure was thought to be produced by its recessive allele. In these studies the propositi were patients with high pressure, and their relatives were classified according to the alternative phenotypes: affected and nonaffected. The validity of this classification was questioned by Pickering and his colleagues. Measurement of the arterial blood pressure of individuals in a given population shows a continuous distribution of both systolic and diastolic pressures—not a separate group with high pressures and another with low (Fig. 186, A). To divide a population into two groups, calling those with pressures below a certain figure "normal" and above "hypertensive," is a somewhat arbitrary procedure. By choosing an appropriate figure, it is possible to make almost any data fit a preconceived 1:1 ratio of normal to affected and, if there is evidence of genetic control, to conclude that a dominant gene is involved. In reality, the blood pressures of close relatives of propositi (parents, sibs, and children) that are above the often-used dividing line 150/100 (systolic/diastolic pressure in mm mercury) range nearly as widely as the pressures of relatives of propositi that are below the line (Fig. 186, B). Although this excludes simple single-locus interpretation, inheritance does seem to be involved, since the mean blood pressure of relatives of propositi with high pressures is significantly higher than that of relatives of propositi with low pressures (Fig. 186, C). This suggests polygenic inheritance.

There are all gradations of the control of traits, from those whose phenotype is altered by alternative alleles at a single locus to those which require allelic substitutions at many loci. The expression of a "major locus" may be dependent on minor modifying loci, or a number of loci with unequal or equal effectiveness may have to collaborate to produce a given phenotype. The control of penetrance and expressivity of a gene by the genetic background (pp. 394ff.) is only a special case of polygenic determination. It is therefore possible to revert to the interpretation of blood-pressure genetics in terms of a single pair of alleles, if one admits the existence of modifying loci and, in addition, of environmentally conditioned variability. Such admittance may include the assumption that presence of a sufficient number of "minor genes" may result in an effect even if the "main gene" is absent.

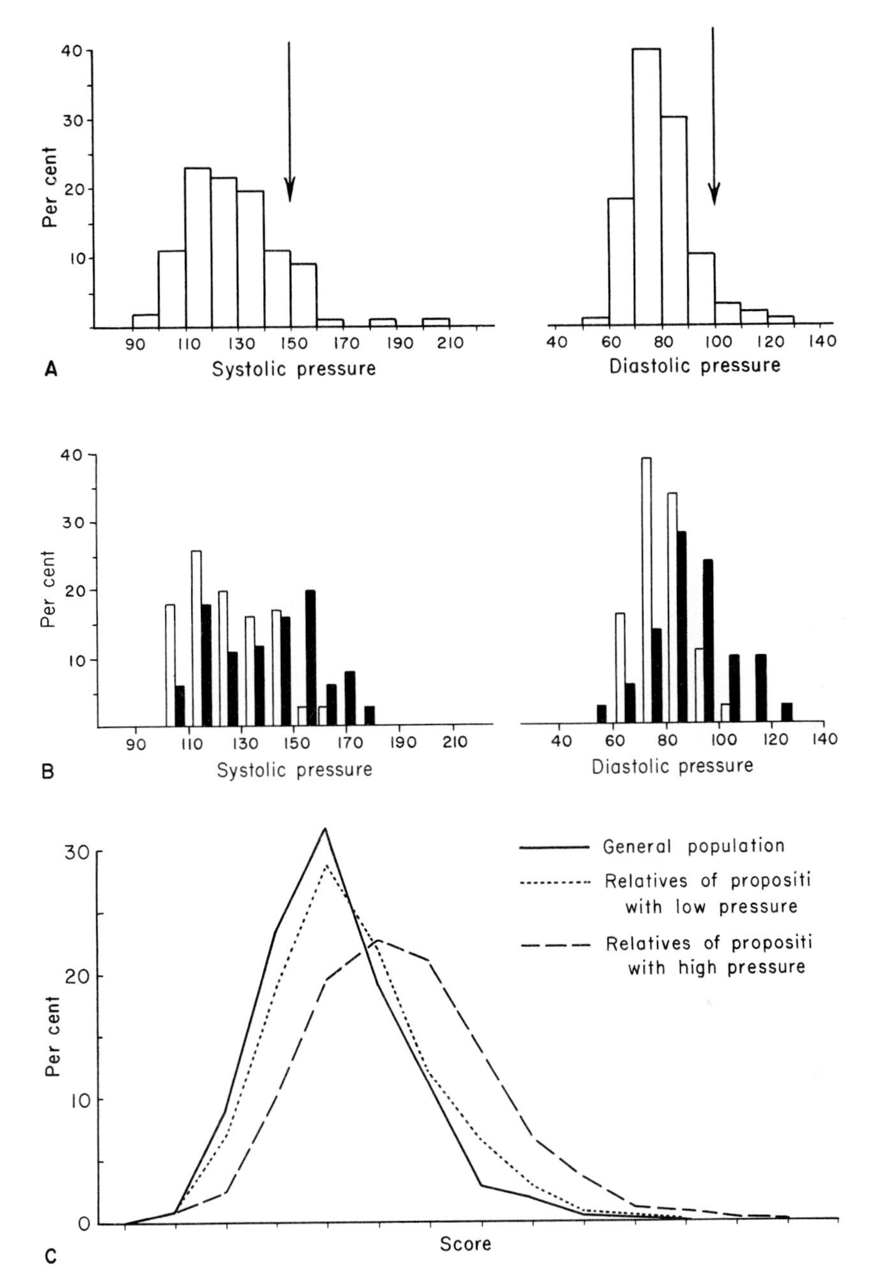

FIGURE 186
Arterial blood pressures. **A.** Sample of 227 women, 30–39 years old. The arrows point to the pressures often used to separate groups with normal and high blood pressure. **B.** Forty-six female relatives of propositi with low pressures ("controls," light columns) as compared with 41 female relatives of propositi with high pressures ("hypertensives," dark columns); ages 30–39 years. **C.** Frequency distributions of diastolic pressures for 867 persons from the general population, 371 relatives of "controls," and 1,062 relatives of hypertensives; males and females, 10–79 years old. Since different age groups as well as both sexes have different mean pressures, the curves are adjusted for age and sex. (After Hamilton, Pickering, Roberts, and Sowry, *Clin. Sci.,* **13,** 1954.)

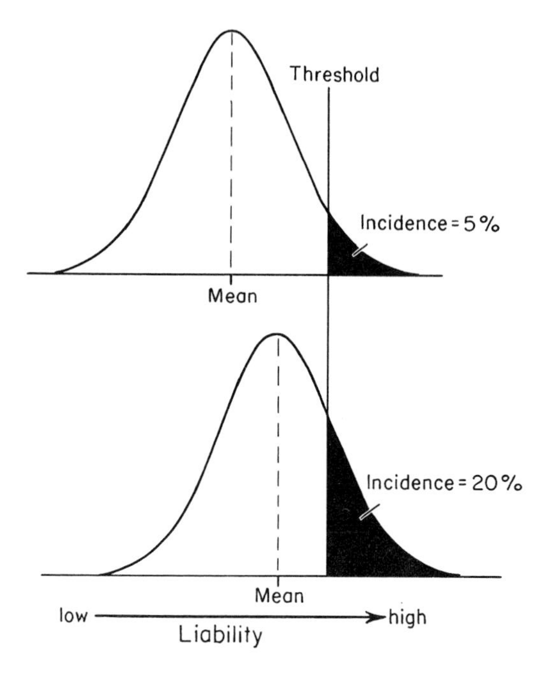

FIGURE 187
Above: A general population in which 5 per cent of the individuals surpass a threshold of liability to a disease and are affected. *Below:* A group comprising certain relatives of those affected in the general population. The distribution of liability of the relatives is so shifted toward greater liability that the threshold is surpassed by 20 per cent of the individuals. (After Falconer.)

A second type of quasi-continuous variation is that in which phenotypes fall into two clear-cut classes. For disease this takes the form nonaffected, healthy versus affected, ill. In many diseases there is an increased incidence of affected among the relatives of affected but the frequencies do not conform to expectations for single-gene transmission. In such cases the assumption has proved fruitful that there is a continuously and normally distributed liability to the disease in question such as resistance to infection or liability to organic weaknesses, and that the observed discontinuity between health and illness is the result of a threshold effect. The disease occurs if a certain threshold has been surpassed. By definition the affected individual represents a genotype for high liability. To some extent his relatives share the predisposition to the disease and thus show an increased incidence (Fig. 187).

The concept of threshold-controlled liability to express specific phenotypes has proved particularly valuable in Carter's studies of pyloric stenosis, a benign tumor of the circular muscle layer of the pylorus, the junction between the stomach and intestine. The disease is present in infants and can be treated surgically with relative ease. In the general population the incidence is about 3 per 1,000 live births. This figure, however, is not as meaningful as the incidences given separately for the two sexes. As the ratio of affected males to affected females is about 5:1, the incidence in males is 5 in 1,000, but that in females is only 1 in 1,000. The incidence of affected among sibs and children of propositi is greatly increased but at first sight the details are complex. The frequencies of affected relatives (sibs and children) of male propositi are

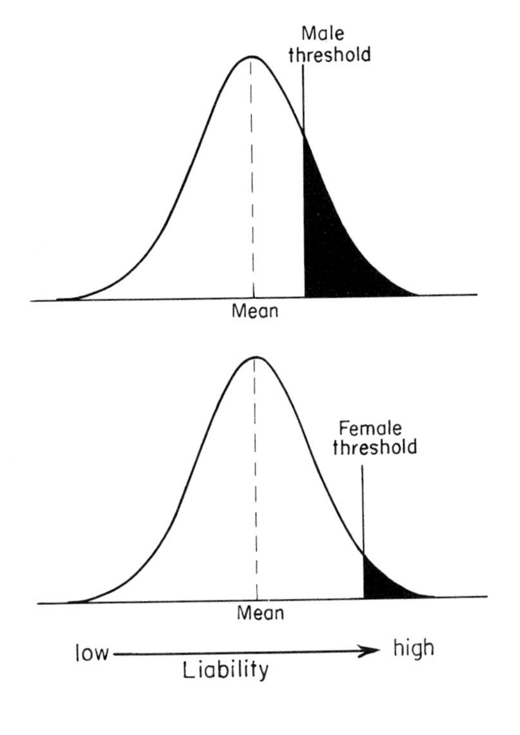

FIGURE 188
A population in which the threshold for liability to a disease is higher in females than in males, resulting in a 5:1 ratio of affected males to females.

much lower than those of female propositi and in both groups the frequencies of affected males are higher than those of females. These findings can be accounted for if it is assumed that there are different thresholds for liability to pyloric stenosis for males and females (Fig. 188). If in the general population the threshold for males is lower than that for females, more males than females will be affected. Affected males, however, will have a mean lower liability than affected females. Since relatives of males will therefore have a mean lower liability than relatives of females, fewer of the relatives of males will surpass the male or female thresholds than of the relatives of females. Frequently there are differences between males and females in the thresholds for liability to diseases or other quasi-continuous traits. They are a form of sex-control.

Among other criteria favoring polygenic inheritance of a trait the following may be mentioned, following Carter. In polygenic inheritance the frequency of further affected sibs rises with the number of already born affected sibs. This is so since a high liability for appearance of a trait—i.e., a genotype well above the threshold—shows itself in a higher number of affected sibs than a lower liability, which corresponds to a polygenic combination that is just above the threshold. This is in contrast to the situation with recessive single genes, for which the probability of an *aa* child being born to $Aa \times Aa$ parents remains 1/4 regardless of the genotypes of the preceding sibs and also in contrast to dominant single-gene inheritance, for which the probability of a *Dd* child being born to $Dd \times dd$ parents remains 1/2. Another consequence of polygenic inheritance is that the more extreme expression of a trait implies a greater frequency of affected relatives than a weak expression—again in contrast to what holds for single-gene inheritance.

Notwithstanding the reasonableness of and the evidence for many cases of polygenic determination, it is often not easy to distinguish beyond doubt a single-gene determination from a polygenic one. The schematic pedigree of a rare defect, Figure 189, may be either interpreted as a case of simple dominance of a gene that is not penetrant in the mother of generation IV, or in terms of additive polygenic inheritance. For the sake of discussion the particular polygenic scheme may be described as four pairs of genes with one of each pair being potentially defect-causing in that an individual who has at least seven such alleles among his total eight is affected. Only large-scale pooled data may permit a decision between a dominant single-gene determination and a polygenic one. Such data may be used for the following comparison. If a single dominant gene is responsible, there will be more-or-less the same ratio of affected to nonaffected offspring among the children of two nonpenetrant parents as among children having at least one penetrant parent. But if the system of inheritance is polygenic, more children having an affected parent will be affected than children having two nonaffected parents.

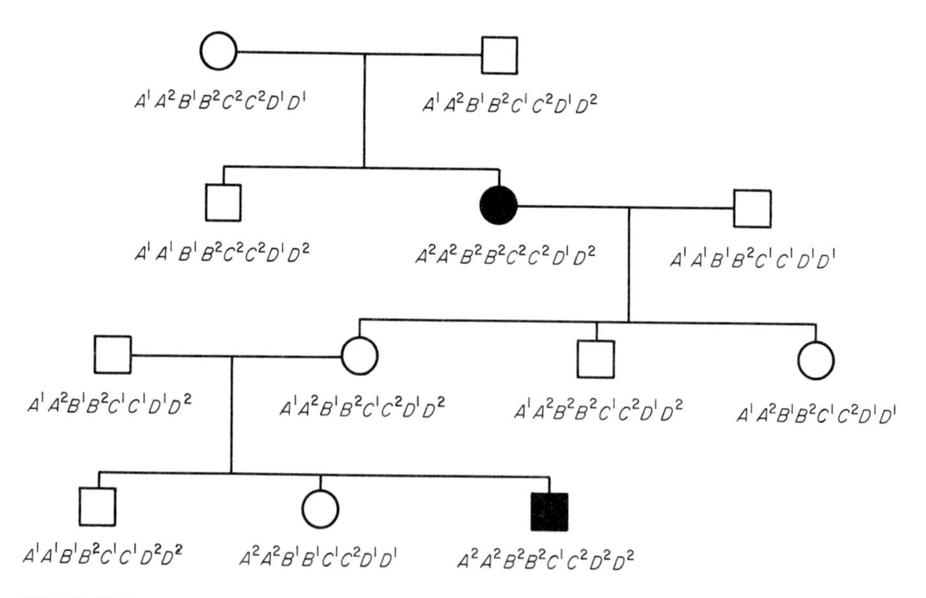

FIGURE 189
Artificial pedigree showing the appearance of a rare defect caused by the presence of at least seven additive alleles. A^1, B^1, C^1, D^1 do not lead toward development of the abnormality, but A^2, B^2, C^2, D^2 do.

It must be stressed that the concept of polygenic inheritance is a statistical one which is an important tool in the analysis of variability, but which at a different level of analysis may be reinterpreted in terms of individual genic action. In the future, detailed studies of the biochemistry of polygenic traits should lead to the recognition of the individual, qualitatively different, single-locus-controlled reactions whose compound effect is now measured on a single quantitative scale. Thus, one locus among the many governing blood pressure may be found to control properties of the kidneys; another locus, those of the adrenal glands; a third, those of fat metabolism; and so on. The recognition of separate components can form the basis for specific controls of abnormal phenotypes.

In this chapter, it has been shown that the concept of polygenic inheritance provides insight into the causes of variability of numerous human traits—variability that cannot be explained in terms of individual genic loci. In later chapters we shall find, particularly when dealing with heredity and environment, that there is evidence for genetic control of many traits for which it is not possible to give a specific genetic formula. Polygenic inheritance can explain both the heritability of these traits and the absence of knowledge regarding the action of the individual loci involved.

PROBLEMS

144. Assume the correctness of Davenport's two-gene-pair theory of skin-color inheritance. What are the types and proportions of colors among the offspring of the following marriages:
    (a) First-generation mulatto × black?
    (b) White segregant × white?
    (c) Light mulatto × black?
    (d) Light mulatto × dark mulatto?
    (e) Assign all possible genotypes to the marriages whose partners have the following phenotypes: white × light mulatto; light mulatto × light mulatto; mulatto × mulatto.
    (f) What phenotypes, and in what proportions, are expected among the offspring of the marriages listed in Part e?

145. Assume that the difference in skin pigmentation of two racial groups I and II is due to three pairs of equally additive genes.
    (a) What range of pigmentation would you expect in a pooled group of children from marriages of F_1 hybrids?
    (b) How many pigmentation types would be present among the children from these marriages?
    (c) What would be the expected frequency of the lightest type?

146. Assume that *aabbcc* leads to 150-cm height and that any "capital letter" allele will "add" 5 cm to the initial height.
    (a) List all genotypes of individuals of size 165 cm.

(b) If only people of 165-cm height married, list the genotypes of all possible combinations of spouses. (For Parts c and d, assume that each marriage type, Part b, consisting of several nonrelated families, is restricted to a single community.)
(c) Give the range of adult size among the children in each community.
(d) What would be the shortest adult height possible among the grandchildren in each community? The tallest?

147. Assume that a defect is caused by the additive action of pairs of genes at three loci. All individuals with three or more normal genes are not affected, one-half of those with two normal genes and all with only one or no normal gene are defective.
(a) Each of two parents has the genotype $A^1A^2B^1B^2C^1C^2$, where $A^1B^1C^1$ are the normal genes. List the genotypes and frequencies of all possible children. What is the probability of their having a defective child?
(b) If the normal and defect-causing alleles at each locus are equally common in a population, what is the frequency of the defect?
(c) If the normal genes have a frequency $p = 0.9$, what is the frequency of the defect?

REFERENCES

Barnes, I., 1929. The inheritance of pigmentation in the American Negro. *Hum. Biol.*, **1**:321–381.

Bresler, J. B., 1970. Outcrossings in Caucasians and fetal loss. *Soc. Biol.*, **17**: 17–25.

Carter, C. O., 1965. The inheritance of common congenital malformations. *Progr. Med. Genet.*, **4**:59–84.

Carter, C. O., 1969. Genetics of common disorders. *Brit. Med. Bull.*, **25**:52–57.

Davenport, C. B., 1913. *Heredity of Skin Color in Negro-White Crosses.* 106 pp. Carnegie Institution, Washington, D.C. (Carnegie Inst. Wash. Publ. 188.)

Edwards, J. H., 1960. The simulation of mendelism. *Acta Genet.*, **10**:63–70.

Edwards, J. H., 1969. Familial predisposition in man. *Brit. Med. Bull.*, **25**:58–64.

Falconer, D. S., 1960. *Introduction to Quantitative Genetics.* 365 pp. Oliver Boyd, Edinburgh.

Grüneberg, H., 1952. Genetical studies of the skeleton of the mouse IV. Quasi-continuous variations. *J. Genet.*, **51**:95–114.

Hamilton, M., Pickering, G. W., Roberts, J. A. Fraser, and Sowry, G. S. C., 1954. The aetiology of essential hypertension. 4. The role of inheritance. *Clin. Sci.*, **13**:273–304.

Harrison, G. A., and Owen, J. J. T., 1964. Studies on the inheritance of human skin colour. *Ann. Hum. Genet.*, **28**:27–37.

Holt, Sarah, 1968. *The Genetics of Dermal Ridges.* 195 pp. Thomas, Springfield, Ill.

Hulse, F. S., 1958. Exogamie et hétérosis. *Arch. Suisses Anthropol. Gén.*, **22**: 103–125.

Mather, K., 1949. *Biometrical Genetics.* 158 pp. Dover, New York.
Pearl, Raymond, and Pearl, Ruth DeWitt, 1934. *The Ancestry of the Long-lived.* 168 pp. Johns Hopkins University Press, Baltimore.
Roberts, J. A. Fraser, 1964. Multifactorial inheritance and human disease. *Progr. Med. Genet.*, **3**:178–216.
Schreider, E., 1967. Body-height and inbreeding in France. *Amer. J. Phys. Anthropol.*, **26**:1–4.
Stern, C., 1970. Model estimates of the number of gene pairs involved in pigmentation variability of the Negro-American. *Hum. Hered.*, **20**:165–168.
Vogel, F., and Krüger, J., 1967. Multifactorial determination of genetic affections. Pp. 437–445 *in* Crow, J. F., and Neel, J. V. (Eds.), *Proceedings of the Third International Congress of Human Genetics.* Johns Hopkins University Press, Baltimore.
Weninger, Margarete, 1964. Zur "polygenen" (additiven) Vererbung des quantitativen Wertes der Fingerbeerenmuster. *Homo*, **15**:96–103.
Wolanski, N., Jarosz, Emilia, and Pyzuk, Mira, 1970. Heterosis in man: Growth in offspring and distance between parents' birthplaces. *Soc. Biol.*, **17**:1–16.
Wright, S., 1934. The results of crosses between inbred strains of guinea pigs differing in number of digits. *Genetics*, **19**:537–551.
Wright, S., 1952. The genetics of quantitative variability. Pp. 5–41 *in* Reeve, E. C. R., and Waddington, C. H. (Eds.), *Quantitative Inheritance.* Her Majesty's Stationery Office, London.

19

CONSANGUINITY

Marriages between relatives—"consanguineous marriages," as they are often called—are important genetically. Since closely related individuals have a higher chance of carrying the same alleles than unrelated individuals, the children from consanguineous marriages are more frequently homozygous for various alleles than are children from other marriages.

In some societies, consanguineous marriages have been encouraged. For example, the ancient Egyptians and the Incas favored marriages of brothers and sisters of the reigning dynasty—"royal blood" being considered worthy only to mix with other "royal blood." Although marriages between relatives are preferred to marriages between unrelated persons in some societies, customs or laws of most societies discourage or prohibit marriages between close relatives. This restriction is probably the result of sociological considerations, perhaps strengthened by biological observations of ill effects of consanguineous marriages on the offspring.

In the United States, all states prohibit marriages between sibs and between parent and child. Most of the states also declare that marriages between a person and his parent's sib—that is, between niece and uncle, or nephew and aunt—are illegal, and marriages between first cousins are prohibited in more than one-half of the states. In some states, even more distant degrees of consanguinity—for instance, the unions of second cousins—are prohibited. In many other parts of the Western world, however, no objection is raised against consanguineous marriages, with the exception of those between sibs and between parent and child. The Catholic church does not permit first-cousin marriages without special dispensation.

It is not uncommon for custom or law also to prohibit marriages between certain classes of individuals who are not related to each other by close common descent. Marriages that have at various times and places fallen under a ban, are those between a person and his step-parent or between a person and his deceased uncle's or aunt's spouse or his deceased wife's sister. Such prohibitions are genetically unjustified. A slightly different example or genetically unjustified discrimination against certain unions was a Chinese regulation which prohibited marriages between first cousins who are the children of two brothers but not between children of a brother and sister or of two sisters (Fig. 190). This differentiation was a consequence of the social custom that assigned a woman to the family of her husband and thus regarded children as "not belonging" to the biological family of the mother. On the other hand, the children of two brothers, considered to be of the same family, fell under the ban of consanguinity, although their genetic endowment is parallel to that of children of two sisters or of brother and sister. Similar considerations are still influential even when no legal obstacles to specific types of cousin-marriages exist. Thus, among 1,805 recent cousin-marriages in Japan, 311 belong to type 1 as designated in Figure 190, 619 to type 2, 378 to type 3 and 497 to type 4—clearly a preference for types 2 and 4 over 1 and 3. For another example of preference for certain types of cousin marriages data from rural regions of the Indian state Andhra Pradesh may be cited. Among a total of 6,945 marriages, 2.1 per cent were of type 2 and 31.2 per cent of type 3. No marriages of types 1 and 4 were permitted, although 2 cases of type 4 did occur.

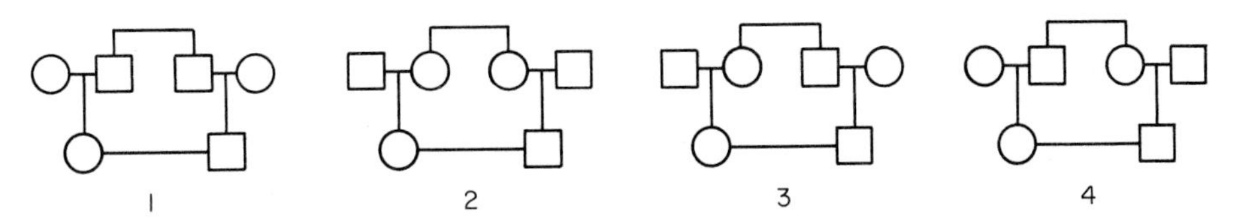

FIGURE 190
The four types of first-cousin marriages according to the sex distributions of the parents of the cousins. Types 1 and 2, between the children of two brothers or two sisters are called parallel cousin marriages, types 3 and 4, between the children of a sister and a brother are called patrilateral and matrilateral cross cousin marriages, respectively.

It may be useful to define a few of the more common types of close consanguinity. First cousins are the offspring of sibs married to unrelated spouses. Double first cousins come from sibs married to unrelated sibs. Second cousins are the children of first cousins who married unrelated spouses. The terms "first cousins once removed" or "one and a half cousins" apply to the relation between an individual and the offspring of one of his first cousins. Other degrees of consanguinity, such as third cousins or second cousins once removed, specify more distant relationships which can be easily derived from the foregoing examples.

In the civil law, consanguineous relations are defined according to the number of "steps" within a pedigree which lie between two related individuals. First-degree consanguinity is that between parent and offspring; second degree is that between grandparent and grandchild or between sibs. Examples of other consanguineous relations are those between uncle or aunt and nephew and niece (third degree), between first cousins (fourth degree), and between second cousins (sixth degree). The canon law defines consanguineous relations by the number of steps from a common ancestor to only one of the related individuals, namely the one more remote from him.

The Extent of Consanguinity. Consanguinity and genetic nonrelatedness cannot be sharply distinguished from each other. Undoubtedly, there are many people descended from common ancestors who are unaware of the fact that they are relatives. In most geographical regions, mankind does not reproduce within pedigrees completely isolated from one another but, rather, in a complex network of relationships which joins all, or most, strains together in a single reproductive unit. This is true even in many places where custom or society favors a separateness of population layers, of castes, or of different races inhabiting the same territory. In the course of generations, such barriers against intermarriages prove ineffective, particularly since legitimate and illegitimate unions lead equally to an interchange of genes between officially separate groups. Therefore, a careful tracing of the pedigrees of any group of apparently unrelated individuals of similar territorial origin will show many of them to possess common ancestors within the last few centuries. It follows that if two such "unrelated" people marry, they contract, in reality, a distant consanguineous marriage.

The magnitude of consanguineous unions can also be demonstrated in a different way. If no consanguinity occurred, then the number of ancestors of any one individual would be 2^n, where n represents the number of ancestral steps removed from the individual. The individual would have 2 parents, 4 grandparents, 8 great-grandparents, etc. Going back one thousand years and estimating three to four human generations per century, we find that an individual would have from 2^{30} to 2^{40}, or from approximately one thousand mil-

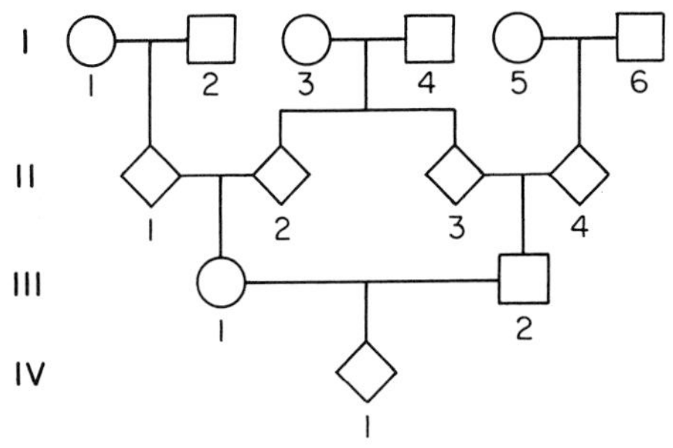

FIGURE 191
A pedigree showing a marriage between first cousins.

lions to one million millions ancestors. Although we have no reliable data on the number of men alive one thousand years ago, available estimates of the total human population three hundred years ago yield figures of between four hundred and five hundred millions. It may safely be assumed that the population of the world was larger in the seventeenth century than in the tenth century. Thus, a thousand years ago, the whole of humanity comprised only a fraction of the theoretical number of ancestors which every now-living individual should have had, provided no consanguineous marriages had occurred. It follows, obviously, that every individual possesses fewer ancestors than the maximum possible number, 2^n. Such "loss of ancestors" can be due only to consanguineous marriages in past generations. For example, the offspring of a marriage between two first cousins—that is, between the children of sibs—have not 8 but 6 greatgrandparents, since 2 of these are common ancestors from both parental sides (Fig. 191).

From the long-range point of view of past evolution, the brotherhood of all mankind is not only a spiritual concept but a genetic reality.

CONSANGUINITY AND HOMOZYGOSITY

What effect do consanguineous unions have in producing offspring homozygous for an autosomal allele *a*? Specifically, what is the frequency with which consanguineous unions lead to the appearance of homozygous recessive phenotypes? In answering this question, unless noted otherwise, primary consideration will be given to marriages between first cousins. The quantitative treatment of this problem depends only on the mechanism of allelic transmission and thus applies equally well to recessive, intermediate, dominant, or codominant alleles. For the last three types of alleles, the problem is of minor interest. The presence of these alleles in the parents is phenotypically observable, and the consequences of close-relative marriages can be easily derived from the appearance of the parents. When the allele is recessive and thus the phenotypes *AA* and *Aa* are indistinguishable, it becomes of great importance to judge the result of consanguinity in terms of the production of *aa* children.

It is obvious that consanguinity has no genotypic influence on the sex-linked genes of males, since males are hemizygous for sex-linked loci, which they receive solely from their mothers. The popular impression that hemophilia in males of the royal families of Europe was due to "inbreeding" is therefore not valid. It would have existed in these men regardless of whom their mothers had married. For females, an evaluation of the effect of cousin marriages on sex-linked genes requires knowledge of the four possible kinds of cousin marriages as pictured in Figure 190.

Cousin Marriages of Heterozygotes. As a first approach, let us oversimplify the problem by assuming that there is an allele *a* of very low frequency in the population. Let us assume, also, that a person is heterozygous for this allele (*Aa*) and ask: What is the probability of his or her first cousin being likewise *Aa*?

Referring to the diagrammatic pedigree of Figure 191, we may consider III–1 to be the *Aa* individual and III–2 the cousin whose genotype is to be determined. Since we assume that the frequency of *a* is very low, we shall disregard the possiblity of *a* entering the pedigree more than once. This reduces our question to the following: What is the probability that III–2 has inherited the *a* allele from the same ancestor as did III–1? Now, III–1 could have obtained *a* from either one of his parents—from II–1 with a probability of 1/2, or from II–2 with an equal probability. Of these, only II–2 is derived from the common line of descent of the two cousins and thus relevant to our question. If II–2 carries *a*, he must have inherited it from one of his parents, I–3 or I–4, who, in turn, had a 1/2 chance of transmitting *a* to II–3, the parent of cousin III–2. Since II–2 carries *a* with only 1/2 probability, there is only 1/2 probability of either I–3 or I–4 carrying it, and, thus, 1/4 probability of II–3 containing *a*. There is a final step from II–3 to the object of the inquiry, III–2. If II–3 carries *a*, III–2 has 1/2 chance of inheriting *a*; and since II–3 carries *a* with the probability 1/4, III–2 has 1/8 chance of carrying it. Thus, the chance of both cousins being carriers of *a*, when one of them is certain to be a carrier, is 1/8.

The significance of this remarkable fact is perhaps best understood if we compare the 1/8 chance for heterozygosity of the cousin of a heterozygote with the chance for heterozygosity of an unrelated individual. The chance of a heterozygous person marrying at random a spouse heterozygous for the same allele depends on the frequency q of this allele. If it is as rare as, for instance, 1 in 50, the chance of its being present in an individual chosen at random is 2pq, or close to 1/25; or, if the allele is as rare as 1 in 200, the chance of its being present in a random individual is only about 1/100. In both cases, however, the chance of the allele's presence in a first cousin is the same, namely, 1/8. This chance is about three times as great as the random expectation 1/25, and more than twelve times as great as 1/100.

The constancy of the figure 1/8 implies that the *relative* probability of an *Aa* person marrying a cousin who is likewise *Aa* increases with rarity of the allele *a*. That this is so is obvious without recourse to arithmetic: the rarer the allele *a*, the less chance of meeting it in a random partner, in contrast to the fixed chance, determined by the fixed-descent relation, of meeting it in a cousin.

The primary purpose of the question of heterozygosity of two spouses is in determining the expectation of homozygous affected children. This expectation is 1/4 for any one child from two carrier parents. Consequently, the probability of a child, IV–1, being affected is $1/4 \times 1/8 = 1/32$ if a heterozygote has married an unaffected first cousin, in contrast to $1/4 \times 2pq$ if he married at random a nonaffected person.

Cousin Marriages of Homozygotes. If the propositus is himself affected (that is, he is homozygous for the rare recessive allele), the probability of his first cousin being a carrier is not 1/8, but 1/4. This is so, since a homozygous individual, III–1, derives one of his *a* alleles from II–2 with certainty instead of with a probability of 1/2. Consequently, there is certainty of either I–3 or I–4 carrying *a*, 1/2 chance of II–3 inheriting it, and 1/4 chance of it being transmitted to III–2. This example may serve as a warning against accepting the fraction 1/8 as an invariable expression of the probable presence of a specific allele in a first cousin. Different genetic situations may lead to different values.

Homozygotes from Cousin Marriages. In the preceding treatment of the genetic implications of cousin marriages, we neglected the possibility that an allele *a* in two cousins may have come from more than one source, except when the propositus is homozygous *aa*. Following Lenz and Dahlberg (1893–1956), we shall now generalize our question. Instead of assuming that one of the cousins is heterozygous, we will make no specific assumption. Let us ask: What is the probability of *aa* offspring from marriages of first cousins? Again we refer to Figure 191. The child IV–1, the offspring of the cousin marriage, can be homozygous *aa* for two different reasons: (1) because of consanguinity, which can bring together two *a* alleles of common origin; and (2) because of the coming together of two *a* alleles of independent origin. These two different ways of acquiring an *aa* genotype will be treated separately.

1. If an *a* allele is carried by one of the two common ancestors of IV–1, for instance, his great-grandparent I–3, then the probability of it being transmitted to IV–1 over all three steps I–3 → II–2 → III–1 → IV–1 is $(1/2)^3$. Likewise, the probability of the *a* allele being transmitted over the alternate three steps I–3 → II–3 → III–2 → IV–1 is $(1/2)^3$. The probability of *a* having descended to

IV–1 over both paths is $(1/2)^3 \times (1/2)^3 = 1/64$. This, then is the probability of IV–1 being homozygous *aa* as a result of two *a* alleles both descended from the same *a* allele in I–3. The probability value 1/64 was obtained under the specific assumption that I–3 carried *a* on one chromosome. Actually, the probability of *a*, and not *A*, being carried is given by the allele frequency q, so that the probability of *a* being carried in I–3 on one chromosome is q and of it becoming homozygous in IV–1 is 1/64 q. The same probability 1/64 q of homozygosity for an *a* allele of common origin exists if the other chromosome of I–3 had carried an *a* allele or if either of the two chromosomes of the great-grandparent I–4 had carried it. Therefore, the total probability of homozygosity *aa* of IV–1 owing to consanguinity is $4 \times 1/64\,q = 1/16\,q$. We have thus derived an expression for homozygosity due to the meeting of *a* alleles of identical origin derived from either one or the other of the four chromosomes in the two common great-grandparents.

2. If 1/16q is the probability of any one of the four possible *a* alleles "meeting itself" in IV–1, then 15/16q is the probability of it not doing so. In the latter case, a given *a* allele can meet either an *A* allele or an *a* allele of independent origin (either from another chromosome of the two common great-grandparents or from any one of the four other great-grandparents). The chance of meeting an *a* allele of independent origin is q, so that the probability of IV–1 being *aa* owing to the coming together of different *a* alleles is $15/16\,q \times q = 15/16\,q^2$.

Altogether, the probability of *aa* offspring from a first-cousin marriage is

$$\frac{1}{16}q + \frac{15}{16}q^2 = \frac{q}{16}(1 + 15q). \qquad (1)$$

Once more, and in a general way, this formula indicates how much greater is the probability of *aa* offspring from cousin marriages (more than q/16) than from random marriages (q^2, which takes on a very low value if q is small). It also indicates that the probability of *aa* from the two kinds of marriages decreases at a different rate with decreasing q: for random marriages, the decrease is steep, being proportional to the square of q; for cousin marriages, the decrease is moderate, being proportional primarily only to q itself.

Cousin Marriages among the Parents of Homozygotes. In a population, individuals of the genotype *aa* are derived from two kinds of marriages, consanguineous and unrelated. The preceding considerations enable us to answer this important question: What proportion of the *aa* individuals in a population comes from cousin marriages? Or, stated differently, how often do *aa* individuals have parents who are first cousins? The answer to this question will obviously depend on the gene frequency q and also on the frequency of cousin marriages in the population. In different populations, cousin marriages take place with frequencies varying between the two theoretical extremes of non-occurrence and exclusive occurrence.

From equation (1) we found that the probability of *aa* children from cousin marriages is

$$\frac{q}{16}(1 + 15q).$$

If cousin marriages occur in a fraction c of all marriages, then the frequency of *aa* children from cousin marriages is

$$\frac{c \times q}{16}(1 + 15q).$$

The frequency of *aa* children from noncousin marriages is

$$(1 - c)q^2.$$

Therefore, the proportion of *aa* children from cousin marriages to *aa* children from all marriages is

$$k = \frac{\frac{c \times q}{16}(1 + 15q)}{(1 - c)q^2 + \frac{c \times q}{16}(1 + 15q)} = \frac{c(1 + 15q)}{16q + c(1 - q)}. \qquad (2a)$$

This relation, which was first derived by Weinberg and later again by Dahlberg, tells us how often it can be expected that individuals homozygous for a recessive allele have parents who are first cousins. From the formula, the factor k, expressed in per cent, has been calculated and listed in Table 54

TABLE 54
Proportions, k, of Recessive Homozygotes *aa* Deriving from Cousin Marriages. Columns 2–4 calculated from formula (2a), page 475, thus neglecting the mean coefficient of inbreeding, F, in the population; columns 5 and 6 calculated from formula (2b), p. 480, thus including consideration of F. ($q =$ allele frequencies; $c =$ frequencies of cousin marriages.)

Allele Frequency (q) *of a*	k *in Per Cent*			k *in Per Cent*	
	$c = 0.1\%$	$c = 0.5\%$	$c = 1\%$	$F = 0.0005$, $c = 1\%$	$F = 0.015$, $c = 20\%$
0.001	5.98	24.17	39.05	42.29	79.30
0.005	1.33	6.32	11.95	12.22	67.19
0.01	0.71	3.48	6.77	6.84	57.50
0.1	0.16	0.78	1.55	1.55	27.17

for a number of different allele frequencies q of *a*, and a number of different possibilities for the frequencies c of cousin marriages in the population at large (for a discussion of the last two columns of the table, see p. 480). Studying the first row of the table, we see that with *a* being as rare as 1 in a 1,000 and c = 0.1 per cent, the proportion of first-cousin marriages among the parents of affected individuals is nearly 6 per cent. In other words, the parents of *aa* individuals should be first cousins nearly 60 times more often (5.98:0.1 = about 60) than the parents of individuals taken at random. This striking fact is, of course, an expression of the genetic situation which makes an *aa* individual, in many cases, a signpost for a preceding cousin marriage. Certainly, it does not mean that a carrier of an *a* allele is more inclined to marry a cousin than is a non-carrier. To say it differently, the fact that a person is a carrier will more often remain unknown when he marries a nonrelative than when he marries a cousin.

Moving to the right in line 1 of Table 54, we see, even more dramatically, the same relation of increased frequency of cousin marriages among the parents of *aa* individuals as compared to that of cousin marriages in the general population. Thus, with a frequency of c of 1/2 per cent, more than 24 per cent of the parents of *aa* children should be first cousins. And at a frequency of 1 per cent, more than 39 per cent of all *aa* children should have parents who are first cousins.

Following the columns of Table 54 downwards—that is, considering the consequences of increasing gene frequencies—we see that the proportion of cousin marriages among the parents of *aa* children decreases. This is to be expected. The more common the *a* allele, the more frequently will a carrier marry an unrelated carrier, and, therefore, the relatively more rarely will the spouse be a cousin.

THE COEFFICIENT OF INBREEDING

When unrelated individuals marry, the resulting children are called outbred. When the parents are related to one another their children are inbred.

Both Wright and Bernstein have devised measures of the degree of inbreeding for individuals and populations. Their coefficients of inbreeding—F and α, respectively—are equivalent. We shall use the symbol F. It may be defined as the probability that an individual is not only homozygous at a given locus, but that the two alleles are "identical" in the sense that they were both derived from an allele present in a certain ancestor. This is obviously possible only if the parents of the individual had the ancestor in common.

When used in measuring inbreeding in populations, F indicates the frequency of homozygotes in a specific population as compared with that in a panmictic population. Being a measure of probability, F values vary between 0 and 1.

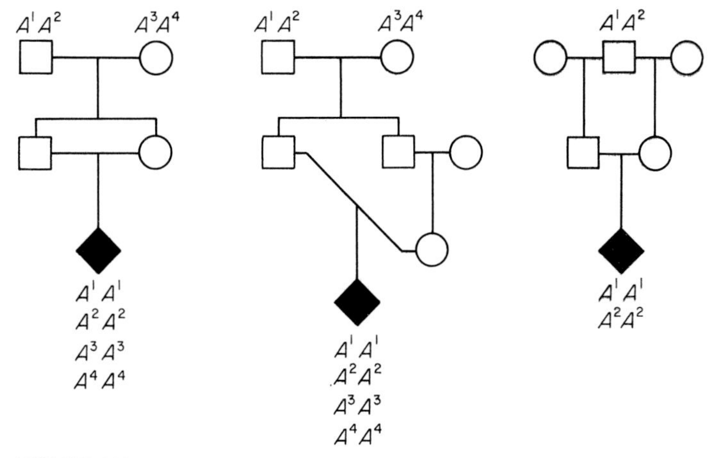

FIGURE 192
The possibilities of homozygosity for a locus in the offspring of unions between brother and sister (*left*), uncle and niece (*middle*), and half-brother and half-sister (*right*).

Thus, F = 0 signifies absence of inbreeding, that is, panmixis; F = 1 signifies complete inbreeding. This value of F = 1 would be obtained in a population containing the alleles A^1 and A^2 if each of the alleles were derived from a single ancestral A^1 or A^2 and if A^1 gametes fused only with A^1 and if A^2 gametes fused only with A^2. The value of F is different when derived for autosomal loci than for X-linked ones. This is obvious since by definition hemizygous males cannot be homozygous for X-linked genes. In the following pages only autosomal F values will be considered.

F can be calculated for any specific individual by tracing his lines of descent to the common ancestor of his parents. Consider as one of the theoretically simplest relations the offspring of a brother-sister union (Fig. 192, *left*). Let the alleles of a specific locus in the two grandparents be A^1, A^2, A^3, and A^4. What, then, is the probability F of the grandchild being either A^1A^1, A^2A^2, A^3A^3, or A^4A^4? In order for the grandchild to be A^1A^1, the grandparental allele A^1 must be transmitted a total of four steps, two on each parental side. Since the probability of transmission at each step is 1/2, the probability of transmission over all four steps is $(1/2)^4$. This same probability also applies, separately, to the other three cases, A^2A^2, A^3A^3, and A^4A^4. Therefore, the total probability F for the product of a brother-sister union is $4 \times (1/2)^4 = 1/4$. For the offspring of uncle-niece or aunt-nephew marriages (Fig. 192, *middle*), there are five steps in the path along which a given ancestral allele can be traced: two steps through the uncle or aunt to the common ancestor and three through the niece or nephew. To each of the four alleles belongs a separate path, so that the coefficient of inbreeding becomes $F = 4 \times (1/2)^5 = 1/8$. For offspring from first cousins,

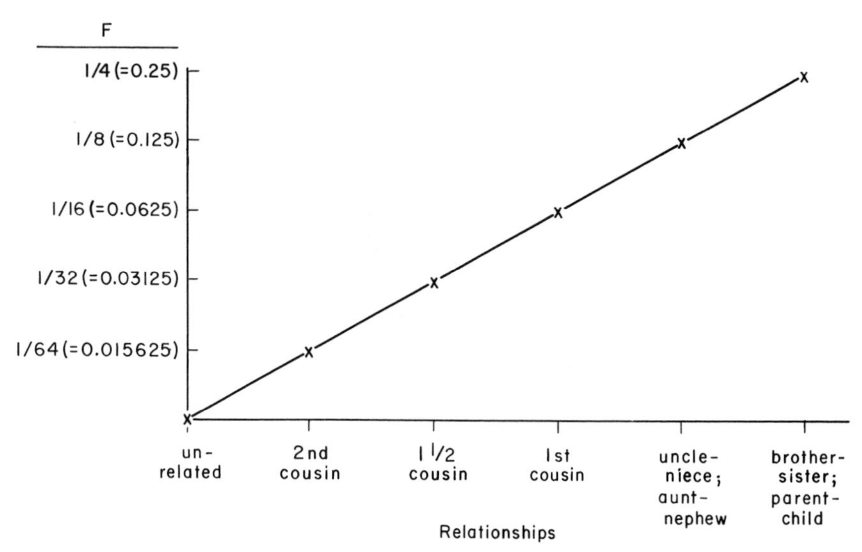

FIGURE 193.
The coefficient of inbreeding, F, for various relationships of the spouses.

F = 1/16. For offspring from first cousins once removed, F = 1/32; and from second cousins, F = 1/64 (Fig. 193).

If the parents of an individual have not two but only one ancestor in common, then the number of possible paths is not four but two (Fig. 192, *right*). If we designate the number of steps from the inbred individual through each of his parents to the common ancestor as n_1 and n_2, respectively, his inbreeding coefficient can be written as

$$F = 2 \times (1/2)^{n_1 + n_2} = (1/2)^{n_1 + n_2 - 1}.$$

Since the number of *ancestors,* N, on the path from the individual through one of his parents and back through the other is one less than the sum of the numbers of *steps* $n_1 + n_2$, we can write

$$F = (1/2)^N.$$

This expression refers to a single common ancestor, who himself is not inbred (ancestral inbreeding coefficient $F_a = 0$). Should the ancestor be of consanguineous origin, his inbreeding coefficient F_a is larger than zero. Under these circumstances, the inbreeding coefficient, F, of his descendant is obviously larger than indicated above. Specifically, as Wright has shown, the general formula for the inbreeding coefficient of an individual in relation to a single common ancestor who is himself inbred is

$$F = (1/2)^N(1 + F_a).$$

For an individual with more than one common ancestor, the total inbreeding coefficient is the sum of the F's as derived from each of the ancestors.

If, instead of single individuals, a population is considered, a mean coefficient of inbreeding can be calculated by averaging the value for not-inbred individuals (F = 0) with the values for groups of different degrees of relatedness. For example, if a population of 1,000 contained 909 individuals of random parentage, 1 individual from an uncle-niece marriage, 30 from first-cousin marriages and 60 from second-cousin marriages, it would have the following mean coefficient of inbreeding:

$$F = \frac{909 \times 0 + 1 \times 1/8 + 30 \times 1/16 + 60 \times 1/64}{1,000} = 0.00294.$$

Mean coefficients of inbreeding have been determined for many populations. Those listed in Table 55 vary between 0.00044 and 0.032.

Earlier (p. 475) we derived a formula for k, the frequency of cousin marriages among the parents of homozygotes. At that time no consideration was given to the mean coefficient of inbreeding in the populations. If consideration of F is included, formula (2a) must be refined. The new formula (no derivation will be given here) is

$$k = \frac{c(1 + 15q)}{16[(1 - F)q + F]},$$

TABLE 55
Percentage Frequencies of Cousin Marriages and Coefficients of Inbreeding (F) in 12 Populations.

Population	*Period*	*No. of Marriages*	*First-cousin Marriages (%)*	F
1. Brazil, Rio de Janeiro (urban)	1946–1956	1,172	0.42	0.00044
2. India, Bombay (Caste of Parsees)	1950	512	12.9	0.0092
3. India, Andhra Pradesh (rural)	1957–1958	6,945	33.3	0.032
4. Japan, Nagasaki (urban)	1953	16,681	5.03	0.0039
5. Japan, Kawajima (village)	1950	414	16.4	–
6. Netherlands (national statistics)	1948–1953	351,085	0.13	–
7. Portugal (national statistics)	1952–1955	276,800	1.40	–
8. Spain, Salamanca (urban)	1920–1957	21,570	0.59	–
9. Spain, Las Hurdes (rural)	1951–1958	814	4.67	–
10. Sweden, Pajola (rural)	1890–1946	843	0.95	0.0008
11. Sweden, Muonionalusta (rural)	1890–1946	191	6.80	0.0058
12. United States, Baltimore (urban)	1935–1950	8,000	0.05	–

Source: After Böök, *Amer. J. Hum. Genet.*, **8**, 1956; Freire-Maia; Neel et al.; Sanghvi, *Eugen. Quart.*, **1**, 1954; and Valls, unpublished.

or, since $(1 - F)$ is approximately equal to 1,

$$k = \frac{c(1 + 15q)}{16(q + F)}. \tag{2b}$$

When F and c are small (e.g., $F = 0.0005$; $c = 1$ per cent) the k values differ only moderately from those obtained with formula (2a) (Table 54, columns 4 and 5). When F and c are large (e.g., $F = 0.015$; $c = 20$ per cent) the absolute k values are much higher than those obtained with formula (2a) (columns 4 and 6) but their relative increase over the large c value in the population is less than when formula (2a) is applied.

Observed Consanguinity in Various Populations. The expectations for k, which have been derived above, will now be compared with some actual data. As a first step in this comparison, we must find the observed frequencies of cousin marriages in various populations. Sources for this information are official data from those few countries, such as France, whose census reports routinely include questions on consanguinity; the records of the Catholic church in various dioceses if made available for demographic purposes; the records of the Mormon church; and the reports of special investigations conducted in specific regions or villages in various countries, on hospital populations, and on other groups.

Table 56 lists not only first-cousin marriages but also other types of consanguinity in various populations. The total frequencies of consanguinity vary from less than 1 per cent to more than 40 per cent. Other data, restricted to first-cousin marriages and to the inbreeding coefficient F, were given in Table 55. In the populations reported in Table 55, cousin marriages vary from a minimum of 0.05 per cent, from Baltimore, to 33.3 per cent in an Indian population. Both Tables 55 and 56 show that only about 1 per cent, or much less, of all marriages in Western populations are between cousins. There are, however, outstanding exceptions: in a rural group in Spain, 4.7 per cent; in a district in northern Sweden, 6.8 per cent; and in an alpine community in Switzerland, 11.5 per cent of the marriages were between first cousins. These exceptional frequencies are not the expression of special systems of mating in which cousin marriages are preferred. They are, rather, the consequence of the fact that these populations constitute unusually small isolates, owing either to geographical or sociological isolation, or to restricted mobility. If the group from which a mate must be selected is small, the proportion of first cousins in it is higher than in larger populations, and random mating will more often include consanguinity.

Consanguinity in small isolates may be complex and manifold. For instance, the pedigree of a Navaho Indian, No. 17 in Figure 194, shows that his parents were simultaneously first cousins, first cousins once removed, and third cousins!

TABLE 56
Percentage Frequencies of Various Types of Consanguinous Marriages.

Population	Number of Marriages	*Consanguinous Marriages*							Total
		Uncle-niece Aunt-nephew	First Cousins	First, Once Removed	Second Cousins	Second, Once Removed	Third Cousins	Third, Once Removed	
Andhra Pradesh (India, rural)	6,445	9.2*	33.3	–	–	–	–	–	42.5
Nagasaki	33,319	–	4.8	1.2	2.0	–	–	–	8.0
Hiroshima	3,283	–	3.9	1.2	2.1	0.5	0.2	<0.01	7.9
Utah, Nevada (9 small communities)	625	–	0.6	1.4	3.7	2.1	1.1	–	9.9†
Rio de Janeiro	1,172	0.1	0.4	0.2	0.1	–	–	–	0.8
England, Wales (a hospital population)	49,315	<0.01	0.6	<0.01	0.1	–	–	–	0.7

Source: After Sanghvi; Schull, *Amer. J. Hum. Genet.*, **10**, 1958; Neel et al.; Woolf et al., *Amer. J. Hum. Genet.*, **9**, 1957; and Bell.
Note: A dash (–) signifies absence of data, not necessarily nonoccurrence of the type of marriage.
The data cover different periods.
*These marriages are between a woman and her maternal uncle.
†Includes 0.96 per cent "other consanguinity."

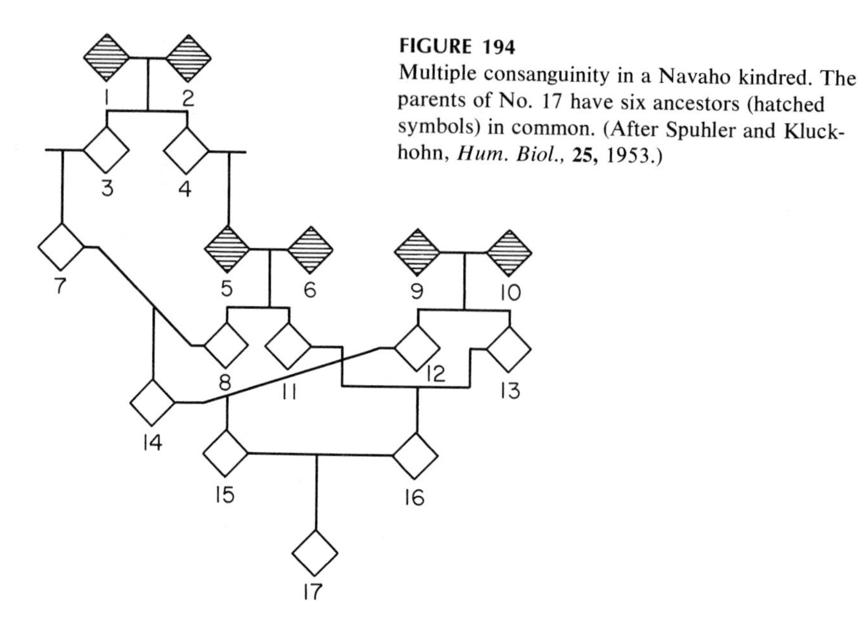

FIGURE 194
Multiple consanguinity in a Navaho kindred. The parents of No. 17 have six ancestors (hatched symbols) in common. (After Spuhler and Kluckhohn, *Hum. Biol.,* **25,** 1953.)

Detailed genealogies of whole isolated villages in Switzerland and in Japan have been worked out among others by Hanhart and Yanase, respectively. The pedigrees often show a bewildering series of multiple, interrelated marriage lines. In Toksenoedao, in Japan, for example, the members of four sibships A, B, C, and D are related to one another as first cousins. In addition to marriages with unrelated or distantly related persons, the following consanguineous unions are recorded: (1) man from sibship A to cousin from B; (2) another man from A to cousin from C; (3) man from B, two marriages, both to cousins from D; (4) woman from B to cousin from D. Furthermore, the woman from C in marriage (2) was herself the offspring of a cousin marriage, as was her father.

Besides isolation, differing attitudes influence the frequency of consanguineous marriages. The high frequencies of such unions in India and Japan are expressions of social preferences. In contrast to attitudes that favor consanguinity are some state and religious regulations that disapprove of such unions. Personal fear of unhealthy offspring may also limit consanguineous marriages. These differing attitudes may possibly depend on the frequency of easily recognizable abnormalities caused by recessive genes. If such genes are present, it may be noticed that cousin marriages produce more abnormal offspring; if such genes are absent, cousin marriages lack serious consequences. Furthermore, if such genes are rare, cousin marriages will bring them to light more often than nonrelated marriages; but if the genes are common, offspring of cousin marriages will not stand out more conspicuously than offspring of nonrelated marriages.

An interesting point may be raised concerning the figure for cousin marriages listed in Table 56 for a hospital population England and Wales. If an appreciable number of individuals who enter a hospital do so on account of being homozygous for a recessive condition, then it is to be expected that more of them are the offspring of cousin marriages than are found in a similar-sized sample of individuals who have not gone to a hospital. This consideration would suggest that the over-all frequency of consanguinity in the British population, as a whole, may have been less than 0.6 per cent. On the other hand, there are factors which may work in the opposite direction. The population studied was from "general" hospitals; it included only a few patients with chronic diseases, since these patients seldom find their way into the wards of general hospitals. Many chronic diseases—for instance, many diseases of the nervous system—have a hereditary basis. If recessive genes are responsible for some of these, we might expect a high rate of consanguinity among the parents of patients with chronic diseases. The noninclusion of these patients in the general-hospital population would have the effect of giving too low a frequency of consanguinity for the population as a whole. It is indeed difficult to avoid unsuspected bias in sampling, and to draw valid conclusions even from carefully collected data.

Observed Consanguinity among Parents of Homozygotes. If we turn to specific records of consanguinity among the parents of persons affected with rare hereditary conditions, we find many observations incidental to descriptions of the traits. Many pedigrees show the parents of individuals affected with various conditions to be closely related. There is a simple means of identifying consanguinity in the pedigrees in the published literature. Consanguinity means partly common descent of two spouses. Graphically, this results in the appearance of a "closed field" in the pedigree (see Figs. 194–197): the marriage line connecting the spouses usually providing the base of the field, the two vertical lines leading to their ancestors forming the sides, and the upper limit of the field the sibship line which joins the two lines of descent of the parents to the common ancestor. Sample pedigrees, selected for their unusual features, are reproduced in Figures 195–197. The first pedigree (Fig. 195,A) is of a man who married twice. His two wives were sisters and also his first cousins. Of the 9 offspring of one marriage, 1 was an albino; of the 3 offspring of the other marriage, 2 were albinos. The second pedigree (Fig. 195,B) also pictures two first-cousin marriages, but here there were four different individuals. Two brothers married cousins: one marriage resulted in 10 offspring—7 normal children and identical girl triplets who lost their hair at six months of age (hypotrichosis); the other marriage resulted in 1 normal and 1 affected child. A third pedigree (Fig. 196,A) includes a marriage of first cousins once removed. The original ancestor I–2 was affected

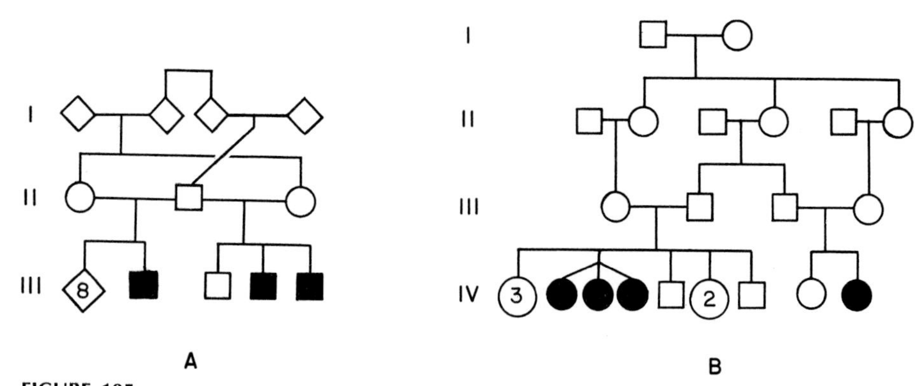

FIGURE 195
Pedigrees showing cousin marriages. **A.** Recessive albinism. **B.** Hypotrichosis. (A, after Pearson, Nettleship, and Usher, pedigree No. 455; B, after Gates, *Human Genetics*, 1946, Macmillan.)

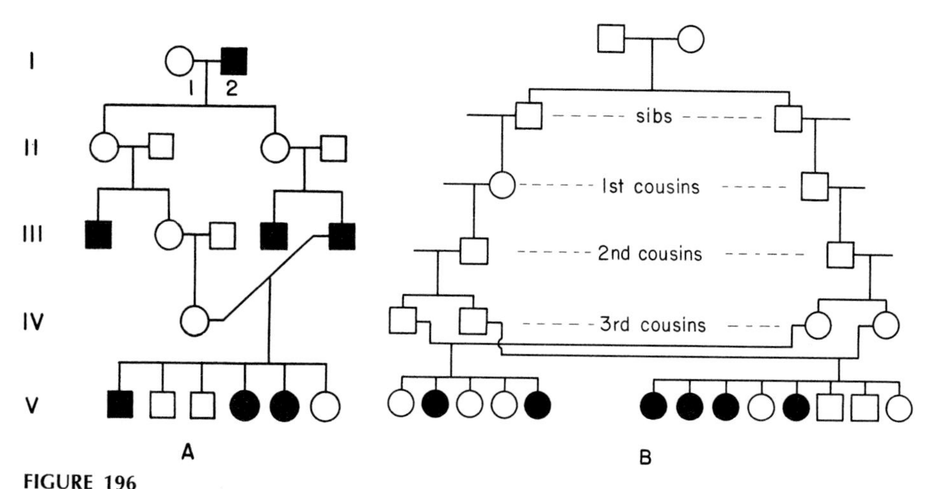

FIGURE 196
Pedigrees showing consanguinity. **A.** Recessive X-linked ichthyosis vulgaris. **B.** Juvenile amaurotic idiocy in two sibships from third-cousin marriages. (A, after Gates, *Human Genetics*, 1946, Macmillan; B, after Sjögren.)

with a skin disease (ichthyosis vulgaris), which, in this family group, was transmitted in recessive X-linked manner. Consanguinity has nothing to do with the appearance of affected males, since they show the trait if it is transmitted by the mother, irrespective of the father's genotype. Females, however, can be affected only if they receive an abnormal allele from each parent. This happened to two of the daughters from the consanguineous marriage. One more pedigree (Fig. 196,B) shows the appearance of the rare juvenile am-

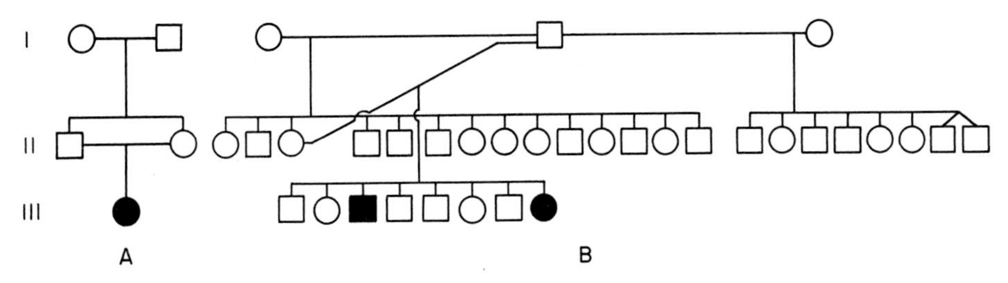

FIGURE 197
Pedigrees showing incestuous unions. **A.** Albinism. Brother-sister union. **B.** Recessive neurological disorder (hereditary ataxia). Two affected individuals from a father-daughter union. (A, after Baur, Fischer, Lenz; B, original by courtesy of Drs. F. E. Stephens and F. H. Tyler, University of Utah.)

aurotic idiocy in the offspring of marriages of two brothers of one sibship to two sisters of another. Here, the parents are only distantly related, being third cousins. Nevertheless, it was presumably one of the two common ancestors who had transmitted, over three generations, a recessive allele to each of the four parents. The last two pedigrees demonstrate the appearance of rare homozygous recessives from incestuous unions. Fig. 197,A, shows an albino child from a brother-sister union; B represents the offspring of a man by his two unrelated wives and by one of his own daughters. The legitimate offspring consisted of 22 normal children (of whom two were identical twins), but among the 8 illegitimate children, 2 suffered from a hereditary neurological disorder.

Although these pedigrees illustrate the production of homozygous recessive individuals from heterozygous carrier parents who presumably obtained their rare recessive allele from the same common ancestor, they cannot serve as a basis for comparison with the quantitative expectations regarding consanguinity cited earlier. Pertinent data on a few specific conditions are available from collections of many pedigrees. Thus, for Friedreich's ataxia, a disease in which muscular coordination is impaired by degeneration of tissue in the central nervous system, among the families of twenty-two propositi, Hogben and Pollack found two in which the parents were first cousins and one in which they were second cousins. This is a rate of about 10 per cent consanguinity. More extensive material concerning another disease of the nervous system is known from Sjögren's studies in Sweden. Juvenile amaurotic idiocy is a condition which begins with failing sight and blindness in children of from four to seven years and leads to progressive loss of sensory, mental, and physical powers and death some ten or twelve years later. The incidence of first-cousin marriages among the parents of those affected is about 15 per cent. Alkaptonuria, the metabolic abnormality resulting in the presence of

TABLE 57
Percentage, k, of Affected Individuals from Cousin Marriages.

Trait	*Caucasians*	*Japanese*
Albinism	18–24	37–59
Tay-Sachs disease (see p. 182)	27–53	55–85
Ichthyosis congenita (skin disease)	30–40	67–93
Congenital total color blindness	11–21	39–51
Xeroderma pigmentosum (see pp. 598f.)	20–26	37–43

Source: Neel et al.

an oxidizable substance in the urine, shows 30–42 per cent ancestral consanguinity among pedigrees collected in different countries.

For phenylketonuria, frequencies of 5–15 per cent consanguinity among parents of affected individuals have been reported. Although the rates of consanguinity among parents of deaf-mutes in Ireland (3–6 per cent) and congenital blinds in the United States (4–5 per cent), as reported in older surveys, are lower, they still exceeded the average of consanguinity in these populations. Data on five other traits, given separately for Caucasian (predominantly European) and Japanese populations, are assembled in Table 57. The percentage k of first-cousin unions among the parents of affected offspring is very high. The higher percentages in Japanese data are to be expected since cousin marriages (c) are more common in Japan than in the Western world, and k is nearly proportional to c (see formulas [2a] and [2b] and Table 54). A detailed comparison of the frequencies k in the two groups would also require a knowledge of the allele frequencies, q, for the five traits, since k (as we have seen) depends not only on c and F but also on q.

The high consanguinity rates among the parents of affected individuals were first recognized in the study of alkaptonuria by Garrod (1902), and the interpretation in terms of the then newly rediscovered Mendelian concepts was given by Bateson (1861–1926). Since that time a higher frequency of consanguineous marriages among parents of individuals affected with a given trait than in the population as a whole has come to be regarded as important genetic evidence for a recessive basis of the trait. This relation holds for familial occurrences of the trait including the chance appearance of only a single affected person in sibships from heterozygous parents. It does not hold for sporadic appearances of a trait due to such causes as mutation, a phenocopy, or unusual polygenic combinations. Thus the frequency of cousin marriages among the parents of affected individuals may serve to distinguish simple recessive inheritance of a phenotype from its sporadic occurrence due to other causes.

Observed and Expected Consanguinity Frequencies. Is the general agreement between observed and expected high rates of consanguinity also quantitative? In order to answer this question, we must have information on the gene frequency q of each condition and on c, the specific rate of consanguinity in the general population concerned. Moreover the degree of consanguinity must be known since the formulas (2a) and (2b) apply to first-cousin marriages only, and other degrees of consanguinity, as well as different inbreeding coefficients F require specific treatments. Such information is only partly available. Therefore, at best we may be able to show that observed and expected values agree in a general way with one another—or that this is not the case. Consider, for instance, alkaptonuria. This rare trait is said to have an approximate frequency of 1 in 1,000,000 (q^2), making $q = 0.001$. Assuming c to be 1 per cent and, as an alternative, 0.5 per cent and entering these values in formula (2a) of page 475, we obtain $k = 39$ and 24 per cent, respectively, which fit into the range of observations. For juvenile amaurotic idiocy, a similar treatment of the data leads to less satisfactory results. Dunn has shown that it is misleading to calculate the gene frequency q from the ratio of homozygotes to the whole Swedish population. The isolates in which the disease occurs represent perhaps 40 per cent of the population. In these isolates, q is higher than when "diluted" by the part of the population which is free from the allele. In the opposite direction, the incidence of cousin marriages, c, in the isolates at the time when the affected children were born, was probably closer to 2 per cent than to the present over-all estimate for a Swedish population of about 0.5 per cent. Only when these corrections for true gene and consanguinity frequencies in the isolates are made, do observation and expectation lead to agreement.

In albinism it has been found that the observed frequencies of parental consanguinity are considerably higher than the expected ones. This had led Roberts to the suggestion that there is more than one locus for recessive albinism. If, for instance, albinos originated in consequence of either one of two equally frequent homozygous genotypes a_1a_1 and a_2a_2, then the frequency of albinos, 1 in 20,000, must be partitioned into that of the a_1a_1 kind,

$$1 \text{ in } 40{,}000 = q^2_{a_1}$$

and that of the a_2a_2 kind,

$$1 \text{ in } 40{,}000 = q^2_{a_2}.$$

The gene frequencies, q_{a_1} and q_{a_2}, then, are both 1/200, which is lower than $q = 1/141$, the value applicable if there were only a single albino locus with $q^2 = 1/20{,}000$. The lower value of 1/200 when entered in equation (2a), page

475, results in a higher consanguinity percentage k, which is closer to observation. Years after Roberts' ad hoc suggestion, independent evidence from pedigrees and biochemical studies led to confirmation of the hypothesis that there are at least two different loci for albinism (see p. 168).

The Recent Decrease of Consanguinity. An examination of percentages of consanguineous marriages in different periods for Utah, Prussia, and the French departments (administrative areas) reveals a general phenomenon (Figs. 198, 199). Over the past hundred years, there has been a continuous decrease in the frequency of near-relative marriages. It is unlikely that this decrease has resulted from a conscious avoidance of such marriages. It is, rather, due to two different changes in population structure that have taken place in recent history. The first is the breakup of isolates. Modern transportation and the large-scale migration from country to country and, even more important, from farm to town have afforded a wider range of human contacts. Formerly, the choice of a mate was largely restricted to the members of a relatively sessile home community. Now, however, people from neighboring districts have an opportunity to meet, and, particularly in the larger cities, people from widely separated regions are brought together. Consequently, the proportion of marriages between people of the same community has decreased, while that between people of different communities has increased (Fig. 200). Since most relatives used to live near one another, they formed

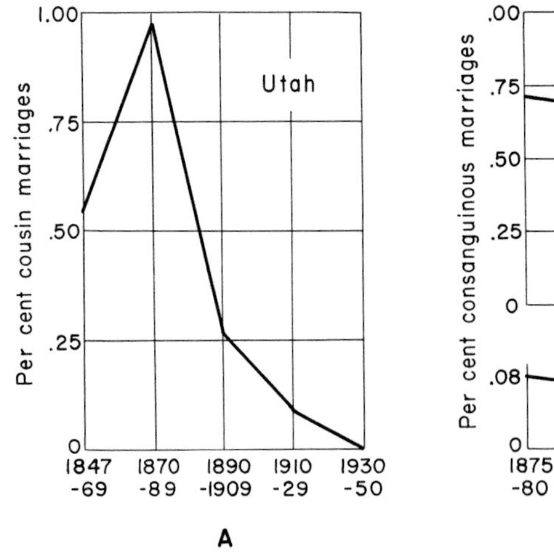

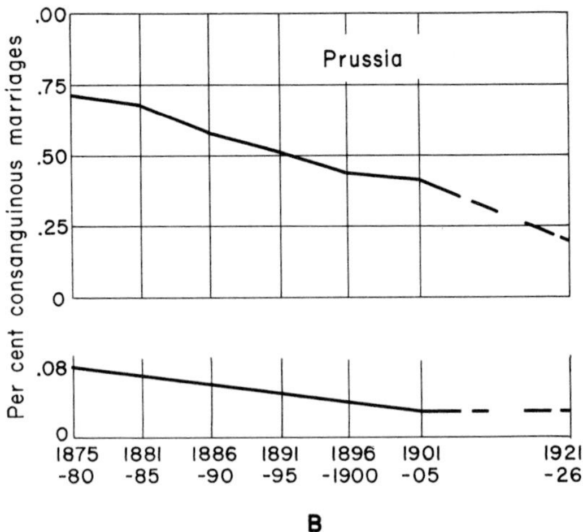

FIGURE 198
Frequencies of consanguineous marriages in Utah and Prussia, during various periods. **A.** Marriages between first cousins. **B.** Marriages between first cousins (upper curve) and uncle and niece (lower curve). (A, after Woolf, Stephens, Mulaik, and Gilbert, *Amer. J. Hum. Genet.*, **8,** 1956; B, after Dahlberg, 1929.)

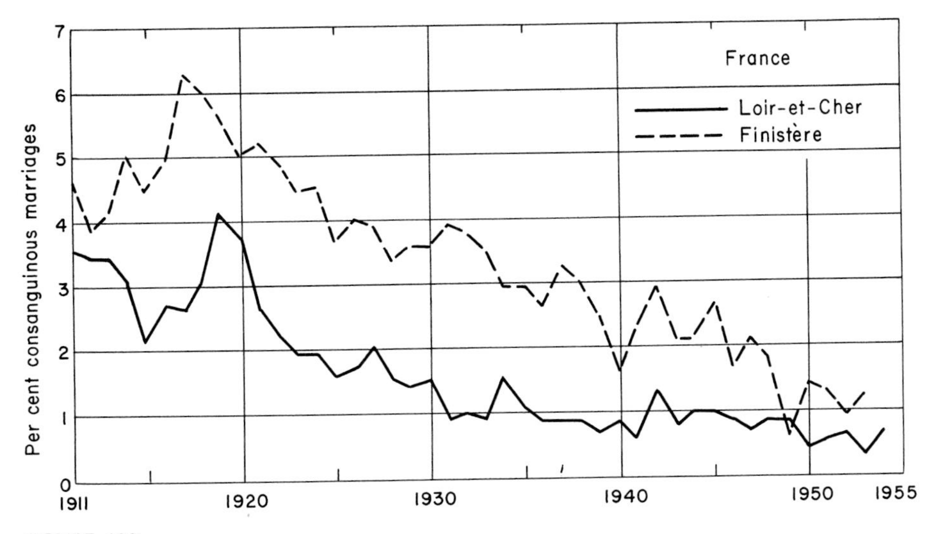

FIGURE 199
Total frequencies of consanguinous marriages, up to those of second cousins, in two French departments, 1911–1954. Finistère, a more isolated district than Loir-et-Cher, had a higher frequency of consanguinity, but the downward trend led to very similar recent frequencies. (Sutter and Tabah, *Population,* **10,** 1955.)

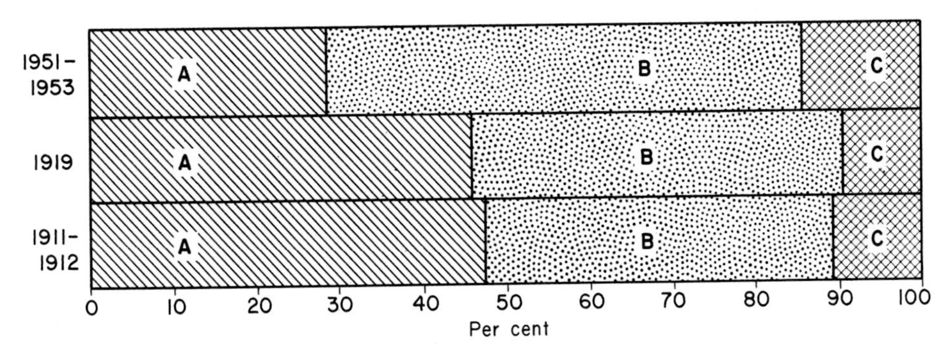

FIGURE 200
Relative frequencies of different kinds of marriages in the Department Finistère of France according to the origin of the spouses. **A.** From the same community. **B.** From different communities of the same department. **C.** One spouse from a community in the Department Finistère, the other from another department. (Sutter, *Population,* **13,** 1958.)

a large fraction of an individual's acquaintances. With the extension of the sphere of acquaintances, the fraction of relatives among them decreases, quite aside from actual separation of relatives by migration. This comparative decrease of relatives among marriageable acquaintances naturally leads to a decrease in consanguinity.

The second cause of decreased consanguinity is due to an absolute decrease in the number of close relatives. This is simply a consequence of the modern

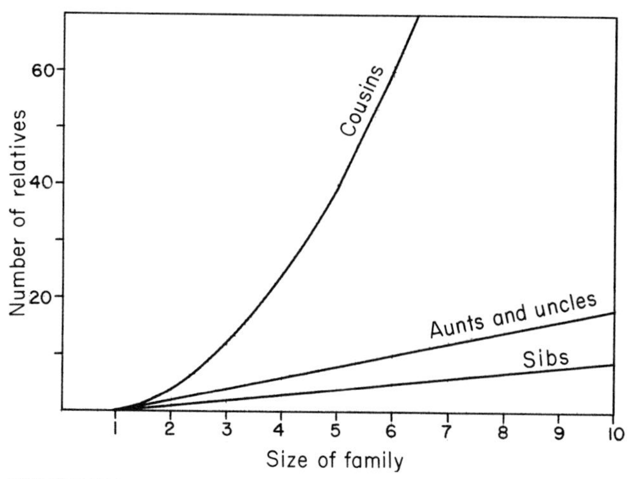

FIGURE 201
Relation between size of family and number of close relatives.

trend toward smaller families. If b is the average number of children in a family, then an individual has on the average (b − 1) sibs. To give some examples, if there are 7 children, each child has 6 sibs; if there are 4 children, each child has 3 sibs; and if there are 2 children, each has only 1 sib. Clearly, not only does the number of sibs become smaller with smaller family size, but the decrease in sibs is also greater than that of sibship size (Fig. 201). This is even more pronounced in more distant relationships. An individual whose parents come from sibships of size b will have (b − 1) uncles or aunts on each parent's side, or a total of 2(b − 1). If each of these uncles or aunts has an average of b children, these children, his cousins, will number 2(b − 1)b. And if one-half of these are of the sex opposite that of the propositus himself, he will have b(b − 1) potentially marriageable first cousins. With this formula, if b = 7, the potential mates will number 42; if b = 4, the number decreases to 12; and if b = 2, it is only 2. In a stationary population b, on the average, is 2, the two children replacing their parents. Although mankind at present is expanding its population, a time will come when the population will only replace itself and thus provide only small cousin groups. In the future the break-up of isolates and small family size may well reduce first-cousin consanguinity to 0.1 per cent or even less.

THE CONSEQUENCES OF CONSANGUINITY

It is clear from a genetic discussion of consanguinity that inbreeding, by itself, is not responsible for the appearance of unfavorable phenotypes. Inbreeding tends to bring into the open recessive alleles present in heterozygous carriers, but, as has been aptly said, inbreeding is no more responsible for the presence of the alleles than a detective is responsible for the crimes which he lays bare. The genetic facts afford an understanding of the often contradictory effects of inbreeding, which sometimes result in undesirable phenotypes and, at other times, in normal or even better-than-average constitutions.

Unfavorable homozygous phenotypes are usually more obvious than favorable ones. Nevertheless, it is to be expected that recessive alleles exist which, in the homozygous state, endow their carriers with better-than-normal characteristics. To cite outstanding examples, the painter Toulouse-Lautrec and the writer John Ruskin were the products of cousin marriages. Yet it cannot be proven that the special accomplishments of these individuals were related to homozygous recessive genes nor, if this were the case, that these genes had become homozygous as a result of consanguinity. Toulouse-Lautrec's multiple skeletal abnormalities were probably caused by homozygosis of the recessive gene for a rare condensing bone condition (pycnodystosis). It may well be that the gene became homozygous as a result of the consanguinity of his parents. It has been suggested that the homozygous presence of this developmentally very undesirable gene kept its bearer from physical activities and indirectly led him to become an eminent painter.

There is one class of genetic situations for which consanguineous marriages would be at an advantage over marriages between unrelated spouses. These situations exist in cases of antigen-antibody incompatibility between mother and fetus as, for instance, when the mother is Rh-negative, *rr*, and the fetus positive *Rr* (see pp. 434ff.). No unfavorable reactions result when mother and child are alike genetically. Since related spouses more often carry the same allele for an incompatibility locus than unrelated spouses, more children of consanguinous than of unrelated marriages should have the same genotypes as their mothers. Therefore, a lower frequency of fetuses incompatible with their mothers is to be expected among consanguinous parents. No observations are yet available to verify this predicted beneficial effect of consanguinity.

Occurrence of Several Unfavorable Traits in the Same Kindred. Not infrequently, more than one unfavorable inherited trait occurs in family groups. The opinion is often expressed that these groups form a "degenerate strain." This view is not compatible with genetic knowledge. The hereditary make-up of individuals consists of particulate, independent units, the genes, while the concept of a "degenerate constitution" vaguely implies a homogeneous basis of heredity. How, then, can the simultaneous occurrence of more than one unfavorable phenotype, either in one individual or in different individuals of a family group, be explained? There are various possible explanations.

Abnormalities are not so rare that two or more cannot occur in one individual or group occasionally. To some extent, then, family strains containing more than one type of abnormality may only be selected cases. Furthermore, a family with one abnormality may have more chance of being scrutinized closely enough to lead to discovery of other abnormalities.

Another explanation is that one and the same gene may express itself in various ways, because of different types of expression in the heterozygous and the homozygous state, because of different genetic backgrounds, or because of different environmental conditions. In other cases, multiple symptoms may be common developmental effects of one and the same gene.

Finally, there exists a true genetic cause of positive correlation between the appearance of two or more rare homozygous recessive traits. This cause is consanguinity. If a person happens to be heterozygous for two different rare recessives, $AaBb$, the chance of his marrying an unrelated partner heterozygous for one or the other of these recessives will be low, and there will be virtually no chance of his marrying one heterozygous for both. Therefore, in a non-consanguineous marriage, at most one trait, aa or bb, will appear among the children. However, should an $AaBb$ person marry a first cousin, the probability of the spouse being heterozygous for a is 1/8, for b is 1/8, and simultaneously for a and b is 1/64. Since, in a consanguineous marriage of $AaBb \times AaBb$ parents, the expectation for $aaB-$, $A-bb$, and even $aabb$ children is rather high, sibships may be produced that contain some individuals affected with one recessive trait, others with another trait, and, occasionally, even an individual affected with both.

The Risk of Consanguineous Marriages

The advisability of cousin marriages is often questioned. If 20 per cent of albinos are the products of such unions, should two cousins to be married fear that they might have albino children? Obviously, this would be wrong reasoning. Although many albinos come from cousin marriages, most cousin marriages do not produce albino offspring. But if the specific fear of albinism is not justified, nor that of deaf-mutism, Friedreich's ataxia, or any one of many conditions caused by homozygous recessives, it is a different matter to wonder about the occurrence of any one unfavorable trait. What are the expectations and the facts?

Theoretical Risks. Let us assume that an individual carries heterozygously one rare recessive gene whose homozygous detrimental effect causes premature death, a major congenital malformation, or some other serious defect. Since the probability of his cousin carrying the same gene is 1/8, the probability of any one of their children being affected is $1/4 \times 1/8 = 1/32$, or about 3.1 per cent (and consequently, of being nonaffected, 96.9 per cent). If the propositus carries a second different gene with detrimental effect, the probability of a child being affected on account of homozygosity for the second gene is likewise 3.1 per cent, and of being nonaffected, 96.9 per cent. Any one of the children of the cousins may be either free from both defects with a probability of $(0.969)^2$; affected by one or the other trait, $2 \times 0.969 \times 0.031$; or affected by both traits, $(0.031)^2$. If the propositus carries n different genes, the probability of a child not receiving any of them in homozygous state is $(0.969)^n$; or conversely, the probability of the child being affected by 1 or 2 or 3 or all n traits is $1-(0.969)^n$. Thus, if $n = 2$, the risk is 6.1 per cent; if $n = 3$, it is 9 per cent; and if $n = 8$, it rises to 22.4 per cent. If it is assumed that the cousin of the propositus also carries, in heterozygous state, a rare recessive gene (different from that postulated for the propositus) there would be an additional of 1/32 or 3.1 per cent of homozygosity for this gene, and correspondingly increased chances in cases of more than one gene in each parent.

Obviously, the risk of having an affected child increases with increasing relatedness of the parents. This is shown by Figure 202, which relates to mortality from stillbirth to 21 years of age. It may also be illustrated by an extreme theoretical example. As we have seen, if a person is heterozygous for a recessive gene, the chance of his having a homozygously affected child is 3.1 per cent when he is married to his first cousin. If, however, he has a child from an incestuous union ($F = 1/4$) between him and his daughter or sister the chance of an affected child is four times greater, i.e., 12.5 per cent. This follows from the fact that the chance of his daughter's or sister's carrying the same gene is 1/2 so that the probability of a homozygous child is $1/4 \times 1/2 = 1/8$ or 12.5 per cent.

These relations between the number of deleterious recessive genes heterozygously present in an individual and the frequency of affected children predicted if he marries a relative were first discussed by Russell. More refined

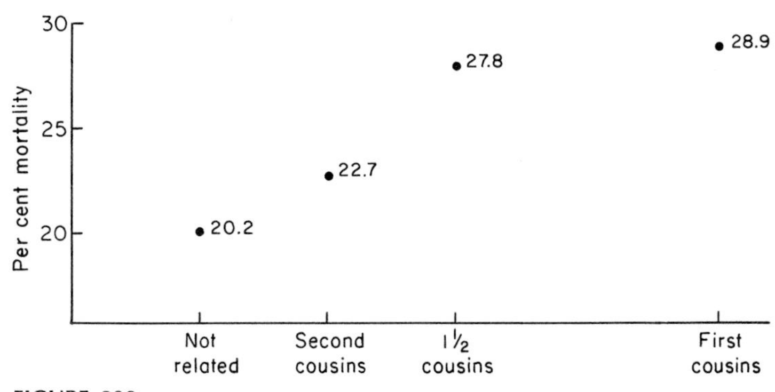

FIGURE 202
Mortality up to 21 years of age of the offspring of various types of marriages registered during the period 1903-1907 in Italian populations. (Data from Serra and Soini, *Proc. 2nd Int. Congr. Hum. Genet.*, **1**, 1963.)

treatments have been worked out by Slatis, Böök, Penrose, Freire-Maia, Morton, Crow, Muller, and others.

Observed Risks. How do these calculated predictions compare with observed results of cousin marriages? There are many reports, dealing with various traits, in a variety of populations. Critical analyses of these reports often reveal biases in ascertainment such as more careful study of the offspring of consanguineous marriages than of the nonrelated controls. This would tend to lower the frequency of affected offspring in the controls. Another possible source of distortion is the finding in some populations that the frequency of consanguinity varies in different socioeconomic groups. As shown by Schull and Neel for Japanese populations, cousin marriages are significantly, albeit slightly, more frequent in lower socioeconomic groups. If, as it turns out to be true, the children of consanguineous marriages attain lower scores in various performance tests than the control children, at least part of the difference appears to be due not to the genetic effect of consanguinity but to the correlated effects of socioeconomic environment. In general, the results obtained by different investigators, at different times and places and with different methods, are not strictly comparable. Some of the data are assembled in Tables 58 and 59 and in Figures 202 and 203.

Seven relevant comparisons on mortality are listed in Table 58. In every one, the percentage of deaths is greater among the offspring of first cousins. All

TABLE 58
Mortalities among Offspring of Marriages Between Unrelated Persons and Between First Cousins.

Deaths	*Period*	*Unrelated*		*First Cousins*	
		N	Mortality (%)	N	Mortality (%)
UNITED STATES					
Young children	Before 1858	837	16.0	2,778	22.9
Before age 20	18th–19th centuries	3,184	11.6	672	16.8
Children 0–10 years	1920–1956	167	2.4	209	8.1
FRANCE*					
Stillbirth, neonatal	1919–*c.* 1950	2,745	3.9	743	9.3
Infantile, juvenile, and later	1919–*c.* 1950	515	9.6	674	14.3
JAPAN*					
Within first year	1948–1954	12,077	3.5	4,947	5.8
Between 1 and 8 years	1948–1954	544	1.5	326	4.6

Source: Bemiss, *Trans. Amer. Med. Ass.*, **11**, 1858; Arner; Slatis, Reis and Hoene, *Amer. J. Hum. Genet.*, **10**, 1958; Sutter and Tabah, *Population*, **7**, **8**, 1952–53; Schull, Amer. J. Hum. Genet., **10**, 1958, Schull and Neel.
*Deaths of children without visible major malformations.

TABLE 59
Frequencies of Diseases and of Physical and Mental Defects among Children of Unrelated and of First-cousin Marriages.

Population	*Unrelated*		*First Cousins*	
	N	Aff. (%)	N	Aff. (%)
France*	833	3.5	144	12.8
Japan†	3,570	8.5	1,817	11.7
Sweden‡	165	4	218	16
United States§	163	9.82	192	16.15

Source: Sutter and Tabah, *Population*, **9**, 1954; Böök, *Ann. Eugen.*, **21**, 1956; Slatis, Reis, and Hoene. *Amer. J. Hum. Genet.*, **10**, 1958; Schull and Neel.

*Morbihan and Loir-et-Cher. Children in completed families from marriages 1919–1925.

†Hiroshima, and Nagasaki. Children born 1948–1954.

‡Three parishes, North Sweden. All cousin marriages registered 1947. The percentages of affected are estimates after various corrections applied to the data.

§Chicago. Children born 1920–1956.

differences are significant and often striking. Thus, 22.9 per cent of the children from cousin spouses in the United States study (Bemiss report) died early versus 16.0 per cent from unrelated parents; 9.3 per cent of the French children from cousins died at birth or during the first month versus 3.9 from unrelated parents; and 4.6 versus 1.5 per cent of children in the Japanese sample died between the ages of one and eight years. For a different Japanese population the cumulative mortality is depicted in Figure 203. It is seen that early mortality for children of first cousins was greater than that for children of unrelated parents. During the second half of the six-year period of study the infant mortality of the two groups of familes did not differ appreciatively.

The increased risk of consanguineous marriages for the offspring is also indicated by a correlation between the frequency of consanguinity and the frequency of perinatal death (stillbirths and death within one month of birth) in the 90 French governmental departments. The coefficient of inbreeding varied from 0.0021 to 0.0236 and the mortality from 38 to 66 per 1,000 births, revealing a high and significant correlation, +0.72.

Data in Table 59 include frequencies for malformations, mental and physical defects, and tuberculosis in the French, Swedish, and United States groups; but only "major congenital malformations" for the Japanese group. Again the differences between children from unrelated parents and from first-cousin spouses are highly significant, indicating a greater risk for the offspring of consanguineous unions.

Children of consanguineous parents often score lower for various traits, while remaining within the normal range, than children of unrelated spouses.

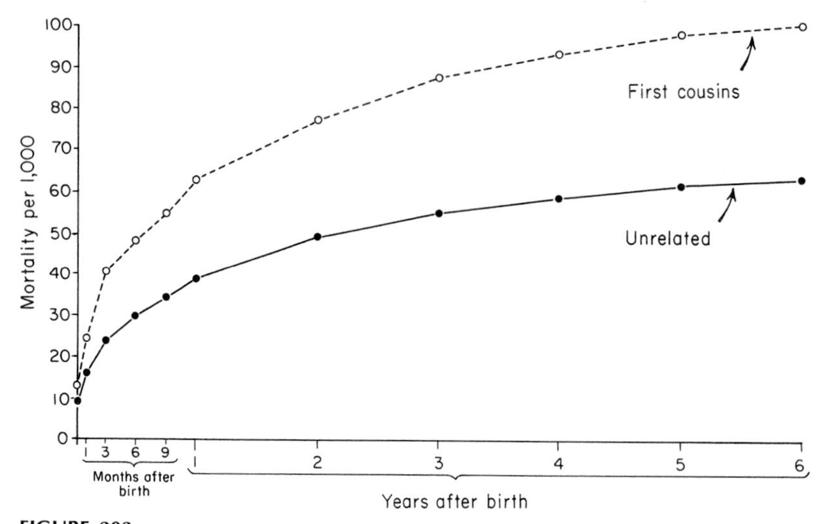

FIGURE 203
Cumulative postnatal mortalities among 5,224 children from unrelated parents and 3,442 children from parents who were first cousins. City of Fukuoka, Japan, births between 1929 and 1956. (Yamaguchi, Yanase, Nagano, and Nakamoto, *Amer. J. Hum. Genet.*, **22,** 1970.)

Thus, various anthropometric measures such as head girth, chest girth, head length, and sitting height have been found to be depressed in Japanese children of consanguineous origin. Although the depression usually is small, it seems over all that inbreeding results in reduced size. On the functional side it may be quoted that, on the average, inbred Japanese children begin to walk several days later than controls. The same is true for the time the children begin to talk. A slight depression in performance also applies to mental tests. The test scores of Japanese children from first-cousin marriages were about 2.5 units lower than those of outbred children who, in the particular tests used, had a mean score of about 55. In Chicago 87 children from first-cousin marriages had a mean I.Q. of 101.5 as compared to a mean score of 104.1 of 72 control children. However, the numbers of children here are too small to make the depression of 2.6 points a statistically significant one.

It is tempting to compare the effects of consanguinity in different racial populations. This has been done by Freire-Maia for Caucasians and Africans in a Brazilian district. Initially, these studies suggested that inbreeding had a quantitatively much greater effect on mortality in blacks and mulattoes than in whites. More extensive later data, analyzed by the same author and his associates, did not confirm the original finding but showed that inbreeding had very similar consequences in the two major racial stocks.

Inbreeding that leads to homozygosity results in selection against detrimental genes and could be a mechanism for cleansing a population from such genes. Sanghvi in India has studied a large population for the frequency of congenital malformations in children of first cousins and of unrelated parents. He found 1.49 per cent affected offspring of consanguineous parentage as compared to 1.39 per cent among controls. The difference was not significant. Sanghvi compared his findings with those of Neel and Schull in Japan who observed 1.42 and 1.02 per cent, respectively, the difference between the last two values being highly significant. Might the Indian population be lower in number of detrimental recessives than the Japanese?

The Genetic Effects of Incest. Carter in England and Adams and Neel in the United States have obtained information on 31 children born to father-daughter or brother-sister unions. Six of the children died early; 12 were severely affected, physically or mentally; and 13 were normal at times of early follow-up. Altogether only 42 per cent of the children were normal. This contrasts with 11 normal out of 12 children (92 per cent) whose parents were not consanguinous and who had been matched as a control group with the 12 American children from incestuous unions. These facts agree with the theoretical expectation of greater frequency of homozygous detrimentally affected children from marriages of very close consanguinity than from cousin marriages. Specifically, extrapolation of findings on effects of lesser degrees of parental consanguinity to the high degree among the parents of the 31 children under consideration yields frequencies of offspring mortality and defects that are compatible with the observations. When adoption is planned for newly born children from incestuous unions, the high risk of homozygosity for detrimental genes must be considered.

The Individual's Load of Detrimentals. As has been pointed out on page 492, for each recessive gene carried by an individual, 1/32 of his offspring from a first-cousin marriage will be homozygous. If, therefore, we wish to know the approximate number of detrimental recessive genes carried by the average individual, it is only necessary to determine the frequency of homozygous recessive genotypes from cousin marriages and to multiply it by 32 (and to divide it by 2 to take account of the fact that both parents contribute to the total percentage of homozygosity in their offspring). For example, and taking the data on face value, 8.1 per cent of the American consanguineous group in Table 58 died before the age of ten years; 2.4 per cent of the control group died during this period. Thus consanguinity accounts for the difference of 5.7 per cent. If it is assumed that each death ascribable to consanguinity is the result of the presence of a single homozygous gene pair—an assumption which will be redefined below—then the value $0.057 \times 32/2 = 0.9$ signifies that, on the average, each parent was heterozygous for 0.9 recessive genes causing childhood mortality. Bemiss' older United States data, with the much higher juvenile mortalities of 22.9 per cent for consanguineous and 16.0 for nonconsanguineous offspring, lead to a very similar figure. The difference in percentage is 6.9, or multiplied by 32/2, a mean of 1.1 recessives carried by each parent. The French, Swedish, and United States data for physical and mental defects (Table 59) place the number of genes for these traits carried per parent in the range of from one to two.

Any detailed estimates must take into account various factors, such as nongenetic social differences between consanguineous and nonconsanguineous groups, the allele frequencies, the degree of inbreeding in the population, and the possible effect of the detrimental genes when present in the heterozygotes. Muller and others have made such estimates. For genes causing premature mortality (from stillbirth to early adult death), the estimates were first expressed in terms of recessive lethals, but later, in terms of "lethal equivalents." A lethal equivalent is either a single gene that is fully lethal in a homozygote, or a group of genes each of which, when homozygous, may cause death in a fraction of cases and which, if separately homozygous in different individuals, would on the average account for one death. The number of lethal equivalents carried heterozygously by the average person seems to be between three and five. There must be additional lethal equivalents per person which cause early abortion and adult, but still premature, death. Another four to five nonlethal equivalents per person may be responsible for anatomical and other abnormalities. The total number of genes per person which are detrimental in homozygotes is likely to be much larger than the ± 10 mentioned here. In Chapter 24, we shall discuss the nature of this load of detrimentals, and point out that it may be in part the result of constantly occurring spontaneous mutations and in part an expression of selection for heterozygotes which may be superior to homozygotes.

Counseling in Consanguineous Marriages. From the preceding data and discussion it is clear that the risk of detrimentally affected offspring is greater for related than for unrelated spouses. Even for related couples the chance of producing a normal child is very much higher than that of producing an affected, but the latter, by no means negligible, possibility must be faced. A study of the families of prospective consanguineous parents affords relatively little helpful information. Most of the rare recessive detrimentals are carried unknown for many generations, and absence of homozygotes for such genes is hardly an indication of absence of the genes in a heterozygous carrier state. Conversely, although knowledge of a relative affected by a specific homozygous genotype may help in identifying one of the several detrimental genes presumably present in the prospective parents, it does not necessarily indicate a heavier-than-normal genetic load.

If all marriages between close relatives were avoided, the effect would be a decrease in homozygous recessives in the following generation. Considering that, on the average, the frequency of consanguineous marriage is low and therefore the number of homozygotes resulting from such marriages small, the decrease in homozygotes, from the point of view of the population as a whole, would be of relatively small consequence. On the other hand, from the point of view of affected individuals and their relatives, any reduction in the number of abnormals is of great significance.

If we consider the fate not only of the first generation after elimination of close consanguinity but of future generations, then the problem takes on a new aspect. Changing the system of mating does not change the frequency of an allele but only its distribution over heterozygous and homozygous genotypes. A decrease in consanguinity, with its resulting decrease of homozygotes, will lead to an increase in heterozygotes. In the course of generations, these heterozygotes will marry unrelated heterozygotes and produce homozygous affected offspring. In other words, the homozygotes prevented from appearing in the immediate future will appear in the more distant future. This is particularly so if the trait is so unfavorable as to exclude the affected individual from reproduction. In this case, consanguinity, which may lead to homozygous individuals, may result in the elimination of two recessive alleles per affected person. One generation undergoes a sacrifice and thus frees the later generations. On the other hand, if consanguinity is avoided, the two recessives will come together in some later generation and will then be eliminated. Thus, exclusion of consanguinity in one generation transfers the load of affected individuals to later generations. It might appear selfish to want to shift the load from one's own to future generations but it may be said in defense that future discoveries may permit normal development of detrimental homozygotes although present knowledge is insufficient for the task.

PROBLEMS

148. A man is homozygous for a very rare recessive gene *a*. What is the chance of a child being *aa* if the man marries his first cousin? (Assume that *a* enters into the pedigree from only one of the common grandparents.)

149. The frequency of a recessive gene is 1 in 90. What is the probability of affected offspring from the marriage of two cousins who are of normal phenotype?

150. A normal-appearing man is heterozygous for a very rare recessive gene that produces an anatomical defect. What is the probability of a defective first child if the man marries
    (a) his first cousin,
    (b) his niece? (Assume that the gene enters the pedigree from only one source.)

151. In a population, the frequency of homozygous recessives for the gene *a* is 1 in 6,400.
   (a) If the general frequency of cousin marriages in this population is 0.008, what is the expected per cent of cousin marriages among the parents of *aa* people?
   (b) If among the parents of 500 unrelated *aa* people 32 are first cousins, would you consider this proportion as significantly different from expectation?

152. How much greater than the frequency in the general population is the frequency of cousin marriages among the parents of homozygous affected persons if the general frequency of cousin marriages is c, and the frequency of homozygotes is f, for:

   (i) $c = 0.005$; $f = 0.0004$.
   (ii) $c = 0.005$; $f = 0.000081$.
   (iii) $c = 0.005$; $f = 0.000025$.
   (iv) $c = 0.002$; $f = 0.000025$.

153. If the general frequency of cousin marriages in a population is c, what will be the frequency of cousin marriages among the parents of color-blind men?

154. Double first cousins are people whose parents are two pairs of sibs. If a woman is heterozygous for a rare recessive *a*, what chance has her double first cousin to be heterozygous for *a*? (Assume that *a* enters the pedigree from only one source.)

155. Determine the coefficient of inbreeding F for the offspring of the marriage of:
   (a) double first cousins, and
   (b) half-brother and sister. Is there a genetic justification for laws which permit marriages of type (a) but prohibit those of type (b)?

156. A color-blind man has parents and grandparents who had normal vision.
   (a) If the man marries his first cousin, who is the daughter of his father's brother, compare the probability of color-blind offspring from this cousin marriage with that of noncousin marriages.
   (b) If the man marries the daughter of his mother's sister from a marriage to a normal-visioned man, what is the probability of color-blind offspring?

157. Assume that a population of one million consists of two generations, of which the parents' generation has an average of b children. If no deaths have occurred, how many individuals belong to the older and how many to the younger generation if b equals:
   (a) 1? (b) 2? (c) 3? (d) 6? (e) 8?

REFERENCES

Adams, M. S., and Neel, J. V., 1967. Children of incest. *Pediatrics,* **40**:55–62.

Bell, J., 1940. A determination of the consanguinity rate in the general hospital population of England and Wales. *Ann. Eugen.*, **10**:370–391.

Böök, J. A., 1948. The frequency of cousin marriages in three North Swedish parishes. *Hereditas*, **34**:252–255.

Böök, J. A., 1957. Genetical investigations in a North Swedish population: The offspring of first cousin marriages. *Ann. Hum. Genet.*, **21**:191–223.

Carter, C. O., 1967. Risk to offspring of incest. *Lancet*, **1**:436.
Centerwall, W. R., Savarinathan, G., Mohan, L. R., Booshanam, V., and Zachariah, M., 1969. Inbreeding patterns in rural south India. *Soc. Biol.*, **16**: 81–91.
Dahlberg, G., 1929. Inbreeding in man. *Genetics*, **14**:421–454.
Dahlberg, G., 1938. On rare defects in human populations with particular regard to inbreeding and isolate effects. *Proc. Roy. Soc. Edinburgh*, **58**:213–232.
Dunn, L. C., 1947. The effects of isolates on the frequency of a rare human gene. *Proc. Nat. Acad. Sci.*, **33**:359–363.
Farrow, M. G., and Juberg, R. C., 1969. Genetics and laws prohibiting marriage in the United States. *J. Amer. Med. Assoc.*, **209**:534–538.
Freire-Maia, N., 1964. On the methods available for estimating the load of mutations disclosed by inbreeding. *Cold Spring Harbor Symp. Quant. Biol.*, **29**:31–40.
Krieger, H., Freire-Maia, N., and Azevedo, J. B. C., 1971. The inbreeding load in Brazilian Whites and Negroes: Further data and a reanalysis. *Amer. J. Hum. Genet.*, **23**:8–16.
Lasker, G. W., 1969. Isonymy (Recurrence of the same surnames in affinal relatives): A comparison of rates calculated from pedigrees, grave markers and death and birth registers. *Hum. Biol.*, **41**:309–321.
Lenz, F., 1919. Die Bedeutung der statistisch ermittelten Belastung mit Blutsverwandschaft der Eltern. *Münch. Med. Wochenschr.*, **66**:2. 1340–1342.
Morton, N. E., 1958. Empirical risks in consanguineous marriages: birth weight, gestation time, and measurements of infants. *Amer. J. Hum. Genet.*, **10**: 344–349.
Morton, N. E., 1961. Morbidity of children from consanguineous marriages. *Progr. Med. Genet.*, **1**:261–291.
Morton, N. E., Crow, J. F., and Muller, H. J., 1956. An estimate of the mutational damage in man from data on consanguineous marriages. *Proc. Nat. Acad. Sci.*, **42**:855–863.
Neel, J. V., Kodani, M., Brewer, R., and Anderson, R. C., 1949. The incidence of consanguineous matings in Japan. *Amer. J. Hum. Genet.*, **1**:156–178.
Reed, S. C., 1954. A test for heterozygous deleterious recessives. *J. Hered.*, **45**: 17–18.
Sanghvi, L. D., 1966. Inbreeding in India. *Eugen. Quart.*, **13**:291–301.
Schreider, E., 1969. Inbreeding, biological and mental variations in France. *Amer. J. Phys. Anthropol.*, **30**:215–220.
Slatis, H. M., 1963. Problems in the study of consanguinity. Pp. 236–243 *in* Goldschmidt, E. (Ed.), *The Genetics of Migrant and Isolate Populations.* Williams and Wilkins, Baltimore.
Steinberg, A. G., 1962. Population genetics: special cases. Pp. 76–91 *in* Burdette, W. J. (Ed.), *Methodology in Human Genetics.* Holden-Day, San Francisco.
Sutter, J., 1958. *Recherches sur les effets de la consanguinité chez l'homme.* 103 pp. M. Declume, Lons-le Saunier.
Woolf, C. M., Stephens, F. E., Mulaik, D. D., and Gilbert, R. E., 1956. An investigation of the frequency of consanguineous marriages among the Mormons and their relatives in the United States. *Amer. J. Hum. Genet.*, **8**:236–252.
Wright, S., 1922. Coefficients of inbreeding and relationship. *Amer. Natur.*, **56**: 330–338.

20

SEX DETERMINATION

The discovery, early in this century, of the chromosome difference between females (XX) and males (XY) afforded a basic, and surprisingly simple, solution of the problem of sex determination. All mature eggs produced by a woman are alike in their possession of one X-chromosome (plus one set of autosomes), but there are two kinds of sperm produced by a man—those with an X-chromosome and those with a Y-chromosome (in both cases, plus one set of autosomes). Fertilization of any egg by an "X sperm" results in an XX zygote, which is destined to develop into a female, and fertilization by a "Y sperm" leads to an XY zygote destined to become a male.

Sex, in man and all other organisms with a similar chromosomal mechanism, is determined at the moment of fertilization and depends on whether an X sperm or a Y sperm unites with the egg. Before the moment of this union, the future of the egg is not fixed and either a female or a male may develop from it. The future role of any individual sperm cell, if it should fertilize an egg, depends on whether it contains an X- or a Y-chromosome. And since both types of sperm are produced simultaneously by every man, the sex of his future offspring is not decided until either one of his X or one of his Y sperm has joined with the egg.

The Male-determining Role of the Y-chromosome. Not until 1959 was it known that it is the presence of a Y-chromosome in a fertilized human egg which causes it to develop into a male, and that it is the absence of the Y-chromosome which causes it to develop into a female. Although this would seem to have been obvious, the matter was not as clear-cut as it appears at first. Genetically, a male is distinguished from a female not only by his having a Y-chromosome, but also by his having only one X-chromosome instead of two. It was rightly wondered whether his sex is the result of the singleness of the X-chromosome rather than of the presence of the Y-chromosome. Moreover, it had been established that the Y-chromosome of *Drosophila* does *not* contain sex determiners. This is shown by the fact that exceptional flies with one X- but no Y-chromosome (XO) are males, and flies with two X- and one or even two Y-chromosomes (XXY and XXYY) are females. Contrary to *Drosophila*, in the silkworm *Bombyx mori* chromosomally similar abnormal individuals made it clear that the Y-chromosome carries sex determiners. Does man's Y-chromosome resemble that of *Drosophila* or of *Bombyx*?

This question was answered in the same way as that concerning *Drosophila* and *Bombyx*. Humans were found who have 45 chromosomes (instead of 46) because they have only one X-chromosome and lack a Y-chromosome. Such XO persons are females! Other individuals were found who have 47 chromosomes because they have two X-chromosomes and a Y-chromosome. Such XXY persons are males! The Y-chromosome, then, determines maleness, normally in XY individuals, abnormally in XXY. And the absence of the Y-chromosome leads to femaleness, normally in XX individuals, abnormally in XO. Thus, man's Y-chromosome, in sex determination, resembles that of *Bombyx* and not that of *Drosophila*.

Man's Y-chromosome is not peculiar among mammals. It shares its male-determining function with that of the mouse. It needs to be added that the known exceptional human XO and XXY persons show certain abnormal traits. These will be discussed in more detail in the section on variations in sexual differentiation.

The Neutral Role of the X-chromosome in Sex Determination. If the Y-chromosome is male determining, does another chromosome or other chromosomes carry a female-determining gene or genes? In *Drosophila* it is the X-chromosome that is female determining, as shown by the fact that "addition" of an X-chromosome to either an XO or XY constitution, thus changing them to XX or XXY, results in "making a female out of a male." In man, on the contrary, the X-chromosome seems neutral in its effect on sex determination: XO are females as are XX, and XY are males as are XXY. Moreover, XXXY, XXXXY and even XXXXXY individuals are known and all are males. (Each one of the mentioned chromosomal constitutions except XX and XY lead to infertility and a variety of abnormal physical and mental traits; the sex as judged by internal and external genitalia is male for all those constitutions including a Y-chromosome and female for those in which it is absent.)

The Balance Theory of Sex Determination. Experiments with plants and animals, and particularly Goldschmidt's experiments on the gypsy moth *Lymantria*, have shown that the sex of an individual is not the result of either pure male or pure female tendencies. Rather, in the development of either sex, both male and female determiners are at work—stronger male than female ones in the origin of males, and the reverse in the origin of females. We may assume that the same "balance theory" of sex holds for man. Presumably, a human male has genes for maleness not only in the Y-chromosome but also in another chromosome or chromosomes; and, in addition, genes for femaleness in some other chromosome or chromosomes. He is a male because the male determiners "outweigh" the female ones. Conversely, although a human female presumably has genes for both maleness and femaleness, in the absence of the strong male-determining factor carried by the Y-chromosome, the female determiners outweigh the male ones.

In *Drosophila* with its female-determining X-chromosomes, the autosomes carry male determiners. This is shown by the fact that "addition" of a set of autosomes to a female constitution (2X + 2 autosomal sets) resulting in a (2X + 3 autosomal sets) constitution shifts the phenotype away from femaleness to intersexuality. In man, where the Y-chromosome directs a tendency to maleness and the X-chromosome is neutral in sex determination, the autosomes must have an over-all female tendency. In females this tendency expresses itself fully, while in males the Y-chromosome outweighs the feminizing action of the autosomes thus resulting in maleness.

It is not known whether the genic control of femaleness is exerted from one locus in one autosome or from many loci in many autosomes. The biochemical action of different genes for femaleness and maleness is likely to differ from locus to locus. Some loci may control the quantity or the molecular make-up of substances which act as embryonic sex hormones, and thus direct development into male or female direction; other loci may control the developmental response of embryonic tissues to such hormones; and still other loci may determine the sexual development of cells and tissues autonomously. We shall return to these possibilities in the course of this chapter.

In *Drosophila*, abnormal alleles of a sex-determining gene may so reverse normal action that intersexual development may result, or even near-transformation of one sex into the other. In man, as will be seen below, the syndrome of testicular feminization furnishes a parallel to such sex-deviants in *Drosophila*.

Parthenogenesis and Sex. In various animals that reproduce by fertilization of eggs by sperm, very rarely unfertilized eggs have been known to develop. If parthenogenesis occurred in man, could it be identified with certainty, and what would be the sex of a parthenogenetically produced child? The first part of the question has been answered by skin grafting. Since a parthenogenetic offspring would not possess any genes that are not also present in the mother, skin from the child grafted onto the mother should not produce the usual antigenic reaction between host tissue and a genetically different graft—a reaction that leads to sloughing off of the latter (Fig. 258, p. 278). A number of presumed parthenogenetic children and their mothers have been tested by blood-group determinations, and for all but one there was evidence of blood-group genes contributed by a father. In the one case in which fertilization could not be proven on blood-grouping and certain other grounds, recourse was made to skin grafting. The graft did not survive, a fact which made it seem very likely that the child did have a father.

Since a parthenogenetic child could obtain only X-chromosomes and no Y-chromosome from its mother, its sex would almost surely be female. It would have either a single haploid chromosome set or be diploid. The latter is more probable if we are correct in extrapolating from various experimental animals in which parthenogenesis starts with a diploid unreduced egg nucleus or with a dipoid nucleus derived from fusion of haploid egg and polar nuclei.

NORMAL SEX DIFFERENTIATION

The mechanism of normal sex determination—by means of the XX–XY alternative—forms only the genetic basis for the developmental processes that transform a fertilized egg into either a female or a male.

Embryonic Sex Differentiation. A zygote of either XX or XY constitution transforms itself, within six to seven weeks after fertilization, into an embryo which, morphologically, is neutral; that is, neither female nor male in its sexual morphology. The embryo possesses gonads which consist of two parts: an external layer of tissue, the cortex, characteristic of an ovary; and an internal mass, the medulla, characteristic of a testis (Fig. 204). Furthermore, two pairs

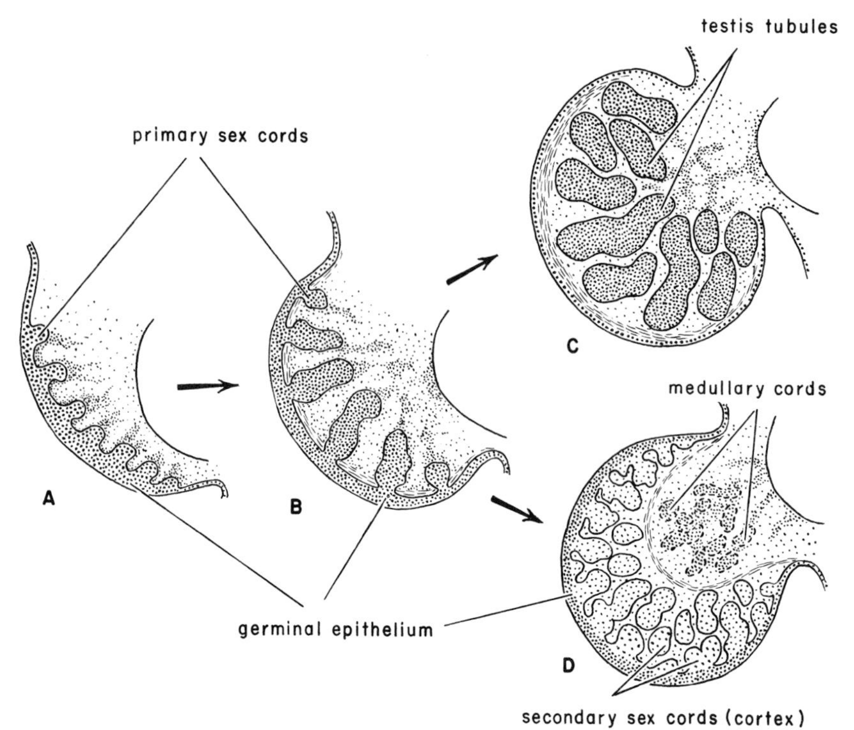

FIGURE 204

Diagrams of gonadal differentiation in mammalian embryos. **A, B.** Indifferent stages. **C.** Development of a testis. The primary sex cords become the testis tubules; the germinal epithelium undergoes involution. **D.** Development of an ovary. Secondary sex cords arise from the germinal epithelium and form the cortex while the primary cords degenerate. (After Burns, *Proc. Nat. Acad. Sci.*, **41,** 1955.)

of ducts are present in the neutral embryo—the Mullerian and the Wolffian ducts—only one pair of which is found in a normal, sexually differentiated individual, the Mullerian ducts persisting in the female and the Wolffian in the male. Finally, the region which is destined to include the openings of the urinary and genital ducts, the urogenital sinus, and the embryonic parts which are to be transformed into the external genitalia are still in such a primitive state that either sexual form may develop from them.

Only after the neutral stage of development of sexual rudiments does the genetic sex constitution of the embryo begin to exert a visible differential effect. In an embryo whose cells are XY, the medullary part of the gonad develops greatly, while the cortical part becomes inconspicuous; in other words, the neutral gonad is transformed into a testis. In an embryo whose cells are XX, the reverse happens: the cortical part of the gonad enlarges and differentiates, while the medullary part disappears, except for remnants, and the neutral gonad becomes an ovary.

While these sexual differentiations are occurring in the gonads, other differentiating processes are taking place in the sexual ducts and the embryonic rudiments of the external genitalia. In an XX embryo, the Mullerian ducts become the oviducts, uterus, and upper part of the vagina of a female, while the Wolffian ducts degenerate, except for small remnants. In contrast, in an XY embryo, the Wolffian ducts become the sperm ducts of a male, while the Mullerian ducts disappear, except for small remnants. The urogenital sinus and the external genital rudiments enter two divergent courses of development. In an XX embryo, the sinus forms the lining of the lower part of the vagina; in an XY embryo, the urethra. In an XX embryo, the external genital rudiments become the clitoris and the labia; in an XY embryo, the penis and the scrotum.

Secondary Sex Differences. In a newborn child, sexual differentiation is not yet complete. Secondary differences between men and women—that is, differences apart from internal and external genital ones—develop during puberty. These are anatomical differences between the larynx of XX and XY individuals, resulting in female or male voice; differences in general body growth and relative growth of various parts of the body, resulting, for instance, in the female or male morphology of the pelvic region and in the development of the female breasts or the undeveloped mammary glands of the male; differences in growth of hair; and many other differences. The development of the secondary differences between men and women proceeds under the influence of a variety of hormones whose production is controlled by secretions of the gonads, the pituitary, and the adrenals, which themselves are under partial control of the hypothalamus, a subdivision of the brain. Actually, the glands of each sex secrete several types of both female and male hormones; and it is, primarily, a matter of male hormones being in excess of female hormones, or vice versa, that controls adult differentiation.

In the embryo the normal alternative sexual differentiation of the internal genital ducts and of the external genitalia is dependent on the type of gonad present. Presence of an embryonic testis is necessary for male-type development of ducts and external organs. The Wolffian ducts are made to persist and to differentiate, while the Mullerian ducts are inhibited and disappear. The neutral primordia of the external genitalia are stimulated toward development of male structures. It is remarkable that female-type development is not the result of the presence of an embryonic ovary, but rather of the absence of an embryonic testis. This has been shown experimentally by Jost in rabbits and mice, where removal or destruction of the embryonic gonad—regardless of whether it was destined to develop into a testis or an ovary—results in disappearance of the Wolffian ducts and in typical female differentiation of the Mullerian ducts, the urogenital sinus, and the external genital organs. It appears that the same course of development is taken by human embryos whose gonads failed to develop.

The concept of hormonal determination of sexual traits is by no means opposed to the theory of genetic sex determination. Hormones are the "tools" by means of which the genetic constitution of the embryo normally directs its sexual development.

VARIATIONS IN SEXUAL DIFFERENTIATION

In almost all embryos, the XX–XY mechanism of genetic sex determination results in the successful formation of individuals of the appropriate normal sex. Occasionally, however, variations in development occur which lead to abnormal sex differentiation. In extreme cases, zygotes of a given "chromosomal sex" may develop into persons of the opposite "bodily sex," or into individuals who possess, side by side, traits typical of both sexes. Although great progress has been made in these fields, much still remains undecided. One fact stands out already: The diversity of abnormal sex types in man cannot be arranged along a simple scale from normal males to normal females. Rather, there is evidence for a diversity of processes which may lead to the different variations in the development of sexual traits.

Certain inherited variations in sexual traits are not caused by abnormal processes of sex determination but represent the specific development of normal sexual attributes. These variations include racial differences, as well as normal differences between individuals of the same population, in size and configuration of external genitalia and breasts, and in type of beard. Some genetically caused abnormalities in the development of sex characters are likewise not part of sex determination as such. Presence of supernumerary mammary glands, in males and females, in some kindreds obviously has a genetic basis, analogous to the inheritance of extra fingers and toes (Fig. 205). The opposite result, absence of nipples and breasts in both sexes, has also been encountered (Fig. 206.) Another abnormality of the mammary gland, gynecomasty, is more interesting, since it leads to female-type devel-

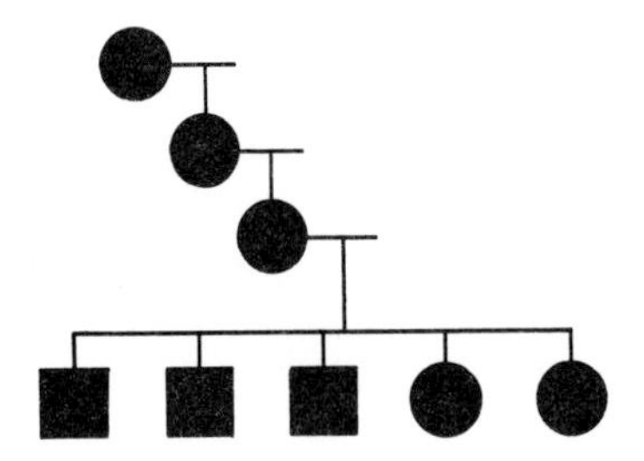

FIGURE 205
Inheritance of supernumerary mammary glands (polymasty). (After Komai, 1934.)

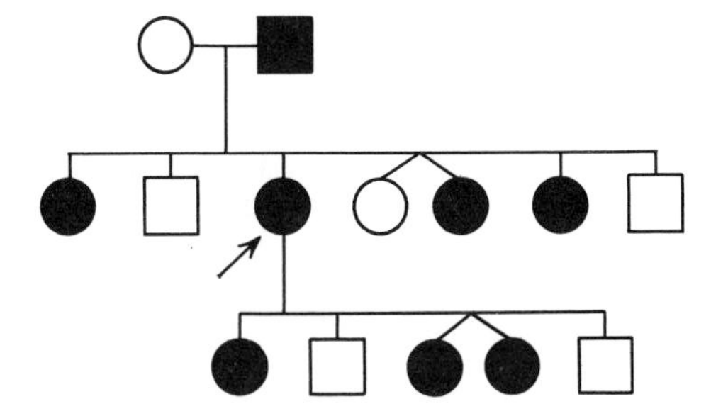

FIGURE 206
Part of a pedigree showing dominant inheritance of absence of nipples and breasts. (After Fraser, *in* Gedda (Ed.),/*Novant' Anni delle Leggi Mendeliane,* 1955.)

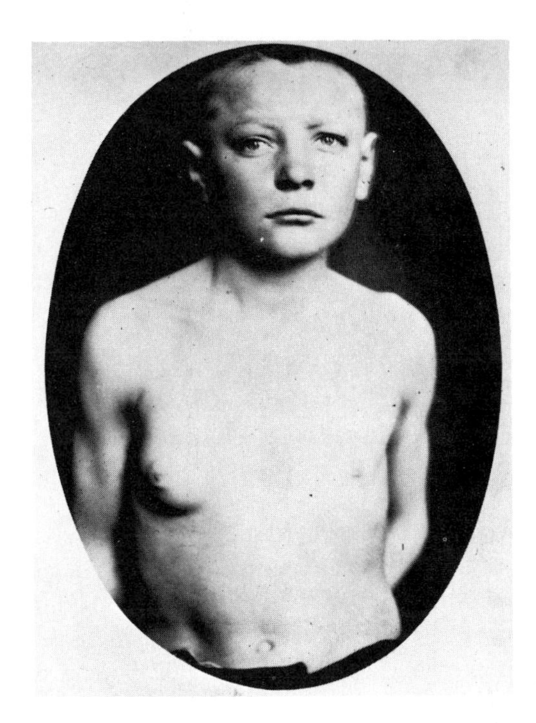

FIGURE 207
Gynecomasty on the right side of a 14-year-old boy. This boy was possibly a mosaic of XY and XO cells (for a possible origin of the mosaic condition, see Fig. 216.)

opment of the organ in otherwise often typical males (Fig. 207.) In some cases of gynecomasty, there is evidence for genetic causation. Perhaps the genotype of the mammary tissue of these males results in a changed response to the normal mixture of male and female sex hormones—a response which leads to well-developed breasts. This possibility is suggested by the existence in races of chickens of roosters that are hen-feathered. Experiments on these roosters have shown that the testicular hormones are perfectly normal, but the feather primordia respond to them in atypical fashion.

Another abnormality of male development which occurs once in every few hundred births is hypospadia, the misplacement of the opening of the urethra from the end of the penis to a lower position. Since, in the female,

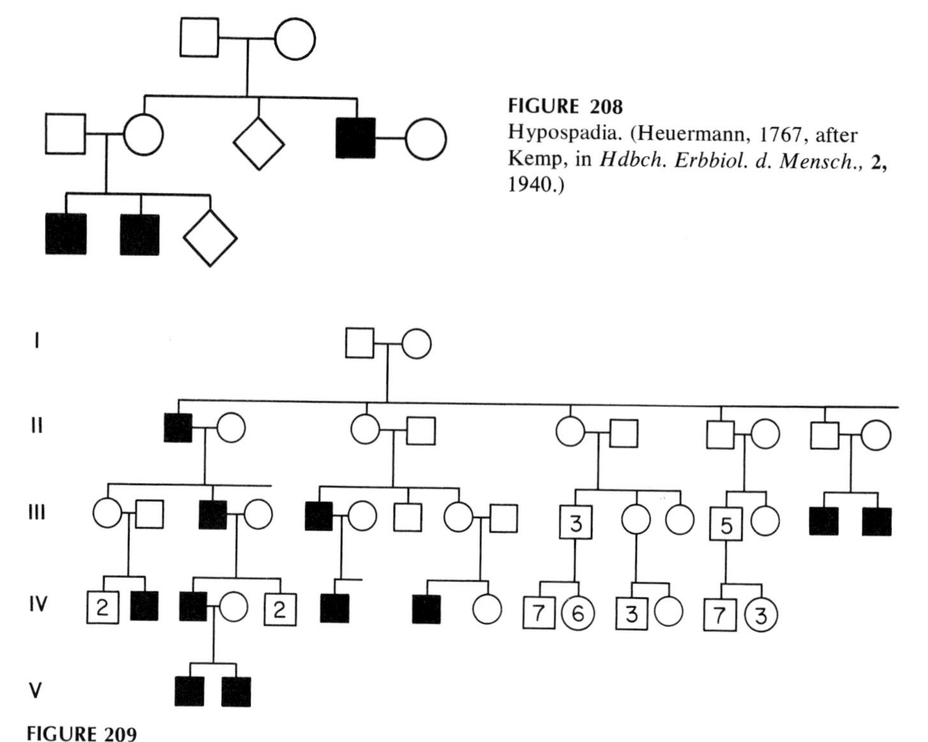

FIGURE 208
Hypospadia. (Heuermann, 1767, after Kemp, in *Hdbch. Erbbiol. d. Mensch.*, **2**, 1940.)

FIGURE 209
Part of a pedigree of precocious male maturity. (After Jacobson and Macklin, *Pediatrics*, **9**, 1952.)

the urethra typically does not enter the clitoris, hypospadia is an approach toward a female trait. Some types of hypospadia are indeed part of fundamentally intersexual development (as discussed on pp. 514ff.), but the common forms are only localized developmental deviations. In some cases there is no evidence for a genetic basis, but in others recessive as well as, more rarely, dominant transmission is encountered (Fig. 208).

As an example of genetically conditioned sexual variations, extreme sexual precocity may be cited. Males in certain kindreds and females in other kindreds develop signs of sexual maturity, such as growth of axillary and pubic hair or development of the breasts, years before the normal ages of adolescence. Figure 209 shows part of a kindred in which altogether 27 males, in four generations, showed precocious maturity, in some as early as two years of age. Inheritance is clearly dominant, through both males and females, with no expression in any female and incomplete penetrance of the dominant in males. A short pedigree of sexual precocity in females, transmitted through an unaffected male, is shown in Figure 210. It is noteworthy that sexually precocious individuals in both kindreds married when they attained adulthood and that they had children.

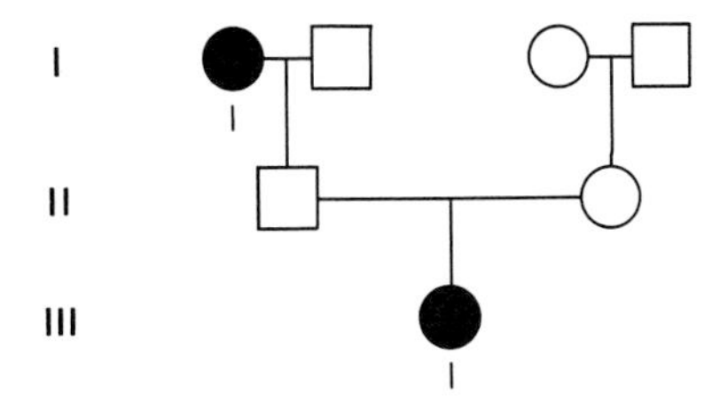

FIGURE 210
Two cases of precocious female maturity. I–1 matured at 8 years, III–1 began to mature shortly after 6 years. (After Wilkins.)

The Barr Body. It had often been wondered whether developmental sex deviants have XX or XY constitutions. However, no knowledge was available on this topic before the discovery of methods that made possible the detailed chromosomal analysis of cell nuclei during mitotic and meiotic division stages. Nevertheless, an unexpected finding bridged the gap between the earlier lack of such material and its later availability. This was the discovery of a small, stainable body in the nondividing nuclei of females and its absence in those of males, the sex-chromatin, or "Barr body," as it came to be called, after its senior discoverer. It can be seen in many tissues of females, including the epidermis and the oral mucosa (Fig. 211), and also in the amniotic fluid surrounding female fetuses.

A variant of the sex chromatin occurs in the so-called neutrophil white blood cells. Here a "drumstick," consisting of a fine stainable thread and a round stainable head, protrudes from a nuclear lobe in a small percentage of female cells but is nearly always absent from male cells (Fig. 212).

The discovery of the sex chromatin initiated a fruitful period, during which the cells of various types of developmental sex deviants were investigated. Individuals whose cells contained sex chromatin were usually interpreted as "chromosomal females," and individuals whose cells contained no sex chromatin as "chromosomal males." Although these interpretations are valid

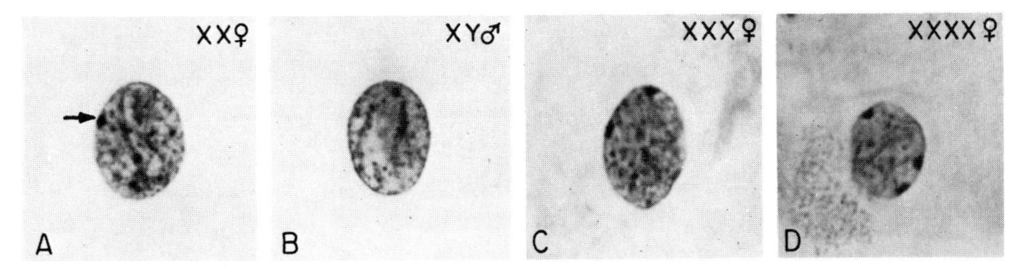

FIGURE 211
Nuclei from oral mucosa smears. **A.** Nucleus from a female showing a Barr body. **B.** From a male; no Barr body. **C.** From an XXX female; two Barr bodies. **D.** From an XXXX female; three Barr bodies. (A, B, Grumbach and Barr, *Rec. Progr. Hormone Res.*, **14,** 1958. C, D, originals from Dr. M. L. Barr.)

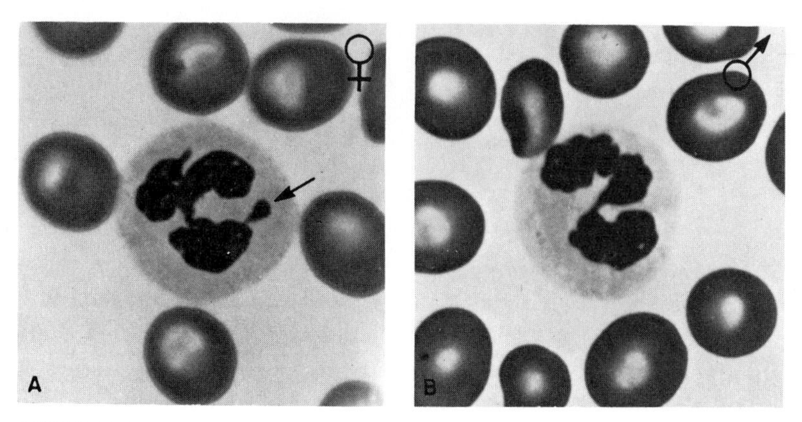

FIGURE 212
Two neutrophil white blood cells. **A.** With an accessory nuclear lobe ("drumstick"). **B.** Without the accessory lobe. (Grumbach and Barr.)

for almost all sexually normal humans, they were only partly applicable to developmental sex deviants. This became known when C. E. Ford obtained and cultivated bone marrow cells from developmental sex deviants and analyzed the individual chromosomes in dividing cells with the use of modern methods (compare Chap. 2, p. 22). Sex-chromatin-positive individuals usually have two X-chromosomes and no Y-chromosome, and thus are truly chromosomal females; but the rare XXY individuals, who are chromosomal males, as judged by the possession of a Y-chromosome, are likewise chromatin positive. Conversely, sex-chromatin-negative individuals are usually XY, and thus chromosomal males; but the rare XO individuals, who are chromosomal females, as judged by the absence of a Y-chromosome, are likewise chromatin negative. It is now clear that decisive determinations of the chromosomal sex require detailed analysis of the numbers and kinds of chromosomes present.

Nevertheless the Barr bodies continue to provide important information. Thus, in surveys of male populations it is easy by means of staining a few cells from the oral mucosa to discover the occasional deviants who possess a Barr body or in female populations those who do not possess it. Following this a karyotype analysis of the exceptional individuals is appropriate. Another application of the Barr body technique consists in the study of cells in the amniotic fluid of pregnant women who are known or likely to be heterozygous for an X-chromosomal recessive deleterious gene, such as that for Duchenne-type muscular dystrophy. Here a test for Barr bodies will indicate whether a girl or a boy has been conceived. If the latter, his 50 per cent chance of becoming affected may be the basis for consideration of legal abortion of the fetus.

For large-scale surveys, the more recent discovery of the striking fluores-

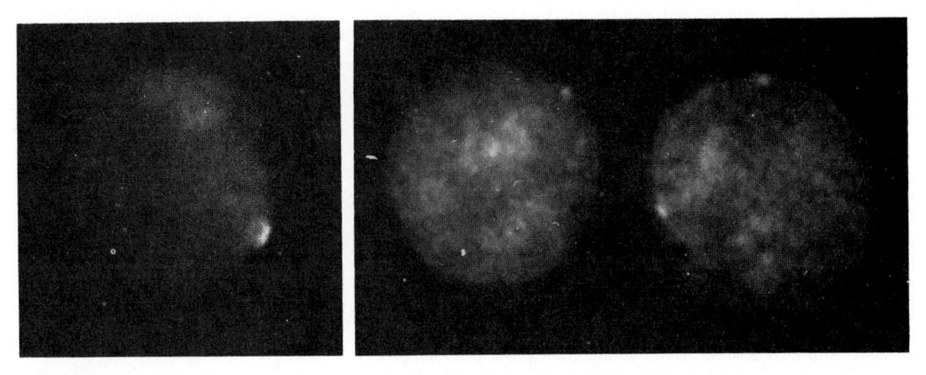

FIGURE 213
Cell nuclei with fluorescent Y-chromosome from the oral mucosa of an XY male (*left*) and an XYY male (*right*), after staining with primaquine. Fluorescent bodies similar in appearance to nonfluorescent Barr bodies occur in a very low percentage of normal female nuclei, in more than two-thirds of normal male nuclei, and in more than four-fifths of the nuclei of XYY males, a high proportion of which have two fluorescent chromosomes. Most of the fluorescent bodies are Y-chromosomes. (Left, original from Dr. Helga Muller; right, originals from Dr. Uta Francke.)

cence after fluorescent staining of the Y-chromosome complements that of the Barr bodies. It is now possible to determine the presence of a Y-chromosome or its absence in nondividing cells such as those of the oral mucosa (Fig. 213) or of the amniotic fluid independent of the more laborious analysis of whole karyotypes in cell cultures. The fluorescent technique even makes it possible to determine the presence of a Y-chromosome in individual sperm cells. (Strangely enough, more than one per cent of sperm cells were seen to possess two Y-chromosomes.)

Anomalies of Sexual Differentiation

The great majority of human beings are normal males or females but somewhat less than one per cent exhibit abnormal sexual differentiation. Such abnormalities vary greatly from person to person but many can be assigned to one or the other of a few well defined categories. These assignments make use of phenotypic characteristics of external and internal anomalies as well as of the sex-chromosomal constitutions of the individuals.

Individuals who exhibit traits typical of both typical male and typical female phenotypes have been called *hermaphrodites*, after the god Hermes and the goddess Aphrodite. They have been subdivided into (1) female pseudohermaphrodites, (2) male pseudohermaphrodites, and (3) "true" hermaphrodites. Group (1) comprises individuals whose gonads are ovaries but who are virilized, having internal or external male features; group (2), individuals whose gonads are testes but who exhibit female traits; and group (3), individuals who have both ovarian and testicular tissues. Sometimes the general term "intersex" is used to designate an individual belonging to any of the three groups, and it has also been applied where the general phenotype is clearly male or female but the presence or absence of a Barr body is typical for that of the opposite sex. Since the Barr body is only a partial reflection of the sex-chromosomal constitution—it reflects the number of X-chromosomes but not the presence or absence of a Y-chromosome—in this book the term intersex will not be applied to morphological males or females with unusual Barr body evidence.

Turner's and Klinefelter's Syndromes. One class of anomalies of sex-chromosomal constitutions consists of lack of normal differentiation of the gonads ("gonadal dysgenesis"). Persons having such anomalies are either phenotypic females or males, but they possess a number of somatic abnormalities. Females of this type have no ovaries. Instead there are streaks of connective tissue but, usually, devoid of germ cells. Among the characteristics of such females are shortness of stature, underdevelopment of primary and secondary sexual features, and webbed skin in the neck region. They are defined as exhibiting "Turner's syndrome." Overall they are mentally normal but often show weakness in some specific mental attributes (p. 702). Typically they do not have a Barr body in their cells, and chromosomal analysis shows that they have only 45 chromosomes, including a single X and no Y (Fig. 214, A).

In addition to the typical Turner's XO constitution several others are known in which a normal X-chromosome plus a fragment of a second X-chromosome

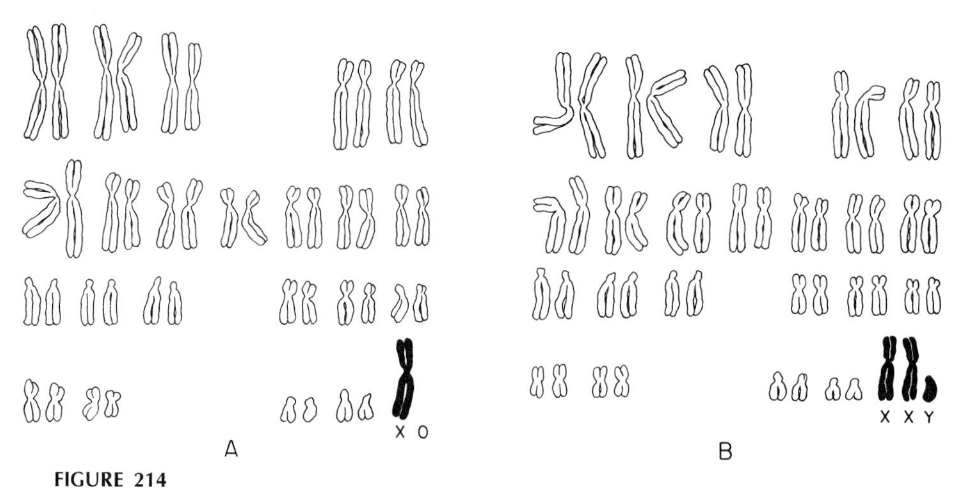

FIGURE 214
A. Karyotype of an XO female (Turner's syndrome). **B.** Karyotype of an XXY male (Klinefelter's syndrome). (After original photographs of Dr. Margery Shaw.)

are present. Among these are XX^L, where X^L stands for the long arm of a normal X-chromosome, and XX^S, where X^S stands for the short arm. Since both types of constitutions lead to absence of ovaries, it may be concluded that both arms of the X in double dose are necessary for ovarian differentiation. XX^L individuals are short and have other traits similar to those of Turner's syndrome, but XX^S individuals have normal stature and in general are not Turner-like. This indicates that stature and other somatic traits that appear abnormally in Turner's syndrome are controlled by genes on the short arm of X.

A fragment of the Y-chromosome is known that presumably lacks the short arm of a normal Y. Individuals who have a normal X plus the Y fragment are females, with only streak gonads, but of normal stature and also without other Turner traits. This suggests that the male-determining genes of the Y-chromosome are located on its short arm and that genes for absence of Turner's traits are on the long arm.

Males with gonadal dysgenesis have small testes in which no spermatogenesis occurs. They are defined as exhibiting *"Klinefelter's syndrome."* These men usually are disproportionally tall. Many exhibit gynecomasty, a trait dependent on abnormal concentrations of sex hormones. Many are below average in mental ability. Typically, they have a Barr body in their cells, and chromosomal analysis shows most of them to have 47 chromosomes, including two X and one Y (Fig. 214, B). In addition, some men with this syndrome have been found to have 48, 49, or 50 chromosomes, including three, four, or five X-chromosomes, respectively, and a Y-chromosome. Still other Klinefelter's males have two X-chromosomes and two Y's.

Most individuals of XXY constitution presumably originate either by fertilization of an exceptional XX egg by a Y sperm or of a regular X egg by an exceptional XY sperm. Persons of XXY constitution are thus "origin opposites" of those with XO constitution, since the latter—except for cases of chromosomal loss in early development—come from exceptional O eggs or O sperm. Klinefelter's individuals occur with a frequency of about 2 per thousand among newborns while Turner's individuals occur with a frequency of only 3 per ten thousand. The difference is in part due to the fact that there is no unusual fetal mortality in Klinefelter's syndrome but that more than 90 per cent of Turner's fetuses die early in development.

Multiple-X Females. Approximately 1 in 1,000 females have two Barr bodies and their chromosomal analysis shows them to be XXX (triplo-X). Many are quite normal but some exhibit subnormal mental capacity. A few females with XXXX and XXXXX constitution have also been observed. They all showed severe mental deficiency.

Triplo-X females are fertile. One would expect their children to be of four types: XX, XY, XXX and XXY, in equal proportions. Actually among more than 30 children born to XXX women all were normal XX and XY individuals except for a single XXY. It is possible to account for the absence of XXX and near-absence of XXY zygotes by assuming a directed meiotic segregation in the mother such that two of her X-chromosomes preferentially segregate into a polar body while the third is assigned to the egg nucleus. "Meiotic drive" of this nature is known from *Drosophila*. A similar unexpected distribution appears among the offspring of human XYY males. Only XX and XY children have been observed and no XYY.

Female Pseudohermaphroditism: Excessive Development of the Adrenal Cortex. Most frequently, female pseudohermaphroditism is caused by an overdevelopment of the fetal adrenal cortex. Although the excess of male hormone released by the cortex results in masculinization, it never leads to a fully developed male. Normal oviducts and a normal uterus are present, but a small vagina may open into the urethra. The external genitalia, although not typically male, approach male differentiation. These female pseudohermaphrodites are sex-chromatin positive. The different congenital types of adrenal cortical virilization are the result of deficiencies—brought about by homozygosity for recessive genes—of specific enzymes concerned with steroid metabolism.

Excessive adrenal cortical secretion, which can also occur in later life, is sometimes caused by a tumor of the gland. Women in whom this happens undergo a change of voice, grow heavy facial hair, and develop other male traits. These symptoms of "virilism," which are less striking than those of congenital female hermaphrodites, disappear after successful treatment of the adrenal cortex. Such virilized women are not included in the class pseudohermaphroditism.

Male Pseudohermaphroditism: Testicular Feminization. A very rare, but rather well understood, anomaly of sexual differentiation is testicular feminization. Affected individuals externally appear as well developed females but internally, except for a small vagina, lack female structures and possess testes and derivatives of the Wolffian ducts. Their chromosomal constitution is XY. Many of these chromosomally male, but phenotypically female, pseudohermaphrodites marry and have normal sex relations with their husbands. Reproduction obviously cannot occur. Part of a pedigree from Switzerland and a pedigree from Sweden are given in Figure 215. The pedigrees show that the intersexuality is transmitted by normal females and that, among the sibs of the 6 male pseudohermaphrodites in B, there were 8 normal sisters and 2 normal brothers, one of whom had, in turn, 6 normal children. From the combined data on fifteen families reported from various countries, it was found that the proportions of females to males to intersexes are 75:20:63. The numbers, and an inspection of the pedigrees, suggest strongly that carrier women are heterozygous for an

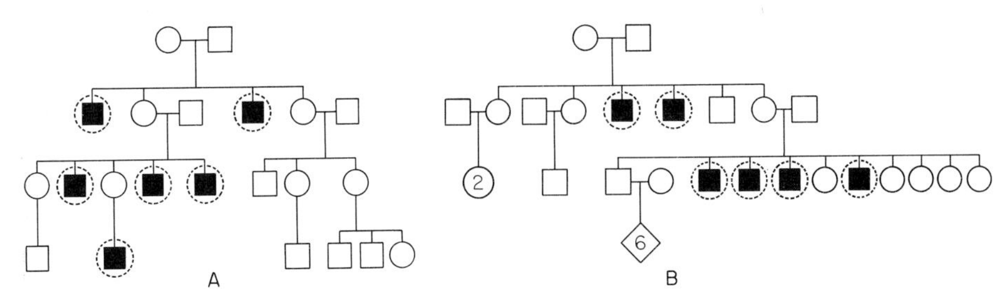

FIGURE 215
Pedigrees of male pseudohermaphroditism (testicular feminization). Solid square inside interrupted circle = affected individual. (A, after Petterson and Bonnier; B, after Burgermeister, *J. génét. hum.*, **2**, 1953.)

abnormal allele which, in an XY individual, results in male pseudohermaphroditism. According to this assumption, some of the 75 normal females are free from the abnormal allele and some carry it. The 20 normal males are free from it, but the 63 intersexes have obtained it from their mothers. Among the children of carrier females, one might expect equality between the number of males and the number of intersexes. The proportion, 20:63, at first does not seem to agree with this expectation. The disagreement is, however, spurious, because it depends on the method of ascertainment (pp. 202ff.): Since the sibships were ascertained through the presence of the intersexes, and since the compilation included only those pedigrees in which at least 2 affected individuals were reported, all those sibships which contained only normal females and normal males from carrier mothers escaped notice. Had it been possible to include these sibships in the total, a 2 female:1 male:1 intersex ration presumably would have resulted. Some investigators have failed to appreciate the fact that, under the method of ascertainment, a deficiency of males, as compared to intersexes, is an expected feature of the simple genetic interpretation. This has led to some ingenious, but unnecessary, genetic or developmental hypotheses.

It may be added that this type of inheritance—of an allele which is without effect in females but which causes pseudohermaphroditism in males—is compatible with either autosomal or sex-linked transmission. If the allele is called *Tr* and assumed to be in an autosome, then XX *Tr tr* = female, XY *Tr tr* = intersex. If *Tr* is in the X-chromosome, then $X^{Tr}\ X^{tr}$ = female, $X^{Tr}\ Y$ = intersex. It may be possible some day to distinguish whether transmission is autosomal or sex linked if the transmission of a suitable X-linked gene *g* can be followed in pedigrees showing male pseudohermophroditism. If *Tr* is autosomal, women who are heterozygous for both *Tr* and the X-linked gene *g* would show independent assortment. On the other hand, if *Tr* is X-linked and within measurable distance from *g* this should become apparent from a preponderance of offspring with noncrossover genotypes. Pedigrees in which both *Tr* and the gene for color blindness, *c*, were transmitted have indicated free recombination of the two loci. This is compatible either with autosomal localization of *Tr* or with localization of *Tr* in the X-chromosome but far away from *c*. Support for one of the alternatives, namely that *Tr* is in the X-chromosome, may be seen in the discovery that a gene for testicular feminization in the mouse is X-linked.

The paradoxical contrast between presence of testes and external feminine appearance is not the result of abnormal hormone secretion by the male gonads but, as first suggested by Wilkins, of inability of the relevant body tissues of persons with testicular feminization to react in a male direction to stimulation by testosterone. In normal males, testostrone, secreted by the testes, is taken up by the cells of such sexual target tissues as the prostate, the seminal vesicles, and the preputial gland. There, it appears, that testosterone does not act directly toward male differentiation, but that it is reduced within the cell nuclei to dihydrotestosterone, which is a powerful determiner of maleness. The conversion of testosterone to dihydrotestosterone is accomplished under the influence of a reducing enzyme. In males with testicular feminization testosterone is secreted by the testes as in normal males, but, as the enzyme is absent, no dihydrotestosterone is formed. The absence of dihydrotestosterone is the reason for the insensitivity of the target organs to testosterone. This insensitivity permits embryonic and adult tissues of the genital organs, the mammary glands, and other parts to develop along a female, instead of a male, pathway. Testicular feminization in man is thus a counterpart to hen-feathering in chickens (p. 509).

The *Tr* gene may also have a slight effect in otherwise normal carrier women, since some of them, like their intersexual offspring, lack axillary and pubic hair, and their sexual maturation seems to be delayed. It is not known whether all cases of testicular feminization are caused by alleles at a single locus. Genetic heterogeneity may account for some of the variations in phenotypes of affected individuals.

The gene for testicular feminization overrides the normal XY sex-determination reactions. It also overrides those of XXY determination, since a pair of twin girls is known who have an XXY karyotype and testes but female bodies.

Genetic Mosaics

It is a basic postulate that—barring somatic mutations—the genotypes of the cells of a normal individual are all alike, having been derived by mitosis from a single fertilized egg. Like any postulate, this one is subject to restrictions. Nondisjunction or loss of chromosomes during cell division may give rise to different cell lines within the same individual and, as we shall see, unusual types of fertilizations or fusions of initially separate embryos will lead to genetic

TABLE 60
Human Sex-chromosomal Mosaics. The mosaics may combine two or three chromosomal constitutions. Phenotypically the mosaics may be female, male, or gynandric.

Female	*Male*	*Gynandric*
XO/XX	XY/XXY	XO/XY
XO/XXX	XY/XXXY	XO/XYY
XX/XXX	XXXY/XXXXY	XO/XXY
XXX/XXXX	XY/XXY/?XXYY	XX/XY
XO/XX/XXX	XXXY/XXXXY/XXXXXY	XX/XXY
XX/XXX/XXXX		XX/XXYY
		XO/XX/XY
		XO/XY/XXY
		XX/XXY/XXYYY

mosaicism. The first chromosomally studied Klinefelter's individual, though showing only male features, contained XXY and XX cells in his bone marrow. Subsequent studies have shown that mosaicism for numbers of chromosome 21 is not infrequent among persons affected with Down's syndrome and that there are numerous varieties of mosaicism of the X- and Y-chromosomal types (Table 60). The left and the middle columns of the table list mosaics consisting of two or even three different chromosomal cell types whose sex is the same. The right column lists mosaics composed of cells of different sex.

Not all chromosomal mosaics are recognizable as such. Usually only bone marrow cells or peripheral blood cells are available for study. They can be complemented by cells from small biopsies of the skin. Biopsies of internal tissues may also be available when surgery is called for in the treatment of abnormalities. It is unavoidable, however, that in some cases only a single cell type is recovered, particularly when the mosaicism originated relatively late in development and includes only a limited part of the whole body. Much variation can be expected from one mosaic to another even of the same sex-chromosomal constitution, depending on the time in development of the initial abnormal mitosis, its place in the body, subsequent cell migrations, and pure chance.

It would be expected that tests for autosomal blood-group genotypes in sex-chromosomal mosaics would show uniformity of the red cells. In the great majority of mosaics this is indeed the case. Very rare exceptions have been encountered in twins. Twin embryos may exchange blood-forming cells and later contain mixtures of red cells with different antigens. If the twins are of opposite sex the blood-forming cells will be mixtures of male and female cells.

The mosaics for blood antigens and sex differ from the majority of mosaics in that their origin is from two separate zygotes, those of the male and female twin. Mosaics originating from more than one zygote are often called "chimeras" after the mythical monster that was compounded of a lion's head, a goat's body, and a serpent's tail. Since it is frequently not possible to decide between single or double zygotic origin of mosaics, the term mosaics is often used to cover both types.

True Hermaphrodites and Gynandromorphism. In insects, we are acquainted with a type of sexual abnormality which is different from intersexuality and is known as gynandromorphism (*gynē* = woman; *anēr*, *andros* = man). Gynandromorphs or gynanders are individuals who are composed, mosaic-like, of typical female and typical male parts. Their chromosomal constitution corresponds to their mosaic appearance. In a *Drosophila* gynander, for instance, the female parts have two X-chromosomes in their cells, the male parts only one X-chromosome. (It should be remembered that in *Drosophila,* the female-male alternative is controlled by the 2X – 1X alternative, the Y-chromosome not being sex-determining.) The two different constitutions in cells which developed from a single egg are the result of some irregularity in the formation of the mature egg or in the distribution of the X-chromosomes in the divisions of the fertilized egg.

Let us consider an analogous hypothetical example in a human egg (Fig. 216). At the left of the figure (A), an egg is shown fertilized by a Y sperm. During the first, or a later, cleavage division (B), one of the daughter halves of the Y-chromosome happens to lag behind on the mitotic spindle and fails to become included in one of the daughter nuclei (C). Consequently, the derivatives, by

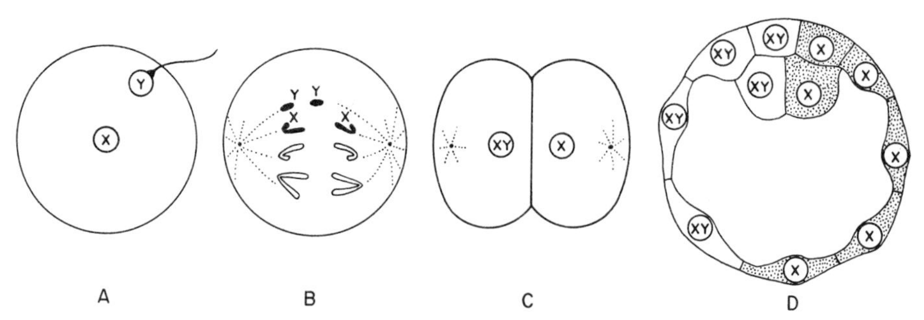

FIGURE 216
One possible mode of origin of a human gynandromorph. **A.** Fertilization of an egg by a Y sperm. **B.** First cleavage division with abnormal distribution of the sister Y-chromosomes. **C.** The resulting two-cell stage consisting of one XY (male) and one X (female) cell. **D.** Cross section through blastocyst with XY and X cells. Variations in the distribution of embryonic cells will not necessarily result in bilaterality of the gynander and the contributions of the two cell types to the whole individual may be very uneven.

further division of the two nuclei, will belong to two different types (D): XY and XO, the first of male and the second of female chromosomal sex. Later, when gonadal tissue is formed in the embryo, the XY cells might form testes and the XO cells ovaries, assuming that the latter genotype does not always result in gonadal dysgenesis. Gynanders with testicular and ovarian tissue could also be produced by embryos which start as exceptional XXY zygotes, but lose, in part of their cells, the Y-chromosome, and thus consist of male XXY and female XX tissues.

Sex hormones, as found in mammals, are absent in most insects, whose sexual traits are determined autonomously according to the genetic constitution of the cells which form each part. This accounts for the sharply mosaic character of insect gynanders. In man, a genetic setup for gynandromorphism would not lead to clear-cut sexual mosaics, since the development of certain tissues and organs is not solely determined by their sexual genotype but also depends on hormonal influences emanating from other tissues and organs. Nevertheless human gonadal mosaics have been discovered in which testicular and ovarian tissues or organs have developed side by side (Fig. 217) and in which secondary sex characters such as a male and a female breast or bearded and nonbearded facial areas likewise give a mosaic appearance. Several of these true hermaphrodites have been shown to be sex-chromosomal mosaics, including, among others, XX/XXY and XX/XY compounds. They are gynanders.

It is unknown whether all true hermophrodites are sex-chromosomal mosaics. In some such hermaphrodites only XX cells were encountered and in others only XY cells. Possibly sex-chromosomal mosaicism existed in all these cases but remained undetected in the necessarily small amount of tissue material that was studied. So much evidence is available in general for the need of the presence of a Y-chromosome for testicular differentiation that we must hesitate to assume that it can also occur without a Y-chromosome. Yet, it is in line with genetic principles that certain genotypes and environmental influences could exist whose effects so override the normal situation that testicular development might occur without a Y-chromosome.

Ferguson-Smith has ingeniously suggested that recombination between X- and Y-chromosomes in a male may occur rarely and result in transfer of male determiners from the Y to the X. In this way an $X^{♂}$sperm will be formed, resulting, upon conception, in an $XX^{♂}$zygote. Inactivation according to the single-active-X theory might lead to activity of the normal X-chromosome in some cells of this zygote and of the $X^{♂}$-chromosome in others, thus providing a basis for true hermaphroditism of a cytologically XX individual.

Sex-chromosomal gynandromorphism is not always based on abnormal mitotic events during development of the zygote. Some human gynanders have been discovered whose body including the gonads were derived from two different sperm, one carrying an X-, the other a Y-chromosome. The first such

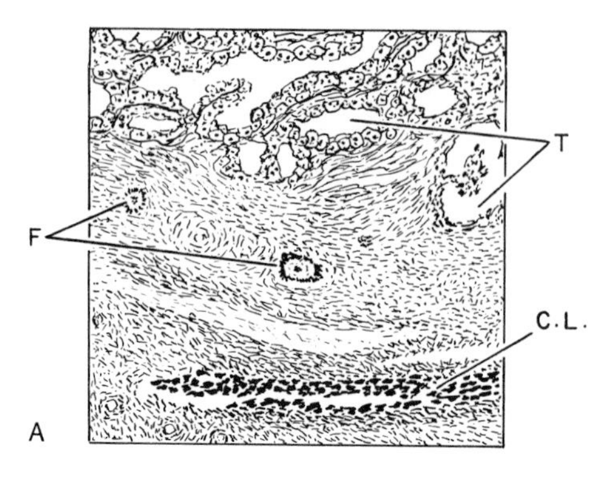

FIGURE 217
Ovotestes in true hermaphrodites.
A. Cross section through part of the ovotestis of a true hermaphrodite. T: testicular tubules; F: egg follicles; C. L: corpus luteum formed by a follicle that had matured and discharged its ovum. **B.** Ovarian follicle from an ovotestis of an XX/XY mosaic individual. **C.** Seminiferous tubules from the same ovotestis as in B. (A, after Schauerte, *Z. Konstitlehre,* **9,** 1923; B, C, originals from Drs. I. Park, H. W. Jones, Jr., and Wilma B. Bias.)

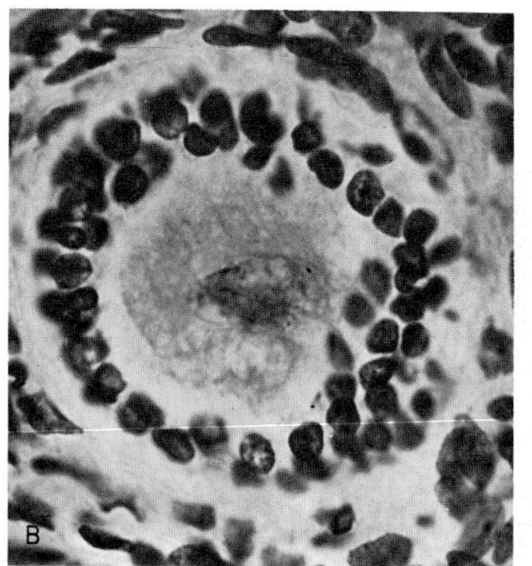

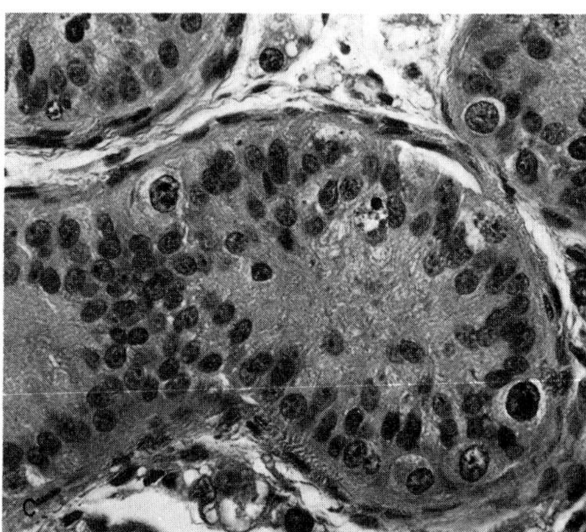

individual was a girl, two years old at the time of the study, who had an enlarged clitoris. She proved to be an XX/XY gynander who had an ovary on one side and an ovotestis on the other. One of her eyes was hazel, the other brown. Her red cells consisted of two different populations, bearing different antigens (Fig. 218); one population was L^{MSu}/L^{MS}; R^1/R^2, the other L^{MSu}/L^{Ns}; R^1/r (see Chapter 12). Her parents' genotypes were as follows: mother L^{MSu}/L^{Ns}; R^1/R^2 and father L^{MS}/L^{Ns}; R^2/r. The child's antigen genotypes must have been derived from two sperm, L^{MS}; R^2 and L^{Ns}; r; the mother's contribution being of only one type, L^{MSu}; R^1.

A second individual who originated from double fertilization had not only two different genotypes from his father but also two from his mother (Fig. 219). He too was composed of XX and XY cells, the latter type of cells being predominant. He was not a true hermaphrodite since his gonads were normal testes that apparently happened to have formed mostly or only from XY cells. A third individual originating from double fertilization was XX chromosomally

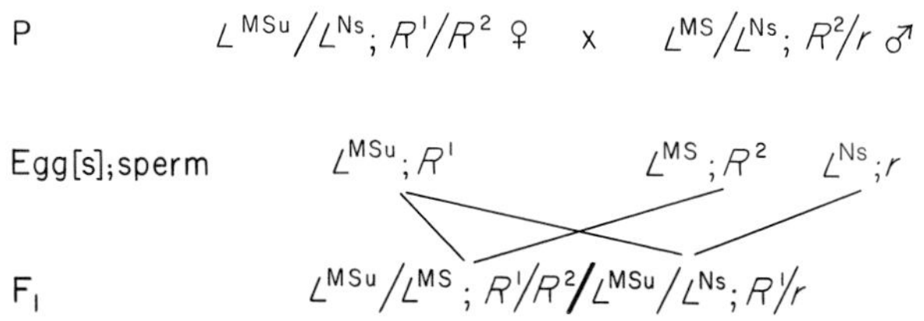

P $L^{MSu}/L^{Ns}; R^1/R^2$ ♀ x $L^{MS}/L^{Ns}; R^2/r$ ♂

Egg[s]; sperm $L^{MSu}; R^1$ $L^{MS}; R^2$ $L^{Ns}; r$

F_1 $L^{MSu}/L^{MS}; R^1/R^2 / L^{MSu}/L^{Ns}; R^1/r$

FIGURE 218

Genotypes of the blood groups of two parents, their mosaic child, and the parental egg nucleus (or nuclei) and sperm cells. One of the sperm carried an X-, the other a Y-chromosome. The heavy bar in the third line separates the genotypes of the two kinds of cells in the mosaic child. (Based on the data of Gartler, Waxman, and Giblett.)

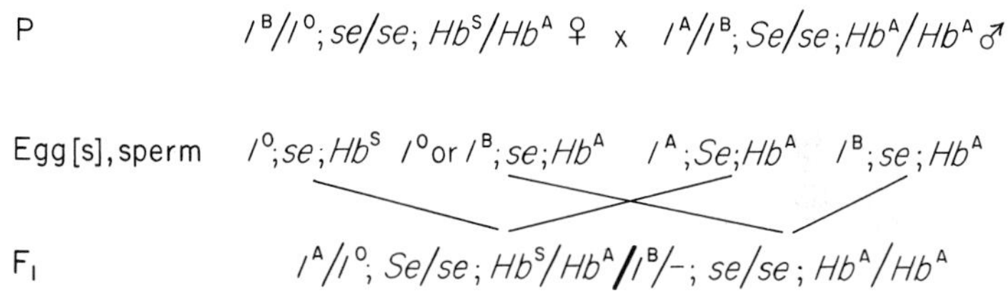

P $I^B/I^O; se/se; Hb^S/Hb^A$ ♀ x $I^A/I^B; Se/se; Hb^A/Hb^A$ ♂

Egg[s], sperm $I^O; se; Hb^S$ I^O or $I^B; se; Hb^A$ $I^A; Se; Hb^A$ $I^B; se; Hb^A$

F_1 $I^A/I^O; Se/se; Hb^S/Hb^A / I^B/-; se/se; Hb^A/Hb^A$

FIGURE 219

Genotypes of two parents, their mosaic child, and the parental egg nuclei and sperm cells. One of the sperm carried an X-, the other a Y-chromosome. (Based on the data of Zuelzer, Beattie, and Reisman.)

throughout and a healthy fertile female. The two sperm that shared in her conception carried different blood-group genotypes but were alike in bearing an X-chromosome.

Several hypotheses have been proposed to explain the origin of a single individual from double fertilization. According to one (Fig. 220, A), two sperm fertilize the two blastomeres of a parthenogenetically dividing egg. In this case the contribution from the mother is the same in both fertilizations. According to another hypothesis (Fig. 220, B), two separate eggs are fertilized by two sperm and the two embryos later fuse to give rise to a single child. In this case each of the parents contributes two different sets of genes. A third hypothesis postulates fertilization of the egg nucleus and a polar-body nucleus of the same egg. In this case too, each of the parents contributes two different sets of genes. It is not known which of the three proposed modes of origin actually occurred in any instance of double fertilization or whether different modes of origin account for different mosaics.

Medical Treatment of Intersexuality. The details of abnormal sexual development are complex, and this account has touched upon only a few of them.

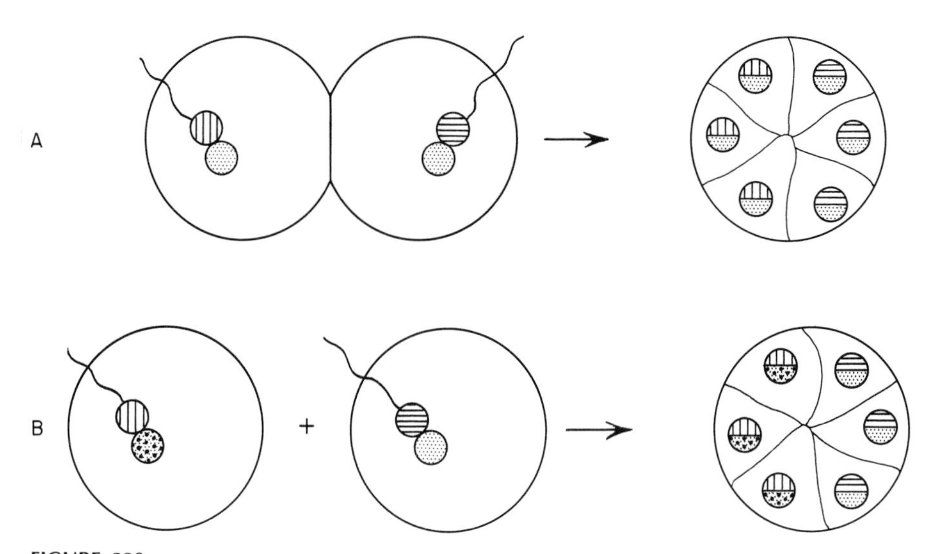

FIGURE 220
Two possible modes of origin of the human mosaics whose genotypes were shown in Figures 218 and 219. **A.** Two sperm fertilize two identical nuclei of the first two blastomeres. **B.** Two sperm feritlize two eggs which later fuse to form a single embryo. (Reprinted by permission from *Genetic Mosaics and Other Essays* by Curt Stern. Cambridge, Harvard University Press. Copyright © by the President and fellows of Harvard College.)

An understanding of the typical hormonal interrelations between gonads and secondary sex characters makes it easy to see that the removal of a gonad or the use of hormonal treatment may, in some cases, help in changing the phenotype of an intersex toward that of a normal person. If, for instance, an ovary and an ovotestis are present, removal of the latter may, through the action of the remaining ovary, result in female development. Only those parts of the body, such as the mammary glands, which may still have enough plasticity to respond to the ovarian hormones can be influenced by such a removal. Surgery may provide further means of restoring a normal phenotype.

Genetic and Psychological Aspects of Sex. We have seen that some individuals are born with developmentally caused deviations from the normal male or female morphologies. Psychologically, most of these persons regard themselves as belonging to that sex which they externally represent, or approach. Two different views have been expressed concerning the origin of psychosexual identification of a normal individual. One view regards the identification with either maleness or femaleness as the result of sociological influences such as being "assigned" a sex by parents and subsequently being reared either as a boy or a girl. This view considers the newborn child as being psychosexually neutral. The other view regards the male-female difference as innate and thus the gender developmentally impressed on the nervous system before birth. According to this view postnatal experiences may at times modify the self-recognition of an individual as male or female, and this particularly in cases of abnormal sexual morphology. Basically, however, the self-assignment of an individual is considered to be a genetically, not a sociologically, controlled trait.

It is not known whether some of the variations in the sexual behavior of men and women have a genetic basis. In other mammals, a genetic basis for variations in sexual behavior is known to exist. Thus, to identical treatment with male sex hormones, different strains of guinea pigs respond with different intensities of sex drive. In cattle, different strains of bulls kept under equal environmental conditions produce very different amounts of sperm, a property which may be related to sexual behavior. In man, it is at present impossible to separate nongenetic from genetic factors and to assign them proportional responsibility for sexual behavior. Cytological studies have shown normal male chromosomal constitution in male homosexuals. In studies of twins, Kallmann and others have found that the genetically identical twin brother of a homosexual nearly always possesses the same tendency, but that a genetically nonidentical twin brother of a homosexual possesses the same tendency in much less than one-half of the cases (see Chapter 27). Although such evidence is regarded by some as indicative of genetic factors, there is still the possibility of an explanation based primarily on environmental influences.

PROBLEMS

158. If it were possible so to change the development of an XX-zygote that a functional male resulted, what would be the sex of the offspring of such an individual?

159. If an XY-zygote could develop into a functional female, what would be the sex ratio among her offspring? (Note: A fertilized egg without an X-chromosome is not likely to be viable.)

160. Assume—though this is very unlikely—that human XXY and XO individuals may sometimes be fertile. Assume further that XXY persons produce only X and XY sperm in a 1:1 ratio.
    (a) If the XXY individuals marry chromosomally normal spouses, what sex ratio would be expected among the offspring?
    (b) If the XXY persons produce also some XX and Y sperm what would be the results?
    (c) If the XO persons marry chromosomally normal spouses, what would be the sex ratio among the offspring?

161. Assume that an XO person is fertile and carries hemophilia. If the XO person were to marry a nonhemophilic spouse, what would be the genotypes and phenotypes of the offspring?

162. Assume that nondisjunction of the sex chromosomes occurs at the first meiotic division in 2 out of 1,000 oocytes and spermatocytes each. If the nondisjunctional

gametes were to fuse with normal gametes, what would be the frequencies of XXY and XO individuals from nondisjunction in:
(a) the mothers, and
(b) the fathers?

163. Consider a chromosomally normal father and mother and their XXY son with the following genotypes:
(1) $Xg^a\ Y$, $Xg^a\ Xg^a$, $Xg^a\ Xg^a\ Y$;
(2) $Xg\ Y$, $Xg^a\ Xg^a$, $Xg^a\ Xg\ Y$;
(3) $Xg^a\ Y$, $Xg\ Xg$, $Xg\ Yg\ Y$.
What is the source of the extra X-chromosome in the son in each case?

164. For inherited testicular feminization, as represented in the pedigrees in Figure 215, it is likely that an X-linked allele *Tr* is responsible for the transformation of male zygotes into intersexes. What, according to this assumption, are the genotypes of all individuals in the pedigrees? If more than one genotype is possible, list alternatives.

165. A pair of identical twins is known who have the constitution XXY and are affected with testicular feminization. The frequency of the XXY constitution has been estimated roughly as 1 in 500, that of identical twin births as 1 in 250, and that of testicular feminization as 1 in 60,000. What is the probability of the combination of traits found in the twin pair?

166. The nucleus of a fertilized egg contains two X-chromosomes, one with the gene for normal color vision, the other with an allele for color blindness. Assume that the first X-chromosome is eliminated during an early cleavage division and that approximately the right half of the developing individual is formed from the descendants of the cell which had received only one X-chromosome. In this individual, what type of color vision would you expect in:
(a) the left eye, and
(b) the right eye?
(c) What would be the sexual type of the individual?

167. A man has the genotype *AA'BBCC'*, where each allele expresses itself as an antigenic property of the red cells. His wife is *AABB'CC'*. They have a son who is a striking mosaic for eye, hair, and skin pigmentation. He has two populations of cells, *AABBCC* and *AA'BB'C'C'*. Discuss the origin of the son's mosaic genotype.

REFERENCES

Armstrong, C. N., and Marshall, A. J. (Eds.), 1964. *Intersexuality in Vertebrates including Man.* 479 pp. Academic, London.

Bardin, C. W., Bullock, L., Schneider, G., Allison, J. E., and Stanley, A. J., 1970. Pseudohermaphrodite rat: end organ sensitivity to testosterone. *Science,* **167:** 1136–1137.

Diamond, M., 1965. A critical evaluation of the ontogeny of human sexual behavior. *Quart. Rev. Biol.,* **40:**147–75.

Edwards, J. H., 1966. Monozygotic twins of different sex. *J. Med. Genet.*, **3**:117–123.

Ferguson-Smith, M. A., 1966. X-Y chromosomal interchange and aetiology of true hermaphroditism and of XX Klinefelter syndrome. *Lancet*, **2**:475–476.

Ford, C. E., 1969. Mosaics and chimaeras. *Brit. Med. Bull.*, **25**:104–109.

Ford, C. E., Jones, K. W., Polani, P. E., de Almeida, J. C., and Briggs, J. H., 1959. A sex-chromosome anomaly in a case of gonadal dysgenesis (Turner's syndrome). *Lancet*, **1**:711–713.

Gartler, S. M., Waxman, S. H., and Giblett, E., 1962. An XX/XY human hermaphrodite resulting from double fertilization. *Proc. Nat. Acad. Sci.*, **48**:332–335.

Goldschmidt, R., 1931. *Die sexuellen Zwischenstufen.* 528 pp. Springer, Berlin.

Hamerton, J. L., 1968. Significance of sex chromosome derived heterochromatin in Mammals. *Nature*, **219**:910–914.

Inhorn, S. L., and Opitz, J. M., 1968. Abnormalities of sex development. Ch. 16 *in* Bloodworth, J. M. B. (Ed.), *Endocrine Pathology.* Williams and Wilkins, Baltimore.

Jacobs, P. A., and Strong, J. A., 1959. A case of human intersexuality having a possible XXY sex-determining mechanism. *Nature*, **183**:302–303.

Jacobs, P. A., and Ross, A., 1966. Structural abnormalities of the Y chromosome in man. *Nature*, **210**:352–354.

Jones, H. W., and Scott, W. W., 1958. *Hermaphroditism, Genital Anomalies, and Related Endocrine Disorders.* 456 pp. Williams and Wilkins, Baltimore.

Jost, A., 1961. The role of fetal hormones in prenatal development. *Harvey Lectures*, **55**:201–226.

Kallman, F. J., 1952. Twin and sibship study of overt male homosexuality. *Amer. J. Hum. Genet.*, **4**:136–146.

Lyon, Mary F., and Hawkes, Susan G., 1970. X-linked gene for testicular feminization in the mouse. *Nature*, **227**:1217–1219.

Mittwoch, Ursula, 1969. Do genes determine sex? *Nature*, **221**:446–448.

Money, J., 1963. Psychosexual development in man. Pp. 1678–1709 *in* Deutsch, A. (Ed.), *Encyclopedia of Mental Health.* Franklin Watts, New York.

Money, J., Ehrhardt, Anke A., and Mascia, D.N., 1968. Fetal feminization induced by androgen insensitivity in the testicular feminizing syndrome: effect on marriage and paternalism. *Johns Hopkins Med. J.*, **123**:105–114.

Moore, K. L. (Ed.), 1966. *The Sex Chromatin.* 474 pp. Saunders, Philadelphia.

Morris, J. M., 1953. Testicular feminization. *Amer. J. Obst. Gynecol.*, **65**: 1192–1211.

Northcutt, R. C., Island, D. P., and Liddle, G. W., 1969. An explanation for the target unresponsiveness to tetosterone in the testicular feminization syndrom. *J. Clin. Endocrinol. Metab.*, **29**:422–425.

Overzier, C. (Ed.), 1963. *Intersexuality.* 563 pp. Academic, London. (English translation of *Die Intersexualität*, 1961.)

Petterson, G., and Bonnier, G., 1937. Inherited sex-mosaic in man. *Hereditas*, **23**:49–69.

Polani, P. E., 1970. Hormonal and clinical aspects of hermaphroditism and the testicular feminizing syndrome in man. *Phil. Trans. Roy. Soc. London, B*, **259**:187–206.

Race, R. R., and Sanger, Ruth, 1968. Blood groups in twinning, chimerism and dispermy. Ch. 23, pp. 467–494 in *Blood Groups in Man.* Blackwell, Oxford.

Sohval, A. R., 1963. "Mixed" gonadal dysgenesis: a variety of hermaphroditism. *Amer. J. Hum. Genet.,* **15:**155–158.

Stern, C., 1968. *Genetic Mosaics and Other Essays.* 185 pp. Harvard University Press, Cambridge, Mass.

Taillard, W., and Prader, A., 1957. Etude génétique du syndrome de féminisation testiculaire totale et partielle. *J. génét. hum.,* **6:**13–32.

Van Wyk, J. J., and Grumbach, M. M., 1968. Disorders of sex differentiation. Ch. 8, pp. 537–612 *in* Williams, R. H. (Ed.), *Textbook of Endocrinology,* 4th ed. Saunders, Philadelphia.

Welshons, W. J., and Russell, L. B., 1959. The Y-chromosome as the bearer of male determining factors in the mouse. *Proc. Nat. Acad. Sci.,* **45:**560–566.

Young, W. C., 1957. Genetic and psychologic determinants of sexual behavior patterns, Pp. 75–98, *in* Hoagland, H. (Ed.), *Hormones, Brain Function and Behavior.* Academic, New York.

Zuelzer, W. W., Beattie, K. M., and Reisman, L. E., 1964. Generalized unbalanced mosaicism attributable to dispermy and probably fertilization of a polar body. *Amer. J. Hum. Genet.,* **16:**38–51.

21

THE SEX RATIO

The XX–XY mechanism not only accounts for the occurrence of XX and XY individuals among the offspring of a couple, but also should be responsible for the production of approximately equal numbers of the two sexes. Two X- and two Y-containing spermatozoa develop from each germ cell that goes through meiosis. The expectation, therefore, would be that fertilization with equal numbers of X and Y sperm would take place, and the numbers of women and men would be equal.

SEX RATIOS IN POPULATIONS

Sex Ratio at Birth. Contrary to the expectation of equality, male and female babies are not born in a 1:1 ratio. Among United States whites, the ratio is approximately 106 boys to 100 girls ($p = 0.5146$, $q = 0.4854$). The data on which this and most other sex ratios are based are so large that the deviations from equality become statistically quite significant. In the literature, sex ratios are given either as the fractions of males (or females) in the total population or as number of males per 100 females.

Among the United States blacks, the sex ratio is less biased in favor of males than it is among whites, but there are still 102.6 males for every 100 females born. In some other countries, the ratio is higher, in still others, lower, than in the United States. Birth data from Korea show values as high as 113 and 117 males per 100 females; the preponderably black populations of several islands in the West Indies are reported to have as few as 100, or even fewer, males per 100 females.

It must be remembered that the observed sex ratios are empirical values which are subject to chance variations. This is forcefully brought out by Table 61, which gives the "95 per cent confidence limits" for a variety of sex ratios obtained in populations of different sizes. Thus for a "true" sex ratio of 106 (stated as numbers of boys per 100 girls), the table shows that, in a population of 100, ratios as high as 158.1 and as low as 71.4 would be within the range in which 95 per cent of all observed values would be found or, expressed differently, the probability of observing a deviation from 106 as large as 158.1 or 71.4 or larger would be 5 per cent. With a population of 10,000 the range would be less, though still as large as from 101.9 to 110.2, and for a sample of one million the range would be from 105.5 to 106.4.

Sex Ratios of Stillbirths. The sex ratio at birth, called the *secondary sex ratio,* does not necessarily equal the *primary sex ratio,* that is, the ratio at the time of fertilization. During the interval between conception and birth, a considerable number of fetuses die. If the prenatal mortality affected the two sexes differentially, then the secondary sex ratio would differ from the primary. The primary sex ratio, in man, is not accessible to direct study, since it would require determination of the sex-chromosomal constitution of large numbers of newly fertilized eggs. It is possible, however, to investigate the sex ratio among pre-

TABLE 61
The 95 Per Cent Confidence Limits for Three Sex Ratios, Stated in Numbers of Boys per 100 girls, Given Various Numbers of Births.

Number of Births	*Sex Ratios*		
	104	105	106
100	70.0–155.0	70.7–156.5	71.4–158.1
10,000	100.0–108.2	100.9–109.2	101.9–110.2
1,000,000	103.6–104.4	104.5–105.4	105.5–106.4

Source: After Visaria, *Eugen. Quart.*, **14**, 1967.

natal deaths. Could it be that the surplus of live-born males is the result of a surplus of female deaths before birth?

Four different methods are available for assigning male or female sex to aborted or stillborn conceptuses: (1) sexing by inspection of the external genitalia; (2) by histological study of the gonads; (3) by presence or absence of Barr bodies and (4) by karyotype analysis, including determination of the sex-chromosomal constitution. Method (1) fails to be accurate for all embryos less than 20 weeks old and is quite unreliable for those younger than 12 weeks; method (2) is inapplicable to embryos younger than about 7 weeks, since the gonads are then not yet differentiated as ovaries or testes. Method (3), "nuclear sexing" by means of the Barr bodies, has been used with embryos as young as 4 weeks, and method (4), study of the sex chromosomes, is theoretically applicable from the very time of fertilization on, although for obvious reasons the earliest embryonic stages can be studied only very rarely. A fifth method, using the recently discovered specific staining of the Y-chromosome, had not yet been applied at the time this was written in 1972.

When the sex ratios of spontaneously aborted fetuses and of stillborns were determined by means of the first method, as reported by the United States Bureau of the Census and from other sources, a surprising result was obtained (Table 62). There was no excess of prenatal female deaths over that of male deaths that would have led to a lower prenatal sex ratio than that at birth. On the contrary, many more male than female prenatal losses were reported.

Much, however, of this surplus of male losses comprises fetuses of the first trimester of pregnancy and is almost certainly the result of misclassification of female embryos, whose clitorises at this stage are very large, as males. Physi-

TABLE 62
Sex Ratios of Stillborn Reported by the United States Bureau of the Census (1925–1934).

Month Pregnancy	*Sex Recognized (n)*	*Sex Unknown (%)*	*Sex Ratio (Males per 100 Females)*
<2	82	89.9	228.0
2	563	72.2	431.1
3	2,388	36.8	361.0
4	6,401	10.5	201.2
5	12,541	2.6	139.6
6	17,857	1.0	122.7
7	23,109	0.6	112.4
8	28,903	0.4	124.7
9	68,932	0.2	134.6
10+	2,671	0.4	133.2

Source: After Ciocco, 1940.

cians and midwives inadequately schooled in embryology could easily make this mistake. In addition, the census data on early abortions include high percentages of "sex unknown," which makes the remaining percentages—those classified as male or female—subject to uncertainty: female embryos may have been preferentially registered in the class "sex unknown" rather than in their correct class.

The very high excess of males among registered prenatal deaths does not agree with the sex ratio of specimens in the collection of the Department of Embryology of the Carnegie Institution of Washington. This collection consists of somewhat less than 6,000 embryos and fetuses contributed over a number of years by various physicians. The sex ratio of the whole sample as determined by the experts of the department is 107.9 males to 100 females. Sexing was based on morphology of external genitalia for fetuses 4 months old or older (method 1). Three month-old abortuses were sexed by histological studies of sections of gonads (method 2). In no month of prenatal life does the sex ratio deviate significantly from the average (Table 63; note that the greatest ratio deviation, in the third month, is based on only 120 embryos).

The findings on prenatal deaths from the census data and from the Carnegie collection may not really be in conflict. Not only are the census data intrinsically less reliable than the Carnegie tabulations but more recent census data do not show as great a disparity of the sexes as that reported earlier, and are similar to the Carnegie data. Nevertheless, a significant excess of male over female fetuses continues to show up in reports of prenatal losses. Independent evidence for a preponderance of male losses comes from application of method (3), nuclear sexing, to fetuses spontaneously aborted over the whole age range from about 4 weeks to term. Stevenson, who made a comprehensive analysis of this kind, found more male deaths for every age group except one. This latter consisted of the earliest abortuses, those of about 4 weeks of age. Here a slight excess of female deaths was registered.

TABLE 63
Sex Ratios of Stillborn in the Collection of the Carnegie Institution.

Month of Pregnancy	*No. of Males*	*No. of Females*	*Sex Ratio (Males per 100 Females)*
3	58	62	93.5
4	598	555	107.7
5	771	667	115.6
6	801	730	109.7
7	775	770	100.6
Total	3,003	2,784	107.9

Source: Tietze, *Hum. Biol.*, **20**, 1948.

Induced early abortions also have yielded excesses of female losses. Mikamo found only 42 males among 108 induced abortuses of less than 8 weeks of age, in contrast to older groups in which males were preponderant. Similar results were obtained when early induced abortuses were sexed by method (4), chromosome analysis: in pooled data from various investigators of both spontaneous and induced abortions a slight excess of female losses was found.

Nuclear sexing entails sources of error which should be mentioned. Absence of Barr bodies is not only characteristic of XY but also of XO genotypes. Since spontaneous abortions include several per cent of XO fetuses, which are assigned to the male category as determined by absence or presence of Barr bodies, the resulting ratio is not strictly one of males to females. Another factor that may inflate the sex ratio determined by this method is the poor stainability of the Barr bodies in tissues that are not excellently preserved. This may lead to misclassification of females as males.

It appears then that higher male mortality, for which there is evidence during most of the prenatal period, may not characterize the very early stages (Fig. 221, broken line). If female mortality is truly significantly higher than that of the males in the very early stages—including even those that are unavailable for sexing—it may well be that the primary sex ratio is one of unity—100 XY : 100 XX.

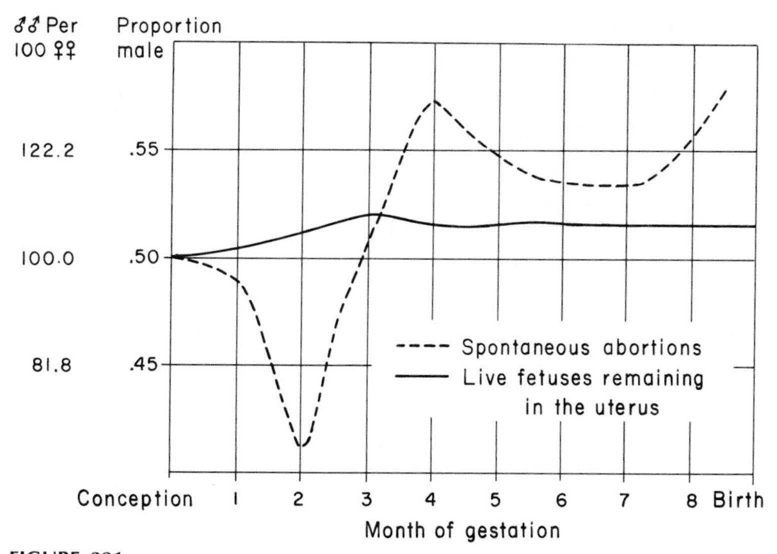

FIGURE 221
Sex ratios of spontaneous abortuses and of live fetuses remaining in the uterus, at different months of gestation. The intrauterine sex ratios are extrapolated from stillbirth and spontaneous abortion sex ratios. The sex ratios of the earliest abortions are so chosen as to fit an assumed 1:1 sex ratio at conception. (After Keller, Ph.D. Thesis, University of California, Berkeley, 1969.)

Making use of the data represented by the broken line, it is possible to construct a curve that gives the sex ratio in utero throughout the pregnancy period (solid curve). Going backward from birth the broken curve shows a variable excess of male deaths down to about the third month of pregnancy, where it shows an excess of female deaths. If we then take account of the probability of prenatal deaths by month of gestation as listed in a "fetal life table," we can calculate the sex ratios of live fetuses remaining in the uterus at any given time. These ratios are then plotted to obtain the curve shown by the solid line. In spite of the considerable fluctuations of the curve for the sex ratio of prenatal losses, the calculated sex ratio of live fetuses remaining in the uterus varies rather little. This is the result of the fact that at most stages abortions represent only a relatively small fraction of the fetuses present.

The hypothetical nature of the curves in Figure 221, particularly their extrapolation to the primary sex ratio, should be emphasized. Stevenson, in 1967, wrote that "it cannot be stated with any certainty that (the primary sex ratio) is above or below unity, or that it is not exactly unity."

It may be useful to comment on the possibility that the primary sex-ratio deviates from 1:1 in the direction of more XY than XX zygotes. If this were true the question would arise why a Y sperm is more able to fertilize an egg than an X sperm. The answer has often been expressed in terms of speed of sperm locomotion. It is reasoned that Y sperm may be lighter than X sperm because of the smaller size of a Y-chromosome, as compared to an X-chromosome. The travel of spermatozoa upward in the uterus and oviducts is sometimes pictured as a race in which the lighter Y sperm are speedier than the heavier X sperm. This comparison, however, is at best of limited value. The upward motion of spermatozoa is not primarily due to their own irregular motility, but to muscular contraction perhaps accompanied by ciliary action within the longitudinal grooves of the female ducts. In order to account for a greater ability, by means of differential motility, of Y sperm than of X sperm to reach the egg, we would have to assume that the contractions and the currents and countercurrents produced by the cilia either aid or hinder one kind of sperm to some degree. (In the rat, approximately twenty million spermatozoa are present in the uterus after mating, but only five to ten spermatozoa are recovered from the part of the oviduct nearest the ovary. In the rabbit, also, very few spermatozoa are found in the corresponding segment of the oviduct. In man it has been estimated that only a few thousand of an ejaculate of several hundred million spermatozoa survive the journey up the oviduct to the egg.)

The speculations of the preceding paragraph imply that immature and mature X and Y sperm have equal survival rates in the testes and sperm ducts, and therefore that, after mating, they are present in equal numbers in the female genital tract. Here three different possibilities may influence their fate: (1)

the environment of the female ducts may be less favorable to the survival of X sperm than of Y sperm; (2) Y sperm may be intrinsically more capable of reaching the egg than X sperm; and (3) the egg surface may react more readily to the approach of a Y sperm than of an X sperm, so that fusion of egg and sperm would be due to preferential, or selective, fertilization.

Postnatal Sex Mortality. The higher mortality of males than females, just discussed as prevailing during most of prenatal life, is also found through all postnatal age classes (Fig. 222). The effect of the postnatal differential mortality is that the secondary sex ratio of live births gives way to tertiary ratios that are different for each age group. The surplus of males at birth is progressively diminished, until, during a certain age period, the sex ratio is equal. Still later, however, more females than males are alive, since more males than females die during every age period. At what age the tertiary sex ratio reaches equality and then again deviates from it in favor of a higher proportion of females depends on the specific mortality figures. Because of medical knowledge and care, and other factors, mortality figures have changed considerably over recent decades. Assuming mortality rates as they have been projected for the different age classes of a cohort of individuals born in 1960, a United

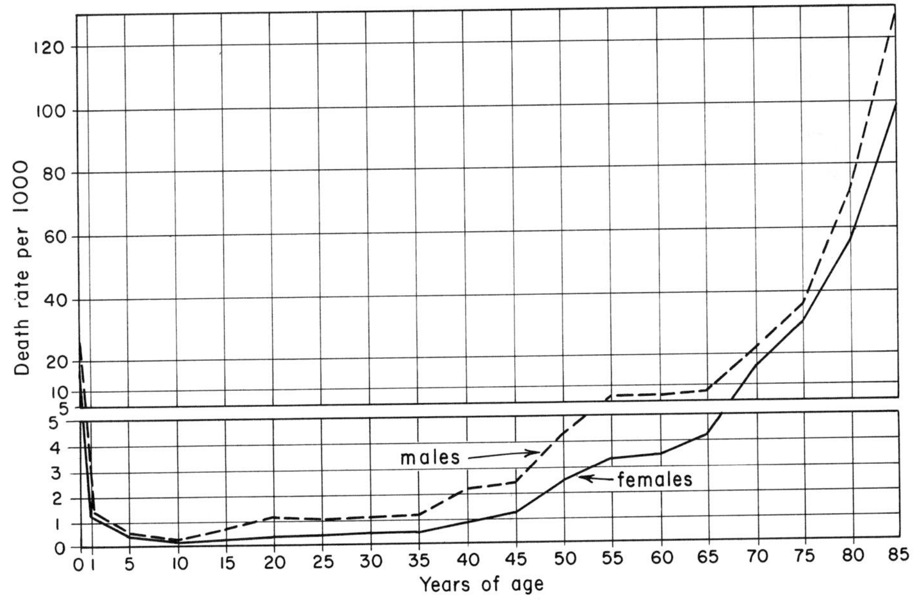

FIGURE 222
Mortality rates of white males and females, United States, born in 1960. Rates corresponding to calendar years 1963 and later are projections. (Data from Jacobson, *Milbank Mem. Found. Quart.*, **42**, 1964.)

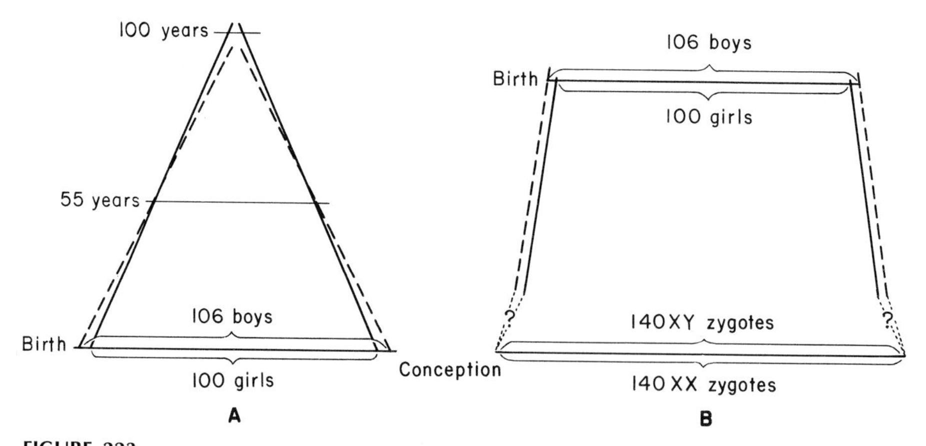

FIGURE 223
A. Diagram of changing sex ratios from birth to 100 years of age. Postnatal mortality reduces the number of males at a higher rate than the number of females. **B.** Prenatal ratios assuming a 1:1 sex ratio at conception, and a prenatal mortality of 29 per cent for females and 24 per cent for males.

States white population that started out with the nation's typical secondary sex ratio would reach equality of numbers of females and males around the age of fifty-five years (Fig. 223). As in many other countries, there is a surplus of males during the marrying ages.

Causes of Differential Sex Mortality. Why the male, at nearly every stage, is less resistant to death than the female is not understood. A study of the causes of death shows that there are a few diseases, such as whooping cough and, for anatomical reasons, gonococcal infections, which are fatal more often to females than to males; but, for most other diseases, the reverse is true. For some causes of death, the greater biological weakness of the male sex is, at least partly, understandable. Thus, the higher rate of mortality at birth of males may be connected with their somewhat larger size, which may make the process of being born more hazardous for them. In most cases, however, no obvious reason can be given for the higher mortality of the males.

It has been suggested that the greater constitutional weakness of males may be due to their having only one X-chromosome. If X-chromosomes carry recessive alleles for lower viability, sublethality, or even lethality, then many hemizygous male zygotes would be subjected to the influence of these alleles, while most female zygotes would be heterozygous and not be affected. Unfavorable X-chromosomal alleles would be eliminated rapidly from a population by their exposure to selection in the hemizygous males. To counteract this disappearance of alleles, which the theory assumes to be present at unchanged frequencies from generation to generation, new mutations from normal alleles to unfavorable ones would have to occur constantly in a large number of X-chromosomes. Stevenson has shown that it is unlikely that X-linked mutations make an appreciable contribution to prenatal death and the same is probably true for postnatal losses. A possible exception is the X-linked Xg^a gene. The sex ratios of children from marriages of Xg(a+) × Xg(a+), Xg(a−) × Xg(a−), and Xg(a+) ♀ × Xg(a−) ♂ fall within the normal range but those from Xg(a−) ♀ × Xg(a+) ♂ are significantly higher than normal. This may be the result of an incompatibility reaction, often leading to fetal loss, between the Xg(a+) zygotes and their Xg(a−) mothers. This hypothesis is strengthened by the finding that the sex ratio is still higher in those children born after the first Xg(a+) daughter. Here, antibodies induced by the first Xg(a+) pregnancy might well lead to greatly increased loss of subsequent female fetuses. Jackson, Mann, and Schull, who discovered the underlying facts, suggest that Xg^a incompatibility may help to explain the lower frequency of female than of male births in general. Earlier Renkonen suggested that the higher rates of prenatal death of males may in part be due to incompatibility reactions between the male XY genotype of a fetus and the female XX genotype of the mother. It may be remarked, that if support for the relation between sex ratio and type of mating in respect to Xg^a increases, we may predict no other discovery of strong incompatibility loci on the X-chromosome. If they existed, their population-genetics random association with the Xg^a alleles would tend to blunt the differences of sex ratios between the various marriage combinations relating to Xg^a.

Variations of the Secondary Sex Ratio. Some data have already been quoted which show that the secondary sex ratio is different in different countries and within the same country in different racial groups. There are other normally slight but significant variations in the sex ratio, and their exploration has been a fascinating pastime of numerous students. For many years and for different countries it used to be true that the ratio of males to females was higher among legitimate than among illegitimate babies, but more recently this difference has apparently disappeared. Similarly, the often reported higher ratio of males to females for rural, as compared to urban, populations has now apparently changed to near equality. Sometimes the upper socioeconomic groups had a higher ratio of males to females than the lower groups, as shown by older data from England and Wales; the number of males per 100 females were: upper and middle classes 106.1, skilled workers 105.7, unskilled workers 103.4.

During times of war, or shortly afterwards, the ratio of males has been found to be higher than during times of peace, not only in the warring countries but also in neutrals. This was first determined for European countries, but also holds for the United States. The change in the sex ratio was not

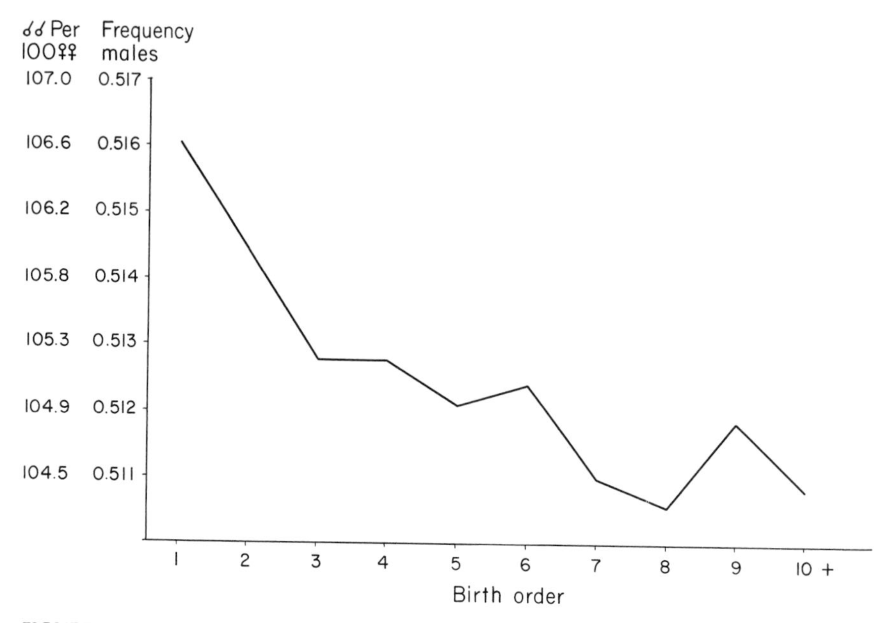

FIGURE 224
The secondary sex ratio of live born according to birth order. White population, United States, 1947–1952. (Novitski and Sandler, *Ann. Hum. Genet.*, **21**, 1956.)

related to a change in birth order or in age of parents, but possibly to what we may call early-fertile, as compared to late-fertile, marriages. From the German *Wer ist's* (*Who's Who*) Marianne Bernstein found that: marriages contracted between 1900 and 1918 in which the first child was born within 18 months gave a 124:100 sex ratio of first born; but those in which the first child was born later gave a ratio of 99:100. In wartime, when many husbands leave home soon after a marriage is contracted, the early-fertile couples may contribute a higher proportion of male children to the population. Whether or not this is the real explanation for the higher ratio of male babies at times of war, it is at least more rational than the popular belief that Nature compensates for war losses by a higher male ratio, which in any case could not benefit the generation in which the losses occurred.

A well-established variation of the sex ratio is correlated with order of birth. In nearly all populations, the ratio of males is highest for the first births and decreases with successive births (Fig. 224). Such a correlation may have various components. It may be directly related to birth order itself; or to ages of parents, which obviously increase with increasing birth order; or to more complex interrelations. Analysis has shown that the age of the mother is not correlated with the sex ratio; but, curiously enough, the age of the father may possibly be, to a very slight degree.

It is impossible, at present, to draw final conclusions from these facts. In some cases, a correlation has been observed between high sex ratio and low prenatal mortality. In several European countries, for instance, there are more stillbirths to unmarried, as compared to married, mothers. Since more males than females are stillborn, it has been reasoned that the stillbirths among illegitimate conceptions eliminate a higher number of males among them, and thus account for the reduced sex ratio at normal birth. In other studies, no persistent correlations could be demonstrated between variations in the secondary sex ratio and the rate of stillbirths.

Altogether, the variability of secondary sex ratios may be due both to different frequencies in prenatal mortality and to differences in the primary sex ratio, which can be thought of as both genetically and environmentally conditioned. If the primary sex ratio should differ from 1:1 then it would be likely that genetic factors, either in the sperm itself or in the female, may have affected the relative survival of X or Y sperm or the response to selective forces in fertilization. It is equally possible that environmental factors, nutritional or metabolic, for example, may exert an influence on the frequency of survival of X versus Y sperm or their relative ability to fertilize. However, no such environmental factors, if they exist, are known; and it is important to realize, in any case, that most of the observed variations in the sex ratio of different population groups are small.

SEX RATIOS IN SIBSHIPS

Unisexual Sibships. Everyone is acquainted with families in which the children are all the same sex. It is an understandable reaction to suspect that a peculiar mechanism is determining the production of only one sex. A consideration of the statistical nature of sex determination should make us cautious about accepting such a conclusion too readily. If the probability of a male birth is p and of female birth is $q(= 1 - p)$, then the probability in a family of n children that all are boys is p^n, or that all are girls is q^n. If we assume p or q to be 1/2, which is, for the present purpose, a sufficiently close approximation to the secondary sex ratio, the probability of an all-boy sibship, as well as of an all-girl sibship, is $(1/2)^n$. If n, the number of sibs, is as large as 10, this probability becomes 1/1,024. This means that, in sibships of 10, chance alone will give rise, on the average, to 1 sibship in 1,024 of all boys and another of all girls, or to 1 unisexual sibship, either male or female, in 512 sibships.

Similar considerations apply to sibships in which both sexes are represented, but in which there is a preponderance of one sex, such as 8 boys to 2 girls or 3 boys to 7 girls.

The assumption that a specific unisexual or very uneven-sexed sibship is the result of an unusual mechanism of sex determination is thus unnecessary, since the normal mechanism yields a certain number of such sibships purely by chance. This statement, however, does not exclude the possibility that, superimposed on chance, may be agents causing unisexual or very uneven-sexed sibships. In *some* families, there may be special mechanisms acting to cause such sibships. A test of this hypothesis can be made in two ways: by studying the relative frequencies of the various types of sibships among all sibships of population samples, and by study of individual pedigrees.

Statistical Studies. An analysis of sibships in populations is statistical in nature. Basically, it rests on the question: Do the observed frequencies of the various types of sibships agree with those expected from chance alone? If there is agreement between observed and expected frequencies, then there is no need for assuming special mechanisms effecting unisexual or uneven-sexed sibships. If, however, there should be disagreement, then a search has to be made for the responsible mechanisms or agents.

A statistical inquiry can be applied to families with different numbers of children. If p and q again represent the probabilities for male or female births as determined by the observed secondary sex ratio, then the inquiry consists of comparing the observed fraction of 2-children sibships with 2 boys, 1 boy and 1 girl, and 2 girls with the expected fractions p^2, 2 pq, and q^2; or of comparing the observed fraction of 3 children sibships with 3 boys, 2 boys and 1 girl, 1 boy and 2 girls, and 3 girls with the expected fractions p^3, 3 p^2q, 3 pq^2, and q^3. In general, the observed fractions of n-child sibships of all possible sex distributions can be compared with expectations according to the binomial $(p + q)^n$, as discussed in Chapter 10, Genetic Ratios.

Before reporting on the results of several relevant studies, a factor which might produce a bias in the observed data must be mentioned. In populations in which birth control is practiced, parents may limit the size of their families not simply after a certain number of children are born, but after some desired sex distribution has been attained. Some parents may desire to have a daughter and may terminate the procreation of children after the birth of a girl; others may wish to have a son and terminate procreation after the birth of a boy. Still other parents may want children of both sexes and may limit their families after at least one of each sex is born.

An inquiry by Dahlberg, in a Swedish population, has shown that most parents like to have both sexes represented among their offspring. The great majority of expectant parents who already had children all or most of whom were of the same sex stated that they wanted their unborn child to be of the opposite sex. The hypothesis that parents to whom children of both sexes were born have a tendency to terminate procreation has been tested in several populations by Gini and others. They investigated whether the last-born child made the sex ratio of the sibships more equal or more unequal. The results indicate some family planning as to sex, mainly for representation of both sexes, but the degree of this planning was low.

It may be thought that the sex ratio in a population might depend on family planning, but a simple consideration shows that this is not true, provided all families are basically alike in their sex-determining mechanism. If, for instance, all parents desired the birth of a son and terminated their families after a boy is born, the first child would be a boy or a girl in the proportion p:q. If only the parents whose first child is a girl have a second child, this second child again will be a boy or a girl in the proportion p:q. In the same way, each next child, in families which have had only daughters, will be a male or a female in the proportion p:q, so that the sum of the sex ratios of all children in the population remains p:q, regardless of family planning. The same conclusion follows from the fact that a decision to have no more children cannot change the sex ratio of children already born. These considerations imply, however, that family planning may lead to a distribution of types of sibships which deviates from chance. If all parents terminated their families after the birth of a boy, and if all parents who have had no boy succeeded in having another child, then all one-child sibships would consist of a boy, all two-children sibships of a girl and a boy, all three-children sibships of two girls and a boy, and so on. If, to use another example, all parents limited their families to the minimum number which gave them at least one child of each sex, then all two-children sibships would be either boy–girl or girl–boy, all three-children sibships boy–boy–girl and girl–girl–boy, and so on. Actually, these schemes of family planning according to sex of children are artificial extremes.

The bias in the distribution of different types of sibships that is introduced by parents who terminate their families after the birth of a child of a specific sex can be reduced by disregarding the last (n-th) child of each sibship and considering the distribution of the sibships with $(n - 1)$ children.

Observed and expected distributions. In an Ohio community, Rife and Snyder collected data on the sex distribution in 1,269 families with from one to five children and compared their observations with chance expectations. They showed that the small deviations between observed and expected distributions were statistically insignificant and that, therefore, chance alone could account for the occurrence of preponderantly, or exclusively, unisexual sibships. A similar conclusion, by Edwards and Fraccaro, was based on a study of the sex distribution in sibships from 5,477 Swedish ministers of religion during the period 1585–1920. The total number of births was 26,037, of which 13,400 were male and the proportion of males therefore was $p = 0.51465$, and that of females $q = 0.48535$. The sibship size ranged from 1 to 18. Let us follow the statistical treatment of one representative set of the data, that of families with

TABLE 64
Frequencies of Various Sex Ratios in a Sample of 565 Sibships of Six.

Ratio per Sibship		*Expected Sex Distribution*		*Numbers Observed*
♂	♀	*Probability*	*Numbers*	
6	0	p^6	10.5	11
5	1	$6p^5q$	59.4	64
4	2	$15p^4q^2$	140.1	129
3	3	$20p^3q^3$	176.1	188
2	4	$15p^2q^4$	124.6	130
1	5	$6pq^5$	47.0	41
0	6	q^6	7.4	2

Source: Based on data in Edwards and Fraccaro, *Hereditas*, **44**, 1958.

6 sibs (Table 64). There is, on the whole, very good agreement between the observed and expected frequencies as seen in the table and particularly in the plotted data (Fig. 225) and as can be shown numerically by calculating the χ^2 value. Excellent agreement between observation and expectation is also obtained when all sibships from 1 to 14 children are considered together. The sibships from Swedish ministers thus conform to the theoretical expectation that the sex of children in sibships is solely determined by chance.

It may be noted that the data in Table 64 show a considerably asymmetry in the number of those sibships which are unisexual or contain all but one child of the same sex. For example 11 all-male sibships were observed but only 2 all-female; 64 five-male sibships but only 41 with five females. This asymmetry is, of course, due to the higher ratio of males at birth. Although the probabilities for male and female births differ only slightly, raising them to a power of 5 or 6 gives strikingly different values for the comparable types of sibships. For sibships of more than 6 children this asymmetry becomes still more pronounced.

The most extensive tabulations of the sexes at birth in sibships were made by Geissler in Saxony during the last quarter of the nineteenth century. These data, while agreeing in general with the expectations of chance determination of sex, show a number of minor but significant deviations from these expectations and observations. Many attempts at analysis of Geissler's tabulations have been made, but it has become apparent that internal inconsistencies in the data are such that it is necessary to discount them.

There are some studies which show that sibships with only one sex occur in excess of chance expectations. For example, Hewitt, Webb, and Stewart found a significant excess of all-male sibships (but not of all-female sibships) in

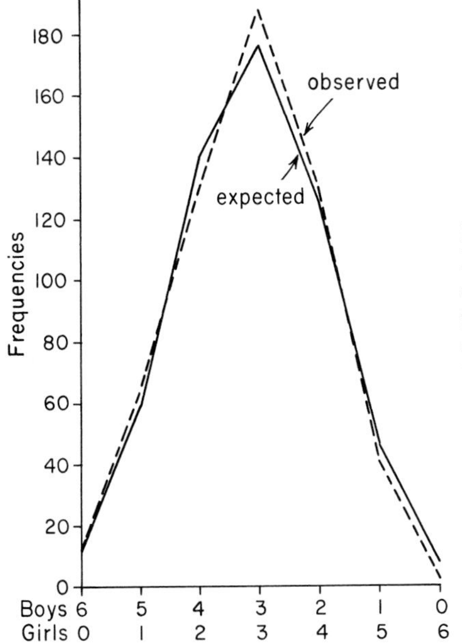

FIGURE 225
Frequencies of various sex ratios in a sample of 565 sibships of six. Solid line: expected frequencies. Broken line: observed frequencies. (Based on data in Edwards and Fraccaro, *Hereditas*, **44**, 1958.)

certain English populations. There were 19 sibships with 5 or more children of like sex, of which 16 were all-male, and all of the 9 sibships with six or more children of like sex consisted of boys. It thus appears that in some sibship samples of sufficiently large size there is a group of sibships of all-male composition that owes its existence to causes other than chance. It is possible to account for an excess of unisexual sibships by assuming that in a small proportion of parents sex of offspring is determined by some specific mechanism which yields children of one sex only. There is indeed evidence for rare families of this type (Fig. 226). Another interpretation, proposed by Gini and Edwards, of a surplus of unisexual sibships does not assume that there is a sufficiently large number of families with a special mechanism leading to unisexual sibships but that the probabilities p and q for male and female births vary between families according to a modified binomial distribution.

This will be shown by a simplified example, using 4-children sibships. Assume first a population A with a constant sex ratio of $p = q = 0.5$. Then $(0.5♂ + 0.5♀)^4$ yields the expectations for the different sex distribution (Table 65, first line). Now assume a population B with $p = 0.6$ and $q = 0.4$, and another one C with $p = 0.4$ and $q = 0.6$. For B the expectations for the different sex distributions are derived from $(0.6♂ + 0.4♀)^4$ and for C from $(0.4♂ + 0.6♀)^4$, as given in lines 2 and 3 of Table 65. If a population consisted of two subpopulations of the types B and C, and of equal sizes, the mean probabilities p and q

TABLE 65
Frequencies of Sex Distributions in Sibships of Four in Four Populations.

Population		*Frequencies of Sibships with Sex Ratios*				
Type	p	0:4	1:3	2:2	3:1	4:0
A	0.5	0.0625	0.2500	0.3750	0.2500	0.0625
B	0.4	0.0256	0.1536	0.3456	0.3456	0.1296
C	0.6	0.1296	0.3456	0.3456	0.1536	0.0256
(B + C)/2	0.5	0.0776	0.2496	0.3456	0.2496	0.0776
obs. (B + C)/2 − exp. A		+	−	−	−	+

would be 0.5, as are the constant probabilities in A. The mean expectations for the frequencies of the different sex distributions would not be like those in A, but rather be the averages of the frequencies for B and C (line 4, Table 65). If we compare these average frequencies with those in line 1, we find too many all-male and all-female sibships and too few of the others (line 5). The assumption that the probabilities for male and female births vary in a population is, as discussed earlier, in agreement with observations on the sex ratio in different socioeconomic groups, different occupations, and other subpopulations.

Variability of the probabilities for male and female births should lead to excesses of both all-male and all-female sibships as well as to excesses of not exclusively but preponderantly unisexual sibships. No such excesses were present in the data from Swedish ministers perhaps because these were an unusually homogeneous group. Since most other tabulations of sex ratios in sibships are based on relatively small population samples, the differences between observed and expected numbers do not reach significance.

A direct way of demonstrating differences in the probabilities for the sexes in different families of a population is to compare the sex ratio of children born after a preceding male birth with that of children born after a preceding female birth. In France, Schützenberger found a small but significant positive correlation between the sex of successive births ($r = 0.029$). Correspondingly, Marianne Bernstein, in a United States population, found a probability for males after a male birth of $p = 0.5387$, but after a female birth of only $p = 0.4988$, the difference being highly significant. Renkonen, in Finland, similarly determined $p = 0.5132$ and $p = 0.4955$, respectively. There is then a slight but real trend in these populations for the sex of a child to be the same as that of a preceding sib. Yet, this phenomenon is not always encountered. In over 100,000 sibships from Mormon records, Greenberg and White found no correlations between the sexes of adjacent sibs.

Genetic Factors in the Variability of the Sex Ratio

A finding that different families vary in the probabilities of having boys or girls could be explained by either genetic or nongenetic differences, or combinations of such differences. Light can be thrown on this question when sibships from related parents are combined. This was done by Nichols who collected genealogical data, mostly for the period 1640–1800, on the sex ratio among New England families. There were 40 groups of sibships, all sibships in a group descended in the male line from a common ancestor, with the sibships selected so as to contain 6 or more children. The sex ratios in the different groups, based on from 63 to 428 children, varied between 177 and 72 males to 100 females. Although in most of the groups the sex ratio did not differ significantly from the mean, there were a few in which the great preponderance of either males or females was probably statistically significant. It seems that the tendency to rather extreme sex ratios was shared by a number of related sibships in successive generations.

The nature of the variation in the probabilities of producing males or females is obscure. Different families may have either different primary sex ratios or different prenatal survival rates. A high number of abortions and early stillbirths could shift the secondary sex ratio, provided the mortality acted differentially on the two sexes, more against the males in some families than in others. It is perhaps likely that more than one mechanism is involved in the variations of the probabilities to produce a given sex.

One source of the variability of the sex ratio at birth is suggested by tabulations according to the ABO blood groups. As shown by Sanghvi and others, in white populations the male ratio among O children is significantly higher than among A children, both in general and from specific types of mothers; and the same is true for the male ratio from AB mothers as compared to the rest. If these findings can be accepted as conclusive, it would seem likely that the immunologic mother-fetal interactions in the ABO groups serve to modify the secondary sex ratio.

Sex-ratio studies on mice by Weir provide experimental evidence for significant genetic variability of the probability of male and female births. Of two different strains, one has a high male ratio, $p_H = 0.556$, and one has a low male ratio, $p_L = 0.437$. Reciprocal matings between the two strains have shown that it is the male which is responsible for the sex ratio of the litter. Artificial inseminations, with sperm removed from the seminal duct of one strain and suspended in sperm-free seminal fluid of the other strain, gave the sex ratio of the strain from which the sperm was obtained. It is likely that the two different sex ratios at birth are due to differences in the primary sex ratio, but the possibility of differential mortality of the sexes in very early stages of development has not yet been excluded.

Pedigrees with Unusual Sex Ratios. A few pedigrees with many births show such a great deviation from the expected sex ratio that the always desirable caution against selection for oddity seems to be unnecessary. Two such pedigrees are reproduced in Figure 226. The first pedigree (A), from England, shows thirty-four births (counting identical twins as one) over ten generations and dates back to the early seventeenth century. Thirty-two of the births were male and only two female. One of the latter died early and nothing specific is known about her; the other female had masculine traits and may have been an intersex. It is somewhat disturbing that the only two females recorded appear in the next-to-the-last generation, since it raises some doubt regarding the reliability of earlier entries in the family Bible on which the pedigree was based. Probably these doubts are not really justified and the pedigree may be accepted as bona fide. In this pedigree, the production of male progeny is clearly transmitted through the males from generation to generation,

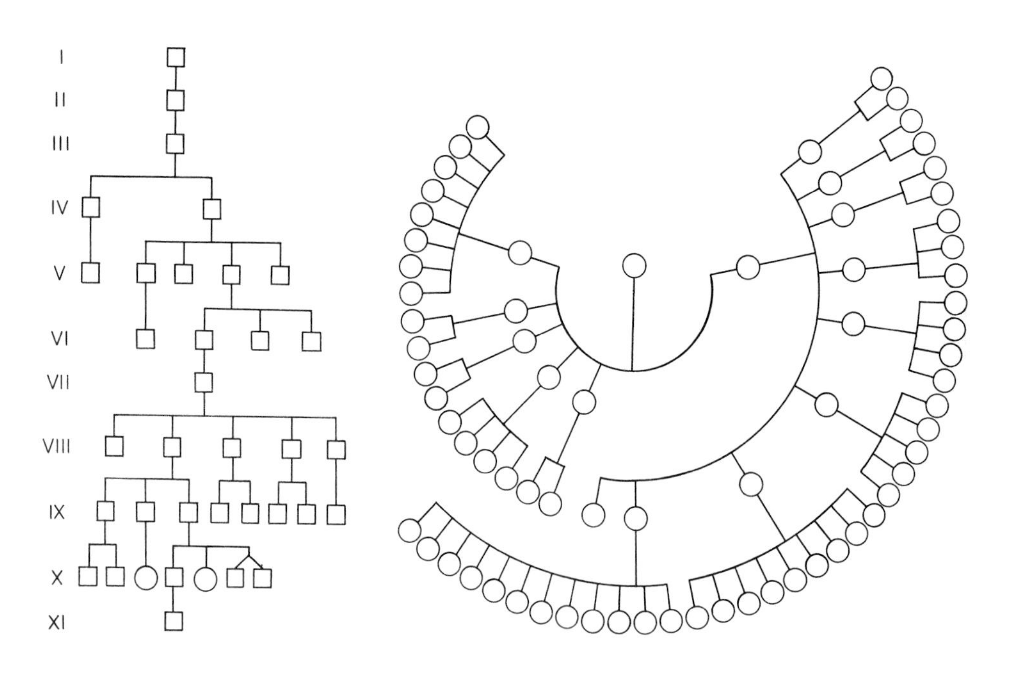

FIGURE 226
A. Nearly exclusively male pedigree. **B.** Exclusively female pedigree. (A, after Harris, *Ann. Eugen.*, **13,** 1946; B, after Lienhart and Vermelin, *C. R. Soc. Biol. Paris,* **140,** 1946.)

irrespective of the diversity of their wives. It may be assumed that the functional sperm produced by these men are almost exclusively Y sperm. Since all these men derived their Y-chromosome from a common ancestor, this specific chromosome may, itself, be the cause of the near nonexistence or nonfunctioning of the X sperm.

The second pedigree, Figure 226, B shows, still more strikingly, a case of sex tendency opposite to that in A. It represents a French kindred with 72 births in three generations, exclusively female. This kindred has been rather ungallantly referred to as "a monstrous regiment of women" (a phrase originally intended, in the title of the pamphlet by the sixteenth-century reformer John Knox from which it was borrowed, to refer to three reigning women of his period). This pedigree, if reliable, points to an inherited property of the females to permit fertilization of their eggs by X sperm only or to cause embryonic death of XY zygotes. In order to have this property transmitted by the original female parent to all fourteen of her daughters and granddaughters whose offspring has been recorded, extranuclear inheritance or some complex chromosomal mechanism would have to be assumed.

Mechanisms Underlying Unisexual Offspring. Whatever the reason for these two remarkable human pedigrees may be, there are similar examples in other organisms, and, in some cases, the mechanism is known. In several species of *Drosophila*, in natural populations males have been found which produce nearly all daughters. The males carry a gene called "sex ratio" that seems to result in inactivation of the Y-bearing sperm. Furthermore, meiotic drive of the Y-chromosome has been observed in mosquitos. In contrast to the "sex ratio" gene, which acts through the male parent, another mechanism, which acts through the female parent, for the production of almost exclusively female offspring has been found, again in *Drosophila*. Certain females carry a spirochete that enters into all eggs. Its presence is fully compatible with the development of XX zygotes, but it kills, at an early embryonic stage, XY zygotes. Perhaps the women in the all-female pedigree (Fig. 226, B) likewise harbor an agent that kills, discriminately, their potential sons. A genetic disposition for "male sex ratio," similar to that shown in the human pedigree in Figure 226, A, has also been found in *Drosophila*; affected males produce mostly sons.

Control of Sex Determination

Someday, the determination of sex in man will be subject to willful control. Theoretically, this might be accomplished by overriding the XX or XY zygote constitution by means of hormones so as to leave the desired sex unchanged but transform the undesired one into its opposite. This approach is not likely to be successful. The intricate processes of early development in an embryo a few weeks of age may well prove to be beyond practical reach for such sex reversal. The selection of X or Y sperm for fertilization seems to be much more promising. If physicists have been able to separate isotopes of atoms, biologists should find it possible to separate two classes of bodies whose sizes and differences are immense in comparison to those of the atoms.

Several methods of sperm separation have been tried, and more will be tried in the future. These methods are of two general types. In one type, the attempt is made so to control the environment provided for sperm in the female body as to favor fertilization by one or the other of the sex-determining kinds of sperm.

In the other type, which has been used with experimental animals, sperm is treated outside of the body. There, separation of X from Y sperm is attempted, and the separated portions are then used for artificial insemination. One method uses sedimentation, another centrifugation, of animal sperm in attempts to separate X- and Y-bearing portions. In a third method an electric current is passed through a sperm suspension, which causes some sperm to migrate to one pole and others to migrate to the other pole. In spite of a long history of experimental attempts of the types just listed, no reproducible success has yet been achieved. It hardly seemed audacious many years ago to predict a solution of the problem in the not-so-distant future but the fulfillment of the prediction is still wanting.

There is another way of controlling the sex of newborns. Now that the sex of unborn children may be determined by sexing the cells of the amniotic fluid (p. 809), it is possible selectively to abort fetuses of an "undesired" sex and to permit development to term of the "desired" sex only. This type of sex control, however, raises ethical questions much more serious than those impinging on other methods. Controlling the kind of sperm, X- or Y-bearing, which is to have access to egg cells will not be objectionable to many persons to whom termination of development of a healthy fetus of undesired sex would be immoral.

Very likely, willful sex determination will first be accomplished in animal husbandry, where artificial insemination is widely practiced. Man's control of the sex of children will be one more step in his efforts to control nature and himself, which has constituted so much of the history of civilization. Such control is likely to lead to changes in the human sex ratio and to the sequence of male and female births within sibships. These effects will raise new social problems, but it should not prove too difficult to cope with them.

PROBLEMS

168. The sex ratio of fetuses stillborn during the fourth month of pregnancy was found to be 107.7, and that of fetuses stillborn during the fifth month 115.6. Using the data given in Table 63, determine whether the difference is statistically significant.

169. In the United States (1925–1929) the sex ratios at birth, according to order of birth, were as follows: first birth 106.32; second 105.99; third 105.90; and fourth 105.33. What would be the sex ratios in a population in which there were only sibships with from 1 to 4 children, and these were in the percentages:
(a) 25, 40, 25, 10; or
(b) 10, 25, 40, 25?

170. The chance of a newborn child being a boy is (approximately) 0.52. What proportions of sibships of four children do you expect to be:
(a) Only boys?
(b) Only girls?
(c) Two of each sex?

171. During a certain period, 14 boys and 27 girls were born to the wives of workers in a radar factory. During approximately the same time, the wives of draftsmen in the offices had 22 boys and 19 girls.
(a) Is the difference between the sex ratios of the two groups significant?
(b) Approximately, what is the probability of the first ratio being a chance deviation from the expected U.S. sex ratio of 106 boys to 100 girls?

172. Assume that, in a number of families, a special system of sex determination results in a preponderance of boys, while, in an equal number of families, another system of sex determination results in an equal preponderance of girls. If parents were to terminate their families after the birth of a boy, which sex would be preponderant in the pooled data?

REFERENCES

Allan, T. M., 1959. ABO blood groups and sex ratio at birth. *Brit. Med. J.*, **1:** 553–554.

Barlow, P., and Vosa, C. G., 1970. The Y chromosome in human spermatozoa. *Nature*, **226:**961.

Bernstein, Marianne E., 1967. Techniques of stratified sampling in the study of variation of the human sex ratio. *Eugen. Quart.*, **14:**54–59.

Ciocco, A., 1938. Variation in the sex ratio at birth in the United States. *Hum. Biol.*, **10:**36–64.

Cohen, Bernice H., and Glass, B., 1956. The ABO blood groups and the sex ratio. *Hum. Biol.*, **28:**20–42.

Colombo, B., 1957. On the sex ratio in man. *Cold Spring Harbor Symp. Quant. Biol.*, **22:**193–202.

Dahlberg, G., 1948. Do parents want boys or girls? *Acta Genet. Stat. Med.*, **1**:163–167.

Edwards, A. W. F., 1962. Genetics and the human sex ratio. *Advan. Genet.*, **11**:239–272.

Edwards, A. W. F., 1966. Sex ratio data analyzed independently of family limitation. *Ann. Hum Genet.*, **29**:337–348.

Edwards, A. W. F., and Fraccaro, M., 1958. The sex distribution in the offspring of 5477 Swedish ministers of religion 1585–1920. *Hereditas*, **44**: 447–450.

Etzioni, A., 1968. Sex control, science, and society. *Science*, **161**:1107–1112.

Gini, C., 1951. Combinations and sequences of sexes in human families and mammal litters. *Acta Genet. Stat. Med.*, **2**:220–244.

Greenberg, R. A., and White, C., 1968. The sexes of consecutive sibs in human sibships. *Hum. Biol.*, **39**:374–404.

Harrison, G. A., and Peel, J. (Eds.), 1970. *Biosocial Aspects of Sex*. 164 pp. Blackwell, Oxford. (Proceedings of the Sixth Annual Symposium of the Eugenics Society, London. Journal of Biosocial Science, Supplement 2. See especially Edwards, A. W. F., Genetic variability of the sex ratio, pp. 55–60.

Harvey, E. N., 1946. Can the sex of mammalian offspring be controlled? *J. Hered.*, **37**:71–73.

Hewitt, D., Webb, J. W., and Stewart, A. M., 1955. A note on the occurrence of single-sex sibships. *Ann. Hum. Genet.*, **20**:155–158.

Jackson, C. E., Mann, J. D., and Schull, W. J., 1969. Xg^a blood group system and the sex ratio in man. *Nature*, **222**:445–446.

Jacobs, Patricia A., and Ross, A., 1966. Structural abnormalities of the Y chromosome in man. *Nature*, **210**:352–354.

Jaquiello, Georgianna, and Atwell, J. D., 1962. Prevalence of testicular feminization. *Lancet*, **1**:329.

Jost, A., 1961. The role of fetal hormones in prenatal development. *Harvey Lectures*, **55**:201–226.

Kang, Y. S., and Cho, W. K., 1962. The sex ratio at birth and other attributes of the newborn from maternity hospitals in Korea. *Hum. Biol.*, **34**:38–48.

Keller, C. A., 1969. Embryonal sex ratios in animals and man. Ph.D. thesis, University of California, Berkeley.

Lindahl, P. E., 1958. Separation of bull spermatozoa carrying X and Y chromosomes by counter streaming centrifugation. *Nature*, **181**:784.

Mikamo, K., 1969. Female preponderance in the sex ratio during early intrauterine development. *Jap. J. Hum. Genet.*, **13**:272–277.

Moran, P. A. P., Novitski, E., and Novitski, C., 1969. Paternal age and the secondary sex ratio. *Ann. Hum. Genet.*, **32**:315–316.

Nichols, J. B., 1905. The sex composition of human families. *Amer. Anthropol.*, **7**:24–36.

Pohlman, E., 1967. Some effects of being able to control sex of offspring. *Eugen. Quart.*, **14**:274–281.

Renkonen, K. O., Mäkelä, O., and Lehtovaara, R., 1962. Factors affecting the human sex ratio. *Nature*, **194**:308–309.

Rife, D. C., and Snyder, L. H., 1937. The distribution of sex ratios within families in an Ohio city. *Hum. Biol.*, **9**:99–103.

Sanghvi, L. D., 1951. ABO blood groups and sex ratio at birth in man. *Nature*, **168:**1077.
Sohval, A. R., 1963. Chromosomes and sex chromatin in normal and anomalous sexual development. *Physiol. Rev.*, **43:**306–356.
Stevenson, A. C., and Bobrow, M., 1967. Determinants of sex proportions in man, with consideration of the evidence concerning a contribution from X-linked mutations to intrauterine death. *J. Med. Genet.*, **4:**190–221.
Teitelbaum, M. S., Mantel, R., and Stark, C. R., 1971. Limited dependence of the human sex ratio on birth order and parental ages. *Amer. J. Genet.*, **23:**271–280.
Turpin, R., and Schützenberger, M. P., 1949. Sur la détermination du sexe chez l'homme. *Semaine des Hôpitaux Paris,* **25:**2544–2545.
Visaria, P. M., 1967. Sex ratio at birth in territories with a relatively complete registration. *Eugen. Quart.*, **14:**132–142.
Weir, J. A., 1962. Hereditary and environmental influences on the sex ratio of PHH and PHL mice. *Genetics,* **47:**881–898.
World Health Organization, Geneva Conference, 1966. Standardization of procedures for chromosome studies in abortion. *WHO Bull.*, **34:**765–782.

22

OCCURRENCE OF MUTATIONS

The appearance of a trait, subsequently inherited, among one or more members of a family in whose ancestry the trait was unknown, suggests to the naïve observer the origin of genetic newness. The geneticist, however, realizes that such a conclusion is not always justified. If a character is based on either an autosomal or a sex-linked allele that is dominant and fully penetrant, then, indeed, the first appearance of the character signifies the origin of the allele in the immediately preceding generation or in a body cell of the bearer of the trait himself. If, however, the character is due to a dominant but incompletely penetrant allele, then lack of penetrance in early generations, and not absence of the allele, may account for the unexpected phenotypic expression of the character in a later generation. Or, if an autosomal recessive allele is responsible for a character, then, of course, carriers may, unknown to themselves, transmit the allele heterozygously for numerous generations, and the homozygous occurrence of the allele, with its resulting specific phenotype, may be far removed from the origin of the allele itself. Finally, should the allele be X-linked and recessive, it may be carried in heterozygous females for several generations without becoming known, revealing itself by its characteristic phenotype only if transmitted to a male.

THE DETECTION OF GAMETIC MUTATIONS

A new allele always arises from one already in existence, and the process that either transforms the allele itself or causes it to produce a new allele is called *gene mutation.*

Mutations may occur in somatic as well as in germ cells. A mutation that occurs in a germ cell can be discovered when this cell or its descendants become gametes, thus transmitting the mutant gene to the next generation. We shall first discuss such gametic mutations, and later take up somatic mutations.

In experimental organisms, various breeding procedures have been worked out in order to reveal gametic changes from one allele to another and whether the changed allele is dominant or recessive to the first. In relation to dominant mutations the original allele is, of course, recessive, and in relation to recessive mutations it is dominant. The detection of mutations of recessive to dominant alleles is simple, since it is only necessary to interbreed homozygous recessives and to watch for the occurrence of dominant mutant phenotypes among the offspring. Mutations of dominant autosomal alleles to recessive ones can be ascertained only after special breeding procedures involving self-fertilization in hermaphroditic animals or plants, or sequences of brother-sister matings in bisexual organisms. In man, brother-sister matings are so rare that mutations to recessive autosomal alleles cannot be ascertained directly; only mutations to dominant alleles—whether autosomal or sex linked—and to sex-linked recessives can be recognized.

Dominant Mutations. An example of a dominant autosomal mutation is presented in a pedigree from England, in which a pathological, severe blistering of the feet occurred for the first time in 1 out of 6 children of unaffected parents (Fig. 227). The condition reappeared in three successive generations in a manner which agrees with full penetrance of a dominant gene. The only reasonable basis for exception to the statement that the first appearance of the character was the result of a mutation of an allele from normal to abnormal is the assumption that the sire of II–5 was not the legal father but an unknown, affected man. The circumstances surrounding this case make illegitimacy so unlikely that the first appearance of blistering in the pedigree may be regarded as the result of a true mutation.

Two other pedigrees showing the first occurrence of autosomal dominant alleles—one for cataract and one for brachyphalangy—are given in Figure 228. In each case, the ancestors of the first affected individual were normal, and the generation or generations derived from him give evidence of dominant transmission. In still another example of this nature, referred to in an earlier chapter, a son with dwarfism similar to achondroplasia appeared among the offspring of normal parents and the trait was transmitted to all succeeding generations (Fig. 229).

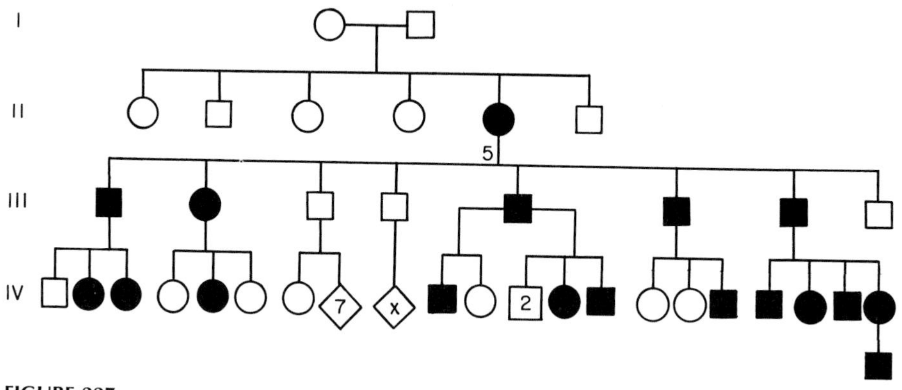

FIGURE 227
Dominant mutation. Pedigree of severe blistering of the feet. (x = several individuals, exact number unknown.) (After Haldane and Poole, *J. Hered.*, **33**, 1942.)

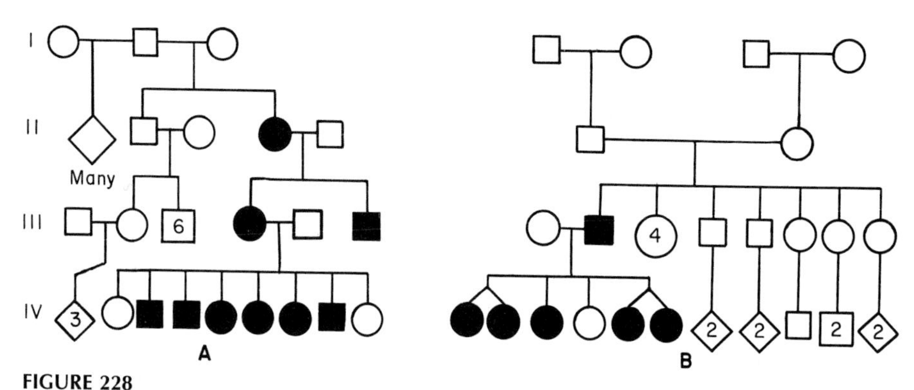

FIGURE 228
Dominant mutation pedigrees of: **A.** Cataract. **B.** Combination of brachyphalangy and hypophalangy. (**A,** after Danforth, *Amer. J. Ophth.*, **31,** 1914; B, after Liebenam, *Zeitschr. Konstitl.*, **22,** 1938.)

Among the numerous known cases of the unexpected appearance of a trait in a child whose ancestors were free from it, an appreciable number may be regarded as dominant mutations, even if proof of inheritance in later generations is lacking. In many, no proof is available simply because the affected individual has not yet reached maturity or has not become a parent. Even if he has produced children, their normality may be due not to lack of heritability of the trait, but, rather, to chance deviation from the 1:1 ratio expected from a heterozygous parent. Attribution of such an appearance of a trait to a dominant mutation is more justifiable if the trait has been shown to be dominant and fully penetrant in other pedigrees than if it has not.

Mutations of normal alleles may result in alleles whose effects are neither better nor worse than those of the original allele or, they may lead to abnor-

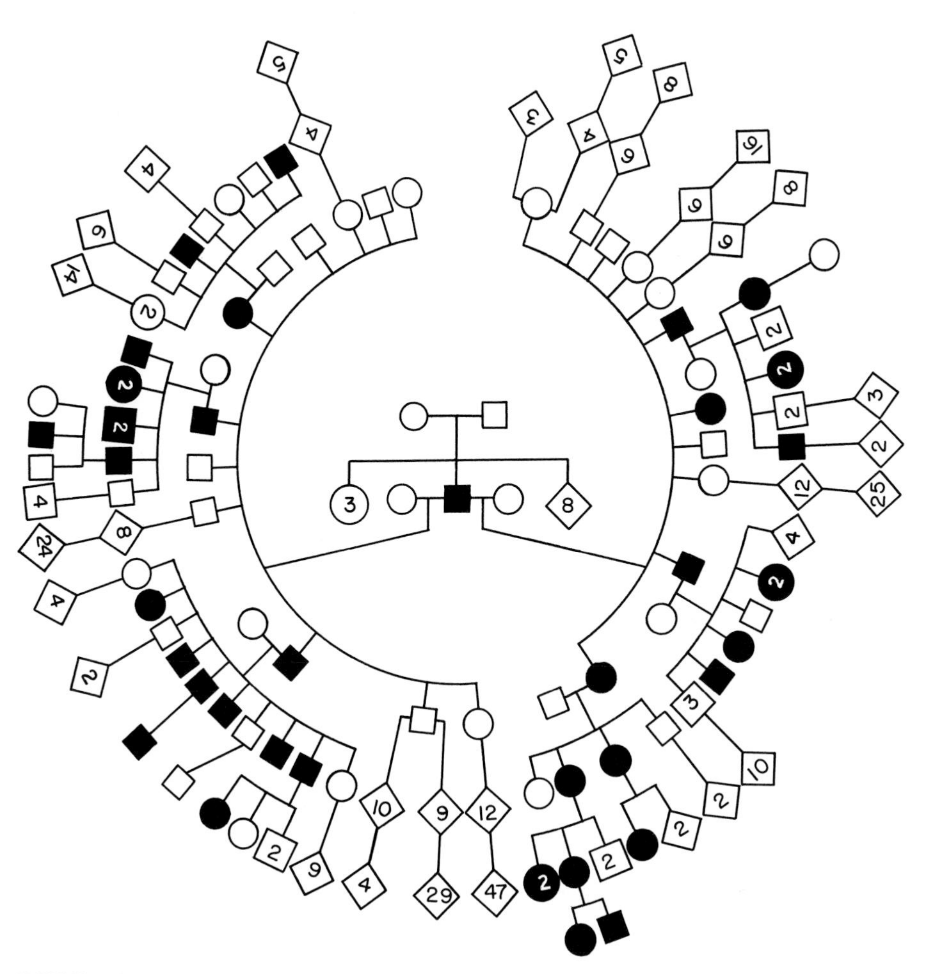

FIGURE 229
Dominant mutation. Abbreviated pedigree of a kind of dwarfism (spondyloepiphysial displasia), similar to achondroplasia, from Utah. (After Stephens, *J. Hered.*, **34,** 1943.)

mal effects such as those referred to in preceding paragraphs. Alleles with abnormal effects may themselves mutate back to alleles with either fully normal or at least less harmful action. No such *reverse mutations* have yet been observed in man, but their occurrence could be demonstrated only under exceptional circumstances.

Sex-linked Mutations. Mutation from an X-linked normal to an abnormal recessive allele has been suggested by several pedigrees, particularly some of hemophilia. For an example, we shall examine one such family (Fig. 230). After four generations of only nonhemophilic persons, a sibship with 6 hemophilic sons, plus 1 normal son and 1 normal daughter, appeared. Considering

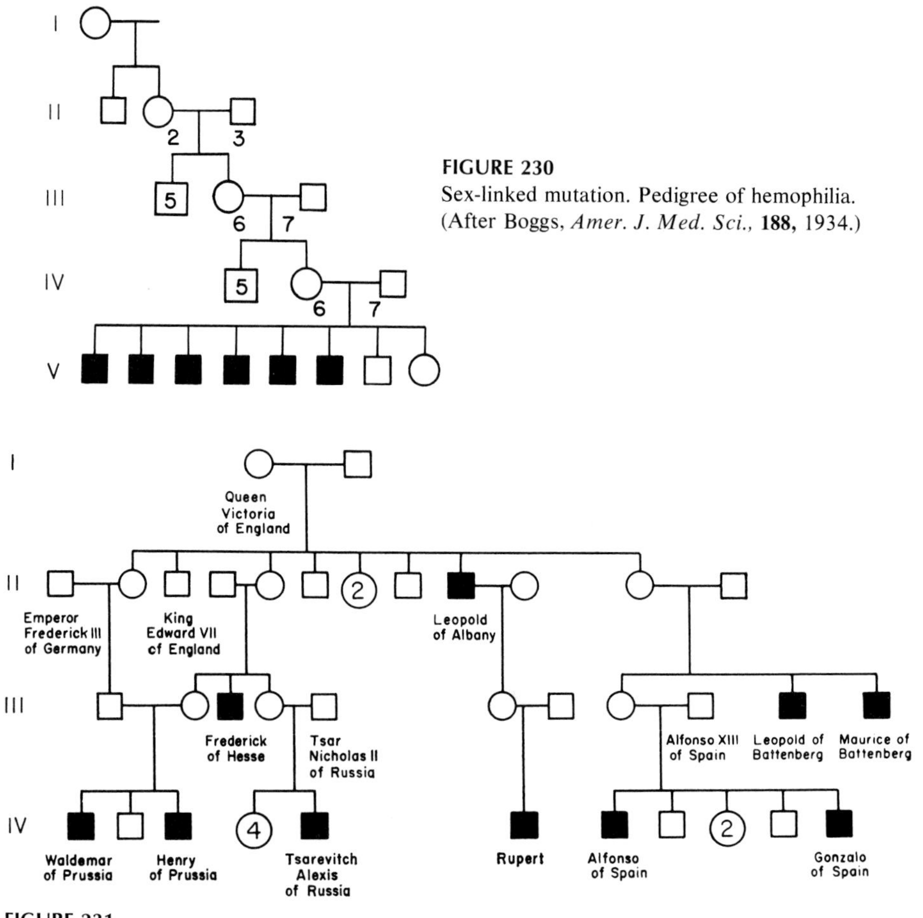

FIGURE 230
Sex-linked mutation. Pedigree of hemophilia. (After Boggs, *Amer. J. Med. Sci.,* **188,** 1934.)

FIGURE 231
Sex-linked mutation. Pedigree of hemophilia in the royal families of Europe. All of Queen Victoria's children are entered. Later generations comprise many more individuals than are indicated here.

the hereditary nature of hemophilia, and that it is usually X-linked, it must be assumed that the mother of this sibship, IV–6, was heterozygous for the recessive allele *h*. She could have either inherited this allele from a parent who carried it, or her *Hh* genotype could have been the result of a mutation in one of the two gametes which gave rise to her. If there was regular inheritance, her father, III–7, was not involved, since he was not a bleeder. But her mother, III–6, from whom she must therefore (still assuming normal inheritance) have received the allele, does not betray heterozygosity for *h*, since none of her five sons was affected. While this does not rule out the possibility that III–6 was *Hh*, there is also no evidence, from her own ancestry, of the presence of the *h* allele. Her normal father, II–3, was certainly *H*, and if her mother, II–2, was a carrier, why did none of her five brothers receive the *h* allele? It is thus very likely that the mother, IV–6, of the affected children owed her heterozygosity to a mutation, even though the analysis cannot eliminate with certainty the possibility of regular transmission of an *h* allele through several generations of the pedigree, with chance having led to the production of only normal sons from *Hh* mothers, except in the last generation.

A similar analysis makes it seem extremely likely that the hemophilia occurring in the royal families of Europe during the nineteenth and twentieth centuries owed its origin to a mutation (Fig. 231). All affected individuals trace their ancestry to Queen Victoria of England, who, undoubtedly, was heterozygous *Hh*. Her father was normal, and nothing suggests that her mother was a carrier. Consequently, Queen Victoria seems to have received a new mutant allele *h* from one of her parents.

Determination of Mutation Frequency—Direct Method

Not only is there evidence of the independent occurrence of the same mutation in different individuals, but the frequencies of specific mutations have been estimated. These frequencies are those of mutant changes that result in clearly recognizable, and usually harmful, phenotypes. They do not include the mutational changes from normal alleles to other normal isoalleles. Presumably, many mutational alterations of the DNA polynucleotide chain of a gene do not lead to obvious phenotypic effects.

Two different methods have been devised for deriving estimates of the frequencies of specific mutations. The first, the direct method, is primarily applicable to dominant mutations. It is based simply on a census of the frequency of children with well-known dominant traits who are born to parents without these traits. Underlying assumptions are: (a) that there is always full penetrance, (b) that the trait is never produced by recessive alleles, (c) that the trait is never produced by nongenetic agents, and (d) that dominant alleles at only one locus produce the trait.

In most cases the validity of these assumptions has not been verified, and full proof of their correctness may be difficult to obtain. Assumptions (a), (b), and (c) should be supported by evidence from numerous pedigrees in which affected individuals have produced offspring. The ratio of affected to nonaffected among their pooled children should be compatible with that expected, 1:1, and the frequency of individual sibships without affected individuals should not be greater than expected in small families as an extreme chance deviation from the 1:1 ratio. Assumption (a), that there is always full penetrance, and (b), that the trait is never produced by recessive alleles, would be invalidated if it were found that the trait skipped some generations. Assumption (c), that the trait is never produced by nongenetic agents, or, in other words, that it does not appear as a phenocopy (see p. 84), would be shown to be false if incomplete penetrance and recessive determination could be excluded but if significant numbers of affected persons had neither affected parents nor children. Assumption (d), that dominant alleles at only one locus produce the trait, can seldom be tested in man, though in the future linkage studies might provide the necessary data.

If any one of the four assumptions does not hold, an overestimate of the mutation rate in terms of change at a single locus would be obtained. Thus, the estimate will be high if the appearance of an affected offspring from non-affected parents is considered to be due to a dominant mutation, although it actually results from expression of an already present incompletely penetrant gene, of homozygosity for a recessive gene, or of a phenocopy. Or, if dominant mutation at more than one locus is responsible for a trait, the mutation rate estimated for one locus will be too high, because it is really a composite of two or more rates. If, for instance, an abnormal dominant trait is caused by mutation in 1 out of 50,000 gametes, then the mutation rate is 1 in 50,000, provided a single locus, P, is concerned. But if two equally mutable loci, P and Q, can give rise to dominant alleles P' and Q', either of which would produce the abnormal trait, then the mutation rate per locus is obviously only 1 in 100,000. Since, in man, it is seldom possible to prove that the same locus is involved in the appearance of a specific trait in different kindreds, the figures to be given below for mutation rates of human genes may be too high, although some mutation rates may be underestimated. If, for example, certain mutations result in recognizable phenotypes in some individuals, but embryonic death without recognizable genetic cause in others, only the recognized occurrences will be available for an estimate of the mutation rate.

Achondroplastic Dwarfism. A direct determination of the rate of mutation has been made in the case of achondroplastic dwarfism. The data were furnished by records on 94,075 children born in Lying-in Hospital in Copenhagen. Ten of these children were achondroplastic dwarfs, 2 of them from an affected parent. This leaves 8 mutant dwarfs, or 1 in nearly 12,000 births. Since pedigree studies in Denmark show simple dominant inheritance of the condition, the occurrence of 1 new case in 12,000 births has been regarded as evidence of mutation.

The rate of mutation is best expressed in terms of the number of alleles involved, not births. Since each individual has two alleles at the assumed single locus for achondroplasia, each birth represents a sample of two alleles, so that 1 mutant birth in 12,000 signifies one mutant allele in 24,000, or a rate of about 4×10^{-5}.

Later studies in Germany modify this conclusion. Achondroplasia occurs in several phenotypically different forms, which do not appear together in the same kindreds. Although most of these are inherited as dominants they are presumably due to different genes; recessive inheritance is also involved occasionally; and the occurrence of phenocopies cannot be excluded. The individual mutation rates for the different genes must, then, be lower than the earlier estimate, which was based on the assumption that a single gene controlled the abnormality.

Retinoblastoma. Another direct determination of a dominant mutation rate has been made for the allele causing the disease retinoblastoma, which consists in the development of tumors of the retina: these appear early in childhood and result in death unless the eye or eyes are removed by surgery. In Michigan, from 1936 to 1945, there were 49 isolated cases of the disease among 1,054,985 children born to normal parents. From these numbers, Neel and Falls derived a rate of 2.3×10^{-5} for mutation of a normal to an abnormal dominant allele—an estimate similar to, though higher than, one made by Sorsby from British data. Subsequently, Vogel demonstrated that approximately three-quarters of all individuals with retinoblastoma who have normal parents do not transmit the disease to any of their children, so that in these affected individuals the trait seems to be either a phenocopy or an expression of a somatic mutation (pp. 574ff.). In any case the frequency of mutation from normal to alleles for retinoblastoma in the gametes must be lower than formerly believed, since only about one-quarter of new occurrences are caused by such mutation. On the other hand, not all mutations are expressed phenotypically, since the penetrance of the abnormal allele is probably below 80 per cent: former unawareness of this tended to lower the estimate of the mutation rate. Taking these facts, as well as some others, into account, the gametic mutation rate of the gene for retinoblastoma has been estimated to be about 1.8 per 100,000 gametes.

Knudson has developed a hypothesis according to which retinoblastoma is a cancer caused by two independent, successive mutational events. In the transmitted cases one of the mutations occurs in the germ cells and the other in somatic cells. In the nontransmitted cases both mutations occur in somatic cells. The rates of mutation of either kind are similar to those calculated on the assumption of a single mutation.

Determination of Mutation Frequency—Indirect Method

Before the direct method of estimating the rate of mutation of a human gene had been applied, an indirect method had been devised by Danforth and later, independently, by Haldane and by Gunther and Penrose. It is based on principles derived from population genetics. Most abnormal human traits, the reasoning goes, decrease the likelihood that the individual possessing them will have an average number of children. Some traits cause early mortality of affected individuals, eliminating them before they reach maturity; other traits reduce the prospects of marriage; still others cause the number of children produced to fall below the average for the rest of the population. Consequently, alleles causing abnormalities are not transmitted to as many individuals as are normal alleles, and this should lead to a decrease in the frequencies of abnormal alleles from one generation to the next. If, for instance, a dominant abnormal allele leads to death of its carriers at an early age, so that, on the average, they have only half as many children as the normal population, then the frequency of the abnormal allele would decrease by 50 per cent in each generation, and in the course of a few centuries the allele should have practically disappeared. With a reproductive rate that is half the normal, as assumed above, in ten generations—about 300 years—there would have been a reduction to $(1/2)^{10}$, or less than 0.1 per cent of the original relative frequency. Even with a reproductive rate as relatively high as 9/10, ten generations would lead to a decline to $(9/10)^{10}$, or less than 4 per cent of the original relative frequency. Applied to such traits as achondroplasia or the dominant trait epiloia, this reasoning would suggest that the present low frequency of these conditions represents only the leftovers of frequencies that were hundreds of times higher a few centuries ago!

This is an absurd deduction, since we know from historical records that the relative frequencies of various inherited abnormalities could not have been strikingly higher in former times. But, since this is true, why are they not lower now?

Equilibrium for Dominant Mutations. An answer is provided by the hypothesis that recurrent mutations from normal to abnormal must have constantly replenished the steadily diminishing store of abnormal alleles of low reproductive fitness. If it is true that the frequencies of such alleles have not changed much in the course of many generations, an equilibrium must have existed between loss and gain; that is, the rate of mutation must have balanced the rate of loss.

This relation can be stated in terms of an equation. Let N be the total number of individuals in one generation, x the frequency of the abnormality among them, and f the reproductive fitness of the abnormal gene, that is, the frequency, relative to that of its normal allele, with which it is transmitted to the next generation. Finally, let u be the frequency per gamete of the mutation from normal to abnormal. Then, for a rare autosomal dominant, the following holds true: The number of mutant births depends on the total number of normal alleles and the frequency, u, with which they can mutate. There are xN affected parents who have xN abnormal and xN normal alleles, and (N − xN) normal parents who have 2(N − xN) normal alleles. The sum of the normal alleles is, therefore,

$$xN + 2N - 2xN = 2N - xN = N(2 - x),$$

and the number of mutations giving rise to mutant births,

$$\text{new cases} = uN(2 - x).$$

Since x, the frequency of the abnormality, is usually very small (e.g., 1/10,000 to 1/100,000), we can neglect x, and, thus, simplifying the last expression, obtain,

$$\text{new cases} = 2uN,$$

which is a good approximation if one makes the reasonable assumptions that the mutant gene is rare and its mutation rate low.

We must now obtain an expression for the number of "lost" cases. The number of abnormal individuals eliminated from the population is equal to the number of abnormalities present, xN, times the fraction lost because of reduced reproductive fitness. If the latter is f, then (1 − f) is the lost fraction (sometimes called the coefficient of selection, s). Thus,

$$\text{eliminated cases} = (1 - f)xN.$$

Postulating an equilibrium between new and eliminated cases means that

$$2uN = (1 - f)xN,$$

which, solved for the mutation rate, u, yields

$$u = 1/2\,(1 - f)x. \tag{1}$$

Here, then, is a means of estimating the frequency of mutations on the basis of two observable data, the reproductive fitness and the frequency of the abnormality.

Achondroplastic Dwarfism. We may apply this indirect method for estimating mutation frequency to the achondroplastic dwarfs of Denmark. The total number of achondroplastics known was 108, living and dead. These 108 dwarfs produced 27 children. In order to judge the relative reproductive fitness of the dwarfs, they were compared with their 457 normal sibs. These had a total of 582 children. Consequently, the relative fitness of the dwarfs is

$$f = \frac{27/108}{582/457} = 0.1963.$$

In other words, relative to the normal alleles only 19.6 per cent of the abnormal alleles present in one generation were transmitted to the next—80.4 per cent, $(1 - f)$, were eliminated.

The best estimate of the frequency of the abnormality comes from the Lying-in Hospital data that have been cited. There were 10 dwarfs in 94,075 births, which yields

$$x = \frac{10}{94{,}075}.$$

Substituting in formula (1) the values found for $(1 - f)$ and x, we find

$$u = 1/2\,(1 - 0.1963)\,\frac{10}{94{,}075} = 0.0000427.$$

This estimate of the mutation rate derived by the indirect method—namely, 4.27 in 100,000 gametes (4.27×10^{-5}) or 1 in about 23,400—agrees well with the estimate obtained by the direct method, 1 in 24,000 (p. 558). However, not too much weight should be given to this agreement. If, as mentioned above, not all achondroplastic phenotypes are genetically equivalent, applications of the direct and indirect methods may each carry the same biases. Some investigators have given reasons for believing that the mutation rate for achondroplasia may actually be only one-fifth of that calculated here from Danish data; others have found evidence for rates two or even three times as high in Sweden and Japan.

Rates of Other Dominant Mutations. Application of the indirect method, and more rarely the direct method, has yielded information on the rate of mutation from normal to dominant abnormal of a variety of genes. These mutation rates are all rather similar, varying only from less than 0.1 to 10 per 100,000 gametes (Table 66). The latter, the highest known rate in man, applies to neurofibromatosis (von Recklinghausen's disease), a syndrome characterized by spots of abnormal pigmentation ("*café au lait*") of the skin and by numerous tumors that arise in association with the central or peripheral nervous systems (Fig. 232). The disease develops in approximately 1 out of every 3,000 births. The relative reproductive fitness of affected individuals has been calculated with particular attention to sources of error in the determination of the relative fertility of affected and normal sibs. For males, $f = 0.41$, for females, 0.75, these low values being in part due to a marriage rate lower than normal and in part to a smaller mean number of children per marriage. Both the direct and the indirect method suggest a mutation rate of approximately 1 per 10,000 gametes.

The lowest mutation rate listed in Table 66 is that for Huntington's chorea. As serious as this disease is, its relatively late onset is compatible with high

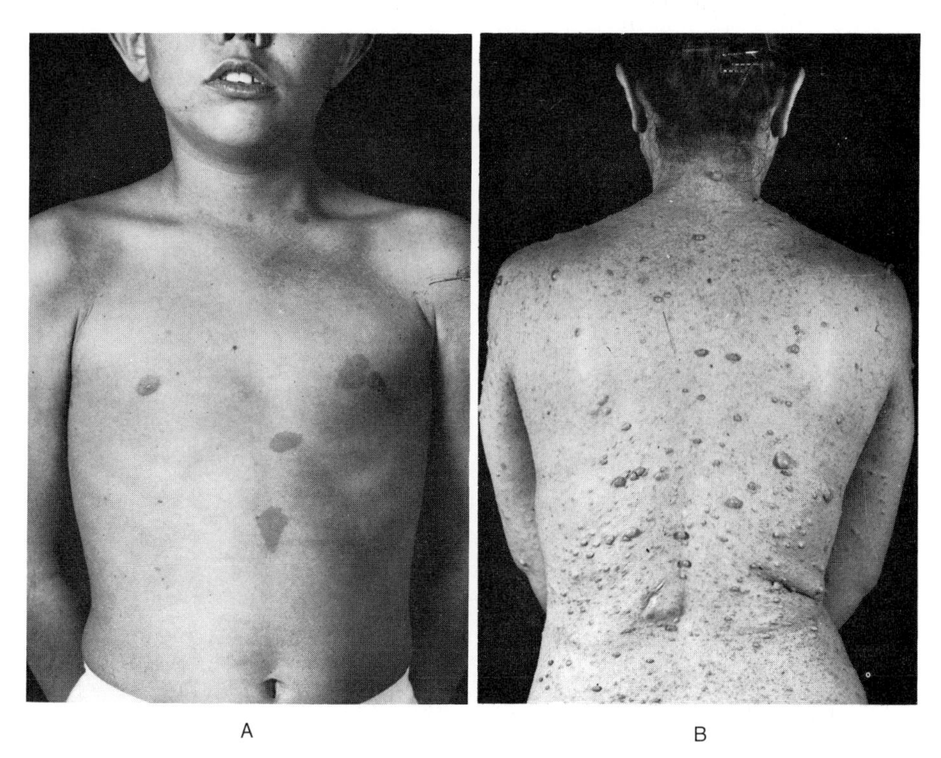

FIGURE 232
Multiple neurofibromatosis. **A.** Pigmented skin areas on a child. **B.** Tumors of varying size on a woman. (Originals from Dr. V. McKusick.)

TABLE 66
Estimates Made by Various Authors of Mutation Rates of Certain Human Genes from Normal to Abnormal.

Trait	*Mutant Gene per 100,000 Gametes*
AUTOSOMAL DOMINANTS	
Huntington's chorea	<0.1
Nail-patella syndrome	0.2
Epiloia (type of brain tumors)	0.4–0.8
Aniridia (absence of iris)	0.5
Retinoblastoma (tumor of retina)	0.6–1.8
Multiple polyposis of the large intestine	1–3
Achondroplasia (dwarfness)	4–12
Neurofibromatosis (tumors of nervous tissue)	13–25
X-LINKED RECESSIVES	
Hemophilia A	2–4
Hemophilia B	0.5–1
Duchenne-type muscular dystrophy	4–10

Source: After Neel, *Proc. Nat. Acad. Sci.*, **43**, 1957; Burdette, *Methodology in Human Genetics*, 1962; and others.

reproductive fitness. Indeed, for a group of choreic families and a selected large kindred in Minnesota, S. C. Reed and Palm reported that the number of children from parents who bore the abnormal *Ht* gene was considerably higher than the number of offspring from the parents' normal (*htht*) sibs. A different result was obtained in studies of Japanese choreics reported by Kishimoto: their fitness was greatly below that of normals. T. E. Reed and Neel concluded that the relative fertility of *Htht* individuals in an unselected large group of families in Michigan, compared to that of Michigan's general population, was about 0.8 or higher. For this population the calculated mutation rate to genes for Huntington's chorea seems to be less than 1 in 1,000,000.

Equilibrium for Recessive Mutations. It should be easy to adjust formula (1) to determine mutation rates for recessive abnormal alleles. Since equilibrium exists when the number of mutations equals the number of eliminated alleles and since, in a recessive trait, *two* alleles are lost with each individual eliminated, the mutation rate, u, for recessive mutations must be twice that of dominant ones: that is,

$$u = (1 - f)x. \qquad (2)$$

It is, however, not advisable at present to use this formula to estimate recessive mutation rates. First, the formula assumes that only the reproductive fitness of homozygotes is affected, but it is now known that many recessive alleles which reduce fitness in homozygotes produce an effect even in heterozygotes—an effect that may increase the fitness of some heterozygotes (heterosis; see p. 455), but decrease that of others. Because heterozygotes for rare recessive alleles are so much more common than homozygotes, even a slight effect on the fitness of heterozygotes would completely invalidate an estimate based solely on the fitness of homozygotes. Second, and even more important, present populations are not in equilibrium. The breakup of isolates in recent times is leading to a decrease in homozygotes and an increase in heterozygotes. Consequently, the number of unfit homozygotes that can be eliminated is decreasing. Unless this is counteracted by increased unfitness of heterozygotes, the unchanged rate of mutation must increase the number of abnormal recessive alleles. Haldane has estimated that this increase of allele frequency may continue for several thousand years, until a stabilized breeding system in mankind results again in a constant frequency of homozygous recessives.

Equilibrium for Sex-linked Mutations. An adaptation of formula (2) can be used for estimates of sex-linked recessive mutation rates. Rare sex-linked recessive alleles will result mainly in elimination of males, and only of a negligible number of females. Since males have one-third of the X-chromosomes of a population, one-third of the sex-linked abnormal alleles are exposed to reproductive unfitness. In an equilibrium between mutation and elimination, the relation is

$$u = 1/3\,(1 - f)x', \tag{3}$$

where x' is the frequency of the abnormality among males. This formula has been applied to two X-linked conditions for which extensive data are available. One of these is the Duchenne type of muscular dystrophy. Since affected individuals hardly ever reproduce, $f = 0$, and the mutation rate becomes simply equal to one-third of the frequency of affected males, namely, 4 to 10 per 100,000 gametes, as calculated from different population samples from Utah, Northern Ireland, and England (Table 66). The second X-linked trait for which the mutation rate has been estimated is hemophilia. Assuming the reproductive fitness of bleeder males to be $f = 0.39$, a combined mutation rate of between 2 and 3 per 100,000 from normal to any one of the alleles causing the more severe types of hemophilia has been calculated for both Danish and Swiss populations. We now know that there are several X-linked types of this disease. It seems that the mutation rate of the gene for classical hemophilia A is several times greater than that for "Christmas disease," hemophilia B (Table 66).

The Distribution of Mutation Frequencies of Different Genes. The observed rates of mutations from normal genes to dominant or to X-linked harmful alleles vary from less than 0.1 to a maximum of 13–25 among 100,000 gametes with an average somewhere in the range 1–5 per 100,000. The average, however, is biased by a tendency toward selection of genes that mutate relatively frequently. This refers particularly to the use of the indirect method, which works best when the fitness, f, is low and the frequency of affected individuals is relatively high, and thus yields a mutation rate, u, that is high. In order to find out how important the bias is toward selection of high mutation rates, Stevenson and Kerr have studied the 54 genes for which the monograph of McKusick lists X-linkage as "possible or very likely." After rejection of 5 of these as not being X-linked, they made estimates of the mutation rates of the remaining 49 genes. These estimates were based in part on specific studies and in part on the frequency with which the investigators had encountered the traits controlled by the mutant alleles in an area, including Oxford, England, whose population includes nearly 900,000 males. The estimated mutation frequencies varied from less than 0.1 per million to 50 per million and their distribution was exceedingly asymmetrical (Fig. 233). Twenty-four—i.e., nearly one-half of the total—of the 49 genes had mutation frequencies of less than 0.1 per million and eleven more had frequencies between 1 and 0.1 per million. Nine others had frequencies of about 1 per million

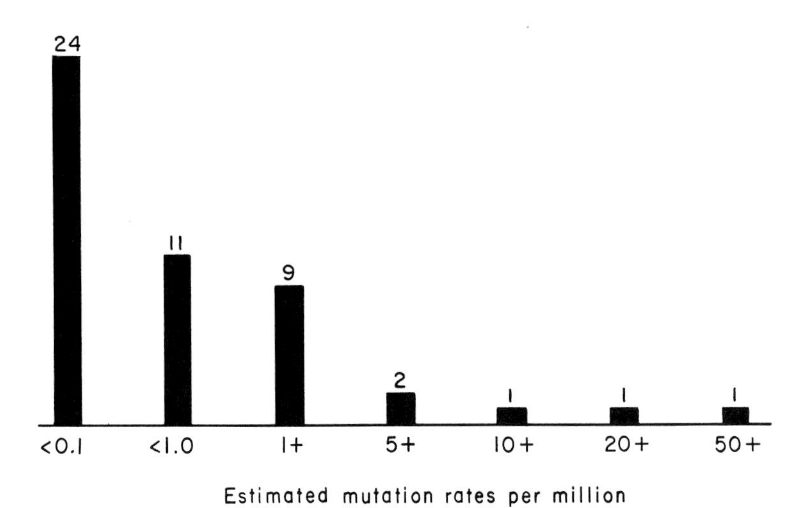

FIGURE 233
Estimated mutation rates of 49 genes. The number of the genes having the given rate appears above each bar. (Based on data of Stevenson and Kerr.)

and the remaining five genes had mutation rates of 5 to 50 per million. "It is difficult to avoid the conclusion," Stevenson and Kerr write, "that the mean mutation rate . . . is not more than 10^{-6}." They add that there is no reason to suppose that such a rate is not also representative of autosomal genes.

The Equilibrium Level. Readers not used to considering equilibria of the type under consideration which are not static but are maintained by a balance between inflow and outgo, may ask the question: Why do mutation and elimination balance each other? Is this balance just a fortunate chance? To answer this question, let us consider a specific example. Assume a population of 1,000,000 per generation, a rate of mutation to a dominant allele of 1 in 100,000 and a reproductive fitness of 0.4, and assume further that there are no abnormal alleles yet present (Table 67). In the next generation, however, 20 such alleles have appeared. One generation later, these alleles have reproduced only 8 of their kind, but 20 new ones have been added, making a total of 28. In the third generation, the 8 alleles from the first generation have dwindled to 3.2, the 20 new alleles of the second to 8, and 20 new alleles have been added by mutation, so that the total is now 31.2. In the fourth generation, the total of abnormal alleles has gone up to 32.48, and it can be seen that an equilibrium is approached in which the total number of mutant alleles is 33.33 (Fig. 234, curve A). Now let us assume that the rate of mutation is only one-tenth as great, making it 1 in 1,000,000. The course of events is shown in Table 68 and Figure 234, curve B. In spite of the great decrease,

TABLE 67
Frequencies of an Abnormal Dominant *A* Allele in Various Generations. (Mutation rate 1 in 100,000. Size of the population 1,000,000. Reproductive fitness of *A*: 0.4.)

Generation	*Normal Alleles a*	*Abnormal Alleles A*		
		Left Over from Former Generations	Newly Mutated	Total
0	2,000,000	–	–	–
1	2,000,000*	–	20	20
2	2,000,000*	8	20	28
3	2,000,000*	8 + 3.2	20	31.2
4	2,000,000*	8 + 3.2 + 1.28	20	32.48
.	.	.	.	.
.	.	.	.	.
.	.	.	.	.
∞	2,000,000*	8 + 3.2 + 1.28 + 0.512 + · · ·	20	33.33

*More accurately, the number of normal alleles would be 2,000,000 minus the total number of abnormal alleles. This small correction is omitted as well as the small error which is introduced by this omission in the expectation for new mutants.

TABLE 68
Frequencies of an Abnormal Dominant *A* Allele in Various Generations. (Mutation rate 1 in 1,000,000. Size of the Population 1,000,000. Reproductive fitness of *A*: 0.4.)

Generation	*Normal Alleles a*	*Abnormal Alleles A*		
		Left Over from Former Generations	Newly Mutated	Total
0	2,000,000	–	–	–
1	2,000,000*	–	2	2
2	2,000,000*	0.8	2	2.8
3	2,000,000*	0.8 + 0.32	2	3.12
4	2,000,000*	0.8 + 0.32 + 0.128	2	3.248
.	.	.	.	.
.	.	.	.	.
.	.	.	.	.
∞	2,000,000*	0.8 + 0.32 + 0.128 + 0.051 + · · ·	2	3.333

*More accurately, the number of normal alleles would be 2,000,000 minus the total number of abnormal alleles. This small correction is omitted as well as the small error which is introduced by this omission in the expectation for new mutants.

an equilibrium is reached, though at a lower frequency of the abnormal allele: the decreased inflow of new alleles is balanced by decreased outgo. Diagrammatically, this state of affairs is illustrated by Figure 235, in which varied rates of inflow into a tank are balanced by arrangements for overflow at different levels.

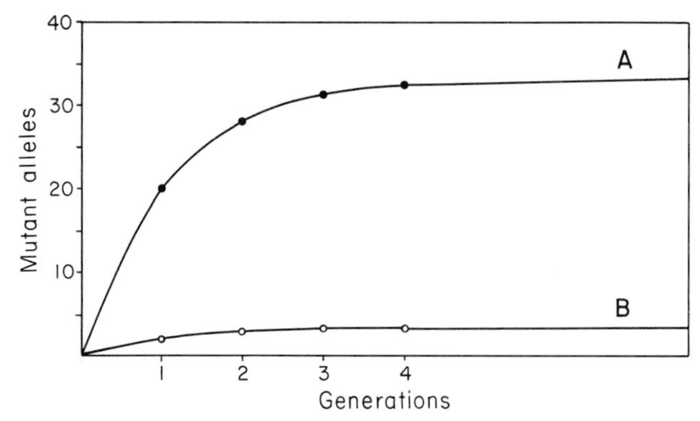

FIGURE 234
Frequency of an abnormal dominant *A* allele in the course of generations after onset of mutations from *a* to *A*. (Size of population: 1,000,000. Reproductive fitness of *A*: 0.4.) **A.** Mutation rate 1 in 100,000. **B.** 1 in 1,000,000.

Keeping the mutation rate constant but varying the rate of loss of abnormal genes also influences the frequency of the allele. If reproductive fitness is zero—that is, if bearers of the allele have no offspring at all—the frequency of the dominant allele at any time will be equal to its rate of production by immediately preceding mutations. The higher the degree of relative fitness produced by the allele, the higher will its frequency rise. Should the reproductive fitness of persons carrying the mutant allele become equal to that of those carrying its normal partner, then recurrent mutation would lead to a continuous increase in the frequency of the mutant. The result of such a process will be discussed in the final chapter of this book. Obviously, the alleles involved cannot be labeled "abnormal," since full fitness contradicts such a term.

Frequency of Mutation in Relation to Age and Sex. Mutation rates derived by the indirect method of estimation are average rates and cannot indicate possible differences of rates in male or female gametes, or differences between different groups of individuals in a population. That such differences may exist is known both from extensive plant and animal studies and from human evidence. It is true, for instance, that rates of sex-linked lethal mutations are higher in *Drosophila* sperm than in eggs, that young *Drosophila* males produce sperm with a higher number of mutant alleles than older ones, and that stored *Drosophila* sperm, like stored plant pollen, accumulate mutant alleles.

In man, as first suspected by Weinberg, the frequency in the offspring of sev-

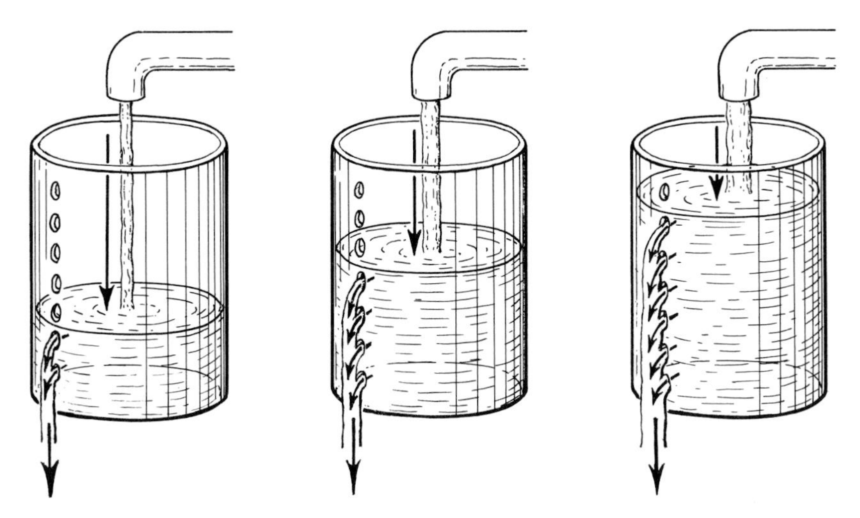

FIGURE 235
The three vessels, each with a different content level, are analogous to three populations, each with a different number of mutant alleles. The level in each vessel remains constant—a result of the balance between input through the pipes above and outgo through the overflow holes. Similarly, the number of mutant alleles in each of the three populations remains constant—a result of the balance between "input" of new mutant alleles and "outgo" due to low reproductive fitness of the bearers of the mutant alleles.

eral types of dominant mutations rises with age of parents. This is shown in Figure 236 for achondroplasia and age of mother, and for the same trait plus two others and age of father. Whether the age of mother, or of father, or of both is the decisive factor cannot be judged by the curves in Figure 236 since older wives, on the average, have older husbands and vice versa. It is noteworthy, however, that in achondroplasia and in myositis the fathers of mutant children are disproportionately older than the mothers. In Danish families with mutant achondroplasia the difference between the mean age of the fathers and that of the mothers was 6.6 years as compared to an estimated difference in the general population of 4.7 years, and in English families 3.4 years as compared to 2.3 years. In myositis, according to data mostly from Germany, the age difference between fathers and mothers of mutant children was 6.2 years as compared to a control estimate of about 4 years. These facts indicate strongly that it is the age of the father that is responsible for the increasing frequency of mutations with age of parents. The same conclusion follows from the fact that if we so select data as to keep age of mother constant, we obtain a positive correlation between age of father and mutation frequency, but if we keep age of father constant, we find that age of mother

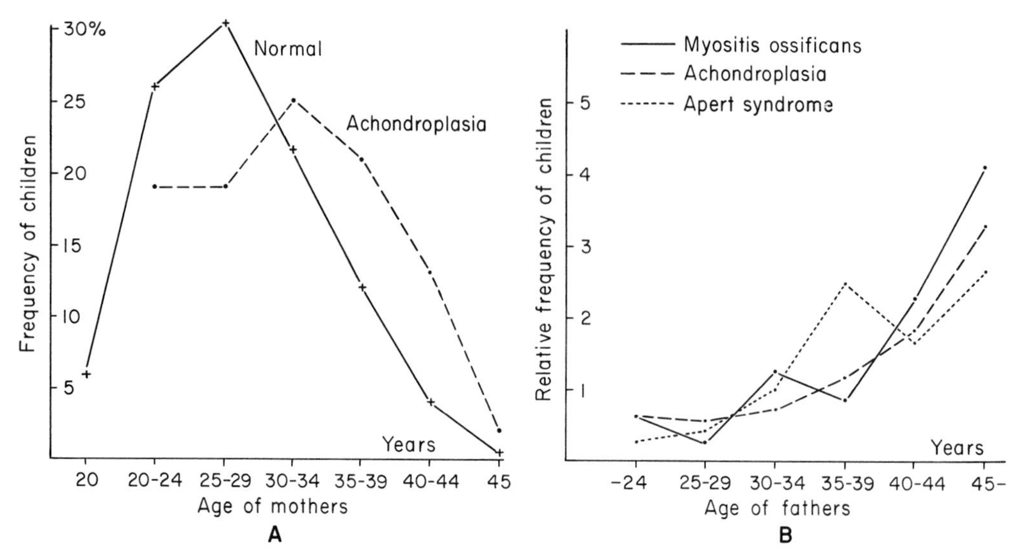

FIGURE 236
A. Age of normal mothers, whose husbands are also normal, and percentage frequency of normal and achondroplastic births. **B.** Age of normal fathers, whose wives are also normal, and relative frequencies of births of children affected by any one of three dominant abnormal traits due to mutation. Myositis is characterized by a special pattern of ossification, congenital deformities, and deficient sexual development. Apert's syndrome consists of malformations of the skull and of the hands and feet. (A, Møřch; B, after Tünte, Becker, and v. Knorre, *Humangenetik*, **64,** 1967.)

and mutation frequency are not correlated. These results also exclude the possibility that number of preceding births, which, of course, rises with increasing parental age, is a factor influencing rate of mutation.

The frequency of mutation to an X-linked hemophilia-A allele also rises with the age of the father independently of the age of the mother. Dependence of mutation frequency on parental age, however, is not a general rule. The rates of mutation for other loci, e.g., those determining aniridia and neurofibromatosis, are not correlated with parental age.

Mutations at X-linked loci provide a particularly suitable material for the study of the relative rates of mutation in the two sexes. If the rates of mutation at an X-linked locus were alike in eggs and sperm, then there would be twice as many new mutant heterozygous women as new hemizygous men, since, after fertilization, half of the mutant eggs would develop into males and the other half into females and, in addition, all of the mutant sperm, being X-chromosomal, would produce females. Were mutations restricted to eggs, then equal numbers of affected males and carrier females of mutant origin would be expected; and were mutations exclusive to sperm, then only carrier females would mark the origin of a mutant allele. Comparing these different expectations with numerous published data on hemophilia, Haldane came to the conclusion that the mutation rate for this condition is much higher in male than in female gametes—possibly ten times as high. Later data do not fully support this conclusion. Given the uncertainties of mutation-rate estimates, the findings are compatible with either equal rates in the two sexes or with an only somewhat higher rate in sperm than in eggs. Similarly, for another X-linked gene, that for Duchenne-type muscular dystrophy, there is no evidence against equality of mutation rates in egg and sperm.

The Total Mutation Rate per Gamete. The estimated frequencies of rates of mutation of given normal alleles into detrimental alleles range from one in ten thousand to one in millions (Table 66). Since each gamete carries many mutable genes, it is important to estimate the frequency with which gametes carry one or more mutated genes at one or another of their many loci. It would be easy to determine this frequency if we knew (a) the mean frequency of mutation per locus and (b) the total number of mutable loci. Simple multiplication of the two quantities would give the frequency of mutant gametes (though uncorrected for the rare event of two or more different mutations being found together in the same gamete).

We have seen that is is difficult to make a satisfactory estimate of the mean rate of detrimental mutations per locus, since the accuracy of the "known" rates is uncertain, and, even if they should be correct, their average may not reflect the true average of all known and unknown individual rates. Somewhat arbitrarily let us assume that this average rate is 1 in 1,000,000. It may well be higher if mutations with rather slight detrimental effects are included, or considerably lower if there should be many genic loci with low mutation rates. The total number of genic loci in man is likewise a matter of uncertainty. Again somewhat arbitrarily it has been assumed in an earlier section that this number lies within the range 10,000–100,000.

If we now use 10,000 as the number of loci in man and multiply it by an estimated rate of 1/1,000,000 mutations per locus we find that 1 out of 100 human gametes carries a new harmful mutation. If alternatively we use an estimate of 100,000 genic loci and multiply this number by the rate of mutation of 1 in 1,000,000 we obtain 10 per cent as the frequency of mutations in the gametes. Since each individual is the product of two gametes, the two estimates would indicate that a minimum of 2 per cent and a maximum of nearly 20 per cent of all humans contain a newly mutated gene that is more-or-less detrimental.

There is another method of estimating the number of newly mutant genes per individual. It is not based on direct determination of mutation frequencies but makes use of estimates of the average number of deleterious genes—

old or new—that are carried by an individual. This number has been derived from studies of the outcome of consanguinous marriages (see Chapter 19). If the analysis is restricted to lethal alleles and the effects of deleterious sublethal genes are combined into "lethal equivalents," an estimate of approximately four lethals or lethal equivalents is obtained. Assumption of an equilibrium between mutational input and selective outgo of mutants suggests that 2 per cent of the 4 lethal equivalents are of new mutational origin. In other words, 0.02×4, or 8 per cent, of all individuals carry one newly mutated lethal equivalent. This figure of 8 per cent lies between the two values 2 per cent and 20 per cent derived in the preceding paragraph. As Neel has emphasized, the interpretation of the results of consanguinity rests on a series of uncertain assumptions. Therefore the estimate of the frequency of newly mutated lethal equivalents must be regarded as very unreliable.

Whatever the actual value is, the estimates of percentages of zygotes with new mutant genes become particularly meaningful when they are considered in terms of the numbers of persons in entire countries or in the world. Even the lowest estimate listed above—namely, 2 per cent—signifies that more than 4 million of the more than 200 million living Americans carry a detrimental gene that was not carried by their parents; in China 16 million; and in all mankind, no less than 70 million carry such genes. We must assume that spontaneous mutations have taken place and poured new genes into the gene pool continuously ever since life existed on the earth. Individually, most of these mutations cause only little harm, and some may even be beneficial, but the harmful mutant genes, accumulated over the generations and in the aggregate, either in heterozygotes or homozygotes or in polygenic combinations decrease the fitness of the population.

THE DOMINANCE OF DETRIMENTAL GENES

We might expect that in order to evaluate the harm done by detrimental genes it would be necessary to know how many of them are dominant and how many recessive. Each single dominant would exert an effect, so that the total effect of dominants in a population would be proportional to their frequency. Recessives would have an effect only in homozygous combination, and the population would suffer in proportion to the square of their frequency. However, detailed studies of *Drosophila* as well as of other species have shown that, in reality, mutant lethals that seem to be recessive are not strictly recessive, but, on the average, reduce the fitness of heterozygotes sufficiently to cause the death of about 2.5 per cent, and may therefore be regarded as dominant lethals with very low penetrance. In any population, heterozygotes for rare alleles are much more common than are homozygotes. Consequently, the reduced fitness of heterozygotes with seemingly recessive lethals will lead to many more eliminations of lethal genes than will the invariable death of the rare homozygotes. If, on the basis of some evidence from *Drosophila*, we assume that not only lethals but also other, less detrimental alleles have an incompletely penetrant dominant action in heterozygotes, then the same conclusion applies to them—namely, that the major part of their effect on a population is exerted in heterozygotes.

The Accumulation of Mutant Genes in a Population. A dominant mutant allele that invariably leads to the death of its carrier before the age of reproduction or that causes sterility, can occur in a population only at the frequency with which it is produced by mutation. By definition, it is eliminated within one generation,as soon as it has been produced. A similar mutant allele that is penetrant in only half of the persons carrying it has a probability of being eliminated in the first generation of 1/2 and correspondingly a remaining probability of being present in the second generation of 1/2. Those mutant alleles that survive the first generation have a probability of 1/2 of being eliminated in the second generation, and so on. On the average, such a gene with its selective disadvantage, s, of 1/2 will persist for two generations. A dominant mutant gene that imposes on its carriers a selective disadvantage of 1/10 will, on the average, persist for ten generations. The general expression for the number of generations of persistence of a gene is 1/s.

Mutant genes that persist for several generations will accumulate in a population, the accumulation varying with the degree of selective disadvantage imposed by the particular gene: The population will contain not only the newly mutated genes but also the leftovers from former generations that have not yet been eliminated (Tables 67 and 68; Fig. 234). The frequency of genes that persist for ten generations will be 10 times their mutation rate and that of genes which persist for 100 generations will be 100 times the mutation rate. It has been estimated (with considerable uncertainty) that about 50 times as many detrimental mutant alleles are present in a population as would be produced by new mutations. This estimate takes into account both the genes lost by elimination of heterozygotes and the relatively small losses from elimination of homozygotes. The estimated frequencies of the homozygotes were not simply based on the Hardy-Weinberg Law but were adjusted for the existence of inbreeding in human populations.

Heterotic Mutant Genes. The foregoing estimates of the persistence and accumulation of mutant genes depend on the estimates of their penetrance in heterozygotes. The estimate of 2.5 per cent for the dominant effect of lethals was an average derived from experiments with *Drosophila* that showed not only higher and lower penetrance of lethality in heterozygotes but also some cases of better-than-normal viability. Some of the mutant genes that were lethal when homozygous increased the chances of survival when present heterozygously! It is not known yet, either in *Drosophila* or in man, how common such heterotic mutant genes are. It is obvious that the accumulation of heterotic alleles has a trend different from that of those with detrimental effects in heterozygotes. A heterotic mutant allele will increase in frequency from generation to generation—an increase which is checked only by the selective disadvantage the mutant imposes on homozygotes. A more detailed treatment of opposing forces will be deferred until Chapter 28, Selection and Genetic Polymorphism, but the discussion of the effect of mutations on human population in Chapter 23 will take account of the significance of heterotic mutant genes.

THE DETECTION OF SOMATIC MUTATIONS

We have treated the process of mutation as if it occurred in the gametes only, but there is experimental evidence indicating that genes are susceptible to mutation in many or perhaps all cells. If one allele in a cell of a very early embryo mutated from A to A', further divisions of the unmutated AA cells and the mutated AA' cell would build up an individual whose tissues would consist partly of AA cells and partly of AA' cells. If A' were a dominant of autonomous phenotypic expression in body cells, the mosaic nature of such an individual might be observed directly. In various mammals, the rare appearance of mosaic coats when uniform ones were expected has thus been attributed to early somatic mutation. This interpretation cannot be tested rigorously when all AA' cells are part of somatic tissues, since proof of presence of a mutation consists of its transmission to later generations of individuals, which is obviously impossible if it exists only in somatic cells. However, in some of the mosaic individuals, proof of somatic mutation was obtainable, since cells descended from the original mutant AA' cell formed part of the germ cell tissue of the gonads. Some of the offspring of these mosaics inherited the mutant allele A', which was not present in the ancestors of the mosaic parents (Fig. 237). Some mosaic human abnormalities have been suspected of being due to mutation during early embryonic development, but there is no definite proof of this. The later in development mutation occurs, the smaller should be the segment of tissue derived from the mutant cell. Thus, some of the fairly common mosaic eye colors—a brown segment in an otherwise blue iris, for instance—are perhaps due to late somatic mutation. Similarly, retinoblastoma in individuals whose parents and offspring do not have it may be caused by somatic mutations.

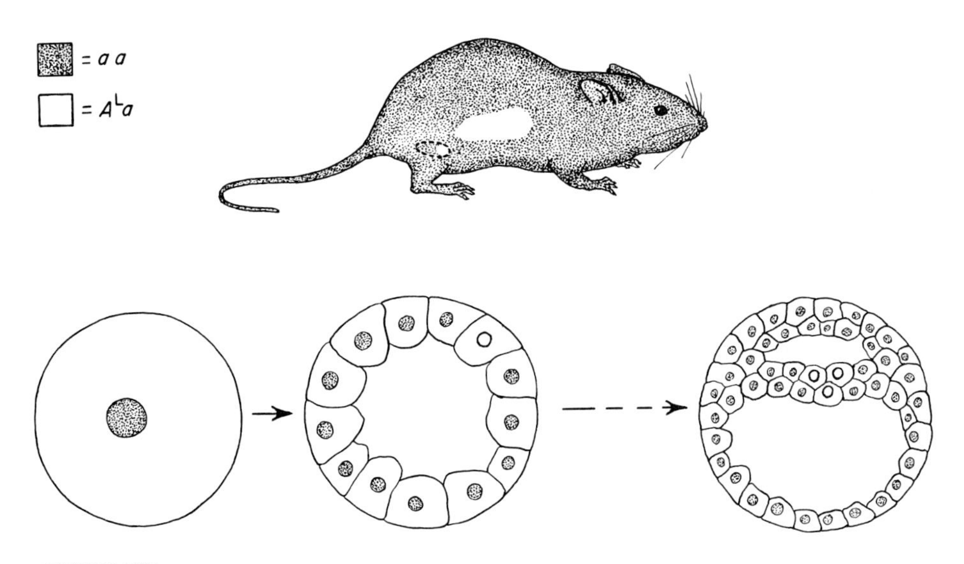

FIGURE 237
Above: A mouse homozygous for not-agouti (*aa*) in most cells, but a mosaic in having an agouti-like ($A^{L}a$) patch. The ovary, too, is mosaic as indicated by the outline above the hind leg. *Below:* The fertilized egg and two later embryonic stages in cross sections. It is assumed that a mutation from *a* to A^{L} occurred in one nucleus of the blastocyst (middle) and that descendants of this cell became incorporated in the embryonic disc (right). (After an analysis by Bhat, *Heredity*, 3, 1949, from C. Stern, *Genetic Mosaics*, Harvard University Press, 1968.)

In the developing individual, mutation may also occur in an embryonic germ cell, which, on division, will produce a "cluster" of gametes of common descent which carry the mutant allele. Such an event may have occurred in a kindred in which a normal couple had 6 normal children and 4 with aniridia (absence of the iris). For three generations this trait was transmitted to 19 descendants in dominant fashion with complete or very high penetrance. It may be concluded either that one of the parents of the first 4 affected individuals carried the gene without penetrance, or that an early mutation had occurred which did not lead to somatic mosaicism but that the descendants of the cell carrying the mutation included germinal tissues, or that the four mutant germ cells were derived from a germinal mutation. Hartl has analyzed the interesting problem of recurrence risks of mutant offspring from a parent whose gonads are mosaic for nonmutant and mutant cells.

Genetic changes in somatic cells can be brought about by various mechanisms of which gene mutation is only one. Thus, if a cell is heterozygous for a pair of alleles, A and A', loss of the chromosome carrying A' will leave a cell hemizygous for A. Even loss of a chromosome or part of a chromosome without a "marker" gene is equivalent to a genetic change, since it upsets

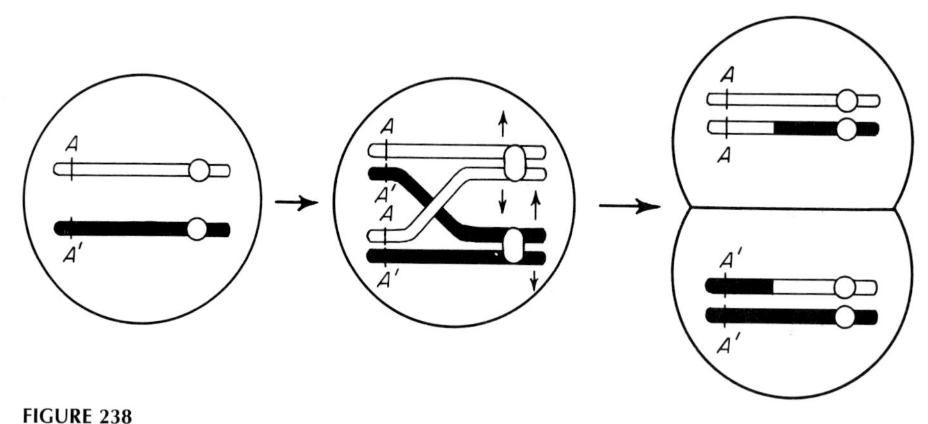

FIGURE 238
Somatic crossing over in a cell of the genotype *AA'* and its possible consequences, the formation of *AA* and *A'A'* cells.

the genic balance. Still another process leading to genetic changes is crossing over between homologous chromosomes in somatic cells. Its occurrence has been demonstrated in *Drosophila*, molds and mice, but no critical genetic studies are available for man. Crossing over in somatic cells, as well as in germ cells, occurs between two of the strands at the four-strand stage (Fig. 238). In a heterozygous *AA'* cell it can lead to the mitotic distribution of *AA* into one daughter cell and of *A'A'* into another. If *A'* is a dominant, then tissue with the recessive phenotype *AA* will develop from the *AA* somatic segregate. If all these genotypes, *AA*, *AA'*, and *A'A'*, are phenotypically distinguishable, then an *AA*–*A'A'* "twin spot" will originate on the *AA'* background. Thus if genotype *AA'* present in the majority of cells of an organism leads to a skin coloration of medium intensity while *AA* is colorless and *A'A'* deeply pigmented then a colorless sector next to a very dark sector will become visible on an intermediate background.

Chromosomal changes other than losses or additions of whole chromosomes comprise deficiencies of parts of a chromosome, duplications of chromosomal sections, translocations, and inversions. Any one of these aberrations can exert a potentially new, transmissible phenotypic effect. In deficiencies and duplications such an effect usually is the consequence of genic inbalance. In those translocation genotypes in which the normal balance has not been disturbed and in inversions in which all genes are present though in a new arrangement, new phenotypes are the result of a *position effect*. It has been shown in a few experimental organisms that genes exert a different effect—either dominant or recessive—when placed in new chromosomal neighborhoods. Genes near a break find a new neighborhood whenever the broken fragment unites with another in such a way as to form a new type of chromosome. A new phenotype produced by a new position of a gene is transmitted in the same way as one produced by a mutated allele that had arisen at the locus, since the rearranged chromosome reproduces itself permanently in its rearranged form. In man such position effects have not yet been discovered, but they have been found in another mammal, the mouse.

Genic Mutation and Chromosomal Change. The existence of chromosomal aberrations is well established in many animals and plants. Yet even in the most favorable material, there is a point at which optical recognition of these chromosomal mutants becomes difficult or impossible. This point is reached when the chromosome section involved is so small as to be undetectable, but since all degrees of ease of detection have been encountered, there is no reason why we should doubt the existence of aberrations below the limits of optical observation. Even very small aberrations could exert a special phenotypic effect by changing the position of a gene or affecting the genic balance of a cell. It is possible that many "mutant" phenotypes ascribed to a typical mutation from one allele to another are produced by chromosomal aberrations too small to be seen. From another standpoint, mutations may often consist of very small chromosomal changes that affect the quantity or position of genes rather than of "real" allelic mutations. Thus, the distinction between genic mutations and chromosomal aberrations must often remain theoretical. This is probable for a type of leukemia, "chronic myeloid," where the appearance of a unique, aberrant chromosome in some leucocytes and bone marrow cells of an individual whose chromosomal constitution had been normal earlier in his life is correlated with the onset of excessive propagation of white blood cells. The pertinent chromosome is the so-called Philadelphia chromosome (Ph^1), observed first in a patient in Philadelphia, a small chromosome that is actually about one-half of the long arm of chromosome 22 (Fig. 239). There is little doubt that the cells containing Ph^1 behave as if they are the leukemic cells, but the correlation between presence of the abnormal chromosome and the disease does not specify what is cause and what effect.

FIGURE 239
The Philadelphia chromosome in a patient with chronic myeloid leukemia. **A.** The four typical small chromosomes 21 and 22 from a chromosomally normal cell. **B.** The three typical and one abnormally short chromosome from another cell. The shortened chromosome, Ph^1, is marked by an arrow. (Drawings after photographs from Nowell and Hungerford, *J. Nat. Cancer Inst.*, 27, 1961.)

In general a somatic mutation hypothesis of cancer can neither be proved nor disproved at present: it is not the only hypothesis that would explain the permanent change in cells that become cancerous. The process of differentiation of genetically identical cells during normal development has always posed a problem. Typical mutational changes in genic composition do not account for embryonic differentiation, and as long as the nature of differentiation is obscure, the characterization of cancer as a somatic mutation must remain tentative. Should, however, such a designation turn out to be correct, it would lead to a specific interpretation of the heritable aspects of cancer. Inheritance of cancer would then be known to consist in an inherited tendency to mutate somatically. There is, as we have said, no evidence for this chain of reasoning, but each single link is based on reasonable or even well-established principles. Thus, inherited tendencies toward somatic mutation, though not to cancerous cells, have been shown to exist in animals and plants.

Some types of cancer, in organisms other than man, are caused by viruses, and the hypothesis that all cancers are virus diseases has been advanced. Should this prove to be correct, then the genetic tendencies to cancer which are known to exist (pp. 672ff.) would be based on genes that control the susceptibility of cells to respond to viruses by cancerous growth.

PROBLEMS

173. In two families, there have been only normal people for many generations. In one family, a brachydactylous boy is born, who later marries and has four children, three normal and one brachydactylous. This brachydactylous offspring marries and has three children, all normal. In the other family, an albino child is born, who later marries and has four children, all normal. What are the most likely explanations for the appearance and disapperaance of the traits in each of these two pedigrees? (Assume only legitimacy.)

174. A child with blue scleras and brittleness of bones is born to normal parents. Does the child represent a new mutation?

175. (a) Assume that the mutation rate of the normal allele x to the dominant allele X responsible for an abnormal trait is 1 in 50,000, and that the reproductive fitness of affected individuals is 0.2. In a population of 100,000,000 per generation which is initially free from X, how many affected persons will be present after 1 generation? After 2, 3, 4, 5 generations? At equilibrium?
(b) What would the answer be if the allele X caused death in childhood?

176. Assume that the rate of mutation of the normal gene for sex-linked color vision to an allele for red-green color deficiency is 1 in 50,000 and that the frequency x' of color-blind men is 0.08. Assuming that the formula for the equilibrium situation is applicable, determine the reproductive fitness of color-blind men.

177. Assume that the mutation rate of the normal allele to that for hemophilia is ten times higher in male than in female gametes. What would be the ratio of new carrier females to new affected males?

178. If the reproductive fitness of heterozygous carriers of a recessive trait is somewhat less than 1, would this call for upward, or downward, or no revision of an estimate of the mutation rate which was originally based on the frequency and fitness of homozygotes only?

179. (a) What is the chromosome number of a boy who has both Klinefelter's syndrome and Down's syndrome?
(b) If he could be fertile, what types of offspring would be expected?
(c) If he formed sperm with XX, Y, X, and XY in the ratio 1:1:2:2, what would be the proportion of the various types of offspring?

180. In different social groups, the age of people entering marriage varies considerably. What effects might this difference have on the incidence of Down's syndrome in these different groups?

181. In a fetus with a genotype for blue eyes, a somatic mutation occurs in a cell, changing its genotype to one leading to brown eyes. If descendants of this cell form part of the iris of the left eye, what will be the appearance of the child after birth?

182. Tissue heterozygous for a histocompatibility locus, H^1H^2, does not usually persist if transplanted to H^1H^1 or H^2H^2 hosts. What would be the results of transplantation if occasionally some cells of the graft become genetically H^2H^2?

REFERENCES

Cheeseman, E. A., Kilpatrick, S. J., Stevenson, A. C., and Smith, C. A. B., 1958. The sex ratio of mutation rates of sex-linked recessive genes in man with particular reference to Duchenne type of muscular dystrophy. *Ann. Hum. Genet.*, **22**:235–243.

Crowe, F. W., Schull, W. J., and Neel, J. V., 1956. *A Clinical, Pathological, and Genetic Study of Multiple Neurofibromatosis.* 181 pp. Thomas, Springfield, Ill.

Danforth, C. H., 1923. The frequency of mutation and the incidence of hereditary traits in man. *Eugenics, Genetics and the Family*, **1**:120–128.

Gunther, M., and Penrose, L. S., 1935. The genetics of epiloia. *J. Genet.*, **31**: 413–430.

Haldane, J. B. S., 1949. The rate of mutation of human genes. *Hereditas*, Suppl. vol.: 267–273. (Proceedings of the Eighth International Congress of Genetics.)

Hartl, D., 1971. Recurrence risks for germinal mosaics. *Amer. J. Hum. Genet.*, **23**:124–134.

Kishimoto, K., Nakamura, M., and Sotokawa, Y., 1957. On population genetics of Huntington's chorea in Japan. *Ann. Rep. Res. Inst. Envir. Med., Nagoya Univ.*, **1958**:84–90.

Knudson, A. G., Jr., 1971. Mutation and cancer: Statistical study of retinoblastoma. *Proc. Nat. Acad. Sci.*, **68**:820–823.

Lejeune, J., Turpin, R., and Gautier, M., 1959. Le mongolisme, premier exemple d'aberration autosomique humaine. *Ann. génét. hum.*, **1**:41–49.

Mørch, E. T., 1941. Chondrodystrophic dwarfs in Denmark." *Opera ex Domo Biologiae Hereditariae Humanae Universitatis Hafniensis*, **3**:200 pp. (see also Popham, R. E., 1953. *Amer. J. Hum. Genet.*, **5**:73–75).

Neel, J. V., 1962. Mutations in the human population. Pp. 203–224 *in* Burdette, W. J. (Ed.), *Methodology in Human Genetics.* Holden Day, San Francisco.

Neel, J. V., and Falls, H. F., 1952. The rate of mutation of the gene responsible for retinoblastoma in man. *Science,* **114**:419–422.

Penrose, L. S., 1955. Parental age and mutation. *Lancet*, **269**:312–313.

Puck, T. T., 1959. Quantitative studies on mammalian cells in vitro. *Rev. Mod. Phys.*, **31**:433–448.

Reed, T. E., and Falls, H. F., 1955. A pedigree of aniridia with a discussion of germinal mosaicism in man. *Amer. J. Hum. Genet.*, **7**:28–38.

Reed, T. E., and Neel, J. V., 1959. Huntington's chorea in Michigan. 2. Selection and mutation. *Amer. J. Hum. Genet.*, **11**:107–136.

Schull, W. J. (Ed.), 1962. 248 pp. *Mutations.* University of Michigan Press, Ann Arbor. (Second Macy Conference on Genetics.)

Steinberg, A. G., 1959. Methodology in human genetics. *J. Med. Educ.*, **34:** 315–334 (also *Amer. J. Hum. Genet.*, **11,** 2 Part 2:315–334), see pp. 330–333.

Stevenson, A. C., and Kerr, C. B., 1967. On the distributions of frequencies of mutation to genes determining harmful traits in man. *Mut. Res.*, **4**:339–352.

Vogel, F., 1963. Mutations in man. Pp. 833–850 *in* Geerts, S. J. (Ed.), *Genetics Today.* Pergamon, Oxford. (Proceedings of the Eleventh International Congress of Genetics.)

23

PRODUCTION OF MUTATIONS

The origin of a new mutant allele from a preexisting gene has been observed in many organisms. Although the first appearance of a given mutant allele may well remain unknown, the presence of such alleles in a population is indirect evidence for mutational input counteracting low fitness outgo. The mutational change may occur "spontaneously"—that is, without exposure to mutagenic agents—or it may be induced by the application of such agents. The causes of mutations obviously are molecular or submolecular. Genes replicate by separating their two strands and by selecting for each nucleotide in the sequence of each strand the complementary nucleotide. Thus, in the replication of an AT pair, the A bonds with a new T and the T with a new A, forming two identical AT pairs. This process functions with accuracy most of the time; but occasionally the A might attract, among other possibilities, a C nucleotide instead of the T, forming an AC combination. At the next replication A and C separate, A resuming its normal complementary pairing with a T, and C pairing in its usual way with a G. In this way a gene with a new CG pair will permanently take the place of the gene with the original AT pair. Mutation in such a case results from an error in replication. Other types of replication errors may lead to insertion or excision of nucleotide pairs. Mutation may also occur in genes that are not undergoing replication such as genes in spermatozoa or in bacterial spores, or, presumably, genes in the oocytes before their meiotic divisions, or genes during the nonreplicating phases of the normal cell cycle. Some of the nonreplicating genes may become chemically modified by a great variety of substances and thus become mutant alleles without further change. Other genes may also react in the nonreplicating state but a full mutant condition may not be attained until replication sets in. In the majority of mutations that have been analyzed in respect to the structure of the polypeptide chains produced by a given gene, the mutant change consists of the substitution of a single amino acid in the original gene product by another amino acid in the mutant product, representing a "missense" mutation. Moreover, this replacement of one amino acid by another is usually compatible with the assumption that there has been only the substitution of a single nucleotide pair among three pairs forming a DNA code triplet. This has been discussed earlier in Chapter 2 for the two mutant hemoglobin genes Hb^{S} and Hb^{C}, which can be derived from the sixth triplet of the normal gene Hb^{A} by a change in one or the other of the base pairs (see Figures 22 and 23, pages 37 and 47):

$$Hb^{\mathrm{A}} = \frac{\mathrm{CTT}}{\mathrm{GAA}};\ Hb^{\mathrm{S}} = \frac{\mathrm{CAT}}{\mathrm{GTA}};\ \text{and } Hb^{\mathrm{C}} = \frac{\mathrm{TTT}}{\mathrm{AAA}}.$$

Altogether, as Vogel has shown for a variety of mutant hemoglobin chains, 34 α chains, 56 β chains, 4 γ chains, and 4 δ chains all may be accounted for by single base-pair replacements at some specific site in the normal sequences. Accountability is, of course, not proof. Given the degeneracy of the code (p. 39), other more complex possibilities exist by means of which the amino acid substitutions could come about but it is highly suggestive that in most cases no contradiction to the assumption of a single base-pair change has been encountered.

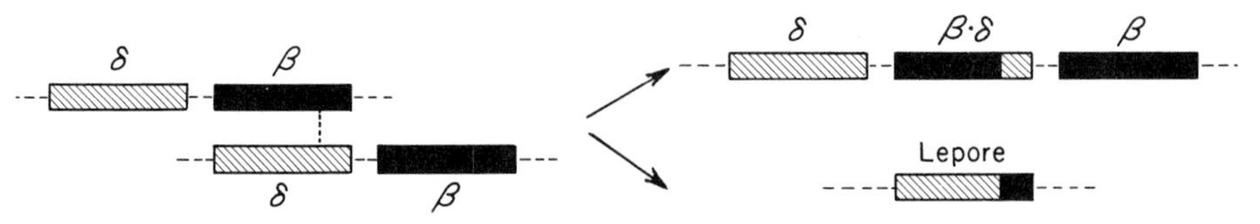

FIGURE 240

Presumed origin of a hemoglobin Lepore gene. *Left:* Unequal crossing over between a Hb_{β} and Hb_{δ} gene. *Right:* The resulting two new kinds of hemoglobin determiners, a hypothetical triplicate *Hb* assembly and a single *Hb* Lepore gene. (After Baglioni, *Proc. Nat. Acad. Sci.*, **48,** 1962.)

There are, however, some real exceptions to the prevalent type of single amino acid change in hemoglobin mutations. Two of these are varieties of the so-called Lepore hemoglobin. They have an amino acid sequence which begins like that of a normal δ chain and ends with a sequence like that of a normal β chain. In Lepore–Boston the length of the δ segment greatly exceeds that of the β segment; in Lepore–Hollandi the reverse is true. These remarkable properties have been explained by Baglioni in the following way. First, it is known that the δ and β chains of normal hemoglobin are highly similar. Second, the genes for the two chains seem to be very closely linked, forming a tandem sequence. Third, pairing of genes is not always between homologous partners, such as pairing of δ with δ and β with β but may occasionally be between nonhomologues, such as β and δ (Fig. 240, left). If the latter process occurs and is accompanied by meiotic crossing over between a β and a δ gene, the resulting recombination products will be either a triplet gene assembly $\delta - \beta \cdot \delta - \beta$ or a single mixed $\delta \cdot \beta$ gene (Fig. 240, right). The triplet is not known but the singleton would be a Lepore gene—of the Boston type if the crossover had occurred in the right half of the original genes and of the Hollandi type if the crossover was in the left half.

The origin of mutant alleles by "unequal" crossing over is not restricted to the hemoglobin genes. Among the haptoglobin alleles some differ from one another by a single amino acid substitution only. The allele Hp^2, however, differs from other Hp alleles by being nearly twice as long. Smithies has postulated that the formation of the Hp^2 allele was the result of a unique event, an "accident," in which an ancestral Hp gene of normal length underwent crossing over with its allele in the homologous chromosome, but at a nonhomologous site. The result would be a partial duplication, in tandem, of the original Hp allele, thus forming the Hp^2 allele. (Fig. 241). In later generations individuals would arise who are homozygous for Hp^2. In the meiotic cells of these individuals the two Hp^2 alleles would have a tendency to pair in two different ways, either "equally" so that the left half of the duplicated gene

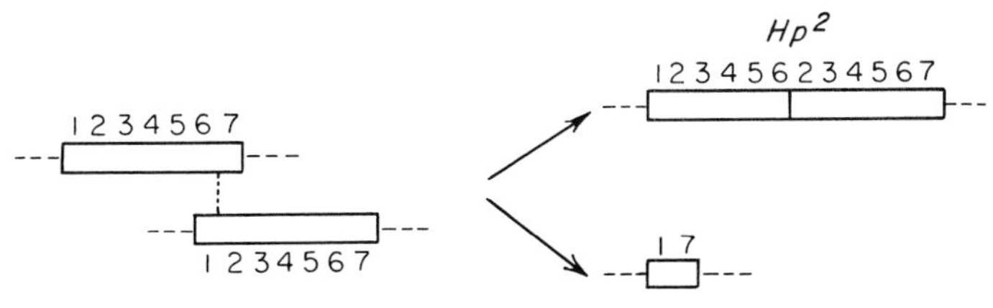

FIGURE 241
Presumed origin of an haptoglobin allele of nearly twice the length of the initial genes. *Left:* Nonhomologous ("illegitimate") recombination between different regions of two alleles of unit length. *Right:* The resulting two new kinds of alleles, one of nearly double length and another, hypothetical, deficient for a large section. The numbers 1, 2, etc., designate the sequences of subunits of the genes.

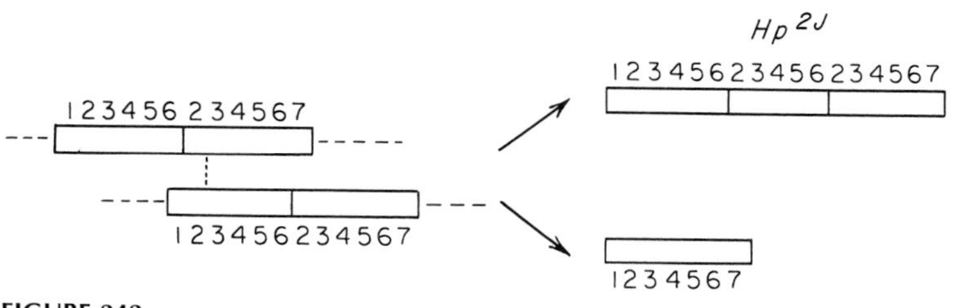

FIGURE 242
Presumed origin of a haptoglobin allele of nearly three times unit length by means of unequal crossing over between two alleles of nearly duplicate length. An allele of unit length is also formed.

in one chromosome pairs with the left half of the other and likewise the two right halves with each other, or "unequally" so that the left half of the gene in one chromosome pairs with the right half of the gene in the other. Both types of pairing would be between homologous sections but in the first case crossing over within the paired sections would result in unchanged alleles in contrast to the second case in which an almost triplicated gene would arise as well as a reverted allele of single length (Fig. 242). A rare *Hp* allele, Hp^{2J}, of triplicated form has indeed been found in several different human populations.

In various organisms, particularly microbes further types of gene mutations have been studied. They consist of insertions of nucleotide pairs into the original sequence of such pairs as well as in deletions of such pairs. These insertions and deletions may involve a single nucleotide pair or more than one. If the number is 3 or a multiple of 3, then, given that the genetic code is written in triplets, insertion or deletion of these pairs will result in production of amino acid chains that differ from the original chain by the insertion or deletion of one or a few amino acids. If the number of inserted or deleted pairs is not divisible by 3, then a "shift in reading frame" usually will result in a very abnormal, "nonsense" polypeptide (see p. 77). There is every reason to believe that these various types of mutations exist in man as well as in microorganisms.

In man, the question of whether mutation occurs when a gene is replicating or when it is not has been approached by comparing the rate of mutation, as related to age, in males and females. The production of mature sperm is preceded all through adult life by a large number of divisions of primordial germ cells, the spermatogonia, but that of mature eggs is restricted to the growth of nondividing cells, the oocytes, laid down in the ovary during the development of the female fetus. If mutations occur only or mostly during gene replication, then their frequency in sperm, but not in eggs, should increase with age. One way of testing this possibility consists in determining the mean ages of fathers and mothers of children with newly mutated genes. If increasing age of the father causes more mutations than increasing age of the mother, then the difference between the two ages should be larger in parents of children with new mutations than in a control group of parents without newly mutant children. Penrose and Vogel have assembled the rather limited data that are available. Some of them were summarized in the preceding chapter (pp. 568–570). No significant increase in the difference between parental ages was found in aniridia and in a number of other mutant traits, but indication of an increase was noted in hemophilia and a definite increase in achondroplasia. Some of these findings are thus compatible with the idea that errors in genic replication may be a cause of spontaneous mutation. It is, however, at least as likely that the great physiological differences between spermatogonia and oocytes are in some unknown way responsible for the different rates of mutation of some of their genes.

It might be thought that a simple comparison of the rate of mutation in men and women, independent of their age, would answer the question of the significance of replication for mutation. Other things being equal, we would expect a higher rate of mutation in sperm than in eggs if replication is an important factor. This indeed has been reported for hemophilia, but some doubts exist concerning the reality of the difference. Even if it is real, more than one interpretation is possible. The phrase "other things being equal" cannot be applied with any certitude to testicular and ovarian cells. They differ not only in the number of cell divisions which precede maturity but also in their differentiation, which undoubtedly is accompanied by differing metabolic events.

The mutations discussed in the last pages are limited to short sections of genic DNA. They are often referred to as "point mutations" although this term is strictly applicable to a specific class, that in which only a single nucleotide pair is changed. Hereditary newness can also be the result of rather large-scale changes in the chromosomes. Trisomy as, for instance, in Down's syndrome consists in the addition of a whole chromosome to the normal diploid set and leads to a mutant phenotype which can be transmitted as if it were due to a dominant gene. Other chromosomal aberrations that may be detected by means of the light microscope are the deficiencies and duplications which lead to the cat's cry and associated syndromes, and translocations which in unbalanced state cause the appearance of other new traits.

Mutant Gene and Mutant Trait. The mutant character which is the external sign of a mutant gene is the result of complex interactions of developmental processes. Although these processes are controlled by genes or their products, they take place largely in the cytoplasm of cells. They may result in movements of groups of cells, as in the folding of cell layers in development, and they include the action of secretions of glands located in one part of the body on the phenotype of another part and the many means by which differentiation and organization of an individual are brought about and maintained.

The "level" of formation of a character and the "level" of the reproduction of any one of the numerous genes which are somehow initially concerned with the development of the character are obviously different. These levels can be at least partly defined by calling them *developmental level* and *genic level.*

In a way, the relation between genes and the characters they control is like that between an assembly line and the automobiles created on it. Changes at any point on the assembly line result in changes in the automobile, but changes in an automobile made by individual adjustment—when it "develops" in the factory, or is repaired—will not affect the assembly line.

Acquired Characters and Genic Stability. If this concept of distant unidirectional relation between the molecular gene and the differentiated character is correct, then it cannot be expected that changes in bodily characters caused by the influence of environmental agents will result in gene mutations. Such somatic changes occur constantly. Skin pigmentation increases or decreases with the amount of solar irradiation received; muscles become more powerful with exercise; learning consists in the acquisition of new skills or new intellectual or emotional attributes. Although these changes occur as a result of reactions in which gene-dependent products play important roles, they take place far from the genic level.

The remoteness of the developmental from the genic level raises a point worth stressing anew: the collaboration of numerous genes in the production of a trait. Many different processes, controlled by alleles at many different loci, are involved in the formation of a muscle of normal size and activity. A mutation at any of these loci may result in a weakened, abnormal muscle. There is neither a specific gene for normal muscle nor one for abnormal muscle. Rather, a large number of genes control reactions that, at early embryonic stages, are not yet part of muscular physiology or structure and whose products are but intermediaries in the interwoven processes that finally lead to the formation of a muscle.

Mutagenic Agents. Evidence that acquired characters are not heritable does not mean that mutations may not be produced by external agents. Such agents have, indeed, been discovered in experiments on various animals and plants. Penetrating ionizing radiation such as X-rays, gamma rays from radium and other radioactive atoms, and neutrons from nuclear reactors are well known mutagens. Many chemicals—mustard gas and various nitrogen- or sulfur-mustard compounds, formaldehyde, nitrous acid, and others—are also mutagenic.

Mutagenic agents produce mutations only in cells on which they act directly: presumably, only if they act directly on the nuclei. Thus, if X-rays are applied to any part of the body, but with the gonads out of their range, no mutations are induced in the gametes, but, if they are applied to the gonads or to the gametes themselves, "artificial transmutations" of genes and chromosomal aberrations result. Similarly, ultraviolet radiation does not cause gametic mutation in most organisms because it cannot penetrate deeply enough to reach the germ cells. But if the gametes are directly exposed to ultraviolet light, mutations are produced. Mutagenic chemical compounds, also, can cause mutation only in cells that they penetrate.

Ever since the discovery of the mutagenic action of radiation in *Drosophila*, the question of artificially increased rates of mutation in man has been debated. It is still unknown whether many of the chemicals that are present in man's surroundings have any noteworthy mutagenic effects on his genes. Though it has been shown that many chemical compounds cause mutations in bacteria, these same compounds, even if present in the blood stream, may never pass through the cellular and nuclear membranes—or if they do, the amounts may be too small to have a similar effect in human germ cells. Only laborious experiments with mammals such as mice can show whether or not we may have reason to expect human mutations to be caused by chemical compounds naturally present in food; substances such as caffeine, nicotine, or alcohol; hydrocarbons from automobile exhausts; insecticides and fungicides absorbed by vegetables and fruit; drugs taken as medicine or for psychic stimulation or relaxation; or drugs contained in meat from animals routinely raised on drug-supplemented feed. It may well be that such substances are now responsible for a significant proportion of seemingly spontaneous mutations and that the mutation rate in man has increased as a result of their presence. Evidence for chemical mutagenesis in man may be the recent increase in the occurrence of cancer of the lung. Since the origin of these tumors is causally related to chemicals in the smoke from cigarettes, and if the tumors are consequences of somatic mutation in the epithelial cells of the lungs, then chemical mutagenesis is active at an ominous rate.

In general, the agents that can cause mutation are not specific for particular loci or groups of loci. This is not unexpected if one considers the structure of genes. All of them are combinations of the same four nucleotide pairs and thus offer relatively little possibilities for specific changes. Nevertheless the sequence of nucleotide pairs varies from gene to gene and some mutagens may well preferentially attack specific combinations within a genic section. Furthermore, some genes are richer in some one base pair than others and certain mutagens interact preferentially with specific bases. For example, ultraviolet light is more highly absorbed by the pyrimidines, thymine, and cytosine than by the purines. Its absorption disrupts certain bonds in these bases with the result that successive pyrimidines in the sequence of nucleotides may become connected with each other to form "dimers," molecular structures formed by two identical single structures, rather than remaining paired with their complementary purine bases at the same levels across the gene molecule. The formation of the dimers may result in a mutant gene.

In any one cell, usually only one or few genes and only one allele of any pair can be made to mutate by irradiation or chemicals. This low yield of mutations per cell is apparently determined by the small amount of irradiation or of the chemical that can be employed without killing the cell; below the lethal level, the mutagenic agent is still relatively "dilute."

ORIGIN OF SPONTANEOUS MUTATIONS

When, in 1927, H. J. Muller discovered the mutagenic effect of X-rays, it was wondered whether the naturally occurring radiation from rare but ever-present radioactive atoms in the soil, air, food, within the body, and from cosmic rays could account for natural, so-called spontaneous mutability. Calculations soon showed that, for *Drosophila* at least, it could not. Although the rate of spontaneous mutation is low, naturally occurring radiation would have to be more than a thousand times as great as it actually is to produce the observed rate of mutation. Consequently, other agents must account for the great majority of spontaneous mutations in *Drosophila*. At present, we know of the two additional classes of such agents mentioned above: chemicals and random energy fluctuation.

We have noted that some experimentally applied chemicals can induce mutations, and there is evidence that radiation exerts most of its effect on the genes indirectly by producing highly reactive chemicals inside cells. It is, therefore, reasonable to suppose that some chemical substances derived from the environment—in food, water, and air—as well as some internal products of normal cellular metabolism, occasionally cause spontaneous mutations.

In contrast to chemical agents as mutagens, energy fluctuations represent physical agents. Since genes are molecules, the laws which govern the stability of molecules should apply to them. From these laws it is known that, given enough energy, bonds between atoms will break and permit the formation of new molecular structures. The high stability of genes seems to signify that a considerable amount of energy is required to effect changes in their molecular structure, but it may be expected that natural fluctuations in energy can be large enough to cause genic changes. These natural fluctuations may be of external origin—for example, temperature oscillations in the immediate environment of a gene—or they may occur inside the gene itself, perhaps as a chance concentration on some specific atomic bond of the interatomic energy that is normally widely distributed over the gene molecule.

Temperature is an important agent affecting the reactivity and stability of molecules. Therefore, a possible source of part of the spontaneous mutation in man may rest on some time-honored customs of human societies which affect the temperature of the gonads. The location of the ovaries within the abdomen makes them largely immune to artificial interference with their temperature, but the testes are exposed to variations in the external temperature which are only incompletely countered by regulatory response of the scrotum and the blood supply. Measurements of the gonad temperatures of clothed and unclothed men have shown those of the former to be more than 3°C higher, and temperatures must rise even more when men bathe in very hot water or take steam baths, as they do in various cultures. There is evidence from *Drosophila* that increased temperature of the testes as well as of mature sperm leads to an increase of mutations. The same may be expected in man.

Are the spontaneous mutations of human genes caused by the same agents which affect the genes of *Drosophila*? We might be inclined to say they are, but a quantitative examination suggests that at least the relative importance of these agents may differ considerably in fly and in man. As stated above, the spontaneous mutation rate in *Drosophila* is far too large to be accounted for by the amount of natural radiation which a gene receives during the life of the fly carrying it. The newly mutant genes found in a mature fly's gametes represent the genetic changes that have accumulated during the brief period from its origin as a fertilized egg to its maturity. Since the life span of a human is much longer than that of a fruit fly, his genes are exposed to natural radiation much longer than those of the fly. Consequently, a much higher proportion of spontaneous mutations in man could be due to natural (or "background") radiation. However, the exact proportion depends not only on the relative amounts of natural radiation received by man and *Drosophila* but also on the relative sensitivity of the genes to radiation. Such simple factors as the oxygen tension within cells and various more complex factors are known to affect the frequency of mutations produced by radiation. We have no knowledge of the radiation sensitivity of human genes. If it is similar to that of the genes of the only mammal for which the results of experiments are available—the mouse—then it is higher than that of the fruit fly. Although it is not impossible that nearly all spontaneous human mutations are due to natural radiation, most geneticists think that only a small proportion of spontaneous human mutation is so caused.

One basis for this opinion is the unlikelihood that the metabolic activity of human cells is so different from that of *Drosophila* cells that it does not occasionally produce chemical mutagens within the cells. Another is that the gonads of a small fly are exposed to external β radiation and perhaps even to some α radiation while those of man are largely shielded from these weakly penetrating particles by overlying tissue. This means that the amount of natural radiation received by human germ cells is less than that received by those of flies in the same period of time.

A further reason for attributing only a small part of the total spontaneous mutation rate of human genes to natural radiation comes from experiments on mice. Extensive tests by W. L. Russell, in Oak Ridge, show that the amount of radiation accumulated during the relatively long life of a mouse cannot account for more than a fraction of one per cent of the spontaneous rate of mutation of seven selected genes. Considering that human beings are exposed to radiation for a much longer time than are mice and assuming that a given amount of radiation causes human genes to mutate as frequently as those of mice, it would seem that the percentage of spontaneous human mutations caused by natural radiation is higher than the percentage for mice. A simple estimate of this percentage can be derived from (1) the rather well-established amount of natural radiation received by human gonads in thirty years and (2) the less well-established dose of artificial irradiation required to produce as many additional mutations as occur spontaneously ("doubling dose").

All people do not receive the same amount of radiation. It increases with the altitude above sea level since exposure to cosmic rays is greater at high altitudes. It depends on whether the rock or soil is rich (e.g., many granites) or poor (e.g., sedimentary rocks) in radioactive atoms of the uranium or thorium series. It is influenced by the type of houses in which people live, since those built of rock containing much radioactive material will expose the inhabitants to higher doses than will houses built of wood or other less radioactive material.

A measure of the genetically relevant dose of natural radiation is the total amount received by an individual during the period from conception until the time when one-half of his children have been produced—a period whose average length in Western populations is about thirty years. In the United States and many other countries this average background dose is approximately 3–4r (Table 69); the r unit is a measure of ionizing radiation (see p. 592). In a relatively small area in Kerala, southwestern India, the presence of highly radioactive monazite sands along the coast increases the exposure ten- to fifteen-fold.

In *Drosophila* and in mice, experiments suggest that approximately 200r units of artificial irradiation administered chronically, i.e., at low intensity over a prolonged period, to spermatogonia and oocytes are required as the doubling dose. The value of 200r is chosen as a reasonable estimate within a probable but not definitively established range of 120–320r. If it were as-

TABLE 69
Average Doses of Natural Radiation Received by the Gonads during Thirty Years. (The doses are given in rem units which correspond closely to r units.)

Source	*Rem*
EXTERNAL	
Cosmic rays	1.5
Terrestrial radiation	1.5
INTERNAL	
Natural radioactive atoms (K^{40}, C^{14}, etc.)	0.78
Total	3.78

sumed that in man the same chronic does of 200r would also lead to a double dose of mutations, then we would have to conclude that only 1.5–2 per cent of spontaneous mutations are caused by natural radiation, since 3–4r, the average background dose, is 1.5–2 per cent of 200r.

It must be stressed that the assumptions underlying these considerations are not strongly supported by facts. For instance, the assumption that the doubling dose in man is like that in *Drosophila* and the mouse may be far from the truth. In addition, evidence regarding the mutagenic effects of radiation on the seven mouse genes studied shows that the mutagenic effects of radiation do not parallel those observed in spontaneous mutations. Many more of the induced than of the spontaneous mutant alleles are recessive lethals. Also, the relative frequencies of mutations at the seven loci are different for induced and spontaneous mutations. In other words, radiation induces a different spectrum of mutations from that arising spontaneously. Calculating a quantitative value for the doubling dose does not take account of the qualitative differences between spontaneous and induced mutations. "We may be talking about doubling the frequencies of apples when we are measuring oranges" (Russell). It is important to keep this possibility in mind. Nevertheless, the concept of a doubling dose remains useful. To modify Russell's comment we may be talking about doubling the frequencies of fruits regardless of their being apples or oranges. In any case, the doubling dose has been found to be different for different genes in the mouse and it is not known whether the average doubling dose for the seven loci is indicative of the average of all loci.

Instead of postulating that the chronic doubling dose for human genes is the same as for *Drosophila* and mice, it would be possible to obtain an independent estimate if the mean rate of spontaneous mutations per locus and the mean mutagenic effect of an r unit were known. We have earlier seen that the mean rate of spontaneous mutation per locus in man is perhaps one per million gametes. In the mouse the mean rate of mutation induced by chronic application of 1r is perhaps about 6 per 100 million (6×10^{-8}). If this value should also apply to man, then it would require 16.7r to induce one mutant allele in a million ($16.7 \times 6 \times 10^{-8} = 1 \times 10^{-6}$)—i.e., as many as occur spontaneously. The value of 16.7r for the chronic doubling dose is considerably lower than the often used value of 200r. A reason for the difference lies in the difference between the formerly prevalent use of the estimate of 1.5 spontaneous mutations per human locus in a hundred thousand gametes and the newer estimate of only one in a million. If the spontaneous mutation rate is actually lower than assumed in the past a lower number of r units or a lower mutagenicity of the r unit is required to match the spontaneous rate. If the radiation-induced rate per r unit differs in man and mouse then the doubling dose may be either smaller or larger than 16.7r, including the high and somewhat conventional estimate of 200r, which assumes that the chronic doubling dose is the same in *Drosophila*, mouse, and man. Arbitrarily, in agreement with an inclination of other geneticists, we shall use the value of 200r for the chronic doubling dose in man. It should be emphasized, however, that even the low value of 16.7r would mean that the natural background would account for only a minority of spontaneous mutations, 3–4r being 18–24 per cent of 16.7r.

The concept of a doubling dose has had its critics. Not only is there no particular merit in selecting a factor of 2 relative to the spontaneous rate, instead of a factor of 1/10, 5, or 10, but the concept implies that the proportions of various kinds of mutations induced by radiation are the same as the proportions of mutations occurring spontaneously. There is evidence, from studies on the mouse, that this is not so.

RADIATION GENETICS

In order to discuss the significance of radiation as a cause of induced mutations, some general results of radiation genetics will first be presented, neglecting the minor differences within and between X-rays, γ-rays, and neutrons.

The amount of radiation received by irradiated tissues (called "dose") is measured in terms of one of several units. The "r," after Roentgen, defines the dose of X- or γ-rays by the amount of ionization produced. In tissue, 1r causes about two ionizations in a cubic micron (μ^3) or, stated differently, about 1.6×10^{12} ion pairs per cubic centimeter. Although this is a large num-

TABLE 70
Doses of Radiation Received by the Gonads during X-ray Examination of Various Parts of the Body.

Parts Examined	*Dose in Thousandths of r (= mr) Received by the Gonads*		
	Male	Female	Fetal
Head	0.8; 0.6–1.0	0.2; 0.2–1.5	0.2
Teeth	4.7; 8	0.8; 2	0.8
Chest, large film	0.4; 1.2	0.1; 0.3	0.1
Chest, mass miniature	0.2; 1	0.1; 3	0.1
Abdomen	69; 200	200; 500	580
Pyelogram (kidney, ureter)	486; 2,000	1,290; 1,200	3,210
Skeleton of pelvis	1,100; 2,000	210; 1,000	800
Skeleton of pelvis-fluoroscopy	6,000	3,000	–
Spine, thoracic	22	15	15
Spine, lumbar	129	713	713
Obstetric examination (in pregnancy)	–	1,280; 260–2,500	2,680; 400–4,000

Source: Selected from Osborn in *The Hazards to Man of Nuclear and Allied Radiation,* 1956; Rep. Int. Comm. Radiol. Protection, *Physics in Medicine and Biology,* **2,** 1957.

Note: The first dose listed refers to the report from one hospital in Britain, the second dose to reports from the United States. If only one dose is listed, it refers to the report from Britain, except for the data on fluoroscopy of the skeleton of the pelvis, which are from the United States. The British data combine estimates for photographic records and fluoroscopic examination, the United States data refer to photographic records unless otherwise stated.

ber, it is very small indeed when compared with the number of atoms in 1 cc of tissue, which is of the order of 10^{23} or 100,000,000,000 times as large. The "rad" is a unit of absorbed dose, equivalent to 100 ergs per gram: this amount of energy is slightly larger than that absorbed after irradiation of 1r.

Genetic damage to future generations is produced only when the gonads or gametes or, in very young embryos, the cells destined to produce gametes are irradiated directly. Irradiation of various parts of the body with X-rays for medical purposes may lead to some degree of exposure of the gonads. Such exposure can often be reduced by using highly sensitive photographic plates and by protective shielding of all areas not of immediate interest. In general, in such irradiation the ovaries, which are protected by the overlaying pelvic tissues, receive a lower dose than do the testes. Table 70 gives some data on estimated gonad exposures in X-ray photography. Most of these data were obtained in a British hospital where particular care was taken to reduce gonad doses to the minimum. In other hospitals or in the offices of individual physicians the doses may be higher. Exposure to X-rays for direct observation on a fluoroscopic screen always requires much greater doses.

EXPOSURES TO RADIATION

Medical Exposure. For most persons, the total amount of irradiation received during a lifetime, as a result of medical diagnosis and treatment, in any part of the body is less than 50r. Not many persons ever receive as much as 100r, although occasionally much larger doses have been administered.

From the point of view of later generations, only irradiation which hits the gonads is of importance. It is obvious that even exposure of the gonads is of no consequence after an individual has ceased to have children. The average dose to the gonads received in connection with medical purposes varies from person to person according to different practices in different countries and different personal health problems. In the United States, the average dose to the gonads is estimated to be between 1- and 2r per 30-year period. Considerably lower doses are received during 30 years in several European countries. Thus, medical (primarily diagnostic) exposures account for about one-third again as much irradiation as that received from natural background radiation. There is an important difference between natural background irradiation and medical irradiation. The former is of low intensity and is received continuously over a prolonged period. It is chronic irradiation. The latter is administered with high intensity during short periods of exposure. It is acute irradiation. Presumably, the acute exposures are more mutagenic than the chronic ones, as will be seen below.

Occupational Exposure. Exposure to radiation is an occupational hazard to certain groups of persons: radiologists and other physicians; X-ray technicians; investigators using radioactive isotopes; and workers in uranium mines, nuclear-energy plants, and certain factories.

When it became known that radiation can cause harm to individuals, such as burns and anemia, occupational protective measures were introduced to limit the amount of radiation reaching the body. This protection is primarily in the form of lead or concrete shields, which absorb most of the radiation. On the basis of experience, 0.1r per day was generally set as an upper permissible dose, that is, a dose to which the whole body of an individual can be exposed for a long time without showing any ill effects. Since 1956, various international and national organizations have lowered the upper permissible dose to 0.3r in any single week. In addition, a further limiting formula has been proposed, according to which artificial exposure should not begin before the age of eighteen and the accumulated exposure after that age should be limited to a total equal to 5r per year.

It is somewhat doubtful whether even this permissible dose may not gradually lead to physiological damage. Evidence from several mammalian species has shown a shortening of the life span of individuals who have been exposed daily to very low amounts of radiation. Also, it seems likely that leukemia and perhaps other types of cancer may be induced by even very low doses of radiation. These findings have suggested that the definition of the permissible dosage should be reduced, perhaps by a factor of ten. Mutations will be induced even by a very low permissible dose. If a chronic doubling dose of 200r is received, the gametes of a worker who accumulated radiation at a yearly rate of 5r for 20 years would possess 50 per cent more mutations than those occurring spontaneously.

Military Exposure. Nuclear explosions result in the nearly instantaneous production of immense amounts of highly penetrating radiation and may cause contamination by radioactive fallout of large areas. Within the area reached by direct ionizing radiation, great amounts of radioactivity are induced. The dose of irradiation received by a person depends on his distance from the center of the explosion. Approximately half of the individuals who were near enough the center to receive an acute total body irradiation of about 450r or more would not survive. Consequently, the highest rate of induced mutations that could be transmitted by a survivor corresponds to about 450r.

Radioactive fallout following a nuclear explosion above ground potentially can cover with high concentration hundreds or thousands of square miles, or, in lower concentration, a whole hemisphere and even more. The above-ground tests of nuclear weapons by the United States, Russia, England, France, and China have resulted in the production of large amounts of radioactive fission products, much of which are present in the upper layers of the atmosphere. These products slowly fall to earth. The total dose of radiation from fallout will be delivered over many years, particularly that fraction of the dose that is due to the radioactive decay of carbon-14. In 1961, after 8 years of testing and assuming—mistakingly!—that no further above-ground tests would be conducted, the sum of the doses that had been and would be delivered to the gonads of a person, the "dose commitment," was estimated to be about 0.1r. By the year 2000 only about 40 per cent of this dose will have been received. These data on radiation received by the gonads do not, however, tell the whole story. Ingestion of radioactive strontium-90, cesium-137, and carbon-14 will result in genetic changes in somatic cells.

It is possible that the amount of radiation, exclusive of that from military sources, to which man is exposed will increase. The use of radiation for medical purposes will spread to many countries, and peaceful applications of nuclear energy will create new sources of radiation, including those from radioactive waste products. Several proposals have been made to keep the mean exposure of individuals in large populations to, at most, 10r per 30 years, a figure that sometimes includes background radiation and sometimes is given in addition to it. This amount of radiation, as we shall see, has far-reaching

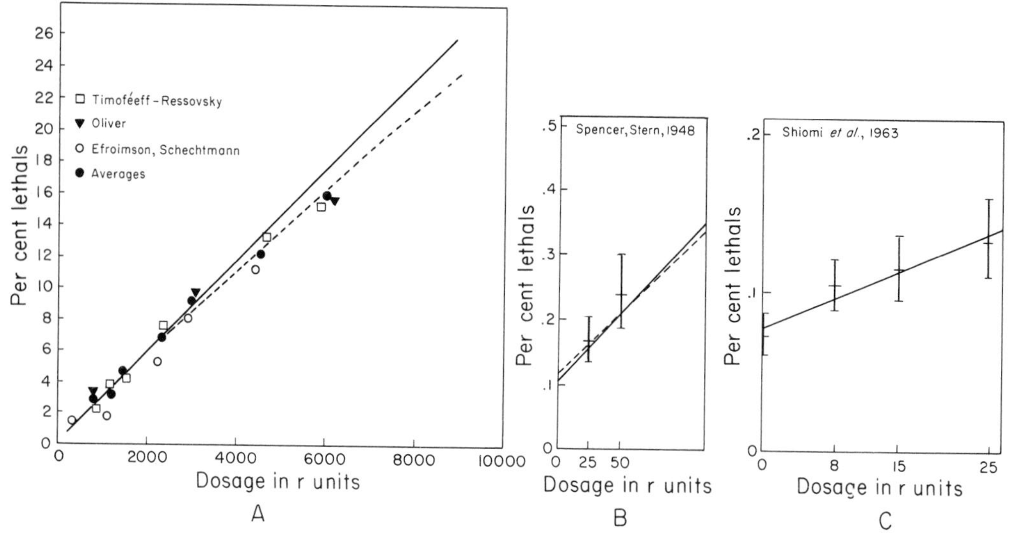

FIGURE 243
The relation between X-ray dosage and frequency of lethal mutations induced in the X-chromosome of the sperm of *Drosophila melanogaster*. **A.** The relation for high dosages according to the experiments of several investigators. The broken line represents the expected curve if some necessary corrections are applied to the straight solid line. **B.** The relation for the low dosages 25r and 50r. The solid and broken lines represent the type of linear relationship obtained if certain data from other experiments with higher dosages are included. The data points for the two dosages are shown as small horizontal marks. The ends of the vertical lines passing through these marks indicate the range within which the observations may vary statistically in a nonsignificant degree. **C.** The relation for the low dosages 8r, 15r, and 25r. The differences between B and C in the control rates and the slopes of the lines are probably due to somewhat different scoring of lethals in the two experimental series. (A, after Timoféeff-Ressovsky, Zimmer, and Delbrück, *Nachr. Ges. Wiss. Göttingen, N. F.* **1,** 1935; B, Spencer and Stern, *Genetics*, **33,** 1948; C. Shiomi et al., *J. Radiat. Res.*, **4,** 1963.)

consequences, particularly if the exposure occurs at a high level of intensity. We are still considerably below such exposures. In the future it may well be decided that a reduction of this so-called permissible population dose is imperative.

Dose and Intensity of Radiation and Frequencies of Mutation. It has been established for a variety of experimental organisms that the number of mutations induced by radiation is proportional to the dose. This proportionality has been proven to hold over a wide range of dosages. Figure 243 shows that, for *Drosophila*, the relation is essentially linear over the range 8–12,500r (insects, unlike mammals, can survive after exposure to many thousands of roentgens). Scoring for a special dominant mutant phenotype in *Drosophila*, even 5r produced an increased rate of mutation as compared to controls. There is then no threshold amount or radiation below which no mutations are produced. This very important observation is in contrast to those on the biological effects of many chemical substances, including drugs, whose effectiveness is zero at low concentration but which become harmful or lethal to man above a threshold dose. It should be added that for exposure of the spermatogonia of the mouse the relatively few data on induced mutations at doses of 0, 300, and 600r fit the linear relation, but that a dose of 1,000r produces fewer mutations than expected. It is believed that the decreased mutagenicity at 1000r is caused secondarily. If a population of spermatogonia should include cells in metabolic stages that increase both their mutability and their tendency to be killed by irradiation, then the high dose of 1000r may selectively destroy these cells, resulting in lowering of the overall mutation rate.

In most of the experiments on which the proportionality rule is based, irradiation was applied for a short period of seconds or minutes. Irradiation of *Drosophila* sperm has shown that it makes no difference to the yield of induced mutations whether a given dose of radiation is applied in a single acute exposure, is fractionated into two or more exposures separated by days or weeks, or is slowly and continuously given over a long period. Mutations in *Drosophila* were obtained at the same rate per roentgen whether the sperm was irradiated chronically with less than one five-hundredth (1/500) of 1r per minute or acutely within a few minutes at an intensity some 200,000 times higher. It is noteworthy that the low-intensity irradiation resulted, on the average, in a single track of ionization transversing a sperm head no more frequently than once in 30 hours.

In contrast to the finding of equivalence in mutagenic effects of acute and chronic irradiation of *Drosophila* sperm, Russell found (in 1958) that a dose applied chronically, at low intensity, to mouse spermatogonia yields significantly fewer mutations than the same dose applied acutely, at high intensity (Fig. 244). Chronic exposure at a rate of 90r per week produced only one-fourth the number of mutations produced by acute exposure at the 10,000

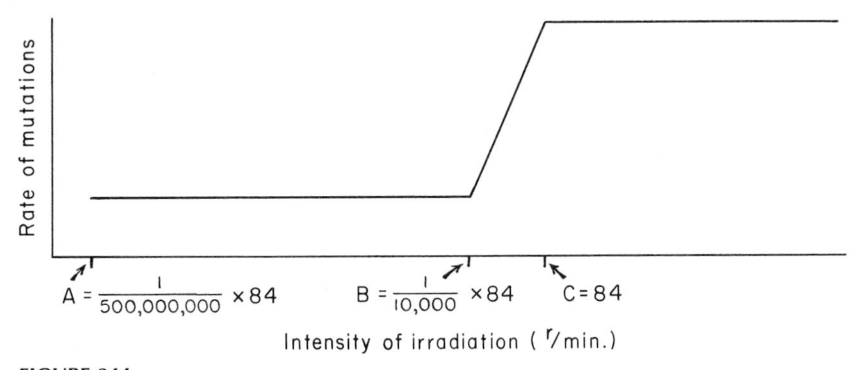

FIGURE 244
Dose-rate effect on rate of mutation in spermatogonia of the mouse. (Based on data of Russell, Russell, and Kelly, 1958.)

TABLE 71
Mutation Frequencies in the Mouse at Various Dose Rates.

Dose rate (r/min)	Mutation Frequency	
	Spermatogonia	Oocytes
90	high	very high
9	intermediate	–
0.8	low	intermediate
0.009	low	very low
0.001	low	–

Source: Russell, 1964.

times higher rate of 80–90r per minute. A similar situation, also in the mouse, seems to prevail for chronic versus acute irradiation of immature female gametes (oocytes). In detail, male and female germ cells, respond somewhat differently to various dose rates. This is shown in Table 71, where the mutagenic effects of the different dose rates are given in qualitative rather than in quantitative terms. Jointly, the lesser efficiency of chronic irradiation in both sexes, is approximately one-sixth of that of acute irradiation. Since, however, the data on male germ cells are more extensive than those on female germ cells, the value of one-fourth for the latter will be employed in the subsequent discussions of the dose-rate effect. Since the discovery of the effect in mice it has also been found in other organisms, notably in the silkworm (by Tazima) and in a wasp species.

Interpretations of the dose-rate effect are centered on recent discoveries concerning the existence of repair mechanisms which involve intracellular enzymes. If radiation has changed one strand of a section of a gene, then the "damaged" piece may be excised and replaced by a normal piece. Low-intensity doses may frequently permit a cell to repair an induced genetic damage while high-intensity doses may cause so many gene-lesions that the repair processes cannot take care of all of them or that the repair mechanisms themselves are damaged. A striking example of a genetically controlled repair mechanism in man is provided by the recessive condition xeroderma pigmentosum (Fig. 245). The skin of affected individuals is highly sensitive to ultraviolet light with the result that severe damage is produced that is followed by development of multiple cancers. When tissue cultures of the skin of normal individuals and of an affected man were irradiated with ultraviolet light, it was found that the DNA of the cells frequently formed bonds between two adjacent pyrimidines resulting in dimers (double pyrimidines; Fig. 246). These abnormal molecular parts were readily excised by normal cells, but cells of the affected man did so at less than one-tenth the normal rate. Further-

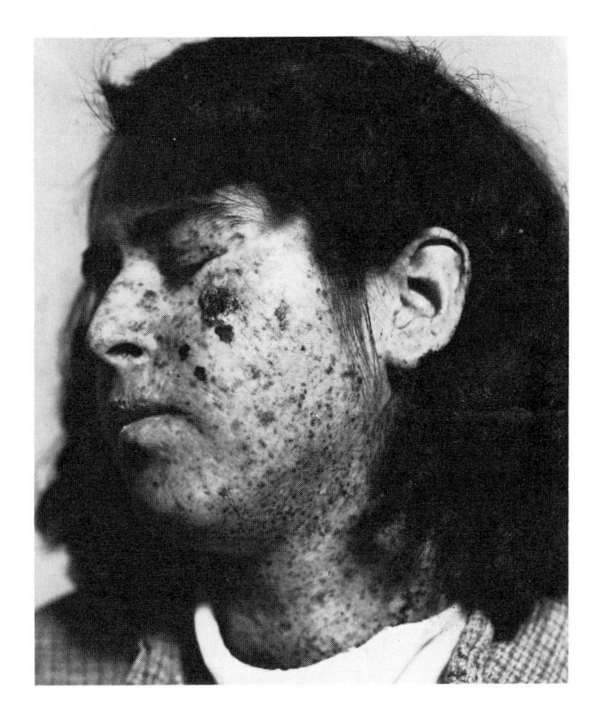

FIGURE 245
Xeroderma pigmentosum. (Original from Dr. F. Ronchese.)

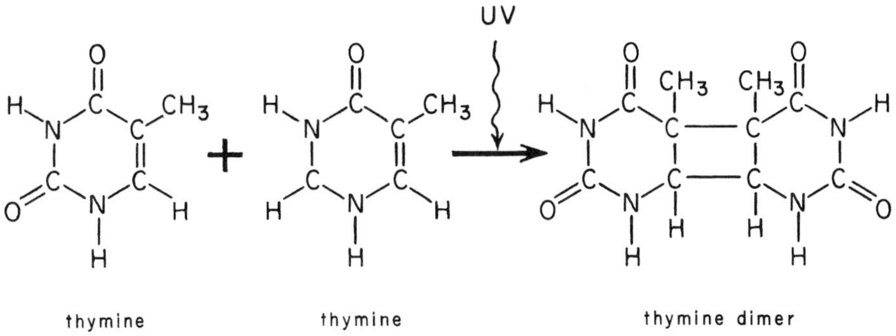

UV

thymine + thymine → thymine dimer

FIGURE 246
Formation by ultraviolet light of a thymine dimer from two thymine bases.

more, after irradiation normal cells exhibit an increase and then a decrease in the number of single-strand breaks in DNA, whereas the affected cells do not exhibit these changes, which are steps in the excision of the dimers. Cells of the xeroderma pigmentosum genotype seem to be defective in an enzyme which is involved in breaking abnormal DNA strands. Without such breakage no excision of the abnormal dimers and no subsequent repair can occur.

If data from mice afford our best clue to the situation in man, then the discovery of mutagenic differences of chronic versus acute radiation in mouse spermatogonia is of greatest importance. Much of human exposure consists in an accumulation of small doses, but until Russell's discovery all considerations of human hazards were based on the assumption that the acute-exposure data were equally applicable to chronic irradiation. It is now necessary to consider the effects of acute and chronic irradiation separately.

Will a further reduction below 90r per week in the intensity of radiation lead, in studies on mice, to a further decrease in the number of mutations for a given dose? Or will the same number of mutations be recovered at all degrees of low-intensity exposures? Experiments with irradiation of lower intensity than 90r per week show that the latter possibility is realized (Fig. 244). Just as the mutagenic effect of irradiation with higher-intensity doses than 84r per minute is independent of intensity so is the effect below one ten-thousandth of 84r per minute. There seems to be no threshold below which low-intensity irradiation ceases to be mutagenic.

Chromosomal Breakage. Chromosomal breakage caused by acute irradiation has been investigated in various plant cells; in embryonic nerve cells of

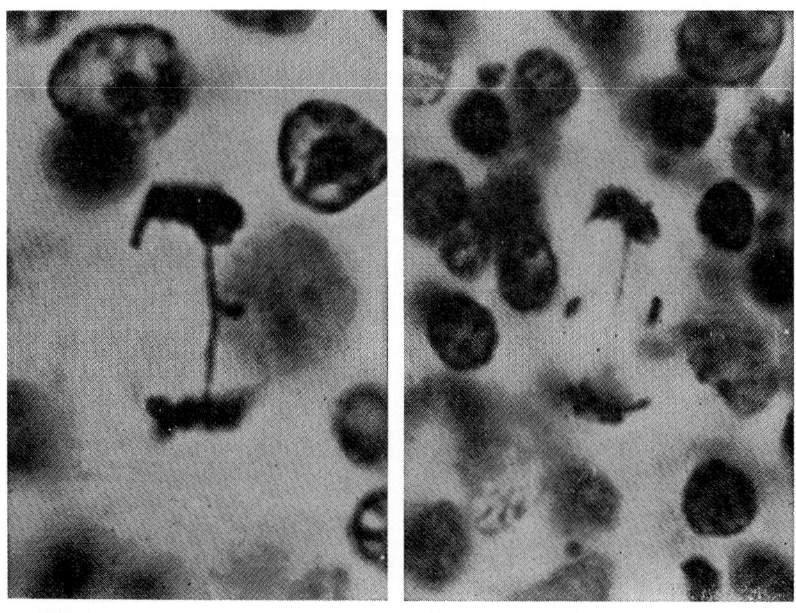

FIGURE 247
Chromosomal breakage, after irradiation with X-rays, of the white-blood-cell-forming lymph tissue of the rat. *Left:* Chromosome bridge and a fragment to right of bridge. (The fragment and bridge are separate from each other and lie in different planes, but the camera makes them appear close together.) *Right:* Chromosome bridge and two fragments, to left and to right of bridge. (Original photomicrographs from B. Thomas Stepka.)

grasshoppers, which can be studied in tissue cultures; in the white-blood-cell-forming lymph tissue of the rat (Fig. 247); and in human tissue cultures. In many cases a simple proportionality between frequency of breakage and dose of irradiation has been found down to doses as low as 3r and 1r. Figure 248 shows, for cells of grasshoppers, that, as is also true for gene mutations, even a dose as low as 8.3r applied for about 2 seconds causes chromosomal breakage, and that at 125r applied for 38 seconds 5.35 per cent of all chromosomes were broken. How great this effect of irradiation is becomes particularly clear if the percentages of chromosomes broken are translated into terms of *cells* damaged. Since each cell has many chromosomes, its chance of being affected is much higher than that of a single chromosome. Accordingly, it was found at 8.3r that 8 per cent of the cells had broken chromosomes, and at 125r that as many as 70 per cent had broken chromosomes. Studies by Puck of human cells in culture suggest that acute exposure to 20r or even less may be sufficient to induce at least one visible chromosome break in each cell.

Dose and Intensity of Radiation and Frequencies of Chromosomal Aberrations. It was indicated in the preceding paragraph that there is proportionality between the frequency of chromosomal breakage and dose of radiation, and independence from the duration or fractionation of a given radiation. It must now be pointed out that this does *not* imply that the frequency of all types of induced chromosomal aberrations follows the proportionality rule. If only one break is induced in one chromosome of a cell, then, obviously, only chromosome fragmentation can be observed. Translocations and inversions cannot be induced, since they require the occurrence of at least two breaks. Since with low dosages the probability of breakage is low, the probability of two breaks—which is the square of the probability of a single break—becomes negligible. An increase in the dose results in a proportional increase

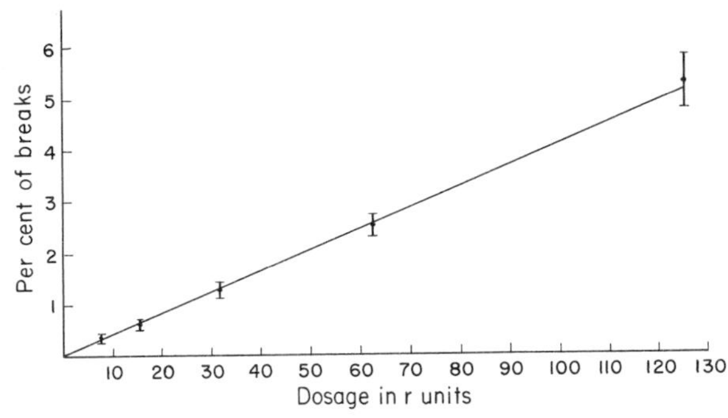

FIGURE 248
The relation between X-ray dosage and frequency of induced chromosomal breaks in cells of a grasshopper. (Carlson, *Proc. Nat. Acad. Sci.*, **27**, 1941.)

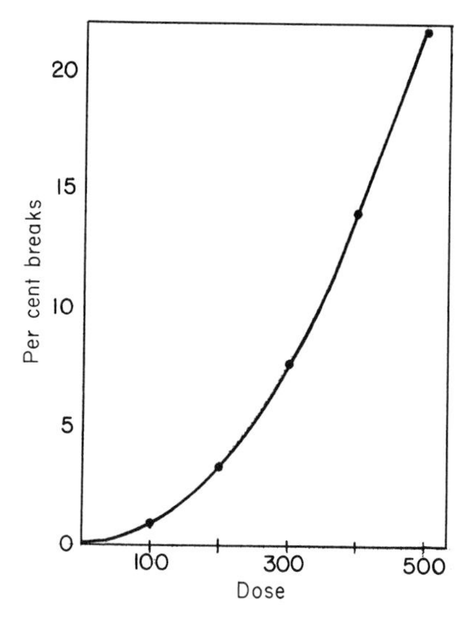

FIGURE 249
The relation between X-ray dosage and frequencies of induced two-break chromosomal aberrations in the plant, Tradescantia. (After Sax, *Genetics*, **25**, 1940.)

of single breaks but in an exponential increase of multiple breaks. For instance, a dose which produces single breaks in 1 per cent of irradiated cells will produce two breaks in approximately 1 out of 10,000 cells. Doubling the dose doubles the total number of breaks (2 per cent) but quadruples the number of cells with two breaks ($[2 \text{ per cent}]^2 = 4$ in 10,000). In agreement with theory, the frequency of translocations or other aberrations resulting from two breaks shows a dependence on dose which is exponential instead of linear; that is, a curve for the frequency of translocations remains close to the zero level for low doses, and for higher doses it rises at an accelerating rate (Fig. 249).

The dependence of translocations and other aberrations on the occurrence of two or more breaks implies that fractionation of a dose, or irradiation over a prolonged period, does not necessarily yield as many of these chromosomal aberrations as short, concentrated radiation. If two breaks are produced more or less simultaneously in a cell, exchange of broken ends may lead to a translocation. However, if two breaks are produced one after the other, a translocation can result only if the chromosome fragment resulting from the first break is still available for union when the second break takes place. From studies of chromosomal breaks induced in mature sperm of *Drosophila*, it has been concluded that broken ends may retain their ability to fuse until fertilization. Other studies on breaks induced in immature germ cells or somatic cells, particularly of plants, show that broken ends soon become unavailable for translocations or similar aberrations. The ends either reunite in the original arrangement, or their duplicated sister strands unite, or they lose the ability to fuse.

These facts are of significance in judging the effect, on human chromosomes, of irradiation concentrated into a short time versus exposure over a long period. While it would perhaps not make a great difference in the frequency of single-hit chromosomal aberrations whether an individual, or individuals, in a population receive 200r within a fraction of a second from say, a nuclear accident or over a period of months from heavy radioactive contamination, the frequency of induced translocations and other multiple-hit events might be considerable in the first instance and negligible in the second.

Estimates of Mutagenic Effects of Radiation. As stated earlier, it is necessary to consider the effects of acute and chronic radiation exposure separately. Obviously, there exists a continuous range of radiation intensities, so it must be specified what intensities are to be called acute and what are to be called chronic. In practice, a distinct separation is possible. Medical exposures are acute and usually require intensities of at least several roentgens up to occasionally thousands per minute. Acute exposures in nuclear-reactor accidents may be of very high intensities, and even higher intensities may result from the immediate radiation of nuclear explosions. In contrast to such acute, high-intensity exposures are legally permissible chronic exposures occurring in certain occupational groups and those resulting from natural background radiation, from radioactive contamination by atomic waste, and by fallout. In general, exposures to these sources of radiation are of very low intensity, consisting of minute fractions of 1r per minute.

Acute Exposures. For many years the only quantitative information on the frequency of induced mutations was based on *Drosophila* experiments. They indicated that a high-intensity dose of 1r applied to spermatogonia causes mutation of about 1.5 out of 100,000,000 genes ($= 1.5 \times 10^{-8}$) from a normal allele to one which either produces an easily discernible different phenotype or acts as a recessive lethal. It was often wondered whether this figure could justifiably be used for human genes. This became more doubtful after it was found that the corresponding frequency for the mouse is several times higher. Russell's data indicate a more than 15-fold greater effect of 1r acutely delivered on the mouse than on *Drosophila*—i.e., 25 mouse mutant alleles in 100,000,000. Other investigators did not find as high a number of induced mutant alleles but all agree that the few selected mouse genes studied are more sensitive to radiation than the few selected *Drosophila* genes.

Estimates of the acute doubling dose in man pose an unresolved problem. On the basis of former estimates, the spontaneous human mutation rate per gene was judged to be approximately 1.5×10^{-5}. Combined with the assumption that the induced rate per roentgen (2.5×10^{-7}), as derived from data on acutely exposed male mice, is applicable to man, a value of 60r for the acute human doubling dose was obtained [$(1.5 \times 10^{-5})/(2.5 \times 10^{-7}) = 60$]. This value is close to those for *Drosophila* and mice. However, if we accept Stevenson and Kerr's more recent estimate of 1×10^{-6} for the spontaneous human rate, a value of only 4r for the acute doubling dose is obtained [$10^{-6}/(2.5 \times 10^{-7}) = 4$]. This seems to be an unlikely small value. It may suggest that the rate per r of induced mutations in man is not like that in mice but is much smaller—more similar to the rate in the insect *Drosophila* than in the mammal mouse.

Both the fly and the mouse data are based on mutations induced in spermatogonia. For the mouse, the available information suggests that the rate per roentgen of induced mutations in the oocytes is similar to that in the spermatogonia. For post-spermatogonial stages, including mature sperm, the rates per roentgen are twice as high. In man, the post-spermatogonial period of the germ cells lasts only a few weeks. This is a very short time in comparison to the period as immature sex cells—a length of time which comprises most of the life of an individual. Therefore, for many purposes, it is realistic to use the spermatogonial rates of mutation. We should remember, however, that the offspring produced from a sperm which, in the mature state, had been exposed to irradiation has a greater likelihood of carrying induced mutations than does the offspring from a sperm which, at the time of exposure, was still in a spermatogonial stage.

If for the sake of the argument it is assumed that a human gamete possesses 10,000 mutable genic loci whose mean rate of induced individually recognizable mutations per roentgen delivered acutely is similar to that of *Drosophila*, namely 2×10^{-8}, then 2 per 10,000 of all gametes ($10{,}000 \times 2 \times 10^{-8}$) would carry a mutation induced by acute irradiation with 1r. About twice as many zygotes (or 4 per 10,000) would have such an induced mutation—assuming for the moment that the sensitivities of male and female germ cells are alike. An irradiation with 10r would raise these figures proportionally to 2 per 1,000 of all gametes and to 4 per 1,000 of all zygotes.

In the next chapter we shall make use of the estimate that an irradiation dose of 10r given at high intensity to both parents of each of 1,000 children causes 4 of the children to carry an induced mutation, visibly or invisibly. It cannot be emphasized too strongly that at the present state of knowledge the estimate is very unreliable.

Chronic Exposures. As stated earlier, the best estimate of the relative mutagenic efficiency of chronic as compared to acute exposure comes from data on irradiated spermatogonia (in the mouse). It is one-fourth. Using this estimate, we can calculate that a low-intensity dose of 1r imposed on all parents would lead to one out of 10,000 zygotes possessing a new induced mutation;

and 10r given to both parents of each of 1,000 children would also cause one mutation.

In this chapter the numerical uncertainties of radiation mutagenesis have been stressed. It should, however, not be overlooked that studies of the induction of mutations by ionizing radiation have given important new knowledge. Rates of induced mutations have become known, differences of effects of chronic versus acute radiation have been discovered, the concept of doubling doses has been introduced. On these bases the following chapter will discuss the genetic hazards of radiation.

PROBLEMS

183. A deficiency is induced in the chromosome of a sperm cell. An individual who receives this chromosome is abnormal. What types of offspring, and in what proportion, may the affected individual produce?

184. Assume that all parents of ten million children had been exposed to a dose of 25r of acute irradiation. Assume further that (i) the frequency of induced mutation from A to a (albinism) is 2×10^{-7} per r, and that (ii) the frequency of the a allele in the panmictic population is $q = 0.00707$.
    (a) How many albinos would have been born if the parents had not been irradiated?
    (b) How many albinos will there actually be among the offspring of the exposed population?

185. Construct a table showing the minimum and maximum percentages of individually recognizable mutations induced by:
    (a) acute exposure to 20r, and
    (b) chronic exposure to 40r. Assume that the mutability of human genes irradiated acutely by 1r is as low as 10^{-9} or as high as 10^{-6}, and that the number of mutable loci is as low as 5,000 or as high as 20,000.

REFERENCES

Crow, J. F., 1961. Mutation in man. Ch. 1, pp. 1–26 *in* Steinberg, A. G. (Ed.), *Progress in Medical Genetics*. Grune and Stratton, New York.

Muller, H. J., 1927. Artificial transmutation of the gene. *Science*, **66**:84–87.

Muller, H. J., 1954. The nature of the genetic effects produced by radiation. Pp. 351–473 *in* Hollaender, A. (Ed.), *Radiation Biology*, vol. 1. McGraw-Hill, New York.

Muller, H. J., 1954. The manner of production of mutations by radiation. Pp. 475–626 *in* Hollaender, A. (Ed.), *Radiation Biology*, vol. 1. McGraw-Hill, New York.

Russell, W. L., 1964. Evidence from mice concerning the nature of the mutation process. Pp. 257–264 *in* Geerts, S. J. (Ed.), *Genetics Today*. Pergamon, Oxford. (Proceedings of the Eleventh International Congress of Genetics.)

Russell, W. L., 1968. Repair mechanisms in radiation mutation induction in the mouse. *U.S. Brookhaven Nat. Lab. Symp. Biol.*, **20:**179–189.

Russell, W. L., Russell, L. B., and Kelly, E. M., 1958. Radiation dose rate and mutation frequency. *Science*, **128:**1546–1550.

Smithies, O., 1967. Antibody variability. *Science*, **157:**267–273.

Uphoff, D. E., and Stern, C., 1949. The genetic effects of low intensity irradiation. *Science*, **109:**609–610.

Vogel, F., 1964. Mutations in man. Pp. 833–850 *in* Geerts, S. J. (Ed.), *Genetics Today*. Pergamon, Oxford. (Proceedings of the Eleventh International Congress of Genetics.)

Vogel, F., 1969. Point mutations and human hemoglobin variants. *Humangenetik*, **8:**1–26.

Vogel, F., and Röhrborn, G. (Eds.), 1970. *Chemical Mutagenesis in Mammals and Man.* 519 pp. Springer, New York.

24

THE GENETIC HAZARDS OF RADIATION

Most spontaneously occurring mutations with clearly noticeable effects must be classified as abnormalities, either of form or of function. This has been demonstrated in extensive animal and plant studies and applies also to the human mutations listed in Table 66 (p. 563). In *Drosophila* the largest group of induced mutations is made up of lethals (this is true also of spontaneous mutations); another large group of "visible" mutations shows a considerably lower-than-normal viability; and only a very small group of visible mutations are as fit as or fitter than the normal. These findings are to be expected, since evolutionary selection in any species is likely to accumulate a majority of genes which work together harmoniously and most substitutions of mutant alleles for these genes result in less-than-perfect functioning of the developmental and metabolic machinery.

The Deleterious Effects of Many Mutant Genes. On the basis of the actual findings on fitness of mutants and evolutionary considerations, it is generally assumed that almost all human mutations are unfavorable. Nevertheless, this may be an exaggerated view. The mutant genes which are easily recognizable are, by definition, relatively striking in phenotypic effect and thus liable to be abnormal. Mutant alleles which produce effects similar to the norm usually escape notice. It is unknown how frequent are mutations to such normal isoalleles. They may form a considerable fraction of all mutations. They may not be disadvantageous to their bearers, and in some cases they may be advantageous. R. A. Fisher has described the probable average deleteriousness of a mutant allele by saying that when its effect is extremely minute, it has "an almost equal chance of effecting improvement or the reverse; while for greater change the chance of improvement diminishes progressively, becoming zero, or at least negligible, for changes of a sufficiently pronounced character."

Selective advantage or disadvantage is not an attribute of an individual gene. The phenotype on which selective properties are based is the result of interaction of numerous genes. Some genes may load their bearers with a selective disadvantage, whatever their genetic background; but many genes may form unfavorable combinations within some genotypes, neutral ones within others, and even favorable ones within still others. Thus, a gene favoring growth within a polygenic system determining height may appear unfavorable when the background tends toward excessive height, neutral when the background tends toward intermediate height, and advantageous when the background tends toward very low height. The very fact of the existence of numerous normal isoalleles at many loci makes the human population continuously segregating. In the sequence of generations a given gene finds itself in ever-new background genotypes. Its fitness, therefore, is not rigidly fixed.

A special class of genes comprises those which are advantageous in heterozygous combination with another allele but less fit in homozygotes. For such alleles, fitness is obviously a property which depends on the allelic partner. In such cases, as well as in others, it is useful to distinguish between fitness of the individual and fitness of the population. The latter takes account of different individual fitnesses and the frequencies of the different genotypes. High fitness of a population is sometimes attained in spite of low fitness of certain, necessarily segregating genotypes.

When we consider the genetic effects of radiation, we may be justified at present in assuming that mutations in general are deleterious to the individual and to the population. Certainly this is true for very many mutations, but in the future we may have to ascertain more accurately the proportion of mutations which are unfavorable to those which are favorable.

Evolution and Induced Mutations. The evolution of species proceeds on the basis of mutations. Without genetic variants among prehuman forms, man would not have evolved. The argument has therefore been voiced that artificial increases in mutation rates may be desirable, since they may serve to speed up further evolutionary changes. This argument neglects the fact that the gene pool of human populations is already extremely diversified and that increases in the frequencies of mutant alleles are not likely to add new types of such alleles. Furthermore, it neglects the fact that evolution, the establishment of genetically new organic systems, is a very different process from mutation, the provision of mutant alleles (see Chapter 32).

Since so many mutant alleles are disadvantageous to their bearers, it is likely that each species builds up a genetic control system which reduces the frequency of mutations as much as is compatible with providing adaptive and evolutionary flexibility. If this is true, then an artificial increase of the human mutation rate is undesirable even from an evolutionary viewpoint. Although mutations have been induced which are useful to man (for instance, mutations in the mold penicillium have resulted in the production of a higher-than-normal amount of penicillin), they are very greatly outweighed by those which are unfavorable. If we wish to produce favorable mutations in man, we must pay a human cost in terms of numerous unfavorable mutations. If someday we should succeed in inducing specific, desirable mutations, only then could we look with satisfaction at the artificial induction of human mutations.

THE PHENOTYPIC CONSEQUENCES OF NEW MUTATIONS AND CHROMOSOMAL ABERRATIONS

Radiation-induced mutations may occur in two basically different anatomical regions: in body cells, and germ cells. Somatic mutations are not transmitted to later generations, but may cause changes in their bearer. Gametic mutations may be transmitted to the offspring, but do not affect their bearer. Therefore, since irradiation of an individual may produce mutations both in his somatic cells and in his germ cells, some effects may be produced in his body, and others, of quite a different kind, may be transmitted to his offspring.

Mutations in Somatic Cells. Somatic mutation effects depend on whether the mutation has been induced in a cell which does not divide any more or in a cell destined to divide in the future. In nondividing cells, as long as the dose is within a few hundred roentgen, it is unlikely that there will be any noticeable mutation damage in the form of either genic mutation or of chromosomal breakage. Broken chromosomes would remain within the nucleus; so no change in allelic quantities would be produced. Position effects, as well as induced genic mutations, might constitute in the most extreme case dominant lethals or, in the male, hemozygous sex-linked lethals, which kill the cell. The death of single nondividing cells would go unnoticed, and even numerous nondividing cells in which radiation may have induced various dominant or hemizygous sex-linked lethals could probably be dispensed with.

In a somatic cell which will continue to divide, a genic mutation will be transmitted to all descendants of the originally affected cell. If a recessive mutant allele is induced, modifying an original *AA* genotype to *Aa*, no phenotypic effect will be observable, but those mutant cells which obtain a dominant or hemizygous sex-linked allele may be phenotypically abnormal. If the descendants of the original mutant cell remain together, as they would in most tissues, a sector of mutant cells will be formed. If the daughter cells become dispersed, as blood cells do, a fraction of the total number of circulating cells will have the mutant genotype.

Since the majority of mutant alleles are recessives, will most somatic mutations remain permanently unnoticed, since they result in heterozygous *Aa* genotypes? There are several possibilities of an *Aa* being converted into an *aa* cell. One is the occurrence, at some later time, of mutation in an *Aa* cell, which transforms *A* into *a*. Another is a chance abnormality in mitosis, which may give rise not to two daughter cells each with an *A* and an *a* chromosome, but to one daughter cell with both *A*'s and the other with both *a*'s. A third possibility is somatic crossing over (Fig. 238, p. 576), and subsequent segregation of two *AA* chromosome sections from two *aa* sections. Also, loss of the *A*-carrying chromosome will lead to a monosomic *a* cell.

Irradiation of part or all of the body may well produce mutations in an appreciable number of the many billions of cells which constitute an individual. If these mutations are recessive, they will remain unknown until perhaps years later, when homozygosity for *aa* may be attained in one of the somatic descendants of an *Aa* cell. Propagation of the homozygous recessive cell may then result in visible effects. It has been wondered whether the proved fact of increased incidence of various types of cancer many years after exposure to irradiation may be explained by these processes.

Induction of chromosomal aberrations in cells destined to divide may have various consequences. Chromosomal fragments without a kinetochore will be lost during mitosis, resulting in cells deficient for chromosome sections. These cells, in many cases, will probably either die or function abnormally. If a breakage-fusion-bridge cycle is induced, the effect on the resulting cells will be either abnormal or lethal. If only some of the body cells suffer chromosomal breakage, the effect on the individual may be small, since it may be assumed that, in many cases, death of abnormal cells is compensated for by increased reproduction of those cells which did not undergo changes. Abnormally functioning cells, too, may not usually be seriously detrimental to the individual but, rather, will be so handicapped that they will be replaced by normal cells. If the whole body is exposed to doses which produce on the average one or more breaks per cell, the consequences to the individual may be lethal. It has been suggested by Puck that chromosomal breakage in dividing cells may be the main cause of death from irradiation in man.

Little is known about the quantitative aspects of radiation-induced changes in body cells. The sensitivity of different kinds of cells to radiation damage varies greatly. In the mouse, during certain early spermatogonial stages, the cells are so sensitive to X-rays that 50 per cent of them are killed by doses of 20–24r, and killing effects are observable with doses as low as 5r. For a variety of human cells kept in tissue cultures, the dose which kills 50 per cent of the cells (LD 50, the dose lethal to 50 per cent) has been found to lie in the range 50–150r. In contrast, brain cells within the intact individual are much more resistant to killing effects of radiation: most of them survive doses of thousands of roentgens.

Very likely, the induction rates of gene mutations are different for different types of body cells and for germ cells. Even if the rates were the same for all types, the effects on the exposed individual and on his offspring would be different. A low rate of mutation in germ cells implies a low expectation that the gametes involved in the creation of a few children will contain a new genetic property. A low rate of mutation in body cells implies a very high expectation that at least one if not many body cells will have a new property. It is therefore possible that irradiation of parts of the body, or of the whole body, even with low doses, may be hazardous.

It is known that heavy doses may have serious effects. Acute whole-body doses of 400r and higher are often fatal, and lower doses, above 100r, cause radiation sickness. These radiation effects occur within hours or days after exposure. Other effects are greatly delayed, often for years or decades. The most significant of these are shortening of the life span, as studied primarily in various laboratory animals, and the occurrence of various types of cancer, and particularly leukemia, a fatal disease of the blood-forming organs. It is not yet clear what the mathematical relation is between amount and intensity of exposure and frequency of induced malignant growth. There are various data on leukemia among American and other radiologists exposed in their occupations, among a special group of British patients who had been heavily irradiated for medical reasons, and among Japanese citizens exposed to the atomic bombs. The data, particularly those from Japan, strongly indicate that there is a correlation between dose and effect and that there is no threshold of radiation below which leukemia, as well as other types of cancer, is not induced. These findings are to be expected if the effect of radiation in inducing malignancies is equivalent to the production of mutations. However, the relation between induced mutations and onset of cancer cannot be predicted. Whether or not a potential cancer cell will become active may depend on reactions between normal cells and the new cell—reactions such as the occurrence or nonoccurrence of immunological destruction of the new cell. Such reactions may themselves depend on a variety of events unrelated to the radiation received. Or, if it is assumed that a potential cancer cell must be homozygous for a recessive mutant gene *a*, then, if irradiation of *AA* cells produce an *Aa* heterozygote, it will require further events such as aberrant chromosomal behavior or somatic crossing over in the descendants of the *Aa* cell to segregate an *aa* cell. Since the frequency of somatic crossing over, if it exists in man, is likely to be variable from person to person and from tissue to tissue, it is impossible to predict the frequency of the origin of *aa* cells.

In general, the expression of a mutated genotype by a body cell poses problems that are different from those of expression by a germ cell. A mutant germ cell that participates in the formation of a new individual will result in the presence of the mutant genotype in all of his cells, while a mutant body cell and its descendants will be surrounded by a nonmutant body. The two types of body cells may interact with each other; or the more vigorous type will replace the less vigorous, either to the benefit or the detriment of the whole; or they may persist together without influencing each other.

Mutations in Germ Cells. Radiation-induced genetic changes in germ cells may become phenotypically apparent in later generations in a way which depends on whether they are autosomal dominant, sex linked, or autosomal recessive, or on the type of chromosomal aberrations.

Gene mutations. Dominant autosomal gene mutations if fully penetrant, will become phenotypically obvious among the children of irradiated individuals. If the mutant alleles cause serious abnormalities, the reproductive fitness of individuals carrying them will be low, and these alleles will die out within a few generations. Sex-linked alleles will appear in an easily predictable manner in the first, or an early, generation, according to whether they are dominant or recessive and to the sex of the irradiated parent. If affected individuals are relatively unfit to reproduce, the sex-linked recessive, too, will soon disappear from the population. Recessive autosomal mutant genes will become visible in the first generation only if two like gametes meet, either if one of the parents happens to be a carrier for the mutant allele, or if gametes with identical mutant alleles produced in irradiated parents meet. The probability that the latter event will occur is negligible. In general, recessive mutant alleles will meet, in fertilization, normal dominants, and will therefore be transmitted unseen to later generations. Sooner or later, depending upon the breeding structure of the population (that is, depending on the amount of inbreeding and of isolate formation or isolate breakdown, and upon similar factors), two recessive alleles will meet and will produce a homozygous individual. Then, low reproductive fitness will, in most instances, reduce the probability of transmission of the recessive alleles.

Chromosomal aberrations. Chromosomal aberrations may be induced in immature and in mature germ cells. Since the fate of these changes is different for each of the two stages, they will be taken up separately.

We shall begin with premeiotic germ cells. In the human male, the testes contain numerous "stem cells" which proliferate mitotically. After their divisions, some of the descendant cells retain the characteristic of stem cells, while the others enter the developmental path toward meiosis. In the female, after birth, the immature germ cells do not divide mitotically any more but are arrested at a premeiotic stage. Meiotic divisions do not occur until ovulation and fertilization.

Chromosomal breaks induced in a stem cell behave in the same manner as breaks in the chromosomes of body cells. After division, some of the daughter cells will have retained all genetic material—although newly arranged—while in others elimination of fragments, or losses and changes due to a fusion-bridge-breakage cycle, will result in unbalanced conditions. Daughter cells with genetic material complete will be capable of entering meiotic processes, but those which suffer important quantitative changes may not undergo further development. If both of a pair of daughter cells become genetically unbalanced as the result of chromosomal breakage, and if this imbalance results in cell death, then both the cell originally destined for meiosis and the stem cell are eliminated. Thus, the divisions of irradiated stem cells serve to purge the gonad of some of the induced genetic changes.

What will happen to those stem cells which continue their premeiotic career or those immature germ cells, in males and females, which have been irradiated and have suffered chromosomal breakage after the last mitosis and during the premeiotic or meiotic stages? It is not necessary to analyze, in detail, the diverse possibilities of distribution and elimination of the chromosomes and their fragments, since a general outline of these events was presented earlier (pp. 124–130). Knowledge derived from experimental organisms indicates that, at least in many cases, both immature egg and sperm cells will probably transform into gametes capable of fertilization, irrespective of whether or not they contain a full and normal set of genetic material. There are two explanations for this phenomenon. Either the developmental processes which transform immature meiotic germ cells into mature gametes are set in motion early under the influence of the intact, unreduced gene content and may proceed successfully, even if an incomplete or abnormal genic complement is produced during meiosis; or the existence of cytoplasmic bridges between adjoining immature germ cells (which has been proved by means of the electron microscope) may permit a sharing of the genotypes so that genic deficiencies in one nucleus may be made harmless by the presence of the genes in a neighboring nucleus. Consequently, mature egg and sperm nuclei may be formed which lack a section of a chromosome, carry a section in duplicate, or have combinations of such abnormalities. Still other mature gametes may have a complete set of genes, some of which, however, are carried in chromosomal fragments without a kinetochore. If there is fertilization, the affected gamete produces an abnormal zygote. If the affected gamete is genetically unbalanced, the zygotic nucleus will be also; if the affected gamete is complete, but includes a fragment without kinetochore, then an early mitosis in the fertilized egg will cause loss of the fragment and thus lead to genetic imbalance in the daughter nuclei. Breakage-fusion-bridge cycles may also be set up by chromosomes with broken ends.

It is known from studies on mice and rats that many, if not all, zygotes with chromosomal imbalance develop abnormally (Fig. 250). Depending on the extent of the imbalance and on the individual effect of the genes concerned, development may cease at an early cleavage stage or at an embryonic stage, either before implantation in the uterine wall or after. It may be said that the gamete which leads to ultimate death of the zygote carries an induced dominant lethal. In man, very early cessation of development and subsequent disappearance of the embryo may cause the mother little trouble or go completely unnoticed. Death of an embryo after implantation may result in abor-

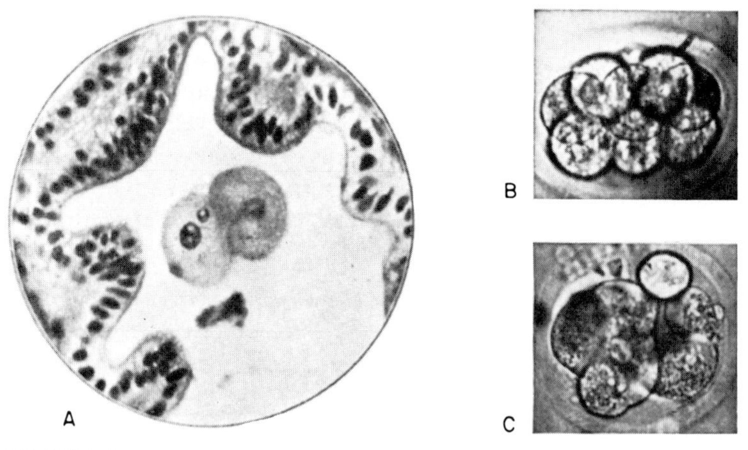

FIGURE 250
Abnormal development of eggs fertilized by unbalanced sperm. **A.** Cross section through a mouse egg after its first cleavage inside the oviduct. In consequence of chromosome breakage, part of the irradiated sperm nucleus was not included in the cleavage nuclei. The egg is destined for early death. **B.** Normal development of a normal rat egg. **C.** Degenerative development of a rat egg fertilized by a sperm produced by a translocation heterozygote (see Fig. 74). (A, after P. Hertwig, *Zeitschr. Abstgs. Vererb.*, **70,** 1935; B and C, after Bouricius, *Genetics*, **33,** 1948.)

tion or in stillbirth. Some genetically unbalanced embryos may develop into viable, but more-or-less abnormal, live births.

Certain types of individuals derived from gametes with chromosomal aberrations may be normal but will produce some abnormal offspring. They are "translocation heterozygotes," having received, from one parent, normal chromosomes, and, from the other, both of the chromosomes which had undergone a reciprocal exchange. The possession of both of these chromosomes insures the existence of all chromosomal sections and thus permits normal development of these individuals. However, during meiosis in their gonads, segregation results in two types of gametes: balanced and unbalanced. If the two reciprocally translocated chromosomes go jointly to one pole and the corresponding nontranslocated chromosomes go to the other, the gametes receive full genic complements (Fig. 74, A, p. 129). But if segregation takes place differently, unbalanced gametes are formed which are deficient for one translocated section and duplicate for the other (Fig. 74, B). These unbalanced gametes fertilize eggs that result in unbalanced zygotes, most of which die in the embryonic stage. Since the numbers of balanced and unbalanced gametes formed in translocation heterozygotes are frequently approximately equal, about one-half of the zygotes develop normally and the other half do not develop to term. Therefore, individuals heterozygous for a translocation are called *semisteriles.* It can easily be seen that they transmit the property of semisterility to one-half of their normal-appearing children. Data on semisterility caused by translocation in the rat are presented in Table 72. Partial sterility due to formation of unbalanced, as well as balanced, gametes may also be caused by an inversion within a chromosome (Fig. 72, p. 128).

In summarizing the genetic effects to be expected from chromosomal aberrations produced in the germ cells of man, it may be said that: (1) Many aberrations initiated in stem cells will be eliminated before meiosis occurs. (2) Many aberrations induced in germ cells which are ready for, or in the

TABLE 72
Semisterility in the Rat Caused by Chromosomal Translocation.

Mating			*Embryos (Mean No.)*			*Degeneration (%)*
			Implanted in Uterine Wall*	Degenerating after Implantation	Normal	
Semisterile ♂	×	Normal ♀	8.7	5.0	3.7	57.5
Normal ♂	×	Normal ♀	9.5	0.6	8.9	6.5

Source: After Bouricius, *Genetics,* 33, 1948.
*The lower number of implanted embryos in the semisterile as compared to the control matings is caused by degeneration of some unbalanced embryonic stages before implantation.

TABLE 73
Early Mortality of Live-born Offspring from Irradiated Males of Mice.

Offspring from Males Irradiated Within	*Death (%) After Birth Within*		
	1–7 Days	7–25 Days	22–75 Days
Two weeks before mating	13	12	18
Two or more months before mating	10	9	3
Not irradiated (control)	8	5	5

Source: After Hertwig, *Biol. Ztrbl.*, **58**, 1938.

process of, meiosis, or which are mature gametes, will lead to gametes which are able to participate in fertilization but which lead to early or late death of the developing zygote. (3) Reciprocal translocations and inversions present in balanced gametes of irradiated individuals will permit normal development of the offspring, but the unbalanced fraction of such gametes will produce inviable zygotes.

The first statement in this summary may contain at least part of the explanation for the fact that, in mice, the frequency of abnormally developing zygotes derived from sperm irradiated when mature is much higher than that from sperm which come from irradiated stem cells (Table 73). It is to be expected, similarly, in man, that genetic effects of irradiation will be found more frequently among individuals who were conceived within a few weeks after irradiation of the mature sperm stored in the male ducts of the father than among individuals who are the products of the sperm produced later from the irradiated paternal stem cells.

We have seen earlier that the frequency of single chromosomal breaks is proportional to the dose of irradiation, but that the frequency of chromosomal aberrations occurring in consequence of two or more breaks is disproportionately very low at low doses, and particularly if the exposure has occurred over a long time. Since only low doses are usually administered to humans within a short period, any but single chromosomal breaks must be rare. Double and multiple breaks are to be expected from severe acute exposures in nuclear accidents or nuclear warfare, or after (the now abandoned) heavy medical irradiation of poorly functioning ovaries to induce ovulation or of well-functioning testes to produce temporary sterility. The total frequency of phenotypically expressed chromosomal aberrations is probably small, since many are eliminated in immature gametes and, by early death, in embryos which receive an unbalanced genotype. If 50r should be the acute doubling dose for gene mutations, then this dosage presumably adds to the phenotypic effect of mutations only a lesser amount due to chromosomal aberrations, and exposure to less than 50r will cause still fewer transmissible chromosomal aberrations.

The Frequency of Affected Offspring from Irradiated Parents

Everyone who expects to become a parent, even though he and his spouse are healthy and of healthy ancestry, must be conscious that it is not in his power to assure the well-being of the children to be born. Besides nongenetic developmental mishaps, the birth of abnormal offspring may be brought about by hidden dominants of low penetrance, recessives carried by both parents, polygenic combinations transgressing a threshold, or new mutations.

It has been estimated that 4–6 per cent of all children either possess or will develop tangible defects, sometimes slight, sometimes severe, of a skeletal, neuromuscular, sensory, physiological, or other nature. In 1–2 per cent of all births the defects are clearly discernible when the children are born. It is probably an underestimate to say that half of all defects, or 2–3 per cent, are genetically caused.

If the germ cells of parents have been exposed to artificial radiation, how many additional defective children will be born as a result of mutations induced in these germ cells? This question may be asked from the "private" point of view of individual parents who want to know how much increased is the probability of their producing a defective child, or from the "public" point of view of the population who wants to know the additional number of defectives to be expected.

As bases of very provisional answers to these questions, we use the following, earlier-derived estimates: (1) number of mutable genic loci = 10,000; (2) number of induced mutations per locus per roentgen = 2×10^{-8} for acute and 0.5×10^{-8} for chronic exposures. An additional requirement for answering these questions is an estimate of the proportion of dominant to recessive mutations in man. This is a ratio whose value is not known. There are rather few dominant, defect-causing genes with complete penetrance, and probably more recessive, defect-causing genes with simple inheritance. Between these extremes, however, lies the whole range of conditionally dominant mutants with incomplete and often very low penetrance, and all those recessive mutants which in heterozygotes produce phenotypic traits detectable in some way even if the heterozygous carriers are normal. If we include all these genes in the estimate, then we may be inclined, following Levit, to believe that dominant, defect-causing human genes are in the majority.

This, however, does not imply that the majority of induced mutant genes will find noticeable expression in the offspring of irradiated parents. It would be necessary to know the mean penetrance of the dominant mutant alleles in order to predict the frequency of their phenotypic expression. Although we lack definite knowledge of both the proportion of dominant mutant genes and their mean penetrance, we may list a number of possibilities (see the first two columns of Table 74). Obviously, the more genes we count as domi-

TABLE 74
Proportions of Induced Mutations Which Become Apparent in the First Generation After Their Induction. (A series of different assumptions are made concerning the proportion of dominant to recessive mutant alleles and their mean penetrance. The allele frequency in the population of those acting as recessives is assumed to be q = .01.)

(a) *Proportion Mutants Dominants* (%)	(b) *Mean Penetrance* (%)	(c) *Dominants Expressed* (%)	(d) *Proportion Mutant Recessives** (%)	(e) *Homozygous Recessives* (q = .01) (%)	(f) *Total Expressed* (c) + (e) (%)
5	40	2	95	.95	2.95
10	20	2	95	.95	2.95
10	10	1	95	.95	1.95
50	10	5	90	.9	5.9
50	4	2	90	.9	2.9
100	4	4	90	.9	4.9
100	2	2	90	.9	2.9

Note: The sum of the dominant and recessives in most rows exceeds 100 per cent since one and the same mutant may produce a dominant effect in the heterozygote and a different, recessive effect in the homozygote.
*Penetrance 100%.

nants, the lower we will have to assume their mean penetrance to be. The entries for the proportion of dominants in the first column, therefore, rise in the arbitrarily chosen range 5–100 per cent; whereas the entries for the mean penetrance in the second column fall from 40 per cent to 2 per cent. In any given case, the percentage of expressed dominant mutant genes (see the third column) is the product of the percentages in the first two columns. It is noteworthy that the values for expressed dominants are all rather similar, being in the range 1–5 per cent.

The values for the expressed mutant dominants must be supplemented by those for expressed recessives. The new values will depend on the proportion of mutant recessives among all mutant genes, the mean penetrance of the recessive homozygotes, and the probability that a gamete with a given induced recessive will form a zygote with another gamete which carries the same recessive. It is assumed, in the fourth column of Table 74, that between 90 and 95 per cent of all induced mutant genes can act as recessives. This assumption implies that some or many dominant genes, particularly those with low penetrance as dominants, will usually be expressed, and in a more striking manner, if present in a homozygous state. The penetrance of the genes in homozygotes is assumed to be complete. This obviously gives a maximum value for the proportion expressed. The last value needed is that for the probability of two recessives meeting in fertilization, for which event two different possibilities exist: The first is the possibility of two independently induced mutant alleles, one from an egg and one from a sperm, meeting each other. The frequency of such an occurrence is negligibly low. Even if each parent had been exposed to a doubling dose, the probability of a specific mutation being induced in one parent would be 1 in 1,000,000, and the probability of the gamete carrying the allele produced by such a mutation encountering a like one from the other parent also only 1 in 1,000,000. In contrast to this very low probability is that of the second type of homozygosis involving an induced recessive mutation. It consists in the meeting of the induced mutant allele with one like it which is already present in the population in consequence of repeated spontaneous mutation. Allele frequencies vary considerably, and only a very few are directly known. It is probably a high estimate if the mean value is regarded as 1 in 100 ($q = 0.01$).

From the three values just discussed—the proportion of mutant genes with recessive expression (fourth column, Table 74), complete penetrance, and an allele frequency in the population of $q = 0.01$—the proportions of recessives expressed in the first generation of exposed parents are given by the products of the values in the fourth column and 0.01. These products are entered in the fifth column. Finally, in the sixth column are entered the sums of the third and fifth columns, which indicate the total proportion of induced mutant genes that may express themselves in the first generation. Depending on the assumptions made for each of the seven horizontal rows, the total proportion of expressed mutant alleles is expected to lie between 1.95 per cent and 5.9 per cent. Since several of the underlying assumptions tend to lead to rather high expectations for expression of expected damage, the figure of 4 per cent chosen for the following discussions is probably still an overestimate. We are now ready to consider quantitatively the expectations of mutational damage to the offspring of individual parental pairs and of populations.

Mutational Damage to the Offspring of Individual Parents. Since the frequency of induced and expressed mutants depends on the exposure received, we shall assume a specific dose, 10r, given to both parents of a pair or to all prospective parents of a population. For lower or higher doses, the expectations for expressed mutant genes are proportionally lower or higher. We have estimated earlier that an acute dose of 10r will induce a mutation in approximately 2 out of 1,000 gametes (postulating that the same figure is valid for eggs and sperm). Since perhaps 4 per cent of all induced mutant genes are expressed in the offspring of a single exposed parent, the probability of affected offspring from both exposed parents is $2 \times 0.002 \times 0.04 = 0.00016$ or 1.6 in 10,000. This is a low probability, particularly in comparison with the general probability of affected children from nonexposed parents, which was earlier estimated as being in the range 4–6 per cent. These last percentages are several hundred times larger than 0.00016. An individual parental pair, acutely exposed to 10r beyond background radiation, has thus a probability of from 0.95984 to 0.93984 of having a normal child, as compared to the probability of from 0.96 to 0.94 per cent from unexposed parents. In view of such a slight decrease, it would seem unnecessary for any individual parental pair to worry about the effect of low-dose exposure. Even with 50r—a considerably higher acute exposure than that assumed in the foregoing discussion—the calculated induced frequency of affected children would be only a small fraction of that for nonirradiated parents. Unless prospective parents are exposed to acute doses much larger than 50r, the probability of their having normal children remains very great. If the parents have been exposed chronically at low intensities, all figures given for high-intensity radiation can be reduced to at least one-fourth of those for acute exposure. Thus, the risk of producing abnormal offspring in consequence of such exposure is correspondingly lower.

What are the prospects for the grandchildren and later descendants of irradiated parents? In general, they are not much different from those of the first generation. Genes with dominant effects of low penetrance have unchanged chances of expressing themselves in future generations, and the same is true for recessives becoming homozygous. A child in whom a dominant effect became expressed is, of course, more likely than a nonaffected child to cause the same effect to appear among his offspring, and, correspondingly, a nonaffected child has better-than-average prospects of having nonaffected descendants. This is true simply because an individual who is phenotypically affected on account of his genotype is known to carry the gene in question; whereas a nonaffected individual may either carry the gene without its being penetrant or, much more likely, be free from it.

There are two exceptions to the statements concerning the appearance of mutant phenotypes in the second and later generations of exposed parents. One exception refers to sex-linked mutant recessives from exposed fathers. Such mutations will first appear among their grandsons and not among their immediate offspring. The second exception refers to consanguinity. If, to take the most significant case, marriages occur between first cousins, one of whose grandparents had been heterozygous for a mutant gene induced in one of his own parents, then the likelihood of both cousins carrying the same mutant gene becomes relatively high. Their children, who are five generations removed from the exposed ancestor, thus have an increased chance of becoming homozygous for the induced mutant allele. However, the increased chance would only slightly enlarge the already greater-than-normal chance of the children from cousin marriages being affected because of spontaneously mutated genes accumulated in the population. The same, to a still lesser degree, would apply in later generations to more distant degrees of consanguinity.

Mutational Damage to Populations. We turn now to the public concern regarding the incidence of phenotypically recognizable induced mutant genotypes in irradiated populations. Again we shall consider separately the effects on the first generation and those on later generations. Let us assume that all prospective parents of a population will be exposed to a mean acute dose of 10r, within the 30 years which on the average precede reproduction. Then the number of affected children will be equal to the probability of such children from individual parental pairs (estimated 0.00016, p. 619) multiplied by the number of births in the population. For example, from 3,000,000 births the number of affected offspring from acutely induced mutations would be 160. In the United States there are at present more than 3,000,000 births annually; so as many as, or more than, 480 affected children would be expected each year. For chronic exposure this number would be reduced to 120. In the world at large the present living generation will be replaced by perhaps as many as 6,000,000,000 people and the total number of affected children among them from parents exposed to 10r of acute radiation would be 960,000. For chronic exposure this number would be reduced to 240,000.

These numbers would be superimposed upon those numbers of affected children whose defects occur independently of the added radiation. Since 4–6 per cent of such children would be produced, their numbers among 1,000,000 births would be from 40,000 to 60,000, and among the world's total population from 240,000,000 to 360,000,000.

In later generations the frequencies of affected children from parents exposed to radiation would depend on a number of circumstances. Among these we may distinguish between the outcome if only the original parental generation, but none of their descendants, had been exposed, and the outcome if a dose of 10r impinges on each successive generation. The outcome of a third type of exposure, in which successive generations are exposed to doses either lower or higher than 10r, can be derived easily from the results of the following discussions.

If exposure is restricted to the parental generation, then the incidence of affected children will decrease in successive generations. The reason for this decrease is implicit, in that affected individuals are, on the whole, less fit than nonaffected ones and therefore reproduce at an average rate which is lower than that of nonaffected individuals. If, in a not-exposed population, there is at least an approach to a balance between spontaneous input of mutations and selective outgo, and if this balance has been temporarily changed by an additional input of induced mutations, selection will tend to reduce the total number of mutant alleles and restore the natural equilibrium. In detail, the selective processes involved will be different for different genes, will depend on whether the genes act as dominants or recessives, and will depend on the

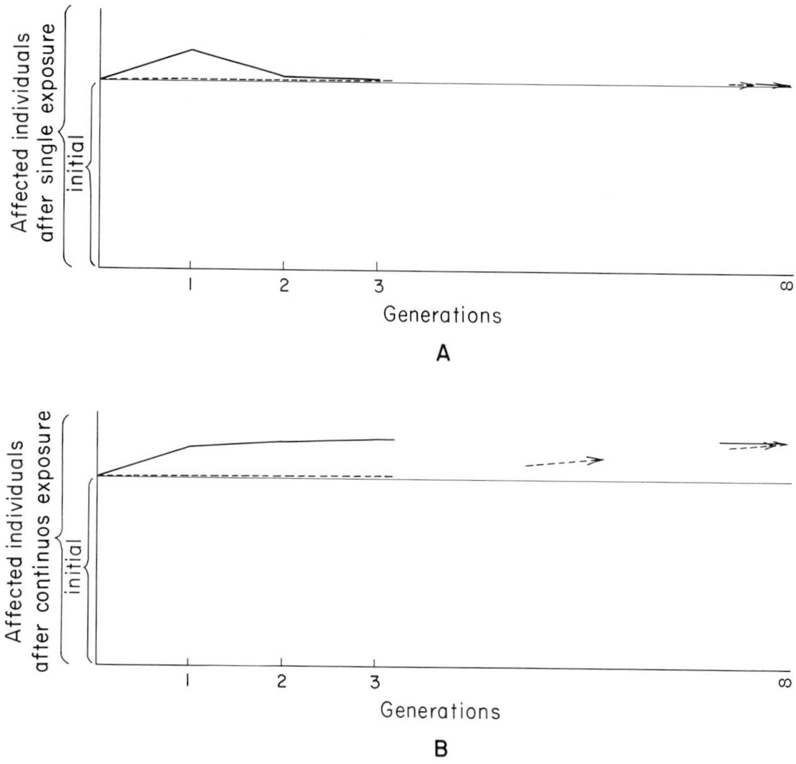

FIGURE 251
The effect of artificially increasing the spontaneous mutation rates by 20 per cent for a dominant and a recessive gene. Solid lines = achondroplastic dwarfism; broken lines = phenylketonuria. **A.** Increased rate of mutation effective for one generation. **B.** Permanently. (Modified from Penrose in Medical Research Council, *The Hazards to Man of Nuclear and Allied Radiations*. 1956.)

selective disadvantage they confer on their bearers. A dominant gene with complete or at least high penetrance and with serious deleterious effects, such as achondroplastic dwarfism, is present in a low frequency in a not-exposed population. With 10r of high-intensity exposure, its incidence among the first generation might rise by 20 per cent (Fig. 251, A). Owing to the low fitness of its bearers, the total frequency of such a gene in the second generation will be no more than 4 per cent above that in a not-exposed group; in the third generation the total frequency will be about 1/2 per cent above that in a not-exposed group; and soon the population will have freed itself from nearly all induced mutant alleles. For a dominant gene with lower penetrance and less serious effects, the lowering of its incidence will proceed in a similar but slower manner. Induced strictly recessive genes in homozygotes will raise the frequency of affected individuals only slightly above the spontaneous rate in the first generation. Thereafter, the number of affected homozygous individuals will gradually decrease, as was true of dominants. However, the course of events will be exceedingly slow, even for mutant alleles that are lethal in the homozygous state. This is so because in each generation only a very small percentage of induced mutant alleles will appear in homozygotes, the great majority being sheltered in heterozygotes. In reality, as we have seen, it is likely that many so-called recessive genes have slightly deleterious effects, even in heterozygotes. Their elimination will tend to follow the course of events characteristic of dominant genes with low penetrance, rather than that of complete recessives. Even so, it may require thirty of more generations until most induced, nearly recessive genes are eliminated. Since the duration of a human generation is about 30 years, this means a very slow decrease over hundreds or even thousands of years.

It may not be superfluous to state specifically that recessive homozygotes, which are rare in the first generation, do not become frequent in F_2. An F_1 plant of the genotype *Aa* which can be self-fertilized will, of course, produce 25 per cent recessive *aa* homozygotes. A first-generation heterozygous human *Aa* will usually marry an *AA* individual and have second-generation *AA* and *Aa* children only. These, in turn, will not intermarry but will mate with *AA* individuals from the general population. Thus, again, no affected *aa* will be produced. Only if two cousins of this third generation marry is there 1 chance in 16 of both being *Aa* and, in this case, of one-quarter affected offspring. If all marriages in a population were of the first-cousin type, then there would be a noticeable increase of affected homozygotes four generations after exposure. However, since in most populations cousin marriages are a small fraction of all marriages, such an increase in the population at large would probably remain insignificant. Lower degrees of inbreeding in later generations would have effects even less noticeable.

In a population that is subjected to irradiation in successive generations, mutations will accumulate until a new, higher equilibrium is reached between mutational input and selective outgo. The rise in affected individuals in the first generation will, in the following generations, continue (Fig. 251, B). Its ultimate level will be proportional to the dose. From that point, with 10r of high-intensity radiation impinging in each generation, and on the basis of the various assumptions made throughout this discussion of induced mutations, there will be produced in each generation 20 per cent more genetically defective individuals than there were before the beginning of exposures. With dominant genes, the rise to the new level will take only a few generations if the penetrance is high, more if it is low. For fully recessive genes, the rise will be very gradual for a long time; finally, the new level, 20 per cent above the old, will be attained.

The reverse trend of events will follow if irradiation should be discontinued. Then, the population will return to its pre-exposed status, speedily for seriously deleterious dominant genes with high penetrance, slowly for less-deleterious or lower-penetrance genes, and particularly slowly for complete recessives.

The preceding discussions have dealt with mutant genes whose effects are deleterious to various degrees. There are other mutant genes whose fitness cannot be classified so simply. Two important classes of these genes will now be considered.

Mutations in Polygenic Systems

Graded characters such as height or intelligence performance depend on the collaboration of many genes, each of which contributes individually small effects. The different alleles at each locus differ only slightly in action, and are thus isoalleles. What effect will irradiation have on these genes? Experiments with animals have shown that irradiation can increase the variability of certain polygenic traits, both in a plus and in a minus direction. Mutations are induced from one isoallele to another, so the range of variation of the trait is widened. It has been shown in experimental organisms that even after considerable exposure the increase in variability is small, as compared to the variability naturally present in populations occurring in nature. It may be assumed that graded traits in human populations, too, would change only slightly and very slowly after exposure. A number of different situations concerning the type of change expected can be envisaged. Among these, following Mather, two may be single out.

The first situation is that in which radiation increases the degree of variation symmetrically on both sides of the average. It may be illustrated by a theoretical analysis of one of many possible examples, the distribution of intelligence performance. Let us assume that one-half of the observed variability of the trait in a population depends on hereditary factors (the other half depending on environmental factors) and that the genetic variability of the population is permanently doubled by continuous irradiation. Such a population may, before exposure, consist of 95.4 per cent of normal persons within the I.Q. range of 70–130, 2.3 per cent of subnormal persons with I.Q. below 70, and 2.3 per cent of persons with I.Q. above 130. After exposure, and in the distant future (when equilibrium has been attained again) only 89.8 per cent would be within the normal I.Q. range and the subnormals as well as the supernormals would have more than doubled, both being present in a frequency of 5.1 per cent.

The second situation, again using an example of intelligence performance, is that in which radiation increases phenotypic variability asymmetrically. Such an effect may be expected if natural selection has, in the past, favored genotypes that result in phenotypes at one side of the average. We might be inclined to think that high intelligence performance has had selective advantages. Therefore irradiation will tend to increase the frequency of alleles which lead to a lower level of performance. As a result, the induced greater variation of the phenotype will lead to a lowering of the mean intelligence, accompanied by a relatively high increase of the subnormal extreme of the population and a low increase of the above-normal extreme.

It is not simple to attach value judgments to the types of change just illustrated. If the variability is increased symmetrically, there will be the regrettable phenomenon of an increase in persons with subnormal intelligence accompanied by a presumably fortunate increase in highly endowed people. Who can say whether these two opposing increases will balance each other in value (for the individuals, for their relatives, for society) or, if they do not balance, which one will be more significant? Even if the variability is increased asymmetrically, it would be a matter of uncertainty how to weigh the large rise of the poorly gifted against the smaller one of the highly gifted. In any case, it is clear that the induction of mutations involved in polygenic, graded traits does not simply fall into the category of mutational damage.

Mutations in Heterotic Situations

Examples are known, particularly for other organisms, but also for man, in which heterozygous individuals A^1A^2 are superior in reproductive fitness to either homozygotes A^1A^1 or A^2A^2. In a population which contains both alleles A^1 and A^2, an equilibrium is attained when the heterozygotes are the most frequent class. (For a more detailed treatment, see Chapter 28, pages 736–740.) Obviously, the terms advantageous and disadvantageous can be applied to either allele, depending on whether heterozygotes or homozygotes are considered. If we consider a series of multiple alleles, it may well be that a whole array of fitnesses exists, with certain heterozygotes being inferior to other heterozygotes and superior to certain homozygotes, and the homozygotes themselves being of varying fitness.

It is likely that any given human population possesses only some of the many possible alleles, of a given series. The consequences of induced mutations in such populations will depend on the types of alleles already present. If a population with heterosis for A^1A^2 contains mostly A^1 and few A^2 alleles, gametes with new induced A^1 mutant alleles will usually meet A^1 gametes. This will lead to relatively unfit A^1A^1 zygotes. Only rarely will the induced A^1 mutant allele meet A^2, thus leading to a more fit A^1A^2 heterozygote. If, instead of an induced A^1 mutant allele, an A^x allele is produced by irradiation, it is conceivable that both A^1A^x and A^2A^x will be superior to A^1A^1

and A^2A^2, and perhaps even to A^1A^2. Here, then, advantageous effects will occur in the beginning. But A^xA^x homozygotes will segregate out in later generations, and, if they are of low fitness, such genotypes will be termed disadvantageous. With heterosis, the sweeping statement that mutations are harmful obviously must be qualified.

The Fitness of Populations and the Load of Mutations

Haldane and Muller have, in different though related ways, considered the impact of mutations in populations. They based their views on a treatment of mutations whose effects lower the reproductive fitness of both heterozygotes and homozygotes, or at least do not increase fitness in heterozygotes. Induced mutant alleles with unfavorable effects in heterozygotes will usually be eliminated from the population before they can become homozygous. Each allele of this type will therefore sooner or later cause one genetic extinction, owing to early or later prenatal death of a zygote that carries the gene, or to pre-reproductive death of an individual, or to relatively lowered fertility or complete infertility of a perhaps otherwise little-or-not-at-all affected person. Completely recessive genes will be responsible for one-half extinction each, since two of them are required to produce their effect. In humans, the load of mutations encompasses more than the number of genetic extinctions might imply. Genes that lower the reproductive fitness of an individual by 10 per cent will, on the average, persist for 10 generations. Some of these genes may remain completely unexpressed before they become extinct, but others may cause suffering in each generation before they are eliminated.

This view of the fitness of a population implies that, ideally, the most-fit genotypes would be those of individuals who are homozygous, at all loci, for long-established normal alleles. Less-than-perfect fitness, which is an attribute of all existing populations, is a consequence of new, spontaneous, harmful mutations that are constantly injected into a population only to be eliminated by selection. The fitness of a population therefore is reduced in proportion to the rate of mutation.

No such proportionality exists for heterotic genes. The heterozygotes are the fittest type. In heterotic populations their mating in each generation leads to segregation of homozygotes whose reduced fitness may result in genetic extinction. In contrast to a hypothetical population consisting of nonsegregating homozygotes exclusively, a heterotic population carries a "segregational load," which lowers its fitness. Induction of mutations to alleles already present will add a mutational load to the segregational one; this mutational load may be far smaller than the segregational. Whatever its relative part the fitness of a heterotic population is not proportional to the rate of mutation.

It is now known that in human populations there are genes of each of the two main types: genes with invariably harmful effects, and genes with harmful effects in homozygotes but beneficial effects in heterozygotes. Admittedly, this distinction draws a sharper dividing line between the two types than is fully justified, since the attributes "generally harmful" or "heterotic" may apply to certain genes in certain genic backgrounds, but may be changed in other backgrounds, and may also vary according to varied environments. Whatever the importance in details is, it is useful to consider separately the significance for harmful and for heterotic genes that induced mutations have for the fitness of a population.

If the existing, total, genetic load of unexposed human populations is due to harmful mutations that occur continuously, then induced mutations will add proportionally to that load. If, for instance, as has been estimated, as many as 25 to 40 per cent of human zygotes are eliminated for genetic reasons, then a permanent doubling of the rate of mutations by radiation would lead to elimination of 50 to 80 per cent of human zygotes. If that were the case, we might justifiably wonder whether the human species would persist or would die out as a result of insufficient reproductive fitness. However, if the total genetic load of unexposed human populations is due largely to segregation of less-fit genotypes from more-fit heterozygotes, then the additional load from induced mutations will only slightly raise the frequency of eliminations. Induced mutation will then constitute a less-serious threat to the survival of the group.

It is not possible at present to assess the relative roles that the mutational and the segregational loads play in human populations. Many genes are known whose effect must be harmful under most if not all conditions. Few examples of definitely heterotic genes can be cited, but it may be that such genes are frequent for graded traits in polygenic systems where their existence cannot be demonstrated readily. Neel has suggested, in harmony with evidence on experimental organisms marshalled by Lerner, that some or many of the sporadic congenital malformations which occur in all populations may be multiple homozygous segregates from normal heterozygotes whose heterozygosis is the very basis of the normality. It may be a long time until the proportion of mutant alleles with generally harmful effects to those with heterotic effects is known. Until then, we cannot confidently evaluate the full effect of induced mutations on human populations. Experiments in which populations of laboratory animals were irradiated have given surprising results. As expected, irradiation of *Drosophila* populations over approximately 24 generations at the rate of 2,000–4,000r per generation have led to the induction of many lethal mutations. Less expected was the fact that as a whole the irradiated populations had increased fitness over nonirradiated controls. Apparently, radiation led to the origin of many more-or-less normal isoalleles which formed the material for new better fitted polygenic combinations and for overdominance effects. Similar experiments with mice and pigs did not yield increased fitnesses but experiments with rats did so. Male rats were treated in each of 8 to 14 sucessive generations with three doses of radiation adding up to 450r. Their female descendants were studied for body weight and age at sexual maturity. On the average, the heterozygotes for induced mutations showed increased body weight and decreased age of maturity as compared with controls; the homozygotes for induced mutations tended in the opposite direction. There was thus overdominance of induced mutations. Irradiation of human populations presumably would likewise be able to increase components of genetic fitness—but at the heavy cost of a multitude of deleterious individual mutations being induced at the same time.

Direct Studies on Man

Certain human groups that have been exposed to unusually high doses of radiation have been the subject of investigations. The offspring of such individuals were scored for frequency of congenital malformations, stillbirths and neonatal deaths, birth weight, and ratio of males to females. The largest groups studied consisted of the offspring of Japanese and American radiologists and radiological technicians, of the offspring of French and Dutch men and women who, for medical reasons, had received relatively high doses in the pelvic region, and of the offspring of Japanese parents exposed to the nuclear explosions in Hiroshima and Nagasaki. Several of these studies were based on answers by the parents to questionnaires. Many parents did not return their questionnaires, as was true also for many of the unexposed control parents. This is unfortunate, since it is difficult to judge whether the persons who did answer were selected samples in the sense that they had special reasons either to report on affected children (and the sex of their children) or to not report on them.

No clear evidence has been obtained on the frequency of congenital malformations. While some data gathered by questionnaire showed a significant increase in malformations, there was no significant difference in frequency of affected offspring from exposed as compared to nonexposed parents in the largest group in which each individual had been carefully inspected: the children born in Hiroshima and Nagasaki during the period 1948–1953. For the exposed parents, 300 congenital malformations are recorded among 33,327 live born (0.89 per cent); for the nonexposed parents, 298 are recorded among 31,904 (0.92 per cent). It would be completely in error to draw from this negative result the conclusion that exposure to the nuclear explosions did not induce mutations. The average dose received by survivors who became parents in later years was not very large, and the expected increase in malformations among their newborn was correspondingly small. Neel and Schull, the leaders of the genetics group of the Atomic Bomb Casualty Commission, submitted their findings to a detailed statistical treatment. It indicated that, had the frequency of malformations been actually doubled, owing to the exposure, there could still have been a 10 per cent chance of the observed frequencies not showing any difference. Furthermore, if many congenital malformations should represent segregational consequences of heterotic systems and not simple consequences of the mutational load, then the absence of an increase in affected children would be even less unexpected. However, the findings also have an important positive aspect. If the sensitivity of human genes were very high—for instance, if the doubling dose were close to its possible minimum of 3r for acute exposures—then there should probably be a significant increase in the malformation frequency of children of exposed parents. The fact that such an increase did not occur suggests that the acute doubling dose in man is at least 15r, and possibly is in the range which led, in this book, to the assumption of the figure of 50r.

A characteristic that has been regarded as an important indicator of induced damage to the genetic material is the sex ratio. If men alone are exposed to radiation, their relatively large X-chromosomes should be broken more frequently than their small Y-chromosomes. Their X-chromosomes, also, should carry induced, lethal mutant genes, some of them dominant or partly so, in contrast to their Y-chromosomes, which should be low in clearly mutable loci. Therefore, more female zygotes from irradiated fathers should die than those from nonexposed fathers, and a higher than normal ratio of males to females would result. This trend toward more male births would be counteracted to an unknown degree by the losses of broken Y-chromosomes in zygotes which, instead of developing into XY males, would develop into XO females. If women alone are exposed, some of their X-chromosomes should be broken and some should carry induced lethals. This would result in greater harm to the hemizygous male zygotes than to the XX female zygotes, and the result would be a relative surplus of female births. These general expectations of changed sex ratios for children or irradiated parents cannot easily be expressed in quantitative terms since we lack sufficient knowledge of the relative frequencies of the various radiation-induced events. Moreover, the sex ratio in nonirradiated populations varies in complex, poorly understood ways in different subpopulations and with time.

The results of the several studies of the sex ratio in children from exposed parents are not uniform. Some studies suggest a shift in the sex ratio in the direction expected from the induction of chromosomal breakage and X-linked lethal mutations, others show sex ratios that diverge in the opposite direction from expectation. Most changes in the sex ratio are small, being of the order of a few per cent. Neel and Schull's findings on the sex ratios in Hiroshima and Nagasaki, which are the most extensive ones, at first gave rather strong support to the validity of the theoretical expectations; but inclusion of additional children, born after 1953, resulted in sex ratios that did not deviate significantly from control ratios of children from nonirradiated parents. On the basis of more than 140,000 births in the period 1956–1962, nearly 74,000 of which were to parents one or both of whom had been exposed to the nuclear bombings, Schull, Neel, and Hashizume concluded in 1966 that "the suggestion of an effect of exposure on sex ratio in the earlier data is not borne out by the present findings."

While the studies in Japan fail to yield evidence for damage to the gametes of exposed individuals, there is unequivocal evidence for damage to somatic cells. The striking increase of most types of leukemia in exposed individuals has been cited earlier; and significant increases in other types of malignancies, particularly of thyroid tumors, have also occurred. Moreover, cytological damage has been observed more than twenty years after the bombings in cells of apparently healthy individuals. Chromosomal abnormalities, such as translocations, inversions, and deletions, have persisted in somatic cells and been transmitted by mitoses to successive cell generations. It may be added that studies in England and other countries seemed to show that irradiation of the pelvic region of pregnant women for diagnostic purposes greatly increases the frequencies of leukemia in the children to be born, even if the dose received by them is as low as less than 3r. However, data on 1,292 children exposed in utero to the atomic bombings do not show a significant excess of mortality due to leukemia or other cancers.

The Social Consequences of Induced Mutations

We have dealt in detail with the mutagenic effects of radiation. As serious as are radiation effects, it should be emphasized that—barring nuclear war—induction of mutations by other environmental agents probably is of greater importance. The list of potential chemical mutagens contained in man's diet, and other ingested or inhaled substances is a long one. It still increases steadily. Such substances may already be responsible for many of the so-called spontaneous mutations. Much of what has been said about radiation-induced mutations also applies to chemically induced ones. Detailed study of specific mutagens in man is needed to provide more than general impressions of their significance.

Although specific values were employed in the estimation of the effects of radiation their uncertainties have been stressed. Whatever specific values will finally be determined, there is no doubt that they will represent many millions of defect-carrying genotypes and that their frequencies will be positively correlated with the amount of radiation received.

It may be expected that future discoveries will provide methods with which to counter the mutagenic effects of radiation. It is already known that fewer mutations are produced when the irradiated tissues are very low in oxygen, and that certain chemicals applied to irradiated animals and bacteria are "antimutagenic." However, no satisfactory antimutagens, harmless to human beings, are yet available.

At present, the major contributions to doses received from man-made devices come from medical radiations that fall into the acute, high-intensity range. The benefit to the health of the individual exposed to such radiations, on the whole, far outweighs the damage to future generations, but damage there will be. Exposure due to radioactive fallout from the now-banned atmospheric tests of nuclear weapons is only a small fraction of that due to unavoidable, natural background radiation. The contribution of fallout to genetic damage is therefore small, particularly since it is due to low-intensity radiation, but fallout will still affect many individuals in the world at large. The foreseeable expanding industrial use of nuclear energy is likely to raise the level of radiation in general at a cost in the well-being of many people in future generations.

Present recommendations of national and international bodies tend to limit the mean exposure of individuals to 10r, excluding background. It would seem that a considerable reduction of this "permissible" dose would be possible since medical exposures, which account for most of the man-made radiation received, can be lowered, perhaps to one-tenth of their present (1972) amount without interfering with medical benefits of diagnosis and therapy.

A relatively new source of radiation exposure that is not included in the permissible dose of 10r derives from nuclear-power plants. Recommendations of various national and international agencies and committees for radiation protection are set at an average maximum dose of exposure per individual of 0.170r per year, thus amounting to close to 5r tor a thirty-year period. Only about 2 per cent of this dose is actually received by the population at present, but this fraction is likely to increase with installation of more and more reactors. The trend of increasing potential exposure should be countered by greater efforts in reducing the amount of radiation escaping the reactors.

Whatever dose will be permissible or tolerated will still cause human suffering and will increase the social burden of mankind in terms of disabled or handicapped members. Such a tragic situation is not new in the history of man. When, in prehistoric times, fire was made to serve human purposes it introduced a new danger which, in spite of extensive safeguards, still kills and maims many persons every year. When modern industry provided the populations of developed countries with a degree of comfort unknown to all past generations, it also created serious physical and chemical hazards to its workers. Likewise, the benefits of peaceful uses of nuclear energy are unavoidably associated with some degree of harmful effects. While every effort must be made to minimize all hazards to the present and to future generations, it must be realized that the moral problems inherent in the use of radiation are not unique.

The wartime use of nuclear weapons poses the moral problems in their starkest form. The carnage of "conventional" warfare is intensified by the long-term, ill effects of radiation on generations still to be born. Nonetheless, the evils of war are sufficiently great without the use of nuclear weapons to require bending all efforts to its permanent abolition.

If mankind were ever to be subjected to unrestricted nuclear warfare, the genetic consequences would be far reaching. Not only would immense numbers of individuals perish from the immediate effects of explosions, but even larger numbers would be exposed to radiation from radioactive contamination of whole continents. Surviving populations living in contaminated areas would be subject to chronic irradiation that might lead to accumulation of higher doses than the 450r which constitutes the acute mean lethal dose. The ill effects would be very severe, from the standpoint of physiological changes and mutations in body cells. In regard to transmissible mutations produced in germ cells, high doses of accumulated irradiation would cause numerous genetic changes, many times above those considered in this chapter. Yet, it is likely that even such high rates would still permit the production of enough descendants with adequate reproductive fitness to avoid the extinction of the human race. Notwithstanding, this likelihood can hardly alter the unimaginable horrors of the catastrophe contemplated here, nor the extreme moral and realistic necessity to ban any thought and means of nuclear war.

PROBLEMS

186. (a) Two employees in a radiological laboratory marry and have an albino child. Since there has been no albino in either ancestry, they ascribe the appearance of albinism to their occupational hazard. Evaluate the genetic aspects of the case.

(b) A man employed for several years in a nuclear-power plant becomes the father of a hemophilic boy, the first case in the extensive family pedigrees of both his own and his wife's ancestry. Another man also employed for several years in the same plant has an achondroplastic dwarf child, the first occurrence of the malformation in his and his wife's ancestry. Both men attribute the abnormality of their respective children to their occupation. They sue for damages. What should be the testimony of a geneticist before the court?

187. If a large group of individuals received 100r of total body irradiation, what prospects exist for determining whether an increased number of mutations were produced? Consider separately: (i) dominant mutations, (ii) incompletely penetrant dominants, (iii) X-linked recessives, (iv) autosomal recessives.

188. A fertilizing sperm cell of an irradiated man contains a reciprocal translocation between two autosomes.
(a) What will be the phenotypic effect on the developing zygote?
(b) If the zygote develops into an adult female, what will be the effect on her reproduction?
(c) If she has offspring, what will be their reproductive attributes?

REFERENCES

Crow, J. F., 1955. A comparison of fetal and infant death rates in the progeny of radiologists and pathologists. *Amer. J. Roentgenol.*, **73**:467–471.

Evans, H. J., Court Brown, W. M., and McLean, A. S. (Eds.), 1967. *Human Radiation Cytogenetics.* 218 pp. North-Holland Publ. Co., Amsterdam.

Glass, B. 1962. The biology of nuclear war. *The American Biology Teacher.* **24**:407–425.

Haldane, J. B. S., 1937. The effect of variation on fitness. *Amer. Natur.*, **71**: 337–349.

Lerner, I. M., 1954. *Genetic Homeostasis.* 134 pp. Oliver and Boyd, Edinburgh.

Macht, S. H., and Lawrence, P. S., 1955. National survey of congenital malformations resulting from exposure to Roentgen radiation. *Amer. J. Roentgenol.*, **73**:442–466.

Medical Research Council, 1956. *The Hazards to Man of Nuclear and Allied Radiations.* 128 pp. Her Majesty's Stationery Office, London.

Medical Research Council, 1960. *The Hazards to Man of Nuclear and Allied Radiations*; A second report. 154 pp. Her Majesty's Stationery Office, London.

Muller, H. J., 1950. Our load of mutations. *Amer J. Hum. Genet.*, **2**:111–176.

Muller, H. J., 1950. Radiation damage of genetic origin. *Amer. Sci.*, **38**:35–59.

National Academy of Sciences, 1960. *The Biological Effcts of Atomic Radiation.* 90 pp. National Academy of Sciences, Washington, D.C.

Neel, J. V., 1963. *Changing Perspectives on the Genetic Effects of Radiation.* 97 pp. Thomas, Springfield, Ill.

Neel, J. V., and Schull, W. J., 1956. *The Effect of Exposure to the Atomic Bombs on Pregnancy Termination in Hiroshima and Nagasaki.* 241 pp. *Nat. Acad. Sci.-Nat. Res. Council*, Washington, D.C.

Puck, T. T., 1959. Quantitative studies on mammalian cells in vitro. *Rev. Mod. Physics*, **31**:433–448.

Russell, W. L., 1963. The effect of radiation dose rate and fractionation on mutation in mice. Pp. 205–217 *in* Sobels, F. H. (Ed.), *Repair from Genetic Radiation.* Pergamon, Oxford.

Scholte, P. J. L., and Sobels, F. H., 1964. Sex ratio shifts among progeny from patients having received therapeutic X-radiation. *Amer. J. Hum. Genet.*, **16**:26–37.

Schull, W. J., Neel, J. V., and Hashizume, A., 1966. Some further observations on the sex ratio among infants born to survivors of the atomic bombings of Hiroshima and Nagasaki. *Amer. J. Hum. Genet.*, **18**:328–338.

Sobels, F. H., 1969. Recent advances in radiation genetics with emphasis on repair phenomena. *Proc. 12th Int. Congr. Genet.*, **3**:205–223.

Sobels, F. H. (Ed.), 1971. Mammalian Radiation Genetics. Special issue of *Mutation Research*, **11**:1–147.

Sonnenblick, B. P. (Ed.), 1959. *Protection in Diagnostic Radiology.* 346 pp. Rutgers University Press, New Brunswick.

Tanaka, K., and Ohkura, K., 1958. Evidence for genetic effects of radiation in offspring of radiological technicians. *Jap. J. Hum. Genet.*, **3**:135–145.

Turpin, R., Lejeune, J., and Rethore, M. O., 1956. Étude de la descendance de sujets traités par radiothérapie pelvienne. *Acta Genet. Stat. Med.*, **6**: 204–216.

United Nations Scientific Committee on the Effects of Atomic Radiation, 1958. [Report.] 228 pp. United Nations, New York.

United Nations Scientific Committee on the Effects of Atomic Radiation, 1962. [Report.] 442 pp. United Nations, New York.

United Nations Scientific Committee on the Effects of Atomic Radiation, 1966. [Report.] 153 pp. United Nations, New York.

United Nations Scientific Committee on the Effects of Atomic Radiation, 1969. [Report.] 165 pp. United Nations, New York.

25

HEREDITY AND ENVIRONMENT: I. TYPES OF TWINS

The interaction of genetic constitution and environment in the production of the phenotype has been discussed at various places in this book, particularly in the chapter dealing with variations in the expression of genes. It was seen that some alleles determine the same phenotype under all external conditions, while others may produce different phenotypes under different external circumstances.

These external circumstances, which are described by the general term "environment," include all nongenetic influences, whether acting before, at, or after birth. The age-old question "How much of a specific trait of an individual is due to heredity and how much to environment?" is meaningless in this form. Since no phenotypic trait is independent of either hereditary or environmental agents, an attempt to divide into two fractions the interrelation of two agents, neither of which alone can produce a phenotype, is futile.

What the question endeavors to ask is: "How much of the variability observed between different individuals is due to hereditary differences between them, and how much to differences in the environments under which the individuals developed?" Even this formulation of the question requires further precision. It is not applicable to "the" phenotype as a whole, but only to well-definable, measurable, or classifiable components. It is easy to measure the weight of a person, the length of any specific structure, the basic metabolism, the speed of reaction in finding his way through a maze, and innumerable other properties, but it is less informative to give quantitative measures to properties such as general body build, constitution, or personality. Indeed, progress in an understanding of these complex attributes is made only by determining and analyzing their separate factors. The question regarding the relative roles of heredity and environment in the determination of differences between individuals must, therefore, be applied separately to measurable components of the phenotype and often leads to a different answer for each separate component.

One other point should be made. It cannot be taken for granted that the effect of an environmental factor on the expression of one genotype can be predicted from the known effect of the same factor on the expression of another genotype. In some cases, the effects on both genotypes may be similar; in others, dissimilar. Thus, increase of food over the minimum necessary for maintenance will result in higher weight both in genetically small and large individuals; but addition of sugar to a sugar-free diet may have serious consequences for a genetically diabetic person, while causing no disturbance in a nondiabetic. Even opposite effects may result from two environments interacting with two genotypes: the use of cow's milk in the diet of a child allergic to it may cause a severe reaction, but in the diet of a nonallergic child is beneficial. For another example, two individuals of different intellectual endowments may both make moderate progress under limited demands from the environment, but the better endowed may thrive under high demands, while the other may fail completely. This example implies that people with varying intellectual endowments may vary for genetic reasons, a matter which will be discussed in Chapter 27.

Haldane has treated the problem of genotypic and environmental interaction in a generalized way. Assume that two genotypes, A and B, exposed to two environments, X and Y, result in four different phenotypes, ranked 1 (best) to 4 (poorest), and that genotype A in environment X gives the best performance. There are, then, six different ways, I–VI, in which the performances of A in Y and of B in X and Y can be ranked:

	I		II		III		IV		V		VI	
	X	Y	X	Y	X	Y	X	Y	X	Y	X	Y
A	1	2	1	3	1	4	1	2	1	3	1	4
B	3	4	2	4	2	3	4	3	4	2	3	2

Thus, arrangement I signifies that genotype A performs better than genotype B in both environments X and Y, and that both genotypes A and B perform better in environment X than Y. This arrangement would fit the earlier example of two genotypes for large (A) or small (B) body size in two environments, providing above-minimum (X) or minimum (Y) food supply. The same example could also be fitted to arrangement II, in which the genotype for large body size performs better than that for small body size in either environment but in which the interaction of genotype and environment leads to a poorer yield of A in Y (rank 3) than of B in X (rank 2).

The other earlier example concerning intellectual accomplishments under conditions of high or limited demands might fit arrangement IV: the "gifted" genotype A performs excellently in the highly demanding environment X (rank 1) and moderately in the less demanding environment Y (rank 2), while the less gifted genotype B fails in environment X (rank 4) but passes in environment Y (rank 3). The reader may begin to appreciate fully the implications of the table if he will formulate in words how an example concerning body size can also be fitted to arrangement III, and one concerning intellectual achievement can also be fitted to arrangements V and VI. And it may prove instructive to invent other examples and fit them to each one of the six arrangements.

If more than two genotypes and more than two environments are considered, the number of possible arrangements in performance increase greatly. Thus, three genotypes, A, B, and C, in three environments, X, Y, and Z, with genotype A in environment X having the highest rank, can be arranged in no less than 40,320 ways!

THE BIOLOGY OF TWINNING

The Use of Twin Studies. It is relatively easy to test the effect of nature and nurture in experimental animals or plants. Different strains can be produced, each of which is more or less isogenic and genetically distinct from the other strains; and controlled environmental variants can be applied to these strains. In some experiments, the environment will be kept constant so that the effect of genetic differences alone can be measured for each type of environment; and, in other experiments, the different genotypes will be kept constant so that the effect of environmental differences alone can be studied.

In man, isogenic strains are not available for such tests; nor is it possible, in most cases, to control at will the environment in which the phenotypic properties develop. Nevertheless, certain phenomena in man approach the ideal arrangements of experimental design. The most significant of these "arrangements" is twinning. Identical twins are isogenic and permit studies of the effect of different environments, while nonidentical twins are genetically different and permit studies of the effect of different genotypes in a similar environment.

Ever since Francis Galton (1822–1911) emphasized the importance of studying twins to obtain information on the nature–nurture problem, twin studies have played a vital role in the development of human genetics. Before we discuss the contributions of twin studies to the nature–nurture problem, let us examine some relevant facts on the biology of twinning.

Frequency of Twins. The frequency of twin births varies in different populations: from as high as 4.51 per cent of all births in a Nigerian population to as low as or lower than 0.8 per cent in several South American populations and among Chinese and Japanese. In the United States in 1964, 0.94 per cent of all white births, or 1 in 106, and 1.37 per cent of all black births, or 1 in 73, were twin births. These figures include only twin births in which at least one twin was born alive.

Identical and Nonidentical Twins. It has long been known that there are two different types of twins. Some are so similar to each other that they are called *identical* twins; others are no more similar than sibs, and they are called *nonidentical* or *fraternal* twins. Identical twins originate from a single egg fertilized by a single sperm, while nonidentical twins come from two eggs, each fertilized by a separate sperm. The occurrence of two-egg, or *dizygotic*, twins depends on the exceptional, more or less simultaneous, release of two eggs from one or both ovaries of a woman and their subsequent fertilization. One-egg, or *monozygotic*, twins are the result of the division of a single fertilized egg into two independent embryonic structures. Monozygotic twins are genetically alike, since mitoses provide the cells of both with descendants of the same chromosomes originally carried by the single zygote. Dizygotic twins are no more alike genetically than two sibs derived from two separate eggs and two separate sperm which matured in the gonads of the parents at different times.

The theoretical question has been raised of whether it is possible that some two-egg twins would be more alike than the usual fraternal, but less alike than identical, twins. Such twins would be produced if, preceding twinning, an unfertilized egg cell divided parthenogenetically and each of its two blastomeres was then fertilized by a different sperm (Fig. 220, A). The differences in genotypes of the two fertilization products would be intermediate between those of one-egg and two-egg twins. Another possibility would be the production of twins following abnormal meiotic divisions of an egg that gave rise, not to a large egg cell and three abortive polar bodies, but to two egg cells, similar in size and fertilizability, and two polar bodies. As first shown by Rosin, twins derived from two such egg cells may on the average be either more or less similar than typical dizygotic twins. In detail the similarities due to sharing like alleles derived from the mother depend on whether the genes considered are located close to, or rather distant from, the kinetochores of their chromosomes. The somewhat complex reasons underlying this statement cannot be given here. They will be found in Rosin's original study. There is no evidence for the existence of twins derived from these various hypothetical abnormal processes; and it is unlikely that more than an insignificant fraction of twins, if any at all, are of these types.

Weinberg's Differential Method. A statistical method, first conceived by Bertillon and later developed by Weinberg, permits us to find how many twin pairs in a twin population are monozygotic and how many are dizygotic. It is based on the fact that sex is determined genetically. Unlike-sexed pairs are undoubtedly derived from two separate zygotes, one XX and the other XY in constitution. Now, the number of like-sexed dizygotic twins should bear a simple relation to that of the unlike-sexed ones. In a population in which the secondary sex ratio is 1:1, dizygotic twins of the ♂, ♂; ♂, ♀; and ♀, ♀ types should occur according to the chance frequencies 1/4:1/2:1/4. This means that the number of like-sexed dizygotic male and female twin pairs, 1/4 + 1/4, would be the same as that of the unlike-sexed twins, 1/2; or, the total number of dizygotic twins would be twice that of the observed number of unlike-sexed twins. The number of monozygotic twins is obtained simply by subtracting the number of dizygotics from the total of all twins. Since the number of monozygotics is represented by the difference between all twins and the dizygotics, Weinberg's procedure is known as the "differential method."

The sex ratio in most populations deviates from equality. Therefore, a more accurate use of the differential method employs, for the population under study, the specific values of p and q for the probability of the male and female sex at birth. The fraction of unlike-sexed twins of all dizygotic twin pairs is

$$\frac{\text{Unlike-sexed twin pairs}}{\text{All dizygotic twin pairs}} = \frac{2pq}{p^2 + 2pq + q^2} = \frac{2pq}{1}.$$

This yields

$$\text{All dizygotic twin pairs} = \frac{\text{Unlike-sexed twin pairs}}{2pq}, \qquad (1)$$

and

$$\text{All monozygotic twin pairs} =$$

$$\text{Total of all twin pairs} - \frac{\text{Unlike-sexed twin pairs}}{2pq}. \qquad (2)$$

By applying these formulas to the white and black births in the United States in 1964, it is found that monozygotics among all twin births were about 40.3 per cent among whites and about 29.4 per cent among blacks. In other words, about 1 out of 2.5 sets of white twin pairs but only fewer than 1 out of 3 black twin pairs were identical. For Japan, with a low frequency of 0.7 per cent, or less, of twin births among all births, the application of the differential method shows that a much larger fraction, namely, more than 60 per cent, of the twins are monozygotic. This means that the difference between the over-all frequencies of twin births among Japanese and Americans is mainly due to differences in the frequencies of dizygotics. The same is true for over-all differences in the frequencies of twin births between a variety of racially or sociologically different groups—e.g., the higher rate of twins among illegitimate than among legitimate births in Scandinavian countries. (The causes of this difference in twinning rates are unknown. Perhaps it indicates different prenatal care resulting in different frequencies of prenatal deaths of legitimate and illegitimate children.)

Age of Mother, Parity, and Frequency of Twinning. An interesting relation exists between age of mothers and frequencies of twin births, both monozygotic and dizygotic (Fig. 252). From the age of 15 to 39 years, the tendency of mothers to bear twins increases, relatively slightly for monozygotics but very strikingly for dizygotics. For mothers 40–44 years of age, monozygotic frequencies continue to increase but dizygotics drop sharply.

Age of mothers is highly correlated with number of pregnancies: younger women have had fewer children than have older women. The variability of twinning frequency is, therefore, correlated not only with age of mothers, but also with parity. It has been shown that the two factors act separately,

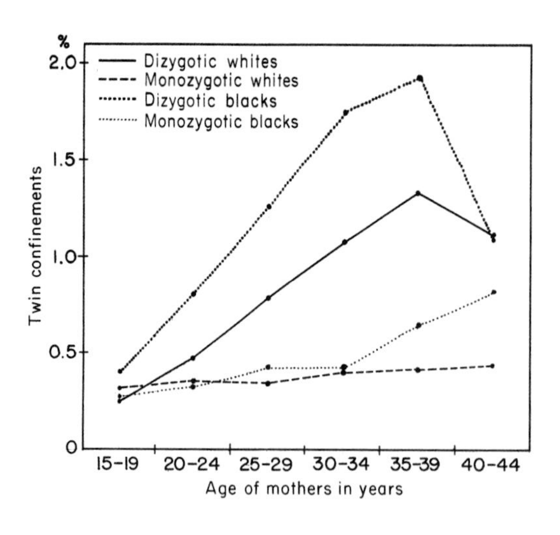

FIGURE 252
Frequency of twin confinements in relation to age of mothers in the United States, 1938. (Enders and Stern, *Genetics,* **33,** 1948.)

and in the same direction, on the frequency of dizygotic twin births. No study has been made regarding their effect on the less variable rate of monozygotic births.

The curves in Figure 252 show that a statement such as "One out of three twin births consists of identical twins" is based on an average and does not take into account the great differences in expectation for different age groups of mothers. A white mother, younger than 20 years, has about equal chances of having either identical or nonidentical twins, while the chance of a white mother between 35 and 39 years bearing nonidentical, rather than identical, twins is 3 to 1. It remains to be investigated whether the somewhat different proportions for black mothers are due to inherent differences in age effect, or to differences in parity for corresponding maternal ages. Since the frequencies of twin births vary with age and parity, changes in the average age of childbearing and of number of children per mother will affect the frequencies of twin births.

There is good evidence that the initial number of twin pregnancies of both types is much greater than the number of twins brought to term. Two embryos growing in a single uterus are in a less favorable environment than a single embryo, and rate of fetal mortality or stillbirth for both or one of twin pairs is several times higher than for single embryos. Estimates of the prenatal deaths of one of a pair of twins have been as high as from 20 to 50 per cent; and, among these deaths, the monozygotics seem to be affected two to three times as often as dizygotics.

Embryology of Twinning. The recognition of monozygotic or dizygotic origin of a given pair of twins is made by means of the *similarity diagnosis*, that is, detailed study of the phenotypes of the twins. This diagnosis is nothing more than a refinement of the inspection method often used by the layman to conclude that a specific pair is so similar as to be called identical, or another one sufficiently dissimilar as to be called nonidentical. The similarity diagnosis will be described later in this chapter.

TABLE 75
Number of Chorions and Amnions, and Similarity Diagnosis in 132 Like-sexed German Twin Births and 581 Like-sexed English Twin Births.

Afterbirth	*German Twins*		*English Twins*	
	Monozygotic	% Dizygotic	% Monozygotic	% Dizygotic
1 chorion, 1 amnion	2	–	3	–
1 chorion, 2 amnions	22	–	17	–
2 chorions, 2 amnions	18	58	8	72

Source: German data from Steiner, *Arch. Gynäk*, **159**, 1936; English data from Cameron, Edwards, and Wingham, *Acta Genet. Med. Gemellol.*, **15**, 1966.

The similarity diagnosis has been used to test another method of twin diagnosis which is based on the analysis of the afterbirth of twins and which has been employed frequently by physicians. As was described in Chapter 3, a human embryo develops from a germinal disc within an amnion, is surrounded by a chorion, and is attached to the uterus by a placenta. Since these structures, referred to as the afterbirth, are ejected from the uterus following the birth of the child, they are accessible to study.

Since it had long been observed that there were two chorions in some twin births but only one in others, it was believed that the number of chorions was diagnostic for two-egg and one-egg twins. Danforth, the first to use the similarity diagnosis as a check of the diagnosis based on the afterbirth, found that the latter diagnosis has only limited validity and often leads to a wrong conclusion. As is shown in Table 75, all twins enclosed in a single chorion are indeed monozygotic; but twins which are enclosed in two chorions are not all dizygotic. Rather, nearly one-fourth of the German and one-ninth of the English twins with two chorions in the table were shown by the similarity diagnosis to be monozygotic. Finally, the table shows that the number of amnions is of little help in diagnosis. Nearly all twins, regardless of whether they are monozygotic or dizygotic, are enclosed in two separate amnions. Only the small minority of twins which have a single amnion can be assigned to monozygosity on the basis of this embryonic finding.

Attending physicians have frequently tried to judge whether a twin birth is monozygotic or dizygotic by means of a single or double placenta rather than by means of the chorionic condition. The reason for this is that it is usually easier to determine the number of placentas than to decide on the number of chorions. Even if two chorions are present, they are often so closely joined that they may be mistaken for a single membrane. (Actual fusion of two chorions, according to Corner, "is to say the least uncommon.") The similarity diagnosis indicates that diagnosis by means of the placenta is even more unreliable than by means of the birth membranes (Table 76). All monochoric twins have only one placenta; but, as the table makes clear,

TABLE 76
Number of Placentas and Chorions, and Similarity Diagnosis in 236 Twin Births.

Afterbirth	*Twins*	
	Monozygotic	Dizygotic
1 placenta, 1 chorion	32	–
1 placenta, 2 chorions	8	80
2 placentas, 2 chorions	16	100

Source: Steiner, *Arch. Gynäk,* **159,** 1936.

a single placenta may also be found in dizygotic twin births, and two placentas may be found in the birth of either type of twins.

The various types of afterbirth provide information on the early stages of development of twins. Since dizygotic twins begin as separate fertilized eggs, they of course form two separate chorions and amnions. When implantation in the uterus occurs, the two embryonic structures sink into the uterine wall, but not in any particularly well-defined region: sometimes they become implanted apart from each other, sometimes in close proximity. In the former case, two separate placentas form (Fig. 253, A); in the latter, the two placentas which started separately grow together into a single one (Fig. 253, B).

Monozygotic twins and monozygotic births of higher multiples owe their origin to the remarkable ability of a single egg to form two or more separate embryos. Students of experimental embryology have succeeded in greatly clarifying the origin of one-egg twins by studies of the developing eggs of such animals as sea-urchins, salamanders, and rabbits. After the fertilized egg of a sea-urchin or salamander has divided once, it is possible to separate the daughter cells from each other, either mechanically, by means of a fine glass needle or a hair loop; or chemically, as by placing sea-urchin eggs into calcium-free sea water which changes the cell surfaces at the region of their contact so that the two cells do not remain together. The two separated cells may develop into whole animals instead of into half animals. In rabbits, Seidel has shown the possibility of monozygotic twinning the following way. He removed newly fertilized eggs from the oviduct and, when they had divided just once, killed one of the two cells by puncturing it with a needle. When the egg was then returned to an oviduct, the unharmed cell proceeded to develop, not into a half embryo, but into a whole rabbit—whose twin mate had been sacrificed (Figs. 254 and 255).

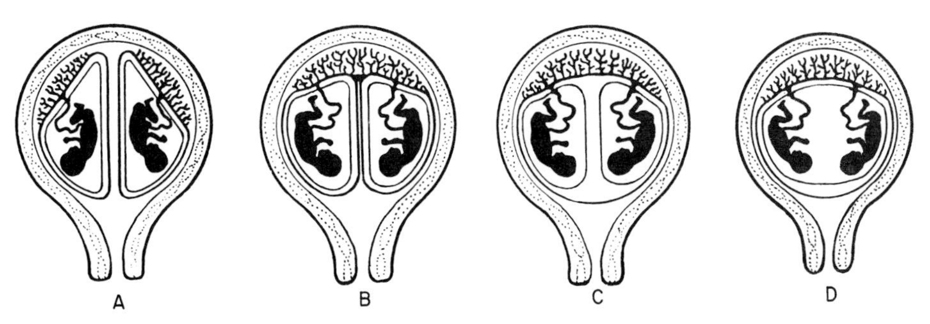

FIGURE 253
Diagrams of twin pregnancies enclosed in the uterus. **A.** Monozygotic or dizygotic twins with separate amnions, chorions, and placentas. **B.** Monozygotic or dizygotic twins with separate amnions and chorions, and fused placenta. **C.** Monozygotic twins with separate amnions, and single chorion and placenta. **D.** Monozygotic twins with single amnion, chorion, and placenta. (A–C, Potter, *Fundamentals of Human Reproduction*, McGraw-Hill, 1948.)

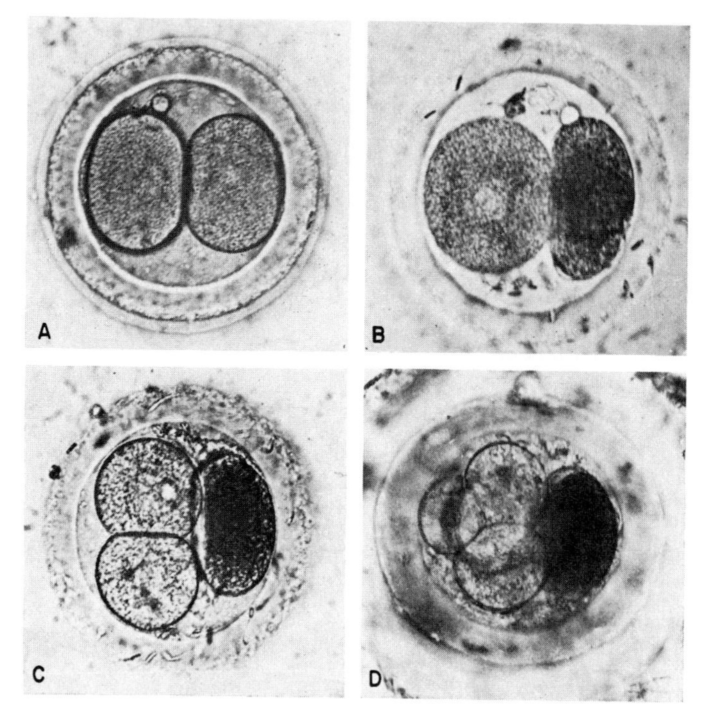

FIGURE 254
Development of a rabbit from one of the first two cleavage cells of an egg. **A.** Fertilized egg, at the two-cell stage, which has been removed from the uterus. **B.** One of the cells (right) has been killed by being pricked with a fine glass needle. **C.** The surviving cell has divided into two cells. **D.** Further division has resulted in four living cells. (Seidel, *Naturwissenschaften*, **39**, 1952.)

FIGURE 255
An operated egg cell was transplanted into the uterus of the large gray female. The transplanted egg developed into the fully formed rabbit shown in the foreground. (Seidel, *Naturwissenschaften*, **39**, 1952.)

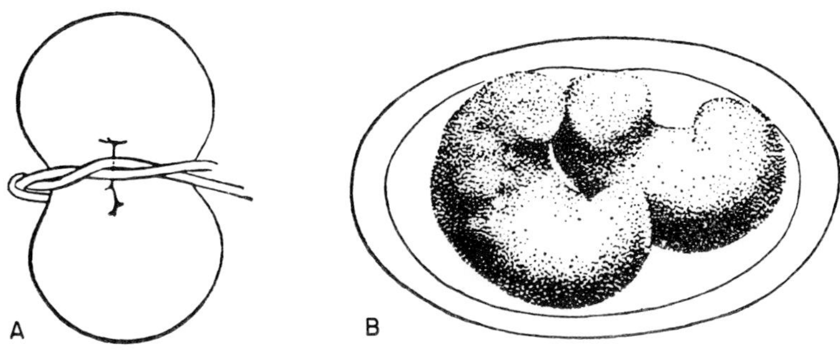

FIGURE 256
Experimental production of monozygotic twins in a salamander. **A.** Constriction of gastrula with a fine hair loop. **B.** The resulting twin embryos. (After Spemann, from Hartmann, *Allgemeine Biologie*, 2nd ed., Fischer, 1933.)

Twins can also be artificially initiated at a later stage in development of a single zygote. This is seen in Figure 256 where, in a multicellular stage of a single egg, salamander twins have been produced by constriction of the egg with a hair. In the nine-banded armadillo, a mammal, the single fertilized egg develops normally into a hollow cell ball which, instead of forming a single embryonic streak, forms four. Consequently, identical quadruplets are born regularly.

Apparently, spontaneous twinning of a single human egg can likewise set in at various stages. Few direct observations of these occurrences have been made, but it is not difficult to reconstruct what may happen. If the two cells of a zygote which has just divided should become separated, or if the cell ball divides during an early stage, each half will probably be able to develop into a blastocyst and differentiate an outer chorion and an inner embryonic shield and streak. Two separate chorions are thus present; and, as in dizygotic twins, implantation of the embryos in the uterus, close together or distant from each other, determines whether one or two placentas are formed. If monozygotic human twinning, similar to the process in armadillos, is initiated after the formation of a chorion, further development may occur in one of two ways. Instead of a single amniotic cavity, two such cavities may appear, each having a separate embryonic shield; or a single amnion and a single shield may appear but two embryonic streaks may form (Fig. 257). In the first case, a pair of twins will be enclosed in a single chorion but each twin separated in one of the two amnions (Fig. 253, C); in the latter case, the single chorion will contain a single amnion with its twins (Fig. 253, D). The different types of afterbirths of twins are thus consequences of early embryonic happenings.

Multiple ovulation may lead to polyzygotic triplets and higher polyzygotic multiple births. Furthermore, any combination of monozygotic and polyzygotic multiplication may occur. For instance, human triplets may be children

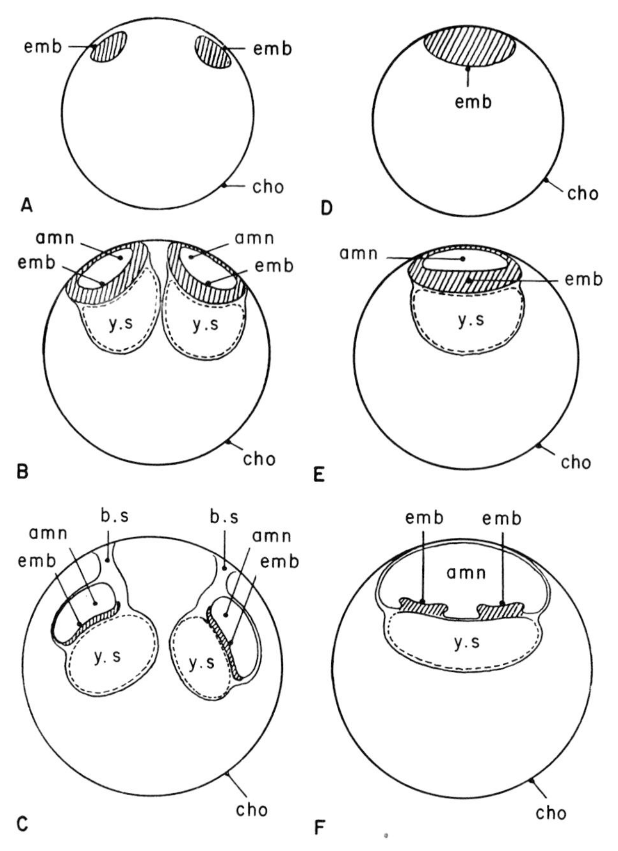

FIGURE 257
Two types of single-ovum twinning in man. **A–C.** By formation of two inner cell masses. **D–F.** By formation of two embryos on a single germ disc. emb = inner cell mass (A, D), embryonic shield (B, E), embryo (C, F); cho = chorion; amn = amniotic cavity; **ys** = yolk sac; **bs** = body stalk (Corner, *Bull. Johns Hopkins Hosp.*, **33,** 1923.)

derived from a single egg; or from two eggs, one of which underwent twin formation; or from three separate eggs. The two popularly known sets of quintuplets, in which all members survived birth, represent two extremes: the Dionnes came from one egg, and the Argentine children came from five.

Very rarely, the embryonic subdivision which normally leads to monozygotic twinning remains incomplete. Experimentally, such incomplete twinning has been produced with salamander eggs by means of incomplete artificial constriction with a hair loop. The results of these partial separations are two-headed, single bodied, or one-headed, two-bodied salamanders. In humans, also, there are cases of incomplete twinning, ranging from duplication of some parts to nearly complete duplication of the whole body. The latter types of conjoined individuals are often referred to as Siamese twins, after a famous Siamese-born pair of the nineteenth century, whose mother was half Siamese and half Chinese and whose father was Chinese.

Heterokaryotic Monozygotic Twins. By definition, monozygotic twins are descended from a single fertilized egg. Basically their somatic nuclei are chromosomally alike, since they are all related to each other by mitotic descent. Occasionally, however, abnormalities in the distribution of the chromosomes result in differences of the karyotypes of different cells, or mutation of genes result in genetic differences between cells. When these events occur in a developing zygote which gives rise to a single individual, mosaicism in regard to chromosomal or genic constitution is produced. When the events occur in a zygote destined to form a pair of monozygotic twins, the two twins may be formed by cells of different genotypes and thus will be heterokaryotic. The first discovery of such twins was made by Turpin, Lejeune, and their associates. Careful analysis of the blood groups, serum types, and other traits of a pair of twins and of their parents and sibs showed that the twins were monozygotic. Yet, one of them was female, the other male. The solution of the apparent paradox lay in chromosomal analysis: The male was XY, the female XO. Apparently, the fertilized egg had been XY, but early in development—perhaps at the first division of the egg—the Y-chromosome was lost from a cell. Subsequently, twinning occurred with one twin being formed by XY cells, the other by XO cells. Among the few other heterokaryotic monozygotic twins known is a pair of XX–XO female twins and a pair of male twins of whom one is normal with two chromosomes 21, the other an individual with Down's syndrome, having three chromosomes 21. Another heterokaryotic monozygous twin pair displays chromosomal constitutions that are not unexpected, given the explanation of the origin of such twins: each is mosaic for XX and XO cells but XX cells predominate in one and XO cells in the other.

Inheritance of Twinning Tendency. In many sibships, more than one pair of twins is found; and in numerous pedigrees, twin births appear in successive generations or in the offspring of close relatives. The meaning of such recurring twinning is not obvious. If accumulation of twin births within a kindred was argued to be evidence of an inherited tendency, it would be easy to point out that selection for oddity may give an erroneous impression. Since twin births occur in nearly 1 per cent or more of all births, by chance alone, sibships or pedigrees which include two, or several, twin births should not be too rare. If only those pedigrees which show numerous twin births are singled out for attention, then no more weight should be attached to them than to selected pedigrees in which only males or only females have been born.

Again, the statistical approach is sounder. One of the first to study the question in this manner was Weinberg. He determined the frequency of twins among the sibs of twins or, as he called it, the *repeat frequency*. Mothers of monozygotic twins did not show a higher frequency of twin births among the rest of their children than did individuals of the general population, but mothers of dizygotic twins were found to have about twice as high a chance of having another twin birth.

A high repeat frequency, by itself, does not prove heritability of twinning; but it shows that a mother who has had twins is, in some way, different from mothers who have had only single births. The frequency could be either due to a special genetic constitution of the mothers of twins or to unknown environmental agents which make these women more susceptible to twin pregnancies than other women. To determine whether particular hereditary or environmental factors are responsible for recurring twinning, studies have been made of the frequencies of twin births in successive generations and among the relatives of parents of twins. The evidence generally indicates that monozygotic twinning has no, or no greatly increased, tendency to reappear in pedigrees in which either monozygotic or dizygotic twin births occurred, but that there is a decidedly increased tendency to recurrence of dizygotic twinning. A recent study by White and Wyshak, using data of the Mormon Genealogical Society, found an over-all twinning rate of 1.16 per cent for their population. In familes with dizygotic, unlike-sexed twin pairs the frequency of twins born to the female partners of the pairs and to their female sibs was 1.70 per cent, considerably higher than in the general population. In contrast, the frequency of twins among the offspring of the male partners of the index twin pairs and their male sibs was significantly lower than 1.70 per cent, namely 1.10 per cent. These findings are consistent with the view that the genotype of the mother affects the frequency of dizygotic twinning, most likely by way of control of multiple ovulations. Obviously, the genotype of the father is unable to express itself in the frequency of twins among his children. It should be added, however, that the data do not fully rule out the possibility of nongenetic, recurrent, external conditions that favor twin births.

The role of the mother in the frequency of twinning is also shown in studies by Morton and his associates on the genetics of interracial crosses in Hawaii. They found that the frequency of dizygotic twinning is independent of the father's race but is highly correlated with the race of the mother.

The genetic hypothesis of the cause of variations in dizygotic twinning rates finds support in studies of families which are distinguished by significantly high frequencies of such twins. Such families were studied in Norway. In a valley near the Trondheim Fjord, Kristine Bonnevie and Sverdrup noted great variations in the frequencies of twins born within different family lines of several isolated agricultural communities. Reliable church registers and other documents provided information on births over the past 250 years. In one pedigree, no twins occurred among 800 births; in another, 101 (or 2.85 per cent) appeared among 3,645; in still another, 107 (or 3.91 per cent) among 2,840. Averaging the twin frequency of all families in which twins occurred gave 3.25 per cent among 10,485 births. This contrasts, on the one hand, with a general twin frequency of about 1.24 per cent in these communities, which included the families with high frequencies. On the other hand, among 1,618 births of selected twin lines within the twin families, the frequency was as great as 8.23 per cent. It is likely that the differences in frequencies of twins in these Norwegian families were based on differences in frequencies of dizygotics.

A curious relation between high monozygotic twinning frequency and chromosomal abnormality, described by Nance and Uchida and others, may be mentioned at this point. In sibships containing individuals with Turner's syndrome the frequency of twins is several times greater than in the general population, with the increment probably consisting of monozygotics. There is evidence from the mosaicism for XX and XO cells in Turner's individuals that loss of the X-chromosome, resulting in the XO constitution, occurs in many of the Turner's zygotes during an early cleavage division. It is possible that the increased twinning frequency is caused by some egg defect that also may cause loss of an X-chromosome.

As a statistical phenomenon, the recurrence of dizygotic twins is different from occasional excessively frequent twin pregnancies in any one particular woman. Bonnevie and Sverdrup report a case, unrelated to their family data, of a woman who had four successive twin births, followed by four single births, then by four more twin births, and, finally, by a single birth. Most likely, the twins were all dizygotic. No other twin pregnancy occurred in the three preceding generations of this woman's ancestors. This, as well as similar cases, is apparently due to special physiological conditions causing, as a rule, double ovulations. Whether or not such an anomaly is hereditary is not known.

If the genetic basis of dizygotic twinning is a simple one, it must have very low penetrance. More likely, perhaps it is polygenic, with incomplete penetrance as a result of nongenetic factors such as maternal age and parity. In this connection, it may be worth pointing out that even a doubling or tripling of the normal frequency of twinning, in twin kindreds, raises the probability for a twin birth by only a few per cent and still leaves an overwhelming probability that any specific birth will be a single one.

TWIN DIAGNOSIS BY MEANS OF THE SIMILARITY METHOD

Weinberg's differential method enables us to determine the number of monozygotic and dizygotic twins in a population of twins. In order to classify a particular twin pair in regard to their one- or two-egg origin, the similarity method of diagnosis, as elaborated particularly by Siemens and von Verschuer, is employed. The general reasoning underlying it is as follows: Two parents are, in general, heterozygous for numerous genes. If for a certain locus, one parent is heterozygous, AA', and the other parent homozygous, AA, the probability for two dizygotic twins being genetically alike is the sum of both of them being either AA or AA', which is $1/4 + 1/4 = 1/2$. If both parents are heterozygous, AA', the probability for genetic identity of dizygotic twins becomes 1/16 (both AA) + 1/16 (both $A'A'$) + 1/4 (both AA') $= 6/16 = 3/8$. These probabilities for genotypic identity of the twins are also valid for phenotypic identity if the heterozygote has an intermediate or codominant phenotype and if penetrance of all genotypes is complete. If A' is dominant, then $A'A'$ and $A'A$ twins will appear alike; and the probability for phenotypically like dizygotic twins from two heterozygous parents becomes 1/16 (both AA) + 9/16 (both having at least one A' allele) $= 10/16 = 5/8$.

Since all these probabilities, 1/2, 3/8, and 5/8 are high, finding twins who appear alike in regard to the A locus will be fully compatible with their being regarded as dizygotics. If it is taken into account that the two parents are both heterozygous not only for one locus, but for many, then phenotypic alikeness of two twins for many characters takes on a different aspect. For example, if ten loci with intermediate or codominant expression of heterozygotes are involved and one parent is homozygous for some loci and heterozygous for other loci, while the other parent is heterozygous for the first and homozygous for the second group of loci, then the probability of two dizygotic twins being phenotypically alike becomes $(1/2)^{10}$. If both parents are heterozygous for ten loci with intermediate or codominant expression in heterozygotes, the probability is $(3/8)^{10}$, or in dominance $(5/8)^{10}$. If, for another example, two loci with intermediate or codominant heterozygotes are homozygous in one parent and heterozygous in the other parent, and four loci with intermediate or codominant heterozygotes and four loci with dominance are heterozygous in both parents, the probability of dizygotic twins being phenotypically alike is $(1/2)^2 \times (3/8)^4 \times (5/8)^4$. This is a very low probability and it becomes still lower with every additional locus taken into account. To classify as dizygotics those twins who are alike in numerous genetic traits for which their parents were heterozygous would, therefore, entail an improbable assumption. On the other hand, if the twins were classified as monozygotics, then their genetic identity would correspond with their single-egg origin.

The main argument described in the preceding paragraphs can be applied even if the genotypes of the parents are unknown. If, for instance, two twins whose parents are not available for study are both recognizable as being homozygous for a pair of alleles, $A'A'$, then the parents must have represented one or another of the following combinations: $A'A' \times A'A'$, $A'A' \times AA'$, $AA' \times AA'$. The relative frequencies of the three types of marriages can be calculated if the allele frequencies p_A and $q_{A'}$ are known. Thus, if $p_A = 0.9$ and $q_{A'} = 0.1$, the probabilities of the types of marriages are 0.0001, 0.0036, and 0.0324, respectively, and the proportions of each marriage to the total of all three ($0.0001 + 0.0036 + 0.0324 = 0.0361$) are

$$\begin{array}{ll} A'A' \times A'A' & 0.0001/0.0361 = 0.0028 \\ A'A' \times A\,A' & 0.0036/0.0361 = 0.0997 \\ A\,A' \times A\,A' & 0.0324/0.0361 = 0.8975 \\ & \text{Total} = 1.0000 \end{array}$$

Now, the probability of an $A'A'$ child from these marriages is 1, 1/2, and 1/4, respectively. Multiplying these probabilities with the proportion of the relevant marriages, we find the joint probability of an $A'A'$ child from either one or another of the marriages as being

$$0.0028 + 1/2\ 0.0997 + 1/4\ 0.8975 = 0.2870.$$

The probability in these marriages of two $A'A'$ children from separate eggs is

$$0.0028 + 1/4\ 0.0997 + 1/16\ 0.8975 = 0.0838,$$

so that, among all those pairs of children which include at least one $A'A'$ child, the relative probability of two children from separate eggs both being $A'A'$ becomes

$$0.0838/0.2870 = 0.292.$$

This, of course, is also the probability of both partners of a dizygotic twin pair being $A'A'$. In contrast to this, the probability of both partners of a monozygotic pair being $A'A'$ if one is $A'A'$ is 1.

Let us assume that the twins are alike in four other loci and that calculations similar to those given for the A locus would yield probabilities of dizygotic twins being alike in these four loci of 0.2, 0.9, 0.3, and 0.5, respectively. Then the combined probability of dizygotic twins being alike for all five loci would become

$$0.292 \times 0.2 \times 0.9 \times 0.3 \times 0.5 = 0.007884,$$

while the combined probability of monozygotic twins being equally alike would remain 1. The odds are therefore highly in favor of monozygotic origin of the twins.

These calculations could still be improved by considering the probability of any randomly selected pair of twins being of dizygotic origin, which is approximately 0.652 (for United States whites using somewhat different data than those of p. 640). Furthermore, since the similarity diagnosis is applied to like-sexed twins only—unlike-sexed twins are always considered to be dizygotic—the probability should be entered of a dizygotic twin pair being like-sexed. This probability is about 0.5. Combining the probabilities 0.652 and 0.5 with that calculated from the five assumed loci, the over-all probability of the twins being dizygotic and alike in sex and five genotypes becomes

$$0.652 \times 0.5 \times 0.007884 = 0.00257.$$

In comparison, the probability of a randomly selected pair of twins being monozygotic is 0.348, with certainty of such a monozygotic pair being alike in sex and five genotypes. Thus the probability that the twins are of dizygotic origin, expressed in terms of their being either of dizygotic or monozygotic origin, is

$$\frac{0.00257}{0.00257 + 0.348} = 0.0073.$$

This is a low probability, and the probability of the pair being monozygotic is greater than 99 per cent ($1 - 0.0073 = 0.9927$).

In accord with these considerations, the similarity diagnosis ideally uses a large number of phenotypic characters based on: (1) numerous loci for which the population is highly heterogeneous, and (2) loci which have complete penetrance and uniform expressivity under all known environmental conditions. Identical, or very close, resemblance between twins in all relevant traits is taken as evidence of monozygosity, while difference in even a single trait is taken as proof of dizygosity. The diagnosis is best made after the twins are at least several years old, because many traits are not well differentiated in infants.

The choice of traits used in the similarity diagnosis presents a special problem. Rather few known loci express themselves in a simple way and have various alleles sufficiently common to be useful for the classification of most twin cases. Loci which are suitable for twin diagnoses are those determining the various bloodgroups, serum constituents, and tissue enzymes. Making use of the known frequencies of bloodgroup alleles in the English population (which may be regarded as similar to many other white populations), Sheila Smith and Penrose have published probability tables for any one of the possible combinations of two sibs in respect to eight different blood-group systems. The use of these tables may be demonstrated by means of an actual example.

For blood groups, two female twins had the following phenotypic identity: B; MS; Rh_1(CDe); P_1; Le^a negative; K negative; Lu^a negative; Fy^a negative. Since the relevant table gives the frequency of B children as 0.084509 and the frequency of two sibs being B as 0.040062, the probability if one of two sibs is B that the other is also B becomes $0.040062/0.084509 = 0.4741$. The probability of a dizygotic twin pair being both B is equally 0.4741. Other tables provide probabilities of dizygotic twins being alike for the other blood groups. These probabilities are listed in the Part A of Table 77, together with the initial probability of a twin pair being dizygotic and that of a dizygotic pair being like-sexed. Multiplication of all these independent probabilities yields a combined probability of 0.0183 for finding twins being dizygotic and alike in all listed traits. Since the probability for finding twins who are monozygotic and, of course, alike in all listed traits is 0.3480, the probability of the two twins being dizygotic is $0.0183/(0.0183 + 0.3480) = 0.0499$. This is close to 5 chances in 100 of the twins being dizygotic, or 95 chances in 100 of their being monozygotic. These probabilities, while highly suggestive, are not sufficient to establish the origin of the twins with near-certainty. To approach this goal, additional traits must be considered.

Two such traits relate to the blood groups Yt and Do discovered after the analysis of the two female twins just discussed had been made. Part B of Table 77 gives the probabilities of likeness of two dizygotic twins in regard to specific phenotypes of the Yt and Do groups. Had these blood groups been analyzed in combination with those in Part A, the combined probability of A and B would have been 0.0061 and the total combined probability of dizygosity of the twins would have been reduced to 0.0172. This is less than two chances in 100 of the twins being dizygotic, or over 98 chances in 100 of their being monozygotic.

Determination of blood groups and other clearly definable genetically variable constituents has displaced former procedures which based the diagnosis of type of twins on such genetically complex traits as number of finger ridges, hair color, eye color, type of lanugo, and shapes of different facial parts. Great similarity in all such traits was taken as evidence for monozygosity, while striking dissimilarity in one or more traits was regarded as evidence for dizygosity. In general, the results of such inspection led to high degrees of agreement with independently made determination of zygosity by means of studies of blood groups. Race and Sanger recount that "for many years Mr. James Shields has been sending us samples of blood from twins." They found "that the blood groups practically never contradict the opinion of such a skilled observer of twins."

TABLE 77
The Similarity Diagnosis for a Pair of Female Twins Who Are Identical in Ten Blood Groups.

Trait	*Probability*
A. Dizygotic origin	0.6520
Likeness in sex	0.5000
Likeness in B	0.4741
Likeness in MS	0.5161
Likeness in Rh	0.5400
Likeness in P positive	0.8489
Likeness in Le^a negative	0.8681
Likeness in K negative	0.9485
Likeness in Lu^a negative	0.9614
Likeness in Fy^a negative	0.6319
Combined probability of dizygotic origin and likeness in above characters	0.0183
B. Likeness in Yt (b+)	0.5356
Likeness in Do (a−)	0.6241
Combined probability of A and B	0.0061
Total combined probability of dizygosity of the twins	0.0172

Source: A, after Smith and Penrose.

Tissue Grafts Between Twins

In an earlier chapter, we described how pieces of skin taken from one person and grafted to another do not remain healthy but slough off after a few weeks. This histo-incompatibility is due to the different genotypes of the tissues of the donor and the recipient. Permanent and successful grafts can be obtained, however, if the donor and recipient are monozygotic twins. Since identical twins have all genes in common, grafts from one twin to the other are "accepted," as are grafts from one part to another part of the same body. If diagnosis of identity or nonidentity is important enough, tissue grafts can be used as the deciding criterion. For example, they were employed in the establishment of zygosity of the first pair of heterokaryotic monozygotic twins studied.

The skin-graft test was also decisive in a famous case of disputed twin identity in Switzerland. Six years after the birth of a pair of clearly nonidentical twins, Victor and Pierre, it was observed that another boy, Eric, of another family bore a striking resemblance to Victor. The suspicion arose that a mistake had been made in the assignment of the babies, all of whom were born on the same day in the same hospital. It seemed likely that Eric was

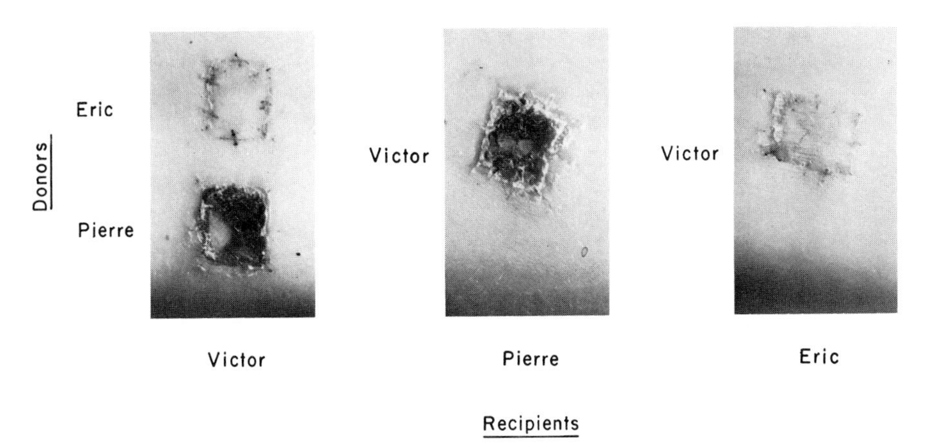

FIGURE 258
Skin grafts and their fate. Pieces of skin from the identical twins Victor and Eric, grafted from one to the other were accepted by the recipients. Grafts between Victor and Pierre, children of different parentage, did not take. (Franceschetti, Bamatter, and Klein, *Bull. Acad. Suisse Sc. Médic.*, **4,** 1948.)

actually the identical twin of Victor, and that Pierre was a single-birth child belonging to the second family. This conclusion was proved correct when skin grafts from Victor to Eric, and vice versa, were permanently accepted (Fig. 258). (On the basis of the skin-graft results it was decided to re-exchange Eric and Pierre and thus to restore the two biological families—a decision that was not without psychological stresses.)

Because of the compatibility of tissues of identical twins, it has also been possible to transplant successfully a thyroid gland and, many times, a kidney from a healthy individual to his identical twin whose corresponding organ had begun to fail.

PROBLEMS

189. Outline three possible sets of situations, each representing effects of your choice, of three genotypes, A, B, and C, in two environments, X and Y, which fit the following ranking arrangements, I, II, and III (1 = best, 6 = poorest):

	I		II		III	
Environment	X	Y	X	Y	X	Y
Genotype A	1	2	1	6	1	3
Genotype B	3	4	3	4	2	4
Genotype C	5	6	2	5	6	5

190. If among 10,000 twin births in Japan, there were 4,350 male pairs, 4,150 female pairs, and 1,500 pairs of mixed sex:
    (a) What were the numbers of monozygotic and dizygotic twins?
    (b) How does the ratio of monozygotics to dizygotics in this population compare with that in a white population?

191. Assume a sex ratio of 1:1.
    (a) What sex combinations of separate-egg quadruplets do you expect, and in what proportions?
    (b) Among a total of 48 American quadruplets, there were 13 with four boys, 6 with three boys and one girl, 12 with two of each sex, 7 with one boy and three girls, and 10 with four girls. How do these data compare with the expectation derived in Part (a)?
    (c) What, most likely, accounts for the deviation?

192. The afterbirth of three pairs of twins consists of:
    (a) 1 amnion and 1 placenta;
    (b) 2 amnions and 1 placenta;
    (c) 2 amnions and 2 placentas. Discuss the problem of monozygosity or dizygosity for each of the three pairs of twins.

193. What is the probability that:
    (a) 2 identical twins from MN parents will both belong to type MN?
    (b) 2 nonidentical twins from MN parents will both belong to type MN?
    (c) 2 girl twins are not identical if they are alike in being *B*, *MS*, and Lu^a negative? (To obtain the answer, use the information given in Table 77.)

194. A man (X) has a father who is an identical twin. If X is heterozygous for a very rare recessive gene, what is the probability that his offspring will be defective if he marries a first cousin who is the daughter of his father's twin brother? (Assume that all individuals mentioned have normal phenotypes and that the recessive gene enters the pedigree only once.)

195. A few marriages are known in which identical twin brothers have married unrelated identical twin sisters. Which relationship in typical families is genetically equivalent to marriages between cousins from these twin marriages?

REFERENCES

Bulmer, M. G., 1970. *The Biology of Twinning in Man.* 205 pp. Clarendon, Oxford.

Corner, G. W., 1955. The observed embryology of human single-ovum twins and other multiple births. *Amer. J. Obst. Gynecol.*, **70**:933–951.

Galton, F., 1874. *English Men of Science: Their Nature and Nurture.* 270 pp. MacMillan, London.

Gedda, L., 1951. *Studio dei Gemelli.* 1381 pp. Ediz. Orizzonte Medico, Rome.

Gedda, L., 1961. *Twins in History and Science.* (Translation, slightly revised, of Chapters I to VIII and X of the Italian edition, *Studio dei Gemelli.*) 240 pp. Thomas, Springfield, Ill.

Haldane, J. B. S., 1946. The interaction of nature and nurture. *Ann. Eugen.*, **13**:197–205.

McArthur, Norma, 1953. Genetics of twinning. A critical summary of the literature. *Australian Nat. Univ. Soc. Sci. Monogr.*, **1**:1–49.

Nance, W. E., and Uchida, Irene, 1964. Turner's syndrome, twinning, and an unusual variant of glucose-6-phosphate dehydrogenase. *Amer. J. Hum. Genet.*, **16**:380–392.

Newman, H. H., Freeman, F. N., and Holzinger, K. J., 1937. *Twins: A Study of Heredity and Environment.* 369 pp. University of Chicago Press, Chicago.

Parisi, P. (Ed.), 1970. *Advances in Twin Studies. Acta Genetica Medicae et Gemellolgiae*, vol. 19. 381 pp. (Proceedings of the First Symposium on Twin Studies.)

Rosin, S., 1947. Theoretisches zur Frage der ovozytären Zwillinge. *Arch. Julius Klaus-Stift.*, **22**:73–84.

Scheinfeld, A., 1967. *Twins and Supertwins.* 292 pp. Lippincott, Philadelphia.

Smith, Sheila M., and Penrose, L. S., 1955. Monozygotic and dizygotic twin diagnosis. *Ann. Hum. Genet.*, **19**:273–289.

Turpin, R., and Lejeune, J., 1969. Monozygotic twinning and chromosome aberrations (Heterokaryotic monozygotism). Ch. 13, pp. 248–261, in Turpin and Lejeune, *Human Affliction and Chromosomal Aberrations.* Pergamon, Oxford.

Verschuer, O. V., 1939. Twin research from the time of Francis Galton to the present day. *Proc. Roy. Soc. London*, **128**:62–81.

Weinberg, W., 1901. Beiträge zur Physiologie und Pathologie der Mehrlingsgeburten beim Menschen. *Arch. f. Ges. Physiol.*, **88**:346–430.

White, C., and Wyshak, Grace, 1964. Inheritance in human dizygotic twinning. *New Eng. J. Med.*, **271**:1003–1005.

26

HEREDITY AND ENVIRONMENT: II. PHYSICAL TRAITS

We are now ready for the data which twin studies contribute to the nature-nurture problem. They are of three main types: First, a phenotypic study of the two members of identical twin pairs who are reared, as is usual, in the same family may afford information on the influence of environmental differences which exist even for such twins in the expression of identical genotypes. Second, a comparison of the degree of the environmentally conditioned differences between the two members of identical twin pairs with the degree of differences between the two members of nonidentical pairs who develop under both environmental and genetic differences may provide a clue to the relative stregnth of nature and nurture in the production of differences in nonidentical twins. Third, a comparison of differences between the two members of identical twin pairs reared apart in different homes with the differences between the two members of identical pairs reared together may illuminate the part played by the greater variation in environment provided by different homes as compared to that of a single home. We shall draw on material from all three types of studies in discussing various points.

A brief reference should be made to the method of "co-twin control." It is customary in the testing of methods for the improvement of health, educational, and social procedures to use both an experimental and a control group of individuals. Applying such methods to population samples, which are nearly always *selected* in one way or another, makes it difficult to ascertain whether the new methods offer advantages or disadvantages over old ones, or whether their effect is neutral. However, if groups of identical twins are used, and if one of each pair is exposed to the new procedure and the other "kept as a control," unequivocal information can be obtained.

The material for twin studies in the nature-nurture problem consists partly of data on large samples of twins and partly of individual descriptions. From such descriptions, particularly of identical twins, facts emerge which sometimes highlight an astonishing similarity between twins (Fig. 259), and sometimes show up a striking difference (Figs. 260, 169). For a more general analysis, statistical data derived from a study of many twins are necessary. Such data provide some measure of the average difference or similarity between the two members of numerous identical pairs as compared to the average between numerous nonidentical pairs. Although the averages of the differences and correlation coefficients are among the measures used, much more refined statistical methods have also been applied to the problem.

Detailed comparisons involving identical and nonidentical twins must be based on *unselected* samples—a necessity that has only been realized in more recent years. Earlier, the data were usually based on the sum of cases culled from the scientific or medical literature. In these reports, there is often a proportion of identical over nonidentical twins in excess of that expected in a random sample. The reason for this excess may be that it seemed more worthwhile to publish cases of striking likeness of twin pairs than of unlikeness—and likeness is more frequent in identicals. But it is also true that an author may be more astonished by a striking difference between identical twins than by lack of a difference, so that he may describe cases of differences rather than of similarities. Many studies also leave some doubt about the identity or nonidentity of twin pairs. It is likely that in some cases an author classified certain pairs as identicals or nonidenticals because they agreed or differed in the trait in which he was interested, instead of diagnosing the type of twinning independently of the trait.

CONCORDANCE AND DISCORDANCE

A particularly simple way of scoring differences between twins is to evaluate traits which are either present or absent. Thus, twins may be either *concordant*, that is, both possess or both are free of a particular trait; or

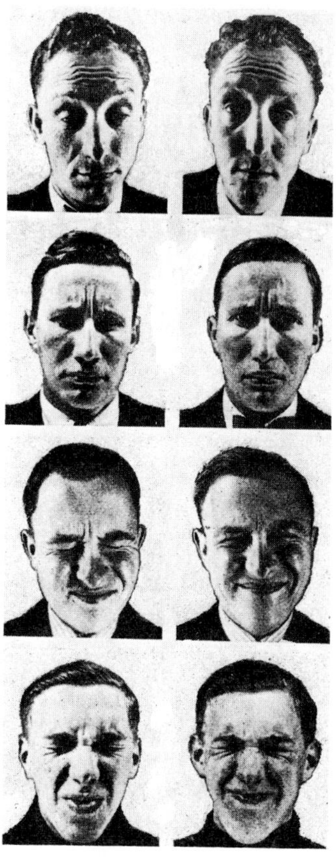

FIGURE 259
Similarity in facial folds of four pairs of identical twins. (After Buehler, from Abel, *Hdbch. Erbbiol. d. Menschen,* **I,** Springer, 1940.)

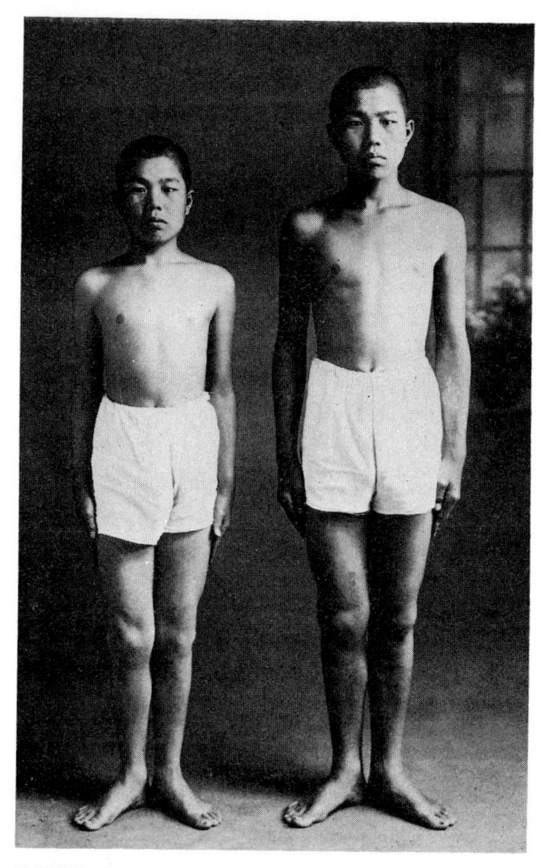

FIGURE 260
A pair of identical twins, aged 15. Until these twins were 5 years old, they were very similar in size. The left twin then became retarded in growth, probably from abnormal functioning of the pituitary gland. (Komai and Fukuoka, *J. Hered.,* **25,** 1934.)

discordant, in that only one of the pair possesses the trait. The terms concordance and discordance can also be applied to twin pairs who have been classified into the two groups, "similar" and "dissimilar" for some specific trait. The expression of many traits is different in males and females. In order to exclude the influence of sex on dissimilarity of twins often only like-sexed twin pairs are used in comparisons of twins. The usual procedure in obtaining concordance rates in twin pairs consists of the ascertainment of affected twin individuals (probands) followed by the determination of whether the twin partners are also affected (++) or are not affected (+−). In detail, concordance rates depend on the method of ascertainment used, which differs in unselected as compared to selected series. The rates given below are taken from various authors and are not always free from ascertainment bias.

Anatomical Traits. Of the exceedingly numerous traits studied, only a few can be discussed. Results obtained by Newman (a biologist), Freeman (a psychologist), and Holzinger (a statistician) on height, weight, head length, and head width are listed in Table 78. (Data on mental traits obtained in the same study will be presented in Chapter 27.) These three investigators studied fifty pairs of identical twins reared together, fifty (sometimes fifty-two) pairs of nonidentical twins reared together, nineteen pairs of identical twins who had been separated very early in life and reared in different homes, and for purposes of some comparisons, a group of fifty-two regular sibs.

Identical twins reared together (second column of the table) showed differences in all four physical traits. This is an effect of nongenetic agents—that is, the environment—which prenatally or postnatally gives rise to phenotypic differences in spite of identical genotypes. In Figure 261, these differences, for height, are shown graphically. There, the curve for the "intrapair" differences—that is, between twins—shows that 35 of the identical pairs were less than 2 cm different in height; 12 were between 2 and 3.9 cm different; 1 was between 4 and 5.9 cm different; and 2 were between 6 and 7.9 cm different. By themselves, neither the fact that there are environmentally caused differences in physical traits of identical twins reared together nor the specific distributions of these differences are very informative. They become significant, however, when comparisons are made between these findings for identical twins reared together and the corresponding findings for the other groups of individuals. The third and fourth columns of Table 78 show that the average intrapair differences are larger in nonidentical than in identical twins for each of the four traits and also larger in sibs for the two traits for which data are available.

TABLE 78
Average Differences Between the Two Members of Identical Twins, Nonidentical Twins, and Pairs of Sibs, Reared Together; and Identical Twins, Reared Apart.

Difference in	*Identical*	*Nonidentical*	*Sibs*	*Identical (Reared Apart)*
Height (cm)	1.7	4.4	4.5	1.8
Weight (kilogram)	1.9	4.5	4.7	4.5
Head length (mm)	2.9	6.2	–	2.20
Head width (mm)	2.8	4.2	–	2.85

Source: After Newman, Freeman, and Holzinger.

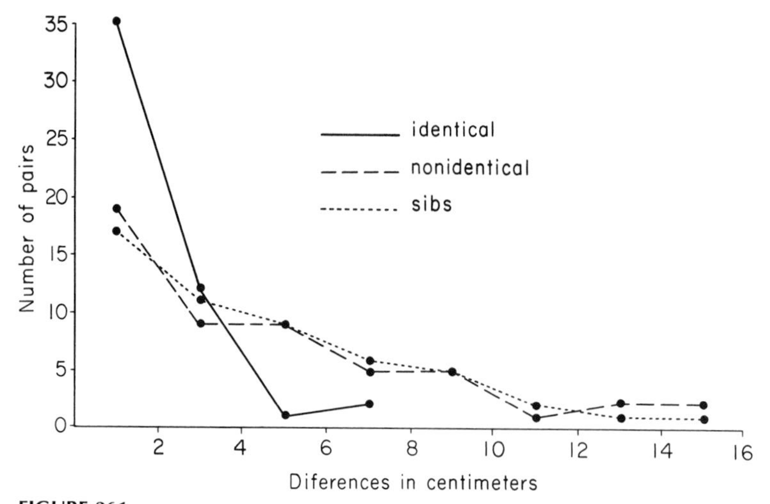

FIGURE 261
Curves of distribution of differences in standing height of 50 identical twins, 52 nonidentical twins, and 52 pairs of sibs. (After Newman, Freeman, and Holzinger.)

Moreover, as shown for height in Figure 261, the ranges of these differences and their distributions were different from those for the identical pairs reared together, but very similar for nonidentical and sib pairs. Among other details, the graphs show that only 19 nonidentical and 17 sib pairs differed less than 1.9 cm in height, as compared to 35 identical pairs reared together; and that 10 nonidentical and 9 sib pairs differed by more than 8 cm, but not a single identical pair. The much greater dissimilarity of nonidentical twins, as compared to identicals, must be ascribed to their different genotypes, in addition to environmental differences which act on them as they do on the identicals.

It is interesting to note that the nonidentical twins and the sibs are practically alike in the degree and distribution of the differences in height and weight, which were the only traits measured in pairs of sibs. Since nonidentical twins correspond genetically to ordinary sibs, their similarity in these studies not only reflects their genetic correspondence but, further, suggests that the average environmental differences which influence height and weight are not larger for two sibs born at different times than for those born together.

The fifth column of Table 78 contains the important average intrapair differences for identical twins reared apart. More will be said later about the differences in home environment in which the two members of each of these pairs lived. For most pairs, the home environments were not strikingly different. It is seen that the weight difference, 4.5 kilograms, is not like that for the identical twins reared together, but is very similar to that for nonidentical

TABLE 79
Correlation Coefficients for Four Physical Traits in Pairs of Twins and Sibs. (Compare with Table 78.)

Trait	*Identical*	*Nonidentical*	*Sibs*	*Identical (Reared Apart)*
Height	0.932	0.645	0.600	0.969
Weight	0.917	0.631	0.584	0.886
Head length	0.910	0.691	–	0.917
Head width	0.908	0.654	–	0.880

Source: Newman, Freeman, and Holzinger; Woodworth.

and sib pairs. We may conclude that weight is a highly modifiable trait, since the different environments in which the separated twins grew up made them as unlike in weight as two genetically unlike sibs reared in the same home. Environmental differences had little effect on the other three traits, height, head length, and head width: the intrapair differences of the separated identicals are as small as those of the nonseparated.

These relations are expressed once more in Table 79, but as correlation coefficients instead of mean differences. All coefficients in the table are positive, signifying that, in general, both partners of a pair deviated from the average in the same direction. The correlations are higher between identical twins reared together than between nonidentical twins or between sib pairs. For separated identical twins, the coefficients are similar—even in the case of weight—to those for nonseparated identical twins.

Physiological Traits. People differ not only in obvious external ways but in innumerable details of both external and internal traits. Interesting pilot studies on metabolic differences have been instigated by the chemist R. J. Williams. He tested various individuals for 31 physiological traits, including taste sensitivity to five different substances, the amounts of twelve substances in the saliva, and the presence of fourteen substances or properties of the urine. The findings were graphically represented by the length of 31 lines which radiate from a center. A hypothetical "average" individual's metabolic pattern would be one in which all 31 rays were drawn of equal length (Fig. 262, A). Actual individuals differ greatly from the average as well as from one another (B and C). Identical twins were found to be strikingly similar in their metabolic patterns (D and E). Yet, in addition to this similarity, which probably has a genetic basis, nongenetic influences can be discerned. Thus, lysine, an amino acid, was present in relatively large amounts in the saliva of one twin but in only small amounts in the other (D and E, lines 12); and the concentration of citrulline, another amino acid, was low in the urine of the first and high in that of the second twin (lines 30).

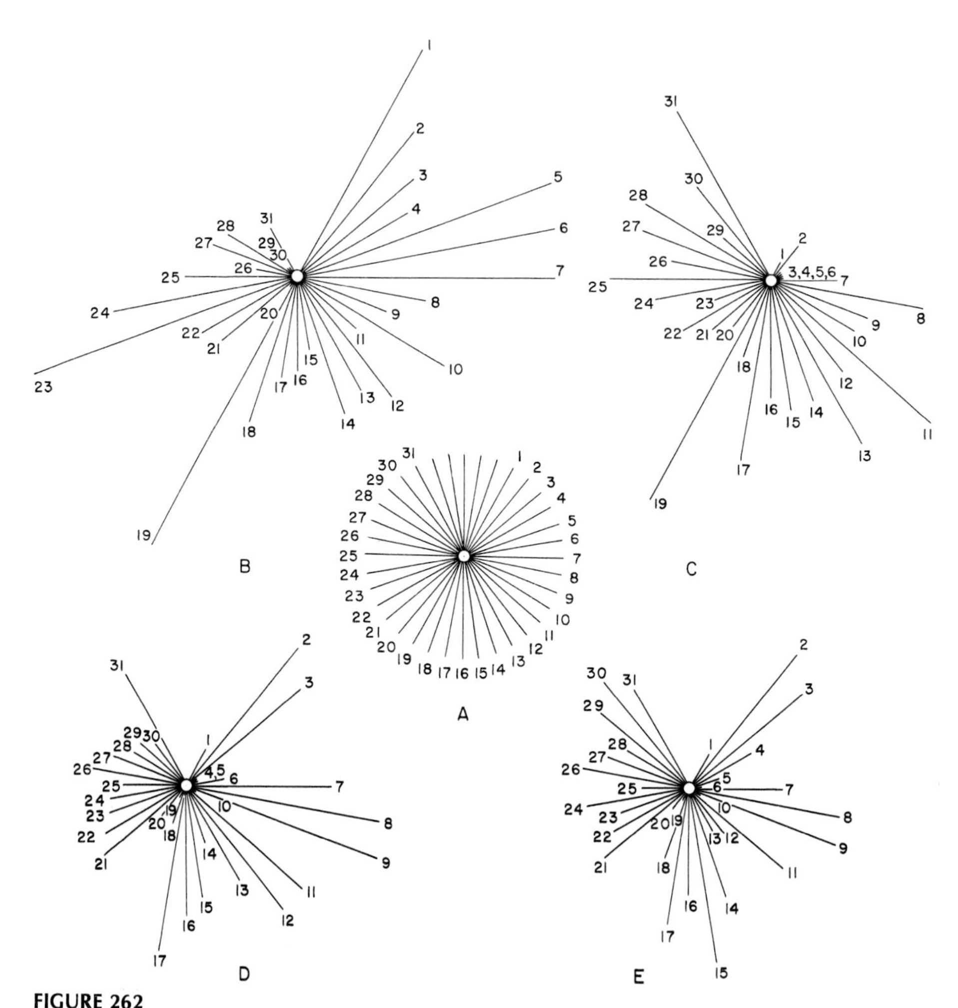

FIGURE 262
The "metabolic patterns" of a hypothetical average individual **(A)**, of two unrelated persons **(B, C)**, and of a pair of identical twins **(D, E)**. The 31 traits are taste sensitivity for: creatinine (1), sucrose (2), KCl (3), NaCl (4), HCl (5); salivary constituents: uric acid (6), glucose (7), leucine (8), valine (9), citrulline (10), alanine (11), lysine (12), taurine (13), glycine (14), serine (15), glutamic acid (16), aspartic acid (17); urinary constituents: citrate (18), an undefined "base Rf. 28," (19), an undefined "acid Rf. 32," (20), gonadotropin, a hormone (21), pH (22), pigment/creatinine ratio (23), chloride/creatinine (24), hippuric acid/creatinine (25), creatinine (26), taurine (27), glycine (28), serine (29), citrulline (30), alanine (31). (Williams.)

A more detailed comparison of a physiological process in twins concerned changes in blood sugar after intake of 50 grams of glucose. By analyzing samples of blood taken at intervals over a period of four hours, a curve was plotted for each individual showing the initial blood-sugar concentration, its rise, and its subsequent decline. For pairs of identical twins, most of the

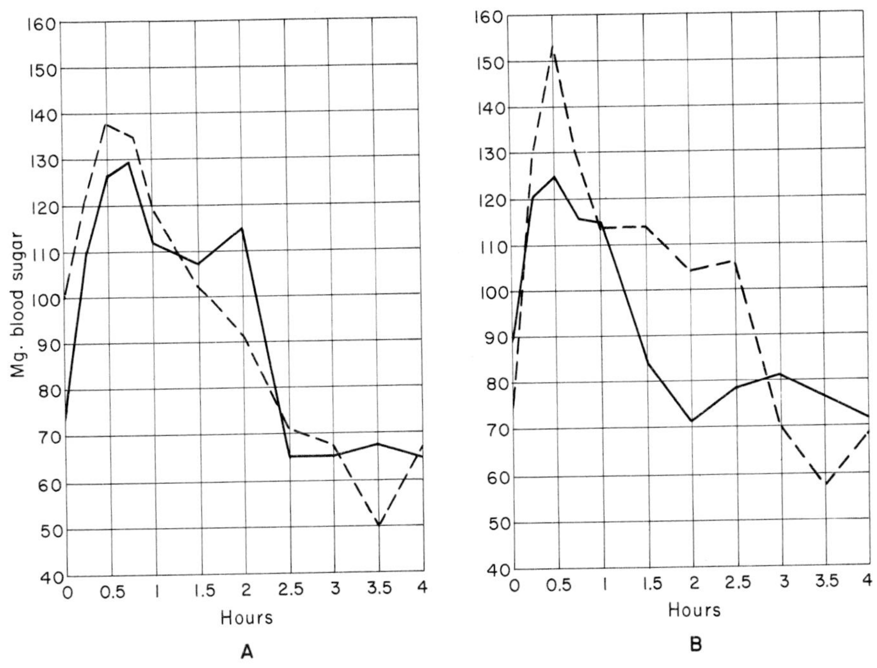

FIGURE 263
Curves of blood-sugar concentration in a pair of identical (**A**) and a pair of nonidentical (**B**) twins. (After Werner, *Ztschr. f. Vererb.*, **67**, 1934.)

pairs of curves were very similar; for pairs of nonidenticals, they differed considerably (Fig. 263, A and B). Nevertheless, as seen in Table 80, for unknown nongenetic causes, even identical twins often differed in their reactions.

Two other physiological traits studied in twins are blood pressure and pulse rate. Concordance in such continuously varying traits is defined as similarity within a specified range. Thus, in the study of blood pressure, concordance meant agreement of the twins within a pressure difference of less than 5 mm mercury. It occurred in 63 per cent of identical twins (Fig.

TABLE 80
Relations of Blood-sugar Curves in 30 Pairs of Identical and 32 Pairs of Nonidentical Twins.

Twins	*Very Similar*	*Intermediate*	*Very Dissimilar*
Identical	10	14	6
Nonidentical	3	9	20

Source: After Werner, *Deutsch, Arch. f. klin. Medizin*, 1936.

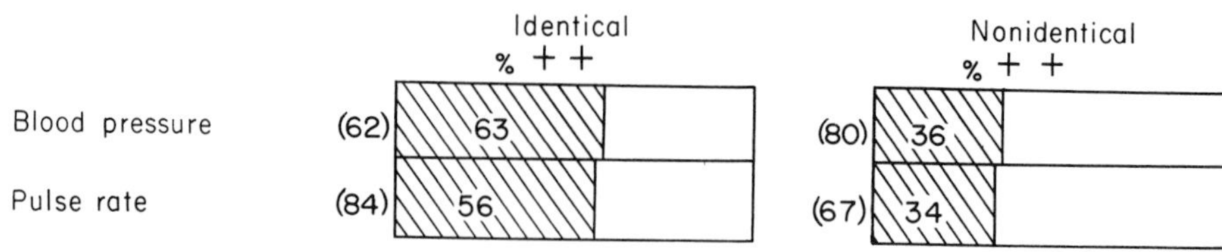

FIGURE 264
Concordance and discordance in twins for blood pressure and pulse rate. The total width of each bar is equal to 100 per cent. The diagonally lined section represents the per cent of concordance, the white section the per cent of discordance. Numbers in parentheses signify the total number of twin pairs investigated. (After Malkova, *Proc. Maxim Gorki Med.-Biol. Res. Institute,* **3,** 1934.)

264, upper half). This shows environmental plasticity of the trait, since, in spite of genetic identity, the rest (37 per cent) were discordant. Again, taken by themselves, these two percentages do not permit further deductions. The higher concordance than discordance figure could be due either to a relatively strong effect of genetic identity or to the similar environment in which the twins grew up. Here, the percentages for the nonidentical twins become significant. They are nearly reversed (as compared to the identical twins): 36 per cent concordant pairs and 64 per cent discordant pairs. Clearly, the genetic dissimilarity of the nonidentical twins has contributed its share to the discordance, thus making it larger than for the identical twins.

In addition to the concordance determination for blood pressure, data are available on the mean intrapair difference of this trait as expressed in mm of mercury: 5.1 for 112 identical and 8.4 for 82 nonidentical twins. These figures are in agreement with the information on concordance and with the interpretation that heredity has a share in the determination of differences in blood pressure. Similar conclusions hold for the pulse rate (Fig. 264, lower half).

TABLE 81
Mean Difference, in Months, in Time of First Menstruation.

No. of Pairs	*Relationship*	*Difference*
51	Identical twins	2.8
47	Nonidentical twins	12.0
145	Sibs	12.9
120	Mother-daughter	18.4
120	Unrelated women	18.6

Source: Petri, *Zeitschr. Morph. u. Anthropol.,* **33,** 1934.

TABLE 82
Intrapair Differences in Life Span of Deceased Twin Pairs Who Were More Than Sixty Years Old at Time of Death.

Twins	*No. of Pairs of Twins*	*Difference in Time of Death (Months)*
Identical	18	36.9
Nonidentical	18	78.3

Source: Kallmann and Sander, *J. Hered.*, **39.** 1948.

Another physiological trait whose dependence on differences in both nature and nurture has been clarified by the twin method is the beginning of female puberty as defined by the first menstruation. Here, the data are particularly satisfactory, since they permit various comparisons (Table 81). Not only is it obvious that identical twin sisters have a very much smaller mean intrapair difference than nonidentical twins, but differences between sibs, between mothers and daughters, and between unrelated women compared in pairs indicate greater environmental differences acting in different families and in different generations than those acting within the same family.

Rate of aging and longevity are collective terms which, undoubtedly, cover a great variety of physiological properties whose details and interrelations are but poorly understood. In spite of the complexity of these two traits, the twin method gives evidence of important hereditary elements in their determination (Table 82). See also page 454.

Other Traits. Twin comparisons have been carried out on so many traits that even a simple listing would be a lengthy task. In order to suggest the variety of properties investigated, Figures 265 and 266 have been abstracted from a very broad survey of Japanese twins (children and students). Figure 265 shows the similarities and differences in form of internal organs and in physiological reactions. Figure 266 shows the intrapair deviations, in percentage of the means, of some traits relating to lung area and blood cells. Identical twins usually score closer to each other than nonidentical twins, but the difference between the two groups is great for some and smaller for other traits.

To obtain information on nongenetic factors affecting the variability of traits, some of the Japanese twins have been separated into two groups: (1) those partners who had similar birth weight, were reared together, and had a similar history of illnesses; and (2) those differing in these respects. The differences between identical twin pairs of group (1) were smaller than between those of group (2), and the same held for nonidenticals. This demonstrates the influence of environmental agents. Nevertheless, the mean

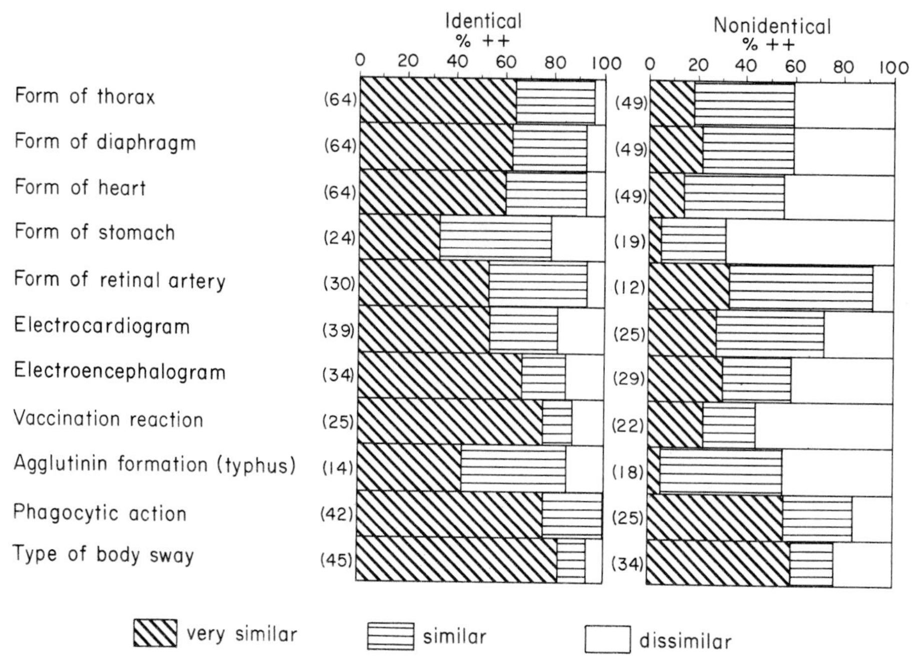

FIGURE 265
Similarity and dissimilarity of identical and nonidentical twin pairs for a variety of traits. The numbers in parentheses signify the numbers of pairs considered. (After Osato and Awano, *Acta Genet. Med. Gemellol.* **6**, 1957.)

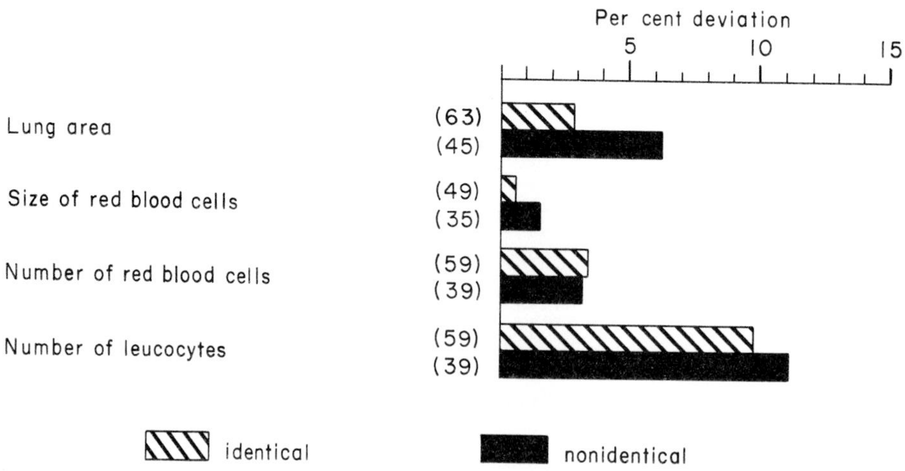

FIGURE 266
Mean percentage deviation of four traits in identical and in nonidentical twin pairs. The numbers in parentheses signify the numbers of pairs considered. (After Osato and Awano, *Acta Genet. Med. Gemellol.*, **6**, 1957.)

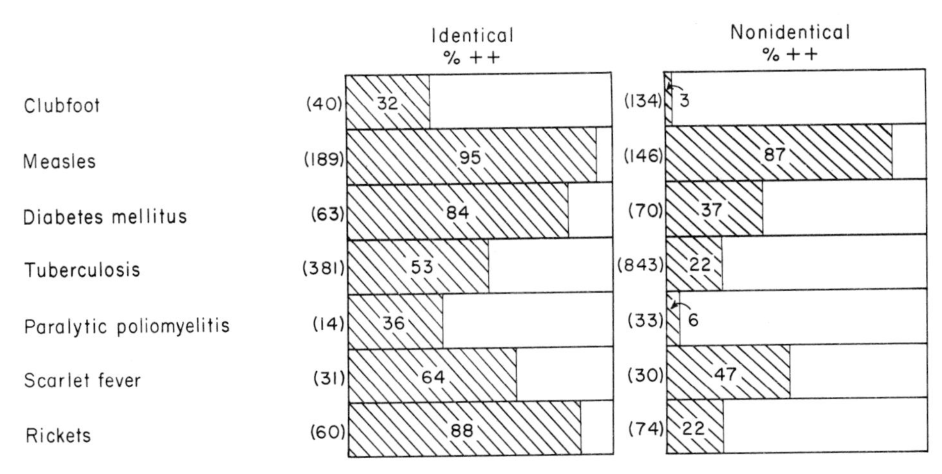

FIGURE 267
Concordance and discordance in twins affected by various pathologic conditions. (After v. Verschuer, *Ergebn. Allg. Pathol.*, **26**, 1932; *Beiträg. z. Klinik d. Tuberkul.*, **97**, 1941; and *Acta Genet. Stat. Med.*, **6**, 1956 [tuberculosis]; Idelberger, *Ztschr. Orthop.*, **69** Suppl., 1939 [clubfoot]; and Herndon and Jennings, *Amer. J. Hum. Genet.*, **3**, 1951 [poliomyelitis].)

differences between identicals with different history were smaller than those between nonidenticals with similar history—an indication of the strength of hereditary influences.

Pathological Conditions. Various pathological conditions have been studied in twins. In collecting the case material, the appropriate procedures consisted of taking a sample of affected individuals and then finding out who of these had twin partners. These twin partners were then investigated in order to determine (a) whether they were identical or nonidentical twins of the propositi and (b) whether or not they were, or had been, affected by the same condition as the propositi.

The results of some of these studies, assembled in Figure 267, are striking. For the congenital deformity clubfoot, 32 per cent of identical twins are concordant. This rather low incidence might suggest that special prenatal circumstances are the only ones which decide whether a child will be born with, or without, the deformity. Although the importance of such prenatal variations in the outer or inner environment of the embryo is, of course, obvious from the frequent discordance of newborn identical twins, there is still the question whether every genotype will respond to such conditions or whether a special genotype or genotypes are necessary for the formation of clubfoot. The validity of this latter interpretation is borne out by the incidence of clubfoot in the nonidentical twins: concordance here is only 3 per cent. Since concordance is so much lower in nonidentical than in identical twins, genetic nonidentity is probably the decisive factor which determines that one twin will be normal in most cases when the other is affected. Studies on clubfoot also serve to emphasize that the frequency of this trait, as well as that of many other abnormal traits, is no higher in nonidentical twins than in single-birth sibs: among the 1,525 sibs of 624 propositi with clubfoot, 46 (equal to 3 per cent) were likewise affected. Family studies suggest that clubfoot is a quasi-continuous trait. It occurs more frequently in males than in females, and its frequency among the relatives of female propositi is greater than among those of male propositi.

Concordance in identical twins is higher for the other conditions listed in Figure 267 than for clubfoot, but, as before, only limited conclusions on the nature-nurture problem can be drawn as long as data of one kind alone are considered. Thus, concordance of 95 per cent in susceptibility of identical twins to measles does not necessarily imply a special hereditary inclination, with very high penetrance, for being affected by this disease. It might mean only that any two children of equal age in the same home will probably get measles if one of them comes down with the disease. That this is close to the truth is suggested by the nonidentical pairs, who have, in spite of their genetic diversity, a concordance nearly as high as that of the identical twins. Yet, there may be some special hereditary element involved, since the difference in concordance between identical and nonidentical twins is significant.

The diseases studied represent several different types of pathological conditions: diabetes mellitus is an abnormality of metabolism; tuberculosis, poliomyelitis, and scarlet fever are caused by microorganisms; and rickets is caused by a deficiency of vitamin D. What, may be asked, has the heredity-environment problem to do with the last four diseases, which so clearly seem externally conditioned? Robert Koch's discovery of the tubercle bacillus in the last century seemed, to many, to dispose forever of the idea that tuberculosis is hereditary; and the early reports of rickets "running in families," which had been interpreted as proof of the hereditary nature of the disease, seemed later to be explainable on the basis of continuous existence of poor nutritional conditions in successive generations. There is, however, no contradiction between the fact that an external agent whose presence (as in tuberculosis) or absence (as in vitamin deficiency) leads to a disease and the fact that different genotypes may make different individuals more or less susceptible to infection by microorganisms or to nutritional deficiencies.

Although many of the concordance figures clearly indicate hereditary differences in susceptibilities to diseases or to vitamin D deficiency, they give only an incomplete picture. This may be shown for tuberculosis and tumors.

Tuberculosis. To no other pathological condition has the twin method been applied as extensively as to tuberculosis. Studies in different countries and covering different periods are available from Germany, Switzerland, the United States, Argentina, Japan, England, and Denmark. Although the frequencies of concordance vary considerably from study to study, perhaps depending on the state of public health and therapy, they all agree in showing a higher frequency of concordance in identical than in nonidentical pairs. A summary treatment as given in Figure 267 of the 1,224 affected twin pairs obscures the fact that concordance included much greater similarity in the expression of the disease for identical twins than for nonidentical twins. In the pioneer studies of Diehl and von Verschuer, concordance for identical twins frequently signified not only that both partners were affected, but also that if one twin had a particular lobe of a lung affected, the other twin was affected in the same part and not anywhere else, or if one was infected in his kidneys the other would be attacked likewise. Concordance in nonidentical twins does not show such similarity.

Conversely, discordance in identical twins was rarely as striking as in nonidentical twins. Only one extreme case of discordance, in which one twin was normal but the other died of tuberculosis, occurred in 16 identical pairs, but such extreme cases occurred in approximately one-third of 68 nonidentical pairs. Two particularly striking examples of similarity in identical twins and dissimilarity in nonidentical twins were: (1) concordance in a pair of identical girl twins, separated for years, one of whom became tuberculous on a farm at the same time that the other became affected in the city; and (2) discordance in a pair of nonidentical boys who lived under slum conditions and slept in one bed, and yet one remained healthy while the other succumbed to tuberculosis. In good agreement with these findings, Kallmann and Reisner's American data show that the degree of resistance to tuberculosis is much more similar in identical than in nonidentical pairs: defining degrees of similarity by somewhat arbitrary quantitative values, they found that identical twins were nearly six times more similar in resistance and susceptibility than were nonidenticals.

Both the German and the American tuberculous twins have been reinvestgated after the lapse of twenty and ten years, respectively. Such "longitudinal" studies are valuable for, among other things, the information they provide on any later development of the disease in the twins. Thus, among the 15 identical and 18 nonidentical German pairs who were concordant for tuberculosis in 1935, both partners of 4 identical pairs, but not a single pair of nonidentical pairs, had died from the disease by 1955. The 1955 survey also furnished important information on the possibility of complete recovery from tuberculosis. In addition to the 4 pairs of identicals who died, there were 6 initially concordant pairs in which by 1955 one twin had died but the other was in good health. In good agreement with these findings on recovery are the results of the restudy of the American twins. There is, however, no agreement on the persistence of the initially greater similarity of identical

TABLE 83
Percentage Frequencies of Tuberculosis in the Families of 308 Tuberculous Twins.

Relationship to Affected Twin	*No. in Group*	*Percentage Affected*
Unrelated general population		1.4
Spouses	226	7.1
Parents	688	16.9
Half-sibs	42	11.9
Sibs	720	25.5
Nonidentical twin	230	25.6
Identical twin	78	87.3

Source: Kallmann and Reisner, *Amer. Rev. Tuberculosis*, **47**, 1943.

than of nonidentical twin partners. The initial differences in degree of similarity between identical and nonidentical pairs were striking in both the German and American twins. The follow-up investigations revealed that these initial differences had persisted in the German group, but had disappeared in the American group. The causes of this disagreement have not yet been resolved.

Studies on the frequencies of a trait in twin partners as well as in other members of the families of affected twins have been particularly informative. Table 83 gives such twin-family data on the tuberculous twins of the American sample. Many instructive comparisons are possible, such as comparison of the frequencies of affected identical and nonidentical co-twins, of nonidentical co-twins and sibs, of sibs and half-sibs, of sibs and their parents and spouses, and of all these with frequencies in the general population.

The conclusion of these studies that susceptibility to tuberculosis is partly under genetic control is supported by the results of experiments in both man and animals. It has been found that the local reaction of the skin to tuberculin injection—that is, the size of the red area—varies from person to person. The difference in diameters of the spots is always small for identical pairs. The differences vary greatly for nonidentical twins and, on the average, are several times larger than for identical twins. There are also strains of rabbits which show strikingly different responses to controlled infections with human-type tubercle bacilli. In some strains, death regularly ensues; in others, the disease soon becomes localized, and the restricted area of infection disappears after a time.

Tumors. There are some genes which are clearly the basis for specific types of cancer. The recessive gene for xeroderma pigmentosum leads to tumors of the skin, and a simply inherited dominant gene for finger-like growths in the large intestine (polyposis) typically leads to cancer of the colon which can be prevented only by timely surgery. The dominant gene for retinoblastoma causes tumors of the retina, and the gene for the complex Zollinger-Wermer syndrome is responsible for tumors of various endocrine glands, namely the pituitary, the parathyroids, the adrenals, and the islets of Langerhans in the pancreas. Certain syndromes, named after Bloom and Fanconi, respectively, are known in which a recessive gene leads to frequent breakage and rearrangements of the chromosomes as seen in cell cultures. Often these syndromes include leukemia. For the great majority of other types of malignant growths, no simple genetic basis is apparent, even though some individual pedigrees show a great concentration of affected individuals in contrast to most pedigrees, in which only one or few affected individuals occur.

Twin data from unselected series show that nongenetic factors are of overwhelming importance in the development of most cancers or their lack of development. This may be seen from the results of a large Danish study by Harvald and Hauge in which 164 identical twin pairs had been ascertained by a twin affected with cancer. In 87 per cent of the twins there was complete discordance in respect to the presence of cancer, and in 8 per cent of additional twins there was discordance in respect to the type and localization of the tumors, e.g., one twin having carcinoma of the stomach, the other of the breast. In 5 per cent only did the twin partners suffer from cancer at the same site as the propositi. The concordance rate for cancer of any type was not significantly different in identical and nonidentical twins, indicating that there was no constitution for the development of "cancer in general." There is evidence, however, for specific genotypes responsible for type and localization of some cancers. The clearest example is that of cancer of the stomach. In pooled data from a variety of sources its rate of concordance in regard to type and localization was 30 per cent in identical twins ($n = 48$) as compared to only 4 per cent in nonidenticals ($n = 73$). The difference is highly significant and suggests, but does not prove, genetic determination of cancer of the stomach. Similarly, uterine cancer was concordant in 15 per cent of identical twins ($n = 39$) as compared to 0 per cent of nonidenticals ($n = 73$).

The specificity of sites affected with cancer has also been found in population studies. Among the mothers and sisters of 200 propositi with cancer of the stomach studied by Woolf, there were 67 individuals afflicted with other types of cancers, not significantly different from the 72.8 individuals expected according to the frequency in a control population. In contrast, the number of stomach-cancer patients in the families of the propositi was 44, significantly higher than the expected 25.3. Similarly, the mothers and sisters of 200 propositi with breast cancer included 59 individuals with other types of cancers where 63 were expected, but 12 with breast cancer where only 5.3 were expected. (There is a report, by Rotkin, according to which there are two different sites at which a liability for a specific cancer is expressed. Here the relatives of patients with cancer of the uterine cervix have an increased frequency both of cervical and skin cancer.)

It is tempting to interpret the increased frequencies of relatives with specific cancers in terms of genetic predispositions. If such predispositions exist they would have low penetrance as shown particularly by the high discordance rates of identical twins. Given the facts that many external variables such as socioeconomic level, geographical distribution, reproductive history of women, dietary habits, and sexual patterns are correlated with incidence of specific types of cancers, we must consider that variations in these factors play decisive roles in the causation of tumors. Added to this is the discovery that at least some cancers are caused by viral infections. Could not, therefore, the increased frequency of specific cancers in the relatives of propositi be due not to shared genotypes but to shared environments? Woolf has attacked this question by studying the incidence of cancer in the genetically unrelated spouses of propositi, who obviously share a greater degree of environmental similarity with the propositi than the rest of the population. He found no increased frequency of cancer in the spouses but carefully pointed out that this finding is compatible with the possibility that the sensitive period for initiating induction of cancer may exist at young ages, before marriage and thus at a time when the future marriage-partners live in different environments. It must be added that the increase in cancer frequencies in the close relatives of propositi has not been encountered in several other studies. At least in part, the different findings depend on biases difficult to avoid in the selection of suitable control populations.

In summary, the evidence, both from twin data and from population studies, is compatible with the existence of a weak genetic component in the occurrence of cancers in addition to the undoubtedly powerful, nongenetic influences. In absolute numbers, the increase over a control population in incidence of cancers of the stomach and of the mammary gland in relatives of propositi is slight. In Woolf's data, expressed in per cent of deaths due to cancer of the stomach, the increase is from 1.5 per cent in the controls to 2.7 per cent in the families of the propositi, and in mammary cancer from 0.7 per cent in the controls to 1.6 per cent in the families of the propositi.

LIMITATIONS OF THE TWIN METHOD

It is noteworthy that the twin method affords a recognition of a hereditary basis for phenotypic differences, but not an analysis of the genotypes responsible. The method provides no answer to the problem of whether dominant or recessive genes, autosomal or sex-linked alleles, single-gene or polygenic combinations, or one or many different genotypes are responsible for the appearance of a trait. Nor does the recognition of "nurture" as having a share in the variability of man involve, by itself, a recognition of what kinds of environmental factors play a role. These are natural limitations of the method which deals with "heredity" and "environment" primarily as collective terms.

Only occasionally can twin data be used to test some specific genetic hypothesis. If, for instance, a given trait is believed to be the result of a simple dominant gene A, then, according to Rife, the following relation should hold for the frequency of discordance among pairs of nonidentical twins. Such pairs can come from two kinds of matings: (1) $Aa \times aa$, and (2) $Aa \times Aa$. These should occur with the frequencies $4pq^3$ and $4p^2q^2$, respectively, where p and q are the allele frequencies. The frequency of discordant pairs Aa, aa from mating (1) will be $2 \times (1/2)^2 \times 4pq^3 = 2pq^3$; and of discordant pairs AA, aa and Aa, aa from mating (2), $2 \times (1/4)^2 \times 4p^2q^2 + 2 \times 1/2 \times 1/4 \times 4p^2q^2 = 3/2\ p^2q^2$. The sum of discordant pairs from both types of matings is $2pq^3 + 3/2\ p^2q^2$. Obviously, the value of this sum varies with the allele frequencies. It can be shown that it never rises above slightly more than 27 per cent. Therefore, the finding of discordance frequencies significantly greater than 27 per cent is, by itself, evidence against simple dominant determination of a trait.

In the preceding pages, a smaller mean intrapair difference, a greater intrapair correlation, or a greater concordance in identical, as compared to nonidentical twins, has been considered the result of genetic identity of the monozygotic twins. Various students of human genetics have gone beyond this statement and have attempted to assign specific quantitative values to the shares which heredity and environment hold in the determination of the observed differences. A particularly simple way of treating the problem has been to assume that the environments under which two identical twins grow up are, on the average, neither more nor less different than the environments of two nonidentical twins. The average intrapair difference, or the discordance, of a trait for identical twins is, therefore, regarded as the measure of environmental differences in identical as well as nonidentical twins. Whenever nonidentical twins have greater intrapair differences or discordance than the identical twins this excess has been attributed solely to the greater difference in genotype of nonidentical twins. Objections have been raised to this interpretation. It has been asked: "Are the environmental differences for identical twin partners really of the same magnitude as for nonidentical twins?" The answer is that frequently they are not. On the whole, identical twins have very similar habits. If one is robust and inclined to outdoor activities, the other shares this inclination; if one is weak and physically inactive, so is the other. If one finds pleasure in social activities, so does the other;

if one prefers solitude, the other does likewise. With nonidentical twins, the situation is different. Their unlike genotypes may give a strong physique to one and a weak one to the other; and the environments they seek may differ greatly, to conform to their genetic differences. In general, it may be said that the environments in which nonidentical twins live are more different that those in which identical twins live. The greater discordance of nonidentical twins may therefore be in part environmentally conditioned. Neglect of this possibility would lead to an exaggerated estimate of the share of heredity in the discordance of unidentical twins.

Another objection to regarding identical and nonidentical twins as equivalent except for heredity is that identical twins, at birth, frequently exhibit larger differences in size and vigor than would be expected from their genetic identity. For instance, prenatal death of one of a pair of identicals is considerably more frequent than of one of a pair of nonidenticals. Such inequalities are probably caused by the often unequal placental blood supply, which particularly affects monochoric twins. The prenatal inequalities might also increase the intrapair differences between identical twins in typical twin investigations. Consequently, the differences in environment (prenatal) responsible for differences between identical twins may be larger than those for nonidentical twins. Neglect of such "prenatal biases" (Price) would attribute too small a share to heredity as a factor in the diversity of nonidentical twins.

Comparisons between identical and nonidentical twins do not take into account the fact that identical twins are embryologically of three different kinds, as shown by their birth membranes: 2 amnions and 2 chorions; 2 amnions and 1 chorion; and 1 amnion and 1 chorion. It would be important to correlate observations on concordance and discordance of identical pairs with the specfic embryologic situation to which the twins were subjected before birth, and to compare them separately with nonidentical twin differences.

Related to these considerations are the comments by Darlington, who stresses that identical twins may receive different kinds of the cytoplasm of the single-celled zygote which splits at some stage to produce twins. These cytoplasmatic differences are partly of the type which call forth the right-left asymmetries observed in single individuals and partly those which are due to irregularities during the embryonic stages of development. In contrast to nonidentical twins, each of which develops from an egg whose cytoplasm remained wholly in each zygote, identical twins may thus differ in kind of cytoplasm received at the splitting of the single zygote.

Doubts concerning the validity of comparisons between identical and nonidentical twins have also been based on the assumption that identical twins are possibly, on the average, constitutionally inferior to nonidentical twins.

TABLE 84
Incidence of Specific Congential Malformations per Thousand Births, Birmingham, England 1950–1954.

Malformation	*Single Births*	*Twins*
Generalized defects	5.0	6.3
Anencephalus	1.3	1.2
Spina bifida	1.6	1.6
Hydrocephalus	1.0	3.1
Cleft lip or palate	0.8	0.4
Cardiac malformation	2.8	6.3
Clubfoot	4.3	4.7

Source: Edwards, *Proc. Roy. Soc. Med.*, **61,** 1968.

This possibility is suggested by the higher prenatal mortality of one partner in identical twin pairs as compared to nonidentical pairs, as well as by the fact that the frequency of both twins being stillborn is also significantly higher for identical than for nonidentical twins. Moreover, it is true that in some populations twins affected with specific types of malformations occur slightly more frequently than singly born infants (Table 84). (The results of different studies, however, vary considerably.) The question has been raised whether or not identical twins who survive birth show the aftereffects of their often unfavorable prenatal environment. If so, then the higher concordance in abnormal traits found in identical twins might be based on their common inferior constitution.

To many students of these problems, the two objections that were described on pp. 675–676 do not seem to be very serious. In most ways, identical twins differ so little that the result of special prenatal inequalities in environment would not seem important enough to have much weight in a general evaluation of nature and nurture. And a constitutional inferiority of no more than a minority of identical twins is apparent in the records of those twins who have survived, unharmed, the hazards of prenatal life, birth, and infancy. Much more significant is the first objection: that the postnatal environments of identical twins are more alike than those of nonidentical twins. It is difficult to estimate the effect of this difference in variability of environment. Many investigators believe that the effect on most traits is not great enough to change the conclusion that the hereditary differences and not environmental ones between nonidentical twins are the main causes of their greater trait differences.

Two factors possibly contributing to the greater similarity of the environments of identical, as compared to nonidentical, twin partners should be distinguished. One factor may be the tendency of identical partners to remain together just because they are alike. The second factor itself may be an expression of the specific, identical genotypes, which may lead the identical partners to select, independently of each other, similar environments; in contrast, the differences in genotypes of nonidentical twin partners may lead them more frequently to select different environments. As far as the second factor is involved, the differences in types of environments would themselves be expressions of the genetic situation. Therefore, instead of weakening the validity of assigning the excess in trait differences of nonidentical twins to heredity, the justification of this procedure would become strengthened.

This evaluation of the limitations of the twin method may be summarized in terms of a formula which plays a basic role in scientific plant and animal breeding. Hereditary and environmental factors are of great importance in determining the yield of, for example, corn per acre, milk per cow, or eggs per chicken. The variability of these traits is expressed in terms of a statistical concept, "variance," whose symbol is V. By referring, for a human example, to the studies on height in twins (Fig. 261), it is obvious that the variability of the differences is greater for the nonidentical than for the identical twins. Computation yields approximate values of $V_{nonid} = 15.6$, and $V_{id} = 2.1$.

The total variance of a population is the sum of three components: genetic variability, environmental variability, and the variability due to interaction of genetic and environmental factors—

$$V_{total} = V_{genetic} + V_{nongenetic} + V_{interaction}. \tag{1}$$

In twin studies the variance of the differences between partners of pairs is determined. For identical twins, this intrapair difference has no genetic component, and for any given twin pair it is due solely to nongenetic influences. For a comparison of the intrapair differences of different identical twin pairs, two components of the variance must be considered: the nongenetic intrapair differences and, in addition, an interaction variance. This is necessary since the same range of environmental agents impinging on a twin pair with the genotype A and causing its partners to differ may cause a different intrapair difference in another twin pair with the genotype B. The total variance of the differences between identical twin partners, therefore, is

$$V_{total\ id} = V_{nongenetic\ intrapair\ id} + V_{interaction\ id}. \tag{2}$$

The variance in nonidentical twins is

$$V_{total\ nonid} = V_{genetic\ nonid} + V_{nongenetic\ intrapair\ nonid} + V_{interaction\ nonid}. \tag{3}$$

Only when it is justifiable to equate the nongenetic components of the total variances in identical and nonidentical twins and also to regard the interaction components as equal could equations (2) and (3) be combined thus:

$$V_{\text{genetic nonid}} = V_{\text{total nonid}} - V_{\text{total id}}. \tag{4}$$

By substitution of $V_{\text{genetic nonid}}$ in equation (3), we obtain

$$V_{\text{nongenetic intrapair nonid}} + V_{\text{interaction nonid}} = V_{\text{total id}}. \tag{5}$$

Stated in words, the last two equations would signify a partitioning of the total variance of nonidentical twins into two measurable components: that due to their genetic differences alone, as expressed in (4); and that made up jointly of nongenetic differences and interactions, as expressed in (5). If we had reasons for assuming that interaction factors were negligible—that is, that the different environments had similar effects on the expression of the different genotypes—then the partitioning of the total variance would yield simply the contributions of genetic and nongenetic causes. In such a situation, it would be justifiable to say that, for instance, half of the variance of a trait in nonidentical twins is due to heredity and the other half to environment.

Even then, however, this partition applies, as stated, to the differences between nonidentical twins only, and not necessarily to differences between other sibs or between individuals in a general population. For the variance in individuals of a general population, the relative shares of nature and nurture may be very different from that for nonidentical twins, since both genetic and environmental differences from one person to the next will greatly exceed those for these twins.

These critical considerations are not intended to discredit the importance of twin investigations. They not only show that—as so often happens—a problem initially believed to be simple has turned out to be complex, but also suggest additional methods of approaching the problems of nature and nurture. By a comparison of the variance of a trait difference in pairs of sibs with that in pairs of nonidentical twins, it can be seen whether the possible greater mean difference in environment for children born to the same parents successively instead of simultaneously expresses itself in greater phenotypic variability. Similarly, a comparison of the variance in identical twins reared apart or in sibs brought up in different homes can provide information on the strength of different environments in molding a trait. Another approach to the same problem consists of measuring the variance in unrelated individuals reared in the same home instead of in different homes. Examples of some such comparisons have already been given, and more will appear in the next chapter.

HERITABILITY

It has been stressed earlier that the question about the relative shares of hereditary and environmental components in the expression of a trait does not refer to specific individuals but implies a population concept. It has been worded somewhat as follows: How much of the variability for a given trait observed between the different individuals of a population is due to genetic differences between them and how much is due to differences in the environments under which the individuals developed? An answer to this question does not yield attributes of the trait in terms of fixed proportions of the genetic and environmental fractions. For the same trait the proportions may vary widely in different populations as well as between different generations of the same population.

The ratio of the genetic to the total phenotypic variance is called *heritability*. It is a concept that is most useful in the analysis of polygenic, continuously variable characters which depend on the additive effects of numerous genes. In its simplest form we can write

$$V_P = V_G + V_E,$$

where the subscripts stand for phenotypic, genetic, and environmental. Dividing through by V_P we obtain

$$1 = \frac{V_G}{V_P} + \frac{V_E}{V_P}, \text{ or } \frac{V_G}{V_G + V_E} + \frac{V_E}{V_G + V_E} = 1.$$

Calling the first term of the last equation h^2, and the second term e^2 we have $h^2 + e^2 = 1$ where h^2 stands for heritability (in the narrow sense; see below). It may be defined as the fraction of the total variation which would still exist had the environments of all individuals in the population been identical. Similarly, e^2 represents the fraction of the total variation which would still be manifested had all individuals identical genotypes. Because h^2 is a fraction, any change in either the numerator, V_G, or the denominator, $V_G + V_E$, will change its magnitude. For instance, if the mating system of the population changes by increasing assortative mating, V_G becomes smaller and the estimate of h^2 goes down.

Heritability in the narrow sense as just defined is also known as additive heritability. It is basically the fraction of the phenotypic variance which may be attributed to the differences between average (additive) effects of genes. But genotypic variance has still other components: for instance, those due to interactions between alleles at the same locus (dominance) or between genes at different loci (epistasis). In addition, the magnitude of the phenotypic variance is affected by such parameters as assortative mating, genotype-environment interaction (i.e., differential responses of a given genotype to different environments), or genotype-environment correlation (determination of the environment to an extent by the genotype, as for instance, when a parent of high intellectual genotypic value affects the environment in which his offspring's intellect develops). If these components of phenotypic variation are included in the heritability estimate, we have h^2 in the broad sense.

The distinction refers to the different uses to which h^2 estimates can be put. In the broad sense heritability essentially measures the totality of the genotypic (additive, some dominance, and all of epistatic variance) contribution to the phenotypic variation. But nonadditive effects are broken up by segregation in each generation. Hence, only h^2 in the narrow sense, that relating to additive effects, can be useful in predicting changes between generations due to selection.

Heritability in human genetics is mainly derived from observed correlations between relatives. Theoretically, we can calculate the fraction of genes that the individuals of a given population have in common: e.g., one-half in parent and offspring, or in sibs; one in identical twins. If the heritability is complete, the observed correlations within the groups would be the same as the theoretical ones. For a very simplified illustration let us consider incomplete heritability, e.g., 0.94 for height as determined from data on identical twins reared apart; the value of h^2 would be the fraction of the observed correlation divided by the theoretical: $0.94/1 = 0.94$. This is heritability in the broad sense because the identity of monozygotic twins applies to all aspects of gene action—additive, dominant, and epistatic, as well as the consequences of assortative mating of the parental pairs. In another example from a different population based on resemblances between half sibs, whose theoretical correlation due to shared genes is 0.25, an observed correlation of 0.10 would signify a heritability of $0.10/0.25 = 0.4$.

In human populations other estimates than h^2 have been often made for various measures of intelligence and personality. They are derived from partitioning variability between and within families or by comparing resemblances between monozygotic and dizygotic twins. The statistical technicalities of such estimations are too complex to be discussed here. The important point to remember is that such estimates reported in the literature under the designation H and often referred to as heritabilities are not exactly equivalent to h^2 (see Jinks and Fulker).

Heritability, then, provides us with a yardstick for measuring the partition of genetic and environmental components in the total phenotypic variance of a trait. It can furnish information on the effectiveness of differential reproductive processes which may be going on unintentionally or are envisaged in the future. And it deserves repeating that heritability is not a fixed attribute of a trait but may vary widely in different populations and with time in the same population.

PROBLEMS

196. Using the data from Figure 267, test the significance of the difference in concordance between identical and nonidentical twins for the following traits: (a) clubfoot; (b) measles; (c) scarlet fever.

197. If a certain congenital human trait is concordant in 60 per cent of nonidentical twins:
   (a) What conclusions regarding the heredity-environment problem can be drawn from this fact?
   (b) What would you conclude if this trait is 93 per cent concordant in identical twins?

198. Analyze the following data and draw conclusions, whenever possible, regarding the roles of heredity and environment:
   (a) In a certain city, symptoms of lead poisoning are concordant in nearly 100 per cent of all pairs of identical twins; in another city, only 80 per cent of identical twin pairs are concordant.
   (b) The frequency of concordance for measles is practically alike in identical and nonidentical twin pairs.
   (c) Concordance for a disease was found in 31 out of 33 pairs of identical twins and in 1 out of 16 pairs of nonidentical twins.

199. For a certain genetic trait which has full penetrance, one-third of the affected nonidentical male or female twin pairs are concordant, but the female co-twin of an affected male is rarely concordant. What is the explanation?

REFERENCES

Allen, G., 1952. The meaning of concordance and discordance in estimation of penetrance and gene frequency. *Amer. J. Hum. Genet.*, **4**:155–172.

Allen, G., 1965. Twin research: Problems and prospects. *Progr. Med. Genet.*, **4**:242–270.

Allen, G., Harvald, B., and Shields, J., 1967. Measures of twin concordance. *Acta Genet.*, **17**:475–481.

Anderson, V. E., Goodman, H. O., and Reed, S. C., 1958. *Variables Related to Human Breast Cancer.* 172 pp. University of Minnesota Press, Minneapolis.

Busk, T., 1948. Some observations on heredity in breast cancer and leukemia. *Ann. Eugen.*, **14**:213–229.

Harvald, B., and Hauge, M., 1963. Heredity of cancer elucidated by a study of unselected twins. *J. Amer. Med. Ass.*, **186**:749–753.

Hay, Sylvia, and Wehrung, D. A., 1970. Congenital malformations in twins. *Amer. J. Hum. Genet.*, **22**:662–678.

Jarvik, L. F. A., and Falek, A., 1962. Comparative data on cancer in aging twins. *Cancer*, **15**:1009–1018.

Jinks, J. L., and Fulker, D. W., 1970. Comparison of the biometrical, genetical, MAVA, and classical approaches to the analysis of human behavior. *Psych. Bull.*, **73:**311–349.

Lurie, M. B., 1953. On the mechanism of genetic resistance to tuberculosis and its mode of inheritance. *Amer. J. Hum. Genet.*, **4:**302–314.

Lynch, H. T., 1967. *Heredity in Carcinoma.* 200 pp. Springer, Berlin.

Macklin, M. T., 1959. Comparison of the number of breast-cancer deaths observed in relatives of breast-cancer patients, and the number expected on the basis of mortality rates. *J. Nat. Cancer Inst.*, **22:**927–952.

Miyao, S., 1966. On the morbidity in twins with special regard to malignant tumor and metabolic diseases. *Jap. J. Hum. Genet.*, **11:**188–207.

Myrianthopoulos, N. C., 1970. An epidemiologic survey of twins in a large, prospectively studied population. *Amer. J. Hum. Genet.*, **22:**611–629.

Newman, H. H., Freeman, F. N., and Holzinger, K. J., 1937. *Twins: A Study of Heredity and Environment.* 369 pp. University of Chicago Press, Chicago.

Penrose, L. S., MacKenzie, H. J., and Karn, M. N., 1948. A genetical study of human mammary cancer. *Ann. Eugen.*, **14:**234–266.

Planansky, K., and Allen, G., 1953. Heredity in relation to variable resistance to pulmonary tuberculosis. *Amer. J. Hum. Genet.*, **5:**322–349.

Price, B., 1950. Primary biases in twin studies. A review of prenatal and natal difference-producing factors in monozygotic pairs. *Amer. J. Hum. Genet.*, **2:**293–352.

Rotkin, I. D., 1966. Further studies in cervical cancer inheritance. *Cancer*, **19:**1251–1268.

Selvin, S., 1970. Concordance in a twin population model. *Acta Genet. Med. Gemellol.*, **19:**584–590.

Spranger, J., and v. Verschuer, O. F., 1964. Untersuchungen zur Frage der Erblichkeit des Krebses. *Z. menschl. Vererb. -u. Konstitutionsl.*, **37:**549–571.

Symposium: Twins, genetics, and constitutional medicine, 1966. *Jap. J. Hum. Genet.*, **11:**162–229. (Japanese, with English summaries.)

Takei, K., 1968. Statistical investigations on genetic aspects of gastric carcinoma. *Jap. J. Hum. Genet.*, **13:**67–80. (Japanese, with English summary.)

Verschuer, O. v., 1956. Tuberkulose und Krebs bei Zwillingen. *Acta Genet. Stat. Med.*, **6:**103–113.

Williams, R. J., 1951. Biochemical Institute Studies IV. Individual metabolic patterns and human disease: an exploratory study utilizing predominantly paper chromatographic methods. *University of Texas Publications*, **No. 5109,** pp. 7–21.

Woodworth, R. S., 1941. *Heredity and Environment.* 95 pp. Social Science Research Council, New York. (Bulletin 47.)

Woolf, C. M., 1955. Investigations on genetic aspects of carcinoma of the stomach and breast. *Univ. Calif. Publ. Public Health*, **2:**265–350.

Woolf, C. M., 1956. A further study on the familial aspects of carcinoma of the stomach. *Amer. J. Hum. Genet.*, **8:**102–109.

Woolf, C. M., 1961. The incidence of cancer in the spouses of stomach cancer patients. *Cancer*, **14:**199–200.

World Health Organization, 1965. The use of twins in epidemiological studies. *Acta Genet. Med. Gemellol.*, **15:**111–128.

27

HEREDITY AND ENVIRONMENT: III. BEHAVIOR GENETICS

Mental activities, expressed in human behavior, are intimately related to physical activities in the brain and nervous system. Destruction of small or large parts of the brain may result in mental changes. Alterations in the biochemistry of nerve cells and their interactions are often accompanied by changes in behavior. Behavior also depends on the organization of the sense organs which relay to the nervous system signals from the outside world and from within the body. Moreover, it also depends on the reactions of the tissues to nervous stimuli which may elicit hormonal secretions and muscular contractions that express themselves in mental states, in movements, and in speech. Since genes exert their actions on the development of the most various types of biological structures and the functioning of innumerable kinds of biological processes, it is to be expected that they also influence those structures and functions on which mental traits depend, and that differences in genotypes may express themselves in differences in behavior.

Proof of genic effects on human behavior comes from various sources. Most striking are the effects of chromosomal abnormalities. Individuals affected with Down's syndrome, which is based on trisomy of chromosome 21, have small brains, with certain brain parts, such as the frontal lobes, the brain stem, and the cerebellum, disproportionately small. Typical individuals with Down's syndrome, according to Penrose, have cheerful and friendly personalities. Trisomy-13 or trisomy-18 also entail gross neuropathology, which in the former often includes absence of the olfactory bulbs. Microscopically, striking changes in the cerebellum are found. The over-all-development of these trisomics is retarded; as they die when very young, it is not possible to make a detailed analysis of behavior. Absence of part of chromosome 5 is responsible for the plaintive cry of children who exhibit the cri-du-chat syndrome, a cry which seems to be generated by nerve action rather than by anatomical abnormalities of the larynx.

Apart from chromosomal aberrations, various single-gene effects are responsible for the breakdown of normal behavior. The Tay-Sachs disease, for instance, is based on homozygosis for a recessive gene. It causes abnormal lipid metabolism of the brain cells, which results in mental degeneration of infants that at first appear to be normal. Or, as another example, the dominant gene for Huntington's chorea leads to degenerative processes of the nervous system and causes distressing changes in the personalities of many patients.

The examples cited in the preceding paragraphs are extreme, but we may expect that a whole spectrum of genetic influences on behavior exists, from such abnormal behavior to all kinds of variations within the range of normality.

Features of human behavior separate it widely from the behavior of even man's closest animal relatives. Nevertheless, because of the basic similarity of nervous organization in man and animals, particularly mammals, it will be helpful to survey briefly the genetics of animal behavior relevant to human genetics in order to point out both agreements and differences.

GENETICS OF ANIMAL BEHAVIOR

Animal species differ not only in appearance but also in behavior. The genotype of a fertilized spider egg determines both the form of the developing spider and the activities which will lead to the spinning of the web characteristic of the species. A chicken's egg of female genotype will transform itself into a bird which, without ever having heard the sound of another hen, will produce the notes typical of its kind. A female rabbit which has never seen another rabbit prepare a home for its young will when pregnant build a rather complex nest and pluck hair from her body to line it. But not all individuals of a species will act alike. For example, birds of the same species often show variations in their song, and different breeds of dogs are known for either friendliness or hostility, ability or inability to retrieve, and many other specific behavioral traits.

Intraspecific differences could be acquired by specific contacts with other animals or with human trainers, or they could be consequences of different genetic constitutions of different individuals. They could also be the results of combined action of specific genotypes and specific outside stimuli—just as the ability of a child to learn the language and behavior of his culture depends both on the human genotype which permits such complex achievements and on the specific culture in which this genotype expresses itself.

To distinguish genetic from nongenetic aspects of behavior, we must observe different individuals under identical external conditions. In higher animals, this may require separating at birth from their mother the young to be studied. In order to repeat behavioral tests on more than one specimen, it is desirable to have available strains of animals whose members are highly alike in genotype. And to find the specific mechanisms which govern genetic differences in behavior, we must study descendants of various crosses between different strains.

The results of such studies have shown that not only nervous diseases but also differences in behavior within the normal range of variability are inherited in mice, rats, and other mammals. Among the abnormalities of behavior, one of the most intensely studied traits is a type of epileptic seizures caused by high-frequency sounds. Different genotypes include those that are immune to these "audiogenic seizures," others that succumb to them early in life, and still others that succumb to them on repeated stimulation. Other inherited behavioral abnormalities include no less than six genetically different types of shaking, "waltzing," and circling in mice, a simply inherited recessive tremor and cramp disease in rabbits, and similar afflictions in deermice and chickens.

Among normal behavioral traits, one of the easiest to measure is the amount of running of mice or rats in an activity wheel. (An activity wheel records the number of revolutions enforced upon it by the efforts of the animal.) Male rats from a highly active strain averaged nine thousand revolutions per day, but males from an inactive strain averaged only one thousand revolutions. As long as certain strains of wild-type and domesticated laboratory rats were fed adequately, the wild type was no more active than the laboratory animals. On a starvation diet, however, the wild rats were much more active than the domestic specimens.

Mice of 15 strains have been studied for exploratory activity, which was measured by the number of maze sections without dead ends that the mice entered. The mean scores differed from strain to strain and were distributed in a normal manner. The activity of the highest strain was 23 times that of the lowest.

"Emotionality" in mice and rats has been assessed by the number of fecal and urinary eliminations occurring under unaccustomed conditions. In a strain of rats selected for high emotionality, approximately seven eliminations occurred in 12 trials; in a strain of low emotionality, a mean of only one-half elimination occurred under like circumstances.

Hoarding of pellets of food was likewise found to be dependent on the genotype. Rats of one strain collected about 44 pellets during a 12-day period; rats of another strain assembled fewer than 11 pellets. This hoarding occurred under conditions of food shortage, but differences in intensity of hoarding remained when a pile of pellets was placed in each rat cage and kept there at all times. Animals of the formerly low hoarding strain virtually ceased hoarding after a few days; whereas animals of the formerly high hoarding strain still collected at least 5 pellets a day for 12 days.

Particularly interesting behavioral differences between strains of rats concern preferences for either plain drinking water or for water containing morphine. Both morphine-addiction-susceptible and addiction-resistant strains have been bred. These strains also showed differences in the amounts of consumption of 10 per cent alcoholic solutions.

Strains with different behavior are sometimes distinguished by different physical traits, for instance, fur color. In the epileptic rabbits of the White Vienna genotype, for example, it has been established that the specific behavior is at least partly controlled by the gene which determines coat coloration. Another interesting example in which the appearance of a special morphological character has been shown to be due to a peculiar type of behavior is in a strain of obese mice. The cause of obesity is not simply one of an unusual food utilization but also of an inability to regulate the food (that is, calorie) intake. On a palatable diet on which other mice thrive, the obese mice overeat; on an unpalatable fare to which other mice adjust, the obese mice undereat and become emaciated. In other cases, it seems that chance and not intrinsic factors are responsible for associations between behavioral and nonbehavioral traits.

Genetic differences in learning ability have been demonstrated in the famous experiments of Tryon by letting rats run through a maze in which alternate pathways lead either to a food supply or into blind alleys. Some genotypes enabled their carriers after a period of training to make few errors on their way toward the goal; other genotypes enabled their carriers to succeed only after many mistakes. The original designations of such strains as either "bright" or "dull" have given way to the more cautious terms "maze-bright" and "maze-dull," since animals with high scores in some types of maze tests may be low scoring in others; and, conversely, low-scoring rats may be higher in other tests. Learning is a highly complex phenomenon and, in rats, may be dependent on such genetically controlled variables as general activity, exploratory activity, and emotionality.

A genetically controlled trait involving social interrelations concerns fighting in mice. When kept under special identical conditions, males of two different strains were opposite in aggressiveness toward a third mouse introduced into their environment, one type being strongly aggressive, the other pacific. It is noteworthy, however, that males from either strain could be easily trained to be either aggressive or peaceable. Genetic differences in sexual drive in male guinea pigs have been mentioned in the discussion of sex determination (Chap. 20).

Traits analogous to human language are the songs of birds. It is well known that different species of birds have different songs, and that within one and and the same species regional variations of the song may be found. Experiments with hand-reared English chaffinches which have been kept separate from birth in soundproof rooms have shown that both heredity and learning are involved in their song. The normal song has three phases, but an isolated chaffinch is able to produce only a song of about the normal length and consisting of a crescendo series concluded by a single note of relatively high pitch. All typical refinements of the song which result in the regional "dialects" have to be learned by contact with other birds of their species. Interestingly, the innate basis of the song is selective enough to insure that the bird does not normally acquire songs from any species other than its own. In contrast to this, certain mockingbirds seem to be genetically endowed with an ability to learn a whole range of songs of other species. The call notes of many birds, however, seem to be fixed genetically.

Final examples of carefully studied genetic control of behavior are furnished by observations made on dogs at the Jackson Memorial Laboratory at Bar Harbor, Maine. Animals of four different breeds were brought up under standardized conditions and rated for such traits as sensitivity to noises, orientation in space, reaction to the approach of the human handler (as measured by change in heart rate), learning behavior, and degree of activity developing in puppies during a period of mild restraint by the handler (Fig. 268). Striking specific differences between breeds were found, some in degree of trait expression (for instance, length of response delay), and others of a nature which might be called alternative. Thus, predominant submissiveness was observed in a sample of cocker spaniels consequent upon restraint, in contrast to predominant aggressiveness developing under the same circumstances in a sample of wire-haired terriers.

These studies on animal behavior have shown the existence of genetic controls. They have also given evidence that genetically controlled behavior can be modified by varying the external conditions. Wild and domestic rats are about equally active when well fed, but not when undernourished; although differences in hoarding do not disappear completely when the nests are well

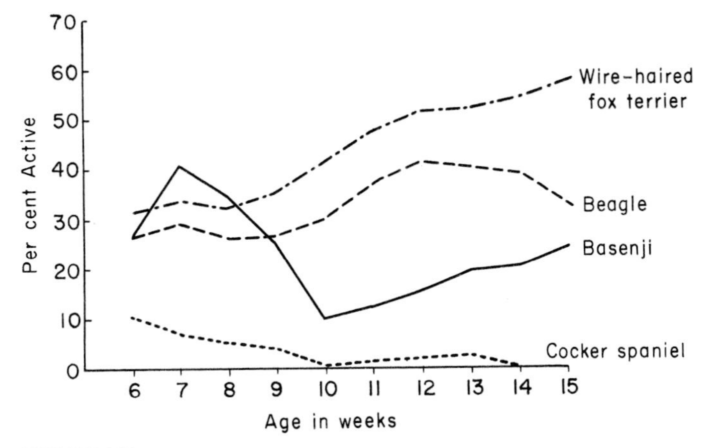

FIGURE 268
The proportion of dogs, of four different breeds, which were rated as "active" when required to remain quiet during weighing on a scale. Note the initial differences in young puppies and the partial increase in these differences with age of dogs. (After Scott and Charles, *J. Genet. Psych.*, **84**, 1954.)

stocked, they become diminished; and mice innately aggressive under one type of environment can be made pacific under another.

Apart from extreme cases, proof of genetic influences on behavior in man is usually more difficult to obtain than in animals. With the latter, more-or-less isogenic strains may be bred whose differences in behavior can then be analyzed by genetic studies involving hybridization, segregation, and selection. In man, such experimental procedures are not applicable. More important still, the study of human behavioral differences within the normal range requires a control of the environment which is often impossible to attain.

The degree to which genetic differences in man may express themselves in mental attributes will vary from case to case, as it does in animals. However, the immensely greater learning potentialities of man may serve as a means of overcoming the effects of different genotypes. The same result may be accomplished by man's power, already large and yet still rudimentary, to manipulate the environment. Beyond such equalizing tendencies, man's special abilities may also be employed so to develop the expression of different genotypes in a diversified manner that each genotype, furnished with its "appropriate" environment, can make a maximum contribution to the well-being of the individual and society. It is obvious that the goals outlined in the preceding two sentences involve discussions and evaluations which lie far beyond the proper realm of human genetics.

The basic attitudes underlying the interpretation of human behavioral differences vary greatly. Some have assigned rigid preformation to mental traits; others have regarded the mind of the newborn as a blank slate which is gradually specified under the impact of the environment. One or the other extreme view is still held by some, but it has become clear that behavioral traits, like physical traits, develop under the joint influence of inborn and environmental agents. There is not an either-or alternative. It is a matter of the relative influences of both factors and their interaction. The following pages are mainly a detailed description of some of the evidence that has led to this recognition.

MENTAL DISORDERS

The nature–nurture problem arouses unusual interest in mental attributes. We shall discuss first some pathological traits and then the particularly important differences within the normal range of human mental capacities. Since geneticists have found genetic bases for many illnesses of the body, they are inclined to look for hereditary factors in the development of mental diseases. They postulate that mental phenomena have a physical basis, and they follow with high expectations the work of physiologists and biochemists who attempt to find physical or chemical properties which characterize patients with specific mental diseases. Such findings might, it is thought, reveal the underlying action of genotypes which physiologically cause abnormal states of the brain and express themselves also in abnormal mentality. A model for such correlations is the well-established case of mental deficiency (not mental disease) caused by the recessive gene for phenylketonuria, which does not permit normal metabolism of phenylalanine and, correlated with this, causes brain damage and low intelligence. Other aspects of the interrelationship between bodily and mental states are demonstrated by the use of certain compounds which when taken by normal persons induce symptoms similar to those found in persons with psychoses, and by other compounds which when taken by mentally ill persons change them temporarily toward normality. Of course, the consideration of genetic factors in the origin of mental diseases does not exclude awareness of the possibliity, or even high probability, that both genetic and environmental factors may contribute to the outbreak of mental diseases.

If the geneticists may have been guilty of prejudice in favor of the importance of hereditary components of psychoses, psychiatrists, and particularly those of the psychoanalytical schools, have often been too easily inclined to consider only the environmental stresses in the lives of their patients.

Yet, even convincing proof of harmful psychological experiences preceding a psychosis would not necessarily rule out the presence of genetic components.

Theoretically, the joint efforts of geneticists and psychiatrists could yield the conclusion (1) that mental disease is the invariable fate of specific genotypes; or (2) that it is equally likely in all known genotypes; or, finally, (3) that nongenetic influences lead to illness in some, but not all, genotypes. The first alternative is easily seen to be false, since identical twins are not always concordant in respect to mental illness. It is more difficult to decide between alternatives (2) and (3), complete versus partial nongenetic determination. The occurrence of increased frequencies of mental illness in various types of relatives of propositi suggests that these frequencies are due to a common genetic basis. But this interpretation can be countered by the argument that the relatives may have been exposed more frequently than unrelated individuals to precipitating external causes of the disease.

Schizophrenia. This mental disease is characterized by a cleavage in the personality structure that, in extreme cases, necessitates permanent institutionalization of the patient. It is the most frequent mental disease, the chance of a random person becoming affected some time in life being nearly one per cent. The high frequency and the serious nature of the disease have made schizophrenia an object of very numerous studies. Since in many of these the nature–nurture problem has played an important part, the study of this disease may be regarded as a model applying to other mental illnesses as well.

The frequency of schizophrenia among the relatives of affected persons is higher than that of the general population (Table 85). The table shows the risk probabilities that specific relatives of a propositus will, at some time during life, have the disease. Such risk figures are usually obtained from data on relatives, which include some nonaffected individuals who have died before or during the age period when onset of the disease occurs. Obviously, the full inclusion of nonaffected individuals will bias the actual risks; thus the raw data have to be "age corrected" by the use of tables which give the age distribution for the disease. Weinberg and others have worked out the necessary methods. The frequency values of Table 85 are often called the "empiric risk" or the "morbidity risk." (The risk figure is called "empiric" because it is not based on any theory of hereditary or environmental causation of the disease, but on the statistical facts of experience.) The very much higher incidence of schizophrenia among the relatives of affected persons than in the general population is most impressive. Among the adult children of patients, for instance, the incidence is nearly twenty times higher than in general. In addition to schizophrenia itself, the frequency of similar, less serious, mental abnormalities is also greatly increased.

TABLE 85
Schizophrenia among Relatives of Affected Individuals. A and D, Data from Germany; B, from United States; C, ranges of risks as found in a variety of studies.

Relation to Propositus	*Morbidity Risk* (%)			
	Schizophrenia			Psychopathic Condition Similar to Schizophrenia
	A	B	C	D
Unrelated	0.85	–	–	2.9
Step-sibs*	–	1.8	–	–
Half-sibs†	–	7.1	–	–
Sibs	10.8	14.2	7–15	9.7
Parents	–	10.3	5–10	–
Children	16.4	16.4	7–16	32.6
Grandchildren	3.0	4.3	3–4	13.8
Cousins	1.8	–	2	10.2
Nephews, nieces	1.8	3.9	3	5.1

Source: Luxenburger, *Fortschr. Erbpathol.*, 1, 1937; v. Verschuer, *Erbpathologie*, 2nd ed., Steinkopf, 1937; Kallmann, 1946, 1953; and Zerbin-Rüdin, 1967.
*Step-sibs are genetically unrelated. They are the offspring of two spouses from their marriages to other partners.
†Half-sibs have only one parent in common.

In detail there is considerable variation in the risk figures as derived from different studies. Apart from chance, the variation is due to differences of ascertainment, differences in diagnostic criteria for the clinical definition of schizophrenia with its variable phenotypic expression, and perhaps differences in different populations in the frequency of environmental or genetic agents which may be responsible for the outbreak of the disease.

Twin studies from various countries almost uniformly show high degrees of concordance for schizophrenia in identical twins and low degrees in nonidenticals (Table 86). There are, however, wide ranges in concordance figures: from more than 80 down to 6 per cent in identicals and from 3 up to 18 per cent in nonidenticals. Concordance tends to be higher in earlier studies than in more recent ones, and although the reasons for this variation are not clear, it seems that at least two phenomena are involved. (1) The earlier studies, particularly those of Kallmann, may have had more severely affected propositi than the later ones, and severity may be correlated with high concordance. (2) The low concordances all come from Scandinavian studies, and it is possible that the criteria for ascertainment vary from those in non-Scandinavian countries. For a general statement, we may cite Shields, who lists seven reasons that might account for concordances being too high, and seven other reasons for their being too low!

TABLE 86
Concordance (+ +) and Discordance (+ −) for Schizophrenia in Twins.

*Authors, Country, Year**	*Identical*			*Nonidentical*		
	+ +	+ −	% + +	+ +	+ −	% + +
Luxenburger, Germany, 1930	14	7	67	2	58	3
Rosanoff et al., U.S., 1934	28	13	68	10	91	10
Kallmann, U.S., 1952	231	37	86	99	586	14
Slater, Great Britain, 1953	31	10	76	16	99	14
Inouye, Japan, 1961	33	22	60	2	9	18
Kringlen, Norway, 1966	12	28	30	4	70	5
Gottesman & Shields, G.B., 1966	10	14	42	3	30	9
Harvald & Hauge, Denmark, 1965	2	5	29	2	29	6
Tienari, Finland, 1968	1	15	6	–	–	–

Source: After Kallmann, 1953; and Gottesman and Shields, 1966.
*Year of analysis or of publication.

Whatever the true concordance for identical twins, it falls far short of 100 per cent. The absence of concordance in perhaps more than 50 per cent of all identical pairs is evidence for nongenetic factors in the origin of the disease. The much greater concordance between identical than nonidentical twins is compatible with the existence of a genetic predisposition for schizophrenia.

Is the high concordance of schizophrenia in identical twins due to the mental shock which the onset of the disease in one twin causes in the other? The data on nonidenticals show that this shock, in most cases, is not sufficient to cause the disease if the genotypes of the twins are different from each other. It is, of course, true that the mental reaction of an identical twin to the illness of his twin partner is probably different from that of a nonidentical twin, who knows that his fate is dissimilar in many other ways. It seems unlikely, however, that this difference in psychological attitude can account for the observed difference in concordance.

A few cases are known of identical twins who were early separated from one another and who both became schizophrenic. In each pair there was concordance for the disease. Two of these cases will be cited here. Identical twin sisters, described by Kallmann, were adopted into different homes soon after birth. The sisters had hardly any contact during the first ten years, and little contact later. At the age of fifteen, one, a factory worker, gave birth to an illegitimate child, while the other lived as a domestic servant in the sheltered home of a private family. Nevertheless, both became schizophrenic—the first shortly after the birth of her child, the second about one and a half years later. Another pair of identical twin sisters was studied by Craike and Slater. After spending the first eight and a half years in the home of a violent, often drunken father, one sister then lived in a chilren's home until she was nineteen; the other sister was adopted after the death of the mother, at the age of nine months, by an affectionate aunt in whose home she stayed until the age of twenty-four. The two sisters had had no personal contact or correspondence before they were twenty-four years old, though each knew of the other's existence. They were rarely together afterwards. Mutual suspicion and dislike dominated their relations. Despite very different upbringings, there was concordance in childhood neurosis, in various aspects of personality and life story, in slight deafness, and in schizophrenic illness. The psychosis of the twin with a harsh childhood was chronic and progressive, while that of the other was less severe and showed its symptoms only at intervals.

The similarities of the early separated twin sisters are highly suggestive of a constitutional background of schizophrenia, but they are special cases which need to be supplemented by more comprehensive, unselected series of similar types. Such series have become available since 1966, and have established that there is a strong genetic component in the liability to schizophrenia. These new series are not concerned with twins raised in separate homes, but are based on single-born propositi who early in life were placed in foster (or adoptive) homes and later became schizophrenic. It was thus possible to compare the frequency of schizophrenia in the biological relatives of the propositi with that in the unrelated families of the foster homes.

The first study of this kind was made by Karlsson in Iceland. He found 8 persons who had been placed in foster homes and had become schizophrenic. They had 29 biological sibs and 28 foster sibs. None of the foster sibs, but 6 of the biological sibs, were affected. Clearly, it was genetic factors, not the home background, that differentiated between the biological and foster sibs.

The second study is that of Heston on 47 persons in Oregon who had been born to mothers hospitalized with schizophrenia and had been placed in foster homes. At the time of the investigation, the propositi had a mean age of 36 years. Their mental and social status was compared with that of 50 carefully matched controls, who had been born to nonschizophrenic mothers and placed in foster homes. Five of the 47 offspring of affected mothers had later become schizophrenic, but none of the 50 controls had. Other mental or social abnormalities were likewise more frequent in the offspring of affected mothers than in the controls: e.g., 9 propositi as against only 2 controls were diagnosed to have sociopathic personalities, and 8 propositi as against only one received psychiatric or behavioral discharges or rejection from the armed services.

The third study, by Kety, Rosenthal, Wender, and Schulsinger, was based on Danish data. Thirty-three persons were found in vital statistics records of Greater Copenhagen who as infants had been placed into adoptive homes and who later developed schizophrenia or related mental anomalies. These propositi were matched by a like number of controls consisting of unaffected

TABLE 87
Schizophrenia or Related Disorders among the Biological and Adoptive Relatives of Affected Propositi and Controls. Numerators = number of affected relatives; denominators = total number of relatives.

Sample	*Biological Relatives*	*Adoptive Relatives*
33 propositi	13/150	2/74
33 controls	3/156	3/83
19 propositi	9/93	2/45
20 controls	0/92	1/51

Source: Kety et al. *in* Kety and Rosenthal.

adopted individuals. The propositi had 150 biological relatives (parents, sibs, and half-sibs) and 74 adoptive relatives, the controls had 156 biological and 83 adoptive relatives. There were significantly higher frequencies of affected biological relatives of the propositi than of the controls (13 out of 150 as against 3 out of 156) but no significant difference between the frequencies of affected persons among the adoptive relatives of the propositi and those of the controls (2 out of 74 as against 3 out of 83). The facts are listed in Table 87, upper rows. Since some of the propositi had lived with their biological families for periods of from one month after birth to one year or even more, a subsample was singled out of 19 propositi who had been placed within one month of birth, in most cases within a few days. Even in this subsample the frequency of affected biological relatives was much higher for the propositi than for the controls; in this subsample, as in the whole, there was no significant difference between the two sets of adoptive relatives. (Table 87, lower rows.) "The conclusion seems warranted that genetic factors are important in the transmission of schizophrenia . . ." (Kety et al.).

The demonstration of a genetic component in schizophrenia does not reveal the specific type of this component. Various theories have been proposed, such as causation by a dominant gene with different degrees of penetrance in homozygotes and heterozygotes, or by a recessive gene with incomplete penetrance in homozygotes, a two-gene pair scheme, and, following the trend of the times, polygenic models. The last seem to fit the facts better than the others, particularly if account is taken of the likelihood that the schizophrenia phenotypes may result from a variety of genotypes.

Manic-depressive Psychosis. Manic-depressive illness has an incidence in the general population of somewhat less than 0.5 per cent. The disease is characterized by alternating emotional states of exaggerated elation and depression. Empirical data on its incidence in relatives suggest inheritance

TABLE 88
Concordance (+ +) and Discordance (+ –) for Manic-depressive Psychosis in Twin Pairs.

Authors	*Identical*		*Nonidentical*	
	++	+–	++	+–
Luxenburger (Germany)	3	1	–	13
Rosanoff, Handy, Plesset (U.S.)	16	7	11	56
Kallmann (U.S.)	25	2	13	42
Slater (England)	4	4	7	23
Mean concordance	77%	–	19%	–

Source: After Luxenburger, from v. Verschuer, 1937; Rosanoff, Handy, and Plesset, *Amer. J. Psych.*, **91**, 1935; Kallmann; and Slater, *Proc. 42nd Ann. Meeting Amer. Psychopath. Ass.*, 1952.

with dominance involved, but no definite knowledge of the genetic details is available. Twin studies again show a strikingly higher concordance in identical as opposed to nonidentical genotypes (Table 88).

INTELLIGENCE

As important as abnormal mental properties are, both to the individual and to society, they play a minor role as compared to the variations within the normal range of mental characteristics. One of the difficulties of making studies within this range is the dearth of devices with which to measure objectively the differences in normal mental traits between different individuals. The most commonly used devices for this purpose are the so-called intelligence tests. The concept of intelligence is a complex one. It includes the assumption of inherited psychological capacities, as well as the knowledge of the behavioristic activities made possible by the acquisition of the cultural tools which society makes available. *Intelligent behavior* is regarded as behavior which, on the basis of inherited capacity, makes good use of the social inheritance, such as language and numbers or scientific and moral concepts. Intelligence tests do not directly measure intelligent behavior, but measure a particular type of behavior as defined by a given type of test. The tests are so constructed that the essentials of the test behavior will closely resemble significant elements of life behavior. How well intelligence tests do this is still a matter for dispute. Different intelligence tests score the behavior of tested individuals in different ways. The most widely used measure is the *intelligence quotient*, or I.Q.—as usually determined by the "Stanford Binet test"—so-called because it is the quotient (multiplied by 100) of the mental age of the individual as defined by the test and his chronological age. It assigns a score of about 100 as the mean value of the population and is so constructed that higher and lower scores are distributed approximately in a normal curve. Accordingly, the number of individuals who score higher or lower than average decreases with the degree of deviation of the score from 100.

Psychologists have subdivided the mental abilities of man into distinct, so-called primary, abilities, such as ability to visualize objects in space, to memorize, or to reason inductively. In addition to such primary abilities it has been proposed that a general ability factor underlies measured intelligence. Since the primary abilities vary independently of each other, individuals with the same over-all I.Q. may be very different in their specific mixtures of primary abilities. There are some newer studies on genetic-environmental interaction in the expression of primary abilities, but most data on test intelligence are given in terms of single scores only.

The nature–nurture problem regarding intelligence is immediately posed by the statement that intelligent behavior implies both inherited capacities and making use of the social inheritance. Differences in intelligence-test behavior among individuals may be due to differences in their genetic endowment, to differences in their opportunities for acquiring the tools of society, and to complex interrelations between the genetic and environmental differences. In general, the designers of intelligence tests have endeavored to make them independent of environmental differences within a given society—none of the tests can be applied to different societies—so that differences in test behavior would be due to differences in genotypes.

A complete independence from nongenetic influences is impossible, however. Human intelligence always operates within a cultural organization, so that the idea of an absolutely "culture-free" intelligence test has intrinsic limitations. What this idea really implies is the desirability of "culture-fair" tests—that is, tests which minimize the effect of cultural differences of groups within a given society. Thus, it would be desirable to reduce or to remove altogether the influence on test scores of the fact that the upbringing of some individuals may motivate them less than others to score as high as they can; or to reduce or remove the influence on test scores of the verbal skill required to score well, since much of this ability depends on home background. One difficulty in such procedures is that there may possibly be some genetic control of motivation or of verbal skill, and efforts to eliminate nongenetic factors may also eliminate genetic ones.

The opportunities for acquiring information helpful in obtaining high scores are not simply different in different homes. Subtle differences have been detected in individuals' attitudes toward gaining knowledge and making use of it in intelligence tests. A certain eagerness for the intellectual success of their children, as is frequently found among middle-class parents, may provide an attitude favorable to test success; the lesser stress on intellectual achievement that is often characteristic of lower socioeconomic families may fail to provide the emotional background on which a high test performance must be based. However, bright and dull children are often found in the same family, and in any layer of the population. Socioeconomic level and cultural status, though important, are not the only factors interacting in intelligence-test performance.

There are examples of improvement or deterioration in intelligence scores of individuals under the influence of particularly favorable or unfavorable circumstances (Skeels). Such examples are important because they may show which factors in the environment influence the expression of the genotypes involved in test behavior, but they have no bearing on the question of whether there are genetic differences among individuals that determine such behavior.

Mental Defect. Before going into the evidence on the nature–nurture problem within the normal range of intelligence, let us briefly discuss mental defect. Intelligent behavior of different individuals, both in life and in tests, varies greatly and continuously from very high, through average and low, to very low. The individuals in the lowest group on the the scale of behavior, which includes the so-called imbeciles and idiots, have to be permanently cared for in institutions. The I.Q. of these persons is below 50. A somewhat higher group, which grades into low "normal" intelligence, still contains individuals incapable of independent life. They need special schools or institutions in which their limited capacities can be trained and put to use. Persons belonging to this group are called "feebleminded," or "mentally retarded." They comprise several per cent of the population of Western societies. In terms of I.Q., feebleminded individuals are, with some variations, defined by scores in the range from 50 to 70.

The causes of mental defect are manifold. As shown in Figure 269, they comprise both environmental and genetic agents. Apart from a large class of unknown causes, external influences, such as prenatal and birth injuries, may be responsible for idiocy and feeblemindedness. The same defects are also often the result of abnormal genic or chromosomal constitution. The outstanding example of chromosomal aberration is Down's syndrome; that of single-gene causation is phenylketonuria. Estimates of the number of loci occupied by recessive alleles responsible for severe mental defect have been as high as several hundred, but individually they are rare and in aggregate they account only for a small fraction of mental defect. Contrary to earlier belief, much of the genetics of feeblemindedness fits a polygenic interpretation. Like stature, measured intelligence in a population shows continuous

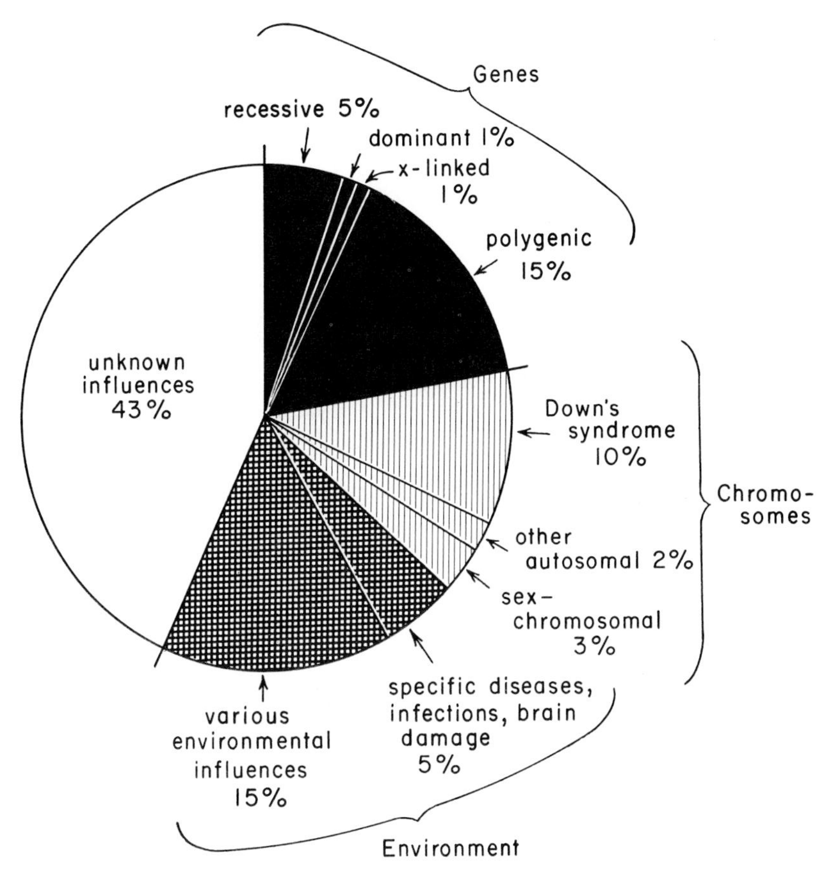

FIGURE 269
Estimates of partitioning of the causes of mental defect in patients of all age groups. (After Penrose, *in* Wendt, (Ed.), *Genetik und Gesellschaft,* Wiss. Verlagsges., Stuttgart, 1971.)

variation from very low to very high. Feebleminded individuals in general seem to possess an accumulation of unfavorable alleles at many loci, and it is this accumulation that assigns these persons to the lowly endowed "tail" of the frequency distribution of mental ability.

Evidence for this interpretation of mental defect comes from family data in which the mental abilities of children from different types of parents are compared (Tables 89 and 90). Certainly, the positive correlations between frequencies of defective offspring and defective parents are not free of the environmental factor. Defective parents are likely to provide less favorable opportunities for mental development than nondefective. Nevertheless, it is true that the great majority of mental defectives comes from the great majority of parents who are not defective (Table 91). Although this does not exclude the possibility that the environment provided by "normal" parents

TABLE 89
Offspring of Various Types of Parents Who Have Had a Mentally Defective Child.

	Children			
Parents	No.	Average or Above (%)	Inferior* (%)	Defective* (%)
Average × Average	18	72	5	22
Average × Inferior	59	64	33	3
Inferior × Inferior	252	28	57	15
Inferior × Defective	89	10	55	35
Defective × Defective	141	4	39	57

Source: After Halperin, *Amer. J. Ment. Defic.*, **50**, 1945.
*Approximate I.Q. range of "inferiors," 70–85; of "defectives," 50–70.

TABLE 90
Empiric Risks That a Mentally Retarded Child (I.Q., 69 or less) Will Be Born to Various Types of Parents.

Parents	*No. of Children*	*% Retarded Children*
BEFORE ANY CHILDREN ARE BORN		
Normal sib of retarded × retarded	71	23.8
Normal sib of retarded × normal	2,996	2.5
Normal sib of normals × normal	9,476	0.5
AFTER ONE OR MORE AFFECTED CHILD HAS BEEN BORN		
Retarded × retarded	76	42.1
Retarded × normal	317	19.9
Normal sib of retarded × normal	139	12.9
Normal, no retarded sib × normal	104	5.7

Source: Reed and Reed.

may in some cases be responsible for the mental defect of their children, such an assumption hardly qualifies as a general explanation. It is also unlikely in view of the following observations. When the mental defectives are separated into two groups, the feebleminded and the imbeciles and idiots, it is found that the mean I.Q. of the sibs of the feebleminded is lower than that of the sibs of the imbeciles and idiots (Fig. 270). If feeblemindedness were the result of poor nurture and not of poor nature, it would have been expected that this nurture was less deficient than that leading to imbecility and idiocy. Therefore the sibs of the feebleminded should on the average be scoring higher than those of the extreme deviants, the imbeciles and the idiots.

TABLE 91
Parental Origin of 1,194 Mentally Defective Children.

*Parents**	*Proposition*	
	Dull to Feebleminded	Imbeciles and Idiots
Superior × Superior	–	1
Superior × Normal	4	5
Normal × Normal	316	481
Normal × Dull	126	70
Normal × Feebleminded or Dull × Dull	73	40
Normal × Imbecile or Dull × Feebleminded	38	16
Dull × Imbecile or Feebleminded × Feebleminded	11	13

Source: After Penrose, 1938.
*The mean I.Q. of the superior parents was estimated as 122, that of the normal, dull, feebleminded, imbeciles and idiots as 100, 78, 56, 34 and 12 respectively.

The fact that the opposite is true does not fit a predominantly environmental expectation. It leads one to agree with Roberts and other investigators in concluding that both types of mental defectives often have abnormal genotypes, but that imbecility and idiocy are primarily caused by single recessive genes, and that feeblemindedness is largely the result of polygenic genotypes.

Twin studies illuminate aspects of the nature–nurture interrelation in feeblemindedness. Some discordance is found in identical twins (in spite of their like genotype), and much more discordance is found in nonidenticals.

Reading Disability. A considerable number of individuals of normal or superior intelligence experience difficulty in reading and writing, sometimes called word blindness, or specific dyslexia. In some, this defect can be traced to special illness; in others, no specific cause can be determined for it. Pedigree data have been interpreted in terms of a single dominant gene, with nearly

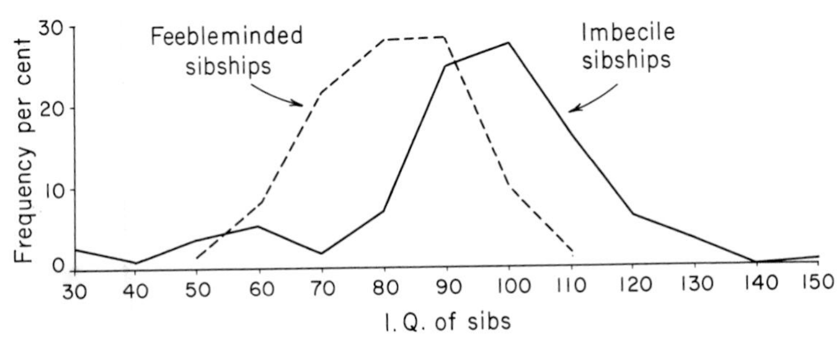

FIGURE 270
Frequency distributions of the I.Q.'s of 562 sibs of 149 feebleminded individuals and 122 imbeciles of the I.Q. range 30–68. (Roberts, 1952.)

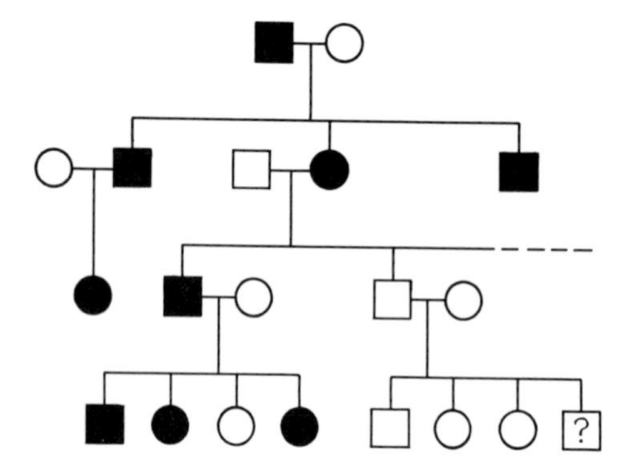

FIGURE 271
Part of a pedigree of word blindness. The youngest child in the last sibship had not yet reached the age of possible manifestation. (After Hallgren, *Acta Psych. Neurol., Suppl.* **65,** 1950.)

complete penetrance (Fig. 271). Of three identical sets of twins, all were concordant, in contrast to two discordant pairs out of three nonidenticals. A trait such as dyslexia is frequently not easy to diagnose with certainty—a fact which leaves some doubt regarding its genetic basis. But the existence of specific mental inabilities caused by certain genes is by no means unlikely.

Space-form Perception. People differ in their ability to orient themselves in space and to perceive forms and perspectives. Two types of evidence show that a genetic element is in part responsible for the observed variation. One type comes from the study of twins, the other from the study of cytogenetically aberrant individuals. The twin method has been applied by Vandenberg to data obtained from answers to questionnaires designed to test spatial orientation ability. He found significantly higher correlations in this ability between identical twins than between nonidentical twins. In earlier studies an interesting test had been devised for the ability to orient oneself toward a specific known place after one has been led, blindfolded and with cotton plugs in his ears, in various directions within an experimental room. In a series of observations on forty pairs of identical twins and forty pairs of nonidentical twins, concordance in orientation in space was defined as an agreement within seven degrees in the angular deviations from the right direction. Identical twins were found to be concordant at least twice as often as nonidenticals.

Individuals with Turner's syndrome usually are of the XO type. According to Money and associates, these females test within the normal range of total I.Q. and various verbal and reasoning performances; but they are deficient in other abilities, including speed and fluency of lexigraphic association, dealing with numbers and particularly in space-form perception. They have been called space-form blind and direction-sense blind. Their cytogenetic peculiarity of having only one X- and no Y-chromosome is a causal factor in their limitations in space-form and other abilities. Since the space blindness of XO individuals is similar to that of some persons with nongenetic brain damage, it has been wondered whether the XO genotype causes maldevelopment of specific regions of the brain.

Normal Intelligence-test Behavior. The most obvious procedure for studying the genetics of intelligence and achievement is to compare the scores of parents and children and other related individuals. Pedigrees of the Bach kindred show an accumulation of great musicians, of the Bernoullis a series of famous mathematicians, of the Darwins a group of distinguished scientists; but a Beethoven, a Gauss, or a Franklin stands alone in eminence among his relatives. Passing from the study of individual families to statistical data, Galton in his book *Hereditary Genius* (1869) showed that the proportion of outstanding men was much greater among the relatives of outstanding Englishmen than in the English population at large. On a still broader scale, positive correlations were found by many investigators between the school performances of parents and children or other individuals closely related to one another. All these facts could be accounted for on the basis of heredity. Since it is possible that some unusual achievements depend on the presence of single genes, the appearance of the responsible trait in various members of a kindred would be likely. It is possible that other unusual achievements have little chance of being present again in someone else. But, clearly, in no human cultural achievement can the influence of nongenetic factors be denied, and no reliable conclusions can be drawn from the facts even though they may suggest genetic participation. The wide range of "intelligence" among the members of a normal population, as judged by general impressions or by test scores, has long formed a basis for divergent interpretations of the relative roles of heredity and environment. Extreme hereditarian views have been countered by extreme environmental ones. Gradually, however, it has been established that both genetic and environmental factors are responsible for the observed variance in normal, measured intelligence. The evidence is manifold. Much of it is based on comparisons of the I.Q. or primary-mental-abilities performances of groups of individuals of different genetic relationships, such as between parent and child or between sibs. Other comparisons are based on groups varying in environmental circumstances, such as between groups of persons raised under different environmental conditions, e.g., twins having been raised together in the same home or apart in two different homes. Usually the comparisons are expressed in terms of correlation coefficients. These in turn are compared with the theoretical correlations to be expected if there were exclusively genetic determination of intelligence.

In detail, most studies of this kind are open to some criticisms. Often, the population samples are not fully random but represent somewhat selected groups. Often, the number of persons compared are too low to lend adequate statistical significance to the findings. Often, environmental factors vary more than ideally they should, for instance, when in a study of twins "reared apart" the time between birth and separation varies from a few days to more than a year. Also, the theoretical correlation coefficients usually depend on specific genetic hypotheses that are not fully valid for the study being made, since they are based on such assumptions as absence of assortative mating, or an infinitely large population. Notwithstanding these limitations, the over-all trend of the results of many studies is remarkably clear: the degrees of similarity in intelligence increase with increasing degrees of genetic similarity, but environmental factors also have undoubted influences on intelligence-test scores.

Before describing some of the important nature-nurture studies, some special findings may be related. One of these concerns the effect of parental consanguinity on test and school performance of their children. Both in Sweden (Table 92) and Japan (Table 93) inbred children had slightly but significantly lower scores in the performance of various tasks than did noninbred ones, even after the possible influences of socioeconomic factors had been taken into account. This not only demonstrates the existence of genes that influence mental performance but also indicates that they have been accumulated as a consequence of consanguinity of the parents.

The relation of prenatal and neonatal factors to intelligence scores is a complex problem which cannot be dealt with here in depth. However, in a Chilean study of infants that died of extreme malnutrition in the first year of life it was found that their brains were below average in weight and in DNA, RNA, and protein content—an indication that the number of cells in the brain was reduced. Intelligence tests of surviving children from the same environment presumably would show lower than average scores, but it would be difficult to separate the specific effect of the underdeveloped brain from that of other unfavorable environmental components. Nutritional deficiencies that are relatively minor may also play a role in test performance. In a pioneer

TABLE 92
Effect of Consanguinity on the Mental Performance of Swedish Children. The decreased frequency of normal or above-normal children (I.Q., 89 or higher) is statistically significantly different from zero at $P < 0.01$.

Parents	*Children*	
	Below Normal	Normal or Above
Not related	17	88
First cousins	51	115

Source: Böök, *Ann. Hum. Genet.*, **21**, 1957.

TABLE 93
Per Cent Decrease in Test Scores for Various Mental Attributes of Male Japanese Children from First-cousin Marriages as Compared with Those of Outbred Children. All percentages are statistically significantly different from zero at $P < 0.01$.

Characteristic	*% Decrease*
TESTS	
Verbal score	4.7
Performance score	3.6
SCHOOL PERFORMANCE	
Language	3.2
Social studies	2.8
Mathematics	4.0
Music	4.1
Fine Arts	3.2

Source: Schull and Neel.

American study by Ruth Harrell, Ella Woodyard, and A. I. Gates, vitamins were added to the diets of groups of low-income pregnant and lactating women, and the I.Q. scores of their children at the ages of three and four years were compared with those of other low-income-family children whose mothers' diets had not been supplemented. There was no difference in the mean scores of a rural sample of children of treated and untreated mothers, but in an urban sample the children of the treated mothers scored 8 I.Q. points higher. More work along these lines is highly desirable.

Data on variations in birth weight of twins have shown interesting associations with later I.Q. performance. For differences in birth weight less than 300 gm, no significant association was observed. For those identical twins whose difference in birth weight exceeded 300 gm, however, a significant correlation between birth weight and I.Q. has been found. Specifically, the heavier of the two identical twins had a mean I.Q. advantage of about 5 points.

It is noteworthy that the mean intelligence score of twins is lower by about 5 points than that of singly born persons. As Record, McKeown, and J. H. Edwards have shown for a large twin sample from Birmingham, England, this difference is not primarily due to prenatal or delivery conditions associated with maternal age, order of birth, birth weight, or duration of gestation. Instead, it appears to be associated with early postnatal experiences of twins. A study was made of a group of surviving twins whose partners had either been stillborn or had died within 4 weeks after birth: The mean score in a test for verbal reasoning of the surviving, essentially singly raised, twin partners was higher than that for twins raised together and nearly as high as that for children born singly.

These examples of nongenetic influences on intelligence scores (as well as similar examples given below) show that such scores are not fixed by heredity in the same way as blood groups or other traits with invariable expressivity. As a matter of fact, in exceptional circumstances variations in I.Q. that are due to external conditions may be very much larger than those considered above. Effects of extreme deprivation of maternal care have resulted in test scores characteristic of mental defect by children who when conditions were improved scored well within the normal range. Some identical twins raised apart with very different amounts of schooling may differ in test scores by as much as 20 or more points. These findings are not unexpected. If an analogy may be permitted, we may compare the genetic intelligence endowments of different individuals to threads of rubber of different length. Their "phenotypic lengths" will be a product of their "genotypic lengths" and the variable degrees of stretching that represent nongenetic influences.

Figure 272 is a succinct summary of more than 50 studies of correlation coefficients of I.Q. scores within and between different groups of individuals. All of the categories whose correlations were determined comprise the sums of several independent studies. The lowest number of such repeats is two, for sibs reared apart; the highest is 35, for sibs reared together. The results within each category show a considerable spread, but the ranges of the observed correlation values show a clear trend toward an increasing test-score resemblance with increasing genetic relationship. They also show an increased test-score resemblance with increased similarity of environment as represented by subjects being reared either in separate homes or in the same home.

If we make certain simplifying assumptions—such as presence of random mating within very large populations—the expected genetic correlations within groups of unrelated persons and between foster parent and child are 0.0; those between parent and child, between sibs, and between members of nonidentical twin pairs are 0.5; and those between identical twins are 1.0. The observed correlations cover nearly the whole range from 0.0 for unrelated persons reared apart to more than 0.9 for identical twins reared together.

The square of the correlation coefficient defines that fraction of the total variance of the trait which is accounted for by the associations studied. To take two examples, if the correlation coefficient for I.Q. in a given population is $r = 0.2$, then $r^2 = 0.04$; if $r = 0.9$, then $r^2 = 0.81$. Since a fraction of the variance is proportional to the square of the correlation coefficient, low coefficients indicate small fractions of the total variance, e.g., 0.04, and only very high coefficients indicate large fractions, e.g., 0.81.

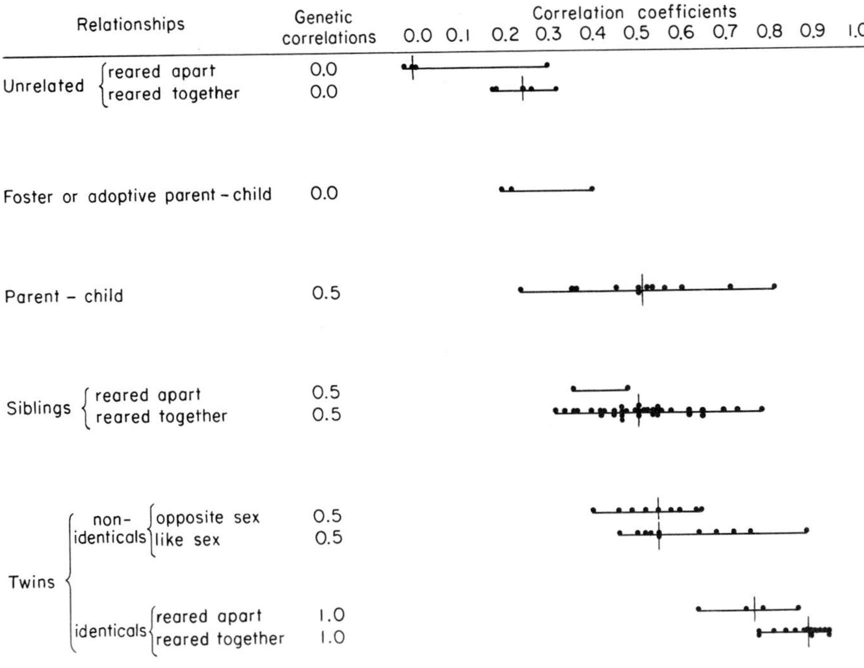

FIGURE 272
A review of the results of numerous studies of the correlation coefficients for "test intelligence" between persons of various relationships. The horizontal lines at right give the range of the coefficients, the dots the individual values, and the vertical lines the averages. (After Erlenmeyer-Kimling and Lissy Jarvik. *Science*, **142,** 1963. Copyright 1963 by the American Association for the Advancement of Science.)

We may now consider specific categories.

Unrelated persons. The first category consists of unrelated persons reared apart. Here, neither genetic nor environmental likeness is involved, and the observed median correlation between the I.Q. scores is indeed close to 0.0. Unrelated persons reared together share a common home environment. The observed median coefficient of 0.23 seems to reflect this environmental similarity, but the observed result is possible due to some degree of selection for similarity of the unrelated persons who were reared together.

Adopted children. The correlations between foster parents and child yield a median of 0.20. This correlation coefficient is probably a reflection of the similar environment of foster parent and child, but it is also possibly due to selection for similarity.

Studies of adopted children deserve some detailed discussion. If a group of children is divided in a random manner into several samples, the average genetic endowment of the children in each sample should be the same, within the limits of statistical error. If the children of one sample are then placed in adoptive homes of one kind, and the children of each other sample into adoptive homes of other kinds, then the effects of different kinds of adoptive homes on the development of the children can be compared. (It should be noted that the original studies usually refer to adopted children as "foster" children. In modern usage, "foster" does not convey the meaning that a child has been legally adopted and permanently placed in an adoptive home.)

It is not easy to find series of adopted children in which no open or hidden selective placement has been practiced; the presumably better-endowed children are usually placed in the better adoptive homes. Selective placement seems to have been minimal in a group of adopted children from Chicago. The sample is small, but agrees in what it shows with other, larger samples which perhaps included some selectively placed children. The adoptive homes were classed as "good," "average," and "poor" (the last term does not mean unfavorable home environment, but one less favorable than the other two). Had there been no influence of home environment, it would have been expected that the average scores of the samples of children in the three classes of homes would have been alike. It was found, however, that the mean I.Q. scores of the adopted children were strikingly related to the quality of the adoptive homes: 45 adopted children in good homes scored, on the average, 112 points; 39 adopted children in average homes scored, on the average, 105 points; and 27 adopted children in poor homes scored, on the average, 96 points. These figures clearly demonstrate the modifying influence of home environment on intelligence-test behavior.

Other studies on adopted children afford an opportunity to judge the hereditary component of the I.Q. score. If heredity has something to do with I.Q., then adopted children should be less similar to their adoptive parents than the "own" (biological) children of a control parent group. For a group of adopted children in Minnesota homes, which were graded in various ways, there was a continuous decrease in mean I.Q. with descent in occupational status of the father from professional to the relatively unskilled occupations (Table 94). This decrease, which reflects the environmental effect on test performance, covered a rather narrow mean range, from 113 to 108. In the control group of own children, there was also a steady decline in mean I.Q. corresponding to the occupational status of the father, but the range was more than three times as great, namely, from 119 to 102. This latter range shows a much more pronounced correlation with the father's occupational status than the narrower range of the adopted children. Specifically, it is significant that for the upper occupational groups the own children scored higher than the adopted children, but for the lower occupational groups the adopted children did better than the own children. It seems reasonable to conclude that the differences between the scores in adopted and in own chil-

TABLE 94
Mean I.Q. of Adopted Children and "Own" Children in Homes of Different Occupational Categories.

Occupation of Father	*Adopted Children*		*Control ("Own") Children*	
	Number	Score	Number	Score
Professional	43	112.6	40	118.6
Business, management	38	111.6	42	117.6
Skilled trades and clerical	44	110.6	43	106.9
Semiskilled	45	109.4	46	101.1
Relatively unskilled	24	107.8	23	102.1

Source: Leahy, *Psych. Monogr.* **17,** 1935.

dren are due to the fact that the latter resemble their parents more than do the adopted children because they inherited part of their parents' genotypes.

Similar conclusions have been reached in other studies on the trend of parent-child resemblance in intelligence during the development of the child. Over a period of twelve years, Skodak and Skeels and, working separately, Honzik, measured the performances of a group of children adopted during their first months of life, and correlated them with measures of the education or actual I.Q. scores of (1) the biological mothers and fathers, and (2) the adoptive mothers and fathers (Fig. 273). As a control, data were available

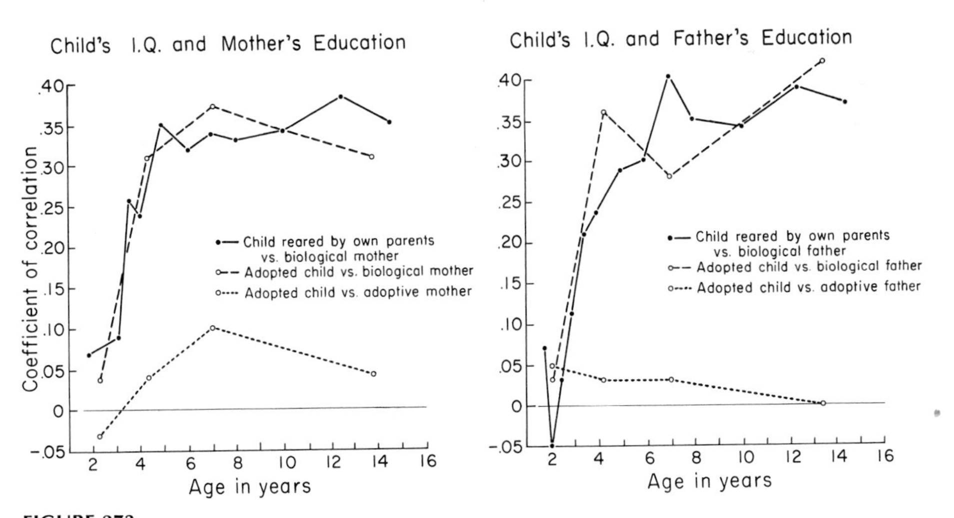

FIGURE 273
Correlation coefficients between education of biological and adoptive parents and children's I.Q. (From Honzik, *Child Development*, **28,** 1957, using data on adopted children from Skodak and Skeels, *J. Genet. Psych.*, **75,** 1949.)

for the correlation between the I.Q. of children and their biological parents, in whose homes they had been reared. Up to the age of two years there was essentially no correlation between the child's performance rating and that of either the biological or the adoptive parents, whether the child was reared by his own or by adoptive parents. With increasing age of the children, there was a steep rise in correlation between I.Q. of the child and the biological parents, whether he was reared by them or by adoptive parents; this correlation reached values of approximately 0.3 around four years of age, after which it increased only slightly more. In contrast to this significant positive correlation between the intelligence of children and biological parents, the correlations between intelligence of the children and their adoptive parents remained insignificantly small. These data strongly suggest that intelligence depends in part on a child's genetic endowment from his biological parents, and is to a high degree independent of the educational status or intelligence performance of his adoptive parents.

Orphanage children. The presence of a genetic component in intelligence is also apparent from a British study on the relation between the occupational status of the parents, particularly the fathers, and the test performances of illegitimate children brought up away from their parents in a relatively uniform environment. These children were under the care of "Dr. Smith's Home," a large and important charitable institution. They had been separated from their mothers before the age of one year, and on the average at six months. Until they were between five and six years old, the children were boarded out individually in the approved and inspected cottage homes of agricultural laborers "of the better type." Thereafter, for about ten years, they lived together in the headquarters of the institution, where they also attended school. The identity and social status of the biological parents remained entirely unknown to the children, and their placement in the cottage homes as well as their treatment in Dr. Smith's Home was unrelated to their parental background. These long-established procedures of the institution provided an opportunity for a study of the relation of the intelligence scores of the children to the status of their biological parents. An ideal way of studying the relationship would have been to compare actual tests of the parents with those of their children. This was not possible, since the parents were not available. Instead, the *mean* scores of children from various occupational layers had to be used. One investigation of this type has already been cited (Table 94), and other studies have borne out the general trend of decreasing test performance with progression from professional to less skilled occupational status of fathers. If the higher mean scores of children from the upper occupational groups were not exclusively due to better environmental opportunities but also to genetic causes, then the scores of the children in Dr. Smith's Home should show a similar relationship to the occupational status of the parents.

TABLE 95
Mean I.Q. Scores, Obtained by Averaging Individual and Group Tests, of Children in Dr. Smith's Home, Classified into Five Groups According to Status of Parents.

	Boys		*Girls*	
Group	Mean I.Q. Score	Number	Mean I.Q. Score	Number
A	99.0	4	105.0	1
B	104.2	15	100.1	18
C	101.1	41	97.1	25
D	96.5	23	92.8	25
E	98.0	1	–	–

Source: After Evelyn Lawrence.

The children were classified, according to occupations of parents, into five groups: group A included not only professional people but also elementary school teachers and farmers (from country gentlemen to working farmers); group B, tradesmen, clerks, and highly skilled artisans; group C, skilled and semiskilled workers; group D, unskilled workers; group E, not regularly employed. Both group A and group E were very small so that the data for some tests were based on exceedingly limited population samples.

The children were tested individually by the Stanford-Binet test and in groups. The scores of both tests, for boys and for girls, showed positive correlations with the occupational status of either father or mother. Examples relating the average scores derived from a combination of the individual and group tests of boys and of girls to the average socioeconomic status of the two parents are given in Table 95. Although there are two minor exceptions to the positive correlation between test performance and status of parents, in general the positive relationship is clearly apparent. In spite of a relatively uniform foster-home and orphanage environment that was unrelated to the status of the parents, the mean performances of the children of parents from different socioeconomic groups showed the same ranking which would have been expected had the children been brought up in the homes of their biological parents. The range of mean scores of children from the five different groups (A–E) in Dr. Smith's Home is narrower than the range of mean scores of children from similar classes living with their parents. Although this greater similarity is a reflection of the less differentiated environment, the fact that, in spite of this, there is still a significant correlation between mean scores and parental classes is evidence that intelligence performance is partly governed by genetic endowment. (A critic may point to the possibility that the I.Q. scores of these children in Dr. Smith's Home were reflections of maternal influences during the prenatal and early postnatal period

TABLE 96
Mean Differences in Stanford-Binet I.Q. and Correlation Coefficients between Twins. The scores of the mean differences in A are corrected for chance errors in measurement.

I.Q.	*Identical (Reared Together)*	*Identical (Reared Apart)*	*Nonidentical (Reared Together)*
CORRELATION COEFFICIENTS			
A	0.88	0.77	0.63
B	0.92	0.87	0.54
MEAN DIFFERENCES			
A	3.1	6.0	8.5
B	–	6.0	12

Source: A, Woodworth, after Newman, Freeman, and Holzinger; B, Burt.

before admission to the home, rather than the result of genetic relations. Only new types of studies can decide the issue.)

Correlations between parents and children, sibs, and nonidentical twins. Theoretical correlations between parents and children, between sibs, and between nonidentical twins are all 0.5. The observed median values as shown in Figure 272, are close to expectation. It is of interest that the correlation coefficients for sibs reared apart fall in the same range as those for sibs reared together and are greater than the correlations for unrelated persons reared together, and that the correlations between nonidentical twins, whether of like or opposite sex, are but slightly higher than those between sibs reared together. These findings show that the differences in environment of sibs reared in different homes or of sibs reared at different times in the same home do not lead to large differences in correlation coefficients.

Identical twins. The median correlation coefficients for I.Q. scores of identical twins are the highest of all the median coefficients shown in Figure 272. For such twins reared together, the median correlation is 0.87; and for those reared apart, it remains very high, being 0.75. For two of the studies of identical twins—one by Newman, Freeman, and Holzinger in the United States the other by Burt in England—the coefficients are listed in Table 96. Burt's data yield a value of 0.92 for identical twins reared together, and the scores of identical twins reared apart are only slightly less correlated. In both studies nonidentical twins reared together had considerably lower correlations than identical twins reared either together or apart. The observed high values for identical twins approach the theoretical genetic value of 1.0.

It is instructive to consider also the mean score differences of the three categories of twins distinguished in Table 96. In the American study (A) the mean difference for identical twins reared together was only 3.1 I.Q. points, 8.5 points for nonidentical twins reared together, and 6.0 points for identical

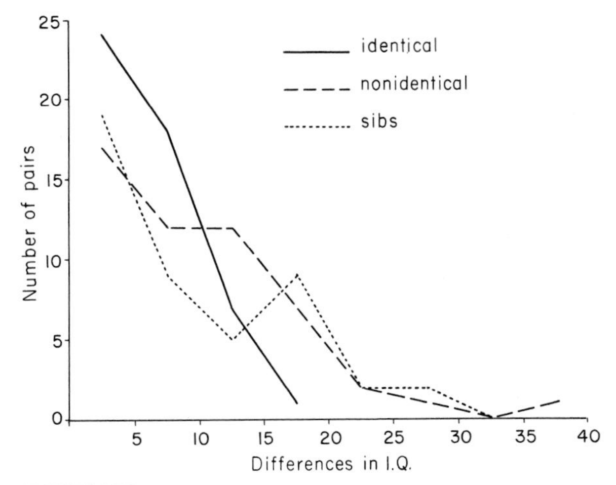

FIGURE 274
Curves of distribution of differences in Binet I.Q. of 50 identical twins, 47 nonidentical twins, and of 52 pairs of sibs. (After Newman, Freeman, and Holzinger.)

twins reared apart. The English study (B) gave the same mean difference of 6.0 points for identical twins reared apart, and a difference of 12 points for nonidentical twins reared together. It is apparent that genetic identity of twins, whether reared together or apart, makes their I.Q. scores more similar than those of nonidentical twins reared together. These findings gain in weight when it is taken into account that the average difference between duplicate tests given the same individual is about 5 points.

Environmental differences, prenatal or postnatal, are not without effect, however, even on the I.Q. scores of identical twins reared together. The mean difference in the scores of identical twins reared together is small but some of the individual values for fifty pairs of identical twins are surprisingly high (Fig. 274). Eighteen pairs had differences between 5 and 10 points, seven between 10 and 15 points, and one pair differed by as much as 17 points. Nevertheless, the differences for nonidentical twins and for pairs of sibs extended over much wider ranges than those for the identicals (Fig. 274).

A detailed consideration of the nineteen pairs of identical twins reared apart, which were described in Newman, Freeman, and Holzinger's work, reveals that, in general the home environments of the separated twin partners did not differ greatly. In most cases, both twins were brought up in either rural or urban communities; and both had similar kinds and lengths of formal schooling. For fifteen of the nineteen twin pairs, it may be said that only moderate intrapair variation existed in the educational, social, and physical environment. The remaining four pairs had been subjected to such differences as "one with two years of regular schooling only, the other with completed college education"; or "one reared on a farm, and completing grade school, the other reared in a town, and completing high school"; or "one reared in the home of a somewhat shiftless, often unsuccessful, man of various semi-skilled occupations, the other in the home of a well-to-do physician." If the intrapair differences in intelligence-test scores of the various separated twins are compared with environmental differences to which they had been subjected, some interesting correlations become apparent. More extensive schooling and general educational advantages resulted in higher scores than those attained by the less favored, genetically identical twin. Much of the average difference is accounted for by only four twin pairs, those for whom the difference in amount of schooling was largest. It seems, then, that it takes great differences in environment to produce a great difference in I.Q. scores of identical twins.

There is some correlation between social advantage and higher I.Q., but it is smaller than that between educational advantage and I.Q. An example is provided by the pair of twins referred to above, one of whom was reared in a low-income home, the other in the home of a well-to-do physician. These boys were thirteen years old when tested and had an equal number of years of schooling. In spite of the estimated social advantages for the adopted son of the physician, his I.Q. score was practically identical with (in reality 1 point lower than) that of his brother. How dangerous it is to generalize from findings in this complex field of human behavior is shown by a pair of English twin brothers who were separated early and who lived under considerably different social environments, though their formal schooling was similar. In this case, the twin from the poorer environment was 19 points higher in I.Q. than his partner. One is inclined to ask with Gates, "Did adverse conditions sharpen his mind?"

In summary, the twin studies on intelligence-test behavior show: (1) modifiability of the I.Q. score under the influence of differences in environment and (2) greater similarity in I.Q. of identical twins, whether reared together or in different homes, than of nonidentical twins reared in the same home. The second fact demands an interpretation in genetic terms: differences in intelligence scores of nonidentical twins are partly due to differences in their environments and partly to hereditary differences.

The same conclusions seem to be justified by twin studies in which some of the primary mental abilities that are part of general intelligence have been analyzed separately. It has been estimated that from less than one-half to more than three-quarters of the variance of scores in tests for different primary abilities in nonidentical twins may be accounted for by genetic influences, and the rest by environmental ones. However, the detailed shares of heredity and environment in the variability of intelligence are still subject to future evaluation.

Differences in the scores of unrelated individuals of a population are, of course, also due to both environmental and hereditary variables. But it is even more difficult to ascertain the specific shares of nature and nurture responsible for degrees of differences between unrelated persons than between nonidentical twins. It is clear, however, that persons with "good" inheritance are more likely to have had parents with "good" inheritance than persons with "poor" inheritance, and that there exists either a "beneficial" or a "vicious" principle of accumulation: "good" heredity of the parents generally provides "good" environment, and "bad" heredity tends to provide "bad" environment.

The share of heredity and environment, whatever it may be, at any stage in history or any locality, is not fixed. Sometimes, the poorer the environment, the more may environmental differences account for differences between individuals. Thus, in societies where only the wealthy can give their children schooling, differences in cultural intelligence between individuals are environmentally conditioned to a much greater extent than where general education is available to everyone, so that differences in inherited capacities are more decisive for score differences than are differences in educational opportunities. The hereditary component of differences between men may sometimes be clearly apparent only because environment has been leveled upward.

Elevated I.Q. test performance. The variations in I.Q. scores in populations of unrelated persons are distributed normally around the score of 100; selected samples of persons having various syndromes are often distributed around a lower mean score. Interestingly, an exception seems to be the adreno-genital syndrome. This congenital condition is produced by dysfunction of the adrenal cortex in fetal and later stages which inhibits the production of cortisone and causes an excess of adrenal male sex hormones. Genetically, the syndrome is the result of homozygosity for a recessive autosomal gene. It leads to an abnormally rapid acceleration of growth; development in a male direction of the fetal external genitalia of both sexes; and, if not treated, preferably beginning at birth, to somatic and sexual precocity. A sample of 70 patients given I.Q. tests by Money and Lewis showed a distribution around a mean of 110. Analysis of the data suggests—but does not yet prove—that the elevated mean I.Q. score is the result of the abnormal hormonal situation.

PERSONALITY CHARACTERISTICS

Intelligence tests are not the only means by which mental traits have been evaluated. Many tests of temperament, emotional behavior, and other personality traits have been devised.

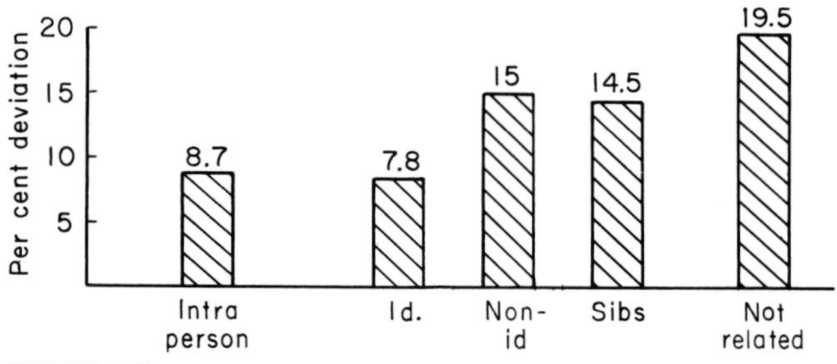

FIGURE 275
Percentage deviations between two tests of a single person and between partners of various twin pairs for a joint index of personal tempo, involving speed of knocking (an act of willing) and speed of metronome ticking (an act of perception). (After Frischeisen-Köhler.)

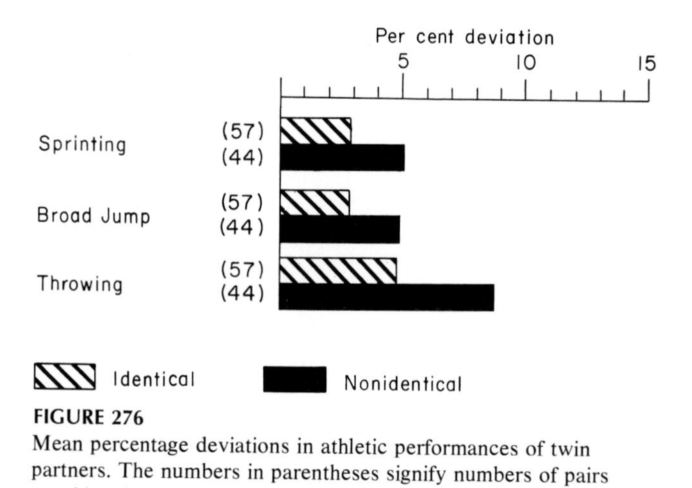

FIGURE 276
Mean percentage deviations in athletic performances of twin partners. The numbers in parentheses signify numbers of pairs considered. (After Osato and Awano, *Acta Genet. Med. Gemellol,* **6,** 1957.)

Personal Tempo. When individuals seated at a table are asked to knock on it at a speed which is "natural" to them, different individuals have been observed to perform at markedly different speeds. Similar differences are found when people are asked at which speed the ticking of a metronome seems to them just right. The "personal tempo" thus determined is rather constant for each individual. On the average, the difference between identical twins is no greater than the difference between repeated tests of the same individual (Fig. 275). Nonidentical twins show more difference, the difference being similar to that between ordinary sibs; unrelated persons differ still more. These tests provide important evidence that the personal tempo, which is

TABLE 97
Concordance (+ +) and Discordance (+ −) for
Beginning of Sitting Up, and of Walking, in Infant Twins.

Beginning of	*Twins*	*Per Cent* ++	*Per Cent* +−	*No. of Pairs*
Sitting up	Identical	82	18	63
	Nonidentical	76	24	59
Walking (v. Verschuer, Germany)	Identical	69	31	39
	Nonidentical	35	65	31
Walking (Bossik, USSR)	Identical	67	33	97
	Nonidentical	30	70	97

Source: Bossik, *Proc. Maxim Gorki Med.-Biol. Res. Institute*, **3**, 1934; and v. Verschuer, *Ergebn. Inn. Med. u. Kinderheilkd.*, **31**, 1927.

one component of what has been called the "natural velocity of the psychological life" is influenced by inheritance.

Psychomotor and Sport Activities. Studies have been made of average differences in athletic attainment of twins (Fig. 276) and of the time when twin infants begin to sit up and when they begin to walk (Table 97). The time of first sitting-up seems little affected by the greater similarity of identical, as compared to nonidentical, twins, since the figure of 82 per cent concordance for the former is not statistically different from the 76 per cent for the latter. However, for the beginning of walking, in two different studies, the concordance of 67 and 69 per cent for identicals outranks concordance of only 30 and 35 per cent for nonidenticals. (It is remarkable how closely the results of the two independent investigations agree with each other.)

A Genetic Component of Language? Different languages make use of widely different sounds, and the variety of sounds that an infant of any race is able to produce is greater than that employed by a child in speaking his native language. Why does any one language use only a specific fraction of the many possible sounds? Darlington has argued that the different genotypes that control the development of the organs of speech must limit the ease with which races and individuals can produce the various sounds, and that a specific language is therefore a reflection of the genotypically controlled phonetic preferences of the people who created it. However, this view has not been substantiated by other than very circumstantial evidence.

Other Personality Traits. Tests have been devised to study such traits as tender-mindedness, general neuroticism, will power, dominance, energetic

TABLE 98
Differences in Concordance from Twin Studies Based on Personality Questionnaires. Left column: Traits that show significantly higher concordances in identical than in nonidentical twins. Right column: Traits whose concordances in the two kinds of twins do not differ significantly.

Significant ($P < 0.01$)	*Not Significant* ($P > 0.05$)
Self-sufficiency	Introversion
Dominance	Sociability
Self-confidence	Hysteria
Social introversion	Hypochonchondriasis
Depression	Socialized morale
Ego strength	Stable temperament
Exhibitionism	Objectivity
Intellectual interests	Motivation
Frequency of drinking	Need for achievement
Average amount consumed	Aggressiveness

Source: Excerpts from findings of various investigators as cited in Vandenberg.

conformity, and submissiveness. Some of these traits seem to be predominantly environmentally determined; others seem to be dependent to different degrees on both nature and nurture (Table 98).

Two examples of performances of identical twins reared apart in the Downey Individual Will-Temperament Test are reproduced in Figure 277. The sums of the twelve scores of the twins Edwin and Fred are 70 and 63 – both higher than those of the twins James and Reece, which are 58 and 57. A different over-all pattern of scores also distinguishes the two pairs. Thus, either Edwin or Fred reaches the top score of 10 for three traits (freedom from load, speed of decision, and resistance to opposition) for which the high scores of either James or Reece are only 3, 6, and 8. On the other hand, James and Reece both score 10 for "finality of judgement," for which Edwin and Fred score 7 and 8.

On the whole, the differences between the responses of James and Reece are smaller than between those of Edwin and Fred. This, however, is in contrast to the life experiences and social adjustments of these individuals. Edwin and Fred have had extremely parallel lives: similar occupations and similar positions in society. James and Reece had very diverse social environments and very different schooling and education. The individual case histories of these and other twins often show a striking similarity in basic personality traits between identical twins, including those reared apart. (The

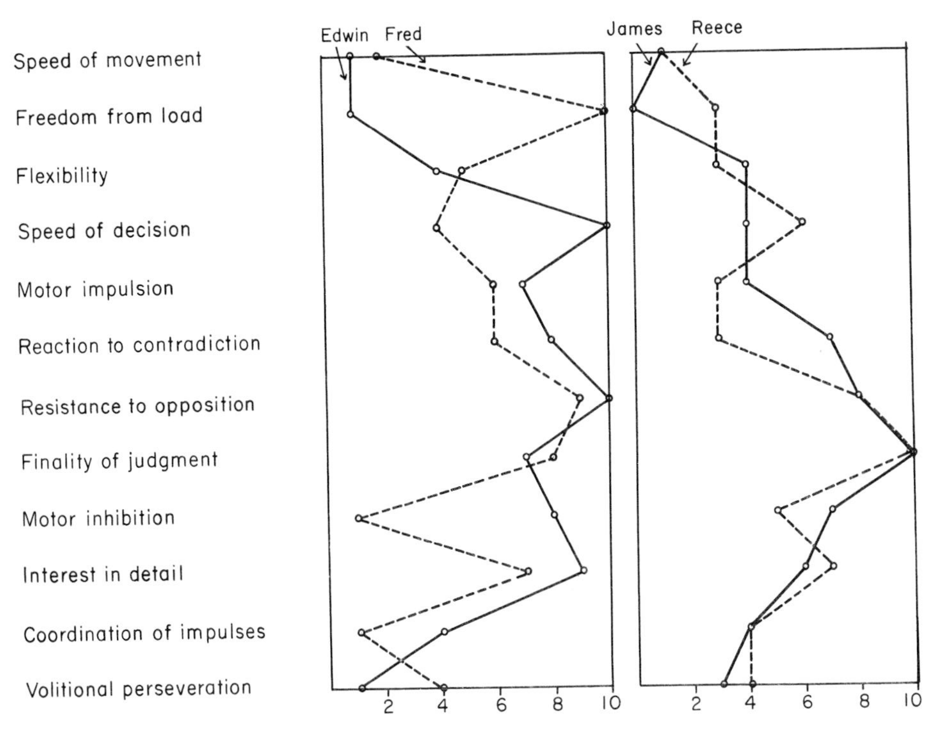

FIGURE 277
"Downey Individual Will-Temperament Test Profiles" of two pairs of identical twins reared apart. The individuals were scored, on a scale from 0 to 10, for their responses to tests which measure aspects of the twelve Will-Temperament factors listed at the left. (After Newman, Freeman, and Holzinger.)

reader will find fascinating descriptions of the life histories of separated twins in the books by Newman, Freeman, and Holzinger and by Shields.

Similarities or differences in personality are sometimes apparent at very early ages. Behavioral concordance of identical twins and discordance of nonidentical twins has been observed by the age of two months (Fig. 278). In contrast to these observations of early genetic influences on personality, similarities in the cries of infants do not correlate with zygosity and thus seem to be less dependent on genetic factors.

One of the most elaborate studies of the over-all personality and development of twins is that of Gottschaldt in Germany. He assembled ninety pairs of twin children in special camps and made detailed observations and tests. Thirteen years after the initial work he was able to find seventy of the twin pairs and evaluate their development during the intervening period, which included the disaster of World War II and its aftermath. Figure 279 shows some of the results. Many mental traits, such as those defined as basic mood, vital tension (which includes the personal tempo), capacity for thinking and

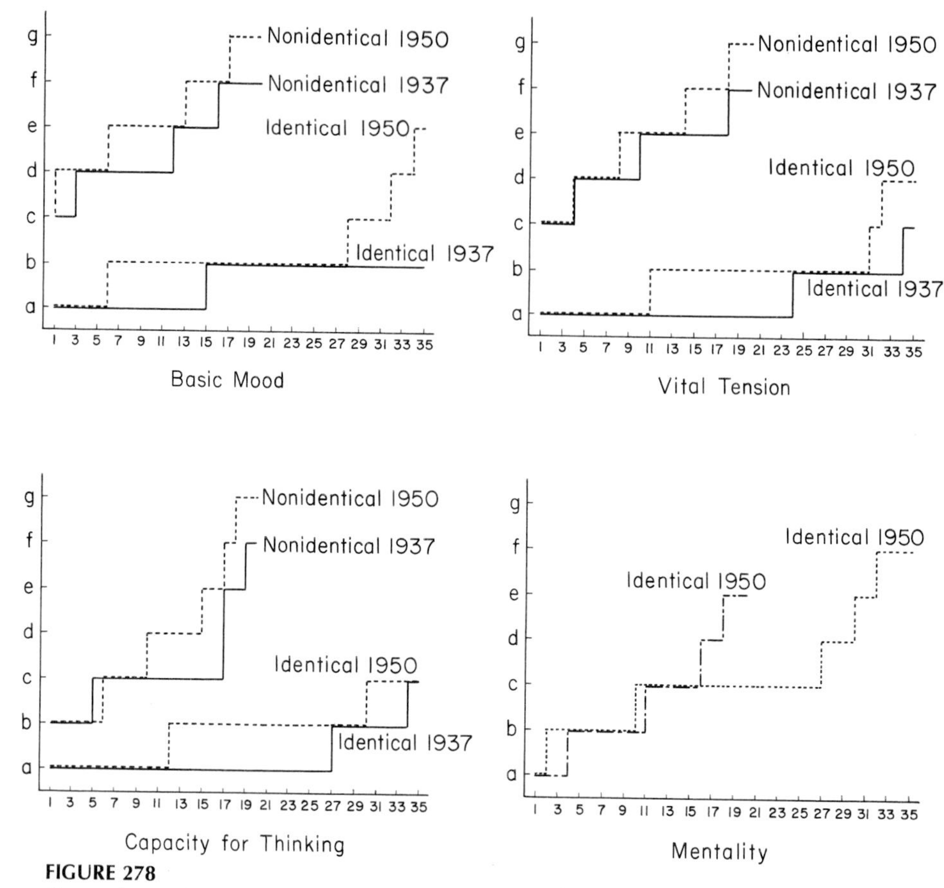

FIGURE 278

Comparisons of concordance and discordance in personality traits of twin partners studied in 1937 and, again, in 1950. The letters from a to g represent a scale from full concordance to striking discordance; the numbers signify the numbers of pairs of a given concordance rating. (Gottschaldt, *Z. Psych.*, **157,** 1954.)

abstraction, show not only greater similarity between identical than non-identical twins, but a persistence of the trends from childhood to adulthood, in spite of separation and, for some, greatly different life experiences. For certain other traits, such as social-personal superstructures of the personality ("mentality") as expressed in judgements of ethical and social values, the similarity between identical twins is not much greater than than between nonidentical partners. Such studies confirm that, by and large, it is not in physical appearance alone that identical twins are much more similar than nonidentical twins. They also provide insight into the shaping of the personality structure by the forces of nature and nurture.

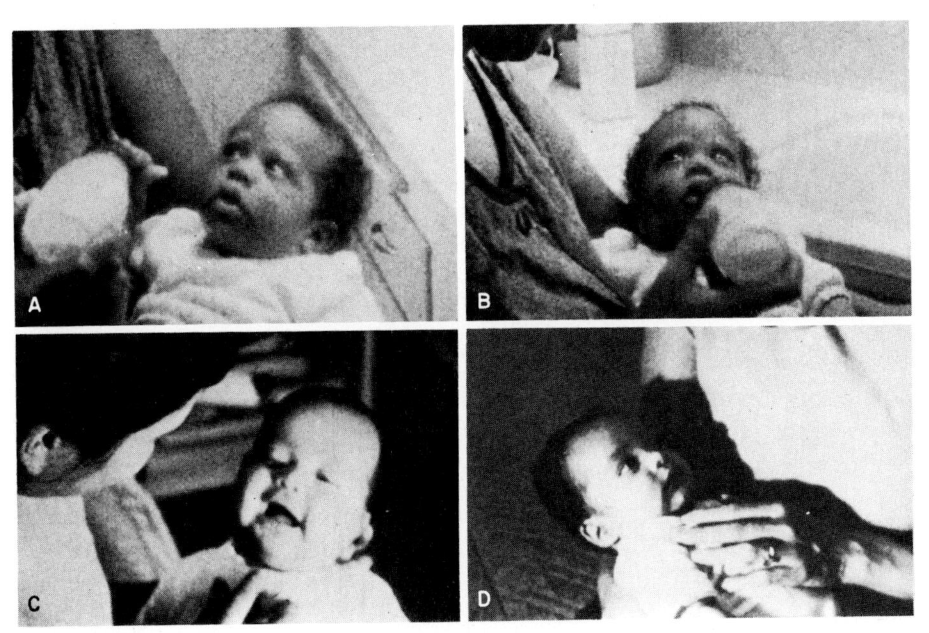

FIGURE 279
A and **B.** Identical twins, exactly 2 months old, with complete concordance in staring into mother's face during feeding. **C** and **D.** Nonidentical twins, 2 months 4 days old, with striking discordance. The twin in C is smiling but drowsy, the one in D is alert but unremittingly sober. (After Freedman *in* Vandenberg (Ed.), *Methods and Goals in Human Behavior Genetics,* 1965.)

Body Types and Personality. Attempts to relate physical appearances to mental traits are of great antiquity. One modern classification of body types is (1) pyknic, with strong development of head and trunk, and inclination to obesity; (2) athletic, with strong skeletal and muscular development; and (3) asthenic, with relative thinness but normal growth in height. A similar classification by Sheldon, based on extensive series of measurements, assigns each individual to a "somatotype"—endomorph, mesomorph, or ectomorph. These are roughly similar to the pyknic, athletic, and asthenic types. Correlations between somatotypes and personality traits, normal and abnormal, have been described. Their interpretation, however, is at present not possible since we lack knowledge of the genetic bases for the physiques defined by somatotyping as well as for the personality traits. Although some investigators regard attempts to correlate over-all mental and physical constitutions as futile, they do expect that fruitful correlations will eventually be made between well-defined "single" characters of behavior and physique or, ultimately, between behavior and biochemical reactions. It is probable that important information will be contributed in the future by both approaches, the study of personality and body types and of biochemical processes.

Criminal Behavior. The case for an inborn tendency toward ill behavior has been strikingly stated by Prospero in Shakespeare's *The Tempest.* He says of Caliban:

> A devil, a born devil, on whose nature
> Nurture can never stick; on whom my pains,
> Humanely taken, all, all lost, quite lost . . .

Such a view is denied by many. In their opinion, since the law is a kind of externally imposed special environment, there is no point in raising the question of whether or not some individuals are genetically criminals. It is certainly true that different social orders will be correlated with different kinds and incidences of crime. Nevertheless, an inquiry into the nature-nurture of criminality is concerned with a legitimate genetic problem—namely, whether in a given social order, with its specific laws, hereditary factors predispose some individuals more than others to criminal acts.

The results of twin studies carried out by different investigators in five different countries are summarized in Table 99. The high concordance of a criminal record in pairs of identical twins is obviously not due to a "bad home background" alone, since concordance in nonidentical twins is comparatively much lower—and nonidentical twins also share a common home background.

TABLE 99
Concordance (+ +) and Discordance (+ –) in Twin Pairs Involved in Criminality.

Authors	*Identical*		*Nonidentical*	
	+ +	+ –	+ +	+ –
Lange (Germany)	10	3	2	15
Rosanoff, Handy, Plesset (U.S.)	35	10	6	21
Legras (Holland)	4	0	0	5
Kranz (Germany)	20	11	23	20
Stumpfl (Germany)	11	7	7	12
Borgström (Finland)	3	1	2	3
Yoshimasu (Japan)	14	14	0	26
Total (number)	97	46	40	102
Total (per cent)	68	32	28	72

Source: Kranz, *Lebensschicksale Krimineller Zwillinge,* Springer, 1936; Lange, 1929; Legras, *Psychose en Criminaliteit bei Tweelingen,* 1932; Rosanoff, Handy, and Plesset, *Psych. Monogr. I,* State of Calif. Dept. of Instit., 1941; Stumpfl, *Die Ursprünge des Verbrechens,* Thieme, 1936; Borgström, *Arch. Rassenbiol.,* **33,** 1939; and Yoshimasu, after Kamide, *Jap. J. Hum. Genet.,* **2,** 1957.

The full degree of concordance is shown only very incompletely by the numbers given in the table. The individual life histories (in the detailed, sympathetic accounts of each pair, by Lange, Kranz, and Stumpfl) disclose that concordance in identical twins signifies not only that both were, at one time or another, in prison, but that the type of crime or crimes was very similar. And discordance in identicals often consisted of only a slight difference in behavior or, when there were great differences, probably resulted from specific birth injuries. Concordance in nonidenticals, on the other hand, usually signified conflicts with the law for rather diverse causes. In this connection, the greater contrast between the identical and nonidentical twins in Lange's study than in Kranz's should be pointed out. Kranz included in his studies many individuals who had only one criminal encounter with the law; Lange included a greater number of repeated offenders. From these two studies it appears that the causes which lead a person to only one criminal offense are less dependent on his genotype—and more on nongenetic "chance"—than those which make him a chronic criminal.

The high concordance of identical twins for many traits which have been determined to have a hereditary basis may seem to suggest that the greater concordance in criminality of identicals as compared to nonidenticals is due to the identity of their genotypes. It may be assumed, however, that the greater similarity in the environment of identical twins plays an important role in this behavioristic trait. If nongenetic chance leads one twin into a criminal offense, would not the similarity of the environmental experiences of his identical partner also lead him to a similar crime? Conversely, if nongenetic chance leads one twin into an offense, would not his nonidentical partner, who is less likely to share his specific environment, be spared the conflict with the law? The facts on criminality in twins actually show only that identity in genes plus the close similarity in social experiences, at least in early childhood, are more likely to bring both identical twins into prison than two nonidenticals who have nonidentity of genes plus less similar social experiences. On the basis of present data, we cannot exclude the possibility that the higher concordance for criminality of identical twins is mainly, or even exclusively, the result of their more similar social experiences; nor can one exclude the opposite possibility that their higher concordance is mainly the result of their identical genotypes. Studies on criminal twins reared apart from birth are needed to clarify this important question.

If the present evidence from twins is insufficient to indicate reliably that criminals are "born," it does point to a genetic component for the type of crime committed, if a crime is committed at all. The personality traits which distinguish people and which seem to have a genetic basis will partly determine whether a criminal is an embezzler, a burglar, or a murderer. This is indicated by the similarity in types of crimes committed by identical twin partners, as well as by the similarity in personalities even of those identical pairs of whom only one partner has had conflicts with the law.

In addition to twin data, a genetic interpretation of some types of criminal behavior has been derived from chromosomal findings. Males of the unusual sex-chromosomal type XYY are very rare in the general population but were found to make up several per cent of the inmates of a British maximum-security prison-hospital. On the average they are very tall men. Some of them were known to be unusually aggressive, and it was thought that it was their XYY constitution which led to excessive aggression and to consequent antisocial conduct. Since the publication of first reports on suspected inclination to criminal activity of XYY men, doubts have arisen about the validity of the presumed association. The early findings were made on selected populations, and it seems necessary to study large unselected samples of both adults and newborns to obtain reliable data on the frequencies of XYY adults and on the social development of XYY infants. Such studies are laborious and have to extend over many years. At the time of this writing (1971) there is not enough evidence on which to base a decision for or against the hypothesis that XYY males are predisposed "to the development of a psychopathic personality and to consequent aberrant behavior and antisocial conduct" (Court Brown).

There is a recessive X-linked gene that condemns its bearers to severe cerebral palsy, mental defect, and compulsive aggressive behavior directed both at themselves and at others. It leads to death in childhood. Affected children mutilate themselves by chewing their lips and fingers and their behavior toward others includes spitting, biting, and hitting. This "Lesch-Nyhan" syndrome results from the gene-controlled absence of the enzyme hypoxanthine-guanine phosphoribosyltranferase required for normal purine metabolism. As a result the children have an excess of uric acid in their blood, producing extreme symptoms of gout. The relation between the biochemical abnormality and the behavior of the children is not understood. This extreme example of antisocial gene-dependent behavior suggests that other genotypes may be instrumental in the causation of lesser degrees and different types of deviant behavior.

NATURE–NURTURE STUDIES

There is nothing unacceptable in the recognition that differences in mental traits between different individuals are due to both heredity and environment. There is no a priori limit which the geneticist can place on the power of specific environments to lead a genotype to its highest expression. The concept of different genetic endowments may even be more helpful in the task of developing a variety of suitable environments than an assumption of lack of genetic differences, which might induce people to search for a single, standardized "best" environment.

Results such as those reported in this chapter frequently justify an optimistic attitude, in spite of the fatefulness which the term heredity seems to imply. If a concordance of 40 per cent among identical twins for schizophrenia, as opposed to only 10 per cent in nonidenticals, emphasizes the influence of heredity, it is, nevertheless, highly important that in 60 per cent of identical twins, one of them was spared by the disease. If we can find out why the same genetic constitution in 60 per cent of the twin partners did not express itself by the symptoms of mental illness, we can hope to use our knowledge in saving still more individuals with these genotypes from breakdown. Instead of regarding the results of nature-nurture studies as static, they can serve the dynamic purpose of continuously fitting more suitable environments to the different genotypes.

The astonishing basic similarity, if not near-identity, of genetically identical twin pairs in both mental and physical traits contrasts with the absence of such close similarity in the genetically nonidentical twin pairs, and remains as strong support for the view that neither in body nor in mind are men born alike.

PROBLEMS

200. Assume that a mental disease depends on a gene A' which is 20 per cent penetrant in heterozygotes and 100 per cent penetrant in homozygotes.
   (a) What frequencies of affected offspring would be expected in the following matings: $AA \times AA'$; $AA \times A'A'$; $AA' \times AA'$; $AA' \times A'A'$.
   (b) If the allele frequency of A' were $q = 0.02$, what would be the sum of the frequencies of all marriages of two affected spouses? What would be the mean frequency of the disease among the offspring of these marriages?
   (c) What would be the answers to Parts (a) and (b), if improved nongenetic circumstances would reduce the penetrance of the heterozygotes to 10 per cent and that of the homozygotes to 60 per cent?

201. Among 278 sibs of criminals, Stumpfl found 103 who had a criminal record. This corresponds to 1 criminal out of 2.7 sibs of criminals. Among 62 nonidentical twin partners of criminals, Stumpfl and Kranz found 30 offenders. This corresponds to 1 criminal out of 2.1 nonidentical twin partners of criminal twins. It has been suggested that the last-named higher frequency of criminals (1 in 2.1) as compared to the first-named frequency (1 in 2.7) is due to the greater environmental similarity for twins than for ordinary sibs.
   (a) What is the statistical significance of the data?
   (b) What bearing has the answer to the preceding question on the suggested explanation for the different frequencies?

REFERENCES

Böök, J. A., 1953. A genetic and neuropsychiatric investigation of a north-Swedish population, with special regard to schizophrenia and mental deficiency. *Acta Genet. Stat. Med.*, **4**:1–100.

Brewster, D. J., 1968. Genetic analysis of ethanol preference in rats selected for emotional reactivity. *J. Hered.*, **59**:283–286.

Brosnahan, L. F., 1961. *The Sounds of Language*: An Inquiry into the Role of Genetic Factors in the Development of Sound Systems. 250 pp. Heffer, Cambridge.

Burt, C., 1963. Is intelligence distributed normally? *Brit. J. Stat. Psych.*, **16**: 175–190.

Burt, C., 1966. The genetic determination of differences in intelligence: A study of monozygotic twins reared together and apart. *Brit. J. Psych.*, **57**:137–153.

Cavalli-Sforza, L. L., and Bodmer, W. F., 1971. *The Genetics of Human Populations.* 965 pp. W. H. Freeman and Company, San Francisco.

Court Brown, W. M., 1969. The development of knowledge about males with an XYY sex chromosome complement. *J. Med. Genet.*, **5**:341–359.

Darlington, C. D., 1969. *The Evolution of Man and Society.* 753 pp. Allen and Unwin, London.

Eells, K., Davis, A., Havighurst, R. J., Herrick, V. E., and Tyler, R. W., 1951. *Intelligence and Cultural Differences: A study of Cultural Learning and Problem-Solving.* 388 pp. University of Chicago Press, Chicago.

Erlenmeyer-Kimling, L. (Guest Ed.), 1972. *Genetics and Mental Disorders.* 230 pp. International Arts and Sciences Press, White Plains, N.Y. (International Journal of Mental Health, **1**: Nos. 1 and 2.)

Erlenmeyer-Kimling, L., and Jarvik, Lissy F., 1963. Genetics and intelligence: a review. *Science*, **142**:1477–1479.

Frischeisen-Köhler, I., 1933. *Das persönliche Tempo. Eine erbbiologische Untersuchung.* 63 pp. Thieme, Leipzig.

Fuller, J. L., and Thompson, W. R., 1960. *Behavior Genetics.* 396 pp. Wiley, New York.

Ginsburg, B. E., 1958. Genetics as a tool in the study of behavior. *Perspectives Biol. Med.*, **1**:397–424.

Hall, C. S., 1951. The genetics of behavior. Ch. 9 *in* Stevens, S. S. (Ed.), *Handbook of Experimental Psychology.* Wiley, New York.

Heston, L. L., 1966. Psychiatric disorders in foster home reared children of schizophrenic mothers. *Brit. J. Psych.*, **112**:819–825.

Heston, L. L., 1970. The genetics of schizophrenic and schizoid disease. *Science*, **167**:249–256.

Honzik, M. P., 1957. Developmental studies of parent-child resemblance in intelligence. *Child Development*, **28**:215–228.

Jensen, A. R., 1969. How much can we boost I.Q. and scholastic achievement? *Harvard Educ. Rev.*, **39**:1–123. (Reprinted in Reprint Series No. 2 on Environment, Heredity, and Intelligence, together with discussions by others, compiled from the *Harvard Educ. Rev.*)

Jensen, A. R., 1970. IQ's of identical twins reared apart. *Behav. Genet.*, **1**: 133–148.

Kallmann, F. J., 1953. *Heredity in Health and Mental Disorder.* 315 pp. Norton, New York.
Karlsson, J. L, 1966. *The Biologic Basis of Schizophrenia.* 87 pp. Thomas, Springfield, Ill.
Lange, J., 1929. *Verbrechen als Schicksal.* 96 pp. Thieme, Leipzig. (Translation: *Crime and Destiny.* 250 pp. 1930. Boni, New York.)
Lawrence, E. M., 1931. An investigation into the relation between intelligence and inheritance. *Brit. J. Psych.*, Monogr. Suppl. No. **16:**1–80.
Marler, P. R., and Hamilton, W. J., III, 1966. *Mechanisms of Animal Behavior.* 771 pp. Wiley, New York.
McClearn, G. E., 1970. Behavioral genetics. *Ann. Rev. Genet.*, **4:**437–468.
McKeown, T., 1970. Prenatal and early postnatal influences on measured intelligence. *Brit. Med. J.*, **3:**63–67.
Meade, J. E., and Parkes, A. S. (Eds.), 1965. *Genetic and Environmental Factors in Human Ability.* 242 pp. Oliver and Boyd, Edinburgh.
Mitsuda, H. (Ed.), 1967. *Clinical Genetics in Psychiatry.* 408 pp. Osaka Medical College, Osaka.
Money, J., and Alexander, D., 1966. Turner's syndrome: further demonstration of the presence of specific cognitional deficiencies. *J. Med. Genet.*, **3:** 47–48.
Money, J., and Lewis, Viola, 1966. IQ, genetics, and accelerative growth: adrenogenital syndrome. *Bull. Johns Hopkins Hosp.*, **118:**365–373.
Money, J., Lewis, Viola, Ehrhardt, A. A., and Drash, P. W., 1967. IQ impairment and elevation in endocrine and related cytogenetic disorders. Pp. 22–27 *in* Zubin, Z., and Jervis, G. (Eds.), *Psychopathology of Mental Development.* Grune and Stratton, New York.
Newman, H. H., Freeman, F. N., and Holzinger, K. J., 1937. *Twins: a Study of Heredity and Environment.* 369 pp. University of Chicago Press, Chicago.
Partanen, J., Bruun, K., and Markhanen, T., 1966. *Inheritance of Drinking Behavior:* A study of intelligence, personality, and use of alcohol of adult twins. 159 pp. Finnish Foundation for Alcohol Studies, Helsinki.
Penrose, L. S., 1963. *The Biology of Mental Defect.* 374 pp. Grune and Stratton, New York.
Polani, P. E., 1967. Chromosome anomalies and the brain. *Guy's Hosp. Rep.*, **116:**365–396.
Record, R. G., McKeown, T., and Edwards, J. H., 1970. An investigation of the difference in measured intelligence between twins and single births. *Ann. Hum. Genet.*, **34:**11–20.
Reed, Elisabeth W., and Reed, S. C., 1965. *Mental Retardation.* 719 pp. Saunders, Philadelphia.
Roberts, J. A. Fraser, 1952. The genetics of mental deficiency. *Eugen. Rev.*, **44:**71–83.
Rosenthal, D., and Kety, S. S. (Eds.), 1968. *The Transmission of Schizophrenia.* 436 pp. Pergamon, New York.
Scarr-Salapatek, Sandra, 1971. Unknowns in the I.Q. equation. *Science*, **174:** 1223–1228.
Schull, W. J., and Neel, J. V., 1965. *The Effects of Inbreeding on Japanese Children.* 419 pp. Harper and Row, New York.
Seegmiller, J. E., 1971. Biochemical and genetic studies of an X-linked neurological disease (The Lesch-Nyhan syndrome). *Harvey Lectures*, **65:**175–192.

Shields, J., 1962. *Monozygotic Twins Brought up Apart and Brought up Together.* 264 pp. Oxford University Press, London.

Skeels, H. M., 1966. *Adult Status of Children with Contrasting Early Life Experiences:* A follow up study. Monographs of the Society for Research in Child Development, **31:** No. 3. 65 pp.

Skodak, M., and Skeels, H. M., 1949. A final follow-up study of one hundred adopted children. *J. Genet. Psych.*, **75:**85–125.

Spuhler, J. N. (Ed.), 1968. *Genetic Diversity and Human Behavior.* 291 pp. Aldine, Chicago. (Wenner-Gren Foundation for Anthropological Research Publications in Anthropology, No. 45.)

Thoday, J. M., and Parkes, A. S. (Eds.), 1968. *Genetic and Environmental Influences on Behavior.* 210 pp. Plenum, New York.

Tsubai, T., 1970. Crimino-biologic study of patients with the XYY syndrome and Klinefelter's syndrome. *Humangenetik*, **10:**68–84.

Vandenberg, S. G. (Ed.), 1965. *Methods and Goals in Human Behavior Genetics.* 351 pp. Academic, New York.

Vandenberg, S. G., 1967. Hereditary factors in normal personality traits (as measured by inventories). Ch. 6, pp. 65–104 *in* Wortis, J. (Ed.), *Recent Advances in Biological Psychiatry*, vol 7. Plenum, New York. (Proceedings of the Society of Biological Psychiatry.)

Vandenberg, S. G., 1967. Hereditary factors in psychological variables in man, with a special emphasis on cognition. Pp. 99–133 *in* Spuhler, J. (Ed.), *Genetic Diversity and Human Behavior.* Aldine, Chicago.

Vandenberg, S. G., 1968. The nature and nurture of intelligence. Pp. 3–58 *in* Glass, D. C. (Ed.), *Genetics.* The Rockefeller University Press and Russell Sage Foundation, New York.

Vandenberg, S. G. (Ed.), 1968. *Progress in Human Behavior Genetics.* Johns Hopkins University Press, Baltimore.

Willerman, L., Naylor, A. F., and Myriantopoulos, N. C., 1970. Intellectual development of children from interracial matings. *Science*, **170:**1329–1331.

Woodworth, R. S., 1941. *Heredity and Environment.* 95 pp. Social Science Research Council, New York. (Bulletin 47.)

Zellweger, H., 1963. Genetic aspects of mental retardation. *Arch. Int. Med.*, **111:**165–177.

28

SELECTION AND GENETIC POLYMORPHISM

In a large population in which each allele present has an equal chance of being transmitted to the next generation, no change in the corporate genetic endowment takes place from generation to generation. If random mating is followed and the reproductive rates of all different genotypes are alike, then not only do the initial allele frequencies remain constant (except for minor chance fluctuations) but also the proportions of individuals who are homozygous or heterozygous for any given locus are constant. There will be, of course, no absolute identity in successive generations. The number of recombinations of the many loci represented by more than one allele is so great that only a small fraction of all possible genotypes is actually found at any given time, even in a large population; thus, different generations contain different samples of the theoretical array of genotypes. By and large, these changes from generation to generation are without trend. Therefore, it may be said that in general the genetic constitution of a large population does not change basically under random mating and equal reproductive rates for different alleles and for different genotypes.

In Chapter 22 on the occurrence of mutations we saw that an equilibrium in the constitution of a population will be preserved even if mutations tend to shift allele frequencies, as long as selective forces balance such mutation pressure. If the frequencies of mutations or the strength of selection change, then alterations in the allele frequencies take place from one generation to the next and, consequently, the genetic make-up of the population changes. Genetic variations in populations without changes in allele frequencies are also possible if the mating structure becomes reoriented. If, in an originally random-mating population, assortative mating sets in, the proportions of homozygous and heterozygous genotypes will no longer obey the Hardy-Weinberg Law but will shift in a predictable fashion. Or, if in a population originally subdivided into genetically different isolates a breakdown of the isolates occurs and random mating becomes established, the alleles will be redistributed until they obey the Hardy-Weinberg proportions.

In the evolution of living forms, selection has played an important role. The preferential survival of certain alleles and certain combinations of alleles has led to the establishment of new types of plants and animals. However, selection is not solely an instrument of change; it is also a means of stabilization. If a species is once genetically well fitted for survival in its environment, the majority of new alleles originating by mutation and of new genic combinations will likely have a lower selective value than those alleles and combinations already present. Consequently, selection will tend to purge the species of genetic novelties. A striking example of the conservative role of selection is the equilibrium between mutation and selection. Recurrent mutations which confer lowered reproductive fitness on their bearers are constantly eliminated from the population. Nevertheless, with recessive mutant genes, even complete lack of fitness of the homozygotes may be compatible with a considerable accumulation in the population of the unfavorable mutant gene in heterozygous individuals.

Examples of selection against genotypes that have low reproductive fitness are numerous. They include such cases as the recessive homozygotes for Tay-Sachs disease, which, leading to death in infancy, confers reproductive fitness of zero, and the dominant heterozygotes for neurofibromatosis, which reduces the likelihood of marriage and reproduction to about 60 per cent of that of normal persons. These phenotypes are rare. We shall discuss aspects of the selective events concerning them in Chapter 29. In this chapter we shall deal with selection for and against relatively frequent alleles.

It has often been considered that, under the influence of civilization, the role of natural selection in human populations has been greatly reduced, and that it is destined to decrease further. Does not civilization, through charitable care and medical skill, reduce or eliminate selection against the less fit by improving their chances of reproduction? Clearly, this question is answered in the affirmative. Nevertheless, there is evidence that natural selection in man is acting at present in more ways than just the most obvious one, that of eliminating extreme deviants.

SELECTION AGAINST HETEROZYGOTES

One of the striking examples of large-scale selective events in man concerns the Rh alleles. As we have seen, erythroblastosis fetalis, which until recently was often fatal, acts against some of the heterozygous children born to Rh-negative mothers and Rh-positive fathers.

Considering only the Rh_0 antigen of the Rh system, and representing its genetic counterpart by *R* and that of its absence by *r*, we may say that with each hemolytic fetus lost, one *R* and one *r* allele are eliminated. This selection against heterozygotes has important consequences. If a population possesses equal numbers of the two alleles ($p_R = q_r = 0.5$), no change in the general make-up of the population will occur, since removal of equal numbers of the two alleles obviously does not alter the allelic frequencies (Fig. 280, A). However, equal frequencies of the two alleles are not found in any human population, and even if such an equality existed at one time, it would soon be disturbed by small chance deviations. Any such inequality would automatically result in still greater inequality since a loss of heterozygotes would mean removal of equal numbers of the two alleles from the unequal numbers to have been reached; for example, the Mongoloids do not possess the *r* allele. No human race is known in which the *R* allele has been lost.
present. Consequently, in each generation the imbalance of the two allelic frequencies would become greater (Fig. 280, B) until, finally, the rarer allele would be completely lost. In most populations the frequency of *R* is considerably higher than that of *r*. Therefore, in these populations the *r* allele should eventually be lost. In some races, provided the assumption that they had both *R* and *r* alleles in the past is correct, such completion of allelic shift seems to have been reached; for example, the Mongoloids do not possess the *r* allele. No human race is known in which the *R* allele has been lost.

In the Caucasoid and in some other races, both *R* and *r* persist. In view of the selection against heterozygotes, the presence of *both* alleles requires a special explanation, since the process of "running out" of the rarer allele should, in the course of evolutionary time, have already been completed.

An explanation, proposed by Wiener and Haldane, suggests that, prehistorically, there were some populations with more *R* than *r* alleles, and other populations with more *r* than *R* alleles. Selection, and perhaps chance deviations, resulted in elimination of the rarer allele, so that nonpolymorphic races, some isogenic for *R* and others isogenic for *r*, became established. Then, in the course of migrations, hybridization occurred between *RR* and *rr* people, which resulted in the appearance of groups such as the present Euro-

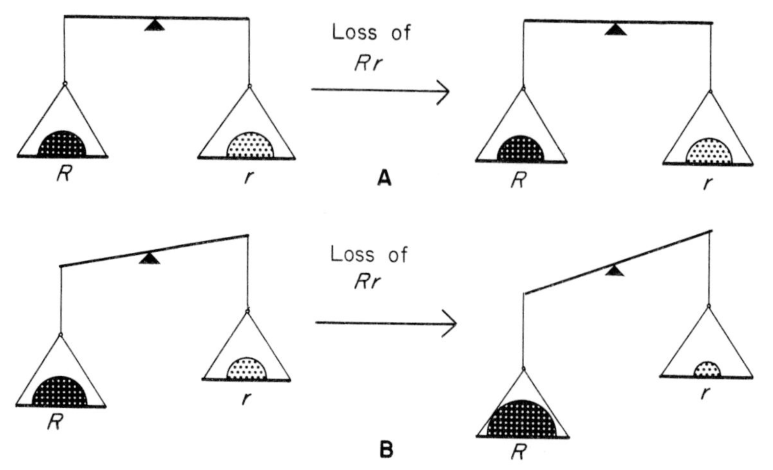

FIGURE 280
Selection against heterozygotes. **A.** In a population in which the alleles *R* and *r* are of equal frequencies, loss of *Rr* children owing to erythroblastosis fetalis will not disturb the equilibrium. **B.** In a population in which *R* is more frequent than *r*, loss of *Rr* children will shift the allele frequencies toward greater inequality.

peans, who are polymorphic for *R* and *r*. Using certain plausible assumptions, Haldane has calculated that selection against heterozygotes in Europeans would require some 600 generations, or 15,000 years, to reduce the frequency of the Rh-negative individuals from the present 14 per cent to 1 per cent. Alternatively, it can also be shown that the present allelic frequencies may be the result of a hypothetical mingling of a large number of *RR* with a smaller number of *rr* individuals about 10,000 years ago, followed by a decrease of *r* alleles due to selection against heterozygotes. The finding that the Basques have more *r* than *R* alleles has given some support to such an assumption. These people have often been regarded as remnants of a very early European race. Possibly, the original European population was isogenic for *rr*, the immigrants belonging to an *RR* race.

These suggestive speculations do not take into account possible population dynamics which might greatly delay or even reverse the decrease of the *r* allele in spite of selection against some of the heterozygotes. One such mechanism might be mutation from *R* to *r*, thus replacing all or part of the *r* alleles lost. This hypothesis will be discussed later in this chapter.

R. A. Fisher and Spencer have suggested another mechanism counteracting selection against heterozygotes. Parents of children lost from erythroblastosis tend to "compensate" for the loss of these children by creating new ones until the number of viable children born is equal to that of the family average in the population. This, of course, is usually possible only if the father is heterozygous *Rr*. The viable children will all be *rr* and will thus replace the loss of *r* alleles caused by the death of their *Rr* erythroblastotic sibs. Unfortunately, the attractive hypothesis of reproductive compensation is not supported by a careful analysis of the Rh data made by T. E. Reed.

A third possibility is that, relative to the homozygotes, selection favors those heterozygotes who come from compatible matings or those who have survived the dangers of incompatibility. Still other possibilities would be selection for *rr* homozygotes, or combinations of the various factors. At present there is no evidence for any of these mechanisms, but this lack of evidence does not mean that such mechanisms may not exist.

Future studies will undoubtedly contribute to our insight into the dynamics of human populations which are subjected to selection against heterozygotes. A fact of general importance is already known and deserves understanding. It concerns the following question: Could opposing forces of selection against heterozygotes and in favor of replacing the rarer of the lost alleles by mutation balance each other so completely that the population would not change in its allele frequencies? For most populations, the answer, as shown by Haldane, is in the negative. In order to follow the argument, we investigate first, in somewhat greater detail than we have up to this point, the fate of *Rr* populations with different allele frequencies.

The Frequency of Mortality from Rh Incompatibility. Obviously, there can be no losses of heterozygotes when either $p = 1$ or $q = 1$. If we neglect slight corrections for the fact that a population with selective loss of some heterozygotes does not fully fit the Hardy-Weinberg Law, then for other values of p and q the proportion of incompatible mother-offspring combinations is p^2q^2 from marriages of *rr* women to *RR* men plus pq^3 from marriages to *Rr* men. The sum $p^2q^2 + pq^3$ is equal to $pq^2(p + q)$, and since $(p + q) = 1$, equal to pq^2. Substituting $(1 - q)$ for p, the proportion of incompatible combinations, expressed in terms of the frequency q for the *r* allele, is $q^2(1 - q)$. Actually, as seen earlier, only a small fraction of offspring from incompatible combinations is lost. If this fraction is taken as 0.05 (1 out of 20)—and owing to medical provisions, at least in the developed countries, the fraction has by now been greatly decreased—we can express the frequency of loss of heterozygotes by a curve, as shown in Figure 281. The percentage of mortality rises from zero for $q_r = 0$ to a maximum for $q_r = 0.67$, then declines rather steeply to zero for $q_r = 1$. An average population of European descent with $q_r = 0.4$ corresponds to point I on the curve. If selection against the heterozygotes alone were determining the future of the *R* allele frequencies, the allele frequencies q_r in successive generations would move toward the left. A Basque population with $q_r = 0.6$ corresponds to point II on the curve. Here the allele frequency q_r would move toward the right.

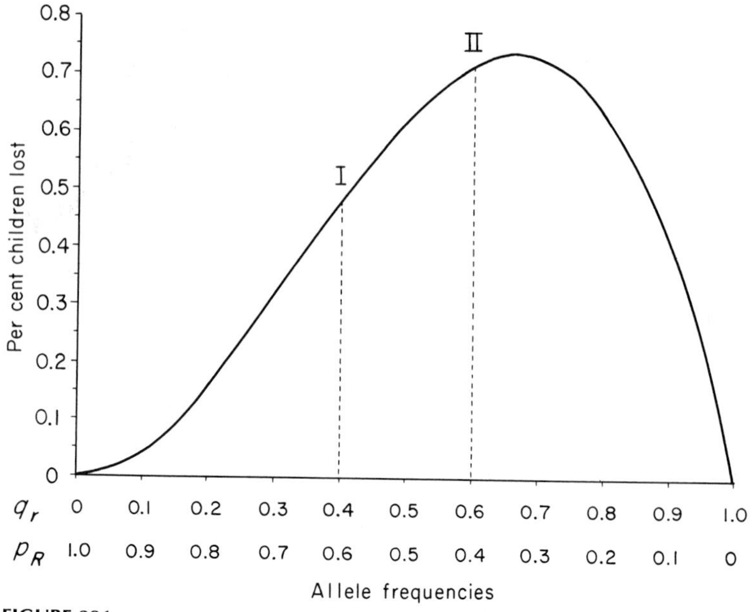

FIGURE 281
The percent mortality of children owing to Rh incompatibility as related to the frequencies p_R and q_r under the simplified assumption of two alleles only. Only 5 per cent of children from incompatible mother-child combinations are assumed to be lethal. I: Position of many Caucasian populations. II: Position of a Basque population. (After Haldane, 1942.)

Selection Against Heterozygotes and Mutation. Let us now assume that the population I in Figure 281 is not subjected solely to selection against heterozygotes, but that mutations from *R* to *r* take place at such a rate as to balance exactly the relative loss of *r* alleles in erythroblastotic mortality. This would require a rather high mutation rate, since the relative loss of *r* alleles in heterozygotes is considerable. Theoretically, however, a mutation rate could be calculated which would furnish the required balancing gain of *r* alleles. *But* the population could not persist with unchanged allele frequencies for more than a generation, and once changed even slightly it would by necessity deviate further in future generations. The reasons for this interesting behavior will be given in general terms.

If the frequency of *r* were q in the first generation and the balancing rate of mutation to *r* were u, chance would bring it about that the allele frequency q_r in a future generation would not be exactly the same as in the beginning. If, for instance, q_r became slightly higher, it can be shown that the erythroblastotic loss of *r* alleles as a fraction of all *r* alleles would be lower than that prevailing at the original allele frequency. Since the rate of mutation, u, would then be higher than the rate of loss, the frequency of q_r in the following generation would increase further. This trend would persist over successive generations until the frequency of q_r became 0.5. At this stage, loss of r alleles would be balanced by loss of R alleles, but mutations to r would continue. Therefore, the frequency of q_r would rise above 0.5. Once r alleles were more numerous than R alleles, the frequency of the former would increase, owing to selection against heterozygotes, even without mutation to r. Given such mutations, the increase would be accelerated, and the population would tend toward the stable equilibrium $q_r = 1$.

Had chance led to an allele frequency q_r of population I slightly below the initial one, then, the detailed treatment shows, the allele frequency would decrease still further in successive generations. Mutations would be insufficient in number to make up for the selective loss of r alleles. The successive shifts of q_r would not tend to lead to the extreme value of $q_r = 0$ but would cease at about $q_r = 0.257$. At this point, there would be a stable equilibrium between the postulated mutation rate and selection, so that the mutation rate which would have supported the (postulated but unstable) initial equilibrium at $q_r = 0.4$ would not only balance the relative loss of r alleles but also tend to overcome any further chance deviations of q_r: a slight increase of q_r is immediately followed by greater selective loss of r alleles, and a decrease by greater gain.

We have seen that the present situation of a typical Caucasoid population, in respect to its frequencies of the R allele, cannot be regarded as one of a stable equilibrium brought about by selection against heterozygotes balanced by mutation. It appears, then, that the frequency of the R alleles is changing among Caucasoid populations. The direction of this change is not obvious. The percentage of Rh mortality of heterozygotes was greater before modern medical help was available than it is at present. With the exception of the Basques, this must have meant a change of R frequencies toward a lower value of q_r. With lowered mortality, the selective loss of r alleles has been slowed. Medical advances are now making it possible to eliminate all losses from Rh incompatibility. In populations in which no more losses occur, no further systematic changes of the allele frequencies would be expected, except for mutation, and unless presently unknown attributes of the different alleles would lead to selection independent of incompatibility.

Selection against heterozygotes resulting from maternal-fetal incompatibility occurs not only for the R allele but also for other blood-group genes. As in selection against Rh heterozygotes, selection against ABO because of mother-child incompatibility should lead to successive decreases in the frequencies of the rarer alleles. In all populations the allele I^0 is more frequent than is any other I allele, so without opposing forces all populations should tend toward $r_0 = 1$. The existence of such opposing forces has been postulated, as will be seen in a later section of this chapter.

It will be remembered that incompatibility for the ABO blood groups protects against the effects of Rh incompatibility. Since the frequencies of ABO incompatible matings vary in different populations, the effect of the strength of selection against Rh heterozygotes must also vary. Thus, in blacks, in whom the frequencies of the I^A, I^B, and I^O alleles lead to potential ABO incompatibility in mother-child combinations more often than they do in whites, more protection against Rh losses would be provided. Nevertheless, it may well be that still other genes tend to change the force of selection against Rh heterozygotes in the two populations, for instance by controlling the severity of the hemolytic disease.

In whites and blacks selection against Rh heterozygotes will simultaneously favor selection for I^A and I^B over I^O alleles, since I^A and I^B are less frequently involved than is I^O in the loss of the ABO-compatible, but Rh-incompatible fetuses. In Mongoloids, in whom most combinations are Rh compatible, the loss of I alleles would be proportional to ABO incompatibility, uninfluenced by Rh incompatibility. The relation between selection against heterozygotes in the Rh and ABO blood groups is a reminder that our treatment of selective forces represents only a first approach to the much more complex situation in the total genic systems of individuals and populations.

At present, selection against heterozygotes in man has been demonstrated only in cases of antigenic incompatibilities. It will be important to investigate whether other situations occur in which heterozygotes are inferior to both kinds of homozygotes. There are indications of such occasional relations in other organisms.

SELECTION FOR HETEROZYGOTES

The vigor of hybrids, as compared to that of their parental stocks, has long been known in animal and plant breeding. The often-quoted example of the mule, the result of a cross between the horse and the donkey, is a case in which the individual vigor of the hybrid is accompanied by nearly complete sterility, that is, absence of reproductive fitness. The most famous, modern example of heterosis is hybrid corn, which combines both greater individual vigor and higher reproductive fitness than its pure-bred parents possess. Part of this high selective value of hybrids is due to the superior fitness of heterozygous constitutions at various loci over that of the corresponding homozygotes. A variety of phenomena in diverse animals and plants has likewise been interpreted as intrinsic superiority of heterozygous genotypes.

A remarkable discovery concerning selection for human heterozygotes has come from studies of the three genotypes Hb^AHb^A, Hb^SHb^A, Hb^SHb^S, which possess normal hemoglobin, a mixture of normal and sickle-cell hemoglobin, and sickle-cell hemoglobin, respectively. Great differences in the frequency, q_S, among different African tribes have been noted, and groups of people with high, medium, and very low frequencies of the Hb^S allele have also been found in southern India, Greece, and Italy. Following a suggestion by Beet, Allison pointed out that the frequency, q_S, of the allele for sickle-cell hemoglobin in different populations is positively correlated with the frequency of malaria in these populations. The hypothesis was therefore suggested that red cells of Hb^AHb^S heterozygotes are more resistant to infection by the malarial parasite than are those of Hb^AHb^A homozygotes. Since Hb^SHb^S homozygotes have a very low reproductive fitness—until recently the majority succumbed early to sickle-cell anemia—the presence of malaria in an area would selectively favor Hb^AHb^S heterozygotes over both kinds of homozygotes. A small experiment, in which volunteers of the genotypes Hb^AHb^A and Hb^AHb^S submitted to infection with the so-called subtertian variety of malaria (caused by the protozoan parasite *Plasmodium falciparum*), showed a decidely greater disease resistance among the heterozygotes than among the normal homozygotes. The superiority of the heterozygotes for the Hb^S allele in a malarial area has since been fully established. They share their resistance to falciparum infection with heterozygotes for alleles at two other loci, those of beta thalassemia and of G-6-PD.

We have seen that selection against heterozygotes typically leads to a reduction of frequency, or a disappearance, of one or the other of the two alleles in the population. An opposite effect is produced by selection in favor of heterozygotes. To start with an extreme model, if both homozygotes, AA and $A'A'$, are lethal or sterile, so that the only parents are AA' heterozygotes, then the population would obviously contain equal numbers of A and A' alleles and would be stable. This is so, since the fertile offspring of the AA' parents would consist of AA' genotypes only. In other words, the population would forever retain both A and A' alleles, and their frequencies would be equal. The same result would be obtained if selection against AA and $A'A'$ were not complete but if both homozygotes could have offspring in equal numbers, though fewer than those of the heterozygotes. At the time of fertilization the proportions of AA, AA', and $A'A'$ zygotes would follow the Hardy-Weinberg Law, $p^2{:}2pq{:}q^2$, in spite of later selection against survival or fertility of the homozygotes. With p initially equal to q, the loss of equal numbers of AA and $A'A'$ parents or their gametes would not change the proportion of A and A' alleles.

An important property of a selective system which favors heterozygotes is: Whatever the initial allele frequencies are in a population, the population will tend toward a single stable proportion of the alleles. The equilibrium point is determined solely by the strengths of selection against the two homozygotes. Two examples will clarify these assertions. Assume that the selective value of both AA and $A'A'$ homozygotes is one-half that of the AA'

heterozygotes, and consider two different populations, I and II, in which the initial allele frequencies are as follows: population I, $p_A = 0.5$, $q_{A'} = 0.5$; population II, $p_A = 0.4$, $q_{A'} = 0.6$ (Table 100). Random mating in population I will produce zygotes in the proportions 0.25:0.5:0.25. After selection, the effective proportions are changed to 0.125:0.5:0.125. The allele frequencies, however, remain as they were before selection, namely,

$$p_A = [(2 \times 0.125) + 0.5]/1.5 \text{ and } q_{A'} = [0.5 + (2 \times 0.125)]/1.5 = 0.5.$$

Population I is thus at equilibrium.

In population II, however, random mating will produce zygotes in the proportions 0.16:0.48:0.36, which after selection become 0.08:0.48:0.18. The allele frequencies are now

$$p_A^1 = [(2 \times 0.08) + 0.48]/1.48 = 0.432,$$

and

$$q_{A'}^1 = [0.48 + (2 \times 0.18)]/1.48 = 0.568.$$

Population II is therefore not in equilibrium. Its allele frequencies have moved from the initial values of $p_A = 0.4$ and $q_{A'} = 0.6$ toward $p_A = q_{A'} = 0.5$. Further progress toward these equilibrium values would be made in successive generations. It is obvious that the reverse change in allele frequencies would

TABLE 100
Selection for Heterozygotes during One Generation.

Population	*AA*	*AA'*	*A'A'*	*Total*	
				Individuals	Alleles
POPULATION I					
Initial proportions ($p_A = 0.5$; $q_{A'} = 0.5$)	0.25	0.50	0 25	1.00	2.00
Selective values	½	1	½		
Proportion after selection	0.125	0.5	0.125	0.75	1.50
Initial proportion in next generation	0.25	0.50	0.25	1.00	2.00
POPULATION II					
Initial proportions ($p_A = 0.4$; $q_{A'} = 0.6$)	0.16	0.48	0 36	1.00	2.00
Selective values	½	1	½		
Proportion after selection ($p_A^1 = 0.432$; $q_{A'}^1 = 0.568$)	0.08	0 48	0.18	0.74	1.48
Initial proportion in next generation	0.187	0.492	0.321	1.00	2.00

occur if population II initially possessed a $p_A = 0.6$ and a $q_{A'} = 0.4$. In one generation, p_A would change to $p_A^1 = 0.568$ and $q_{A'}$ to $q_{A'}^1 = 0.432$ and thus would have tended toward $p_A = q_{A'} = 0.5$. In selection for heterozygotes, the equilibrium point thus serves as a "point of attraction" for deviating allele frequencies, in contradistinction to selection against heterozygotes where equilibrium points usually are "points of repulsion."

Only rarely will the selective disadvantages of the two homozygous types, relative to those of the heterozygotes, be of equal degree, as assumed in the last examples where the fitnesses of *AA* and *A'A'* were both one-half. If the homozygotes differ in fitness, then the equilibrium point will not be at $p_A = q_{A'} = 0.5$ but will be shifted toward a frequency which is higher for that allele whose homozygote is fitter. If, for instance, in a malarial environment the fitness of $Hb^A Hb^A$ individuals were 0.8, relative to a value of 1 for $Hb^A Hb^S$, and if the fitness of $Hb^S Hb^S$ were only 0.1, then a stable equilibrium would be represented by allele frequencies of approximately p_A^{Hb} 0.82 and $q_S^{Hb} = 0.18$. For the simple formulas from which these equilibrium values have been obtained, the reader may be referred to Li's *Population Genetics.*

The heterozygotes for normal and sickle-cell hemoglobins show codominance. Subtle differences in fitness values of heterozygotes for genes that seem to be fully recessive may also exist. They would play important roles in the equilibria conditions of populations. A possible example of such a situation has been analyzed by Knudson and by Danks and their collaborators in their study of cystic fibrosis. This disease affects the pancreas, the respiratory system, and other organs, and, until recently, led to death in infancy or in any event before the reproductive period. The condition occurs with a frequency of about 4 in 10,000 births and is caused by an autosomal recessive gene. Why does this gene have such a high frequency in spite of its high rate of elimination due to homozygosity for it being lethal? One possible answer would be an equilibrium between loss of the allele from the zero fitness of homozygotes and gain of the allele by recurrent mutation from normal *A* to *a*. However, the mutation rate required in Caucasians would be much higher than that usually found for other genes, which seems unlikely. In view of this it has been proposed that an equilibrium is based on selection for the heterozygotes *Aa*. Suggestive evidence for such selection has been obtained by comparing the number of offspring of *Aa* × *AA* matings with the number from *AA* × *AA* matings. The *Aa* × *AA* matings consisted of the grandparents of *aa* children. The mean number of offspring of the *Aa* × *AA* couples was 4.34, and that of the *AA* × *AA* couples 3.43 – a highly significant difference. Calculations show that a heterozygous advantage of only 2 percent would be sufficient to maintain an equilibrium for the observed frequency of the *a* allele. Very similar findings have been reported by Myrianthopoulos for Tay-Sachs disease, which also is due to an autosomal recessive that is lethal in homozygotes (see p. 180). The data give a mean reproductive fitness of $Aa \times AA$ matings as 1.06 (or less), as compared with the value of 1.00 for $AA \times AA$ matings. The observed difference, however, is not "yet" statistically significant. Moreover, it is not certain for either disease whether the existence of some subtle bias leading to preferential ascertainment of the more fertile grandparents does not exist, which, if excluded, would remove the difference.

It has been considered whether selection for heterozygosis may play a stabilizing role in polygenic inheritance. For many traits intermediate phenotypes may be of selective advantage. For example, it is known that the frequency of infant death at or within a month of birth is higher for the lightest and for the heaviest babies than it is for middle-weight babies. Although it is true that birth weight is determined to a large degree by factors other than the genotype of the child itself, its genotype does play a certain role and it is likely that the intermediary classes contain more heterozygotes for the pertinent loci than do the extreme classes. However, as was first pointed out by Fisher, selection for intermediate phenotypes in polygenic traits is not a stabilizer of allele frequencies since such selection may favor both heterozygotes, e.g., $AA'BB'$, and homozygotes, e.g., $A'A'BB$ and $AAB'B'$. Nevertheless, if selection were effective not for intermediate phenotypes in general but specifically for heterozygous genotypes within them, stability may be attained.

The genetic basis for mental abilities is likely to be polygenic. Persons with low and high endowment have been considered to be mostly homozygotes, and those with intermediate endowment have been considered to include most heterozygotes. Since, in many populations, the reproduction of both very lowly and highly endowed individuals is less than that of those with average endowment, it has been suggested that the higher reproductive fitness of heterozygotes would tend toward stabilization of allele frequencies. Such a hypothesis involves heterosis from the standpoint of reproductive but not of mental performance. As an alternative, it might be suggested that the genetic component of high mental performance tends to consist of heterozygous genotypes. In this case the low reproductive fitness of high mental performers would signify selection not for, but against, heterozygotes. As seen earlier, this type of selection is not a stabilizer of allele frequencies. It is likely that the genetic basis for mental performance is different in different individuals, involving both homozygous and heterozygous polygenic combinations. Until we know more about the genetic control of mental performance and its correlated reproductive fitness, we can only work out models from which the trend of the endowment of hypothetical populations can be deduced. The actual processes occurring around us will undoubtedly remain obscure for a long time.

GENETIC POLYMORPHISM

The existence in a population of more than one allele at a given locus leads to the appearance of more than one genotype. In the past such genetic polymorphism was often accepted as a fact which did not pose any unusual problem. Indeed, the existence of rare dominant or recessive alleles which cause the appearance of clearly abnormal phenotypes can be understood as a consequence of mutation of the normal allele and of selection which keeps the abnormal allele at a low equilibrium level. With the coexistence in populations of frequent alleles, none of which seems to have an obviously abnormal effect, the situation is different. Examples such as those of the I^A, I^B, I^O alleles, or those of the alleles for so many other blood groups, or of the presence of the two taster alleles *T* and *t*, the secretor alleles *Se* and *se*, and many others, could hardly be regarded as the result of the opposing mutation–selection processes. How did the high frequencies of the different alleles become established in the different populations? Why does the same allele occur in different frequencies, for instance, I^O as low as, or slightly less than, 0.5 and as high as 1.0, or *t* as low as 0.80 and as high as 0.97? Why, also, did not chance often lead to complete loss of an allele from a population when, as must have happened frequently in the prehistoric past, famine, disease, or war decimated the different, mostly small, human groups? Answers to these questions that were based on such unlikely assumptions as very high mutation rates and striking differences in these rates in different populations were not satisfactory.

There is another set of facts which requires explanation. As widely as allele frequencies may diverge from one population to another, they are usually limited to a range which is narrower than the possible range of 0.0 to 1.0. This is clear from the two examples of I^O and *t* just given.

Polymorphism has been defined as the occurrence together of two or more alleles in a population in such proportions that the least common of them is too frequent to be accounted for by recurrent mutation alone. This leaves room for more specific statements concerning the allele frequencies in polymorphism. Such statement define "common" alleles as having frequencies of at least 0.01.

Polymorphisms are of two types, transient and balanced. The former characterizes a temporary evolutionary state during which an "old" allele in a population is replaced by a "new" allele. In many cases this would be brought about by selection not countered by other processes. In contrast, balanced polymorphism, according to Fisher and Ford, is a stable condition in which the allele frequencies are maintained by a balance of selective agencies. Selection for heterozygotes, originally known only for a few animal cases,

TABLE 101
Polymorphisms Detected from Electrophoretic-variant Enzymes at 7 out of 20 Arbitrarily Chosen Enzyme-determining Loci. Only common alleles are listed.

Enzyme	*Frequencies among Europeans of Alleles*			*Frequencies among Africans of Alleles*		
	1	2	3	1	2	3
Red-cell acid phosphatase	0.36	0.60	0 04	0.17	0.83	–
Phosphoglucomutase–locus PGM_1	0.77	0.23	–	0.79	0.21	–
Phosphoglucomutase–locus PGM_3	0.74	0.26	–	0.37	0.63	–
Adenylate kinase	0.95	0.05	–	1.00	–	–
Peptidase A	1.00	–	–	0.90	0.10	–
Peptidase D (prolidase)	0.99	0.01	–	0.95	0.03	0.02
Adenosine deaminase	0.94	0.06	–	0 97	0.03	–

Source: Harris, *The Principles of Human Biochemical Genetics*. North Holland, Amsterdam.

seemed to be the most obvious mechanism which would account for the retention of two or more common alleles in a population, but it was also foreseen that additional selective forces may be acting on the homozygotes.

There are many balanced polymorphisms in man. One estimate of their frequency, by Lewontin, is based on the expectation that polymorphic loci would have a greater chance of being discovered than nonpolymorphic loci. This would result in a situation in which the frequency of polymorphic loci discovered in the early years of the twentieth century would be biased toward a higher than true frequency of polymorphic loci and that this bias would be reduced by the discoveries of nonpolymorphic loci in later years. Applied to data on red-cell antigens it was found that the proportion of polymorphic loci discovered was indeed high in the early decades but reached a more or less constant value in recent decades. Plotting of the time sequence of the discoveries of red-cell antigenic loci against numbers of polymorphic and nonpolymorphic loci led to the conclusion that the proportion of loci for which the English population is polymorphic is about one-third. This estimate is in close agreement with findings of Harry Harris on polymorphisms at enzyme-determining loci in European and African populations. Of 20 enzymes chosen arbitrarily to be examined for electrophoretically distinguishable variants 7, i.e. one third, were shown to have 2 or 3 common variants and thus 2 or 3 alleles at their loci (Table 101). Most of these polymorphisms occurred both in whites and blacks. About one-quarter of the twenty enzymes were shown to have polymorphic loci in each racial group, which, of course, underestimates the frequency of polymorphisms at enzyme-determining loci since variants that do not differ in electric charge remain undiscovered in electrophoretic experiments.

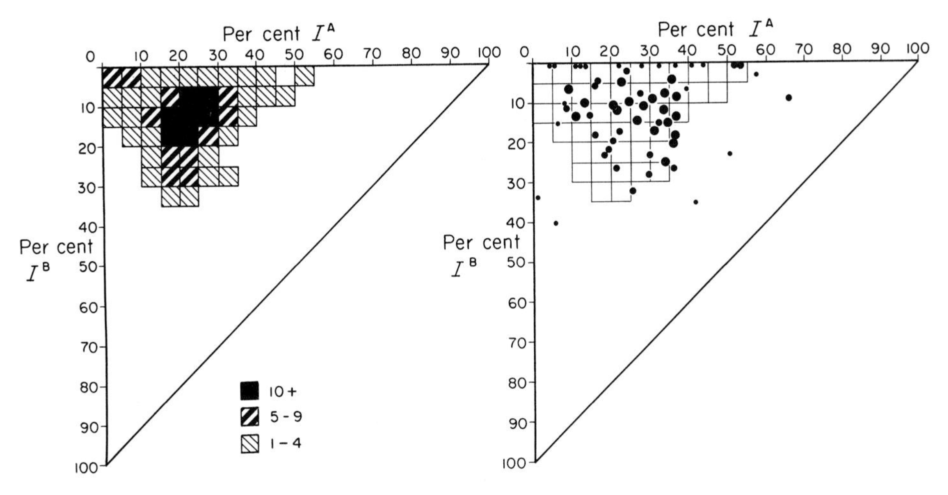

FIGURE 282
Left: Limits of the world range in frequencies of the I^A, I^B, and I^O blood-group alleles, in relation to the total possible range, based on 215 representative populations. *Right:* Computer simulation of the distribution of 60 populations with selection coefficients for I^AI^A, I^BI^B, and I^OI^O homozygotes of 0.74, 0.66, and 0.79, respectively, and for I^AI^O, I^BI^O, and I^AI^B heterozygotes of 0.89, 0.86, and 1.00 respectively. The 60 populations consisted of 3 groups with different numbers of individulas as indicated by the size of the circles: 40–50 adults, 80–100 adults, and 160–200 adults. (A, after Brues, *Amer. J. Phys. Anthropol.*, **12,** 1954; B, after Brues, 1963.)

Selective Attributes of the ABO Blood Groups. A penetrating analysis of the frequency distributions of the I^A, I^B, and I^O alleles has been made by Alice Brues. In Figure 282, A, the allele frequencies of 215 different populations are graphed within the possible ranges of I^A and I^B frequencies (and thus also according to their frequencies of I^O, since $p + q + r = 1$). It is seen that of the whole possible range of gene frequencies, represented by the triangle, only about one-fifth is actually occupied by existing populations. Moreover, even within the occupied range, the populations are not equally distributed; the majority is concentrated in a restricted area.

Using a computer, Brues simulated a number of possible situations in which there were a series of selection coefficients for and against the three *I* alleles. Starting with 60 identical populations with allele frequencies of 0.25, 0.15, and 0.60—frequencies more-or-less centrally located within the world range of allele frequencies—each population was followed for 20 generations. After that time most of the 60 initial populations had moved under the influence of chance processes (pp. 849 ff.: "drift") as well as of selective agents and covered an array of locations on the triangular field of possible allele frequencies. The following assumptions did *not* yield distributions corresponding to those of the natural populations: absence of selective influences; maternal-fetal incompatibility risks; and such incompatibility risks combined with several alternative arrays of strengths of selection coefficients favoring the I^AI^B genotype equally over all others. It was possible, however, to devise a scheme of selection coefficients which assigned highest fitness to the I^AI^B genotype, lowest fitness to I^BI^B, and four different intermediate fitness values to the remaining genotypes and which resulted in a very satisfactory approximation of the simulated distribution to that of the known world pattern of frequencies (Fig. 282, B). This agreement has a rational basis since various selective coefficients were so chosen as to produce a stable polymorphism. Nevertheless, it must be realized that the fit of the model does not prove its correctness.

Blood Groups and Disease. Early attempts were made to demonstrate selective advantages of one or another of the ABO blood groups, particularly by investigating whether some diseases were less frequent in members of one of the groups than in members of the others. None of these studies seemed to yield convincing evidence. In 1958, however, Aird, Bentall, and Roberts reported significant associations between the ABO blood groups and several diseases; and there has been a constant flow of announcements of new findings of such associations since then.

In the studies that have produced these findings, a large number of patients suffering from a specific disease were classified according to blood group. The percentages of the different blood groups among the patients were compared with those among a control group of persons not affected by the disease. For persons with congenital defects such as hydrocephalus, harelip, and cleft palate, and for patients with certain diseases such as appendicitis, kidney stones, and numerous others, the blood-group frequencies were not significantly different from those for the controls. But with certain other diseases there were significant differences. Among more than 7,000 sufferers from duodenal ulcers, 55.5 per cent belonged to blood group O; whereas among more than 83,000 persons of the control groups, only 47.3 per cent belonged to blood group O (Table 102). Correspondingly, the incidence of

TABLE 102
Percentages of ABO Blood Groups in 7,112 Patients with Duodenal Ulcers and in 83,126 Controls, from Eight Different Research Centers.

Group	*Ulcer (%)*	*Control (%)*
O	55.51	47.32
A	34.07	39.81
B	7.93	9.60
AB	2.49	3.27

Source: Roberts, 1956–1957.

TABLE 103
Relative Incidence of Duodenal Ulcers in Patients of Different ABO Blood Groups.

Persons of Group	*Compared with Persons of Group*	*Relative Incidence*
O	A	1.37
O	B	1.42
O	AB	1.54
A	B	1.04
A	AB	1.12
B	AB	1.08

Source: Same data as in Table 102; Roberts, 1956–1957.

the A, B, and AB blood groups was lower in the disease sample than in the control. It is possible to express these facts in terms of relative incidence of duodenal ulcer. In Table 103 it is seen that a person belonging to blood group O had from 37 to 54 per cent more likelihood of developing ulcers than had persons belonging to groups A, B, or AB. It is also seen that there was no significant difference in the relative incidence of the disease among the A, B, and AB patients and controls. Consequently, a single, meaningful comparison gives the incidence of duodenal ulcers in group O relative to the sum of incidences in the three other groups. This relative incidence is 1.40, which implies a 40 per cent increase in the likelihood of the disease in group O as compared to other than O individuals.

A similar disadvantage of belonging to blood group O is seen in the increased incidence of gastric ulcer in this group (Fig. 283). On the other hand, O individuals are favored in regard to cancer of the stomach, sugar diabetes, and pernicious anemia—diseases which have an increased relative incidence in members of blood group A.

Overall, persons belonging to group O seem to enjoy better health than those of group A. This is suggested by a German study of the blood groups of active athletes above 40 years of age, volunteer soldiers selected for good health, and healthy persons above age 75. All three populations had significantly higher frequencies of group O as compared with the general population. The same was true to an especially high degree for a comparison of healthy subjects older than 75 with patients older than 75 who had undergone major surgery.

A particularly interesting association has been observed between ABO groups and certain infectious diseases. It is known that some infectious microorganisms possess blood-group-like antigens in common with human beings. If, for instance, a virus has an A-like substance, group O persons who have anti-A antibodies might partially neutralize the invading virus in contrast

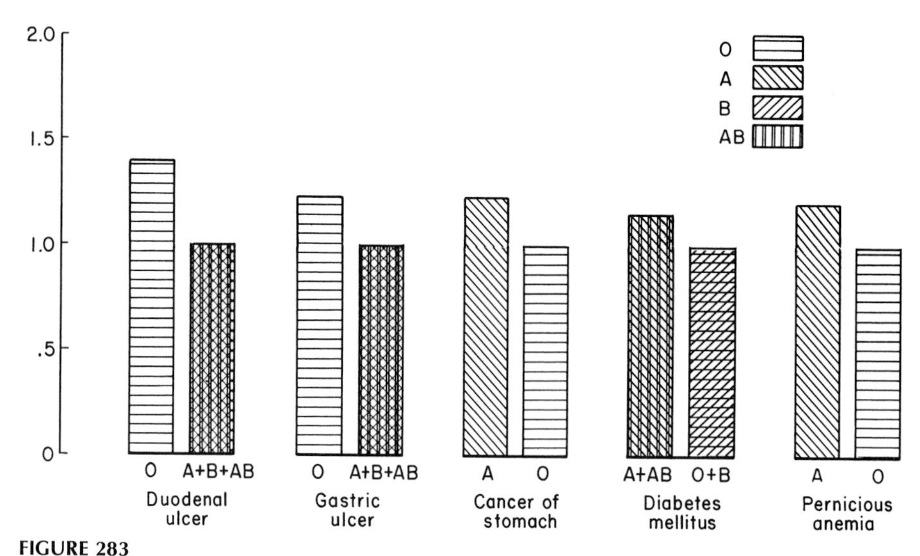

FIGURE 283
Relative incidence of five different diseases in patients of different blood groups. (After Roberts; and Bentall, *Acta Genet. Stat. Med.*, **6**, 1956–1957.)

to group A persons who do not have anti-A antibodies. Thus the incidence of group A persons among patients having the virus infection would be higher than in a group of healthy controls. This has indeed been found in smallpox infections in an unvaccinated Asian Indian population living under primitive conditions. Vogel has pointed out that such association between infectious diseases and blood groups may have played an important role in former times in selection of ABO genotypes during periods of severe epidemics.

It is important to realize that correlations between ABO blood groups and various diseases do not necessarily imply that the possession of a certain blood-group allele increases the likelihood of contracting the illness. Before assuming such a causal relationship, we must exclude the possibility that the patients belong to a subpopulation, i.e., a partial isolate, in which both the specific blood-group allele and the disease happen to be more frequent than in the control group.

A direct approach to the problem has been made by studying the relative incidence of diseases, in respect to the blood groups, not between patients and a general control population but between patients and their own sibs. In such a comparison a genetic stratification in respect to blood groups of patients and controls is obviously excluded unless it is causally related to the disease. So far, the data available have shown that the relative incidence of the disease among patients and their healthy sibs differs according to blood-group frequencies in the same general way it does among patients and unrelated controls, but these differences are less striking and statistically not significant. The weight of these findings is limited, since sib studies rely on much smaller samples than do population studies. This smallness is responsible for the fact that while patients and sibs do not diverge significantly from identity of blood-group frequencies, they also do not differ significantly from distributions predicted on the basis that there are true differences in the relative frequencies of affected and healthy individuals!

The meaning of the relations between ABO blood groups and disease, for the dynamics of human population, is not simple. Some diseases, such as cancer of the stomach, mainly affect persons past the reproductive age and therefore should influence the proportions of alleles in successive generations only slightly. Other diseases, such as duodenal ulcers, frequently affect younger people. Although this condition is rarely fatal, it may possibly have a more pronounced effect on the frequency of reproduction, that is, the transmission of alleles. Duodenal ulcer is also correlated with the nonsecretor allele, *se*, of the secretor locus and thus should be a selective agent concerning this gene.

The incidences of the diseases mentioned depend on complex and partly unknown environmental factors. For instance, in some diseases the type of food consumed seems to play a role, and the occurrence of duodenal ulcers is obviously correlated with psychological stress. These various interactions will have to be fully explored before all aspects of the polymorphism of the ABO blood groups are understood.

The ABO groups are not the only polymorphic genotypes for which relations with diseases have been studied. In addition to mother-fetus-incompatibility interactions, an association has been found between incidence of alleles of the haptoglobin serum groups and leukemia. Furthermore, an interaction between the haptoglobin constitution and ABO incompatibility seems to exist.

In the foregoing pages we have summarized some of the evidence in favor of associations between blood groups and disease. It must now be stated that there are various fallacies pertaining to the collection of data and their statistical treatment that may lead to mistaken conclusions concerning a significant relation between blood groups and diseases. An outstanding blood-group investigator, Wiener, has repeatedly discussed critically the evidence for intrinsic associations and has rejected most of it. A valuable dialogue between Wiener and Vogel—the latter being convinced that there are many intrinsic associations between blood groups and disease—has been published in the *American Journal of Human Genetics* under the heading "Controversy in Human Genetics." While Wiener's critique seems to be valid in various instances, few students agree with him in rejecting the evidence for blood-group associations with duodenal ulcers, cancer of the stomach, and some of the other diseases mentioned in this discussion.

PTC Tasting and Goiter. A curious relation exists between the ability to taste phenylthiocarbamide (PTC) and certain types of the thyroid disease called nodular goiter. Among affected persons a significantly higher frequency of nontasters has been found than among nonaffected. PTC is chemically related to substances which produce goiter, and it is possible to speculate on selective forces involving disease in determining the polymorphism at the *T*, *t* locus. A deeper understanding of this problem is still to be attained.

Selection and Disease Resistance. One of the important selective processes which must have occurred during historical times was selection for resistance to infectious diseases. The accumulation of people in cities and densely populated areas facilitates the spread of disease and the occurrence of epidemics, and it is likely that the great toll of deaths taken by epidemics in former times led to selective elimination of genetically susceptible persons. The many historical incidences in which isolated native populations became seriously endangered by such diseases as measles and tuberculosis, through contact with the relatively resistant Caucasoids, indicate differences in susceptibility which seem to have originated by earlier exposure of, and consequent selection among, the Caucasoids. To account for the low frequency of alleles for disease resistance in unexposed populations, Haldane has suggested that these alleles may have had minor, but unknown and unfavorable, effects on the reproductive fitness of their carriers. If this is true, we might foresee a future in which incidence of alleles conferring resistance to various diseases might decrease again, since modern chemotherapy and antibiotic drugs abolish selective consequences of most infectious illnesses.

Direct evidence of selection in human populations would be provided by data on successive generations of interbreeding populations. If it could be shown that allele frequencies change from one generation to the next, and if chance could be excluded as the cause of such changes, then the existence of "directional" selection would be proved. Absence of allele-frequency changes, however, does not constitute evidence against selection. Selection for heterozygotes is a stabilizing process. The interplay of selection for and against certain homozygotes and heterozygotes (for which examples were given in the associations between blood groups and diseases and mother-child incompatibilities) may likewise result in relative constancy in allele frequencies. Within a single generation the results of selective processes should become apparent through comparisons of allele frequencies of different age groups, of reproducing versus nonreproducing groups, and of the fertility of different subgroups. One such study, by Petrakis, Wiesenfeld, and others, has yielded data suggesting that there is selection against glucose-6-phosphate dehydrogenase deficiency, which is caused by an X-linked allele and is common in black males. The frequency of the allele in a group aged 5-20 years was 12.1 per cent, in a group aged 21-49 years 5.6 per cent, and in a group 51 years old and older 3.8 per cent. Although these findings may be interpreted in more than one way, it is perhaps most likely that mortality selection against G-6-PD deficiency acts all through life and leads to a decline of affected males in the older age groups. This hypothesis finds support in observations according to which G-6-PD deficient black males have a mean elevated blood pressure over controls, an elevated pulse rate, and an elevated level of serum creatinine. It is known that throughout all age groups the risk of dying for blacks is greater than for whites; and it is not inconceivable that this increased risk, in addition to being caused by unfavorable environmental components is accentuated in the genetically enzyme-deficient male. A partition of the existing mortality differential between blacks and whites into genetic and nongenetic components may suggest specific means of improving the health of the affected population.

Mechanisms Underlying Polymorphisms. The existence of polymorphisms at many loci poses questions about their origin and maintenance. Are the polymorphisms due to a mutation–selection equilibrium in which the input of alleles by mutation balances their outgo by selection? Or are the polymorphisms due to superior fitness of the heterozygotes over that of either homozygote, making selection for the heterozygotes the prime agent for keeping different alleles in the population? Or are there other balancing mechanisms, such as alternating selection values of alleles, causing at one time, or in one ecological niche, one allele to be selected and at another time or in another niche an alternative allele? Not much consideration has been given these last possibilities, but the former two have been discussed in great detail, with some investigators preferring a mutation–selection interpretation and others preferring single-locus-heterosis models.

Originally, mutation–selection theories were applied to alleles with severely unfavorable effects. Selection against such alleles, however, keeps them at very low frequencies, with the result that true polymorphism is ruled out for their loci in accord with the definition that the rarer allele at a polymorphic locus must have a frequency of one per cent or more.

If mutation–selection models do not seem adequate, single-locus heterosis also seems unable to serve as a mechanism for creating and maintaining the high frequency of polymorphisms. In a very simplified way, the argument against heterosis may be stated as follows. Assume that there are three genotypes at one locus, AA, AA', and $A'A'$ in the proportions 1/4:1/2:1/4. Assume further that the fitness of both homozygotes is 90 per cent of the fitness of the heterozygote—that is, 10 per cent of the homozygotes that die prematurely or fail to reproduce would not have died or failed to reproduce if they had had the heterozygous genotype. Since half of the zygotes would have homozygous genotypes, the average fitness of the whole population is reduced to 95 per cent of the fitness of the heterozygote. If a second independent locus

B had two alleles with the same attributes as those of locus *A*, additional zygotes would fail to survive because of homozygous *BB* or *B′B′* genotypes; if the two loci act independently, the average fitness of the population would be reduced to 95 per cent of 95 per cent, or 0.9025 of the fitness of the best genotype. For n such loci the proportion of surviving zygotes would be $(0.95)^n$, which for only 100 polymorphisms would mean that more than 99 per cent of all zygotes would fail to survive for genetic reasons. Since a third of the gene loci that have been investigated in man have proved to be polymorphic, and there are almost certainly many thousands of gene loci, there must be well more than a thousand polymorphisms. With balancing selection at each locus, the fraction of surviving zygotes becomes excessively small, even if much smaller differences in fitness than 10 per cent are assumed to apply. If we accept the reasoning outlined, we must conclude that single-locus heterosis at many loci is not able to maintain widespread polymorphism without an absurdly high loss of zygotes—a loss which has been given the name "segregational load."

Possible ways out of these difficulties were simultaneously found in 1967 by Sved; T. E. Reed and Bodmer; J. King; and Milkman. Instead of assuming that the alleles at different loci act independently of each other, these investigators realized that it was much more likely that the different genes interact physiologically. The fitness values of their joint effects are therefore not algebraic products of their individual fitnesses. On this basis the majority of zygotes may have numbers of homozygous loci, which are small enough to be compatible with normal fitness of the individual. Only those rare zygotes which have an unusually large number of homozygous loci will cross a threshold which assigns them to low fitness. "If different loci interact in such a way that their disadvantageous effect is fully expressed only when there is a coincidence of recessive conditions at several loci, the 'affected individuals' will appear very sporadically indeed. Such a situation would exaggerate the effect of single-locus recessiveness by permitting higher equilibrium levels of each of the individual genes involved, and is, in effect, a 'superrecessiveness.'" (Brues). These new insights make it likely that many polymorphisms are based on heterosis at many loci of interacting genes. There is still another theory that will account for polymorphism. It is based on the recognition that many mutations will result in alleles whose effects differ only very slightly or not at all from the most frequent alleles that lead to a normal phenotype. Given the immense number of mutations possible in the DNA, it is to be expected that purely by chance some of them will reach frequencies which fit the definition of polymorphism (see the discussion of "random genetic drift" pp. 849–853). We thus have at least two mechanisms each of which may lead to polymorphism, heterosis and random establishment of selectively neutral or nearly neutral normal isoalleles. Different polymorphisms may depend upon one or the other of these mechanisms.

PROBLEMS

202. (a) What percentage of all pregnancies is potentially subject to selection against *Rr* heterozygotes in each of seven different populations with the following frequencies of *r*: 0.01; 0.1; 0.4; 0.5; 0.6; 0.9; 0.99?
    (b) If selection actually eliminated one-tenth of all children from potentially incompatible mother-child combinations in the above seven populations, what mutations rates, and from which allele to the other, would be required to maintain an (unstable) equilibrium?

203. List all those mother-child combinations which are compatible for ABO and all those which are potentially incompatible.
    (a) From the data of Table 27, determine for various populations the frequencies of the maternal genotypes and of the incompatible sperm types.
    (b) Calculate and compare with one another the total frequencies of potentially incompatible combinations.

204. In a certain population, selection during childhood favors the heterozygotes *AA'* so that the adult population consists of: 10 per cent *AA*, 10 per cent *A'A'*, and 80 per cent *AA'*. Assume that this population intermarries with another one of equal size consisting of *A'A'* individuals only. If the selective processes continue to be active in the mixed population, what will be the allele frequencies after equilibrium?

205. Carry population II of Table 100 through a second generation of selection. What are the allele frequencies $p_{A''}$ and $q_{A''}$? What are the initial proportions of the three genotypes in the following generation?

206. Assume that, in comparison with normal reproductive partners of *AA'* individuals, *AA* are reduced in fitness by a specific disease to 0.9 and *A'A'* to 0.8.
    (a) Starting with a population in which there are 24 per cent *AA* and 22 per cent *A'A'*, what will be the percentages in the following generation?
    (b) What will be the percentages in the second generation?

REFERENCES

Bajema, C. J. (Ed.), 1971. *Natural Selection in Human Populations.* 406 pp. Wiley, New York.

Brues, Alice M., 1963. Stochastic tests of selection in the ABO blood groups. *Amer. J. Phys. Anthropol.*, **21**:287–299.

Brues, Alice M., 1969. Genetic load and its varieties. *Science*, **164**:1130–1136.

Blumberg, B. S. (Ed.), 1961. *Proceedings of the Conference on Genetic Polymorphisms and Geographic Variations in Disease.* 229 pp. Grune and Stratton, New York.

Crow, J. F., 1958. Some possibilities for measuring selection intensities in man. *Hum. Biol.*, **30**:1–13.

Ford, E. B., 1965. *Genetic Polymorphism.* 101 pp. Faber and Faber, London.

Haldane, J. B. S., 1942. Selection against heterozygosis in man. *Ann. Eugen.*, **11**:333–340.

Harris, H., 1969. Enzyme and protein polymorphism. *Brit. Med. Bull.*, **25**:5–13.

Hertzog, K. P., and Johnston, F. E., 1968. Selection and the Rh polymorphism. *Hum. Biol.*, **40**:86–97.

King, J. L., 1967. Continuously distributed factors affecting fitness. *Genetics*, **55**:483–492.

Kirk, R. L., Kinns, H., and Morton, N. E., 1970. Interaction between the ABO blood group and haptoglobin systems. *Amer. J. Hum. Genet.*, **22**:384–398.

Knudson, A. F., Jr., Wayne, L., and Hallett, W. Y., 1967. On the selective advantage of cystic fibrosis hybrids. *Amer. J. Hum. Genet.*, **19**:388–392.

Levine, P., 1958. The influence of the ABO system on Rh hemolytic disease. *Hum. Biol.*, **30**:14–28.

Lewontin, R. C., 1967. An estimate of average heterozygosity in man. *Amer. J. Hum. Genet.*, **19**:681–685.

Li, C. C., 1955. *Population Genetics.* 366 pp. University of Chicago Press, Chicago.

Luzzatto, L., Usanga, E. A., and Reddy, S., 1969. Glucose-6-phosphate dehydrogenase deficient red cells: Resistance to infection by malarial parasites. *Science*, **164**:839–842.

Luzzatto, L., Nwachuku-Jarrett, E. S., and Reddy, S., 1970. Increased sickling of parasitized erythrocytes as mechanism of resistance against malaria in the sickle-cell trait. *Lancet*, **1**:319–322.

Matsunaga, E., 1956. Selektion durch Unverträglichkeit im ABO-Blutgruppensystem zwischen Mutter und Fetus. *Blut*, **2**:188–198.

Milkman, R. D., 1967. Heterosis as a major cause of heterozygosity in nature. *Genetics*, **55**:493–495.

Morton, N. E., Krieger, H., and Mi, M. P., 1966. Natural selection on polymorphisms in Northeastern Brazil. *Amer. J. Hum. Genet.*, **18**:153–171.

Myrianthopoulos, N. C., and Naylor, A. F., 1970. Tay-Sachs disease is probably not increasing. *Nature*, **227**:609.

Neel, J. V., 1951. The population genetics of two inherited blood dyscrasias in man. *Cold Spring Harbor Symp. Quant. Biol.*, **15**:141–155.

Petrakis, N. L., Wiesenfeld, S. L., Sams, B. J., Collen, M. F., Cutler, J. L., and Siegelaub, M. S., 1970. Prevalence of sickle-cell trait and glucose-6-phosphate dehydrogenase deficiency. Decline with age in the frequency of G-6-PD-deficient Negro males. *New Eng. J. Med.*, **282**:767–770.

Price, J. 1967. Human polymorphism. *J. Med. Genet.*, **4**:44–67.

Reed, T. E., 1971. Dogma disputed: Does reproductive compensation exist? An analysis of Rh data. *Amer. J. Hum. Genet.*, **21**:215–224.

Roberts, J. A. Fraser, 1957. Blood groups and susceptibility to disease. *Brit. J. Prevent. Soc. Med.*, **11**:107–125.

Schull, W. J. (Ed.), 1963. *Genetic Selection in Man.* 355 pp. University of Michigan Press, Ann Arbor.

Sved, J. A., Reed, T. E., and Bodmer, W. F., 1967. The number of balanced polymorphisms that can be maintained in a natural population. *Genetics*, **55**:469–481.

Vogel, F., 1970. ABO Blood groups and disease. *Amer. J. Hum. Genet.*, **22**: 464–475.

Wallace, B., 1970. *Genetic Load:* Its biological and conceptual aspects. 116 pp. Prentice-Hall, Englewood Cliffs, N.J.

Wiener, A. S., 1970. Blood groups and disease. *Amer. J. Hum. Genet.*, **22:** 476–483.

Woolf, B., 1955. On estimating the relation between blood group and disease. *Ann. Hum. Genet.*, **19:**251–253.

Workman, P. L., 1968. Gene flow and the search for natural selection in man. *Hum Biol.*, **40:**260–279.

29

SELECTION IN CIVILIZATION

When Charles Darwin, in the middle of the last century, pointed out the great role which natural selection of certain genetic types has played in the evolution of animal and plant species, it was soon realized that modern man may be subjected to similar selective influences. As a result, two slightly different considerations were advanced, complementary to each other. One dealt with changes in natural selection which civilization has brought about. Darwin stressed that, in the ruthless struggle throughout the millennia of evolution, the genetically less fit had a poorer chance of reproducing its kind than had the more fit. The question arose: Had not civilization created an ominous situation in which the survival not only of the fittest but of many unfit was possible, thus leading to an increase in the undesirable genetic constitutions? But, in contrast to this pessimistic view, another idea fired the imagination toward hopeful prospects: Could not man take into his own hands the future genetic fate of his species? Could he not be more efficient and successful than nature, and by the use of his knowledge, improve the genetic qualities of future generations of men? Francis Galton, who early recognized the importance of twin studies for human genetics, also recognized the social implications of genetic changes in man. He coined the word *eugenics*, meaning hereditary well-being, to cover the whole "study of agencies under social control, that may improve or impair the racial [i.e., hereditary] qualities of future generations, either physically or mentally."

Preventive and Progressive Eugenics. In accordance with its two aspects, the field of eugenics has often been subdivided into two branches called "negative and positive eugenics." Instead of these terms, we shall use the designations preventive and progressive eugenics. The first is concerned with combating the increase or the presence of alleles which produce undesirable phenotypes; the second, with furthering the increase of alleles which cause desirable phenotypes or, at least, guarding against the decrease of such alleles. In so far as specific alleles often do not produce undesirable or desirable results in every genetic background, eugenics may also include the combating of undesirable, and the furthering of desirable, allelic *combinations*. In a sense, the two branches of eugenics are identical, since decrease of undesirable genic constitutions implies increase of desirable ones and vice versa. In practice, the distinction between preventive and progressive eugenics rests on a difference in emphasis and on the definition of desirable and undesirable genetic constitutions, in relation not so much to each other but, rather, to an average "norm." In preventive eugenics, attention is concentrated on undesirable, subnormal traits; in progressive eugenics, on desirable, supernormal ones. Trends which improve the genetic endowment of a population are called *eugenic*; and those which entail a deterioration, *dysgenic*.

Frequencies of Defective Traits. Although the frequencies of strikingly subnormal traits are not known precisely, the total number of affected individuals in the United States amounts to millions (Table 104). Of course, only in a fraction of these individuals are the subnormal conditions due to hereditary causes: many physical and mental abnormalities are the result of accidental injuries and infectious diseases, and the group of feebleminded includes an unknown large number of persons whose subnormal intellectual status is partly due to social handicaps. Nevertheless, the sum of genetically defective persons is very large.

If it were possible to decrease the number of afflicted individuals born in each generation, obviously a great reduction in human suffering would be achieved. Besides eliminating the suffering most experienced by those directly involved—the affected individuals, their nearest relatives, and their associates—such a reduction would benefit society at large.

It is customary in this connection to emphasize the monetary public expenditures required for the care of defective individuals. Early in this century,

TABLE 104
Estimates of Prevalance of Common Birth Defects in the United States, 1971.

Trait	*No. Affected Persons Under Age 20**
Mental retardation of prenatal origin	1,170,000†
Congenital blindness and lesser visual impairment‡	300,000
Congenital deafness and lesser hearing impairment	300,000
Genitourinary malformations	300,000
Muscular dystrophy	200,000
Congenital heart and other circulatory diseases	200,000
Clubfoot	120,000
Cleft lip and/or cleft palate	100,000
Diabetes mellitus	80,000
Spina bifida and/or hydrocephalus	60,000
Congenital dislocation of hip	40,000
Malformations of digestive system	20,000
Speech disturbances of prenatal origin	12,000
Cystic fibrosis	10,000

Source: Courtesy of The National Foundation.
*Since many children have more than one type of birth defect, the sum of these estimates exceeds the total number of children disabled by birth defects.
†Includes an estimated 250,000 with Down's syndrome.
‡Includes congenital cataract, strabismus, and certain refractive errors.

a number of studies were published of kindreds in the United States who, in the course of generations, had produced a large number of mentally subnormal and socially undesirable individuals. The "Jukes" and the "Kallikaks," which are literary names assigned to these kindreds, became household words in the discussion of eugenics. The recurrence, generation after generation, of various types of criminality and mental deficiency was taken as proof of the hereditary nature of these traits. It is now recognized that the methods used in gathering these family histories were highly uncritical and that, therefore, these studies give a distorted picture. Moreover, even if the data were unbiased, no valid conclusions about the genetic component of the traits in these families can be drawn, since it is impossible to separate the part played by genetic factors from the influence of the very unfavorable social environment which persisted generation after generation.

It has often been stated that the physical defectives and, especially, the "insane," are on the increase in Western nations. If this implies that the absolute numbers of such persons are increasing, this may well be so, since populations as a whole are increasing. A statement of this kind is meaningful only when it refers to the *relative* frequency of defectives in a population. When the facts are stated in relative terms, it is indeed found that the relative number of patients *in institutions* until recently has steadily increased, but the reasons for this rise are by no means obvious.

A rise in the number of institutionalized individuals is furthered by a change in social attitudes and an increase in opportunities for social care. Whereas, formerly, the mentally ill were kept at home, they are now sent to hospitals. Better diagnosis and better methods of obtaining full reports also result in adding to the census of defectives. In regard to the latter factor, an example may be cited. In Sweden, with its highly developed census system, 4,349 epileptic individuals were registered in 1940. However, a calculation of the number of epileptics in Sweden based upon medical examination of all men of conscription age placed the total of epileptics in that year at about 12,000, indicating that the census had unearthed less than 40 per cent of the actual number of epileptics. Apparently, the accuracy of the census data depended on the willingness of people to divulge the relevant information to the authorities; and this willingness was far from satisfactory. Dahlberg, to whom we owe this example, concluded that "there is plenty of room for an increased frequency through improved registration, even if the actual frequency of hereditary epilepsy were to decrease appreciably."

One more factor may be mentioned, which enters into an interpretation of increased frequencies of certain defects. A number of pathological conditions, among them Huntington's chorea and certain types of mental derangements as well as organic diseases like cancer and diabetes mellitus, tend to make their appearance in the later periods of life. The prolongation of man's average life span through medical and social progress has resulted in a greater number of persons affected with such illnesses.

We have seen that the breakdown of isolates—an aspect of modern civilization—will reduce the frequency of the many types of defects caused by homozygous recessive genotypes. Only a few observations suggesting this genetic trend have actually been made: a decrease in the incidence of juvenile amaurotic idiocy in Sweden from 1 in 25,000 to 1 in 50,000 in 30 years, and decreases in the incidences of mental retardation in the same country over 20 years. Reduction may also be expected of those defects whose frequencies are positively correlated with parental age and birth order if the present tendency to terminate childbearing relatively early continues to characterize large parts of populations. Matsunaga for Japan and Vogel for Germany have discussed the effects of family planning on the reduction in the frequencies of relevant defects. In general such favorable trends are overshadowed by the factors described above, which may simulate an increase in genetic defectives. It must be remembered that a reduction in the frequencies of defective recessive homozygotes due to the breakdown of isolates does not signify a reduction in the frequencies of the recessive alleles. The number of homozygotes per generation is higher within the isolates than in the total population after over-all panmixis. But the absolute number of defective homozygotes produced over many generations is the same for both types of population structure. Furthermore, after the breakdown of isolates, new mutations from normal to recessive alleles will have a lower chance of being counterbalanced by selection against the homozygotes than before. This will lead to increasing allele frequencies until, very gradually, a new equilibrium is reached in which the increased frequency of the recessive allele is matched by an increased frequency of homozygous defectives.

Medical Progress and Rise in Frequency of Abnormal Genes. Apart from factors which lead to increased registration of defects (independently of any change in the actual frequencies), it must be assumed that the frequencies of alleles for certain defective conditions have risen. One of these is diabetes mellitus. Since the discovery of insulin, in 1922, the life expectancy and general health of diabetic persons have improved greatly and their ability to reproduce has been strengthened considerably. This ability has probably led to a higher frequency in the present generation than in earlier ones of the genotypes that control the diabetic status. Direct evidence for this supposition is difficult to obtain, since comparable census data on diabetics in different generations are not available, particularly in view of the varying nutritional and social circumstances which influence the appearance of the defect in an individual.

Other traits which used to reduce the likelihood of reproduction of affected individuals are harelip and cleft palate. These traits often appear together, since they may have a common embryological basis—the failure of lateral anatomic parts to grow together. Formerly, many infants severely affected by cleft palate died soon after birth on account of difficulties in feeding or as consequence of respiratory infections. Of those who survived and had less severe degrees of the defect, a number developed speech defects and minor malformations which reduced their chances of marriage or induced them to abstain from parenthood. Modern surgery has not only succeeded in keeping alive many affected newborns formerly doomed to death but frequently leads to aesthetically highly satisfactory repair of the congenital defects. In all probability, a rise in the frequency of the alleles controlling harelip and cleft palate has taken place in recent times.

Still another example of the fact that progress in medical procedures often results in the propagation of alleles which would otherwise be subject to selective elimination is the disease congenital pyloric stenosis. Occurring in from 2 to 4 of 1,000 live births, this condition is caused by constriction fibers at the opening of the stomach into the small intestine, owing to an extensive overdevelopment of the circular muscles. Through the ages many of these affected children have died and thus not contributed to the gene pool the assembly of genes which gave them their low selective value. In 1912, however, an operation was devised which corrects the condition, permitting the affected individuals to attain adulthood and have children. When these individuals became parents it was found that 70 out of 1,000 of their offspring required the corrective surgery—more than 20 times as many as in the general population.

For final examples let us consider congenital heart disease such as atrial septal and ventricular septal defects and patent ductus arteriosus, the abnormal open channel from the pulmonary artery to the aorta. Again, modern methods enable surgeons to treat successfully affected individuals, the majority of whom would have died early in former times. Now they often have children and the incidence of heart defects among these children is many times as great as in the general population (10–37-fold increases have been reported for septal defects, and 20-fold increases for patent ductus arteriosus).

Diabetes mellitus, harelip and cleft palate, pyloric stenosis, and congenital heart diseases are relatively common defects. There are many others which are rarer when appraised singly, but which together add up to a considerable total for which medical knowledge and surgical skill have made possible not only survival but a normal life, including greatly increased chances of reproduction. This increased fitness of the bearers of abnormal genes, together with a presumably unchanged rate of mutation from normal to abnormal, is bound to lead to a higher accumulation of abnormal alleles responsible for the hereditary classes of such defects.

"Relaxation of natural selection" is not restricted to recent medical advances. The development of different modes of life, such as the replacement of hunting and food gathering by agriculture, may have been accompanied by relaxed selection for traits that were disadvantageous in an earlier culture but became more neutral in a later one. Post has surveyed a number of traits that occur with different frequencies in less highly evolved populations, such as those of hunters, and in more highly evolved populations. He finds increased frequencies of various abnormal traits in the latter populations. Among the traits considered, the frequencies of red-green color blindness are particularly striking. The over-all rate of color blindness in males of hunting populations is 2 per cent, that in populations somewhat evolved from a hunting technology 3.2 per cent, and that in populations furthest evolved from a hunting technology 5.1 per cent. If we assume—without firm evidence—that color vision was important in hunting and food-gathering cultures but became selectively neutral in settled habitats, we may attribute the high frequency of color blindness to relaxation of selection. (Post's suggestive hypothesis as applied to color blindness has been both criticized and defended.) Another possible example of relaxed selection concerns the incidence of deformed nasal septa, which is considerably higher in "civilized" populations than in more primitive ones. Septal deformities often result in mouth breathing instead of nasal breathing, a handicap in hunting, flight from enemies, and other activities requiring sustained physical effort. Mouth breathing also may lead to infection and decrease of the sense of smell. It may be assumed that in "advanced" societies selection against septal deformities is much relaxed, and, as a result, that the frequency of the trait has increased over what it was in the past.

Diabetes mellitus was cited above as probably becoming more frequent in recent decades on account of successful treatment of the disease, which permits increased reproductivity. Why, however, did the diabetic status ever reach the high frequency which it already had before the discovery of insulin? Neel has asked whether diabetes mellitus is "A 'thrifty' genotype rendered detrimental by 'progress'?" The reasoning behind this question is as follows. It is known that after ingesting food persons with a diabetic genotype possess a greater-than-normal amount of circulating insulin. In primitive hunting societies, availability of food often follows a "feast-or-famine" pattern. Under these circumstances the relatively rare occurrence of excess insulin production is instrumental in making maximum use of the food ingested. In modern societies where food is available more regularly the resulting higher frequency of stimulation of insulin production leads to the release of the recently discovered insulin antagonists. These anti-insulins in turn interfere with the normal production of insulin, thus causing the diabetic illness. Whatever the merit of the specific hypothesis, it serves as an example of the relativity of the value assignments for some traits. Very likely the diabetic state was an asset in hunting societies, but it is a liability in modern societies.

DIFFERENTIAL REPRODUCTION AND INTELLIGENCE

One sphere of eugenic concern which requires a special discussion is the future of the intellectual endowment of populations. The topic is closely linked to the facts of differential reproduction (often referred to as differential fertility), within many populations, particularly of the Western world. If, for instance, the population of the United States is subdivided into different categories according to occupation, educational background, income, or in other ways, it is found that the average number of children per family is different for the different groups. There is a high correlation between factors used in the various classifications; for instance, longer periods of schooling are more common in the higher-income groups. The striking fact regarding the number of children in the different categories is that the average decreases as the socioeconomic status increases. Figure 284 depicts the total number

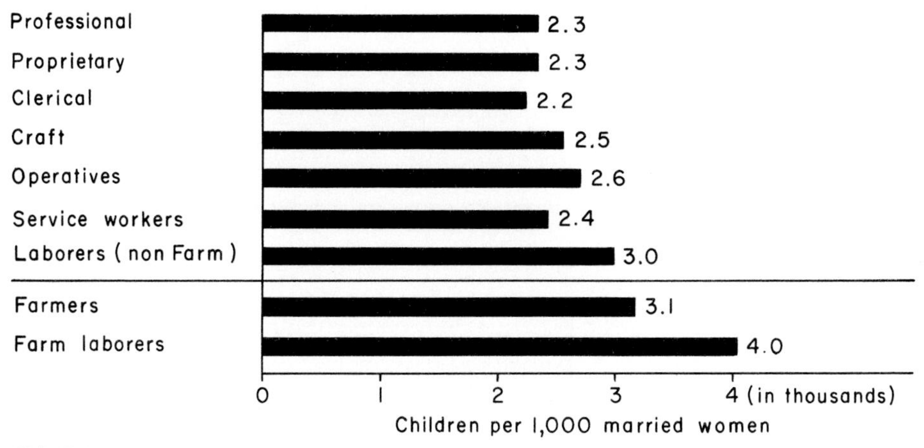

FIGURE 284
Total number of children born to 1,000 married white women, 45-49 years old, by occupational class of the husbands. United States, 1965.

of children born to 1,000 white married women, who were 45-49 years old in 1965, by occupational class of the husbands. By and large the number of children increases with decreasing socioeconomic status of the father. The upper series of bars shows that couples in the professional class had an average of 2.3 children, while the average for laborers was 3.0, with the other occupational classes taking intermediary positions between the extremes. The lower two bars, relating to farmers and farm laborers, demonstrate that the reproductive patterns of these groups differ from those of the other occupational classes; but, again, the reproductive average of the economically better-situated is lower than that of the poorer one. That a similar reproductive differential of different occupational levels occurs in other countries is shown in Figure 285 for Ireland and, in relation to educational level of the husband, for Egypt in Table 105. In Ireland, the reproductive differential among occupational levels is smaller for Catholics than for the rest of the population, but the trend is the same nevertheless, except for the rural groups.

TABLE 105
Family Size and Educational Level of Father in Urban Egypt, 1960.

Education	*Mean No. of Children*
Illiterate	7.0
Elementary	7.1
Secondary	5.9
University	3.9

Source: El-Badry and Rizk, *World Popul. Conf.*, **2**, 1967.

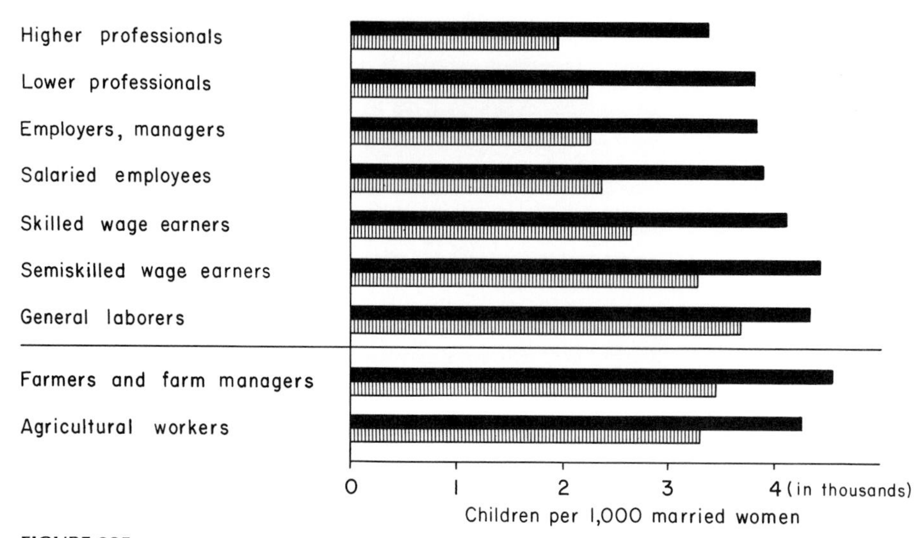

FIGURE 285
Number of children born per 1,000 married women in Ireland, according to occupational classes (1946 census data). Solid bars = Catholics; open bars = others. (After McCarthy, *Proc. World Popul. Conf.*, **1**, 1954.)

Differential reproduction in the United States as related to education of the mothers is shown in Figure 286 both for whites and nonwhites. In 1960, women with completed histories of childbearing showed a steady rise of number of children with decrease of education of the mother both in whites and nonwhites, with only one exception for the lowest educational group. If we consider—in a very rough approximation to a complex problem—that about 2.1 children per family are needed to replace one childbearing woman in the preceding generation it is seen that women in all groups with more than 1–3 years of high school did not replace themselves, while those with less than 1–3 years of high school increasingly tended to have more children than the replacement number. Comparisons of specific reproductive rates of white and nonwhite women are of interest. They show, among other things, that the range of differential reproduction among the latter women is considerably wider than among the former.

While there is no doubt regarding the existence of differential reproduction, is should be pointed out that the data presented do not give a full picture of the situation. The number of children born to married women is of course a prime factor in considering the reproductive attributes of a population—that is, its fertility—but there are others which also enter into the situation, such as death of persons before reproduction, number of unmarried individuals, and length of generation. Information on differentials in these factors is incomplete, and an accurate evaluation of differential reproduction must await the gathering of more data.

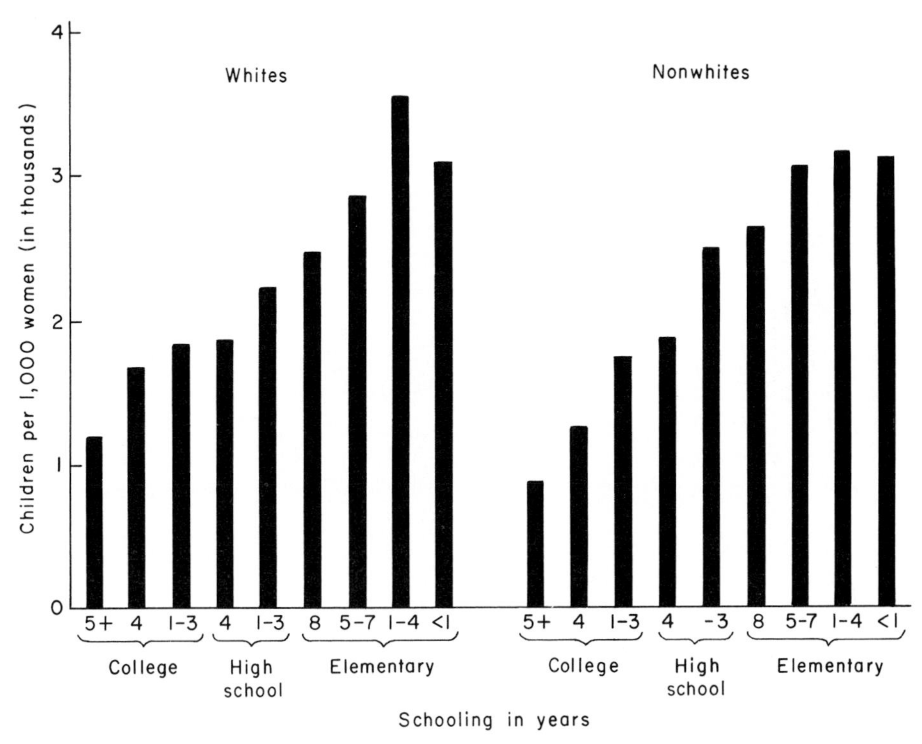

FIGURE 286
Total number of children born per 1,000 women, 45-49 years old, white and nonwhite, by years of school completed. United States, 1960.

The facts of differential reproduction of groups within a population are, of course, independent of the absolute birth rate of a population as a whole. In the period before World War II, the birth rate in many Western countries was insufficient to replace the number of parents. Since that time, the various national birth rates have increased significantly. Whether a population is decreasing, stable, or increasing, differential reproduction of subgroups will determine what proportion of the future population will be derived from each subgroup.

The degrees of differential reproduction of various subgroups are not constant. Although a check on reproduction by control of conception, abortion, and infanticide has been practiced regularly or intermittently for several thousand years by the most diverse peoples, the striking differentials in Western countries among different socioeconomic classes seem to be of relatively recent origin. Birth control became an important social practice in the second half of the nineteenth century, but at first it was restricted largely to the upper and middle socioeconomic layers, resulting in a decrease of their reproductive rates. The differential use of birth-control measures in itself was a cause of differential reproduction. In addition, the survival rate

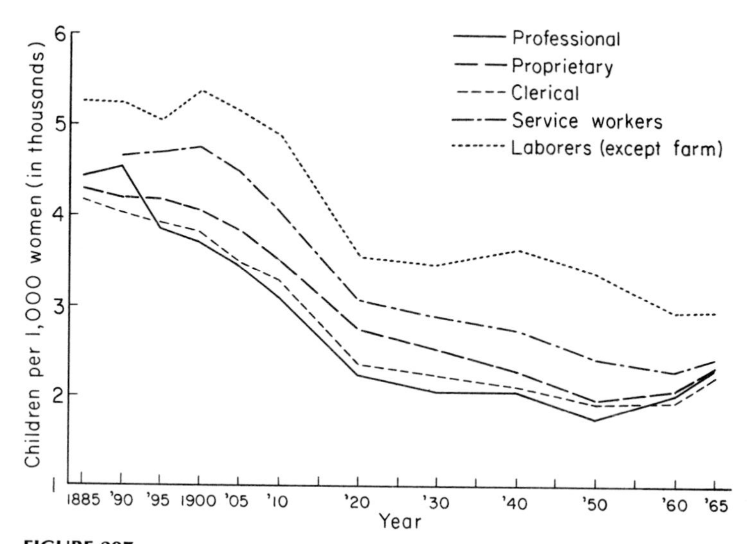

FIGURE 287
Total number of children born per 1,000 married white women, 45-49 years old, by occupational class of the husband. United States, 1885–1965. (After Kiser).

of infants and children was greatly increased by improvement in public health generally, so that the effective reproductive rate increased among the layers which did not practice birth control: more of their numerous children became adults. The changes in relative reproductive rates which occurred in the United States between 1885 and 1965 are evidence of these trends in the use of contraception and in mortality rates (Fig. 287). In addition they reflect changing attitudes in respect to the number of children desired per family. These attitudes include psychological, economic, and social factors whose detailed interplay is only very partially understood.

In recent decades there has been a narrowing of the gaps in reproductive differentials among different groups in several countries. In the United States (Fig. 288) the mean number of children born to a woman with four or more years of college education dropped from 1.2 in 1940 to 1.1 in 1950. Then, however, the trend of fertility was reversed sharply resulting in 1965 in a mean of 1.8 children for women in this educational class. During the same periods the mean number of children born to women with 8 or less years of schooling first dropped from 3.1 to 2.7 (1940–1960) and then rose slightly to 2.8 (1965). The gap in fertility between the two groups of women was 1.9 in 1940 but had been reduced to 0.9 in 1965. These data are based on women with completed fertility and only partly indicate the trends in reproduction of women still in the childbearing ages.

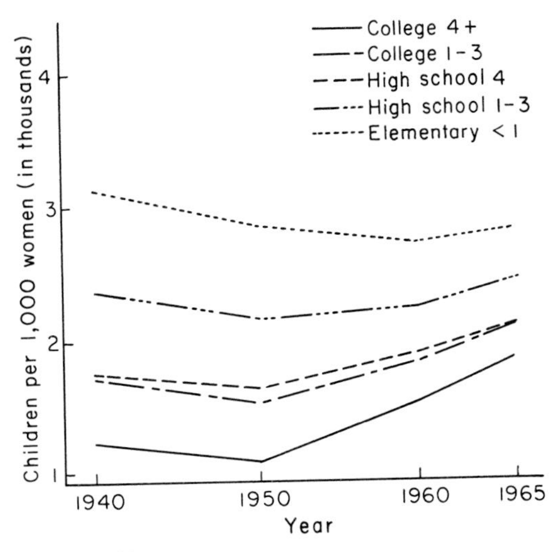

FIGURE 288
Total number of children born per 1,000 white women, 45-49 years old, by years of school completed. United States, 1940–1965. (After Kiser).

The decreases in the differentials of reproduction in recent decades are probably due to a variety of causes. The spread of contraceptive practices to the lower socioeconomic groups has resulted in an over-all lowering of their formerly high rate of reproduction. The increase in reproductivity after World War II among all groups—the postwar "baby boom"—has led to a particularly striking increase of births among the groups which formerly had the lowest reproductive rate. In view of the complexity of the biological and especially the social factors affecting the human reproductive rate, it is hazardous to predict future trends.

Little is known about reproductive differentials within socioeconomic groups. Some data indicate that the most successful members of the upper groups (success being estimated in various ways) are more reproductive than the less successful ones. The reasons for this higher rate may be, at least partly, related to the favorable financial status of these successful families, which permits the bringing up of more children under good conditions without straining unduly the personal and material resources of the family.

Data relating I.Q. performance of parents to their reproductive levels will be reviewed in the next section. Otherwise, no studies have been made which give information on the reproductivity of different subgroups within the groups of middle or lower socioeconomic statuses. It seems possible, however, that the correlation between success and rate of reproduction is negative.

In the upper subgroups of the groups of middle or lower status with their relatively small financial resources, the desire to provide for one's children the most favorable conditions may lead to particularly stringent birth limitation. Moreover, ethical considerations in regard to the threatening overpopulation of the globe may also result in a decrease in reproduction among these subgroups.

Differential reproductivity is a phenomenon not only within individual countries but also among the different racial groups of the earth as a whole. One of the most striking examples on a large scale is provided by the reproductive rate of European Caucasoids and their descendants on other continents relative to that of the rest of mankind. In the seventeenth century they made up approximately 20 per cent of the world's population; in 1940 they represented nearly 40 per cent of all people. It appears that an opposite trend —a relative increase of Asian and African races—has set in more recently.

I.Q. Scores in Various Socioeconomic Groups. The differential reproduction of groups of different socioeconomic statuses would not be of concern to the human geneticist if the genetic endowments of the different layers of the population were alike—that is if, by and large, the allele frequencies at corresponding loci of the different groups were alike. If, on the other hand, different layers differ in their corporate genetic make-up, then differential reproduction would constitute a selective agent in favor of increasing alleles in a population layer which reproduces at a higher rate than the rest. Differential reproduction thus would lead to permanent changes in the genetic constitution of the population as a whole.

It is not possible, at the present time, to state with certainty whether different socioeconomic groups are genetically differentiated. The difficulties of research in this important field are great. The concept of socioeconomic levels itself is subject to various definitions involving occupation, social prestige, amount of income, education, etc.; nor does a simple scale of such levels represent actualities satisfactorily. Although there is a correlation between size of income and occupation or educational level, there are also great overlaps in income among different occupations.

Difficulties of definition are, however, minor in comparison to those involved in finding out whether different groups are, or are not, genetically alike. Mental traits, as we have seen in Chapter 27, differ greatly in expressivity, according to environmental conditions in the widest sense of the term. Undoubtedly, much of the variability in mental traits among different socioeconomic levels is, therefore, attributable to differences connected with the different environments represented by these levels.

Intelligence tests of the children of parents belonging to different socioeconomic levels show a rather consistent phenomenon (Tables 106, 107;

TABLE 106
Mean I.Q. scores of Children, Aged 18 to 54 Months, of Fathers of Different Occupational Classes, United States.

Occupational Class	*Mean I.Q. of Children*
Professional	125
Business, clerical	120
Skilled	113
Semiskilled	108
Unskilled	96

Source: After Goodenough, from Osborn, 1951, *Preface to Eugenics*, Harper, N.Y.

TABLE 107
Occupations of Fathers and Intelligence-test Scores of Children, Aged Eleven Years, Scotland. (Maximum possible score 76.)

Occupational Class	*Mean Score of Children*
Professionals, large employers	51.8
Small employers	42.7
Salaried employees	47.7
Nonmanual wage earners	43.6
Skilled manual wage earners	37.2
Semiskilled manual wage earners	33.2
Unskilled manual wage earners	31.1
Farmers	36.2
Agricultural workers	32.3

Source: Maxwell, in Scottish Council for Research in Education, 1953.

Fig. 289; also Table 94, Control Children): a decline of the mean scores as one descends from groups of higher to those of lower socioeconomic levels. In attempting an interpretation of this fact, we must take into account a general relation between test score and number of sibs. Various studies have shown that a negative correlation exists between these two variables: the more sibs, the lower the test score of a child (Fig. 290). Numerically this correlation is

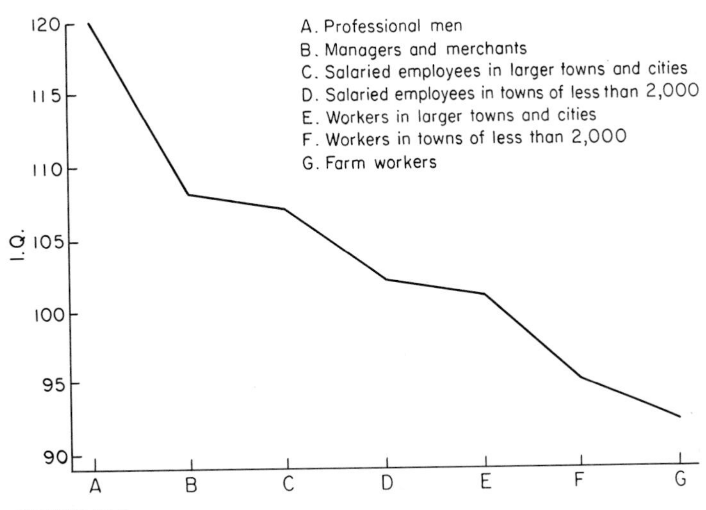

FIGURE 289
Distribution of mean I.Q. scores among singly born French children grouped by occupational class of the father. (After Zazzo, *Les Jumeaux*, Paris, 1960.)

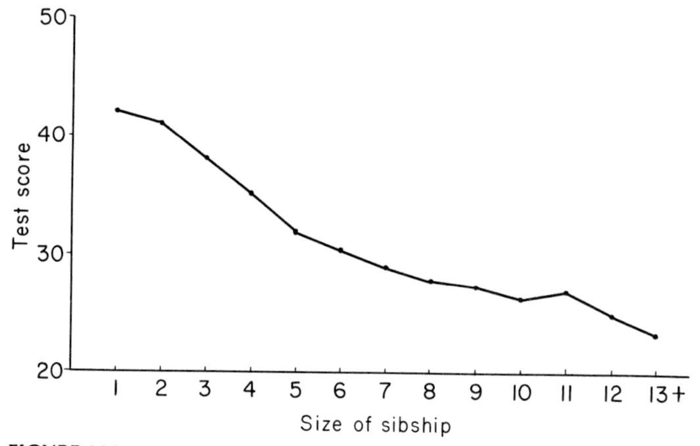

FIGURE 290
Mean scores in a group intelligence test of Scottish school children of sibships of different sizes. (Maximum possible score 76.) (After Thomson, in Scottish Council Res. Educ., 1949.)

rather small, approximately −0.3, thus indicating that whatever is responsible for the inverse relation between I.Q. score and size of sibship contributes only a minor fraction to the total variability. Given the differential reproduction of different socioeconomic groups, the decrease in mean test score as one descends the sequence of socioeconomic levels, therefore, is at least partly an expression of the negative correlation between score and size of sibship.

Notwithstanding its low value, the existence of this negative correlation raises the question of its causes. Do the children in small sibships score higher because their parents have (1) "better-than-average" genes and (2) fewer children? Or is it because their parents are more intelligent for nongenetic reasons and (1) therefore provide more favorable environments and (2) have fewer children? Or is the smaller number of sibs itself solely responsible for the higher scores? If the last two possibilities were actually true, then the negative correlation between intelligence score and size of sibship would have no genetic basis and differential reproductivity no genetic consequences. Parents from different socioeconomic groups would be assumed to have, on the average, the same mental endowment, and their children would have the same average endowment. Only the more favorable social environment of some groups, as expressed either independently of size of sibship or by size of sibship or by both, would be the cause of the higher scores of children in those groups. If, however, the first-named possibility were proved, namely, that endowment of the parents is responsible for their children's scores, then the negative correlation under discussion would involve selection against the better-endowed groups since size of sibship is small.

It has indeed been found that the environment provided in homes of different socioeconomic levels is reflected in differences in test scores. One example was furnished by the mean scores of adopted children and the occupational status of their adoptive fathers: the higher the occupational class, the higher the score (Table 94, Adopted Children). Another example is the fact that mean test score is inversely related not only to size of sibship but also to average number of persons occupying a room. Thus, in a Scottish survey of 1947 (to be described below in detail), the mean scores in a test (whose range of possible scores was between 0 and 76) were 47.3 for children from homes with the lowest occupancy rate (less than 1 person per room), 39.3 for children from homes with from 1 to 2 per room, and 34.2 for children from homes with from 2 to 3 per room; the same trend held for still higher occupancy rates. ("Occupancy rate," which is often dependent on size of family and size of home, is, to some degree, an indication of socioeconomic status; but even within a particular occupancy rate there is a negative correlation between test score and size of sibship [Fig. 291]).

In spite of the undeniable nongenetic component of intelligence-test performance, the data discussed earlier concerning the differences between test scores of "own" and adoptive children (Table 94) and the differences

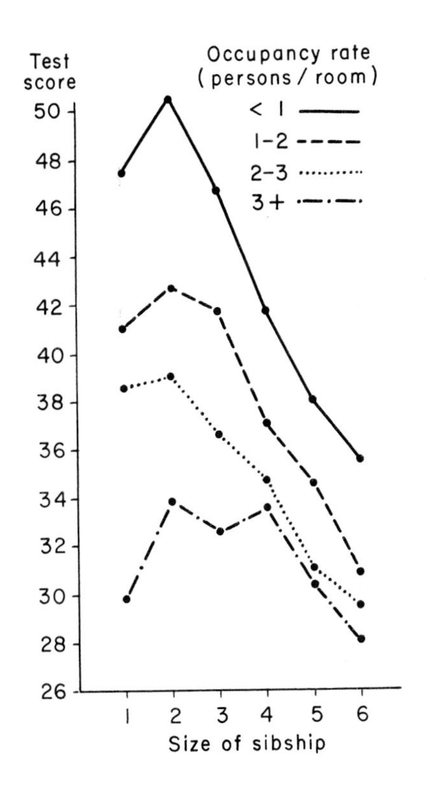

FIGURE 291
Mean group-intelligence-test scores in sibships of different sizes according to occupancy rate. (After Maxwell, in Scottish Council Res. Educ., 1953.)

among the test scores of children placed under common institutional care (pp. 710–712) suggest strongly that environment is not the sole agent—that there *are* mean differences in the genetic endowment of the different socioeconomic groups.

The objection has been raised that the assumption of the existence of genetic differences among different socioeconomic groups is sociologically undesirable and dangerous. It is indeed true that often, in the past, proponents of eugenic measures were biased by class prejudice. Since these individuals usually were members of the upper or upper middle class, they tended to ascribe their favorable socioeconomic position almost exclusively to their assumed good genetic endowment and to regard the plight of the lower classes as an unavoidable consequence of their assumed poor genotypes. In more recent times such extreme opinions have rarely been voiced, but some modern writers, among them Darlington, have continued to advocate the basic thesis. Usually there is no denial any more that much of the difference in intelligence-test performance by members of different groups is due to nongenetic factors and that there is a very great overlap in the range of performance among different groups. If, however, as judged by a variety of data, there are also genetic factors involved in the differentiation of socioeconomic groups, then the existence of such factors must be accepted. A valid interpretation of the consequences of a degree of genetic determination of graded group differences will emphasize the diversity in endowment of persons within each socioeconomic group as well as the need to judge each person as an individual and not as a representative of a fixed type. Misuse of established facts can best be combated when these facts are acknowledged and then become the bases of humane social measures.

I.Q. scores within a socioeconomic group. A detailed analysis of the scores of individuals in a population shows that within each socioeconomic group, the scores vary much more than do the mean scores among the different groups. An example of the overlap of test performance of children from different occupational classes will illustrate this situation. Table 108 lists the percentages and numbers of Scottish children who were in the upper 6.3 per cent of the performance range. It is seen that children with high scores occurred in all occupational classes, even though the percentage frequency decreased with most descending classes, reflecting the decrease in the mean score of each class. Nevertheless, the largest number of high scorers came from the skilled manual wage earners, who formed the largest group of parents, and less than 16 per cent of all high scorers—66 out of 416—came from the class of professionals and large employers, who formed one of the smallest groups.

The fact that there is a wide spread in I.Q. scores within each socioeconomic group undoubtedly mirrors to some extent the great environmental

TABLE 108
Distribution by Occupational Class of Fathers of Scottish School Children, Aged Eleven Years, Scoring 60 or More in a Group Intelligence Test. (Maximum possible score 76.)

Occupational Class	*No. of Children*	*Scoring 60 or More* (%)	*Scoring 60 or More* (N)
Professionals, large employers	221	29.9	66
Small employers	330	10.0	33
Salaried employees	236	18.6	44
Nonmanual wage earners	556	11.3	63
Skilled manual wage earners	2,392	5.6	133
Semiskilled manual wage earners	1,190	2.6	31
Unskilled manual wage earners	1,132	2.1	24
Farmers	142	5.6	8
Agricultural workers	428	3.3	14
Total	6,627	6.3	416

Source: Maxwell, in Scottish Council for Research in Education, 1953.

differences among homes and the different opportunities within any one group. But it must also be assumed that the spread in I.Q. scores is partly due to hereditary differences, which express themselves in varying capacities even within a particular group.

The same conclusion is indicated if the investigation is restricted to the subnormal categories of intelligence. Several studies have shown that the frequency of feebleminded children was considerably higher in the lower socioeconomic groups than in the upper ones. There is thus an association between low mean I.Q. score of a socioeconomic group and high frequency of *very* low I.Q. scores among their children. This association undoubtedly has an environmental component, in that the cultural environment of a lower socioeconomic status may be more prone to relegate a child of low intellectual potentialities to the feebleminded group; whereas the environment provided by a higher socioeconomic status might shift the same child into the range of better I.Q.'s.

It seems unlikely, however, that such environmental factors are solely responsible for the higher rate of feebleminded children in the lower groups. It is more likely that, frequently, certain recombinants of genotypes which are involved in low I.Q. scores in the parents result in genotypes among the children which relegate them to the range of feeblemindedness. In addition, it is probable that the expressivity of genetic constitutions involved in low scores may vary from feeblemindedness on upwards, so that the same genotype which, in a parent, permits a score short of feeblemindedness may cause some of his children to fall into that category.

I.Q. Scores and Reproduction. In the preceding pages various findings were reported according to which socioeconomic or educational levels of parents were inversely related to the number of their children. We have now to consider some relatively recent reports in which the relation between intelligence, as measured by I.Q. performance, of parents and the number of their children was studied. The first of these studies, published in 1962 by Higgins, Reed, and Reed, was based on approximately two thousand parental pairs from Minnesota who had completed their reproductive period and whose I.Q. scores had been determined when they themselves were children. When the parents were divided into six or seven groups according to the I.Q. average of the couple, and the mean numbers of offspring of each group was determined, a large number was found not only for the lowest I.Q. group (0 to 55) but also for the highest (131 and above). It was then realized that reproductive rates ascertained by studying only couples who have children do not characterize the whole population since childless persons are excluded. Such individuals were not numerous in any but the lowest I.Q. group. The investigators decided to add data on the childless sibs of the parents in the original study. When these data were taken into account the reproductive rates of the various groups were not changed greatly except for the lowest I.Q. group. Here the number of childless sibs, the majority of whom were unmarried, was so large that the mean number of offspring for this group dropped from the high value of 3.8 to 1.4. Altogether, the Minnesota study shows the following striking features (Table 109, second column): a high rate of reproduction for the highest parental I.Q. group (those couples with an I.Q. average of 131 or above), a lesser rate for a group with I.Q. 86 to 130, a slightly increased rate for a group with I.Q. 56 to 85, and a very low reproductive rate for the lowest group (I.Q., 55 or less). Thus the Minnesota population sample does not show an inverse relation between I.Q. and reproductive rate, but rather a bimodal relationship. Similar findings were recorded in 1963 by

TABLE 109
Mean Numbers of Offspring per Individual in Relation to Mean Parental I.Q. Scores.

I.Q. Range	*Minnesota Sample*	*Michigan Sample*
> 130	3.0	3.0
116–130	2.4	2.5
101–115	2.3	2.1
86–100	2.2	2.3
71–85	2.4	2.0
56–70	2.5	0.0
< 56	1.4	0.0

Source: Data of second column by Higgins, Reed, and Reed, *Eugen. Quart.*, **9**, 1962; data of third column by Bajema, *Eugen. Quart.*, **10**, 1963.

Bajema for a sample of about one thousand individuals who, as children, had been given I.Q. tests in the Kalamazoo school system in Michigan (Table 109, third column). Here too, the highest I.Q. group was found to have the highest reproductive rate, an 86–115 I.Q. group had a somewhat lower rate and the lowest I.Q. groups had the lowest reproductive rates.

The results of census tabulations and of the Minnesota and Michigan studies differ from each other in important ways. The census data pertain to socioeconomic and educational categories; the studies use I.Q. ranges as bases for group designations. The census data show peak reproductivity for the lowest socioeconomic and educational level and minimum reproductivity for the highest; the Minnesota and Michigan data show peak reproductivity for the highest I.Q. group and minimum for the lowest. These opposite findings are not easily understood, since a positive correlation exists between socioeconomic or educational level and I.Q. score. Perhaps the Minnesota and Michigan populations are not fully representative of the United States population as a whole but come from areas in which class differences are smaller than in the country at large. Perhaps the correlations in the Minnesota and Michigan samples between social and educational class and I.Q. score are so low that close agreement in the findings of the different studies should not be expected. It would be of value to subdivide the Minnesota and Michigan populations according to socioeconomic or educational status and to compare those findings on reproductivity with the ones already determined using I.Q. scores.

The reproductivity of subclasses of a population is not solely dependent on the numbers of children born. If there are differences from subclass to subclass in mortality rates for persons in the age range of the childbearing period, in the mean age of mothers when the first child is born, or in the parental age distribution over which successive children are born, the relative reproductive rates of the subclasses may be affected. These factors determine the average length of a generation. In the Michigan study generation length was determined to be 29.4 years for the highest I.Q. group and from 28.0 to 28.9 years for the other groups. Thus, the high mean number of children of the highest I.Q. group is somewhat counteracted by its relatively long generation length.

Genetic Differentiation and Social Mobility. To reach the probable conclusion that there are genetic differences in intellectual endowment from one socioeconomic group to another is one thing, but to arrive at a specific determination of the type and magnitude of these differences is quite another. Undoubtedly, the differences are not absolute in the sense that any socioeconomic group in a population is in the exclusive or even near-exclusive possession of specific alleles that control intelligence. There are no sharp boundaries between the different groups, since an appreciable number of individuals in each generation rise from a lower to a higher socioeconomic status, while another number move from a higher to a lower one. Some of these shifts may be thought of as being due to genetic segregation—of better genotypes leading to upward movement and of poorer genotypes leading to downward. But even if this interpretation should be true, there is a lag in mobility caused by the socioeconomic environment into which individuals are born, which keeps many with better genetic endowment from rising, and others with poorer endowment from falling.

Given a wide range of genotypes within each layer of a population, it is to be expected that, in the course of generations, a society with high social mobility will favor within each socioeconomic group an accumulation of specific, similar genotypes. In contrast to this, a society with little mobility should retain a wider range of genotypes in each of its "castes." There are no methods available at present to test these expectations in detail, but there is evidence, for various physical and mental traits, that social "sifting" has brought about differential gene flow and thus has contributed to the creation of genetically divergent socioeconomic subpopulations. Detailed data have been provided for Germany by Ilse Schwidetzky, for France by Schreider, for Belgium by Cliquet, and for England by Gibson. For the French sample Schreider finds that only 40 per cent of children belonging to the uppermost social layer maintain themselves at that level, while 60 per cent of them drop down in the social scale. In this connection it is worthy of some consideration that a society which tends to give equal opportunity to all may by this very endeavor also tend to create genetically differentiated classes.

Tendencies toward a genetic differentiation within a population are strengthened in various traits, including intelligence-test performance, by positive assortative mating. In such matings, the intellectual correlation between spouses has been found to be of the order of +0.55. Opposing tendencies also exist: genetic heterogeneity within each layer of a population is maintained by marriages between individuals belonging to different levels, particularly since such marriages are often entered into without regard to the intellectual similarity or dissimilarity of the spouses.

The Presumed Dysgenic Effect of Differential Reproduction. Whatever the genetic details, and as important as knowledge of them is for complete understanding, one fact is already apparent: If there are genetic differences between the different socioeconomic groups, then differential reproduction will result in selective increase of some, and decrease of other, allelic frequencies in the population as a whole. Many authors, impressed by the facts of differential reproduction during the last 150 years or more, have expressed fears that a deterioration of the genetic endowment of the population must have resulted.

Assuming that the past trends continue into the present and future, attempts have been made to calculate expected changes in I.Q. scores of successive generations. Using the observed mean scores in different socioeconomic groups and the observed reproductive differentials among these groups, Cattell and other investigators arrived at estimates for the decrease in I.Q. for the population as a whole, from one generation to the next. These estimates vary from about one to around five points on the I.Q. scale as the rate of decrease. The uncertainties—such as the degrees of heritability of the mean I.Q. scores characteristic of the different socioeconomic groups—which entered into these calculations are great.

Since these estimates, there have been some actual studies to determine what, if any, decline in mean intelligence scores occurs. The most extensive study is the Scottish Mental Survey. In 1932, more than 87,000 eleven-year-old Scottish school children, constituting more than 90 per cent of all Scottish eleven-year-olds, were given, in groups, a verbal intelligence test; and in 1947 the same test was administered to a similar population of children totaling more than 70,000. In addition, smaller samples of children were given individual intelligence tests of the Binet type. The results of the group test showed an increase—not a decrease!—in the mean score of the 1947 population over that of the 1932 population: in 1932, the mean score was 34.5; in 1947, it had risen to 36.7 points. This difference is equivalent to about two points on the I.Q. scale (Fig. 292).

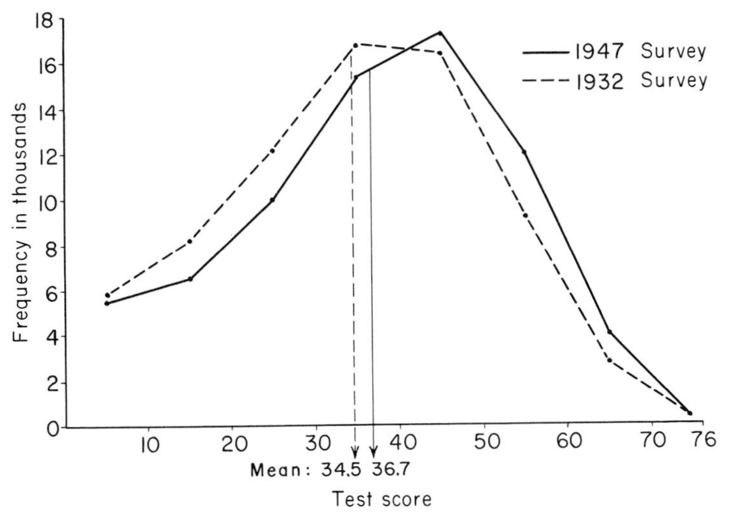

FIGURE 292
Distribution of group test scores for all pupils in the 1932 and 1947 Scottish survey. (Maxwell, in *The Trend of Scottish Intelligence*, University of London Press, 1949.)

Interestingly enough, the improvement in test performance was not the same for boys and girls. The mean score of the former changed only from 34.5 to 35.9, while that of the latter rose from 34.4 to 37.6. No significant change in the Binet test performance had occurred from 1932 to 1947. Figure 292 shows not only that the mean scores of the group tests had improved but that the whole curve of scores in 1947 had shifted to a level higher than that of 1932. Thus, it is clear that the predicted decline in intelligence-test performance had not taken place among Scottish children. Studies elsewhere led to results similar to those in Scotland.

The interpretation of these facts is not obvious. It is unlikely that the observed rise was due to genetic improvement of the population. No selective mechanism for such an improvement is readily envisaged, nor is it likely that such a mechanism would have led to the rather strikingly greater rise for girls. One might therefore be inclined to speculate that some nongenetic factor, such as differences in sibship size of the 1932 and 1947 groups, greater familiarity with mental testing ("test sophistication"), or earlier maturation was responsible for the over-all rise in group test performance.

Do the results of the Scottish Survey, then, disprove the validity of the theoretical forecasts of a diminishing genetic endowment of populations due to differential reproduction? Two different answers have been given by those who accept the likelihood of genetic differences among socioeconomic groups. One answer suggests that the predicted decrease in endowment may actually have taken place but that it was masked by improvement in performance due to nongenetic circumstances. If nongenetic circumstances can lead to higher scores from one generation to the next without a change in genetic endowment, then, it is argued correctly, suitable nongenetic circumstances can lead to higher scores even in the face of genetic decline. Thus, the Scottish Survey may not have led to a decision on the hypothesis of decreasing genetic endowment.

Another answer, proposed by Penrose, denies the need for expecting a decline in innate intelligence. His reasoning has been demonstrated by a greatly simplified model. Assume that intelligence depends on a single pair of genes and that the genotype *AA* makes for superior, *AA'* for intermediate, and *A'A'* for very low performance. Assume further that complete assortative mating of like genotypes occurs and that the *A'A'* individuals are sublethal weaklings who do not reproduce. Assume finally that the proportion of *AA* to *AA'* persons is 90:10 and that their birth rates are 1.89 and 4.00, respectively. Then, as shown in Table 110, the 90 per cent of $AA \times AA$ marriages will yield 170 *AA* children only, while the 10 per cent of $AA' \times AA'$ marriages will yield 10 *AA*, 20 *AA'*, and 10 *A'A'* children. The proportion of the *AA* and *AA'* individuals in the F_1 generation then becomes 180:20, identical with the 90:10 proportion of the parental generation.

TABLE 110
A Hypothetical Population Segregating for a Single Pair of Alleles Determining Intelligence Performance. (Completely assortative mating, and highest reproductive fitness of the heterozygotes.)

Type of Marriage	*Frequency (%)*	*Relative Birth Rate*	F_1		
			AA	*AA'*	*A'A'**
$AA \times AA$	90	1.89	170	–	–
$AA' \times AA'$	10	4.00	10	20	(10)
Proportion of future parental pairs from F_1			90	10	–

Source: After Penrose, 1954.
*Sublethals.

In this model the birth rate of the superior group is lower than that of the intermediate group, but the relative loss of the *A* gene owing to insufficient replacement by the *AA* class is counteracted by the lack of replacement of *A'* by the *A'A'* class. The stability of the allele frequencies in this model population obviously rests on reproductive selection for heterozygotes. While intellectually only of intermediate grade, reproductively the heterozygous state is postulated to confer on its bearers the advantage of heterosis.

The assumption of a compensation for low reproductivity of the superior group by low fertility of the inferior group is not wholly artificial. As earlier illustrated for Minnesota and Michigan populations, many persons of very inferior mental ability remain childless. Furthermore, if the findings on the high reproductive rates of the highest I.Q. groups in these populations can be generalized, we would even expect not a constancy of allele frequencies in successive generations but an increase of those alleles favoring high I.Q. scores. The genetics of mental endowment is of course, vastly more complex than Penrose's ingenious but intrinsically unlikely model suggests. To take account of some aspects of this complexity, Penrose has succeeded in devising elegant schemes in which heterosis for higher fertility of heterozygotes at many polygenic loci yields an equilibrium of allele frequencies. It must be realized, however, that the assumption of an equilibrium situation may not be correct and that selection, upward or downward, may play an important part in determining the intelligence of populations.

An analogy to the pattern of relationship between socioeconomic group and intelligence-test score has come to light in the Scottish Survey. Very similar relations exist between height or weight of children and sibship size and occupational class as between intelligence scores and the latter two categories. Therefore, decline in height and weight was to be expected. On the contrary, the height of Scottish pupils of a given age has increased during the period covered by the Survey. Indeed, it has done so since 1910 and is still increasing. This apparent paradox may be explained by environmental improvement whose results masked an actual decrease in genes for greater height, possibly in combination with a heterotic mechanism.

The fear of a genetically controlled decline of intelligence may have been unwarranted. But even if "gene erosion" is not a serious problem at present, a progressive eugenic program may someday—when we know more than we do now—be concerned with increasing the proportion of favorable genotypes instead of being satisfied with maintaining the current fraction of unfavorable ones. At present, a more pressing problem—one whose solution is bound to yield immediate, tangible results—is the improvement of social circumstances that prevent large numbers of individuals from developing their innate abilities to the highest degree. Lederberg has called the discipline whose goal is to bring about such improvement "euphenics."

The Possible Eugenic Effect of Differential Reproduction. Apart from the problem of whether differential reproduction among different socioeconomic groups has a dysgenic effect on intelligence, the question may be raised whether there are other desirable gene-controlled mental traits whose frequencies may be positively correlated with reproductive differentials. Although the upper socioeconomic groups of Western societies apparently have a relatively high frequency of genetic constitutions favoring intelligence, they seem to be no better off than other strata in their frequency of alleles which lead to idiocy. Could there be yet a third type of genetically controlled mental trait, for which the upper groups are relatively deficient but which constitute assets to the individual and to society? An unequivocal answer to this question cannot be given, because it would involve consideration of many factors whose exact role is unknown and some of which are inherent contradictions of others.

The first difficulty arises in defining a desirable trait. Emotional stability might be one such trait, but its presence in all individuals would eliminate the appearance of many types of genius which, though often characterized by emotional instability, enrich civilization. Altruism may be another desirable trait, but acquisitiveness and the egocentric ambitions of individuals, while often producing misery, have also led to advances which have contributed to the welfare of society at large. It will be hard to agree on definitions of desirable traits—but it is clear that the goal does not lie in uniformity.

Even if some agreement could be reached, a second difficulty is that we have no measures of the genetic component which determines the variability of men in regard to these traits. The social plasticity of man is very great, and different societies and groups within societies mold the attitudes of their members in most diverse ways. Thus, psychological studies have shown that environmental influences can develop cooperativeness or aggressiveness in the same individuals; but this phenomenon does not preclude the possibility that certain genetic components, as yet unknown, may bring out one or the other trait more readily than do certain others.

In this connection, we may refer here once more to the evidence from two genetically different strains of laboratory mice (see p. 688). Under certain conditions, the males of one strain react peaceably to a strange male mouse, while the males of the second strain are highly aggressive toward a stranger. Yet special training can transform both strains of mice, within a few days, either into peaceable or fiercely combative individuals.

Preventive and Progressive Eugenic Measures. Eugenicists have proposed various measures to counteract the decrease in the intensity of natural selection against alleles which cause severe physical and mental abnormalities, and against the presumed dysgenic effects of differential reproduction in Western societies. In order to reduce the frequencies of undesirable genotypes, it has been suggested that genetically defective individuals be prevented from reproducing; while, in order to increase the frequencies of favorable alleles, individuals with better-than-normal endowments should be given incentives to have more children.

Specific measures which would lead to a reduction in the number of undesirable genotypes include dissuasion from procreation, prevention of procreation by segregation of the two sexes, or sterilization (not by castration, but by surgical interruption or removal of part of the egg or sperm ducts—an operation which does not affect the physiology or the sexual drive of the individual but only makes the passage of the gametes impossible). Another preventive procedure is the medically induced abortion of the fetus if the genetic constitution of either or both parents makes it highly probable that a severe, incurable defect would appear, or if direct study of the constitution of the fetus provides evidence for the presence of such a defect.

In addition to these preventive measures, which would apply mainly to certain severe defects, the eugenics program lists provisions for education on the genetic basis of human traits and encouragement of contraception in the lower socioeconomic groups in order to reduce their, supposedly dysgenic, high rate of reproduction.

Progressive measures consist partly in the enlightenment of public opinion by emphasizing the desirability that a larger share of future generations be provided by parents who are best endowed genetically. It is hoped that presentation of the presumed facts will often lead to more children per family in the upper socioeconomic groups. In part, progressive measures consist, also, in improving the social and economic conditions which discourage genetically well-endowed parents from having more children. Sufficiently large subsidies, in the form of higher salaries, greatly reduced taxes, or special bonuses for parents of more than a specified number of children, or community help in the upbringing of the children, are among the methods suggested. However, it is seldom proposed that parents from specific groups be singled out for these premiums, because such a procedure would certainly result in great individual injustices. It is hoped, rather, that the measures would induce those persons who make genetically desirable parents to have more children, and that the less desirable ones who already have a large number of children would not be induced to have still more. May it be noted that there is an inherent conflict between the desirability of increased numbers of well endowed children and the desirability of limiting population growth.

QUANTITATIVE ASPECTS OF SELECTION

We have stressed the tentative nature, if not the often complete lack, of knowledge about the genetic basis of many differences among human beings. Even if the information were more complete, we would still require detailed investigation of the consequences of selective agents on the genetic and phenotypic composition of later generations. In the following pages, a number of theoretical situations will be discussed, in which selection of varying strengths for or against various genotypes is postulated. Some important applications to problems of human genetics will result from these considerations.

Selection Against a Dominant Genotype. The simplest situation is that of selection against a single-factor, autosomal, dominant genotype (*DD* or *Dd*) or, conversely, selection for a recessive genotype (*dd*). In the case of a rare dominant allele, practically all persons carrying it are heterozygous, so that we may restrict our discussion to a population consisting of only *Dd* and *dd* individuals. If the dominant allele is fully penetrant and causes its phenotypic effect to occur before the individual reaches the age of reproduction, then suppression of reproduction of all affected individuals will lead to the elimination of the dominant condition from the next generation (Fig. 293, broken line), except for new mutations from the recessive to the dominant allele. (The possibility of mutations, while significant, will not be considered in the remainder of this chapter.)

If, on the other hand, selection, in terms of suppression of reproduction of *Dd* individuals, acts on only some of these individuals either because penetrance is incomplete or for other reasons, then a fraction of affected persons will appear in successive generations. If, for instance, half of the carriers of a dominant allele are selected against, while the other half reproduces at the rate of the genetically normal population, the number of carrier individuals will be reduced to one-half in the first generation following selection, to one-

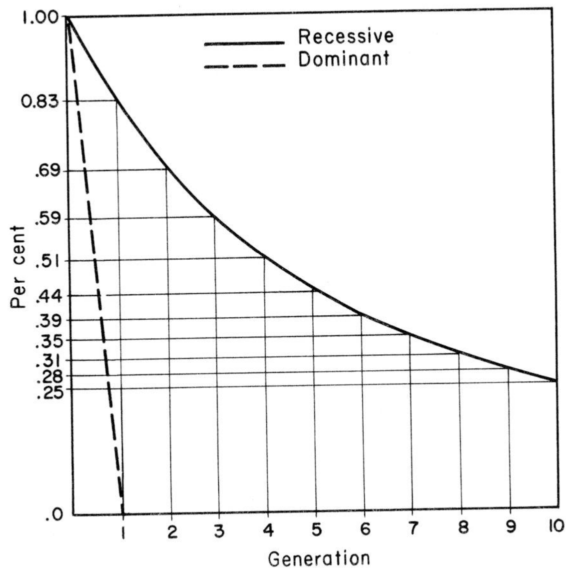

FIGURE 293
Complete selection against a rare dominant genotype (broken line) and against a recessive genotype (solid line). Initial frequency of the selected genotype is 1 per cent. The reduction in frequency of the genotypes is shown for ten successive generations of selection.

quarter in the second, and, in general, to $(1/2)^n$ of the original number, where n equals the number of generations (Fig. 294, broken line). If selection affects not the specific fraction of 1/2 of all affected individuals but a fraction k, then the number of affected individuals in n generations, after the onset of selection, is k^n.

Partial selection in which only a fraction k of the relevant individuals are prevented from reproducing is only a special case of a great variety of partial-selection systems. From the genetic point of view, it makes no difference, for instance, whether the fraction k of the individuals are sterile and the rest fully fertile, or whether all these individuals are fertile but leave only a fraction k of children, as compared to the average quota of genetically normal persons. This latter possibility is often approached when an inherited abnormality results in the death of affected individuals before the age when reproduction normally ceases.

Complete selection against a single dominant factor is thus 100 per cent effective in a single generation, and even partial selection accomplishes much. For example, with a selection factor of one-half, the number of dominants is reduced to a little more than one-tenth of the original number in three generations and has practically disappeared in ten generations. If all dominant

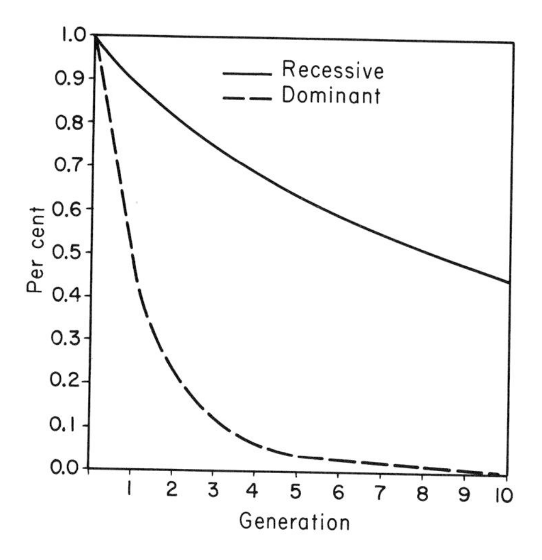

FIGURE 294
Selection of 50 per cent for ten successive generations against a rare dominant and a recessive genotype. (After Koller, *Z. Konstitl.*, **19,** 1935.)

achondroplastic dwarfs or all individuals with neurofibromatosis did not reproduce, then the unhappiness caused by the birth of affected individuals in these families would be eliminated in one generation. For dominant inherited diseases like Huntington's chorea, which often sets in after the reproductive age has been attained, the reduction of the disease through nonprocreation by phenotypically affected persons will follow the exponential decrease represented by k^n. Even with such incomplete penetrance of a dominant allele, a nearly complete elimination in one generation could be accomplished if *all* children who had an affected parent remained childless regardless of whether they were phenotypically normal or affected. This procedure would involve not only the *Dd* individuals, who might later become diseased, but also their *dd* sibs, who are genetically normal. Such a situation is fraught with tragedy. A person who knows for certain that he is the carrier of a genotype which leads to a very serious disease later in life will usually not wish to risk the chance of producing potentially affected children, but the personal sacrifice in remaining childless might appear very heavy to an individual who did not develop the condition later in life, thereby establishing too late for founding a family that he was free from the dreaded allele. The discovery of methods of distinguishing between a *Dd* and a *dd* individual when both still appear normal would obviously be of great benefit, especially to the *dd*'s (see pp. 802–806).

Selection Against a Recessive Genotype. Selection against a homozygous, single-factor, autosomal genotype (*dd*) or, conversely, selection for a dominant genotype (*Dd* or *DD*) is less effective than the type of selection just discussed. This is due to the fact that many *dd* individuals have heterozygous ($Dd \times Dd$) parents who are phenotypically normal and thus not directly subject to selection.

If, with full penetrance, the frequency of *dd* is q_0^2 before the onset of selection, and if *no* affected person reproduces, the number of *dd* individuals in the next generation can be easily calculated from the frequency q_1 of the *d* alleles in the reproducing population, as Figure 295 shows. The total size of the reproducing population is the sum of the *DD* and *Dd* individuals, equal to $p_0^2 + 2p_0q_0$, while the frequency of the *d* alleles in this population is one-half of the frequency of the *Dd* persons, that is p_0q_0. The proportion of *d* among all alleles after disregarding the nonreproducing *dd* persons is thus

$$q_1 = \frac{p_0q_0}{p_0^2 + 2p_0q_0} = \frac{q_0}{p_0 + 2q_0},$$

which, because $p_0 = 1 - q_0$, becomes

$$q_1 = \frac{q_0}{1 + q_0}. \qquad (1)$$

Therefore, the frequency of *dd* in the new generation amounts to

$$q_1^2 = \left(\frac{q_0}{1 + q_0}\right)^2. \qquad (2)$$

The significance of formula (2) becomes apparent if some specific values for the initial frequency q_0^2 of *dd* individuals are used. If, for instance, this frequency is 1 per cent, then, after one generation of complete selection against the affected individuals, q_1^2 amounts to 0.83 per cent. If the initial frequency is 0.83 per cent, selection in one generation will reduce it to 0.69

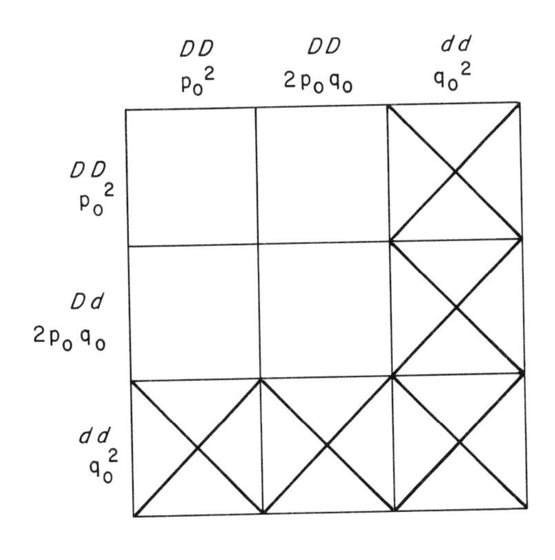

FIGURE 295
Selection against a recessive genotype *dd* in a population which initially is at Hardy-Weinberg equilibrium. Five of the nine possible marriage types are excluded.

per cent; if the initial frequency is 0.01 per cent, the reduction will lead to 0.0098 per cent.

These figures show two main facts: (1) the lowering of the frequency is only a fraction of any initial frequency, and (2) the relative efficiency of selection against recessives becomes less with a decrease of the initial frequency. This latter point is well illustrated by a comparison between the first and last examples. The reduction from 1 to 0.83 per cent represents a lowering of the frequency of *dd* by 17 per cent of the initial frequency, while the reduction from 0.01 to 0.0098 per cent represents a lowering by only 2 per cent.

The decrease in the effectiveness of selection against recessives with a lowering of the initial frequency is of great significance if we consider the results expected from selection continued over many successive generations. In order to calculate the frequency of *dd* after any given number of generations of selection, use may be made of a simple relation which gives the allele frequency q_n after n generations. Since, after one generation of selection, according to equation (1),

$$q_1 = \frac{q_0}{1 + q_0},$$

after two generations

$$q_2 = \frac{q_1}{1 + q_1}.$$

Substituting q_1 by the above fraction, we obtain

$$q_2 = \frac{\dfrac{q_0}{1 + q_0}}{1 + \dfrac{q_0}{1 + q_0}} = \frac{q_0}{1 + 2q_0},$$

and, in general,

$$q_n = \frac{q_0}{1 + nq_0}. \tag{3}$$

If we apply formula (3) for ten successive generations to a population with an initial *dd* frequency of q_0^2 equal to 1 per cent, we obtain the results plotted in Figure 293 (solid line). It is seen that the reduction of frequency of *dd* becomes less in each successive generation and that, after ten consecutive generations of complete selection, it is still one-quarter of the initial frequency of 1 per cent. To reduce it to one-tenth, that is, 0.1 per cent, would require twenty-two generations, and the number of generations to reduce it further by factors of 10 are 68 generations to reduce the frequency from 0.1 to 0.01

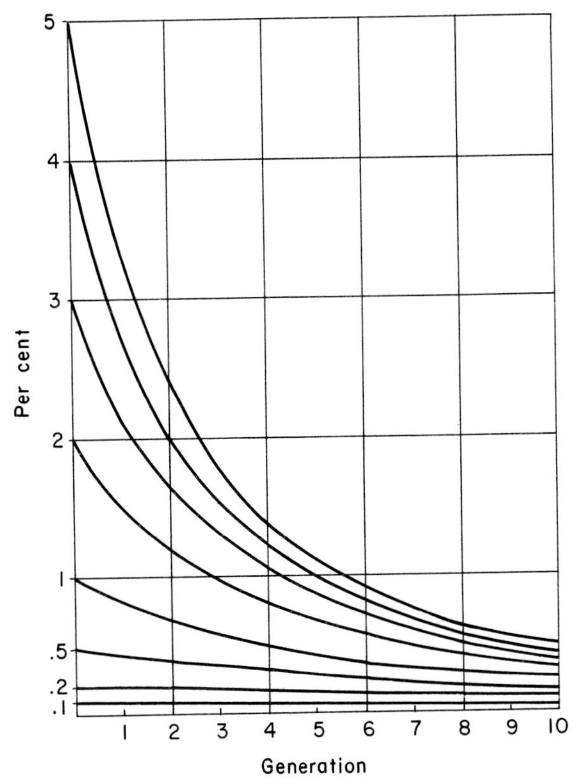

FIGURE 296
Complete selection for ten successive generations against recessive genotypes occurring initially at frequencies of from 0.1 to 5 per cent. (After von Hofsten, *Hereditas*, **37**, 1951.)

per cent, 216 generations to reduce it from 0.01 to 0.001 per cent, and 684 generations to reduce it from 0.001 to 0.0001 per cent.

The dependence of the efficiency of selection against recessive homozygotes on their initial frequency is shown in Figure 296, in which the results of selection continued for ten generations are shown for eight different populations with initial frequencies of *dd* ranging from 5 per cent down to 0.1 per cent. It is easily seen that progress in elimination of *dd* is faster when the initial frequency is higher, and that, because of the slower progress in the population with the lowest initial frequency, the fiftyfold difference in *dd* carriers at the onset has been reduced to less than ninefold.

If selection against recessives is *not* complete but reduces the average reproduction of *dd* individuals to some fraction of the normal rate, then obviously the effect of the selective process is even less than that discussed above. Elimination from reproduction of one-half of the *dd* individuals cuts an initial frequency of 1 per cent down to 0.91 per cent in the next generation, and, in ten generations, a reduction to little less than one-half of the initial frequency is accomplished as compared to the decrease to one-quarter with full selection. These facts may be visualized by a comparison of the two continuous curves in Figures 293 and 294.

Instead of the results of selection being expressed in terms of reduction of phenotypes, they may be given in terms of reduction of allele frequencies. Corresponding to the decrease of homozygous recessives after complete selection in one generation from an initial frequency of q_0^2 to q_1^2 as shown in equation (2), the decrease in allele frequency is from q_0 to q_1. In our first example, where q_0^2 was 1 per cent and q_1^2 was 0.83 per cent, the decrease in allele frequency is from $q_0 = 0.1$ to $q_1 = 0.091$. Thus, the loss of the recessive allele amounts to only about 9 per cent of the initial allele frequency in contrast to the more severe reduction of 17 per cent in the frequency of homozygous recessive individuals.

The application of these calculations to human populations is easily seen. Selection, partial or complete, against recessives, reduces the number of affected persons in future generations; but the amount of this reduction per generation is low even for the comparatively common genotypes with a frequency of only 1 per cent, and it drops sharply for rarer conditions.

Selection would be more effective if it included not only the homozygotes themselves, but also certain of their close relatives whose genotypes may be known. Thus, the children from marriages of an affected and a normal person are all heterozygotes; or, to use another example, two-thirds of the normal-appearing sibs of affected persons are heterozygotes. Reducing the reproduction of such relatives, in addition to reducing that of the affected individuals, would lead to greater reduction of the allele frequency from generation to generation; but the immediate effect would be very small, since, in the case of rare alleles, most marriages of heterozygous persons are with homozygous normal ones. It is true, however, that for any two heterozygotes who do not reproduce, two recessive alleles are eliminated, which, on the average, is equivalent to the nonoccurrence, at some future time, of one affected person.

There are serious limitations to selection against heterozygotes for recessive defect-causing genes. These limitations are due to the fact that probably all human beings carry several such genes. If all heterozygotes were excluded from reproduction, few if any persons would be left to people the earth.

Selection Against a Sex-linked Genotype. We shall omit a detailed discussion of selection against sex-linked alleles, dominant or recessive. Such selection resembles, but is not identical to, selection against autosomal dominant alleles. This property of selection against sex-linked recessives follows from the fact that in the hemizygous male sex-linked recessives have the same phenotypic effects as do dominants. For this reason, selection against sex-linked alleles is rather highly efficient.

In spite of their selective disadvantage, deleterious sex-linked recessives may occasionally spread in a population and be transmitted for many generations. Hemophilia, for instance, which used to eliminate many affected males before they reached the reproductive age, is found in high frequency in several isolates in different parts of the world. It would be possible to eradicate most of the existing alleles for hemophilia in one generation if the following individuals refrained from having children: affected men (h), daughters of affected men (Hh), and sisters of affected men (one half HH, other half Hh). This would leave some h alleles in Hh women whose fathers were normal and who did not have affected brothers. It should be noted, however, that the selection scheme outlined involves the abstention from procreation by some women who do not carry the h allele.

Selection Against a Two-factor Polygenic Genotype. A particularly important group in a consideration of selection are polygenic cases, since they account for the inheritance of quantitative characters. Some examples of selection for and against certain polygenic combinations will be considered.

We shall begin with the assumption that there are two independently inherited pairs of alleles—A^1, A^2 and B^1, B^2—and that the genotype $A^2A^2B^2B^2$ is phenotypically distinguishable from all others. This condition is fulfilled, not only if A^2 and B^2 are recessives, but also under a variety of other conditions, for instance, if A^2 and B^2 have equal and additive effects and the heterozygotes are intermediate between the homozygotes.

The genotypes of the population are $A^1A^1B^1B^1$, $A^1A^2B^1B^1$, $A^1A^1B^1B^2$, $A^2A^2B^1B^1$, $A^1A^1B^2B^2$, $A^1A^2B^1B^2$, $A^2A^2B^1B^2$, $A^1A^2B^2B^2$, and $A^2A^2B^2B^2$. If selection is directed against $A^2A^2B^2B^2$ and if it is assumed, for the first example, that the allele frequencies of A^1, A^2, B^1, and B^2 are alike ($p_{A^1} = q_{A^2} = 0.5$; $p_{B^1} = q_{B^2} = 0.5$), then the initial frequency of $A^2A^2B^2B^2$ individuals is $(q_{A^2})^2 \times (q_{B^2})^2 = (0.5)^4 = 0.0625$, or 6.25 per cent. If selection completely eliminates the $A^2A^2B^2B^2$ class from reproduction, all $A^2A^2B^2B^2$ individuals in the next generation are from the types of marriages, and in the frequencies, indicated in Table 111.

The sum of all new $A^2A^2B^2B^2$ individuals amounts to 4 per cent, a considerable reduction from the initial frequency of 6.25 per cent and the same as that in selection against single-factor recessive inheritance with an initial frequency of aa individuals of 6.25 per cent. This statment can be verified by calculating the value of q_1^2 according to equation (2) on page 783. If $q_0^2 = 0.0625$, q_1^2 becomes 0.04.

TABLE 111
Selection Against $A^2A^2B^2B^2$. Types and Frequencies of Marriages Which May Give Rise to $A^2A^2B^2B^2$ Offspring.

Genotypes of Marriages	*Frequency of Marriages**	*Frequency of $A^2A^2B^2B^2$ Offspring*
$A^1A^2B^1B^2 \times A^1A^2B^1B^2$	$\frac{4q^4}{1-q^4}\cdot\frac{4q^4}{1-q^4}=\frac{16q^8}{(1-q^4)^2}$	$\frac{q^8}{(1-q^4)^2}$
$A^1A^2B^1B^2 \times A^2A^2B^1B^2$	$2\cdot\frac{4q^4}{1-q^4}\cdot\frac{2q^4}{1-q^4}=\frac{16q^8}{(1-q^4)^2}$	$\frac{2q^8}{(1-q^4)^2}$
$A^1A^2B^1B^2 \times A^1A^2B^2B^2$	$2\cdot\frac{4q^4}{1-q^4}\cdot\frac{2q^4}{1-q^4}=\frac{16q^8}{(1-q^4)^2}$	$\frac{2q^8}{(1-q^4)^2}$
$A^2A^2B^1B^2 \times A^2A^2B^1B^2$	$\frac{2q^4}{1-q^4}\cdot\frac{2q^4}{1-q^4}=\frac{4q^8}{(1-q^4)^2}$	$\frac{q^8}{(1-q^4)^2}$
$A^2A^2B^1B^2 \times A^1A^2B^2B^2$	$2\cdot\frac{2q^4}{1-q^4}\cdot\frac{2q^4}{1-q^4}=\frac{8q^8}{(1-q^4)^2}$	$\frac{2q^8}{(1-q^4)^2}$
$A^1A^2B^2B^2 \times A^1A^2B^2B^2$	$\frac{2q^4}{1-q^4}\cdot\frac{2q^4}{1-q^4}=\frac{4q^8}{(1-q^4)^2}$	$\frac{q^8}{(1-q^4)^2}$
Total frequency of $A^2A^2B^2B^2$ offspring:		$\frac{9q^8}{(1-q^4)^2}$

*$q = p_{A^1} = q_{A^2} = p_{B^1} = q_{B^2}$. The frequencies of the different genotypes have been adjusted to the size of the population after the $A^2A^2B^2B^2$ group has been excluded. This size equals $(1-q^4)$. For example, the frequency of $A^1A^2B^1B^2$ individuals is $4q^4$ in a population of $(1-q^4)$, or $4q^4/(1-q^4)$.

It might be thought, after this result, that the effect of selection against a doubly homozygous two-factor combination is identical to that against a recessive single-factor homozygote. This, however, is not true. As will be shown in the following pages, selection against the $A^2A^2B^2B^2$ class continued after the first generation is increasingly less effective than is selection against an *aa* class. The effect of selection in polygenic inheritance cannot be judged adequately by its immediate result.

For a second example that demonstrates the effects of continued selection, it will be assumed that the initial frequency of the $A^2A^2B^2B^2$ class is 1 per cent and that the allele frequencies q_{A^2} and q_{B^2} are alike. The assumed frequency of $A^2A^2B^2B^2$ is equivalent to a q_0^4 of 0.01, from which it follows that $q_{A^2} = q_{B^2} = 0.31623$ and $p_{A^1} = p_{B^1} = 0.68377$. After one generation of complete selection against the $A^2A^2B^2B^2$ class, the frequency of new individuals of this genotype is reduced to 0.83 per cent. In order to calculate the further reduction after another generation of complete selection, we could draw up a table similar to Table 111, in which the frequencies of the relevant genotypes would have to be entered in terms of the allele frequencies, p_1 and q_1, established after the preceding selection. If this tabulation is carried out, it is found that the frequency of the $A^2A^2B^2B^2$ class has decreased to 0.735 per cent in the second generation. This is a smaller reduction than that for a single-factor

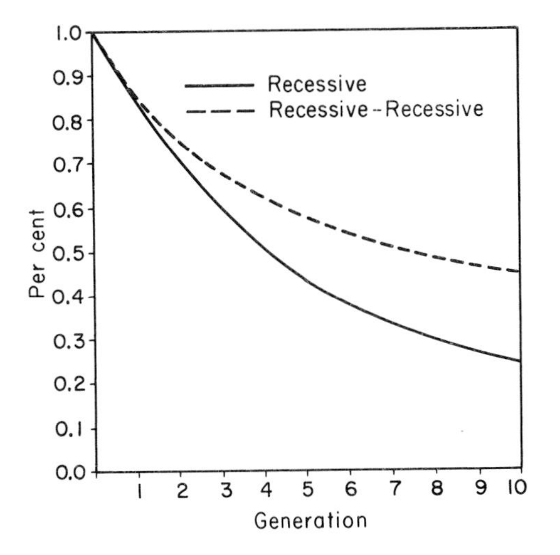

FIGURE 297
Complete selection for ten successive generations against a double homozygote (broken line). Solid curve shows complete selection against a single locus homozygote. (After Koller, *Z. Konstitl.*, **19,** 1935.)

recessive, *aa*, in which the frequency of the affected type by the second generation goes down to 0.694 per cent. The results to ten generations of continued selection against $A^2A^2B^2B^2$ individuals are given in Figure 297 (broken line), as well as the results of selection against a single-factor recessive, *aa*, for comparison (continuous line). It is seen, that after ten generations, the $A^2A^2B^2B^2$ genotype still recurs in 0.455 per cent of the individuals, in contrast to 0.25 per cent of the *aa* class.

In a manner similar to that indicated above, we may calculate, among other examples, the results of selection in two-factor inheritance if the phenotypes against which selection is directed are controlled by *dominant* alleles.

Equilibrium after selection against a Two-factor polygenic genotype. It was stated above that "the effect of selection in polygenic inheritance cannot be judged adequately by its immediate result." In single-factor inheritance with random mating, a change in an allele frequency immediately results in the establishment of a new equilibrium. This is an expression of the Hardy-Weinberg Law. It was Weinberg himself, however, who realized soon after his discovery of the situation relating to one gene pair, that for multiple pairs, if the equilibrium were disturbed, a new equilibrium would be established only gradually (pp. 244–246).

The previously treated $A^2A^2B^2B^2$ class in two-gene-pair genotypes may serve as an example of the contrast between the apparent immediate result of selection in the first generation and a "comeback" in later generations. It was seen that, after one generation of selection, this class is reduced from an initial frequency of 6.25 to 4 per cent, and from an initial frequency of 1 per cent to 0.83 per cent. We shall determine the allele frequencies q for A^2 and B^2 after one generation of selection and from these the frequency, q_1^4 of the type $A^2A^2B^2B^2$, at equilibrium.

Since selection removed all $A^2A^2B^2B^2$ individuals, which occurred with the frequency q_0^4, the new allele frequencies for A and B become

$$q_1 = \frac{q_0 - q_0^4}{1 - q_0^4}. \tag{4}$$

The value of q_1 calculated according to this formula for each of the two examples has been entered in Table 112, as well as the value for q_1^4, which gives the equilibrium frequency of the $A^2A^2B^2B^2$ against which complete selection had occurred for one generation. The last column of the table shows that the initial success of selection, which lowered $A^2A^2B^2B^2$ from 6.25 to 4 per cent, does not continue undiminished since the frequency of $A^2A^2B^2B^2$ rises to the equilibrium value of 4.7 per cent. Similarly, the initial lowering of the $A^2A^2B^2B^2$ class from 1 to 0.83 per cent is partly reversed by the later rise to 0.91 per cent; that is, the comeback in this case is approximately one-half of the originally lost frequency!

Even more significant is a comparison between q_0, the initial, and q_1, the new, allele frequency of A^2 and B^2 after selection (see the second and fourth columns of Table 112). In the first example, the reduction in allele frequency is from 0.5 to 0.47; and in the second, from 0.32 to 0.31. These very slight changes are a consequence of the genetic situation which makes the overwhelming number of all individuals of these populations carriers for at least one of the alleles A^2 and B^2. Selection against these alleles reaches only the fraction q_0^4 of the population (the $A^2A^2B^2B^2$ individuals), while the alleles A^2 and B^2 present in the rest of the population, $(1 - q_0^4)$, remain untouched. With the exception of the class $A^1A^1B^1B^1$ (p_0^4), all genotypes in the population which are not subjected to selection carry either A^2 or B^2, or both. Therefore, the sum of all carrier classes, S, is

$$S = 1 - p_0^4 - q_0^4. \tag{5}$$

In a population in which $A^2A^2B^2B^2 = 6.25$ per cent, $p_0 = q_0 = 0.5$; therefore, S is 87.5 per cent. In a population in which $A^2A^2B^2B^2 = 1$ per cent ($q_0 = 0.316$, $p_0 = 0.684$), S is still as large as 77.1 per cent.

TABLE 112
Immediate Effect and Equilibrium Effect of One Generation of Selection Against $A^2A^2B^2B^2$.

Initial Frequency of		*Frequency After One Generation of Selection of*		*Frequency of $A^2A^2B^2B^2$ After Selection and After Equilibrium*
$A^2A^2B^2B^2$ (q_0^4)	Alleles (q_0)	$A^2A^2B^2B^2$	Alleles (q_1)	(q_1^4)
0.0625	0.5	0.04	0.46667	0.04743
0.01	0.316	0.00826	0.30932	0.00915

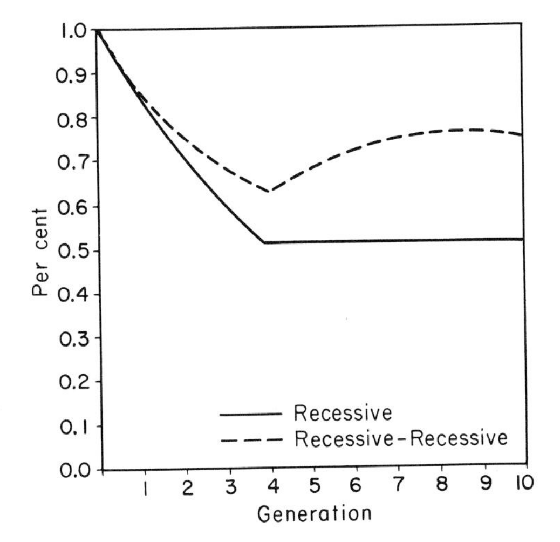

FIGURE 298
Complete selection for 4 successive generations, and cessation of selection during the following 6 generations. For the first 4 generations, the curves relating the frequency of various genotypes to the sequence of generations are identical with the ones in Figure 297. For the ensuing 6 generations, during which selection has ceased, the curves show either no change in frequency of genotype (Rec.) or a rise (Rec.-Rec.). (After Koller, *Z. Konstitl.*, **19**, 1935.)

The comeback phenomenon in two-gene-pair inheritance is obviously not restricted to cases in which selection ceases after one generation. In order to show its existence and extent for another example, Figure 298 is provided; it gives the changes in the frequencies of two phenotypes after complete selection had acted for four generations and then ceased.

Polygenic inheritance is usually based on many more than two pairs of genes. The foregoing discussions, modified accordingly, apply likewise to inheritance involving three or more pairs. Since, in a population, the frequency of individuals who are carriers of at least one of the alleles concerned in selection increases with the number of loci, the speed with which selection permanently accomplishes specific results decreases with increasing number of loci.

The numerical data presented for selected cases of polygenic inheritance are examples of the kind of information needed for a detailed understanding of the effect of differential reproduction in man upon the phenotypic and genotypic composition of later generations. The model examples used in our discussion cannot be regarded as adequate representations of the as yet unknown genetic situation in such traits as performance in intelligence tests or of other genetic components which are believed to participate in the assignment of individuals to different socioeconomic groups. It already seems plausible, however, that genetic changes brought about by selective agents are small from one generation to the next, and that the effectiveness of selection cannot be judged adequately from a consideration of phenotypic changes. The loss of alleles involved in specific phenotypes against which selection acts is smaller than the reduction in frequency of the phenotypes, because the population at large represents a great reservoir of these alleles.

Opposing Forces Influencing The Effectiveness Of Selection

The efficiency of selection has been discussed for populations which, at the beginning of the selective processes, were in equilibrium. It is, however, quite common for specific alleles or combinations of alleles to occur in relatively high concentration in pockets of the population. As we have seen earlier, such conditions are brought about by isolation of subpopulations. Moreover, regardless of the population structure, the allelic distributions will not fulfill the equilibrium demands if assortative mating takes place. Thus, the positive assortative mating which exists in regard to body height and intelligence must result in increased frequencies of homozygotes relative to the heterozygotes.

The efficiency of selection is *increased* if there is a relatively high concentration of alleles in isolates, or if the frequencies of homozygotes or of selected polygenic combinations are higher than in panmixis. This can easily be seen in positive assortative mating, which accumulates, relative to random mating, too many alleles in homozygous or specific polygenic combinations. Consequently, selection against such homozygotes eliminates a larger proportion of the alleles than it does in a population at equilibrium. If alleles are concentrated in an isolate, the situation is similar: more alleles will appear in homozygous or specific polygenic combinations than if the concentration of these alleles is diluted over the whole population. Again, therefore, there is a larger number of individuals who will be subject to selection.

In another respect, the effectiveness of selection may often be *lower* than predicted at first. If an inherited trait has a certain frequency in a random-mating population, it must be ascertained whether the appearance of the trait is due to alleles at the same locus in each affected individual or whether it is caused by different loci. If the former is true, then the observed frequency of the trait enters directly into the calculations of the result of selection, in the manner which has been worked out in the various cases discussed in this chapter. If the trait is the result of two or more different genotypes controlled by different loci, then the effect of selection has to be considered separately for each genotype.

This may be shown by an example. Assume a population in which 1 in 10,000 individuals is blind for genetic reasons, that is, a frequency of 0.0001. If the blindness of all individuals were due to the same homozygous recessive allele, then one generation of selection against the trait would, according to equation (2) (p. 783), reduce its incidence to 0.00009803. On the other hand, if there were ten different and equally numerous genotypes, each of which produces blindness, then the initial frequency of any one of them would be one-tenth of 0.0001, or 0.00001. Selection, for one generation, against one homozygous genotype would reduce it to 0.000009937, so that the frequency of the sum of all individuals affected with any one of the ten genotypes is 0.00009937. If this figure is compared with the one derived under the assumption that hereditary blindness depends on a single homozygous recessive genotype, it is seen that selection against the trait has led to a reduction of only 0.63 per cent of the the original frequency when any one of ten different genotypes is responsible, as compared to 1.97 per cent when a single genotype causes the trait.

This hypothetical example is probably representative of numerous actual situations, since it has been shown that, often, similar or apparently identical hereditary phenotypes are the result of diverse genetic constitutions. If it is true that selection against certain traits is often *more* effective on account of isolate concentration or assortative mating, than would be expected from random mating, it is equally true that selection is often *less* effective than expected on account of the genetic heterogeneity of many traits. Which of the two opposing forces is stronger will vary from case to case. The recognition that selection acts very slowly against all but single-factor dominant genotypes remains the fundamental result of our considerations.

Eugenics and the Slow Action of Selection. The slowness of selective processes against recessive single factors and against polygenes is both a blessing and a curse. It is a fortunate feature, since it forms a powerful buffer against sizable, undesirable effects of selection, as in differential reproduction, when it is directed against intellectually well-endowed genotypes. It is an unfortunate feature, since selective measures aimed at eliminating undesirable traits are rendered relatively ineffective. In either respect, the population at large serves as a huge reservoir for alleles from which the desirable and undesirable genotypes can be reconstituted.

If the hopes and fears of the eugenics movement seem greatly exaggerated in the light of a numerical treatment of the problems, it should not be forgotten that the idealism which concerns itself with the genetic fate of future generations has a sound core. To say that the loss of supposedly desirable genotypes in one or even many generations of differential reproduction is small does not mitigate the fact that is *is* a loss which may be regrettable and, possibly, even have serious consequences. To state that reproductive selection against severe physical and mental abnormalities will reduce the number of the affected from one generation to the next by only a small percentage does not alter the fact that the small percentage may represent tens of thousands of persons. Conversely, even a slight increase of desirable genotypes, through progressive eugenic measures, would be a social gain.

Eugenics and the Control of Environment. Eugenic selection and dysgenic selection are concerned with genetic constitutions which, under present physical, mental, and social circumstances, may lead to desirable or undesirable phenotypes. Since many genotypes express themselves differently under different environmental conditions, it is theoretically possible that new kinds of environments can so influence the developmental reactions of those genotypes which now lead to undesirable phenotypes so that desirable phenotypes result. Thus, certain genotypes formerly led to the serious disease diabetes mellitus by causing a deficiency in internally produced insulin, but now persons with these genotypes can lead an almost normal life owing to insulin injections. Or, it may be assumed that one genotype gives its carriers, in their specific educational and social environment, a lower I.Q. than that which another genotype gives to individuals in a different environment. In spite of the genetic circumstances, it might be possible to bring the carriers of the "lower" genotype to the same high achievement as the carriers of "higher" genotypes, provided that a particularly appropriate environment were furnished. A successful search for environments best suited to the development of desirable phenotypes may accomplish sometimes less, at other times as much, and at still other times more improvement of humanity than does genetic selection in a static environment. Provision of favorable environments involves a task equivalent in importance to selection for genotypes which, under present conditions, lead to better phenotypes than others or selection against genotypes which, under present conditions, lead to poorer phenotypes.

It is often asked if the institution of special environments as compensation for deficient gene function does not necessarily lead to a weakened constitution of mankind. If this question means that man thus becomes more dependent on his environment than he would be if natural selection were allowed to operate—with the eventual elimination of deficient genotypes—the answer is in the affirmative as far as the specific trait is concerned. Such dependence on special environments, however, did not start with the advent of civilization. When, in earliest evolutionary times, "animals" first developed, a then new kind of dependence on environment arose, since animals are incapable of synthesizing their substance from inorganic sources and have to rely on other organisms for food. When, much later, man's ancestors lost most of their mammalian body hair, another, but this time minor, step was taken. Man had to rely on fur from other mammals and on fire to keep his temperature at the necessary physiological level.

Man's greater dependence on specific outside sources has not been equivalent to degeneration. On the contrary, dependence has often resulted in greater freedom from the restrictions of the external world. The loss of ability to use inorganic material for food became correlated with the evolution of nervous systems and sense organs which make possible the many autonomous adjustments of animals. The dependence on clothing and fire enabled man to occupy regions of the globe where formerly he could not have survived. The passing of the primitive stages in which each man was, to a large extent, independent of the help of others gave rise to the complex interdependence of men in modern civilization. This new dependence has released man from the physical and mental starvation of earlier times. It is true, however, that man's freedom from many limitations imposed by the external world and even by his own genotypes can persist and be enlarged only if he retains and extends his control over environment and particularly over himself.

Man's ability to create new environments has its limits. It is important to realize that many undesirable traits, such as severe myopia or harelip and cleft palate, against which selection was formerly strong and which have tended to become selectively neutral, may continue to remain disadvantageous for a long time. Although we can correct rather well for myopia by appropriate eyeglasses, harelip and cleft palate often require difficult and repeated surgery. Too much reliance on medical and surgical progress may lead to an ever-increasing number of persons whose normal functioning is made possible only by the performance of major operations, by the permanent use of artificial organs or other functional aids, by lifelong provision of complex, special drugs or diets, or by regularly repeated blood transfusions. At some stage, it seems, preventive eugenic measures will become truly urgent.

The tasks of human genetics concern the future as well as present generations. Genetic counseling has been largely devoted to individual problems, and the eugenic implications of specific decisions have been considered only rarely. Ultimately it will be necessary to consider both the individual and society. Much of eugenic thinking in the past was based on inadequate knowledge and prejudice, and has been harmful. With increasing insight, wise planning will be possible in the future. Then, genetic and eugenic counseling will become the foundation of man's direction of his own biological evolution. Although eugenic problems are not as immediately urgent as the pessimists believe them to be, their ultimate importance can hardly be overestimated.

PROBLEMS

207. A dominant allele *A* has a frequency of 0.01. A recessive allele, *b*, at another locus, has a frequency of 0.1.
   (a) What is the frequency of persons affected with the dominant trait?
   (b) What is the frequency of persons affected with the recessive trait?
   (c) If none of the affected reproduce, how many dominantly affected persons will appear in the next generation (barring mutation)?

(d) If none of the affected reproduce, how many recessively affected persons will appear in the next generation (barring mutation)?

208. In a population of 100,000,000 people, 40,000 are afflicted with a disease caused by a homozygous recessive gene. If these individuals are kept from reproducing and if the size of the population remains constant, what will be the number of afflicted individuals in the next generation?

209. In population I, 50 per cent of all individuals are heterozygous for a dominant gene D, which is penetrant in one-half of all cases. The other 50 per cent are dd. No marriages of $Dd \times Dd$ occur. In panmictic population II, 25 per cent of all individuals are homozygous for a recessive gene a, which is fully penetrant.
(a) After one generation of complete selection against the affected individuals in populations I and II, which population will have a larger number of affected?
(b) After many generations of selection, which population will have a larger number of affected? (Give reasons for your answers.)

210. With the use of equation (3) (p. 784), determine the frequency of dd individuals after 5, 20, and 50 generations if their initial frequency is
(a) 4, (b) 0.25, and (c) 0.04 per cent.

211. With the use of equation (3) (p. 784), show how numbers of generations required for various reductions in frequencies of dd listed on pp. 784–785 were obtained.

212. Assume that the frequency of a sex-linked affliction in men is 1 in 10,000 and of carrier women 1 in 5,000.
(a) If affected individuals ceased to reproduce, what would be the frequency of affected men in the next generation?
(b) What would be the frequency of carrier women?
(c) What would be the frequency of affected men in the second generation?
(d) What would be the frequency of affected men in the tenth generation?

213. The difference in number of children per 1,000 mothers of two different socio-economic groups decreased from 1,700 in 1910 to 1,400 in 1940 and to 1,200 in 1952. How much greater was the fertility of one group than the other group in each of the three periods? (The less fertile group had 3,600, 2,200, and 1,800 children respectively.) Compare the trend shown by your answer with the trend of decreasing absolute differences between the two groups.

214. Assume that the genotype A^1A^1 leads to an average test score of 90, A^1A^2 to 100, and A^2A^2 to 120. A population consists of two equal-sized absolute isolates, I and II. In I, the frequency of the allele A^1 is 1/2; in II it is 1/4.
(a) What is the average score of isolate I? Of isolate II? Of the whole population?
(b) If all individuals in I reproduce at an equal rate, which is twice that in isolate II, what will be the average score of the whole population in the next generation?
(c) In isolate II, if the A^2A^2 individuals reproduce at a rate which is 50 per cent greater than either the A^1A^1 or A^1A^2 individuals, what will be the average score of isolate II in the next generation?
(d) If the individuals of isolate I reproduce at a rate which is twice as great as either the A^1A^1 or A^1A^2 individuals in isolate II, and if in isolate II the A^2A^2 individuals reproduce at a rate 50 per cent greater than either A^1A^1 or A^1A^2 individuals, what will be the average score of the whole population?

215. In the preceding problem, what will be the average score of the whole population after complete breakdown of the isolates and random mating if:
(a) The initial population becomes panmictic?
(b) The populations defined in Parts (b), (c), and (d), respectively, become panmictic?

REFERENCES

Bajema, C. J., 1963. Estimation of the direction and intensity of natural selection in relation to human intelligence by means of the intrinsic rate of natural increase. *Eugen. Quart.*, **10**:175–187.

Bajema, C. J., 1966. Relation of fertility to educational attainment. *Eugen. Quart.*, **13**:306–315.

Bateson, W., 1919. Common sense in racial problems. *Eugen. Rev.*, **11**. (Reprinted *in* Bateson, Beatrice, 1928. *William Bateson, F. R. S., Naturalist.* pp. 371–388. Cambridge University Press, London.)

Cliquet, R. L., 1968. Social mobility and the anthropological structure of populations. *Hum. Biol.*, **40**:17–43.

Crow, J. F., 1958. Some possibilities for measuring selection intensities in man. *Hum. Biol.*, **30**:1–13.

Dahlberg, G., 1947. *Mathematical Methods for Population Genetics.* 182 pp. Karger, New York.

Darlington, C. D., 1969. *The Evolution of Man and Society.* 753 pp. Allen and Unwin, London.

Gibson, J. B., 1970. Biological aspects of a high socioeconomic group. I. I.Q., education, and social mobility. *J. Biosoc. Sci*, **2**:1–16.

Haldane, J. B. S., 1924. *Daedalus.* 93 pp. Dutton, New York.

Haller, M. H., 1963. *Eugenics. Hereditarian Attitudes in American Thought.* 264 pp. Rutgers University Press, New Brunswick, N.J.

Higgins, J. V., Reed, Elizabeth W., and Reed, S. C., 1962. Intelligence and family size: a paradox resolved. *Eugen. Quart.*, **9**:84–90.

Kiser, C. V., 1968. Trends in fertility differentials by color and socioeconomic status in the United States. *Eugen. Quart.*, **15**:221–226.

Kiser, C. V., 1970. Changing patterns of fertility in the United States. *Soc. Biol.*, **17**:302–315.

Kiser, C. V., Grabill, W. H., and Campbell, A. A., 1968. *Trends and Variations in Fertility in the United States.* 338 pp. Harvard University Press, Cambridge.

Li, C. C., 1970. Human genetic adaptation. Pp. 545–577 *in* Hecht, M. K., and Steere, W. C. (Eds.), *Essays in Evolution and Genetics in Honor of Theodosius Dobzhansky.* Appleton-Century-Crofts, New York. (A supplement to *Evolutionary Biology.*)

Lunde, A. S., 1965. White-Non White Fertility Differentials in the United States. *U.S. Health, Education and Welfare Indicators.* Sept, 1965: 23–38.

Mather, K., 1964. *Human Diversity.* 126 pp. Free Press, New York.

Matsunaga, E., 1966. Possible genetic consequences of family planning. *Amer. Med. Ass.*, **198**:533–540.

Maxwell, J., 1969. Intelligence, education and fertility: a comparison between the 1932 and 1947 Scottish surveys. *J. Biosoc. Sci.*, **1**:247–271.

Osborn, F., 1968. *The Future of Human Heredity*. 133 pp. Weybright and Talley, New York.

Penrose, L. S., 1955. Evidence of heterosis in man. *Proc. Roy. Soc. London, B.*, **144**:203–213.

Penrose, L. S., 1963. *The Biology of Mental Defect*, 2nd ed. 374 pp. Grune and Stratton, New York.

Post, R. H., 1962. Population differences in red and green color vision deficiency: A review, and a query on selection relaxation. *Eugen. Quart.*, **9**:131–146. (For corrections of errors, see *Eugen. Quart.*, **10**:84–85.)

Post, R. H., 1969. Population differences in tear duct size. Implications of relaxed selection. *Soc. Biol.*, **16**:257–269.

Post, R. H., 1971. Possible cases of relaxed selection on civilized populations. *Humangenetik*, **13**:253–284.

Ray, A. K., 1969. Color blindness, culture, and selection. *Soc. Biol.*, **16**:203–208.

Schreider, E., 1967. Possible selective mechanism of social differentiation in biological traits. *Hum. Biol.*, **39**:14–20.

Schwidetzky, Ilse, 1950. *Grundzüge der Völkerbiologie*. 512 pp. Ferdinand Enke, Stuttgart.

Scottish Council for Research in Education, 1949. *The Trend of Scottish Intelligence*. 151 pp. University of London Press, London.

Scottish Council for Research in Education, 1953. *Social Implications of the 1947 Scottish Mental Survey*. 356 pp. University of London Press, London.

Sutter, J., 1950. *L'Eugénique*. 254 pp. Institut National d'Etudes Démographiques. Travaux et Documents, Cahier no. 2, Presses Universitaires de France, Paris.

Wilkins, J. C., 1969. Risks to offspring of patients with patent ductus arteriosus. *J. Med. Genet.*, **6**:1–4.

Zazzo, R., 1960. *Les jumeaux le couple et la personne*. Thèse pour le Doctorat ès Lettres, Presses Universitaires de France.

30

ASPECTS OF MEDICAL GENETICS

Medical genetics is that branch of human genetics which is directly concerned with the relationship of heredity to disease. Of course, there is no sharp line of demarcation between basic work in genetics and possible applications to medicine. In fact all Nobel prizes in medicine or physiology awarded to geneticists and molecular biologists were in recognition of basic research with the fly *Drosophila*, the mold *Neurospora*, the bacterium *Escherichia*, the bacteriophages, or with the properties of nucleic acids.

The chromosome theory of inheritance, which was the foundation of pedigree analysis and population genetics, has more recently shown its value in the field of human cytogenetics with the discovery of abnormal chromosome constitutions in inherited syndromes (e.g., Klinefelter's, Turner's, Down's). Radiation genetics has become of utmost importance in the use of nuclear energy, for the diagnosis and treatment of disease and to society in general in the determination of radiation effects on future generations. Biochemical and microbial genetics have likewise furnished concepts and tools for the tasks of the modern physician. Conversely, the medical professions's interest in blood transfusion and immunology has led to the fundamental discoveries of blood groups and their genetics, and the concern with sickle-cell anemia has culminated in the elucidation of the molecular structure of different hemoglobins as determined by different alleles.

The present chapter will briefly survey specifically. medical aspects of human genetics, many of which have been discussed elsewhere in this book. For a treatment of some medicolegal problems, see Chapter 13.

Genetic Heterogeneity of Diseases. Clinical studies of individuals affected with some defect or disease have frequently shown that a presumed single disease is in reality a group of different diseases with similar manifestations.

Mental defect, for instance, may have any one (or more) of many underlying causes, both nongenetic and genetic. The clinician is able to distinguish some of the different kinds of inherited mental defect as part of various syndromes (e.g., Tay-Sachs disease, Down's syndrome, cri du chat) or by biochemical differentiation (e.g., phenylketonuric feeblemindedness). The geneticist has contributed evidence for heterogeneous causes of diseases by finding single-gene recessive inheritance in some kinds of mental defect (Tay-Sachs disease, phenylketonuria), gross chromosomal imbalance in others (Down's syndrome, cri du chat), and polygenic determination in still others ("undifferentiated" mental deficiency). In diabetes insipidus, one of two different physiologic disturbances is responsible for the abnormally high excretion of water: a primary deficiency, or absence, of a pituitary hormone; or an abnormal primary function of the kidney tubules. The geneticist has only partly been able to discover heterogeneity of causes for this disease, since both types of abnormal water metabolism may be inherited in the same X-linked way, though presumably through genes at different loci. However, the genetic heterogeneity of diabetes insipidus goes beyond the X-chromosome, for extensive pedigrees showing dominant autosomal inheritance are also on record.

With other diseases, either the genetic evidence suggests that for apparently the same disease different mechanisms may be operating in different kindreds, or it at least supports the finding of differences by the diagnostician. For example, retinitis pigmentosa, a retinal disease that may lead to blindness, is inherited in autosomal fashion in some kindreds and in X-linked fashion in others. In addition, there are dominant as well as recessive genes for both autosomal and X-linked types. It may be that the pathology of this disease is basically different in the genetically different types, but full knowledge is still lacking. Genetic heterogeneity is also revealed by linkage studies. The heterogeneity of recombination values between the Rh blood groups and elliptocytosis (p. 360) implies the existence of at least two different gene loci involved in the abnormality of the red cells and therefore, presumably, of at least two different, but still unknown, physiological mechanisms which lead to this condition.

The finding of genetic heterogeneity of an abnormality may be fundamental to the discovery of the specific steps in its development. And knowledge of how the defect is inherited may later help in its prevention or cure.

Disease Resistance. In many diseases, external agents are clearly the major causes. This is pre-eminently true of infectious diseases produced by bacteria, fungi, viruses, and animal parasites. But external agents are also instrumental in diseases due to malnutrition, such as degeneration of the liver owing to alcoholism; psychosomatic diseases, such as gastric or duodenal ulcers, which are related in some degree to mental stress; and some types of mental illness. It has become increasingly apparent, however, that humans differ in their genetic susceptibility to nongenetic causes of disease. This is not surprising in view of the successes of plant and animal breeders in raising organisms resistant to various conditions: rust-resistant wheat, cold-resistant fruit trees, wilt-resistant watermelons, typhoid-resistant poultry, and cattle resistant to tick-borne diseases. In man there is evidence for genetic control of susceptibility to infectious agents in general, as well as to specific pathogens. In hypogammaglobinemia, which is caused by an X-linked recessive gene, because there is a deficiency in the production of antibodies against bacterial infections, the invaders cannot be combated by the usual immunological reactions of the host. Another cause of recurrent bacterial infection is based on impairment of the leucocytes' capacity to phagocytose. A dominant gene seems to be responsible for the defect. Susceptibility to specific infections, such as tuberculosis or poliomyelitis, indicates more complex genetic control; and the same holds for susceptibility to ulcers and to mental illnesses. Genetic predisposition to certain types of tumors has also been demonstrated—with full or high penetrance in special types such as polyposis of the colon and neurofibromatosis, or with presumably low penetrance in the more common types of cancer. Such predisposition may ultimately be shown to be susceptibility to an external agent, should the virus theory of the origin of cancer be generally substantiated.

Pharmacogenetics. Genetic differences in susceptibility to drugs are also known. Ordinarily, the X-linked gene for glucose-6-phosphate dehydrogenase deficiency is of no obvious disadvantage to its carriers. When, however, certain antimalarial drugs, such as primaquine, were given to persons who happened to be deficient for G-6-PD, severe anemia resulted. Knowledge of such possible relations between drugs and diseases in certain normal-appearing persons is important if serious consequences are to be avoided for the few persons susceptible to ill effects from drugs that have only beneficial effects in the majority of people. An analogous case is that of the barbiturates. Most normal persons may take these drugs and obtain the desired effect. Persons who carry a dominant gene for porphyria, however, suffer consequences, such as violent abdominal pain when they take barbiturates. Still another important variation in drug response concerns suxamethonium, a usually short-term muscle relaxant that had been administered by anesthesiologists and by psychiatrists to many patients before it was discovered that some individuals react in a dangerously abnormal way to this compound. Instead of the muscular paralysis and associated suspension of breathing lasting for about 2–3 minutes as it does in most persons, in some persons the effect of the drug may last for several hours. The abnormal reaction of the affected individuals is due to their genetically determined deficiency of the enzyme pseudocholinesterase. In these individuals, suxamethonium is broken down too slowly and exerts its effect on the respiratory muscles for an excessively long time.

An inverse situation exists when the genotype of an individual interferes with the desired effect of a substance by tending to make it ineffective. This applies to the dominant gene for exceptional resistance to coumarin anticoagulant drugs. Twenty times the average dose of warfarin had to be administered to patients having this gene to maintain the prothrombin time within the desired range. Similarly, children having the genotype for vitamin-D-resistant rickets require a hundred or more times the normal amount of the vitamin for normal bone development.

An interesting difference in the ability to digest milk (or its sugar, lactose) has also been observed. Infants of all races thrive on milk; but many healthy adults of Mongoloid and African races seem to have a lactase deficiency that makes the ingestion of milk bring on abdominal cramps or diarrhea. Among adult Caucasians the prevalence of lactase deficiency is low. Although the genetic basis of the trait is not known in detail, it appears that—unexpectably—milk intolerance is a frequent trait analogous to those of pharmacogenetic importance.

The Recognition of Carrier States. It would seem that, by definition, individuals who are heterozygous for a "recessive" gene, A^2, are not distinguishable from the homozygous dominant A^1A^1. However, dominance and recessivity are relative terms, in the sense that their recognition depends on the ability to distinguish the three phenotypes corresponding to A^1A^1, A^1A^2, and A^2A^2. When A^1A^2 is indistinguishable from A^1A^1, then A^2 is correctly called a recessive; but when refined observations disclose two different phenotypes produced by the two genotypes, then the designation of A^2 as a recessive is no longer fully valid. It is nevertheless customary in human genetics to speak of A^2 as a recessive if A^2A^2 causes a phenotype strikingly different from that of A^1A^1 and if the heterozygote A^1A^2 seems to resemble the normal A^1A^1.

The often arbitrary use of the terms dominance and recessivity is illustrated in the case of sickle-cell anemia. The genotype Hb^SHb^S may lead to a severe, often fatal anemia, while both Hb^AHb^A and Hb^AHb^S produce a normal, nondefective phenotype. In terms of the anemic effect, Hb^S is a recessive. However, a study of red cells which have been subjected to low oxygen pressure reveals absence of sickling in Hb^AHb^A, severe sickling in Hb^SHb^S, and an intermediate degree of sickling in Hb^AHb^S. Finally, at the molecular level, the two alleles in the heterozygote act codominantly, leading to the production of a mixture of hemoglobins A and S.

Many geneticists have attempted to discover differences between normal homozygotes and apparently normal heterozygotes for recessive genes that cause defects in homozygotes. Such discoveries would strengthen genetic hypotheses of recessiveness by providing direct proof of the heterozygosity of both parents of defective children. More important, as emphasized by Neel, these findings would be of general significance for human genetics in various other ways. An understanding of the interplay, in human populations, of mutation and selection for and against heterozygotes and homozygotes would be furthered if direct observation could distinguish between all genotypes. Also, the recognition of genetic factors in malformations would be facilitated if we could identify not only grossly abnormal types which may be based on homozygous genotypes but also transitional stages between normal and abnormal which may have a heterozygous genetic basis. And, in genetic counseling, it would represent a great step forward if it were possible to distinguish between the heterozygous and homozygous normal sibs or other relatives of homozygous recessive affected individuals, since advice could then be given founded on certain knowledge instead of on probability.

The search for recognizable phenotypic criteria of the heterozygous carrier state has indeed been successful in numerous genotypes. For some genes it is now possible to distinguish unequivocally individuals of all three genotypes, but for most genes the overlap in phenotypic expression makes it difficult to distinguish some of the heterozygotes from the normal homozygotes. An example is the X-linked disease, Duchenne-type muscular dystrophy, which is fatal to affected hemizygous males during the first two or at most three decades of life. Heterozygous women are not affected and were formerly impossible to distinguish from normal homozygotes. Sisters of affected males and daughters of heterozygous mothers have equal chances of being either carriers of the abnormal allele or homozygotes for the normal gene. The two genotypes can now, to some degree, be distinguished from each other by level of creatine phosphokinase in the blood, which is somewhat elevated in carriers. There is some

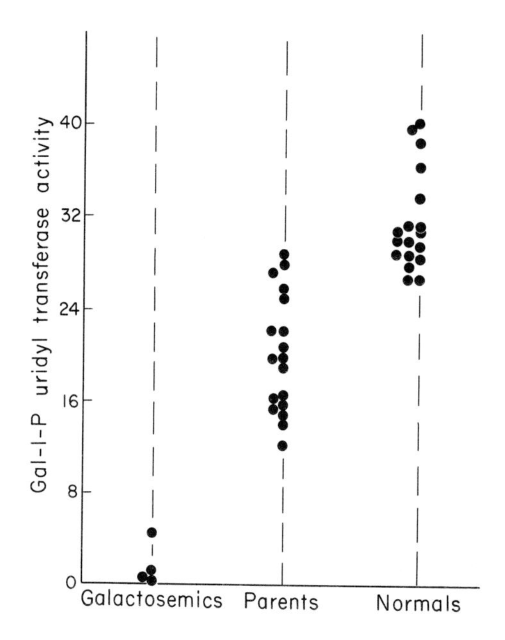

FIGURE 299
Galactosemia. Determinations of the activity of the enzyme uridyl transferase in 18 homozygous normal individuals, 18 heterozygous parents of affected children, and 4 homozygous affected persons. The activities are expressed as microliters oxygen uptake per volume for 0.3 cc packed and hemolyzed red cells. The affected individual whose activity was measured as above zero had been given a transfusion of normal blood some time before the test. (After Kirkman and Bynum, *Ann. Hum. Genet.*, **23**, 1959.)

overlap in the two distributions of creatine phosphokinase level but it is possible, on the basis of an enzyme assay, to tell about 80 per cent of the women who have a chance of being carriers either that they carry the gene or that they do not carry it.

Another example of partial recognition of the carrier state is provided by the rare congenital disease galactosemia. The abnormal homozygotes are fully separable from the heterozygotes, but there is some overlap in the degree of a specific biochemical activity of heterozygotes and normal homozygotes (Fig. 299). Affected individuals seem to lack completely the enzyme α-D-galactose-1-phosphate-uridyl transferase, which is necessary for the normal metabolism of the sugar galactose. Heterozygotes generally have between 60 and 70 per cent of the enzymatic activity of normal homozygotes. The pedigree shown in Figure 300 gives details of the enzymatic state of the parents, sibs, and other relatives of two affected individuals. It may be added that the amount of overlap between homozygotes and heterozygotes depends on the type of test. The galactose "tolerance test" determines the amount of milk sugar lost from the blood after intake of a given amount of the substance. The more direct tests (whose results are listed in Figs. 299 and 300) measure the enzymatic activity either in liver tissue obtained by surgery or in red cells.

Two more examples of autosomal recessives demonstrate success in completely separating the phenotypes of all three genotypes. One of the examples concerns a very rare gene, discovered in Japan by Takahara, which in homozygous persons leads to the absence in the blood and many body tissues of cata-

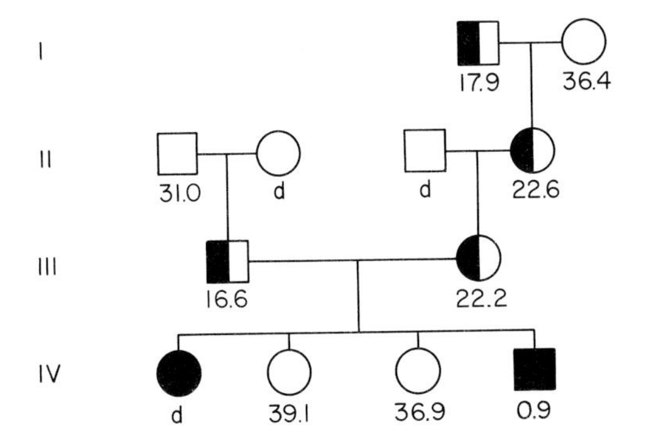

FIGURE 300
Galactosemia. A pedigree giving the enzyme activity for individuals of four generations in terms explained in the legend for Figure 299. Black-and-white symbol = low activity; solid black = galactosemia; d = dead. (After Kirkman and Bynum, *Ann. Hum. Genet.*, **23**, 1959.)

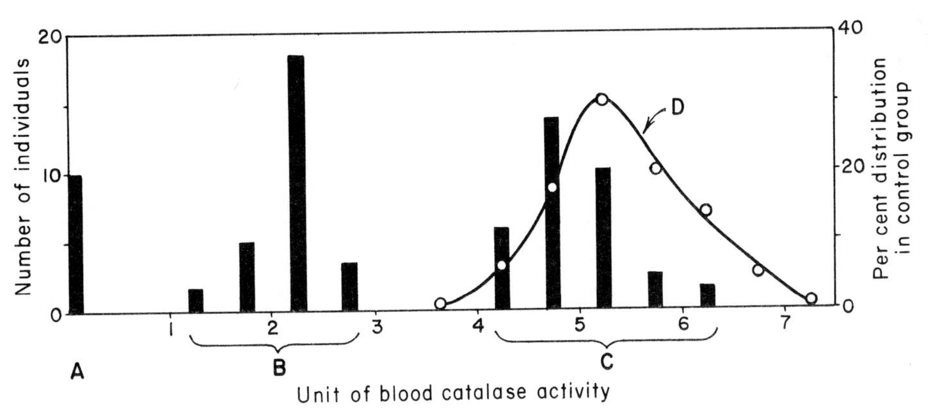

FIGURE 301
Acatalasemia. Activity of the enzyme catalase in (**A**) 10 affected persons, (**B**) 30 relatives with low activity, and (**C**) 36 relatives with normal activity. The curve (**D**) represents the percentage distribution of activity in 206 controls. (After Nishimura, Hamilton, Kobara, Takahara, Ogura, and Doi, *Science*, **130**, 1959.)

lase, an enzyme found in considerable amounts in most other persons. The catalase content of the blood of heterozygotes is intermediate, without overlap, between that of the two homozygotes (Fig. 301). The second example of success in completely separating the phenotypes of all three genotypes relates to Tay-Sachs disease. The serum of homozygous normal individuals contains two enzymes, hexosamidases A and B, while that of homozygous recessive infants shows practically no hexosamidase A activity. The heterozygotes are intermediate in serum activity without overlap with values for either homozygote (Fig. 302). It is, of course, possible in this as well as in studies of other diseases that some overlap may be found as more individuals are tested.

The term "carrier state" has been used in the preceding discussion to designate genotypes heterozygous for "recessive" abnormal genes. It can be extended to cover normal phenotypes made possible by incomplete penetrance

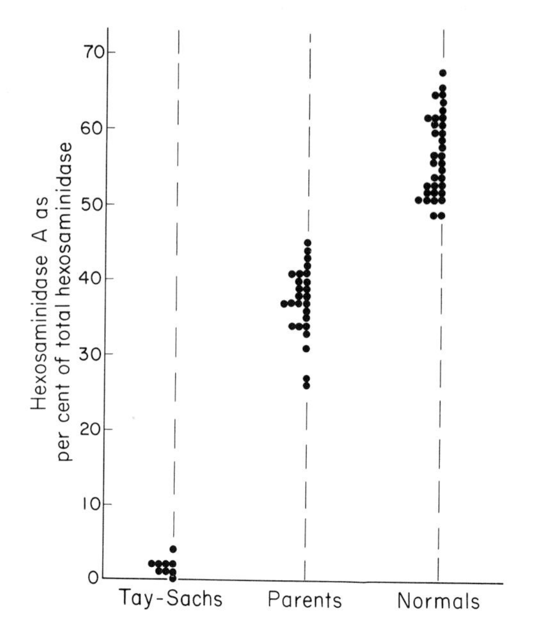

FIGURE 302
Tay-Sachs disease. The percentage of hexosamidase A in the total hexosamidase in serum of 9 affected children, 26 parents of affected children, and 33 normal individuals. (After O'Brien, Okada, Chen, and Fillerup.)

of homozygous recessive abnormal genes or incompletely dominant abnormal genes. Sugar diabetes, for instance, with its polygenic basis may not be expressed openly even if a person is a carrier of a potentially affected genotype. Tolerance tests for sugar can distinguish between genetically normal and genetically susceptible individuals.

For Huntington's chorea, which is caused in heterozygotes by a dominant allele and usually develops in middle age or later, the attempt to discover at an early age which of the children of an affected parent are carriers of the gene has thus far been unsuccessful. It is easy to predict that determining the carrier state will be possible someday, but it is hard to decide how at that time the gratifying certainty of some individuals that they and their future children will remain healthy should be weighed against the disturbing certainty of other individuals that they will develop the disease and against the knowledge that, should they have offspring, half of their children will suffer the same fate.

The phenotypic recognition of carriers of defect-causing genes will often permit taking preventive measures. A homozygous galactosemic infant when fed milk is likely to develop an enlarged liver, cataract, and mental deficiency, but it thrives on a diet free of galactose or lactose. A heterozygous carrier develops normally on the usual milk diet; but we may ask whether, in view of his reduced enzyme activity, a reduction of galactose intake might be beneficial. A person who is genetically susceptible to sugar diabetes may be able to remain nondiabetic by controlling his carbohydrate intake. Frequently, knowledge of the biochemical mechanism by which a specific gene causes a defect will enable the physician to supply the patient with the material which his body is unable to produce in sufficient amounts. The treatment of diabetes with insulin to compensate for the lack of production in the patient's pancreas is an outstanding example of therapy for a genetic defect.

Somatic Cell Genetics. The study of human genetics is not restricted to that of whole persons, pedigrees, and populations but encompasses processes that can be investigated at the cellular level, either in the person himself or in cultures of his body cells, particularly the fibroblasts. Somatic cell genetics deals with a variety of topics: normal and abnormal chromosomal behavior, including somatic segregation, somatic crossing over, somatic nondisjunction, somatic doubling of the chromosome complement; somatic mutations leading to mosaicism in whole individuals or to heterogeneity in tissue cultures; hybridization between cultured somatic cells (including man-mouse hybrids! —pp. 366–367); biochemical pathways; and application to clinical diagnosis and eventual therapy. Several of these topics have been referred to in this book at various places. Some additional findings in somatic cell genetics will now be cited.

Evidence of drug resistance, presumably by spontaneous mutation, has been obtained in the somatic cells of mice and rats. Tumor tissues which are perpetuated by transplantation from one animal to another may be highly susceptible to various chemotherapeutic agents such as amethopterin or 6-mercaptopurine, which are lethal in certain concentrations. But within an originally homogeneous line of tumor cells, some cells may develop that are resistant to, or even dependent upon, the usually destructive chemicals. Normal tissue cells grown in cultures have also been found to develop drug-resistant strains, which result from a spontaneous, presumably genetic change in a single cell that occurs before administration of the drug. The drug itself is solely instrumental in killing the majority of nonresistant cells but selectively sparing the mutant resistant ones.

These striking phenomena had been discovered earlier in microorganisms, particularly in bacteria. Among millions of bacterial cells killed by such agents as sulfa drugs, penicillin, and streptomycin, a few survived—resistant to, or even dependent upon, the usually lethal compounds. For some of these microorganisms, hybridization or equivalent methods have proved that spontaneous genic mutations resulting in drug resistance or dependence, and subsequent selection by application of the drug, were responsible for the new strains.

Abnormalities in the biochemical pathways that lead to defects can often be studied in tissue cultures of cells of affected persons. While experimentation with the affected persons would be neither feasible nor morally justifiable, cell cultures derived from them may obviously be used for a great variety of experimental procedures. Thus, the availability of cultures of cells from boys afflicted with the Lesch-Nyhan syndrome and the fact that the abnormal uric acid metabolism can be analyzed in detail in these cells should greatly improve our understanding of normal uric acid metabolism. Possibly it may lead to the finding of ways to control the disease.

Cell cultures with enzyme deficiencies characteristic of those in the Lesch-Nyhan syndrome have not only been derived from affected males but have also been obtained from cell cultures of normal males. Two different cell lines, one with extreme deficiency of the relevant enzyme and one with a reduction of the enzyme activity to about one-third of normal have been found – both presumably representing somatic mutations at a specific enzyme locus. These cell cultures represent additional systems in which to study human uric acid metabolism.

"Therapy" performed on cell cultures has sometimes been effective in restoring a normal phenotype, and we may hope to extend such "cures" to the whole individual. The cultured cells, for instance, of children with Hurler's and Hunter's syndromes have granules that stain with specific dyes and also have a high mucopolysaccharide content in contrast to cells from normal persons, which lack the granules and have a low mucopolysaccharide content. Danes and Bearn have shown that if the mutant cells are grown in vitamin A alcohol the granules disappear and the amount of mucopolysaccharide is reduced. Another example concerns Gaucher's disease, a defect which in its infantile variety affects liver, spleen, bones, and central nervous system function. The primary biochemical defect seems to consist of a severe reduction in activity of a β-glucosidase. Beutler and his associates have been able, within limits, to control the activity level of the enzyme in cell cultures. In a minimum medium the activity is higher than in an enriched medium. The authors conclude that "the possibility of manipulating the enzyme activity of fibroblasts could lead to therapeutic approaches in this disorder."

Amniocentesis. The use of biochemical "markers" in cell cultures has made possible an important advance in genetic counseling. Much of genetic counseling is based on probability consideration. This is unavoidable in general when the outcome of a future pregnancy is in question, but more certain determinations can be made in some cases when a pregnancy is already underway. A human fetus is surrounded by the amniotic fluid, and this fluid contains cells that are derived from the fetus and its amnion. Cytological and biochemical study of these cells can give information about the presence or absence of specific genotypes for which the fetus is at "high risk."

In order to obtain a sample of the amniotic fluid the procedure of amniocentesis is used. In its most common form it consists of puncturing the anesthetized abdominal wall of the mother with a needle and withdrawing of a small amount of the amniotic fluid from the amniotic sac. The cells in the fluid either are used immediately for diagnosis or are cultivated for later diagnosis. Immediate diagnosis at present is possible for determination of the fetal sex as judged by the presence or absence of Barr bodies and Y-chromosomes. After culturing of the amniotic cells detailed cytological diagnosis can be made on the basis of chromosome studies in dividing cells, and biochemical diagnosis can be carried out by assays for specific enzymes or for other metabolic features.

Some examples of the use of amniocentesis will show its power. Chromosome studies of cultured amniotic cells have been carried out by Nadler and others for pregnancies having a high risk of producing an infant with Down's syndrome. Among fetuses from parents one of whom carried a translocation involving chromosome 21, 15 fetuses were judged normal by amniocentesis and 7 affected. Among 28 pregnancies for which the normal parents had had a previous Down's syndrome child, 27 fetuses were normal and 1 affected. Among 82 pregnancies of women aged older than forty years, 80 fetuses were normal and 2 affected.

Biochemical studies are useful only for those abnormal conditions which are demonstrable in tissue cultures. One of these conditions is the absence of an enzyme that participates in uric acid metabolism, resulting in the Lesch-Nyhan syndrome (pp. 312, 724). The gene responsible for this disorder is X-linked and women who have had an affected child are, therefore, known to be heterozygous for the mutant allele. Each male offspring of a heterozygous woman has a 50 per cent chance of being affected. Amniocentesis has led to early prenatal recogniton of the presence of the defect in several pregnancies. In Tay-Sachs disease, which provides another example, cells homozygous for a recessive autosomal allele lack the enzyme hexosamidase A. A couple who have had an affected infant may now obtain information during a subsequent pregnancy on whether the 1-in-4 chance of their bearing another affected child has been realized or whether the 3-in-4 chance for a normal child has.

One more example of the use of cultures of cells from fetal amniotic sacs demonstrates the application of radioactive labeling of abnormal cells. The Hurler (autosomal recessive) and Hunter (X-linked recessive) disorders of mucopolysaccharide metabolism, as already mentioned, express themselves in tissue cultures by the appearance of numerous mucopolysaccharide granules in the cytoplasm. This was first demonstrated by Shannon Danes and Bearn by means of special staining techniques. Later, Fraccaro, Fratantoni, and their associates confirmed the staining properties and improved the diagnosis further by the use of radioactive sulfate, which is preferentially incorporated into the granules (Fig. 303).

What can be done with the information acquired by means of amniocentesis? If the cytological or biochemical diagnosis is that the child will be normal, the concern of the parents facing a high risk of producing an abnormal offspring

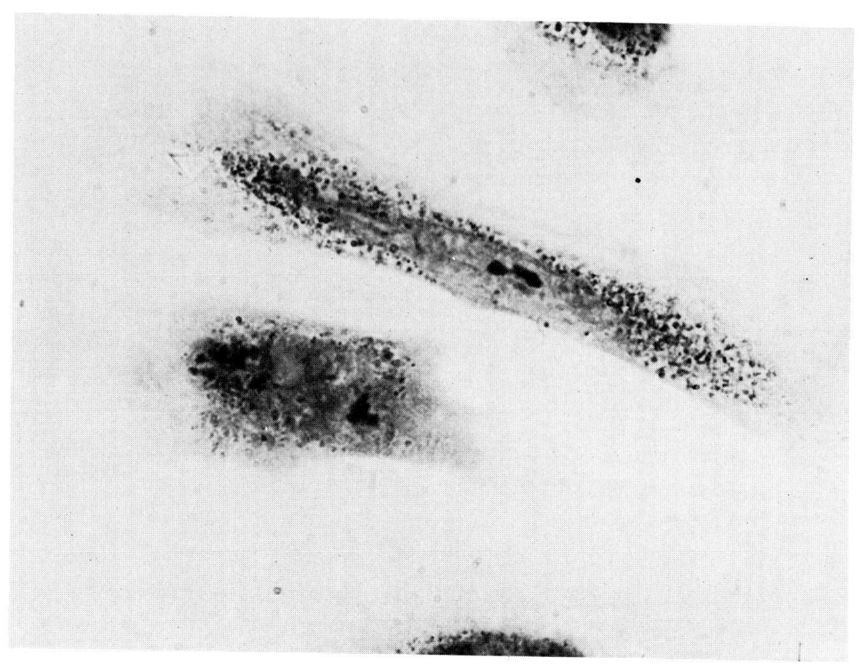

FIGURE 303.
Cultured cells from the amniotic fluid surrounding a fetus that after birth exhibited the phenotype of Hurler's syndrome. Radioactive sulfate has been incorporated into the mucopolysaccharide granules in the cytoplasm. (Original from Drs. Fratantoni, Neufeld, Uhlendorf, and Jacobson.)

can be relieved. If the diagnosis reveals an affected fetus, then a therapeutic termination of pregnancy (abortion) would be advocated by many persons, including many parents and many physicians. However, the laws concerning abortion vary in different states and countries, and there are ethical and psychological considerations which are not judged uniformly by different people. Considering the suffering of the parents of severely abnormal children, of the children themselves, their sibs and other relatives, and the load on society at large, it seems likely that in the future a more general acceptance of therapeutic pregnancy termination will evolve.

Genetic Counseling. The use of genetic knowledge in counseling has been discussed in various places in this book: as in genetic prognosis (Chap. 9), X-linkage (Chap. 14), linkage (Chap. 15), and consanguinity (Chap. 19). Well-established information on simple genetic control of traits, together with the knowledge of gene frequencies in a population, may often be adequate to provide a basis for detailed advice.

In most cases, however, it is necessary to study the specific family pedigree in order to determine the type of inheritance involved. While there are a few inherited defects whose genetics seem to be the same in all known kindreds, the heterogeneity of many genetic traits should prohibit the mechanical interpretation of pedigree data according to simplified textbook statements.

When the genetic situation is not fully understood, or hardly understood at all, counseling must become less specific, although the availability of empiric risk figures may still allow useful predictions. Thus, regardless of what hypothesis is favored on the causation of schizophrenia, the established frequencies of the disease among various types of relatives of a propositus furnish yardsticks for counseling (Table 85, p. 692). It is true, however, that empiric risk figures only report what has happened in the past and do not interpret the events. Different circumstances, genetic and nongenetic, may, in the future, greatly change such figures, particularly if whole classes of defects formerly lumped together can then be separated into individual types. Although it is important to know, for instance, the incidence of congenital malformations in relation to the age of the mother or in sibs subsequent to one or more malformed children, such information is of limited help in predicting the fate of future pregnancies in a particular family.

Genetic counseling goes beyond advising individuals. Besides its fundamental role in evaluating the significance of mutations, both spontaneous and induced, for problems of public health (Chap. 24), genetic counseling contributes information on the constitution of the populations of different areas and of mankind as a whole. We may not now have sufficient knowledge to counsel wisely on genetic problems of entire populations, but the time for such counsel is bound to come.

The fact that even individual counsel involves problems that extend beyond the immediate case may be illustrated by a fatal infantile paralysis due to a rare recessive gene (see p. 193). A heterozygous couple first had two affected children before a normal child was born. Although this child had a normal phenotype, there was a probability of only 1/3 that he would be free from the recessive allele and 2/3 that he would be heterozygous. It may be questioned whether it was advisable to produce a third child, since the probability of his carrying the defective gene was so high. Two considerations would be taken into account: the welfare of the child himself, and that of the population at large. Though the child would remain physically normal, his mental well-being in adult life would be affected by the possibility of his own offspring being paralytic. The chances for such an event, however, are negligible, since the disease is very rare and the probability of his marrying a carrier not great. As to the welfare of future generations of the population: it is true that the undesirable gene may thus be perpetuated, and at some later date, a homozygote may again be produced. Such conflicts between the happiness of one generation and the possible unhappiness of a later one are not rare in practical problems of human genetics. The decision, in such cases, is not the task of the scientist, as such, but of the individual or society. Moreover, the optimist may foresee the discovery of a remedy!

A similar decision, between short- and long-range effects, must be faced in counseling heterozygous carriers of a specific deleterious recessive whether or not to avoid marrying each other. Avoidance would eliminate the appearance of affected children in the first generation. But in time, the frequency of heterozygotes will gradually increase because new mutations from the normal to the defect-carrying allele are not balanced by elimination of mutant alleles in recessive homozygotes. Given sufficient time, the population would tend to consist of a growing majority of heterozygotes, and avoidance of heterozygous marriage partners would become difficult—if not impossible! It may be argued with merit that the time needed for the majority of the people to become heterozygous is so immense that the problems of preventing the development of the defective phenotype from homozygous genotypes may be solved long before this eventuality. The extreme formulation of the problem has been made in order to point out the kinds of effects which any choice of alternatives will entail.

The Problem of Having Affected Children. Prospective parents with the history of a defect in their family often ask the consultant geneticist whether it is genetically advisable to have children. The only objective statement that can be made is in terms of probability. Assume, for example, that the chance of an abnormal child of the type in question being born is calculated to be 1 out of 50. Whether it is deemed advisable to have children, particularly in view of the fact that the chance for a normal child is, after all, 49 out of 50, will depend on the severity of the trait in question and on the inclination of the prospective parents to take the risk for themselves and their prospective offspring.

Secrecy Concerning Genetic Defects. The strong tendency to hide knowledge of hereditary defects in one's family is a reflection of the belief that a hereditary "taint" belongs in the same category as a moral offense. In addition, it is not only a feeling of shame which induces many people to keep genetic facts secret but also the fear that the marriage prospects of their children may be decreased. There is, of course, no reason for being ashamed of genetic constitutions, which fate, and not the individual, has decreed. Perhaps the knowledge that many more individuals than is generally supposed are carriers of some unfavorable genes—indeed, that probably all people are such carriers—will gradually help to reduce the feeling of shame or guilt.

Most important, however, is the fact that the consequences of hiding the truth may be worse than their open discussion. This may be illustrated by a family, known to the author, in which Huntington's chorea occurs. Only after the father of the family developed Huntington's chorea was it revealed to him that his mother had had the disease and had succumbed to it. His adult children realized that there was a 50 per cent chance that they, too, would suffer from it. The psychological load of this knowledge is, necessarily, extremely heavy. Had the father known earlier, he could have refrained from having children, thereby averting great suffering in a later generation. In circumstances like these, secrecy itself is culpable, for it may have more serious consequences than the truth.

Aspects of Genetic Advice. The responsibility of a genetic counselor does not end when he has made as accurate a diagnosis as possible. He must, in transmitting his information to the individual concerned, take into account the psychological effect of unfavorable prospects. He must weigh the possible chance of defect against the probability that other, favorable traits may at least partly compensate for genetic misfortune. He must remind his questioner that, often, weakness itself is made a source of strength by man's ability to conquer difficulties: A man with malformed hands may become a distinguished artist; and a paralytic, a world leader. Even with the most menacing genetic diseases, such as certain types of nervous deterioration which may set in later in life, the adviser may suggest that probably years of normal activity lie ahead, and that medical science may still discover means of preventing or curing the illness. The genetic counselor, then, is in the same position as the physician who has diagnosed the prospects of a patient suffering from a serious, and perhaps at present incurable, disease. It is not sufficient to tell the truth; it is necessary to tell it humanely.

The Rising Importance of Genetics in Medicine. In recent decades the relative importance of genetics in medicine has greatly increased. The incidence of infectious and environmentally caused deficiency diseases, which once formed a very large part of medical ills, has been, and continues to be, sharply reduced in many parts of the world. In proportion, the share of genetically determined defects, susceptibilities, and aging processes takes on more and more significance. Knowledge of the role of heredity in disease not only is compatible with hope for control but is one of the foundations on which such control can be erected.

REFERENCES

In addition to the references below, consult the list of treatises on medical genetics at the end of Chapter 1.

References on genetic counseling include those listed below by Berg, Bergsma, Fuhrman, and Vogel. See also "Panel Discussion: Genetic Counseling." 1952, *Amer. J. Hum. Genet.*, **4**:332–346.

Berg, J. M. (Ed.), 1971. Genetic Counselling in Relation to Mental Retardation, 2nd ed. 44 pp. Pergamon, Oxford.

Beutler, E., Kuhl, W., Trinidat, F., Toplitz, R., and Nadler, H., 1971. β = glucosidase activity in fibroblasts from homozygotes and heterozygotes for Gaucher's disease. *Amer. J. Hum. Genet.*, **23**:62–66.

Bergsma, D. (Ed.), 1970. *Genetic Counselling.* 106 pp. William and Wilkins, Baltimore. (National Foundation–March of Dimes Birth Defects, Original Article Series, **6**: No. 3.)

Bergsma, D. (Ed.), 1971. *Symposium on Intrauterine Diagnosis.* 36 pp. William and Wilkins, Baltimore. (National Foundation—March of Dimes Birth Defects, Original Article Series, **7**: No. 5.)

Carter, C. O., (1970). Prospects in genetic counselling. Pp. 339–349 *in* Emery, A. E. H. (Ed.), *Modern Trends in Human Genetics.* Butterworth, London.

Cleaver, J. E., 1969. Xeroderma pigmentosum: a human disease in which an initial stage of DNA repair is defective. *Proc. Nat. Acad. Sci.*, **63**:428–435.

Dice, L. R., 1952. Heredity clinics: their value for public service and for research. *Amer. J. Hum. Genet.*, **4**:1–13.

Fraser, F. Clarke, 1954. Medical genetics in pediatrics. *J. Pediat.*, **44**:85–103.

Fratantoni, J. C., Neufeld, Elizabeth E., Uhlendorf, B. W., and Jacobson, C. B., 1969. Intrauterine diagnosis of the Hurler and Hunter syndromes. *New Eng. J. Med.*, **280**:686–688.

Gartler, S. M., 1953. Elimination of recessive lethals from the population when the heterozygote can be detected. *Amer. J. Hum. Genet.*, **5**:148–153.

Goodman, R. M. (Ed.), 1970. *Genetic Disorders of Man.* 1009 pp. Little, Brown, Boston.

Haldane, J. B. S., 1949. Disease and evolution. *La Ricerca Scientifica,* Suppl. **19**:3–11.

Hammons, Helen G. (Ed.), 1959. *Heredity Counseling.* 112 pp. Hoeber-Harper, New York. (This is a slightly changed version of: Heredity Counseling Symposium, *Eugen. Quart.*, **5**:3–63, 1958.)

Harris, M., 1964. *Cell Culture and Somatic Variation.* 547 pp. Holt, Rinehart, and Winston, New York.

Harris, H., 1971. The Croonian Lecture. Cell fusion and the analysis of malignancy. *Proc. Roy. Soc.*, London, *B*, **179**:1–85.

Huang, S., and Bayless, T. M., 1968. Milk and lactose intolerance in healthy Orientals. *Science*, **160**:83–84.

Krooth, R. S., 1969. Genetics of cultured somatic cells. *Medical Clinics of North America*, **53**:795–811.

LaDu, B. N., and Kalow, W. (Eds.), 1968. Pharmacogenetics. *Ann. N.Y. Acad. Sci.*, **151**:691–1001.

Lehmann, H., and Simmons, P. H., 1958. Sensitivity to suxomethonium apnoea in two brothers. *Lancet*, **2**:981–982.

Magrini, U., Fraccaro, M., Tiepolo, L., Scappaticci, Susi, Lenzi, L., and Perona, G. P., 1967. Mucopolysaccharidoses: autoradiographic study of sulphate ^{35}S uptake by cultured fibroblasts. *Ann. Hum. Genet.*, **31**:231–236.

McKusick, V. A., 1971. *Mendelian Inheritance in Man*, 3rd ed. 738 pp. Johns Hopkins Press, Baltimore.

Milunsky, A., Littlefield, J. W., Kanfer, J. N., Kolodny, E. H., Shih, Vivian E., and Atkins, L., 1970. Prenatal genetic diagnosis. *New Eng. J. Med.*, **283**: 1370–1381, 1441–1447, and 1498–1504.

Motulsky, A. G. (Ed.), 1970. *Counseling and Prognosis in Medical Genetics.* Hoeber, New York.
Motulsky, A. G., 1971. Human and medical genetics: A scientific discipline and an expanding horizon. *Amer. J. Hum. Genet.*, **23**:107–123.
Nadler, H. L., 1969. Prenatal detection of genetic defects. *J. Pediat.*, **74**:132–143.
Neel, J. V., Shaw, Margery W., and Schull, W. J. (Eds.), 1963. *Genetics and the Epidemiology of Chronic Diseases.* 395 pp. Government Printing Office, Washington, D.C. (U.S. Public Health Service Publ. No. 1163.)
O'Brien, J. S., Okada, S., Chen, Agnes, and Fillerup, Dorothy L., 1970. Tay-Sachs Disease: detection of heterozygotes and homozygotes by serum hexosamidase assay. *New Eng. J. Med.*, **283**:15–20.
O'Reilly, R. A., Aggeler, P. M., Hoag, M. S., Leong, L. S., and Kropatkin, M., 1964. Hereditary transmission of exceptional resistance to coumarin anti-coagulant drugs: The first reported kindred. *New Eng. J. Med.*, **271**:809–815.
Reed, S. C., 1964. *Parenthood and Heredity.*, 2nd ed. 278 pp. (Originally, *Counseling in Medical Genetics.*) Wiley, New York.
Roberts, J. A. Fraser, 1970. *An Introduction to Medical Genetics*, 5th ed. 296 pp. Oxford University Press, London.
Simpson, Nancy E., 1969. Heritabilities of liability to diabetes when sex and age at onset are considered. *Ann. Hum. Genet.*, **32**:283–303.
Stevenson, A. C., Davison, B. C. Clare, and Oakes, M. W., 1970. *Genetic Counselling.* 355 pp. Lippincott, Philadelphia.
Uchida, Irene A., and Soltan, H. C., 1969. Dermatoglyphics in medical genetics. Pp. 579–592 *in* Gardner, L. I. (Ed.), *Endocrine and Genetic Diseases of Childhood.* Saunders, Philadelphia.
Witkop, C. J., Jr., 1962. *Genetics and Dental Health.* 300 pp. McGraw-Hill, N.Y.
World Health Organization Expert Committee on Human Genetics, 1964. Second Report. *Human Genetics and Public Health.* 38 pp. WHO, Geneva.
World Health Organization Expert Committee on Human Genetics, 1969. Third Report. *Genetic Counselling.* 23 pp. WHO, Geneva.

31

GENETIC ASPECTS OF RACE AND RACE MIXTURE

There is no full agreement among students of the systematics of animals and plants as to what constitutes a "species." In general, however, separate groups of organisms are regarded as belonging to different species if they are more-or-less *reproductively isolated*, a term which signifies limitation or lack of interbreeding under natural conditions.

The criterion of reproductive isolation is independent of the more obvious structural differences among organisms. While many groups of structurally different organisms do not cross with each other, there are also numerous examples of reproductive isolation in spite of very small or nonobservable morphological differences, as well as of greatly different populations without reproductive isolation. Such populations are able to retain their differences because of geographical or other isolating factors.

Often, two strikingly different groups of organisms inhabiting different regions are connected by a more or less continuous series of intermediate types. Despite the range of variability, all these organisms form one single unit, differentiated into subgroups which, in different regions, replace each other.

Even members of two subgroups which are most distinct morphologically can, in many cases, interbreed freely if brought into contact. Many systematists call the large group a *species*, and the groups out of which the species is composed its *subspecies*, often called races. There are species which are not subdivided into subspecies; others in which only two or a few subspecies can be distinguished; and still others, sometimes called *polytypic species*, in which a great many subspecific types exist, each of which inhabits a different region.

A taxonomic observer of mankind, using the criteria which have just been described, would classify man as a single species subdivided into numerous subspecies. There is no doubt about the existence of morphologically different populations of mankind; these form the various races studied by the physical anthropologist. There are also many intermediate populations among the different groups, forming either a more-or-less continuous chain of connections or consisting of "hybrid swarms"—the products of interbreeding and consequent segregation of genetic traits derived from different races which were once separated and later came into contact.

As in other polytypic species, the subdivisions of mankind are not all of equivalent rank. There are major races, as the Mongoloids and the Caucasoids, for example, and minor variants, as the Mediterraneans and the Alpines, and many groups of intermediate taxonomic rank. Because all degrees of differences occur between human populations, there is no generally accepted system of classification. As far as is known, members of every human race can successfully hybridize with members of every other. These phenomena—morphological and reproductive—have led the taxonomists since Linnaeus' time, two and a half centuries ago, to assign a single species name, *Homo sapiens*, to all mankind.

GENETICS OF RACIAL DIVERSITY

It is one thing to recognize the existence of racial diversity and another to define it in terms of the underlying genetic facts. We shall attempt the latter after familiarizing the reader with specific examples of types of genetic diversity among human groups. Our discussion assumes that racial diversity has a genetic basis and is not exclusively conditioned by environment. This premise is undoubtedly true for many of the differences among groups which are studied by the physical anthropologist. Color of skin and hair; shape of hair, nose, and lips; amount of body hair; prominence of cheekbones, and large differences in stature and in many other traits are highly independent of the climatic, nutritional, or cultural environments in which human beings are reared. Whether there are mental differences of genetic nature among races will be discussed later.

Genetic analyses of the main morphological differences between human races are few and incomplete. Such differences are probably based on differences at several if not many loci. Some examples are the dark pigmentation of the Negroids and the lighter of most of the Caucasoids; the kinky hair of the former and the straight or wavy hair of the latter; the different configurations of lips, noses, eyelids, and other facial features among Mongoloids, Negroids, and Caucasoids; and the differences in body build of African Pygmies, Hottentots, and Polynesians.

Results of crosses between members of the different races consist mostly of data on first-generation hybrids. The phenotypes of such hybrids are mixtures of dominant and intermediate expression of individual traits. Thus, the hair type of first-generation mulattoes is similar to that of the black parents, while the pigmentation is intermediate between black and white. In general it is not known whether the distinguishing morphological traits of different races are due to differences at single loci or whether they have polygenic bases. The latter alternative is more likely to be true. Thus, while the kinky hair of the black appears in first-generation white-black hybrids, in later generations a whole array of different hair types is found. First-generation children from marriages between American Indians and blacks frequently show the straight hair of the former, but, again, various grades and types of hair occur in later generations. The polygenic determination of pigmentation differences between whites and blacks has been discussed earlier (Chap. 18). The discussion of the inheritance of quantitative characters given there is relevant to the majority of the main racial characters.

The genetic interpretation of phenotypic differences among races must take into account the possibility that different genotypes have resulted in similar phenotypes. Thus, the absence of crosses between such groups as African blacks and the small groups of Negritos in Southern India, Malaya, Java, and the Philippines leaves an open question of whether their similar characteristics of skin color and hair type have the same genetic basis or are the phenotypically similar products of different genotypes. The latter possibility is thought by Gates to be true for the dark skin color of Australian aborigines and that of Africans. Several loci are involved in the pigmentation difference between Africans and Caucasoids but perhaps only one major locus is responsible for the difference between the latter and Australians.

Antigenic Differences. Anthropologists in recent decades have studied racial differences in genetically well-defined characters, particularly blood groups and types of hemoglobin. Most human populations have the same array of alleles at the different loci, even if in different proportions. Table 113 shows that the blood-group alleles I^{A1}, L^{Ms}, L^{Ns}, R^0, R^1, R^2, and probably P and Fy^a are found in all six human races listed. Several other alleles, such as r', r'', R^z,

TABLE 113
Frequencies, in Per Cent, of Various Blood-group Alleles in Six Racial Groups Belonging to Three Major Races.

Genetic Loci	*Caucasoids*		*Negroids*	*Mongoloids*		*Australoid*
	Basques	Others		Asian	Amer. Indian	
I^{A1}	20	20–30	10–20	15–25	0–55	20–45
I^{A2}	3	4–8	5	0	0	0
I^{B}	0–3	5–20+	10–20	15–30	0	0
L^{MS} }	55	20–30	7–20	4	15–30	0
L^{Ms} }		30	30–50	56	50–70	26
L^{NS} }	45	5–10	2–12	1	2–6	0
L^{Ns} }		30–40	30–50	38	5–20	74
r	48–53	30–40	10–20	0–7	0	0
r'	1–3	0–2	0–6	0	0–17	13
r''	1	0–2	0–1	0–3	0–3	0
R^0	1	1–5	40–70	0–5	0–30	9
R^1	38–42	30–50	5–15	60–76	30–45	56
R^2	5–7	10–15	6–20	20–30	30–60	20
R^z	0	0–1	0	0–0.5	1–6	2
P	?	40–60	50–80	15–20	20–60	?
Lu^a	?	2–5	0–4	?	0–10	0
Fy^a	?	40	< 10	90	0–90	?
Di^a	?	0 +	0	1–12	0–25	0

Source: After Boyd; Mourant; and Wiener and Wexler.
Note: The frequencies given are sometimes based on a single or a few samples. The ranges are based on from two to many samples. Altogether, the data are somewhat selected and intended to show general trends only. For details, consult the original sources.

and Lu^a, which are rare wherever they occur, may well be present even in those races for which they have not yet been recorded.

There are some alleles which are common in some races and apparently absent in others: I^{A2}, for instance, is restricted to Basques, other Caucasoids, and Negroids; I^B appears in these groups as well as in Asian Mongoloids but not in American Indians and Australian aborigines (Australoids); and r has a high frequency in Basques, other Caucasoids, and Negroids and a low frequency in Asian Mongoloids but is lacking in American Indians and Australoids.

The major races have long been differentiated from one another on the basis of their external physical characters. It is significant that the study of blood groups leads to a similar classification. Wiener, Boyd, and Mourant have pointed out that most people can be assigned to one of six "serologic races," which coincide fairly well with the differentiation according to external physical characters. Caucasoids are characterized by possession of I^{A2} as well as I^{A1}, high frequency of r, and total or near absence of Di^{a}; Negroids, by possession of I^{A2} as well as I^{A1} and high frequency of R^{0}; Asian Mongoloids, by absence of I^{A2}, low frequency of L^{MS} and L^{NS}, and high frequency of R^{1} and Fy^{a}; American Indians, by absence of I^{A2}, I^{B}, and r and low frequency of L^{NS}; and Australoids, by absence of I^{A2}, I^{B}, L^{MS}, L^{NS}, and r and high frequency of L^{Ns}. The Basques of Spain and southern France belong physically to the Caucasoid race, though they are to some degree culturally and linguistically separate. It is believed that they are the remnant of an early European population which has been largely replaced by later arrivals. Serologically the Basques are distinguished by very low frequency of I^{B} and high frequency of r.

Differences in Hemoglobin Types. There are some striking associations between racial groups and incidences of alleles for certain types of hemoglobin. One of the most significant is the high frequency of the allele for hemoglobin C in Equatorial West Africa and its virtual absence elsewhere. The allele for sickle-cell hemoglobin is found in wider groups of African Negroids as well as in various Mediterranean people and a few Asian groups, and the gene for thalassemia is restricted to Mediterraneans and dispersed populations of Asia. The very rare presence of these alleles in northern and western Europeans and their descendants in the United States is perhaps the result of past intermarriages with Mediterraneans. We will return to the distribution of hemoglobin types in the final chapter of this book.

Other Differences in Allele Frequencies. The genetic polymorphism of blood-antigen types is certainly representative of a great number of other traits. The genetic basis of some of these polymorphic characters is more-or-less well known and there are often extensive data on racial variations. Among these are type of singing voice, "secretor" property, pattern of finger ridges, ability to taste phenylthiocarbamide (PTC), and haptoglobin types.

The proportions of people with basso or soprano voices decrease from northern to southern Europe. Under the assumption that the single-locus two-alleles theory of the inheritance of singing voices in Europeans was correct (see also p. 404), the frequency for the basso-soprano allele in the geographical range studied was calculated. It was concluded to have a maximum of 61 per cent along the northwest coast of Germany and a minimum of about 12 per cent in Sicily. Although this specific interpretation of the polymorphism of singing voice had to be abandoned, there is undoubtedly a genetic basis for it, and differences in allele frequencies must account for the different frequencies of types of voice in different populations.

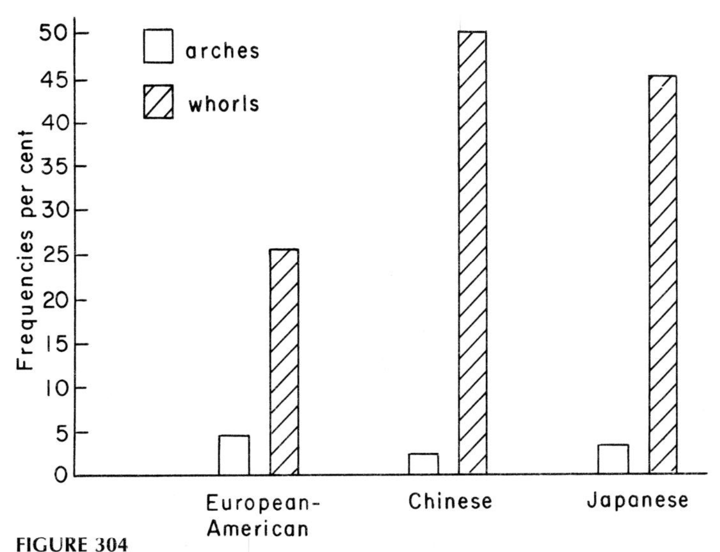

FIGURE 304
Frequencies of two types of dermal finger ridges, arches and whorls, in three racial groups. (after Holt, based on data of several authors.)

The frequency of the gene *Se* (for secretion of the ABH substances into body fluids) is approximately 50 per cent in Caucasoids, only 38 per cent in American blacks and probably less in African Negroids, and rises to nearly 100 per cent in American Indians.

Some racial variations in dermal ridges are illustrated in Figure 304, which gives the frequencies of arches and whorls on the fingers in three populations. The Caucasoid sample had higher frequencies of arches than the two Mongoloid samples, while the latter have strikingly higher frequencies of whorls than the former. The polygenic bases of these traits exclude determination of allele frequencies.

The frequency of tasters for PTC varies from as low as 51 per cent among Australian aborigines to 98 per cent among some American Indians (Table 114). These phenotype frequencies correspond to a range of allele frequencies for the nontaster allele from about 0.7 to 0.14. It is, however, not fully justified to assume that the differences in frequencies of tasters are the result of differences in the frequency of the *t* allele only. The somewhat variable expression of the trait among Caucasians and its sex modification in many racial groups suggest that when the frequency of modifying genes is studied, the interrracial variability may be found to depend, at least in part, on genes at other loci. It may be added that polymorphism for PTC-tasting ability is met in other primates as well as man. Out of 28 chimpanzees in British zoos, 20 showed by their unambiguous reactions that they were tasters, while 7 were

TABLE 114
Racial Variations in Ability to Taste Phenylthiocarbamide.

Racial Group	*No. Tested*	*Tasters (%)*	*Frequency of t Allele*
Australoids	152	51	0.70
American whites	> 6,000	65–75	0.50–0.59
Egyptians	208	76	0.49
American blacks	> 3,000	91	0.30
African Negroes	> 1,000	91–97	0.17–0.30
Chinese	> 200	89–94	0.24–0.32
American Indians	> 1,000	90–98	0.14–0.32

Source: After Valls, 1958.

obviously nontasters (one animal "was too shy to be tested"). Among the 3 orangutans available, 2 were tasters and 1 was a nontaster. Other species were represented by one or two specimens only, some being tasters and some not.

A final example of polymorphism in different races concerns the frequency of the Hp^1 haptoglobin allele. A small selection of the numerous determinations is listed in Table 115. In general Asian populations have the lowest Hp^1 frequencies, Europeans come next, and American Indians and African populations have the highest frequencies. In detail there are many exceptions to this summary statement.

Two important generalizations have resulted from the studies on polymorphisms as related to racial distribution:

1. There are striking differences in the frequencies of various alleles in different races and racial variants.
2. These differences are usually merely relative, "more or less," and not absolute, "all or nothing."

According to Sarah Holt, Galton's conclusion (1892) concerning racial differences in dermatoglyphic features is still true: "The only differences so far observed are statistical, and can not be determined except through patience and caution, and by discussing large groups." This conclusion also applies to many other polymorphisms. Nevertheless, there are some examples of populations possessing solely one specific allele (e.g., I^O) at a locus for which other populations are polymorphic (e.g., I^O, I^A); as well as examples of populations being totally devoid of an allele (e.g., L^{MS}) that other populations possess at some frequency less than 100 per cent. There are no examples of two races being homozygous for different alleles at the same locus, e.g., there are no examples of one race being isogenic for I^O and another isogenic for I^A.

The finding of so many relative differences in the frequencies of alleles suggests that even apparently absolute differences in traits between two races may

TABLE 115
Frequencies of the Haptoglobin Allele Hp^1 in Various Populations.

Population	*Frequency*	*Population*	*Frequency*
EUROPE		NORTH AMERICA	
Norwegians	0.36	Navahos	0.83
Germans	0.38–0.40	Lacandons	0.92
North American whites	0.38		
		SOUTH AMERICA	
AFRICA		Venezuelans (10 tribes)	0.21–0.90
Liberians	0.70		
Zulus	0.53	Peruvians (9 tribes)	0.44–0.73
Egyptians	0.21		
North American blacks	0.55	AUSTRALIA	
		North Queenslanders	0.17
ASIA			
Chinese (Hong Kong)	0.39	PACIFIC ISLANDS	
Japanese	0.23–0.27	New Guinea natives (5 tribes)	0.66–0.77
Pakistanis	0.21		
Indians (Tamils)	0.09	New Britain natives	0.73

Source: Selected from Giblett, 1969.

be dependent on alleles which occur in both of them. If the differences in skin color between Negroids and Caucasoids are based on a polygenic system involving several loci, then it would be possible to account for the dark pigmentation of the African by the presence of alleles for darkness with frequencies of, say, 99 per cent and presence of alleles for lightness of 1 per cent. Inversely, the Caucasoids may possess the alleles for darkness in frequencies of 1 per cent and those for lightness of 99 per cent. With such large differences in allele frequencies, practically all members of one group would be dark and practically all of the other group light.

A GENETIC DEFINITION OF RACE

A genetic definition of race must take into account the fact that all populations consist of individuals who are heterozygous for many loci and many alleles. There is no "pure" race—a designation which, in genetic terms, would signify homozygosity and isogeneity of all individuals.

The study of genetic polymorphisms has shown that races often differ only in the relative frequencies of alleles, and this phenomenon may be true for most loci. A race, then, is a group of individuals whose corporate genic content, called the *gene pool*, differs from that of other groups. The members of a race retain the differences, more or less, over the course of generations because geographic or cultural isolation results in only a small amount of genic exchange between them and members of other races. Considerations such as these may lead to the following definition:

> "**Race.** A geographically or culturally more-or-less isolated division of mankind whose corporate genic content (gene pool) differs from that of all other similar isolates."

Does this definition coincide with the everyday concept of race? The definition includes this concept but goes beyond it. It fits the anthropological characterization of the major groups of mankind as well as of the minor racial types, since they are all endowed with different genic contents and are more-or-less isolated from one another by geographic or social barriers. In addition, the genetic definition considers as "races" different groups of individuals who are more-or-less isolated but whose genic contents are so slightly different from one another that our language has no word to characterize these groups as separate entities. For example, during World War II, it was found in northern Wales that the ABO frequencies of blood donors with Welsh family names differed from those of donors with non-Welsh names (Table 116). There were more O and B individuals and fewer A and AB among the Joneses, Williamses, Robertses, and other people with Welsh surnames than among the people with English names. The differences in allele frequencies for both sexes are significant. Obviously, then, the genic content of the group of Joneses, Williamses, etc., is distinct from that of the rest of the population among whom they live, a difference which can only be due to some genetic isolation between the two groups. Yet, we would hardly call the Joneses, Williamses, etc., a separate race.

The difficulty is inherent in the facts of nature. There exists a continuous series of degrees of difference in the genic content of isolates, from the slight differences between the Welsh isolate and the population of which it is a part to the obvious differences among the major races of the anthropologists. The biological phenomenon is the same, irrespective of the size of the difference. The basic genetic similarity of all differences among isolates results in somewhat arbitrary decisions about when to apply the term race and when to regard

TABLE 116
Percentage Frequencies of ABO Blood Groups among Donors with Welsh and Non-Welsh Family Names.

Men and Single Women Donors	*No. of Individuals*	O	A	B	AB
Welsh family names	909	52.7	35.0	9.7	2.6
Non-Welsh family names	1,091	46.6	42.0	8.3	3.2

Source: Roberts, *Ann. Eugen.*, **11**, 1942.

its use as inappropriate. A problem of this kind is frequently met. Natural bodies of flowing water of different width and depth are called brooks, creeks, streams, and rivers. No sharp definition can be drawn up to separate these different terms from one another, although there is no doubt about which term to apply to the Mississippi or, conversely, to a very small and shallow current.

There is not only a wide range of genic diversity between different groups but also many degrees of isolation. Absolute isolation of different human groups hardly exists, since contacts between such groups have always resulted in some interbreeding. The degree of isolation, itself, is variable not only from group to group but also with time. In historical periods of mass migration, barriers to interchange of genes have always been decreased. Technical developments in modern transportation, which facilitate and increase contact between formerly geographically isolated groups, have led to equivalent results. Wars, too, with their shifting of military forces into foreign areas, have contributed to the breakup of genetic isolation, either through legally sanctioned marriages or from illegitimate unions.

Frequencies of Specific Genotypes in Different Races. Let us assume that two races of equal numbers possess n loci, A^1, A^2; B^1, B^2; C^1, C^2; ... N^1, N^2, each of which occurs in two allelic forms, with the frequencies p_{A1}, q_{A2}; p_{B1}, q_{B2}; q_{C2}; ... p_{N1}, q_{N2} in race I and the frequencies p'_{A1}, q'_{A2}; p'_{B1}, q'_{B2}; p'_{C1}, q'_{C2}; ... p'_{N1}, q'_{N2} in race II. In both races, there are homozygotes for both of the alleles at any one locus and also heterozygotes for each locus. Consequently, an N^1N^1 individual taken from race I cannot be assigned to his race on the basis of his constitution at the one N locus. However, a priori, the probabilities of his belonging to race I or II may be different, the proportion of the expected frequencies of N^1N^1 individuals in the two races being $(p_{N1})^2$ and $(p'_{N1})^2$, respectively.

If we assume that the allele frequencies p and q for every locus are 1/10 and 9/10 in race I, and that the allele frequencies p′ and q′ for every locus are 9/10 and 1/10 in race II, the probability that an N^1N^1 individual will occur in race I becomes $(p_{N1})^2 = 1/100$, and that he will appear in race II $(p'_{N1})^2 = 81/100$. Thus, the probabilities that an N^1N^1 individual will be a member of race I or II differ from each other in the proportion 1:81, so that the probability that an N^1N^1 individual belongs to race I is 1/82 and to race II is 81/82. If an individual's genotype were known to be $A^1A^1N^1N^1$, his chance of occurring in race I would be only $(1/100)^2 = 1/10{,}000$ as opposed to the chance $(81/100)^2 = 6{,}561/10{,}000$ of his occurring in race II. Therefore, the probability of an $A^1A^1N^1N^1$ belonging to race I and not to race II is 1 out of 6,562. Clearly, the more loci known, the more reliable may be the assignment of an individual to his race. However, not only is there a small uncertainty, even in favorable cases, but there are genotypes which have rather similar probabilities or even an equal probability of belonging to either race. Among these genotypes are the complete heterozygotes. The probability of a person being heterozygous for n loci is $(2pq)^n$ in race I, and $(2p'q')^n$ in race II. In our example where pq $(1/10 \times 9/10)$ is equal to $p'q'(9/10 \times 1/10)$, there would be an even chance that an A^1A^2, B^1B^2, ... N^1N^2 individual would belong to either race.

In general, gene frequencies in two races will not be as "symmetrical" (1/10:9/10 *vs.* 9/10:1/10) as in our two assumed populations. Consequently, it will usually be possible to discriminate with a high probability between the two alternatives. It must be emphasized, however, that these calculations apply only to races that are absolutely isolated genetically and not to races between which there is some degree of hybridization. Particularly between the minor anthropological subdivisions, intermarriages are frequent, or have been so in the past. For example, in the course of many centuries, the Nordic, Alpine, and Mediterranean racial variants of Europe have interchanged genes freely. Consequently, allele frequencies for different loci are probably rather similar in these racial subtypes. The result is that it is often difficult to place a given individual in one or another of these groups. Indeed, such an attempt may become meaningless, since segregants from the same parental pair may show phenotypes characteristic of different minor racial divisions.

In view of the imperfections of any definition of the term race and particularly in view of the fateful misuses of the term, some anthropologists have been inclined to strike the word race from modern vocabulary, using exclusively words like "population" or "ethnic group." Since, however, the word race will probably remain in our language, it has been retained in this book and is used, without value judgement, in its scientific sense.

RACE MIXTURE

Whenever history brought together two or more races in the same territory, unions of persons of different races occurred. This interbreeding, often called *miscegenation* (from the Latin *miscere* = to mix, *genus* = race), proceeded sometimes at a slow pace and at other times rapidly. Often, complete "amalgamation" or "assimilation" of the different groups was the end result. Before the particulate nature of the hereditary material was understood, it was thought that such joining of the germ plasms of the races would result in a new, homogeneous population. The recognition of the existence of separate genes has, of course, changed this expectation. Recombination of the different alleles brought into the gene pool of a mixed population results in the production of numerous diverse genotypes in proportions that are predictable if the allele frequencies are known.

Segregation of allelic differences brought into a mixed population should increase its variability, as compared with the variability within each parent race. However, a comparison of the variability of the mixed population with that *between* the parent races shows a decrease in variability, since statistical measurements of variability are less dependent upon the rare extreme variants of a group than upon the distribution of the majority around the mean. Expressed differently, the *intra*racial variability within the mixed group is less than the *inter*racial variability before mixture. The specific application of these considerations to human hybridization on a population scale depends on the number of loci, on the phenotypic expression of polygenic genotypes, and on the relative sizes of the two parent races involved in any specific crossing.

In a population derived from white-black mixtures, a trait like skin color shows a greater variability than is found in the parental races. Other traits may be less variable. In fact, it has often been found that many traits do not vary appreciably more in the mixed population than within the parental groups. There are several ways in which this low variability may be interpreted. Perhaps the most probable explanation, advanced by H. J. Muller, is that if the intraracial variability of a character in each of the two parent races is due to numerous recessives of individually rather low allele frequencies, and if these genes are at different loci in the two races, then the variability of the mixed race would be lower than that of either of the parent races. This is a consequence of the relative lowering of each allele frequency due to "dilution" (by intermarriage) of the concentration of alleles present in one, but not in the other, race. This dilution results in a lower frequency of homozygous recessives in the new population and, thus, in a lower number of phenotypes which vary from the mean. Opposite in effect to this lowering of the variability is the segregation of interracial differences. It is not to be expected that the two factors affecting variability, dilution and segregation, will always compensate for each other completely. Hence, a mixed race may show increase of variability in some traits, decrease in others, and no, or little, change in still others. That tha changes in variability observed in mixed populations of black-white, Tahitian-white, or Hottentot-white intermarriages, as well as those observed by Morton, Chung, and Mi in the wide array of interracial crosses in Hawaii, have been small may be taken as an indication that the differences between individuals within the parental groups are of magnitude comparable with the differences between the groups. Beyond this deduction, data on allele frequencies for protein loci in whites, blacks, and Japanese populations indicate, according to Nei and Roychoudhury, that the differences between these groups are rather small compared with those within the same population.

White-Black Hybridization in the United States

One of the most important examples of hybridization, both because of the number of individuals involved and because of its contemporary historical significance, is that between the white and black populations of the United

States. Neither group was racially uniform, in the sense that neither consisted of a single random-mating isolate, but the whites stemmed mostly from western and central Europe and thus, probably, were less differentiated into genetically different groups than the blacks, who came from widely separated parts of Africa. The slight initial genic isolation between the white subgroups has had a strong tendency to disappear, although geographic and religious factors, as well as new immigration, have retarded their complete amalgamation. The different black subgroups, likewise, underwent a process, by intermarriages, of gradual transformation into a single black race.

Superimposed on these two separate processes of amalgamation within each of the two major races, hybridization has occurred and continues to occur. Whites and American Indians have been involved in this hybridization, but the discussion will be restricted to blacks and whites, since the over-all contribution of the American Indians to the gene pool of the other groups has been very small. This is proven by comparisons of the frequencies of blood-group alleles among the three racial groups.

The Colonial period and the succeeding decades before the emancipation of the black slaves was the most significant period of hybridization. Hybridization, primarily between white men and black women for more than a hundred years, had led to the existence of a very large number of first-generation hybrid mulattoes. Although half their genes were derived from whites and half from blacks, socially the mulattoes were classified together with blacks, as "colored" people. Accordingly, they married either with each other or with blacks who had no white ancestors. The result of this system of mating at first was an infiltration of "white" genes into the black population without an appreciable reciprocal gene flow from blacks to whites. Lately, as a result of the segregation of alleles concerned with the more obvious characters which differentiate whites from blacks, black parents have produced children whose skin pigmentation, hair type, and facial features are similar enough to those of whites for them to "pass" for whites. The segregants undoubtedly carry various combinations of less obviously recognizable genes of African derivation than those responsible for the traits mentioned above. They also contain heterozygously recessive alleles concerned with some of the more obvious differences between whites and blacks, as well as alleles which, in specific combinations with alleles at other loci, determine genotypes characteristic of blacks. Consequently, individuals who have "crossed the color line" and have "white" children represent a channel through which various genes derived from blacks flow into the white population. Thus, the initially one-sided gene flow, from the white to the black population, has developed into a mutual exchange of genes.

The frequency of white or near-white segregants is probably higher than would be expected from random mating within the black population. There is a tendency, perhaps declining recently, for positive assortative mating among black individuals; for example, lightly pigmented persons preferentially marry light ones. This mating preference favors the reconstitution of homozygosity for pigmentation alleles derived from the Caucasoid race; that is, it favors the production of individuals who function in the transfer of genes from the black to the white population.

The process of extraction of alleles for light pigmentation from the black population and the return, by means of "crossing the color line," of these light alleles to the white population from which they were derived is bound to continue—with the result that the black population will be drained of those alleles or combinations of alleles which make for the more striking anthropological white phenotypes, particularly light pigmentation. It may be expected, therefore, that the American black population, whose mean coloration is lighter than that of the original Africans, will gradually darken again due to selective "back-migration" of white genes.

This process will be counteracted by continued hybridization between whites and blacks. The time span of a human generation is not negligible if it is measured against historical periods of one or two centuries. For whites and blacks in the United States, this time span certainly has been much too short to result in a complete breakup of genic isolation. Possibly, in the future, hybridization will increase, or continue long enough, so that a single random-mating population will finally be produced. It is more likely that selective mating of whites with whites and blacks with blacks will uphold the relative genic isolation between the two groups for a long time, but that an inconspicuous gene flow in both directions, as described, will continue.

What will the final consequences of these processes be, in terms of the over-all genic differences which existed between the original whites and blacks? There will be a tendency toward equalizing the frequencies of any one allele in the white and black groups, except for those genes which are concerned with the most obvious racial characters. This equalization of numerous allele frequencies will proceed fastest for loci which are not chromosomally linked to those genes for which racially assortative mating exists, as for genes concerned with pigmentation. Even for loci linked to them, crossing over will tend to establish an equilibrium in which the original racial linkage combinations will occur no more frequently than exchange combinations. The result will be that the white and black populations will become similar for most allele frequencies for which they were different before hybridization and will remain different only in those probably fewer loci which contribute in an appreciable degree to superficial diversity of Caucasoid and Negroid individuals.

The gradual breakup of isolation for most genes has an important bearing on the genetic evaluation of segregants in future generations. Apart from genes causing obvious phenotypic differences, the least African-like members of

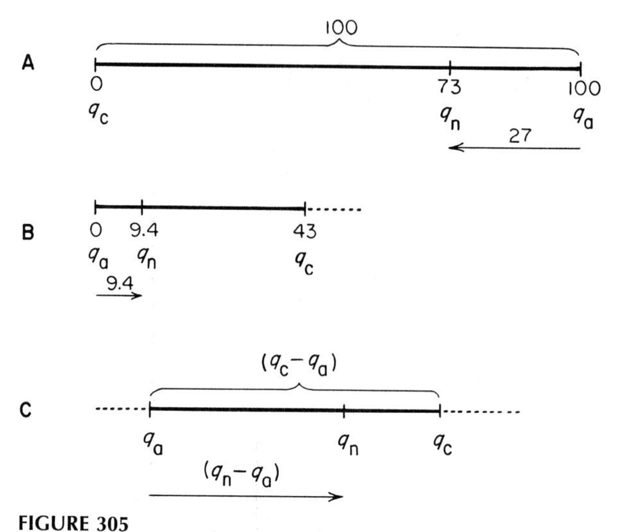

FIGURE 305
Determination of frequencies of specific alleles brought into a population by hybridization with another population. See text.

the white group will belong to the same array of genotypes and with equal probability distribution as the most African-like members of the black group. Even at the present time, the skin pigmentation of white-black segregants is but a poor indicator of their total genotypes.

Allele Frequencies in Hybrid Populations

Populations that originated from hybridization of two racially different populations may yield information on the degree of hybridity. If one of the original races exclusively possessed an allele a^1 and the other exclusively a^2, and if equal numbers of persons from the two races had produced a hybrid population, then the ratio of a^1 to a^2 alleles would be 1:1. Conversely, if a 1:1 ratio of a^1 and a^2 were found in the hybrid population, it could be concluded that equal numbers of genes had been derived from each of the original populations. This reasoning can be extended to cover less restricted situations. It has served in various studies to determine the proportion of "Caucasian genes in American Negroes" (T. E. Reed).

In Figure 305, A three frequencies of the allele $Gm^{1,5}$ which is responsible for certain antigenic immunoglobulin determinants of the blood have been entered on a horizontal line. Among Caucasians this frequency in per cent is $q_c = 0$, among West Africans it is $q_a = 100$, and in a large sample of the black population of Oakland, California, it is $q_n = 73$. Hybridization then has resulted in a "dilution" of the West African allele frequencies of $Gm^{1,5}$ from q_a to q_n by the Caucasian allele. The fraction of admixture, M, is 27/100 = 0.27; in other words, 27 per cent of the allelic content of the hybrid population is derived from Caucasians.

A similar treatment of the frequencies of the Duffy allele Fy^a is as follows (Fig. 305, B): Among Caucasians the frequency is $q_c = 43$, among West Africans $q_a = 0$, and in the hybrid population $q_n = 9.4$. Hybridization has resulted in an intrusion of Fy^a alleles into the West African stem race. The amount of this intrusion is M = 9.4/43 = 0.22; thus, 22 per cent of the allelic content of the hybrid population is derived from Caucasians.

Generalizing for any values of q_c and q_a, and any intermediate value for q_n, it is seen (Fig. 305, C) that the amount of admixture is

$$M = \frac{q_n - q_a}{q_c - q_a}.$$

This formula, which can be applied to mixtures of fluids of different concentration, and to similar problems, was first used in genetic considerations of racial hybridizations by Bernstein.

Provided that the allele frequencies in a hybrid population are solely the result of racial mixture and not also influenced by selection, mutation, or other factors, the value of M should be the same for any allele at any locus for which the ancestral races had different allele frequencies. The two M values derived above, 0.27 and 0.22, differ from each other within the probability of chance only and thus confirm the postulate stated in the preceding sentence. A third M value for the black population has been derived from data on I^A and I^B alleles. Here M = 0.20, again compatible with the postulate (although Workman feels that selective forces do play a role in the frequencies of *I* alleles). A fourth M value derived for the R^0 allele was first estimated by Glass and Li as 0.31 but later, on the basis of better data on the West African allele frequency, was revised to either 0.28 or 0.22 depending on details of the Rh gene system. Estimates of M greater than 0.40 have also been made using the haptoglobin allele Hp^1, the sickle cell hemoglobin allele Hb^S, and others. Here selection for certain alleles is indicated.

The blacks in the United States are not homogenous for their Caucasian genes. Southern blacks, specifically those from isolated areas in Georgia and South Carolina, have a smaller amount of white admixture than non-Southern blacks. Values of M as low as 0.04 and 0.10 have been encountered. Brazilian blacks deviate in the opposite direction. On the basis of the frequencies of the *t* allele for nontasting, Saldanha has estimated African genes in the Brazilian blacks to make up about 50 per cent of all genes (M = 0.5). The other 50 per cent of alleles comes, to varying degrees, from Caucasians and American

Indians. Puerto Ricans seem to be mixtures of about three-quarters Caucasian alleles and one-quarter African, plus a small third component of American Indian genes.

It is to be expected that American Caucasians possess African genes acquired by direct race mixture and by the process of blacks "passing" into the Caucasian population. Numerically this African fraction among Caucasian alleles is small. It has been estimated, with 95 per cent probability, that it is less than 1 per cent.

THE CONSEQUENCES OF HYBRIDIZATION

Many persons regard hybridization as undesirable. So far as it focuses attention on sociological problems arising from unsolved difficulties in the attitudes of races toward one another, the question of the undesirability of hybridization does not fall within the province of the biologist. Nor is it the geneticist's task to evaluate the historical consequences of a gradual disappearance of the diversity of cultures as a possible result of extensive racial intermingling. Nor are the psychological conflicts which may confront an individual whose parents belong to two racial groups with widely different cultures of primary genetic concern. The opinion is often expressed, however, that there are biological reasons for the undesirability of racial hybridization. The most important arguments for this opinion center on views concerning (1) the breakup of well-adapted racial genotypes; (2) the origin, in the first or later generations, of disharmonious gene combinations; and (3) the superiority of certain races over others. We shall take up these arguments in the order given and further divide our discussion by considering physical characters first and mental ones later.

Physical Traits

Racial Adaptations. It is indeed possible that some of the major racial physical characters are the result of selective forces which acted against certain genotypes in one isolate and favored their appearance in others. Such adaptations may be in response to environmental (ecological) factors. With pigmentation, which has often been cited in this connection, two different and opposing selective forces are known: an advantage in geographic regions of strong sunlight, where dark pigmentation provides a shield against sunburn as well as against excessive production of vitamin D, which can be toxic, by natural irradiation in the skin from dihydrocholesterol; and a disadvantage in regions of low sunlight, where it may prevent production of enough vitamin D for good health. The darker pigmentation of the inhabitants of tropical regions thus may be regarded as an adaptive protection against intense sunlight, and the lighter pigmentation of northern people as an adaptation which permits enough of the less intense light rays to penetrate the skin for manufacture of vitamin D. That the establishment of such adaptions may require many thousands of years is suggested by the fact that American Indians living in tropical and subtropical regions have not yet acquired the degree of pigmentation found in African and some Asian populations. Other traits also have been interpreted as ecologically adaptive. The long, narrow nose of Caucasoids is thought by some anthropologists to allow for warming up cold outside air before it enters the lungs, while the shorter and broader nose of Africans seems more suited to their evenly warmer surroundings. The lean physique, with its long arms and legs, of inhabitants of deserts, whether in North or South Africa or in Australia, seems to be an adaptation to the heat of these regions, since the relatively large amount of body surface provides for more cooling evaporation than compact bodies.

It is difficult to prove the correctness of such interpretations, but it seems likely that some, even if not all, racial characters represent adaptations to the specific environments in which the major races evolved. It is questionable, however, whether the value of these adaptations is still as great as it may have been in the past. Technological developments, particularly those of the last 200 years, have so greatly changed man's environment and his ability to cope with it (and greater changes are bound to come) that adaptations to former environments are becoming largely obsolete; i.e., they lose their positive selective value. If this is so, then the breakup of formerly adaptive racial genotypes is of little importance.

Regardless of whether many racial characteristics are of ecological adaptive significance, the proper working of any human body depends on the harmonious adjustment of its different parts and functions. We may speak of this adjustment as *internal adaptiveness*. The blood-pumping function of the heart must be fitted to body size; the size and activity of the different glands of internal secretion must be delicately related to one another; the proportions of limbs to trunk and of various bones to one another must be fitted within the limits of normality. It may be wondered whether different human races contain different gene combinations which provide, within each race, the necessary internal harmony. If this were so, the breakup of former racial isolates would justifiably be a matter of concern. It appears, however, that internal adaptiveness should not be conceived as the ability of an organism to fit together, in a harmonious way, separately determined parts. Rather, the sizes and the degrees of function of organs are genetically provided with a wide range of possible expressions, and the specific expressivity that will result is dependent on developmental interrelations. The marvelous ability of an organism to regulate—that is, to respond in an adaptive way to a great variety of conditions—is responsible to a large degree for internal adaptiveness. If one kidney is removed, the other kidney compensates for the loss by increasing its activity; or, if a bone is broken and heals in an abnormal fashion, it rebuilds internal structure in a new manner best fitted to cope with the different mechanical stresses imposed on it. It may be assumed, similarly, that, in general, a developing human being will form during his embryogeny and later an internally adaptively balanced system, regardless of the gene combinations he inherits from his ancestral race or races.

Disharmonious Gene Combinations. The regulatory abilities, or developmental homeostasis, of organisms makes the occurrence of disharmonious phenotypes rare. It is, however, conceivable that different parts of the body may sometimes be genetically determined in a sufficiently independent manner so that actual incongruities may arise. Such disharmonies occur occasionally when different species of animals or plants are crossed with each other. Disharmonies have also been described within a single species, namely, in crosses of widely different breeds of dogs. Thus, some hybrids between a short-legged, slender-bodied dachshund and a long-legged, heavy St. Bernard have the short legs of the first and the large body of the second breed, so that the body drags on the ground.

As instructive as this example is in showing the limits of regulatory development, it is hardly comparable to human race crosses. The dachshund is a disproportionately dwarfed animal, corresponding to achondroplastic human dwarfs and not to naturally occurring human races. Differences between human races seem to be dependent, not on a few genes which independently determine striking properties of parts, but rather on polygenic combinations of which each single gene affects to a small degree one or more characters, so that the various allelic combinations of the system are able to direct development toward a reasonably harmonious system. This seems to be the explanation for the fact that no well-substantiated examples of disharmonious constitution resulting from human hybridization have been reported.

In this connection, a very special case may be recalled: the Rh incompatibility between pregnant Rh-negative mothers and their Rh-positive fetuses. If two isolated human races existed, one isogenic for the allele R, the other for r, the pathological phenomenon of Rh-determined erythroblastosis fetalis would not be known in either race. If the two races intermarried, many disharmonious mother-child combinations would appear, resulting in a disease which might then be called typical for hybridization. (We neglect here the medical prevention of the disease, which is now possible. See p. 438). There are races—the Mongoloids, for example—which have a frequency for the allele *r* of nearly 0.0. No known race has an *r* allele frequency of 1.0, but intermediate frequencies occur. These differing allele frequencies account for the near absence of Rh erythroblastosis in Mongoloids and its presence in Caucasoids.

Let us imagine the immigration of whites into China and of Chinese into a country inhabited by whites, and the hybridization of the immigrants with the native race. In China, no erythroblastosis fetalis will occur in the first generation of intermarriages between Chinese women and white men, since all Chinese women are Rh positive and not subject to Rh iso-immunization. However, in the marriages between white immigrant women and Chinese men, the disease will affect some of the children, since about 14 per cent of white women are Rh negative and can be iso-immunized by fetuses, all of whom have inherited the *R* allele from their fathers. In later generations, the relative frequency of erythroblastosis in the population of China, now of mixed Chinese-white origin, will be less than in the first generation, since the frequency of the *r* allele and, therefore, of *rr* women will be lower than among the original whites due to the decreased frequency of *r* after dilution with the *R* allele of the original Chinese. Still, the frequency of the *r* allele will be higher than it was before the white immigrants came. Hence, *rr* women will result as expected from random mating, and some of them will have erythroblastotic pregnancies. Thus, from the point of view of the Chinese, and judged purely from Rh incompatibility, hybridization will have had permanently bad effects.

The results will be very different in the "white country" with its Chinese immigrants. None of the Chinese women married to white men will add to the fetal disease, but Chinese men married to white women will cause the appearance of a higher frequency of erythroblastotic children than among marriages of white men and white women. In later generations, the frequency of the *r* allele in the white population, now of mixed white-Chinese origin, will be *lower* than before the *RR* immigrants came; the frequency of *rr* women will be correspondingly less, and the incidence of the disease will be lower. Thus, again judged purely from Rh incompatibility, hybridization will be found to be permanently beneficial.

Racial "Superiority." The argument that certain races are superior to others and that hybridization involving the superior type is bound to destroy its excellence has rarely been used in connection with purely physical characteristics. While it is likely—though not established—that some races have a better genetic endowment than other races in regard to normal eyesight, hearing acuity, endurance of extreme temperatures, longevity, and so on, it is improbable that some races contain many or all of these desirable traits and that others lack most or all of them. It would be difficult to give an objective rating to races, because they form different combinations of genic endowments—

some resulting in one array of excellent phenotypes and others in another array.

The same difficulty in evaluating relative superiority exists in regard to what has sometimes been called *race pathology*, the study of diseases in relation to race. It has been found that certain diseases are nearly unknown in some races and that the frequency of other diseases varies from race to race. Environmental differences may account for many such variations. Moreover, where a genetic basis is well established, differences in the frequencies of the disease-conditioning alleles often account for the observed racial differences. Examples of such differences, cited earlier in this book, are thalassemia, which is mostly restricted to people of Mediterranean racial background, and sickle-cell anemia, which is mostly restricted to blacks. The frequency of erythroblastosis fetalis depends on the highly variable frequency of the different R alleles in different races. Other more-or-less well-substantiated examples are the rare occurrence of scarlet fever among Mongoloids and Negroids as compared to Caucasoids, and the less severe course of the disease when it does appear in the former races; the less frequent occurrence of tuberculosis among Jews as compared to members of many other racial variants; and the more frequent occurrence of infantile amaurotic idiocy (Tay-Sachs disease) among Ashkenazy Jews as compared to Sephardim Jews and to Caucasoid non-Jews.

Public health statistics contain further data on different frequencies of various diseases in different racial groups, and the causes of some of these diseases are known to have a genetic component. Frequently, however, the penetrance of this component is dependent, to an unknown degree, on complex external circumstances, such as social status and its interrelation with housing, nutrition, type of occupation, etc. It would, therefore, be premature to draw general conclusions regarding differences in racial frequency of pathogenic alleles.

The examples given for racial differences in hereditary diseases should not overshadow the large number of diseases for which no striking difference exists. One of the most extensive compilations on the subject is Komais two-volume *Pedigrees of Hereditary Diseases and Abnormalities Found in the Japanese Race*, which comprised probably all such pedigrees ever published in Japan up to 1943. The list is similar to a recounting of a great number of the frequent and the rarer abnormalities well known among Western peoples: albinism, ichthyosis, harelip, arachnodactyly, hemophilia, diabetes mellitus, cataract, blue sclerotics, partial and total color blindness, Huntington's chorea, amaurotic idiocy, deaf-mutism, and many others. In contrast to genes which differentiate the normal characters of Japanese from those of Caucasoids, most of the alleles for pathological traits are found in both groups. Specific allele frequencies are not known, however, and it is likely that quantitative differences exist in the incidence of alleles involved in the different diseases. It is also possible that some hereditary, apparently identical diseases or abnormalities in the two racial groups may be caused by genes at different loci.

If two different races possess different frequencies of certain pathogenic alleles, hybridization will result in the allele having a lower frequency in the mixed population than it had in the original race with the higher incidence. As we have seen earlier, for recessive traits, this will result in a lower *total* frequency of the affected homozygotes. Hybridization, then, like the breakup of isolates within racial groups, will result in a reduction of pathological conditions. This may be illustrated by an example for sickle-cell anemia. If, in a racial isolate of 20,000,000 people, the frequency of Hb^S is 5 per cent, then $20 \times 0.05^2 \times 10^6 = 50{,}000$ individuals are affected by the homozygous disease. Should hybridization occur with another isolate consisting of 180,000,000 people in whom Hb^S is virtually absent, then the allele frequency would be 0.005 so that after panmixis the frequency of the homozygotes would be $(0.005)^2 \times 200{,}000{,}000$, or 5,000.

Mental Traits

There is no doubt that racial differences in psychological traits exist. Attitudes vary among different races, and their study is the object of special sciences, such as social anthropology. It is clearly true that psychological attributes of races are greatly influenced by the particular historical, cultural, and sociological environment. It is difficult enough to define an over-all social psychology, but even when some valid approximation can be made, it seems to apply only to specific historical periods, or, if the race occupies different parts of the globe, only to specific regions. Differences in group psychology are also well known between different social layers of populations presumably genetically rather homogeneous.

Test Intelligence. Numerous studies have been made of the I.Q. performance of white and black Americans. There are great variations within each group and a large amount of overlap. This overlap means that high scoring individuals of each group outrank not only lower scoring individuals of their own group but also lower scoring individuals of the other racial group. On the whole the distribution of I.Q. scores of each group conforms to a normal curve, but the mean scores of the two groups differ from one another: The blacks regularly showing a mean score about 10 to 15 points lower than the mean of a comparative white population sample. What is the cause of the difference? Undoubtedly nongenetic sociological circumstances are responsible at least for part of the difference. Are such nongenetic factors responsible for all of it? We do not know. Even within the white or the black population the interpretation of I.Q. differences is difficult; the least ambiguous interpretations are possible only for such selected groups as twins reared apart and adoptive children. Without special designs for the study of I.Q. differences between the two racial groups, it is no more than unfounded speculation to assign specific shares of nongenetic and genetic factors to the interpretation of the observed differences in I.Q. scores of the two races. Given the fact of polymorphism for numerous human traits and given the polygenic basis of normal intelligence, the geneticist will find it unlikely that two racial populations will have identical mean I.Q. scores. He will, however, be unable to state which of the two races has an intrinsically higher mean genetic endowment. If the mean endowment of the whites is higher, then a fraction of the observed difference between the two racial scores would be due to a lower mean genetic endowment of blacks. If the blacks are endowed with higher genotypes, then their lower scoring is due to nongenetic circumstances so unfavorable as to more than cancel their genetic superiority.

The white-black comparisons in I.Q. scores are not the only ones between different racial types. Some data are available for Americans of Chinese and Japanese ancestry as compared to Caucasian Americans. The mean scores of the Oriental populations were higher than those of the Caucasians. The interpretation of these findings in terms of genetic as against nongenetic factors is as uncertain as in the black-white comparisons.

From a human point of view the problems of interpretation of mean I.Q. scores should not influence the attitude toward an individual. Within each race there is a wide range of I.Q. phenotypes. The mean phenotype of a population is not an attribute of each individual. Rather, the determination of his specific place in a wide range of phenotypes is a task independent of the statistical parameter "mean phenotype." It seems to be difficult to realize the simple fact that we are all unique, and imperfect.

Racial Adaptiveness. Biological arguments against hybridization as it concerns mental traits assume that there are basic mental differences among races and that these are at least partly determined by different genetic endowment. Such discussions emphasize the first two arguments which were raised in connection with physical traits, namely the supposed breakup of well-adapted racial genotypes and the supposed origin of disharmonious gene combinations, as well as the third, the supposed existence of superior races. "The psyche," wrote von Verschuer (1896–1969), "is a more sensitive reagent [than the body]. Disharmony of genes therefore, will probably become apparent more easily in psychological than in physical disturbances." An opposite point of view is held by other writers, who doubt the existence of well-adapted and of disharmonious gene combinations which affect mental traits. This opinion is guardedly expressed by Dobzhansky and Ashley Montagu: ". . . genotypic differences in personality traits, as between individuals and particularly as between races [are probably] relatively unimportant compared to their phenotypic plasticity." These authors, as well as others, stress evolutionary factors which "in all climes and at all times have favored genotypes which permit greater and greater educability and plasticity of mental traits. . . ." The consequences of this point of view for hybridization are, obviously, that there is no specific racial adaptiveness in mental traits, but that mental harmony is potentially present in all human beings.

The two points of view are, of course, not mutually exclusive. Stressing possible genetic factors in racial mental differences does not deny plasticity, and stressing plasticity leaves room for genotypic differences. Even though we lack exact knowledge, we may still be rather confident of the existence not only of great plasticity, which is an obvious phenomenon, but also of genotypic differences in racial endowment. Since genic differences influence all parts of the body and since differences in allele frequencies have been established for various genes in different races, we may expect some genetic influence on mental traits. The important problem is how great this influence is in differentiating races mentally. How does it compare with the inherent plasticity of mental traits, which may diminish or obliterate phenotypic expression of genotypic divergence; and how does it compare with external factors, which cause different expressions of like genotypes in different individuals and different groups?

Cultural Achievements. In an attempt to answer questions concerning differentials in racial achievement the facts of history are often cited. The cultural achievements of different races are very diverse, and such differences in achievement are taken as proof of different genetic endowments of mental traits. In such discussions, particularly, the concept of a scale of racial superiority is employed, the measure being achievements in such fields as mechanical inventions, abstract thought, social and political organization, religious creativity, and accomplishments in architecture, sculpture, music, and other arts. The scientist and the historian alike regard established differences in achievement as very inadequate evidence of genetic causation. Not only each individual but, even more, societies seem to be similar to electronic amplifier systems, in which the relation between intake and output is most complex. The development of any society is intricately conditioned by history. The fortuitous appearance or lack of appearance of some external circumstances which provide stimuli of just the "right" intensity to specific cultural endeavors may, with all probability, determine the most divergent future developments. Similarly, the appearance or lack of appearance of a specific influential individual at a specific moment may possibly decide the course of a culture, although here, as in the case of external circumstances, conviction cannot point to controlled experiments in history.

In this connection an opinion voiced by Lord Raglan is relevant: "It has been said against the African Negroes that they never produced a scientist; but what kind of a scientist would he be who had no weights and measures, no clock or calendar, and no means of recording his observations and experiments? And if it be asked why the Negroes did not invent these things, the answer is that neither did any European, and for the same reason—namely, that the rare and perhaps unique conditions which made their invention possible were absent!" Remote indeed seems the time when a reliable statement can be made regarding a possible genetic component in the cultural achievements of different races.

Apart from the statement itself, it is pertinent to ask whether an objective scale of superiority is possible if we take into account cultural achievements which defy a simple hierarchy of values, such as European Gothic art as compared to Chinese T'ang art; if we consider simultaneously excellence in different fields, such as Roman law and Buddhist thought; and if we include negative values, which stem from ruthless acquisitiveness, coercion, and intolerance. If there are differences in mental endowment among races, they could be of various kinds, including variations in mean endowment, in relative frequencies of specific endowments, and in range. How such variations would be mirrored in racial achievement will be estimated differently by different men. What importance should be ascribed to differences in the average endowment of two groups, as compared to differences in the frequency of a few exceptional individuals? How important are variations in the frequency of a large, better-than-average groups? How great is the drain on a group's achievement if the distribution of grades of endowment is biased too heavily on the poorer side? To ponder questions like these not only shows our inability to answer them rationally, but, even more, brings to light the inadequacy of an approach which attempts to analyze the history of cultures in terms of allele frequencies.

Confronted with the lack of decisive evidence on the genetic consequences of hybridization for physical and mental traits, the conservative will still counsel abstention, since the possible ill effects of the breakup of races formed in the course of evolution will not be reversible; whereas the less conservative will regard the chance of such ill effects as small and will not raise his voice against the mingling of races which, from a very long-range point of view, is probably bound to occur anyway. It should not be forgotten, however, that the problem of race is only partly genetic; men and women will have to consider the biological, sociological, and ethical problems when they attempt to plan for the future.

REFERENCES

Boyd, W. C., 1950. *Genetics and the Races of Man.* 453 pp. Little, Brown, Boston.

Coon, C. S., Garn, S. M., and Birdsell, J. B., 1950. *A Study of the Problems of Race Formation in Man.* 153 pp. Thomas, Springfield, Ill.

Dobzhansky, Th., and Montagu, M. F. A., 1947. Natural selection and the mental capacities of mankind. *Science,* **106:**587–590.

Flatz, G., and Rotthauwe, H. W., 1971. Evidence against nutritional adaption of tolerance to lactose. *Humangenetik,* **13:**118–125.

Garn, S. M., 1971. *Human Races.* 3d ed. 196 pp. Thomas, Springfield, Ill.

Giblett, Eloise R., 1969. *Genetic Markers in Human Blood.* 629 pp. Blackwell, Oxford and Edinburgh.

Glass, B., and Li, C. C., 1953. The dynamics of racial intermixture—an analysis based on the American Negro. *Amer. J. Hum. Genet.,* **5:**1–20.

Goldsby, R. A., 1971. *Race and Races.* 132 pp. Macmillan, New York.

Harrison, G. A., 1969. The Galton Lecture, 1968: The race concept in human biology. *J. Biosoc. Sci.,* Suppl. 1, 129–142.

Harrison, G. A., and Peel, J. (Eds.), 1969. *Biosocial Aspects of Race.* 202 pp. Blackwell, Oxford. (Proceedings of the Fifth Annual Symposium of the Eugenics Society, London. Journal of Biosocial Science, Supplement 1.)

King, J. C., 1971. *The Biology of Race.* 180 pp. Harcourt Brace Jovanovich, N.Y.

Mange, A. P., 1964. Growth and inbreeding of a human isolate. *Hum. Biol.,* **36:**104–133.

Mayr, E., 1963. *Animal Species and Evolution.* 797 pp. Harvard University Press, Cambridge.

Morton, N. E., Chung, C. S., and Mi, M. P., 1967. *Genetics of Interracial Crosses in Hawaii.* 158 pp. Karger, Basel. (Monographs in Human Genetics, 3.)

Nei, M., and Roychoudhury, A. K., 1972. Gene differences between Caucasian, Negro, and Japanese populations. *Science,* **177:**434–436.

Penrose, L. S., 1955. Evidence of heterosis in man. *Proc. Roy. Soc. London, B.,* **144:**203–213.

Reed, T. E., 1969. Caucasian genes in American Negroes. *Science,* **165:**762–768, 1353.

Saldanha, P. H., 1962. Taste sensitivity to phenylthiourea among Brazilian Negroes and its bearing on the problem of White-Negro intermixture in Brazil. *Hum. Biol.,* **34:**179–186.

United Nations Educational, Scientific, and Cultural Organization, 1952. *The Race Concept.* 103 pp. UNESCO, Paris.

Valls, A. Medina, 1958. Estudio antropogenetico de la capacidad gustativa para la fenitiocarbamida. *Fac. Cienc. Univ. Madrid.* 67 pp.

32

THE ORIGIN OF HUMAN DIVERSITY

We have seen that men differ widely in their genetic constitutions. Let us now examine some of the causes of human polymorphism and polytypy. These two kinds of human diversity require separate treatment, since it is one problem to explain why one individual differs from another, and another problem to determine why a racial group has genetic similarities within it which distinguish it from other groups.

THE ORIGIN OF RACIAL DIVERSITY IN MAN

The racial diversity of man is, presumably, the result of evolutionary processes which occurred after the species *Homo sapiens* had evolved from pre-human ancestors. The later divergence of humans into racial groups is a phenomenon of *microevolution*, the evolution of genetically distinct populations within a single species. The nature of microevolution in general has been the subject of many studies since the 1920's, when the theoretical work of

J. B. S. Haldane, R. A. Fisher, and Sewall Wright introduced mathematical genetic concepts into the study of populations, and Tschetverikoff and his followers begain to study experimentally the genetic diversity of populations (rather than of individuals) of *Drosophila* and other organisms.

The prerequisite for any evolutionary change in a population is the presence of different genotypes. This presence depends on the existence of different alleles at various loci. The existence of different alleles is brought about by mutation from one allele to another. Mutation, whether of intragenic nature or of chromosomal-aberration type, is thus the condition necessary for the origin of racial diversity.

The mechanism of Mendelian recombination creates an immense amount of genetic variation from even a small number of mutated genes. For example, if a population is initially homozygous and isogenic at the two loci $A^1A^1B^1B^1$ and if mutations occur from A^1 to A^2 and from B^1 to B^2, then recombination can result in 9 different genotypes: $A^1A^1B^1B^1$, $A^1A^1B^1B^2$, $A^1A^1B^2B^2$, $A^1A^2B^1B^1$, $A^1A^2B^1B^2$, $A^1A^2B^2B^2$, $A^2A^2B^1B^1$, $A^2A^2B^1B^2$, and $A^2A^2B^2B^2$. Four of these genotypes are homozygous, and five are heterozygous. With mutations at m loci from one allele to another, recombination can produce 3^m genotypes, of which 2^m are homozygous; and with n multiple alleles at each of m loci, the total number of combinations becomes $[1/2\ n \times (n+1)]^m$ (see p. 251). Even if m, the number of loci concerned, were as low as 100, and n, the the number of alleles per locus, as low as 5, the term $(1/2 \times 5 \times 6)^{100}$ would be so large that the possibilities of genetic newness become inexhaustible.

The origin of genetic newness produced by mutation and recombination is, in itself, not sufficient for any evolutionary phenomenon. In an industrial society, it is one thing to make a technological invention, and another to insure its widespread use and persistance. Similarly, evolutionary history consists of two different processes: one, the origin of newness; the other, the more-or-less permanent establishment of the innovation in all or many members of a population.

Mutation, Nonrecurrent and Recurrent. As a first hypothesis on the establishment of evolutionary newness, we might consider mutation a method not only for *originating* newness but also for *establishing* it. Without recourse to secondary processes, like selection, could not the mere occurrence of a mutation lead to its permanent presence in the population? In the analysis of this problem, it is well to separate singular—that is, nonrecurrent—mutational events from recurrent mutations.

The chance that a single mutation will become established is very small. Assume that, in a population containing the A^1 allele only, one gamete is produced which contains the mutant allele A^2, so that an A^1A^2 individual appears. Assume, also, that the number of individuals remains the same, so that each individual of one generation is represented, on the average, by one individual in the next generation—though some individuals will leave no progeny at all; others, one or two; and still others, many. If the A^1A^2 individual reaches maturity and becomes one of the parents of the next generation, he may have one or more children, who, themselves, become parents of the succeeding generation. As a result of segregation, he forms two kinds of gametes, A^1 and A^2, in equal numbers. If he has one child, it may or may not receive the new A^2 allele; if he has two children, chance may give A^2 to both, to only one, or to neither. In general, for any given number of children there are specific probabilities of all (n) or (n − 1) or (n − 2), etc., down to no child receiving the A^2 allele. It is thus seen that the new allele A^2 may not occur again in the next generation, or it may recur only once or more than once. If it does not recur, then the evolutionary potential of the mutation has obviously been cut off. If it recurs in single number, the chance that it will be extinct by the next generation is like that of the original A^2 allele. If it recurs in several individuals, its total chances of survival—that is, continued existence in later generations—are, of course, improved, but the likelihood persists that any one of the A^2 alleles present in A^1A^2 individuals will become extinct in the next, or a later, generation. Fisher has calculated the probabilities of survival of a mutation which appears only once and in a single individual (Table 117). It is seen that the chance of survival of a single mutation is very small indeed. For example, there are only 153 out of 10,000 chances—or 1 in 65—that A^2 will exist after 127 generations; and, after 40,000 generations, these chances will decrease to only about 25 out of 1,000,000—or 1 in 40,000.

However, these probabilities are not infinitely small; and if the allele A^2 survives at all, it may be found in small, or even in rather large, numbers. Just

TABLE 117
Probability of Survival of a Mutation Appearing in a Single Individual. (No selective advantage or disadvantage.)

Generations	*Probability of Survival*
1	0.6321
3	0.3741
31	0.589
63	0.0302
127	0.0153
1,000	Approx. 0.0010
10,000	Approx. 0.0001
40,000	Approx. 0.000025

Source: Fisher.

as chance may lead to nonoccurrence or extinction of the A^2 allele, conversely, chance may result in an increase. The survival of any specified number of A^2 alleles derived from the original A^2 can be expressed in terms of specific probabilities. Table 118, which affords some insight into these relations, lists the probabilities that a minimum specified number of survivors will be found, provided that survival has taken place at all. Thus, as the second column shows, there exists a probability of 1 in 10 that 58 or more A^2 alleles will appear in the descendants 50 generations after the occurrence of the original single A^2, provided that at least one A^2 allele escaped extinction. After 100 generations, at least twice as many alleles possess the 10 per cent chance of having become established (again provided that at least one has survived). After 1,000 generations, there is a 10 per cent chance of finding 1,160 or more A^2 alleles, and, as seen in the final column, a rarer but still finite probability of 0.1 per cent of finding 3,480 or more A^2 alleles. After 40,000 generations, the numbers of A^2 alleles corresponding to the various probabilities have increased further.

We have seen that the survival and the establishment in fairly large numbers of a unique mutation are improbable, but not impossible, events. If we assume that many thousands of loci which might give rise to unique mutations exist, it follows that most of these new alleles never become established but that an occasional one may escape extinction. As a general phenomenon, however, it is very unlikely that the present diversity of man is the result of singular, nonrecurrent mutations.

Most mutations are not singular events, but recur at an appreciable rate. Such recurrent mutations have a much better chance of becoming established. For each specific mutational event, in a recurrent change of A^1 to A^2, the

TABLE 118
The Minimum Numbers of Surviving Descendants of a Mutant Allele Which Appeared First in a Single Individual and Did Not Become Extinct. The second, third, and fourth columns give the minimum numbers of surviving alleles determined by the probabilities 0.1, 0.01, and 0.001. (No selective advantage or disadvantage.)

Generations	*0.1*	*0.01*	*0.001*
50	58	116	174
100	116	232	348
500	578	1,160	1,738
1,000	1,160	2,320	3,480
10,000	11,600	23,200	34,800
40,000	46,400	92,800	139,200

Source: Fisher.

prospects are identical to those listed in Tables 117, and 118. Recurrent mutation, however, may greatly increase the number of opportunities for establishment of A^2. If, in each generation, a large enough number of individuals carry the A^1 allele and a large enough number of generations are considered, numerous A^2 alleles will be produced, and the probability that at least one of them will survive and even become established in appreciable numbers will be greatly increased. If mutation from one allele to another occurs recurrently and at numerous loci, most of the single mutations at any one locus will never be established and even none of the many recurrent mutations at any one locus may become established. Some mutations at some loci, however, will escape extinction and may, indeed, even become numerous.

When mutation is unidirectional—from A^1 to A^2, but not from A^2 to A^1—then recurrent mutation may not only initiate the establishment of A^2 but even lead to the complete replacement of A^1 by A^2. In the course of time, all the A^1 alleles will be transformed into A^2 alleles. If, however, mutations in both directions occur, then the A^2 alleles will produce A^1 alleles, and an equilibrium will be reached: the A^1 alleles will give rise to just as many A^2 as the A^2 alleles give rise to A^1. The equilibrium frequency will be determined by the relative rates of the two mutation processes. Although there is no available information on the existence in man of mutations in both directions, evidence from other animals and from plants strongly suggests that such reversible mutations may occur in man as well. They would, however, be very rare, considering that most reverse mutations would have to have undergone reversion of a specific nucleotide pair among several hundreds pairs per gene. In all these cases, mutation pressure alone from recurrent mutations will not lead to isogeneity of the new allele in the population.

Selection. If the primary process of creating genetic variability (mutation) does not, in general, carry the "evolutionary reaction" $A^1 \rightarrow A^2$ to completion, what processes do accomplish evolutionary changes? Two main processes are recognized: natural selection, and genetic drift. Natural selection is the principle emphasized by Charles Darwin. Applied in terms of modern genetics—and given genetic variability—natural selection has retained its status as the most important agent of evolution. If certain genotypes endow their bearers with higher reproductive fitness than is found in others, then the alleles responsible for such relatively "better-adapted" types will be represented in higher proportion in the succeeding generations than the alleles responsible for reproductively less fit—that is, relatively more "poorly adapted"—types. "Better" or "worse" adaptation is used here in the general, and strictly objective, sense of reproductive fitness.

Selection, even to a slight degree, can be a powerful agent for shifting allele ratios from one extreme to another, and thus can lead to the establishment of

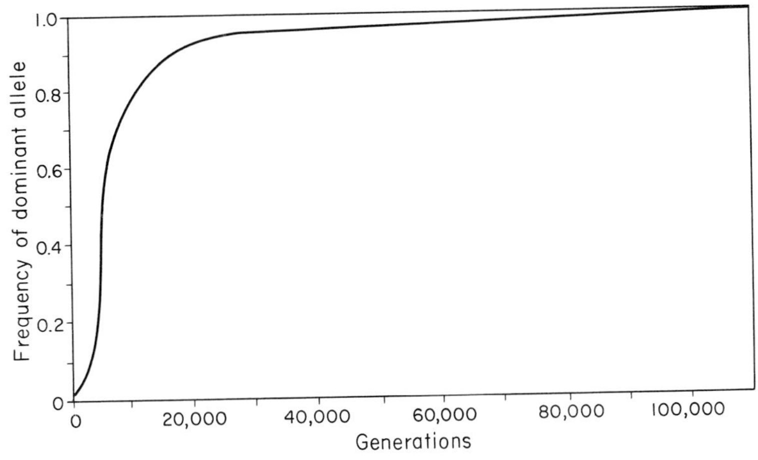

FIGURE 306
Effect of selection for a dominant allele. Selective advantage of bearers of the allele, 1 in 1,000. The allele increases rapidly from an initially low frequency. After having reached a frequency of approximately 0.9, the allele's further increase is very slow and at a decreasing rate.

new genotypes. If a new dominant A^2 allele allows A^2A^1 individuals to produce one-thousandth more offspring in each generation than an equal number of A^1A^1 individuals, then it can be calculated that it takes fewer than 10,000 generations to change a population with only a fraction of a per cent of A^2 alleles to one with 50 per cent A^2 (Fig. 306). To raise the frequency of A^2 from 50 to 90 per cent requires fewer than another 10,000 generations. The final increase toward complete replacement of A^1 by A^2 is, however, very slow, as shown by the curve which approaches 100 per cent A^2 asymptotically. The reason for this lies in the fact that, with high frequency of A^2, most individuals will be A^2A^2 and A^2A^1, and only very few will be A^1A^1. Therefore, the effect of selection against A^1A^1 becomes negligible, while the production of A^2A^1 individuals results in perpetuation of A^1 alleles in the population.

As stated, these figures apply to dominant alleles which endow their bearer with a selective advantage, and no difference is postulated for the selective advantage caused by heterozygous A^1A^2 or homozygous A^2A^2. With recessives, selection is effective only in homozygotes. This means that a rare recessive, which only occasionally appears in homozygous form, is nearly free of selective influences. Only when a recessive has become frequent enough to result in the production of an appreciable number of homozygotes will selection become effective. The effect of selection on the frequency of a recessive allele is illustrated in Figure 307. There is an extremely slow increase in the frequency of the allele *a* as long as it is rare, and a steep rise in its frequency once it has become common.

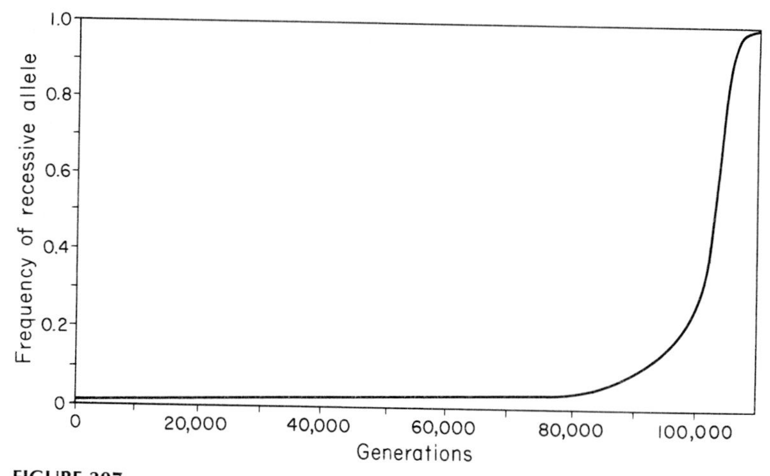

FIGURE 307
Effect of selection for a recessive allele. Selective advantage of homozygotes, 1 in 1,000. The allele increases very slowly from an initially low frequency, but relatively rapidly after it has attained an allele frequency of approximately 0.1.

We have seen that mutation pressure alone is very inefficient in establishing new alleles. We now find that selection is relatively efficient in increasing the number of advantageous dominant alleles and that this is also true for recessive alleles, provided they are present in appreciable numbers. If selective advantages are of the order of 1 in 100 or higher, instead of 1 in 1,000, then the rate of increase in the frequency of the new alleles is greatly enhanced.

Concerning recessive alleles, the efficiency of selection is low as long as they are still rare. But, as we have seen in earlier discussions, many so-called recessives are likely to have a slightly dominant effect in heterozygotes, so that their response to selection resembles that of dominants more than that of complete recessives.

It is remarkable that the coexistence of mutation pressure and selection in nature results in greatly increased speed of evolutionary changes. Where it requires many millions of generations of selection to double the frequency of a from 0.00001 to 0.00002, pressure produced by mutation from A to a at the rate of 1 in 100,000 will accomplish the same result in a single generation.

Thus, both mutation and selection may play roles in the establishment of new recessive alleles. At one stage, mutation is the more important and, at another stage, selection. If details are to be worked out, one must, of course, consider separately autosomal and sex-linked alleles; differentiate between the mutant alleles with dominant, recessive, or intermediate effects; and consider, furthermore, whether a genotype at the locus under discussion has selective advantages in all genetic backgrounds or only in certain specific ones. Still more important is the consideration that a selective advantage in one environment may not be, and often is not, an advantage in another environment; for instance, heavy pigmentation of the skin may be selected for in southern regions and selected against in northern regions.

In general, it must be emphasized that selection affects simultaneously the whole gene pool rather than the alleles at single loci. An individual selected for his reproductive fitness is liable to transmit assemblages of alleles which differ from those of individuals with lower fitness.

Sometimes the relatively high reproductive fitness of an individual may depend primarily on the specific alleles he carries at a certain single locus. In this case, the assemblages of alleles at other loci that he transmits are selected by the chance which brought them together with the specific alleles at the single locus. More often, his relatively high reproductive fitness will be the result of an interaction of the alleles at the single locus with those at many other loci. When this is true, the assemblage of alleles which he transmits has a tendency to be "coadapted" within itself. In either case, the evolution of diversity as a result of selection constitutes evolution of new gene pools.

Genetic Drift. Selection, by definition, is powerless in the establishment of new neutral alleles—that is, those which confer neither reproductive advantage nor disadvantage on individuals who carry them, and which, therefore, leave in later generations, neither relatively more nor fewer alleles than the original allele. Whether or not neutral alleles exist cannot be demonstrated at present but there must be many isoalleles whose selective differences approximate zero. It is of great importance to realize that a process, *genetic drift*, has been recognized which is powerful in leading to the establishment of nonadaptive, neutral characters, or even slightly unfavorable traits, and which, following Wright, must be taken into account, together with mutation and selection, as an evolutionary agent.

Let us introduce genetic drift by the following hypothetical situation. Assume that each of 160 isolated islands with exactly identical environmental conditions is settled by a man and woman, all of the genotype A^1A^2. Let each couple have two children, one of each sex, who, on reaching maturity, will become the ancestors of the subsequent populations of the 160 islands. Assume, finally, that the alleles A^1 and A^2 are of equal selective value. What will be the genic composition of each of the 160 populations? As a first approximation, we might be inclined to state that whatever the specific composition might be, it would be alike in all 160 groups, since it was postulated that the original genotypes were identical, the environments are identical, and the selective values for the two contrasting alleles are the same.

This answer is false, as will now be shown. The two original parents $A^1A^2 \times A^1A^2$ will have children of the genotypes A^1A^1, A^1A^2, and A^2A^2 in the expected proportions 1/4:1/2:1/4. Since we postulated that there were to be only two children, chance will result in:

both A^1A^1	in 1/16,
both A^1A^2	in 1/4,
both A^2A^2	in 1/16,
one A^1A^1 and one A^1A^2	in 1/4,
one A^1A^1 and one A^2A^2	in 1/8, and
one A^1A^2 and one A^2A^2	in 1/4 of all cases.

Any one of these six different situations may arise in each one of the 160 islands. And considering the various probabilities, the most likely result will be that 10 islands will have only two A^1A^1 children, 40 islands only two A^1A^2, 10 islands only two A^2A^2, and the remaining 100 islands will be divided into three different groups each with two children of different genotypes.

The genic constitution of the later populations will obviously depend on the genotypes of their second-generation ancestors. To consider the extreme first, the 10 island populations descended from the two A^1A^1 individuals will contain only A^1A^1 people, while the 10 other island populations descended from the two A^2A^2 individuals will contain only A^2A^2 people. Thus, polytypy for neutral differences in genotype will have arisen, not by mutation pressure or by selection but by the play of chance. This happening, as in these two groups of 10 islands, has been called "chance loss" of an allele (A^2 in the first, and A^1 in the second group) or, conversely, "chance fixation" of the other allele (A^1 in the first and A^2 in the second).

The same process of allelic loss and fixation may occur in the course of later generations in any one of the other 140 islands. On some of the islands, the two children of the original parents still jointly carry two A^1 and two A^2 alleles. On others, instead of two of each kind, a shift by chance will already have resulted in the appearance of three A^1 and only one A^2 alleles. On still others, one A^1 and three A^2 alleles will appear. As long as both types of alleles occur in succeeding generations, chance may shift their ratio back and forth in either direction. Whenever chance leads to fixation of one and loss of the other allele, an irreversible situation will have been established (barring mutation).

The speed with which genetic drift changes the ratio of A^1 to A^2 alleles depends on the size of the population, "N" (more specifically, the part of the population which actually becomes parents in each generation). With two parents and two maturing children, as in the island examples, 1/8 of all populations undergo complete loss and fixation for one locus within one generation. With more parents, the chances of loss and fixation become lower, since it will be less frequent that the children of all of the parents will have lost the same allele. Rather, some of the children will have lost A^1 and others A^2, with the over-all result that both alleles persist in the population. Yet, there always remains a small but finite chance, even with large populations, of complete loss of one allele and fixation of the other.

The size of the population not only is decisive for the speed of drift, which, results in loss and fixation of alleles, but also plays a similar role within the reversible range of shifting gene ratios, that is, within the range where both alleles are still present. When populations are small, the speed of drift is high, and often of striking degree from one generation to the next; but when populations are large, drift proceeds slowly and is mostly of slight degree. This is shown diagrammatically in Figure 308. The upper part shows allele ratios attained in successive generations in a population with large numbers of parents, and the lower part shows the same for a small population.

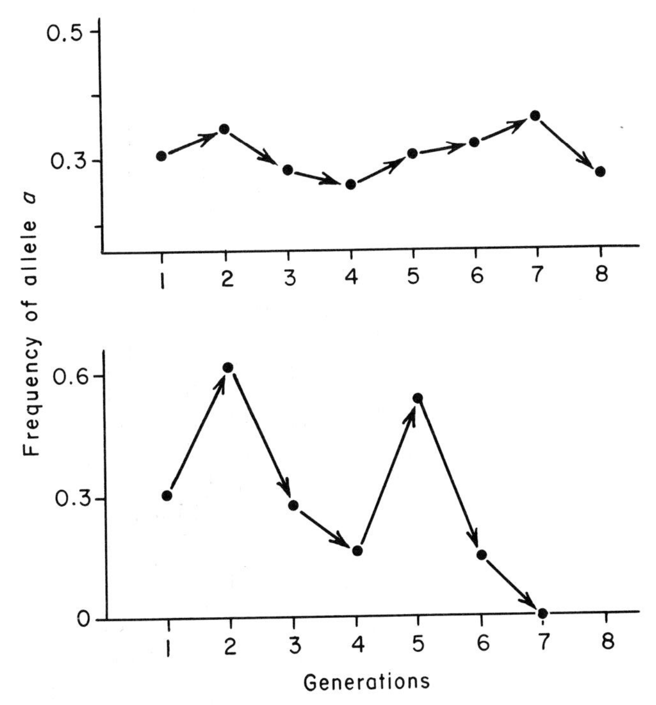

FIGURE 308
Changing allele frequencies under the influence of drift. *Upper part:* Drift in a large population. Starting with the frequency of 0.3, the allele *a* increases slightly after one generation, and its frequency continues to fluctuate moderately in successive generations. *Lower part:* Drift in a small population. Again starting with a frequency of 0.3, the allele *a* drifts violently. After six generations, drift leads to the extinction of *a*.

It might appear that the concept of genetic drift contradicts the Hardy-Weinberg Law, which demands an equilibrium, that is, a constancy of allele frequencies from generation to generation. However, the conflict between the expectation of changing allele frequencies and that of an allelic equilibrium is not real. The equilibrium rule applies to infinitely large populations, but has limitations in smaller populations. Basically, the situation is the same as that regarding Mendelian ratios in small sibships. In Chapter 10 we found that the expectation of a 3:1 or a 1:1 ratio needs modification for small sibships, and that expected deviations from the ideal expectation can be calculated. Similarly, the details of drift are predictable as deviations from equilibrium expected in small populations.

In our discussion of drift, we have made the assumption, among others, of neutral selective value of the various alleles. Drift plays a role even when the selective value of an allele is positive or negative. The interplay of the processes of drift and selection is complex. Their relative strength in changing the genetic composition of a population depends, among other factors, on the population size. Since drift is most effective in small populations, it can overshadow, and even run counter to, the trend which selection alone would favor. Thus, even an allele against which selection would discriminate may become established and fixed by drift as long as it is not so disadvantageous as to lead to the extinction of its carriers, and, conversely, an allele which would be favored by selection may be lost by drift before selection could show its full effect.

Genetic drift, small population size, and isolation are closely interrelated agents of evolution. They may not only be effective in the manner described—that is, acting on individual alleles—but may also be particularly important in the preservation and establishment of special combinations of alleles of different loci. In a large population, two rare alleles A^2 and B^2 will only occasionally happen to form the combination $A^2A^2B^2B^2$. Regardless of the selective value of such a combination, a cross with any one of the prevalent genotypes which is not homozygous for A^2 and B^2 will immediately break up the $A^2A^2B^2B^2$ combination, so that it would not reappear in the next generation. The situation might be different in a small isolate within the population. If chance had given the otherwise rare alleles A^2 and B^2 a relatively large share of the allele frequencies, the danger of loss of the combination $A^2A^2B^2B^2$ would be decreased. The higher allele frequencies of A^2 and B^2 will increase the likelihood of an A^2B^2 gamete from $A^2A^2B^2B^2$ meeting an A^2B^2 gamete produced in the isolate. Drift or selection, or both, may lead to the establishment of the new combination $A^2A^2B^2B^2$ and the disappearance of all other genotypes.

If mankind were, and had always been, a single interbreeding population, then genetic drift with its shifting allelic ratios and potentially complete loss or fixation would have been of minor consequence. In reality, mankind has always consisted of more or less separate isolates, owing to geographical and social barriers. Particularly in prehistoric times, the total number of individuals was small and was distributed discontinuously over many parts of the globe, and very frequently, groups were divided into reproductively segregated tribes. Furthermore, fluctuations in the size of these isolates must have been great, as the result of famines, epidemics, and wars. During each such low in population number, the sample of alleles that passed through the narrow "bottleneck" of the few parents which connected an earlier, relatively large, with a later, relatively large, population provided a striking opportunity for a demonstration of the effect of drift and often must have resulted in greatly different allelic ratios before and after the bottleneck generations.

Similarly, migration of a fraction of a population to a new region constituted opportunities for drift. The migrants may, in many genetic respects, have been a random sample of the original group, but if they were few enough in number, this random sample may not have been a representative one. In other words, the allelic ratios found in the migrants may have deviated considerably from those of the original group, and, in any case, these ratios may often have undergone great changes during the early generations when the migrated population remained small.

Selective Migration and Stratification. Genetic diversity of populations can also be brought about by the separation of subgroups within a common gene pool. It has often been asked whether emigrants form a group which is somewhat genetically distinct from the group which stays at home. It is likely that emigrants often have different mental attributes than the sessile group—for instance, a higher initiative—but the genetic basis of such properties is doubtful. Emigrants as a group conceivably may also have different morphological traits, be it a greater physical strength which might favor the overcoming of physical hardships, or a smaller body size which might permit more individuals to occupy a primitive boat.

Some modern anthropometric measurements seem to indicate that there are no differences between Mexican and Chinese immigrants to the United States and the native populations from which they stemmed, although it is clear that the new environment, if effective early in life, can strikingly modify such traits as stature. In other, historical migrations, differences between migrants and nonmigrants may well have existed. The type of selection involved in such migrations has been referred to as "sifting."

Within a population, different socioeconomic layers may to some degree represent different gene pools. If the stratification of such a population becomes relatively fixed, a genetic diversity of subpopulations is thereby originated.

Stimulated by his findings concerning genetic and sociological factors in the population structures of the Xavante Indians of Brazil, Neel has stressed that allelic frequencies in a migrant subpopulation may frequently have deviated in a highly nonrandom manner from those in the original group. This is because a group of migrants presumably included close relatives, i.e., was a selected subpopulation of genetically similar individuals. The fission of an original group into a migrant and a nonmigrant population may, therefore, have led to striking differences in allele frequencies between the two groups.

The term "genetic drift" has been critically examined by Mayr and shown to cover a variety of meanings. In general it does not imply unidirectional changes in allele frequencies as suggested by the everyday sense of the word "drift" but rather indeterminate up-and-down moving changes in these frequencies. In order to emphasize this point Kimura has suggested that the word "random" should precede the term "genetic drift."

Migrant subpopulations bring into consideration aspects of genetic drift to which Mayr has given the designation "founder principle." In so far as a migrant group consists of individuals who may become the ancestors of a large new population they may rightly be called "founders." In the following pages the words "genetic drift" will cover various situations, including the founder effect.

The Role of Selection and Drift in Specific Cases

In a general fashion, the origin of racial diversity can be explained by the action of the various evolutionary processes which have been outlined: mutation, recombination, selection, and drift. Added to these, the interbreeding of already diverse populations must have led to new types of gene pools, for instance, as in the American black and Australian aborigine (to be discussed below). In most specific instances of racial diversity, it is, at present, difficult if not impossible to say whether selection or drift has been the sole or at least major agent, or whether an interplay of the two has been decisive. What, for instance, is the explanation for the fact that all South American Indians, and many of the North American Indians, possess only the blood-group allele I^{O}? Since the ancestors of the Indians were Asian Mongoloids who presumably came to the Americas across the Bering Strait, they should have brought with them the three alleles I^{A}, I^{B}, and I^{O}. It has seemed likely that the loss of I^{A} and I^{B} was due to drift and happened in one of two different ways: either the probably small number of prehistoric migrants who came to North America all happened to belong to blood group O; or they and their descendants did carry one or both of the other I alleles but ultimately happened to transmit to a small number of surviving children the I^{O} allele only. One or the other of these happenings may indeed constitute the whole explanation of the origin of the isogeneity for I^{O} in the Amerindian. But one wonders whether selection can be disregarded.

Since we know that the populations of the earth fail significantly to cover the whole possible array of allele frequencies for I^A, I^B, and I^O (Fig. 282, p. 743), that maternal-fetal ABO and Rh incompatibitities have complex selective effects, and that selection finds at least one more point of attack on the ABO blood groups in their relation to various diseases, it is an open question whether selection was not important in making most tribes of the American Indian I^OI^O.

Similarly, one may ask: Are the observed relative differences in ABO blood-group frequencies of Egyptians and Central Arabs, who otherwise are racially similar, the result of chance processes that acted during earlier periods when small, isolated populations were forming the ancestors of the contemporary larger groups? Or did ill-defined different modes of life favor selectively different allele frequencies? Conversely, did the observed similarities in ABO blood-group frequencies between the Central Asian Mongoloids and the Hindus, who are racially so different, come about by chance shifting of allelic ratios, or did selective forces play a role? Only future discoveries will permit decisive answers to these questions. In the meantime, we can point to some important examples of diversity in which chance and others in which selection seem to have been of main importance.

Probable Examples of Selection. Drift, as an explanation for the origin of racial diversity, accounts primarily for differences determined by single gene substitution. Polygenically determined traits are less easily changed by chance assortment, since random processes are likely to shift some allelic proportions in one direction and others in an opposite direction. Moreover, if the hypothesis is true that many of the characteristics of different races are adaptations to their specific environments, present or past, then selection must have been instrumental in their establishment.

Among the differences between racial groups that depend on single genic loci are some for which selective forces, and not drift, are clearly responsible. The foremost example of these is that of sickle-cell hemoglobin. The allele Hb^S is nearly absent in most human populations, but occurs in frequencies as high as 10–20 per cent in a broad belt across Central Africa, in Madagascar, and in high or low (but still clearly elevated) frequencies in most North African and southern European countries which surround the Mediterranean. In irregular fashion, hemoglobin S has also been found in certain racial groups in Arabia and India (Fig. 309). The sickle-cell trait is thus a characteristic of many African populations, but its presence in other groups suggests either migrations and separation of subgroups or admixture by interbreeding. It is likely that such processes have served to distribute the Hb^S allele over wide areas. It is also clear that an additional factor must exist which accounts for important, detailed features of the distribution as well as for the unexpectedly high frequencies of this homozygously lethal, or at least severely deleterious,

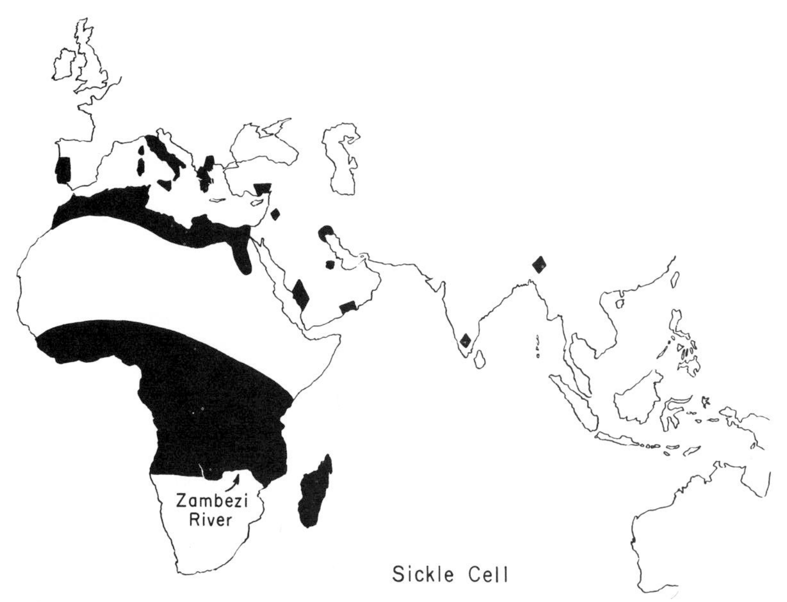

FIGURE 309
Sickle cells. The distribution of the gene for hemoglobin S. (Edington and Lehmann, *Man,* **36,** 1956.)

allele. This additional factor is selection for the $Hb^{A}Hb^{S}$ heterozygote, accomplished by greater resistance to malaria of the heterozygotes than of normal $Hb^{A}Hb^{A}$ homozygotes.

In favor of this theory is the considerable positive correlation between the incidence of malaria and the frequency of the sickle-cell allele in various areas. The correlation breaks down, however, in certain places where African tribes living in malaria-infested regions have only low frequencies of the Hb^{S} allele. Livingstone has analyzed these relations in West Africa. He considers that selection for sickle-cell heterozygotes is a phenomenon which has come relatively recently to mankind in general, and to West Africa in particular. He believes that malaria began to attack man on a large scale only after the invention of agriculture led to the clearing of forests and to the rise in population densities. The clearing of the tropical rain forests, as well as man's refuse and his villages, provided new and abundant breeding places for the malaria-carrying mosquito, and the increasing population density created a favorable situation for the persistance of the parasite. According to this view, in many populations the frequency of the sickle-cell allele is not yet in equilibrium with its selective attributes. Its existence, in minimal numbers, depends on its initial presence in certain tribes, its spread on their migration into new areas, and the "diffusion" of the gene, through hybridization, into other populations. But the actual frequencies of the gene are not determined by the initial state or the degree of admixture. Rather, they are the result of selective forces which, in a malarial environment and an agricultural mode of life (which favors exposure), lead to high frequencies of Hb^S, or which, in a relatively malaria-free environment and a mode of life such as hunting in the forest (which decreases exposure), lead to low frequencies. The Hb^S allele may well have originated by mutation, in other populations of the world. Its near absence in them may be no more than a reflection of the fact that most individual mutations never are established but become extinct by random sampling soon after they appear.

Among various other genes for abnormal hemoglobins are the allele Hb^C for hemoglobin C and the gene for thalassemia. The occurrence of Hb^C is nearly restricted to West Africa, where it overlaps the distribution of Hb^S. It may reach a frequency of more than 10 per cent. Homozygous hemoglobin C individuals are affected by an anemia in a much milder way than are sickle-cell homozygotes. Compound heterozygotes for Hb^S and Hb^C are severely affected, although less so than Hb^S homozygotes. There is thus selective pressure against Hb^C in $Hb^C Hb^S$ heterozygotes and in homozygotes. Its high frequency, in spite of this negative selection, requires the existence of compensating positive selection. It has naturally been wondered whether resistance to malaria, in the case of hemoglobin C, also provides the agent for such selection. This does not seem to be the case, and the nature of the polymorphism for hemoglobin C is not understood.

The frequency of the gene for thalassemia offers problems similar to those of the Hb^S and Hb^C alleles. Thalassemia occurs primarily in the Mediterranean area, but also in various Asian regions particularly in Thailand. Careful studies of its distribution in Italy have been made by Silvestroni and his associates. In Figure 310 it is seen that the thalassemia allele is very unequally distributed, having frequencies of less than 1 per cent in Florence and Bologna, and reaching peaks of 3 and 5 per cent in Sicily and Ferrara, respectively. Since the reproductive fitness of the homozygotes is practically zero, the high frequencies of the gene for thalassemia must be maintained by the increased fitness of the heterozygotes. The nature of this increased fitness is not yet clear, but there are again suggestive correlations between the incidence of the allele for abnormal hemoglobin production and the presence of malarial environments.

Probable Examples of Drift. The Blood and Blackfeet Indians of North America are exceptional in that they contain, in addition to I^O, the I^A allele, in frequencies of more than 50 per cent. This is not only higher than in other North American Indian groups, but also higher than in any other population. It would seem likely that this is the result of shifting allele ratios in initially small populations. Perhaps the immigrant Asian ancestors of these American Indian tribes did bring with them the I^A allele, and drift brought it to a high frequency. Or, perhaps, even the Bloods and Blackfeet were first all $I^O I^O$ and mutation secondarily led to I^A alleles. Even with a low rate of mutation, a new allele, once

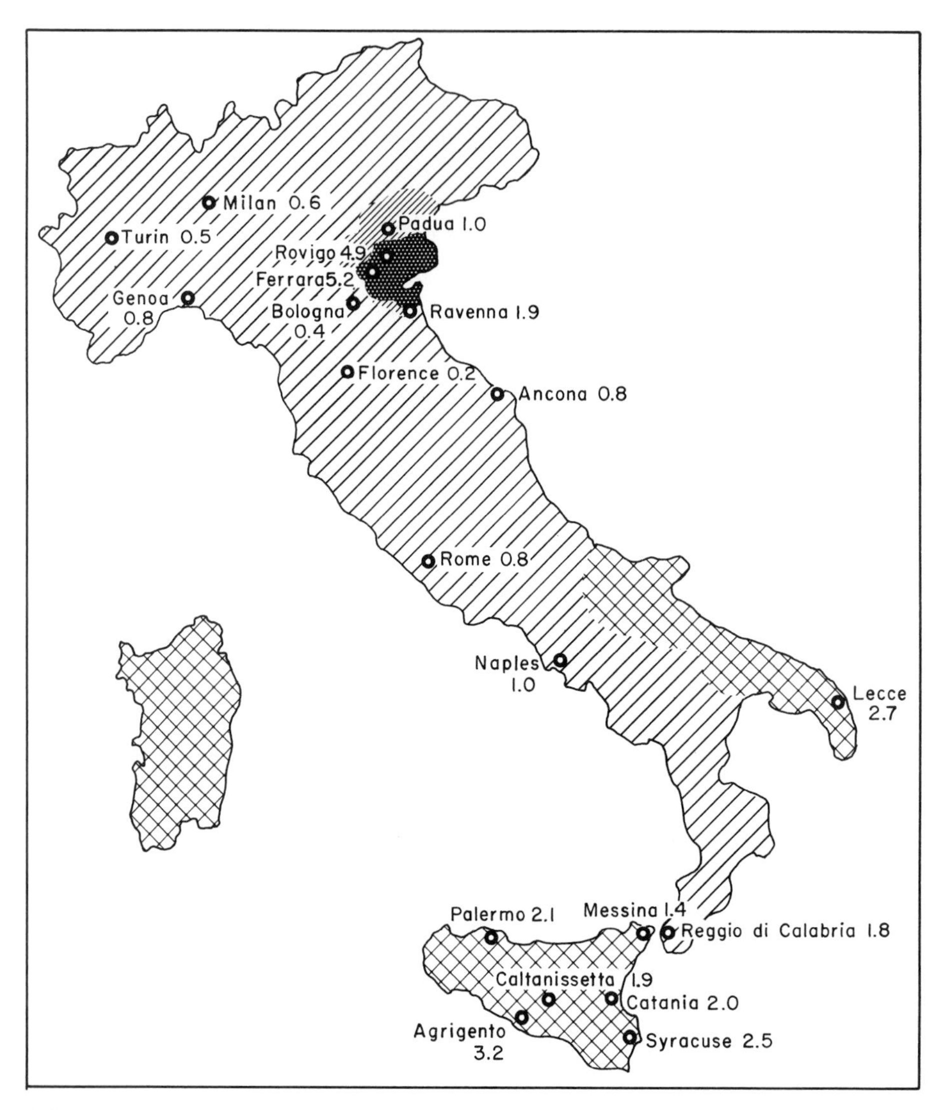

FIGURE 310
The distribution of the gene for thalassemia in Italy. The numbers represent allele frequencies in percent. Frequencies are progressively higher in the more heavily shaded areas. (After Bianco, Montalenti, Silvestroni, and Siniscalco, *Ann. Eugen.*, **16**, 1952.)

originated, may by chance escape extinction and become numerous, particularly in a small population.

Other examples in which it is unlikely that selection was important are the peculiarities in blood-group frequencies of certain Eskimos. The Aleutians and the Eskimos inhabiting the areas from western Alaska to eastern Greenland have allele frequencies of I^A of about 30 per cent and of I^B of about 6 per cent. But small tribes of the Polar Eskimos have only 7 per cent I^A and 1 per cent I^B, and those from Baffin Island and Labrador, while having I^A in high frequency, have almost no I^B. For the MN group, most Eskimos, like most American Indians, have the high frequencies of 78–91 per cent L^M, but again the Eskimos from Baffin Island and Labrador diverge greatly by having only 56 per cent L^M.

These variations in allele frequencies represent special cases. To generalize beyond them, Cavalli-Sforza, Barrai, and Edwards have studied in ingenious ways the distributions of a total of twenty alleles at five blood-group loci in fifteen human populations selected to represent the world population. They have come to the conclusion that random drift during the evolution of human populations has been a major factor in producing the genetic diversity among present-day populations.

A direct demonstration of changes in allele frequencies can be made only by comparisons of successive generations. Glass has studied the blood groups of a very small religious isolate—a community of Dunkers in Pennsylvania. The ancestors of this sect were 27 families who, in the early eighteenth century, came from German Rhineland to North America. When studied, their ABO frequencies differed significantly from those typical for Germans in the Rhineland as well as for Americans. Thus, the frequency of group A among the Dunkers was nearly 60 per cent, as contrasted to 40–45 per cent in the related populations; O group was somewhat rarer in Dunkers than in Germans or Americans; and the I^B allele was nearly absent among the Dunkers. It is not known when these changes in allele frequencies occurred—whether they were a result of initial deviations from the general population or whether they occurred during the 200 years in which the Dunkers remained a small isolate. Although there was no difference, at the time of the study, between the ABO frequencies in three different age groups of Dunkers (representing more or less three generations), there was a significant difference in frequencies of the MN groups. The age group of fifty-six years and older had a L^M frequency of 55 per cent, which was similar to that of West German and American populations. In contrast to this, the middle-age group, from twenty-eight to fifty-five years, contained 68 per cent of L^M alleles. In the youngest group, from one to twenty-seven years, the frequency had risen to 73 per cent. Because of the small numbers of subjects, the differences between adjoining age groups are not statistically significant, but the difference between the oldest and youngest groups is highly significant. It is hard to avoid the conclusion that the shifts in allele frequency in the Dunker isolate have been the result of drift.

Some populations are exceptional in possessing a high frequency of some clearly detrimental gene. In such cases the founder principle provides an explanation. One such gene is that for variegate porphyria, a metabolic disorder in which excessive amounts of porphyrin, an essential part of the hemoglobin molecule, are excreted in the urine. Affected individuals are liable to develop blisters and abrasions on those parts of the skin which are exposed to light. They can become acutely ill, suffering abdominal pain and other symptoms, particularly after the ingestion of barbiturates and certain other drugs. The condition is caused by a dominant gene. It is rare, though not absent, in most parts of the world, but seems to occur in about 4 per thousand of the Afrikaner population of South Africa. These people are descendants of Dutch and French settlers who arrived in Africa during the latter part of the seventeenth century.

Why is the allele for porphyria so frequent among the Afrikaners? The answer seems to be that it is due to chance! The present population of about two million Afrikaners is descended from a rather small number of original immigrants who raised large families. Their descendants too had large families, as may be seen by the fact that the affected great-grandfather of an affected proposita left 478 descendants. A careful search of the genealogies of South African porphyrics makes it probable that all of them are descendants of a single Dutch couple who were married in Cape Town in 1688. Either the husband or the wife seems to have been a carrier for the porphyria allele. Since the total number of original settlers was small, the one allele for porphyria present in the original carrier corresponded to a rather high allele frequency, which is reflected in the unusually high frequency of porphyria in South Africa at the present time. The single gene of the first carrier is now found in more than 10,000 persons.

Similar chance sampling of genes among the small numbers of original ancestors of specific populations probably accounts for unusual incidences of other abnormal alleles. A striking example is the 4–10 per cent incidence of individuals affected with a peculiar type of congenital blindness among the Pingelapese people of the Eastern Caroline Islands. The high frequency of the phenotype is due to a very high frequency of a recessive autosomal allele. Sometime between 1780 and 1790, a typhoon hit Pingelap and reduced the population to approximately nine men and an unknown number of women. By 1970 the Pingelapese numbered about 1500. It seems most likely that one or more of the few survivors of the eighteenth-century typhoon was heterozygous for the allele causing blindness, thus accounting for the present high incidence of the eye defects.

Albinism, although rare in most populations, occurs with relatively high frequencies in others. Again, random drift, in one or the other of its forms, may be responsible for the high incidence of the trait, which has apparent disadvantages, among the San Blas Indians of Panama, in an isolated population of mixed black-white-American Indians in the United States, and in diverse American Indian populations. It is possible, however, that drift is not the sole and perhaps not even the main process responsible for the high frequencies of the albino allele. Since albinos are poorly protected from sunburns and males may remain in their villages at times when the nonalbino men work in the fields, it has been suggested that albino males may sire relatively more children than nonalbinos and thus increase the albino allele frequency.

An interesting example of the founder principle in which the unit of heredity is a chromosome rather than a gene has been traced in kindreds of French Canadian men (Fig. 311). Seventeen men were found to have an unusually short Y-chromosome. All of them were descended in the male line from a man who had been married in the seventeenth century, about 11 generations ago. He must have been the carrier of the short Y-chromosome and thus the founder of the group of males bearing it in the nineteenth and twentieth centuries.

Absence of an allele, or its presence at a low frequency—as well as the high frequency of an allele—in some populations may likewise be due to chance phenomena. Böök has noted the near absence of manic-depressive psychosis, but not that of schizophrenia, in a north Swedish isolate, and Eaton and Weil have found the reverse, namely low incidence of schizophrenics but high incidence of manic-depressives, among the Hutterite isolates of North America. These data on psychoses are subject to not only genetic but also cultural interpretations, and the reference to genetic drift at present cannot be more than a suggestion.

The interpretation of striking differences in the incidence of clearly unfavorable traits as the result of drift presupposes genetic determination of the traits. Sometimes it is difficult to establish such genetic bases even if numerous pedigrees show an accumulation of affected individuals in sibships, successive generations, and collateral families. Although such pedigrees suggest genetic determination, the possibility of unknown external causative agents often remains. This may be illustrated by two fatal diseases of the nervous system.

Amyotrophic lateral sclerosis is a disease in which slowly progressive degeneration of nerve cells in the spinal cord and in the brain stem leads to progressive muscular wasting and other degenerative changes. The disease is found in persons of middle age and older. Death occurs, on the average, within three years after the onset of the disease. In most parts of the world amyotrophic lateral sclerosis is very rare, accounting for only one death annually among one hundred thousand people. In contrast to this over-all low frequency is the very high incidence among the Chamorro people of the Mariana Islands, including Guam. There the incidence of the disease is at least one hundred times higher than in the rest of the world; nearly 10 per cent of all adult deaths are due to it. Among white and black Americans and Europeans, most cases of the disease are sporadic, but a number of kindreds have been found in which the disease occurs in a manner which fits inheritance by means of a dominant, incompletely penetrant gene. The sporadic cases are probably of heterogeneous nature. Some cases may be from kindreds in which there are other affected members who have not been reported, some may be due to new mutations, some may be caused by genes with unusually low penetrance, and

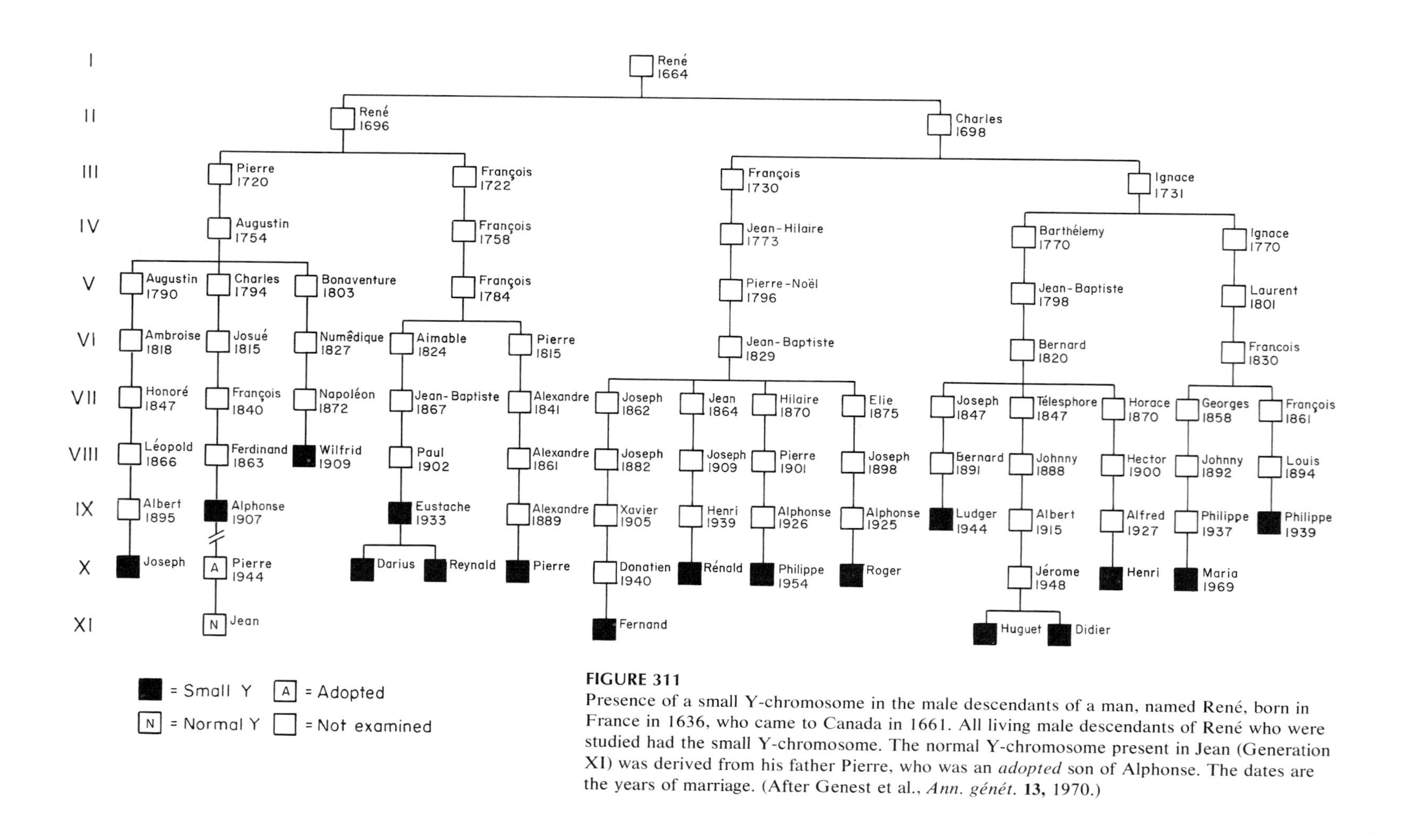

FIGURE 311
Presence of a small Y-chromosome in the male descendants of a man, named René, born in France in 1636, who came to Canada in 1661. All living male descendants of René who were studied had the small Y-chromosome. The normal Y-chromosome present in Jean (Generation XI) was derived from his father Pierre, who was an *adopted* son of Alphonse. The dates are the years of marriage. (After Genest et al., *Ann. génét.* **13,** 1970.)

some may be nongenetic in origin. Among the Chamorro people in Guam, most cases are familial. This could be due to the common presence of a responsible gene, but could also be due to a high incidence of some unknown nongenetic condition. If the pattern of recurrence of affected individuals in pedigrees is analyzed, usually it is relatively easy to distinguish between these two possibilities. But the family histories in Guam are not very reliable since there is a tendency not to reveal the existence of affected relatives, the family relationships are often not clear, and the church and civil records were destroyed during World War II. Nevertheless, Kurland is inclined to interpret the evidence as indicative of dominant inheritance with incomplete penetrance. This interpretation is strengthened by the finding that amyotrophic lateral sclerosis is rare among other people than the Chamorros living on Guam, and by the fact that the disease is common not only among the Chamorros in Guam but also among those Chamorros who have settled in California and live like other Californians of their socioeconomic groups. If the genetic interpretation is correct, the population of Guam was probably very small some two and a half centuries ago, and the occurrence of a mutation, or the presence of the mutant gene, in one or a few individuals may have provided the opportunity for drift, which led to a high frequency of the deleterious allele. On the other side, there are some facts that suggest a nongenetic interpretation of the disease. Possibly an unknown external factor, e.g., a virus, is responsible for the disease, together with a genetically controlled susceptibility to virus infection. These possibilities have come to the foreground since a clarification of the nature of another disease, the Kuru disease of the Fore natives of New Guinea, has been obtained. This fatal illness, which is due to rapidly progressive malfunction of the cerebellum, had, on the basis of extensive genealogical data, been interpreted in genetic terms. This interpretation had to be abandoned when it was recognized that a slow persistent virus infection is the primary agent of the disease and that its peculiar mode of transmission followed lines apparently determined by cannibalistic ritual.

Gene Flow and Gradients in Allele Frequency. Selection and drift lead to diversity in human populations, but such diversity has a tendency toward equalization. Matings between neighboring, genetically diverse populations result in the introduction of genes from one to the other. Migrations of populations into new territories may lead to contacts, accompanied by gene exchange, between diverse populations formerly separated by great distances. Migrations of groups of people may be likened to the coming together of separate rivers which flow into the same sea. The consequent intermingling of the water molecules from different rivers into a homogeneous body of water may be likened to the interbreeding of initially separate populations.

The formation of new gene pools by interbreeding is often a slow process. Populations are known that have lived side by side for hundreds or even thousands of years and yet have retained genetic diversity. Thus, Sanghvi has shown that different castes in India have strikingly different frequencies for ABO, MN, P, and Rh blood groups as well as for taster ability and color blindness. Conversely, the gypsies of Hungary resemble in their ABO frequencies the Indian immigrants who came to Europe from Asia in the fifteenth century and who have remained different from the population by which they are surrounded. Such retention of diversity depends on cultural factors which prescribe endogamy. Sometimes these factors restrict gene diffusion primarily in one direction: the African Negroids brought to North America experienced an influx of genes from Caucasoids; whereas the latter did not acquire any appreciable admixture of genes from Negroids. Even if a social system does not place legal barriers in the way of gene exchange, personal preferences often lead to positive assortative mating and thus counteract tendencies toward panmixis.

If two genetically diverse populations, I and II, occupy different but adjoining areas, gene exchange should result in gradients of allele frequencies. An allele A^1 which is more common in population I than in II would be expected to be relatively frequent in II near the border of I and II and to decrease in frequency in II farther away from the border. Similarly, if there are migrations of individuals of population I into the territory occupied by II, gradients of the frequency of A^1 among II may reflect the number of migrants who penetrated into specific regions and there mixed with the indigenous population. A famous example is the gradient of frequencies of the I^B allele from Central Asia to western Europe (Fig. 312): from a peak frequency of 25–30 per cent in Central Asia, this allele becomes less and less common closer to western territories. Candela has explained this gradient by assuming it to be a consequence of the invasions of the Mongoloids who, from about 500 A.D. to 1500 A.D., pushed westward in numbers which decreased with distance from their region of origin. Hybridization of the invaders with the native populations, in which the I^B allele is supposed to have been absent, led to "diffusion" of I^B from Central Asia to the west. The gradient of I^B concentration, therefore, is a reflection of the diminished contact between Mongoloids and populations farther west. The external features of the Mongoloids are not found strikingly in the present populations of Europe, since polygenic combinations determining racial characters were broken up in hybridization. The singly determined blood-group traits, however, still bear witness to the influx of many other eastern alleles. It should be added that there are pockets of increased frequencies of the I^B allele in some of the most western areas of Europe, making it unlikely that all I^B alleles on that continent are derived from those of the eastern invaders. And there remains the slight possibility that a gradient in a blood-group allele may reflect not only gene flow but also unknown selective influences which follow a geographic gradient.

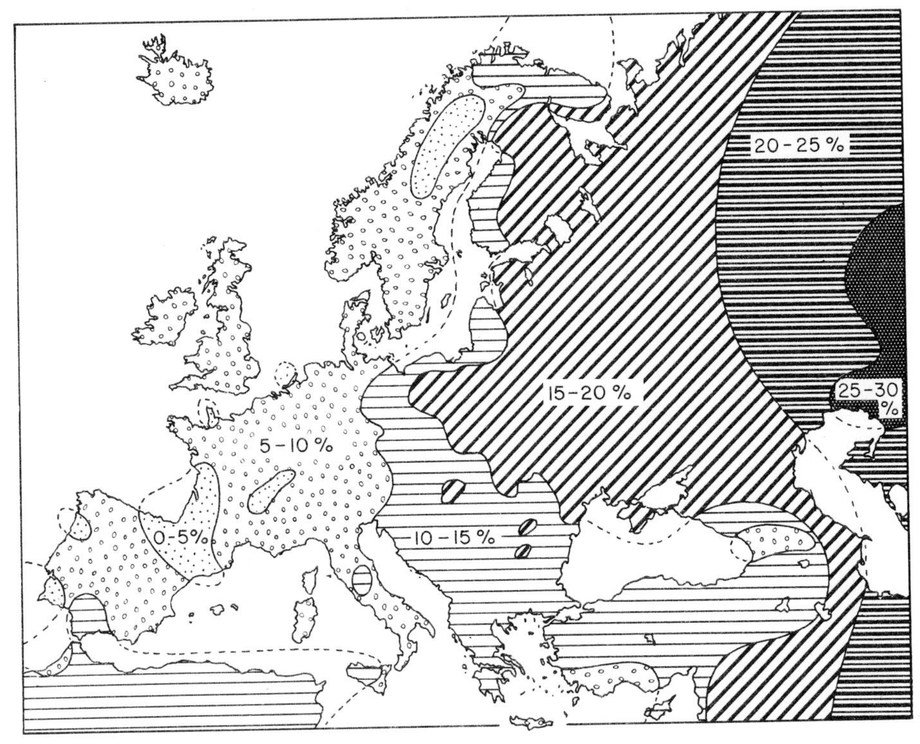

FIGURE 312
The frequencies of the I^B allele in Europe. Note the gradient from high frequencies in the east to low frequencies in the west. (After Mourant, 1954; and Manuila, *Amer. J. Phys. Anthropol.*, 14, 1956.)

A similar penetration of I^B alleles into a region known to have been free of I^B alleles has been noted in Australia. On most of this continent the aborigines possess only I^O and I^A alleles, but tests of individuals from the northern coastal areas have revealed I^B frequencies of 8–20 per cent (Fig. 313). It seems that the I^B allele was introduced from two separate sources: the Papuans of New Guinea, northeast of Australia, and the Malayans, northwest of Australia. It may be expected that in the course of time a gradient toward the south of decreasing frequency of the I^B allele will be formed.

Birdsell, who has summarized the data on I^B in Australia, has also considered various other traits. One of his many interesting uses of population genetics in the exploration of anthropologic problems concerns the question of the racial origin of the aborigines. It is believed (but not generally so) that these people were derived from three different racial groups which reached Australia in successive periods: first, Negritos; second, primitive Caucasoids, related to the Ainus of northern Japan; and, third, a group called Carpentarian, which seems closest to some of the aboriginal tribes of India. Birdsell

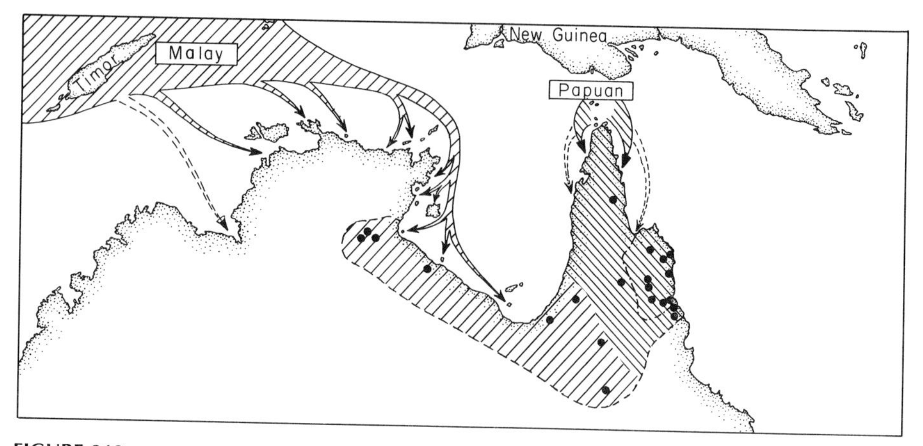

FIGURE 313
Map of northern Australia, showing areas of penetration of the I^B allele from adjacent regions. The black dots signify locations of aborigines possessing I^B (Birdsell.)

has constructed a map, based on a subjective taxonomic evaluation, which presents estimated Carpentarian gene frequencies. They show a general gradient from north to south with some interesting irregularities (Fig. 314, A). Thus, there is a small region of zero frequency in the northeast, a large zone of very low frequency along most of the eastern shores, and a tongue of zero frequency protruding from the south into the otherwise even southern contour of the lowest frequency belt. These irregularities are correlated with physical or geological barriers. Birdsell has worked out two theoretical models to explain, by means of gene flow, the observed gradients of Carpentarian elements. These models take into account the existence of not only physical and ecological barriers but also cultural barriers to gene flow. Foreign alleles arriving in a given tribe will easily become distributed in it, but endogamic rules and preferences will form relative barriers to penetration into adjacent tribes. Thus "genetic space," defined in relation to the number of tribes between two regions rather than to the geographic distance between them, will determine the speed of gene flow along the gradient.

Any detailed model of gene flow will have to make unproved, and at present unprovable, assumptions concerning the genetic space involved. The consequences of gene flow derived from models must therefore be regarded as highly hypothetical. Nevertheless, models are valuable in clarifying concepts and suggesting interpretations.

Birdsell's first model assumes the entry of the Carpentarian element by way of the Torres Strait. Its detailed elaboration results in an expected gradient of estimated allele frequencies that is very unlike that derived from taxonomic observations (Fig. 314, B). The second model assumes a wide area of entrance, along much of the northern coast of Australia. This model

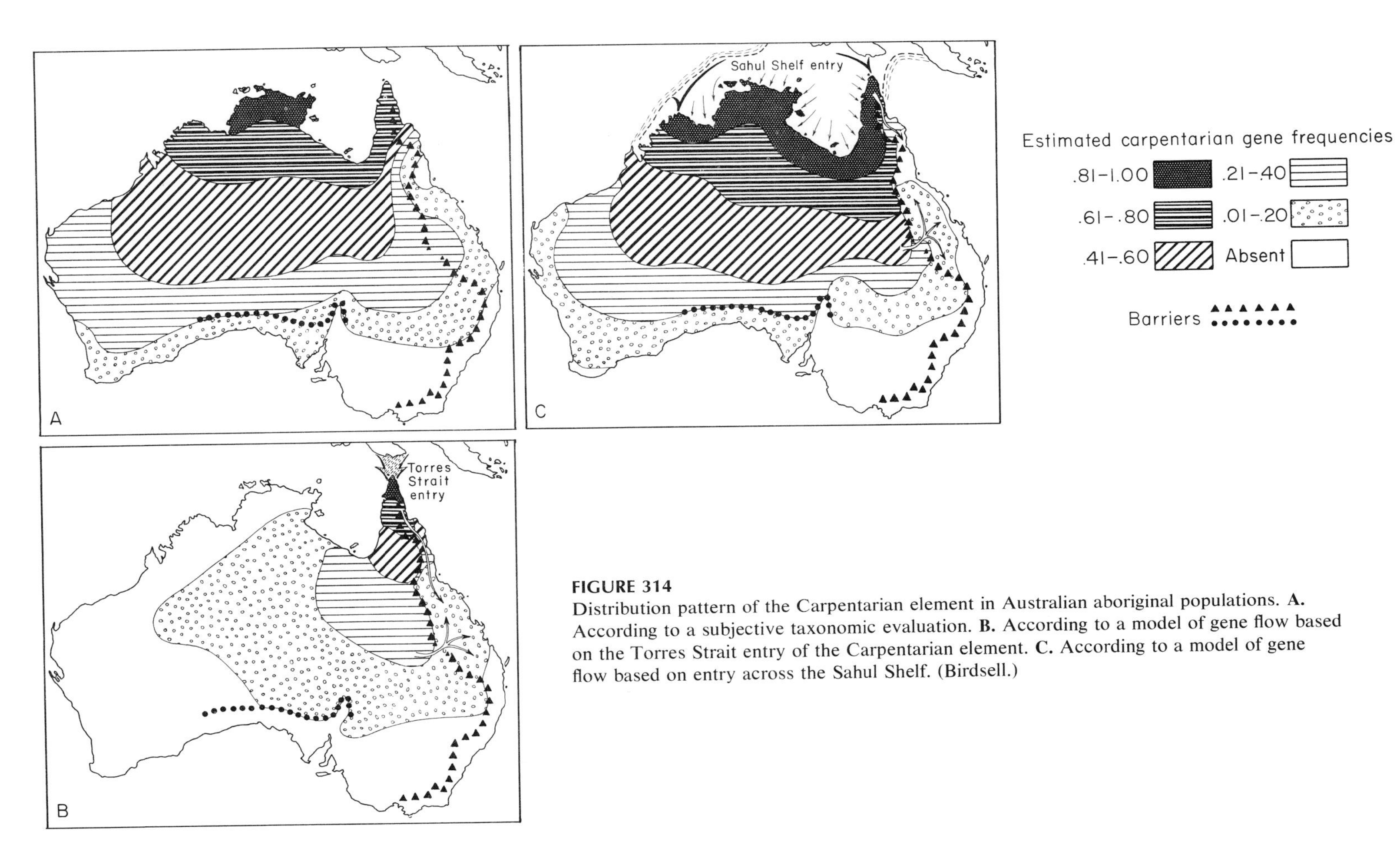

FIGURE 314
Distribution pattern of the Carpentarian element in Australian aboriginal populations. **A.** According to a subjective taxonomic evaluation. **B.** According to a model of gene flow based on the Torres Strait entry of the Carpentarian element. **C.** According to a model of gene flow based on entry across the Sahul Shelf. (Birdsell.)

results in an expectation of the gradient distribution which is surprisingly similar to that based on observation (Fig. 314, C). It therefore seems possible that the second model reflects the prehistoric actuality of Carpentarian gene flow into Australia.

Birdsell's models did not involve direct knowledge of allele frequencies. They made use of taxonomic observations which were interpreted in terms of population genetics. The validity of such interpretations has been tested in certain situations where degree of relationship of hybridizing populations could be expressed in terms of both taxonomic similarities and specific allele frequencies. How closely these two methods lead to the same conclusion can be seen for the example of four populations represented by West African blacks, Caucasoids, and two groups of hybrids between them—a typical American black group, and the blacks of Charleston, South Carolina. Owing to isolating factors, the last group has experienced less gene flow from the Caucasoids than have most other American blacks. The graphic representations of relationships according to morphological and genetic evaluation of subgroups are very similar (Fig. 315).

THE FUTURE

In the attempt to predict the future biological evolution of man, two main facts may serve as guides. The first is biological, the second cultural. Biologically, man became a large species in terms of breeding population. Furthermore, this large species is becoming less and less subdivided into relative isolates of some permanence, but is tending to form increasingly larger interbreeding groups. This by itself should result in a decisive slowdown, if not a cessation, of changes so far as these are caused by genetic drift in its various forms.

Mutation pressure in the future may lead to the further reduction and disappearance of organs or functions which have already ceased to be vital. The loci which are, at present, concerned with the retention of vestigial properties are likely to give rise, by mutation, to new alleles which will result in loss of these properties. These mutant alleles will accumulate in the population, since selection will not act against them. Likewise, the polygenic combinations which result in losses of vestigial traits will accumulate.

Interbreeding of different racial groups will lead to the appearance of many hitherto unknown combinations of alleles which, until now, had been kept isolated from one another. While these new variants may possess selective advantages or disadvantages over the already known types, natural selection in the large interbreeding species *Homo sapiens* will have little effect, since good and bad combinations alike will be broken up in each generation.

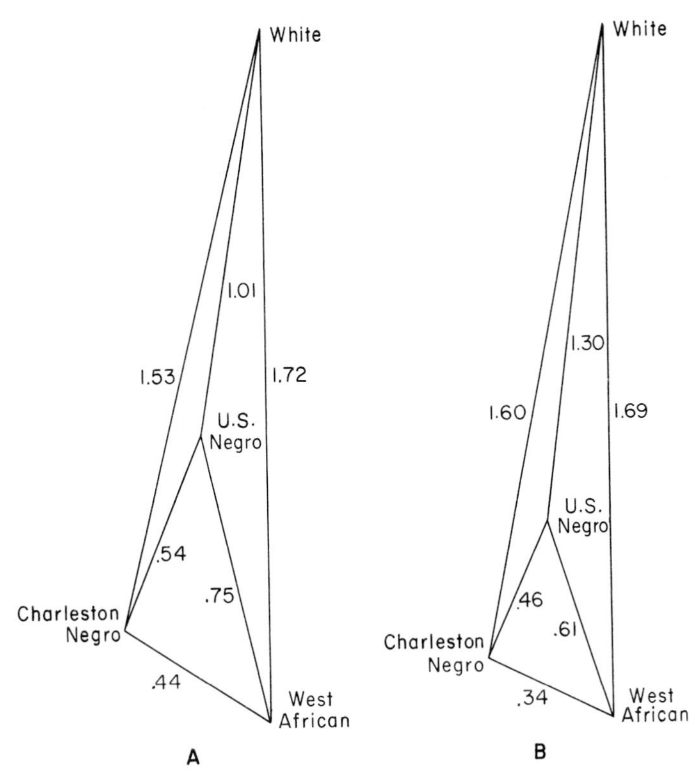

FIGURE 315
Diagrammatic representation in terms of "distance" of four populations from one another. **A.** According to morphological criteria. **B.** According to indexes computed from allele frequencies for blood groups and hemoglobin types. (After Pollitzer, *Amer. J. Anthropol.* **16,** 1958.)

Changes in the physical environment of man will hardly act as stimulants for evolution. Such changes force other organisms to, so to speak, "explore" their genetic potential for necessary new adaptations. Man's techniques have made him so largely independent of physical environmental influences that he will succeed in stabilizing his surroundings and thus escape the need of adapting himself to new ones. Hence, it appears likely that the normal agents for evolutionary biological change will be greatly reduced in efficacy.

Opposed to this conservatism of biological influences are the revolutionary potentialities of cultural ones. At present, differential reproduction of different genetic groups is a cultural phenomenon which produces changes in the allele frequencies of mankind. The reality of such changes in the over-all genetic composition is beyond doubt, in so far as it concerns the differential reproduction of the major races. On changes in allele frequencies within racial groups, the evidence is less decisive, but it is likely that such alterations are also taking place. As we have seen, there is little reason for being too alarmed about such present phenomena, since only long-continued trends will appreciably change the genetic composition of large populations. Human societies have been so unstable that persistent genetic trends have been virtually impossible.

The endeavors of eugenicists to check undesirable trends and to further desirable ones may be the precursors of human actions which will become evolutionary agents of incalculable influence. Natural selection will be superseded by socially decreed selection.

Genetic Engineering. In the future, man may not be restricted to selective processes concerning the genes already present in his gene pool. He may also change the genes themselves. This is done already, even if unintentionally, when radiation or other external agents produce gene mutations. Even if it should not prove possible to be sufficiently specific in "repairing" defect-causing genes by external means it may be feasible to replace them by normal genes. The understanding of pneumococcus transformation provides one of the models whose realization with the genes of man is envisaged. A DNA extract from one strain of pneumococcus applied to another, genetically different, strain may lead to the replacement of genes of the second strain by those of the first. Might it not be possible to bathe human cells having defective genes in a DNA solution derived from cells with normal genes and transform the former genes by incorporation of the latter? The difficulties of this approach, technical and intrinsic, are great; and only beginnings of success have been recorded. Nevertheless, gene therapy of this type does not seem to be utopian. Moreover, gene transformation by means of DNA extracts of whole cells may not be the choice of the future. Possibly, a more specific way of replacement of one by another gene may ultimately prove feasible. The biochemists' triumphs of recent years include the elucidation of the whole sequence of purine and pyrimidine bases in a transfer RNA and, with only a little less certainty, those in the DNA gene which codes for it. The triumphs also include man-made assembly of bases in the right sequence to synthesize a transfer RNA and, implicitly, its corresponding gene. Most human genes consist of much longer sequences of base pairs than those for transfer RNA's, and their synthesis will be more difficult. It may be too early to expect that within the next 10 years physicians will be able to prescribe applications of pure non-PKU genes to their PKU patients or of pure non-Tay-Sachs genes to others; but given a less limited period the prospects appear good.

Perhaps innocuous viruses will play a role in controlled gene transformation. Bacterial viruses sometimes carry small sections of their hosts' DNA from one bacterium to another, where it may be permanently incorporated into the bacterial DNA. In man a case of harmless virus infection is known by means of which a viral gene for a special type of the enzyme arginase has been permanently added to the genotype of somatic cells. Events such as these may well be exploited in future genic therapy. Indeed, hardly had this paragraph been written when the result of the following experiment was announced. Human cell cultures from a galactosemic patient had been infected with a bacterial virus which carried an allele for normal galactose metabolism, derived from the bacterium *Escherichia coli.* Before infection the cells were genetically unable to produce an enzyme necessary for normal galactose metabolism. After infection they were able to do so (see p. 804).

The most radical proposal so far made in regard to genetic engineering does not rest on discoveries of molecular biology but on those of classical experimental embryology. In the frog *Xenopus* unfertilized eggs have been deprived of their nucleus and then been provided with the nucleus of a body cell of another Xenopus. In favorable cases development proceeded in normal fashion. The resulting frogs were identical in genotype with the donor or donors of their somatic nuclei. This method applied to man would result in the development of individuals genetically identical to whatever donor individuals had been chosen: boys genetically exactly like their father, girls like their mother, or individuals like some true or false hero of art, of science, or sports, or like some demagogue, or some saint.

The method of asexual reproduction by means of the nucleus of a somatic cell is called cloning. It is now attempted in mammals. Mice are the subject of experimental trials, and cattle are next as subjects of applied significance. Dairy cattle would not have to be bred and selected in each generation, but a single cow of excellence could donate nuclei to numerous enucleated eggs and thus be the progenitor of numerous cows identical in genotype to her. Beef cattle likewise could be multiplied by cloning a superb specimen.

And if successful in cattle, then on to man?

As long as gene therapy is applicable to somatic cells only, changes in the pool of transmissible genes do not occur. When, however, directed changes in the germ cells can be induced, or when by cloning the genotypes of specific individuals can be multiplied, or when socially directed selection within the established human gene pool becomes a reality, man himself becomes the master of selective actions. Whether such actions will be directed toward human uniformity or toward variability, and what traits will be set up as desirable ones, will not be decided soon. In the course of time, however, the control by man of his own biological evolution will become imperative, for the power which knowledge of human genetics will gradually place in man's hands cannot but lead to action. Such evolutionary controls will be world-wide in scope, since, by its nature, the evolution of mankind transcends the bounds of unrestricted national sovereignty. It will be a difficult problem to direct the actions of individuals toward a socially desired long-range goal, and to retain at the same time the essential aspects of personal freedom. This, however, is not a new problem. The resolution of the conflict between freedom and organization is a permanent theme in human history.

REFERENCES

Birdsell, J. B., 1950. Some implications of the genetical concept of race in terms of spatial analysis. *Cold Spring Harbor Symp. Quant. Biol.*, **15**:259–314.

Brody, J. A., Hussels, Irena, Brink, E., and Torris, J., 1970. Hereditary blindness among Pingelapese people of Eastern Caroline Islands. *Lancet*, **1**: 1253–1257.

Candela, P. B., 1942. The introduction of blood group B into Europe. *Hum. Biol.*, **14**:413–443.

Cavalli-Sforza, L. L., Barrai, R., and Edwards, A. W. F., 1964. Analysis of human evolution under random genetic drift. *Cold Spring Harbor Symp. Quant. Biol.*, **29**:9–20.

Dean, G., 1963. *The Porphyrias. A Story of Inheritance and Environment.* 117 pp. Lippincott, Philadelphia.

Dobzhansky, Th., and Allen, G., 1956. Does natural selection continue to operate in modern mankind? *Amer. Anthropol.*, **58**:591–604.

Eaton, J. W., and Weil, R. J., 1955. *Culture and Mental Disorders.* 254 pp. Free Press, Glencoe, New York.

Fisher, R. A., 1930. *The Genetical Theory of Natural Selection.* 272 pp. Oxford University Press, London. (Also 1959, 2nd rev. ed. 291 pp. Dover, New York.)

Friedmann, T., and Roblin, R., 1972. Gene therapy for human genetic disease? *Science*, **175**:949–956.

Gajdusek, D. C., 1964. Factors governing the genetics of primitive human populations. *Cold Springs Harbor Symp. Quant. Biol.*, **29**:121–135.

Glass, B., 1954. Genetic changes in human populations, especially those due to gene flow and genetic drift. *Advan. Genet.*, **6**:95–139.

Haldane, J. B. S., 1932. *The Causes of Evolution.* 235 pp. Harper, New York.

Hulse, F., 1957. Some factors influencing the relative proportions of human racial stocks. *Cold Spring Harbor Symp. Quant. Biol.*, **22**:33–46.

Livingstone, F. B., 1964. The distribution of the abnormal hemoglobin genes and their significance for human evolution. *Evolution.*, **18**:685–699.

Mayr, E., 1963. *Animal Species and Evolution.* 797 pp. Harvard University Press, Cambridge.

Mourant, A. E., 1954. *The Distribution of the Human Blood Groups.* 438 pp. Thomas, Springfield, Ill.

Neel, J. V., and Ward, R. H., 1970. Village and tribal genetic distances among American Indians, and the possible implications for human evolution. *Proc. Nat. Acad. Sci.*, **65**:323–330.

Scarr-Salapatek, Sandra, 1971. Race, social class, and I.Q. *Science*, **174**:1285–1296.

Schwidetzky, Ilse, 1950. *Grundzüge der Völkerbiologie.* 312 pp. Ferdinand Enke, Stuttgart.

Wiesenfeld, S. L., 1967. Sickle-cell trait in human biological and cultural evolution. *Science*, **157**:1134–1140.

Wright, S., 1932. The roles of mutation, inbreeding, crossbreeding and selection in evolution. *Proc. 6th Int. Congr. Genet.*, **1**:356–366.

INDEXES

AUTHOR INDEX

SUBJECT INDEX